Significant Progress in Geophysical Research

Significant Progress in Geophysical Research

Edited by **Annabelle Turner**

Callisto Reference
New York

Published by Callisto Reference,
106 Park Avenue, Suite 200,
New York, NY 10016, USA
www.callistoreference.com

Significant Progress in Geophysical Research
Edited by Annabelle Turner

International Standard Book Number: 978-1-63239-560-3 (Hardback)

Printed in China.

Contents

Preface

Geophysics is the fundamental study of the earth, its atmosphere and earth's environment in space. This book provides an extensive analysis of the biophysical issues related to earth and the impact of climate change on the earth's environment. Climate change is a phenomenon by which the long-term averages of weather events such as temperature, precipitation and wind speed that define the climate of a region do not remain the same but change over time. Since, the ice age up to contemporary times; climate change has affected the earth and influenced it in certain patterns which are registered in historical or paleoecological records. Geophysics also extends to examine the effects on the flora and fauna in different ecosystems; the ecosystems and landscapes in alpine and arctic regions are expected to be among the most affected, as they suffer more extreme changes.

This book on is dedicated to the progress in geophysical research. It discusses the state-of-the-art research projects involved to explore the earth, the biophysical causes for climate change and the techniques currently being used and developed for studying the earth and environmental changes in several regions of the earth. The book includes some important chapters that discuss the issues of seasonal and solar cycle variations, regional climate variation and monsoon rainfall, groundwater flow systems, the application of GPS systems networks, and also presents some case studies analysis and experimentations.

Several international experts on geophysical researches have contributed to this book. I am grateful to them for their excellent research works. The book would be of immense benefit to students, researchers and scientists involved in the study of geophysics. Finally, I would like to thank my family for supporting me in my research endeavours.

Editor

1

Measurement of the Characteristics of TIDs Using Small and Regional Networks of GPS Receivers during the Campaign of 17–30 July of 2008

Cesar E. Valladares[1] and Matthew A. Hei[2]

[1] *Institute for Scientific Research, Boston College, Chestnut Hill, MA 02467, USA*
[2] *Plasma Physics Division, Naval Research Laboratory, Washington, DC 20375, USA*

Correspondence should be addressed to Cesar E. Valladares, cesar.valladares@bc.edu

Academic Editor: Y. Sahai

This scientific report presents the results of a dedicated experiment that was conducted within the framework of the Low-latitude ionospheric Sensor Network (LISN) observatory to measure the characteristics of medium-scale (hundreds of km) Traveling Ionospheric Disturbances (TIDs) as they transit through the low-latitude ionosphere. A small array of 3 GPS receivers separated by 4-5 km placed in a triangular configuration was installed near Huancayo in Peru possessing several characteristics of a radio-interferometer. During the campaign days, 17–30 July 2008, TIDs were observed daily. On July 20, 2008 between 22 and 24 UT several TIDs moved across the small array of GPS receivers with a velocity near 130 m/s, were directed northward and had wavelengths close to 450 km. Other GPS receivers that were operating hundreds of km away from Huancayo show also similar TEC traces and provide a phase velocity equal to 150 m/s. This value was measured using the GPS at Piura, Cuzco and Huancayo. Based on this positive result, we conclude that small and/or regional arrays of GPS receivers can be used at low latitudes to study the role that gravity waves may have on seeding plasma bubbles.

1. Introduction

Traveling ionospheric disturbances (TIDs) are the ionospheric manifestation of neutral density oscillations called atmospheric gravity waves (AGWs or GWs). AGWs originate in the troposphere or stratosphere, and often propagate to F-region heights, where they induce oscillations in the plasma density via ion-neutral coupling. When AGW winds drive plasma along magnetic field lines via ion-neutral collisions, "classical" TIDs result [1–3]. However, at night the electrodynamics are favorable to the production of "nonclassical" or "electrodynamic" TIDs, such that vertical movement of the F-layer results [4–6]. TIDs that reach F-region altitudes are divided into two classes: medium-scale (MSTIDs) and large-scale (LSTIDs). MSTIDs primarily originate from tropospheric weather systems and mountain turbulence; depending on background wind conditions and other factors, these may propagate in any direction. LSTIDs are produced by auroral energy injection during geomagnetic storms. They typically propagate meridionally from polar regions towards the equator. AGWs have a wide range of observed wavelengths and propagation velocities, frequently having horizontal wavelengths of up to hundreds of kilometers and speeds of hundreds of meters per second. As they propagate from lower altitudes, their amplitudes grow exponentially. At mid-latitudes, AGWs have been observed to have preferred propagation azimuths that change with season [7]; it is logical to assume that similar seasonal patterns exist at low latitudes.

Gravity waves are widely believed to play a critical role in seeding the development of equatorial plasma depletions in the nighttime low-latitude ionosphere. Current understanding holds that both favorable ionospheric conditions [8] and an initial seed perturbation on the bottomside are necessary to create rising depletion channels. If both are present, the Rayleigh-Taylor instability creates intense polarization electric fields within the perturbations, causing the depleted plasma inside to $E \times B$ drift upward into the topside [9, 10]. Gravity waves, [11], as well as penetration electric fields, dynamics associated with the post-sunset vortex [12], and E-region structure mapped to F-region altitudes along magnetic field lines [13] have all been cited as possible seeding mechanisms.

To unravel the importance of the gravity wave seeding mechanism, it is important to observe and characterize both early-evening gravity waves and the equatorial plasma depletions seen later in the evening. However, most current observing techniques have severe limitations. Ionosondes can measure the GW vertical period, but they are not sensitive to other parameters. A steerable-beam radar such as Altair can measure the large-scale zonal wavelength and amplitude, but the GW must be traveling roughly orthogonal to the radar beam to be detected. Phased-array radar systems such as the Japanese MU system [14, 15] are sensitive to GW horizontal wavelength, propagation azimuth, and amplitude, but these systems are expensive to deploy and operate. OI (557.7 nm, emission altitude ~96 km) and OH band (720–910 nm, ~86 km) airglow imagers can measure GW horizontal wavelengths, propagation azimuths, and amplitudes, but at altitudes far below the bottomside [16, 17]. OI (630 nm emission) optical measurements have been used to characterize the seasonal and solar cycle variability of MSTIDs [18, 19]. OI has an emission layer between roughly 200–300 km, which is very close to the bottomside [20]. But even here, airglow imagers operate only during the nighttime and require good seeing conditions (no clouds) to operate.

The radio beacon technique [7, 21–23] overcomes many limitations of other observation methods, enabling simultaneous characterization of key GW parameters. Here, satellite radio beacon signals are measured by stationary ground receivers; TIDs are evident as fluctuations in total electron content (TEC) derived from the beacon signals [24]. Although TEC is an integrated quantity, various processes, including wind-filtering [7] and breaking [25], prevent AGWs from penetrating far into the thermosphere. Observations [26] and model results [27] have shown that most TIDs do not rise above 300 km. Thus, the bulk of the TEC perturbations occur on the F-region bottomside and the upper E-region. Moreover, after the E-region decays in the evening, TEC fluctuations become highly localized in the bottomside region. Thus, while radio beacon TEC measurements are integrated over the whole altitude range of the ionosphere, the narrow vertical localization of GW-induced TEC fluctuations makes this technique a powerful tool for studying early-evening TIDs. In the radio beacon technique, also known as radio-interferometry, phase differences measured at the various stations are used to determine TID velocity, propagation azimuth, and amplitude.

Recently, the radio-interferometric technique has been adapted for use with GPS satellites [28–30]. This new innovation makes it possible to utilize inexpensive, easily deployed GPS receivers to study gravity waves at a wide variety of locations. The large number of GPS satellites in orbit makes it possible to continuously monitor the bottomside region over a given receiver array.

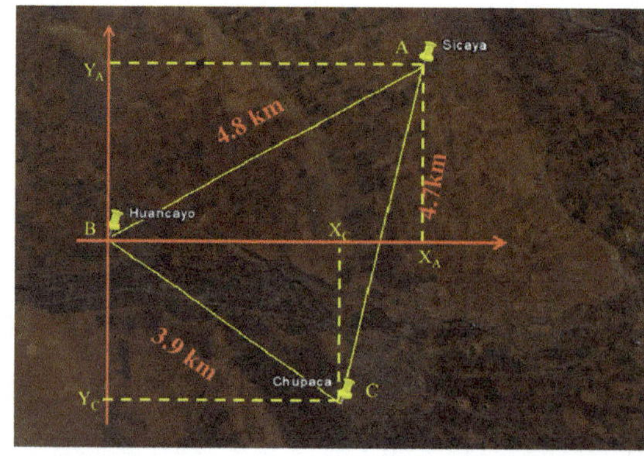

FIGURE 1: Geographic location of the Huancayo radio interferometer test array.

This paper describes measurements observed with a three-station array of GPS receivers that were located near Huancayo, Peru. These receivers constitute the first low-latitude radio-interferometer array dedicated to studying TIDs. The hardware layout and the analysis technique, that was used to derive the TEC perturbation associated with the TIDs, are described in the next section. The phase velocities, scale sizes, and the direction of motion of the TIDs calculated using the SADM-GPS methodology are presented in Section 3. The TIDs wave characteristics derived using a cross-correlation method are introduced in Section 4. The cross-correlation method was also applied to TEC measurements recorded by other low-latitude ionosphere sensor network (LISN, http://lisn.igp.gob.pe/) GPS receivers that are operating hundreds of kms apart from Huancayo. The results of this analysis are described in Section 5. Discussion and conclusions follow in Sections 6 and 7.

2. Observations

Since the late 90s several publications have reported the characteristics of mid-latitude GWs using the GPS radio-interferometry (GPS-RI) method [28, 30]. Based on this success, we conducted a campaign in South America aiming to characterize the local time, duration, and wave properties of GWs that are transiting near the magnetic equator in the Peruvian sector. We installed two GPS receivers near the city of Huancayo (12.042°S; 75.321°W) due to its proximity to the magnetic equator, the existence of a permanent GPS receiver in that city and the presence of nearby towns. The additional GPS receivers were temporarily installed, one at the town of Chupaca (12.061°S; 75.292°W) and the other at Sicaya (12.021°S; 75.282°W). All 3 GPS receivers of the small network of GPS receivers consist of GSV4004B Novatel GPS using geodetic antennas. Similarly, Novatel GPS receivers were also installed in the cities of Piura and Cuzco. However, the receiver in Iquitos consists of a CRS1000 LEICA GPS that uses a choke ring antenna. Figure 1 indicates the geographic locations of the three sites, where GPS receivers operated between July 17—30, 2008 and July 30. As indicated in Figure 1, the distance between GPS receivers was between 3.9 and 4.8 km. They were distributed forming an isosceles

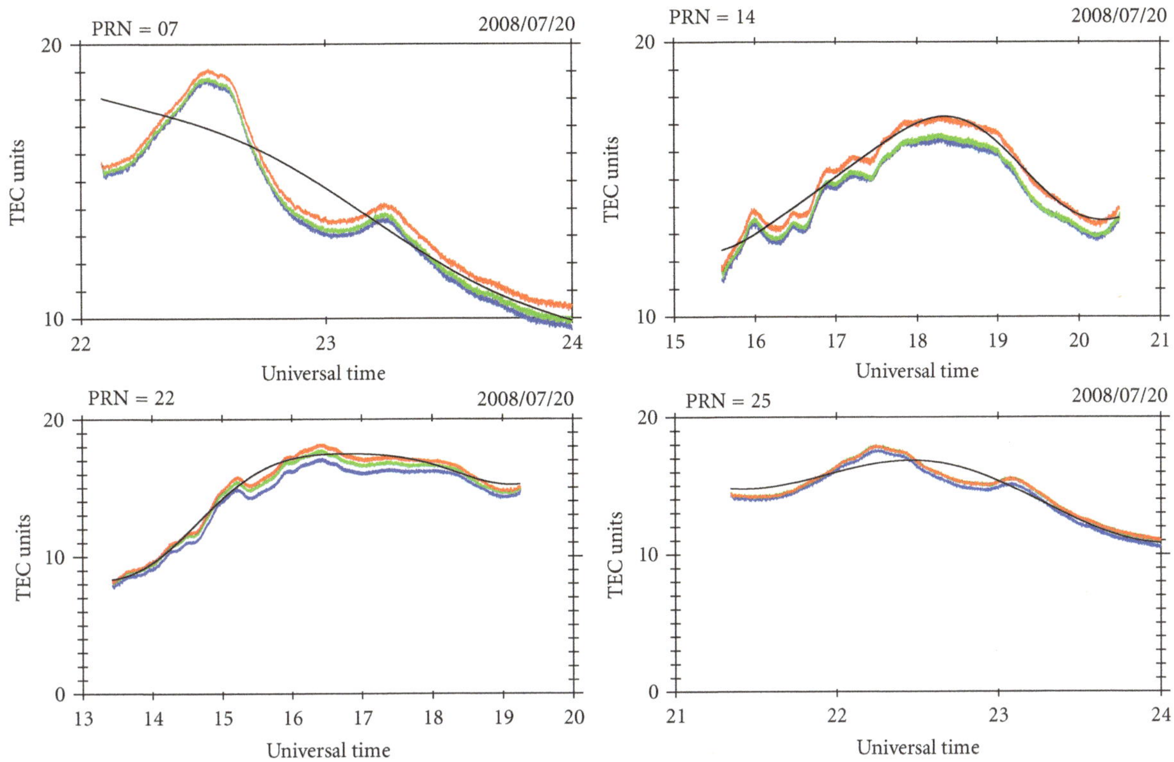

FIGURE 2: TEC versus UT measured by the Huancayo GPS receivers; red, green, and blue traces represent TEC values from the three receivers; black lines represent the estimated background/unperturbed TEC values. Each of the four panels represents signals from one of four GPS satellites (PRN = 07, 14, 22, and 25).

triangle to provide good estimates of the phase velocity as a function of both latitude and longitude. Due to the relatively short distance between receivers and the relatively large phase velocity of GWs that can vary between 100 and 500 m/s [31], we opted to sample the phase and group delay of both L1 and L2 frequencies once per second (1 Hz).

Each of the four frames of Figure 2 present TEC values measured by all 3 GPS receivers that operated near Huancayo on July 20, 2008. The TEC values from different receivers have been color coded to differentiate them and to indicate their similarity. The TEC values were computed following the standard procedure that involves using the differential phase and pseudorange values of both L1 and L2 frequency signals [32] and leveling the differential phase to the average differential code values to eliminate phase ambiguity. The absolute values of TEC were then calculated by introducing the differential satellite biases published by the University of Bern and calculating the receiver bias by a minimization process of the TEC variability between 0200 and 0600 local time. The thin black line, displayed in each of the 4 frames of Figure 2, indicates an estimate of the background or unperturbed TEC values. This estimate takes into account the typical daily variability of the TEC values that is a function of the local time variability produced by solar EUV radiation, changes in tides, presence of external electric fields, and more importantly the TEC variability introduced by the motion of the GPS satellites that continuously samples different latitudes and longitudes. These background TEC values are used to estimate the TEC perturbation, or in other words, derive the contribution produced by the passage of the gravity waves. The TEC curves corresponding to each satellite pass were fitted to a 4th-order-polynomial to filter out the diurnal variability and to remove spatial changes in TEC due to the GPS satellite motion and to evaluate TEC perturbations (TECP) containing time scales of 3 hours or less. Similar analysis was conducted by Valladares et al. [24] to estimate the TEC perturbation attributed to the passage of large-scale gravity waves. Equatorial plasma bubbles can also produce significant perturbations in the TEC traces. However, their effect consists of rapid and deep negative excursions of the TEC traces, commonly called TEC depletions. To avoid any possible erroneous identification, the GW/TID campaign was carried out during a period of minimum bubble activity, in which no TEC depletions were observed. As shown in the following figures, all the analysis methods are based on the difference between the measured and the estimated background "unperturbed" TEC values. This differential value is called TEC perturbations (TECP).

Figure 3 shows TECP values measured on 3 different days. The top panel, corresponding to July 17, 2008, displays perturbations that reach 1 TEC unit between 20 and 22 UT. Smaller amplitude TEC perturbations (±0.4 TEC units) are observed between 13 and 19 UT. The noise level in this and other panels is about ±0.2 TEC units. The TEC perturbations of the center panel (July 20, 2008) exceed 1 TEC unit between 22 and 24 UT and exhibit moderate levels (±0.6 TEC units)

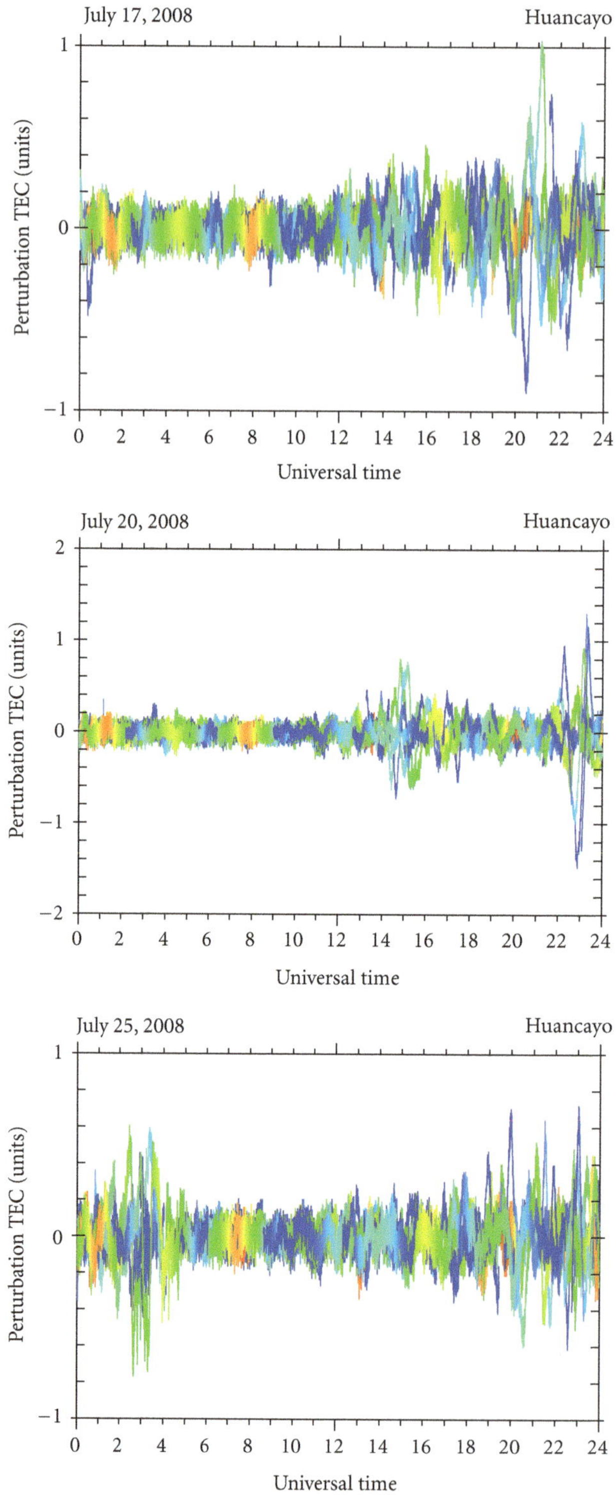

FIGURE 3: TECP versus UT at Huancayo for three days in July. Data from various GPS satellite signals are plotted; each satellite is distinguished by a different color.

of TEC perturbations between 14 and 16 UT. Note the different amplitude scale that is used in this panel. The bottom panel corresponding to July 15, 2008 depicts two large patches with significant levels of TEC perturbations. The first occurs during early evening hours (00–04 UT) and the second in the afternoon hours (19–24 UT). In summary, during the campaign TEC perturbations are observed preferentially a few hours before sunset (24 UT). A second interval of TEC perturbations occurs at different times of the day.

3. Analysis of TID Characteristics Using SADM-GPS Algorithm

We have calculated the TIDs horizontal phase velocity $V_h(t)$ and the azimuthal propagation direction of the phase front $\alpha(t)$ using the statistical angle of arrival and doppler method for GPS interferometry (SADM-GPS) developed by Afraimovich et al. [28–30]. This method assumes that the ionosphere changes in a nondispersive manner due to the passage of the TID or a packet of TIDs. Under this assumption, Afraimovich et al. [28, 30] calculated the velocity and direction of propagation of mid-latitude TID packets during quiet ionospheric conditions. The SADM-GPS represents the TEC perturbations by the form: $I(x, y, t) = \delta(t)\sin(\Omega t - k_x x - k_y y + \varphi_0)$. This expression implies that a monochromatic or a largely dominant TID propagates through the field-of-view of the small receiver network. For events when more than one TID is propagating through the measured volume, this representation will be invalid, and a spectral characterization of the TIDs is more appropriate. Using the equations proposed by Afraimovich et al. [28], it can be demonstrated that the TID wave parameters can be calculated using:

$$\alpha(t) = \arctan\left(\frac{I'_x(t)}{I'_y(t)}\right) = \arctan\left(\frac{G_x}{G_y}\right), \quad (1)$$

$$u_x(t) = \frac{I'_t(t)}{I'_x(t)}, \quad (2)$$

$$u_y(t) = \frac{I'_t(t)}{I'_y(t)}, \quad (3)$$

$$u(t) = \frac{\left|u_x(t)u_y(t)\right|}{\left(u_x^2(t) + u_y^2(t)\right)^{1/2}}, \quad (4)$$

$$V_h(t) = u(t) + w_x(t)\sin\alpha(t) + w_y(t)\cos\alpha(t), \quad (5)$$

$$G_x = \frac{(Y_A(\mathrm{TECP}_B - \mathrm{TECP}_C) - Y_C(\mathrm{TECP}_B - \mathrm{TECP}_A))}{X_A Y_C - X_C Y_A}, \quad (6)$$

$$G_y = \frac{(X_C(\mathrm{TECP}_B - \mathrm{TECP}_A) - X_A(\mathrm{TECP}_B - \mathrm{TECP}_C))}{X_A Y_C - X_C Y_A}, \quad (7)$$

where $I'x(t)$, $I'y(t)$, and $I't(t)$ are the spatial and time derivatives of the TECP values. $u_x(t)$ and $u_y(t)$ are the propagation velocities of the phase front along the x-(east) and y-(north) axes in a frame of reference centered at the Huancayo station. $w_x(t)$ and $w_y(t)$ are the x and y projections of the sub-ionospheric intersection point velocity. Note that G_x and G_y symbols used in Figure 4 are equivalent to the $I'x(t)$ and $I'y(t)$ derivatives. As the velocity of the subionospheric intersection (80–100 m/s) is comparable to the speed of the GWs, it becomes imperative to subtract this effect from the estimated phase velocity $u(t)$. Equation (5) includes this correction in which the component of the satellite velocity in the direction perpendicular to the phase front is subtracted to determine the true horizontal velocity of the TID. The symbols X_A, Y_A, X_C, and Y_C are defined in Figure 1 and correspond to the coordinate distances of Sicaya (A) and Chupaca (C) in a Cartesian system. To eliminate noise fluctuations containing periods less than 5 min, a low-pass filter was applied to all TECP traces shown in subsequent figures.

Figure 4 presents the results of applying (1) through (7) to the TECP values measured on July 20, 2008 between 21 and 24 UT. This figure follows closely the format presented by Afraimovich et al. [28] in Figure 9. Panels (a) and (e) show the time derivative (G_t) of the TECP measured at Huancayo from GPS satellites 13 and 25, respectively. Panels (b) and (f) display the east-west (G_x) spatial derivative using a solid line, and the north-south (G_y) spatial derivatives using dots. Note that (6) (G_x) and (7) (G_y) are a function of the TECP values measured by all 3 GPS receivers belonging to the small array. Panels (c) and (g) present the angle of the wave front, $\alpha(t)$, that was calculated using (1). It is noted that this angle is measured clockwise with respect to geographic north. Panels (d) and (h) introduce the phase velocity of the TID calculated using (5). The large scatter of the $\alpha(t)$ and $V_h(t)$ values is caused by the nonmonochromatic nature of the TECP values. However, the average lines in these frames indicate our best estimate of the TID phase velocity that varies between 3° and 6°. The estimation of the horizontal velocity of the dominant TID falls between 107 and 143 m/s. Comparison of the G_t traces for satellites 13 and 25 (panels (a) and (e)) show largely different magnitudes; this is due to the different look angle direction that varies between these two GPS satellites. GPS satellite 13 passed 200 km away from Huancayo and GPS 25 went over Huancayo almost overhead. Mercier and Jacobson [33] concluded that the largest TECP values were found when the line of sight made a small angle with the phase front. The satellite trajectory of GPS 25 was almost parallel to the propagation direction of the phase front.

Figure 5 shows hourly histograms of the statistics of the phase front angle, $\alpha(t)$, and the horizontal velocity, $V_h(t)$, comprising all GPS satellites that passed near Huancayo between 21 and 24 UT. A total of 6 GPS satellites observed significant TEC perturbations associated with the TID. Each frame of Figure 5 contains a vertical arrow to point out the location of the statistical average. The propagation angle of the TIDs (left frames) indicates a consistent northward direction. The phase velocity of the TIDs varies between 60 and 120 m/s with the maximum value occurring between 22 and 23 UT.

4. Analysis of TID Characteristics Using Cross-Correlation Algorithm

This section describes the cross-correlation method (CCM) and presents our results of the TID phase velocity, its direction of propagation and scale size. The aim of the CCM

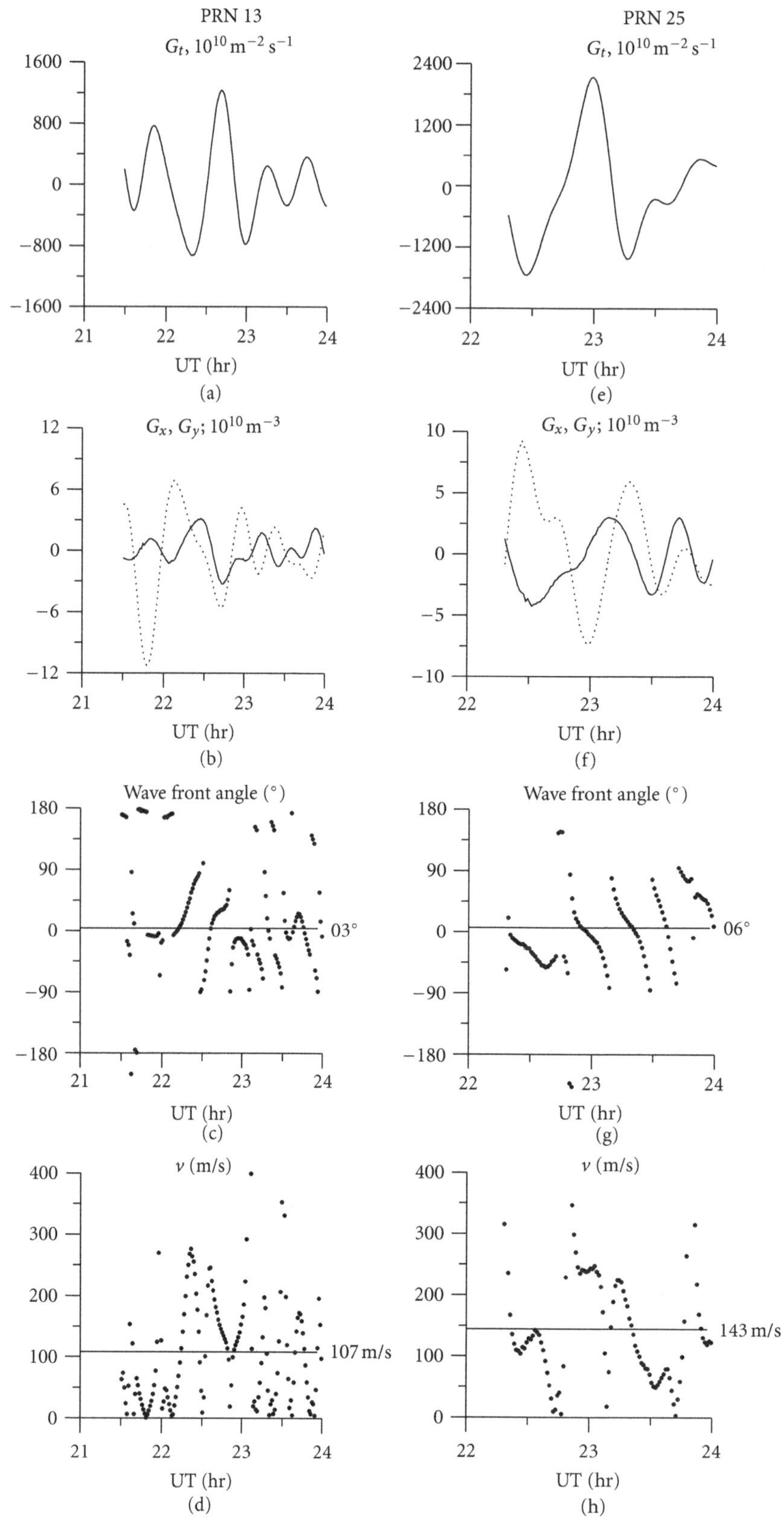

FIGURE 4: Derivation of different parameters used by the SADM-GPS method for GPS satellites 13 and 25. Panels (a) and (e) show G_t; (b) and (f) G_x solid curve and G_x, dots; panels (c) and (g) show the direction of propagation $\alpha(t)$; (d) and (h) present the horizontal velocity, $V_h(t)$.

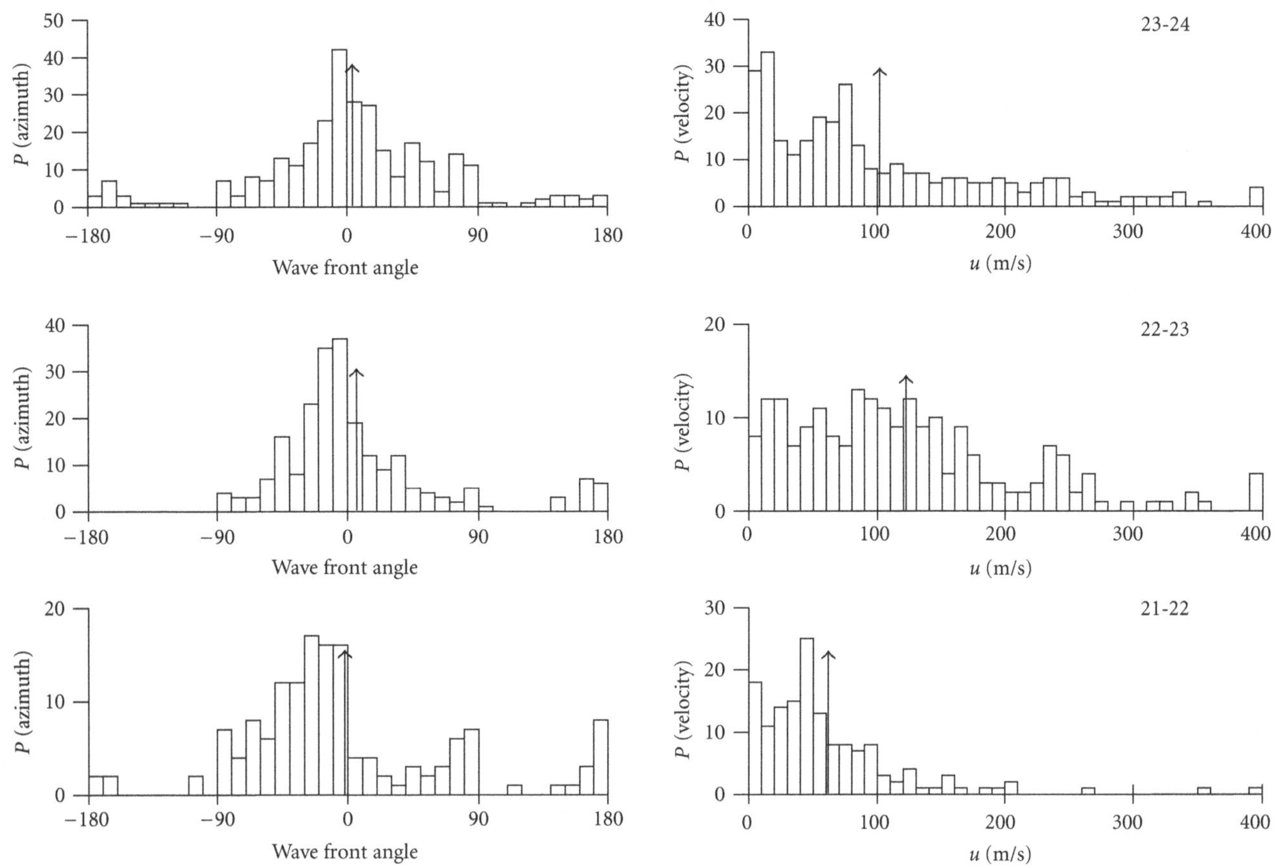

Figure 5: Distributions of the azimuth angle on the left and velocities on the right, corresponding to the calculated dependencies of $\alpha(t)$ and $V_h(t)$. The large arrows indicate the location of the mean value for each histogram.

method is to calculate the running cross-correlation function using TECP traces originated from any of the GPS satellites, but measured by two different receivers. The cross-correlation function is calculated every minute using segments of TECP data one hour long. We use TECP values from the same GPS satellite due to the motion of the satellite that produces the same temporal delay or advance on receivers placed at relatively close distances. This satellite motion effect can be removed after we calculate the phase velocity. If signals from different GPS satellites are used, it would be necessary to remove the effect of the satellite motion before the CCM method is applied. Figure 6 shows the results of the CCM method based on signals from the GPS 13 satellite recorded at Huancayo and Chupaca. The right frame of Figure 6 shows the same analysis using the TECP values measured at Huancayo and Sicaya. The lower panels show the TECP traces for GPS 13 in TEC units observed on July 20, 2008 and between 20 and 24 UT. The lower left frame displays the TECP values measured at Huancayo (blue line) and Chupaca (green line). Due to the coarse resolution of the plot, the small time delay between the traces becomes invisible. The cross-correlation functions, displayed in the upper frames, indicate that the time delay is of order 10 sec. Note the color scale of the cross-correlation functions varies between 0.9 and 1.0 due to the high degree of coherence of both signals. To obtain the scale size of the TECP, we fitted an ensemble of monochromatic sinusoids to each segment one-hour long. The numbers printed in the lower frame point out the scale size in minutes of the best fit for each hour of TECP data. These results are affected by the motion of the GPS satellites and need to be corrected by subtracting the apparent motion of the sub-ionospheric intersection in the direction perpendicular to the phase front. The propagation direction and velocity of the TID was obtained using:

$$\alpha(t) = \arctan\left(\frac{Y_C T_{B-A} - Y_A T_{B-C}}{X_C T_{B-A} - X_A T_{B-C}}\right), \quad (8)$$

$$V_h(t) = \frac{(Y_C \cos(\alpha(t)) - X_C \sin(\alpha(t)))}{T_{B-C}} + w_x(t)\sin\alpha(t) + w_y(t)\cos\alpha(t). \quad (9)$$

T_{B-A} and T_{B-C} are the time delay between the Huancayo and Sicaya TEC perturbations and the offset time between the Huancayo and Chupaca TECP traces. As in the previous section, $w_x(t)$ and $w_y(t)$ are the projections of the sub-ionospheric intersection point velocity. A negative offset, as seen on the upper left frame indicates that the TECP observed at Huancayo lags Chupaca's TECP. A positive delay, as seen in the upper right frame is an indication that Huancayo's

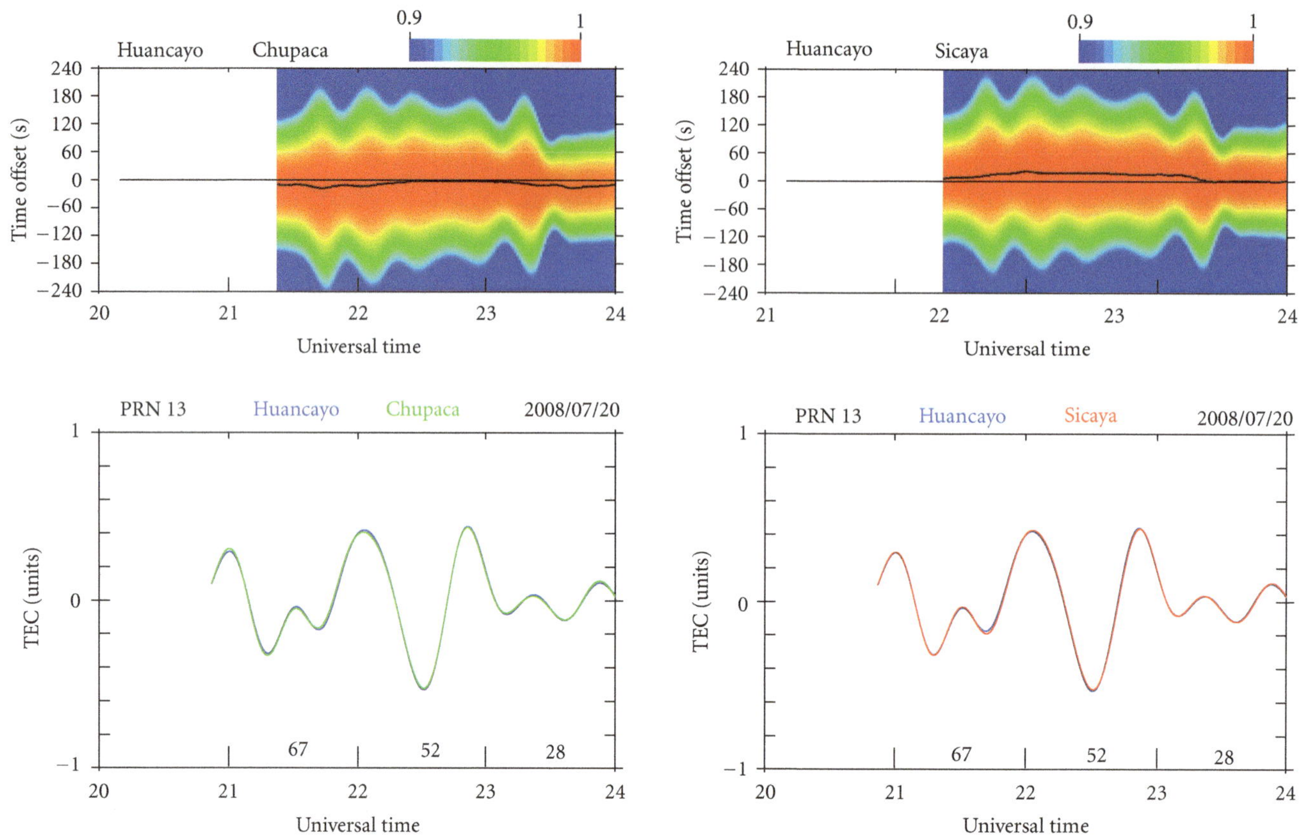

Figure 6: Cross-correlation functions (top panels) and TECP curves (lower frames) for GPS satellite 13. The left frames show the CCFs for TECP signals from Huancayo and Chupaca. The right panels correspond to Huancayo and Sicaya. The cross-correlation functions are plotted using a color scale that varies between 0.9 and 1.0. The black line that crosses near the middle corresponds to the maximum and it is taken as the time delay (T_{B-A} or T_{B-C}).

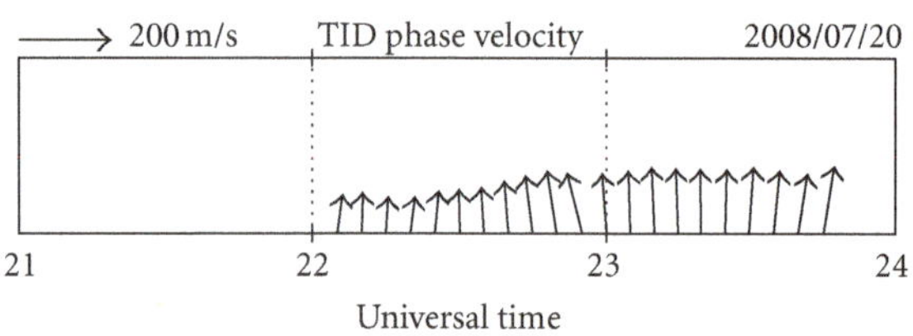

Figure 7: Phase velocity of the TIDs calculated using the CCM method and based on the data collected by the small array that operated near Huancayo.

TECP leads Sicaya's. Based on this timing relationship, it can be concluded that the perturbation associated with the TID was seen first at Chupaca, located further south, then at Huancayo placed near the center of the array, and later at Sicaya placed further north (Figure 1). This sequence of events indicates that the TID motion was in the northward direction.

A more quantitative estimation of the TID phase velocity was obtained using (8) and (9). Figure 7 shows the magnitude and direction of the TID phase velocity represented by arrows. The size of the arrow is proportional to the phase velocity and the azimuthal rotation of the vector is a function of the propagation angle. An arrow pointing to the top indicates a propagation direction toward geographic north. It is evident that between 22 and 24 UT, the TID moved toward geographic north. A wavelength equal to 449 km is obtained using the true phase velocity of the TID and the velocity of the satellite sub-ionospheric intersection point. The phase velocity varied between 84 m/s and 144 m/s near 24 UT. These values are consistent with the results of the SADM-GPS method (Section 3).

5. Analysis of the TID Characteristics Using GPS Receivers Spaced Hundreds of Kilometers

We also report our investigation of the TID morphological properties using GPS receivers that belong to the low-latitude ionospheric sensor network (LISN) that operate at different cities separated by hundreds of kms. LISN is a distributed observatory designed to nowcast the state and dynamics of the low-latitude ionosphere over South America. The LISN observatory is comprised of GPS receivers, flux-gate magnetometers and vertical incidence-pulsed ionospheric radar (VIPIR) ionosondes. A unique feature of the observatory is the ability of all the instruments to relay their observables in a near real-time manner via Internet. At the time of this

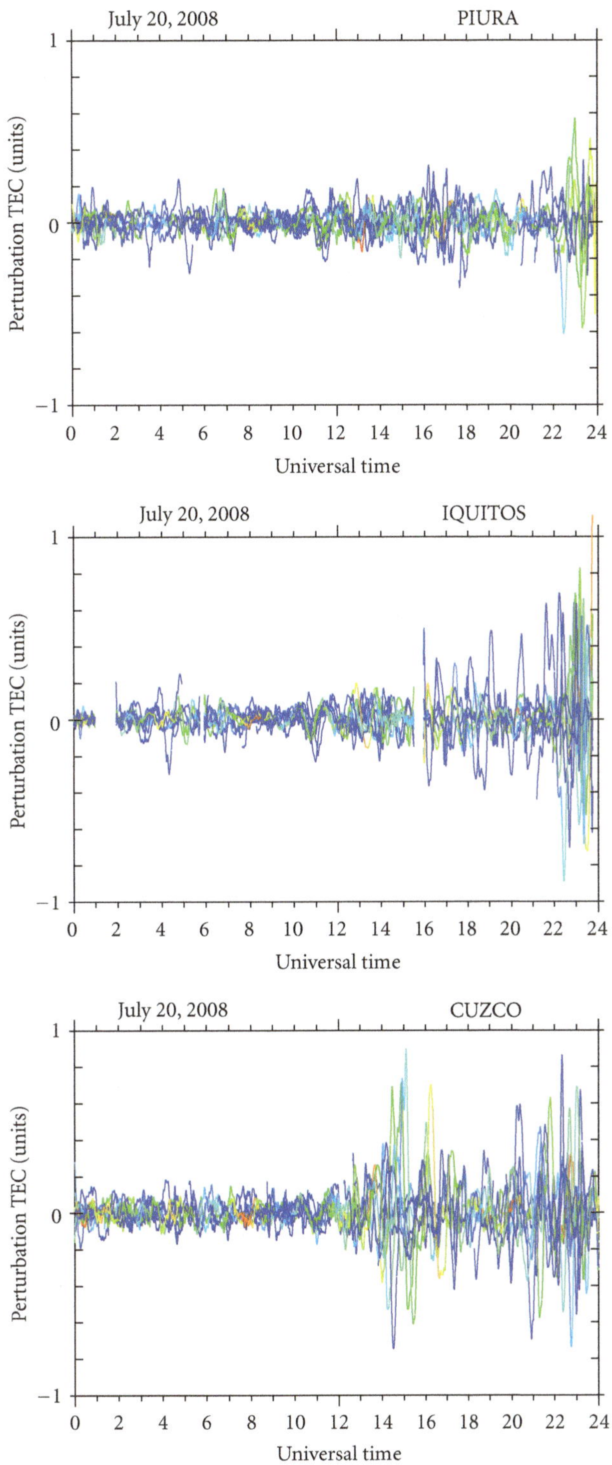

FIGURE 8: Same as Figure 3, but for three of the LISN stations. From top to bottom: Piura, Iquitos, and Cuzco. These TECP values were obtained on July 20, 2008.

paper (November 2011), 44 dual-frequency GPS receivers were part of the LISN network. During the time of the campaign a total of 126 GPS receivers—including the LISN GPS receivers—operated in the South American continent. TEC perturbations from these receivers, especially the ones located close to Huancayo, were used to provide supportive evidence for the TID dynamic characteristics and give a regional view of the presence of TIDs over South America.

Figure 8 shows the TECP values measured on July 20, 2008 at Piura (5.17°S; 80.63°W), Iquitos (3.77°S; 73.27°W) and Cuzco (13.52°S, 71.96°W). The TEC perturbations in Piura were seen mainly between 21 and 24 UT. Iquitos,

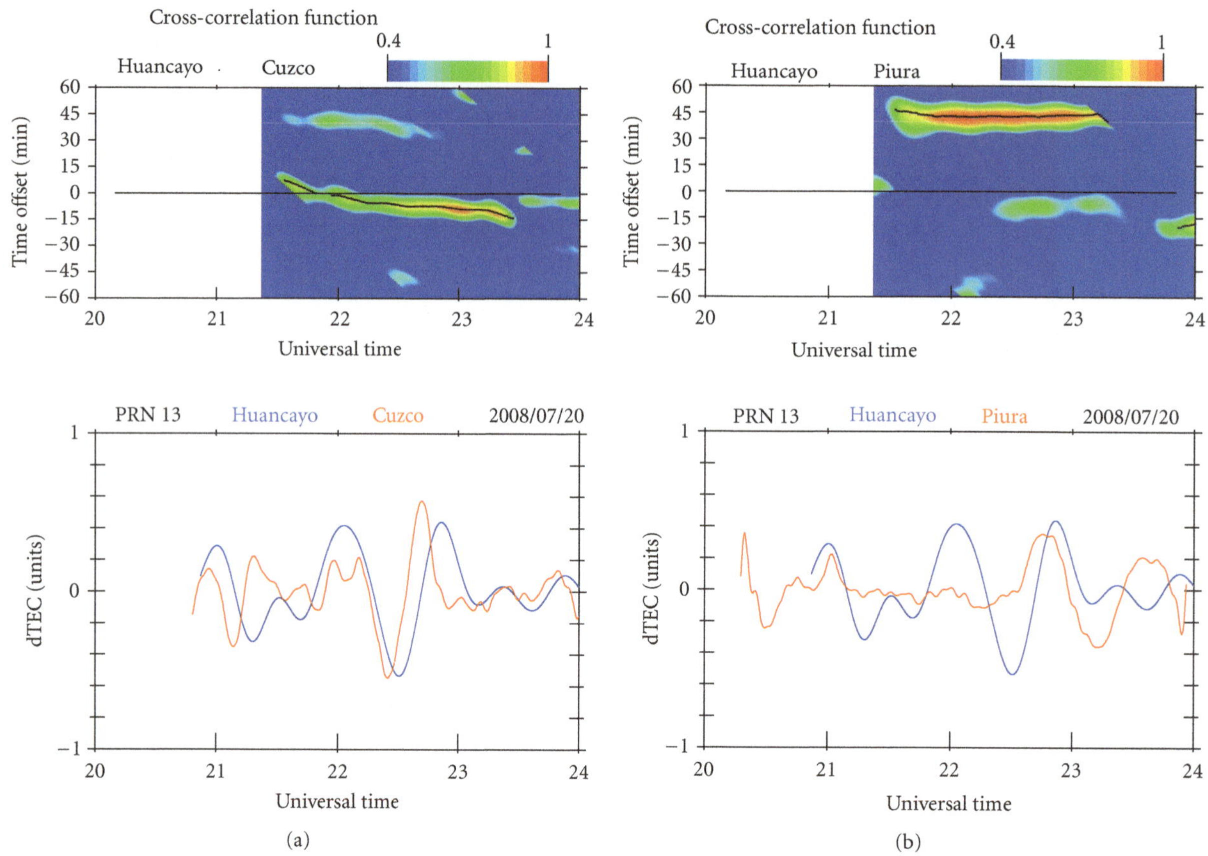

FIGURE 9: Same as Figure 6, but for CCF between Huancayo and Cuzco on the left side and Huancayo and Piura on the right side.

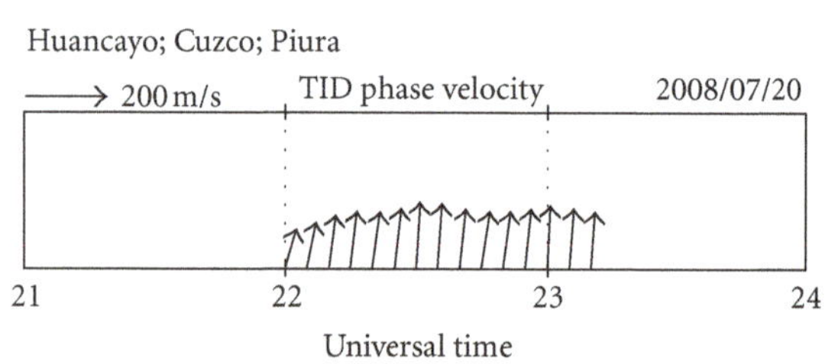

FIGURE 10: Same as Figure 7, but for TID's phase velocities that were calculated using TECP values from Huancayo, Cuzco, and Piura.

located in the north-eastern side of Peru, registered large TECP values between 16 and 24 UT. Cuzco, located close to the magnetic equator and 250 km east of Huancayo, observed large TEC values between 14 and 24 UT. These 3 stations show TECP values much larger than the noise level almost simultaneously. However, the TECP amplitude is different at different stations. Figure 8 suggests that between 21 and 24 UT, TIDs existed in the western region of South America. Below, we discuss further this finding.

The lower panels of Figure 9 display the TECP values for Huancayo (in blue), Cuzco (left panel in red), and Piura (right panel in red). The upper left frame presents the CCFs calculated using the Huancayo and Cuzco TECPs, showing a relatively short delay that varies between +10 and −15 min. Note that due to the large distances between the stations, the time delays between these station's TECP values are of order of tens of minutes. The right upper frame of Figure 9 displays CCFs for Huancayo and Piura. The time delay between the TECP values for these two stations is ~45 min. This value agrees quite well with the time offset that is observed in the lower panel. Visual inspection of the minimum value of the blue curve (Huancayo) observed at ~2230 UT and the minimum value of the red trace seen at ~2315 UT indicates that the Huancayo TECP values lead the Piura TECP values by 45 min. Note that CCFs in the upper panels also show secondary maximum with much smaller amplitudes (~0.6) and spaced roughly 50–55 min away from the primary maximum. In the case of a perfect sinusoidal oscillation, this spacing would be equal to the wavelength of the sinusoid. However, since MS-TIDs are more properly viewed as traveling wave packets [30], this spacing corresponds to the scale size of the TID packet. It is also pointed out that the CCF amplitude scale was limited to values above 0.4 and did not include negative values as they would indicate anticorrelations. If we had used a CCF amplitude scale varying between −1 and 1, a negative peak would have been observed between the primary and secondary maxima.

Figure 10 shows the TIDs phase velocity that was obtained using (8) and (9) and the time delays presented in Figure 9. The similarity of the angle $\alpha(t)$ that was derived based on measurements by GPS receivers separated by

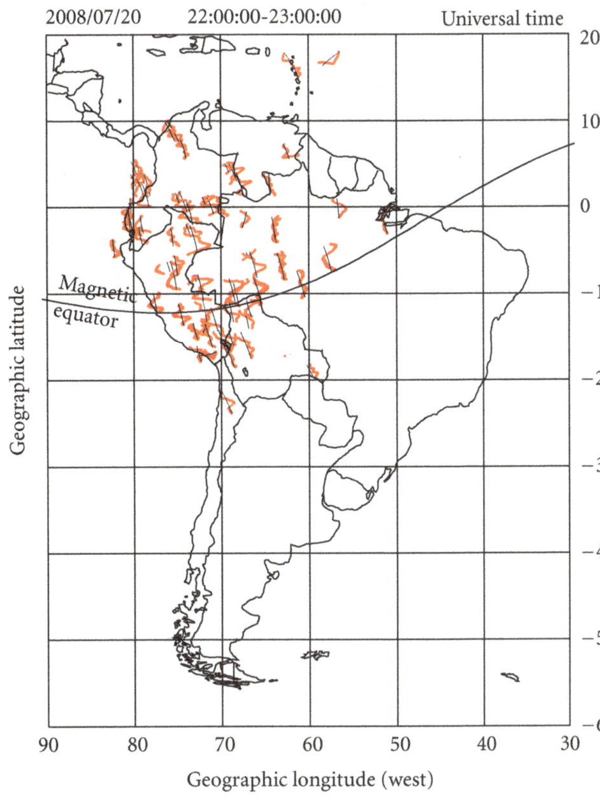

FIGURE 11: TEC perturbations recorded by several GPS sites that operated in South America on July 20, 2008. The red traces correspond to the TECP values after subtracting the daily variability.

hundreds of kms and the $\alpha(t)$ calculations based on GPS receivers that were installed only few kms apart is striking. Figures 7 and 10 show $\alpha(t)$ to be directed very closely in the geographic northward direction. However, the amplitude of the horizontal phase velocity $V_h(t)$, seen at 2230 UT, varies between 100 m/s in Figure 7 and 160 m/s in Figure 10. The difference in the $V_h(t)$ values is reduced at 2300 UT, when 135 m/s is shown in Figure 7 and 147 m/s in Figure 10.

The TEC perturbations for all 126 GPS receivers that operated in South America on July 20, 2008 are presented in Figure 11. These GPS receivers provide a good coverage over the whole South American continent, except for latitudes southward of 40°S. Figure 11 shows TECP values in red, centered along the satellite trajectory (thin black line), that have amplitudes higher than 0.4 TEC units. This baseline value is about 4 times the noise level of TEC fluctuations. The TECP traces of Figure 11 demonstrate that TIDs were restricted to the north-western side of South America between 2200 and 2300 UT. Similar Figures corresponding to earlier times, not shown here, display the TECP traces to be located further south. They indicate that the TIDs were propagating northward and were probably originated in the northern part of Chile (lat. = 20°S).

6. Discussion

We conducted a campaign at Huancayo in Peru to assess the effectiveness of GPS receivers in detecting and measuring the morphology and characteristics of TIDs that circulate at low latitudes. To accomplish this, we temporarily installed two additional GPS receivers near the Huancayo station to form 3 closely spaced GPS receivers having a geometry that can be used as a GPS radio-interferometer. The TIDs that were observed during the campaign can be classified as medium-scale disturbances (MSTIDs). They have horizontal phase velocities between 100 and 250 m/s, wavelengths of several hundred kilometers and periods between 15 and 60 min [34]. We also indicate that the TIDs observed on July 20, 2008 are associated with AGW, that are likely generated in the lower atmosphere.

To derive the TID characteristics we applied two different methods (SADM-GPS and CCM) and used two different datasets. The SADM-GPS is a statistical method that provides the correct phase velocity and the angle of propagation of the TIDs. However, it requires averaging the derived parameters for one hour and for all the GPS satellites with look angles above 35° elevation. The CCM method seems more robust, and it is less affected by the fact that more than one AGW may be circulating across the volume probed by the GPS receivers. A more detailed comparison of the results provided by the SADM-GPS and the CCM methods shows an excellent agreement in the angle of propagation, as both indicate the angle of propagation to be equal to +3° ± 5°. Calculation of the phase velocity with the SADM-GPS method indicates an average velocity of 120 m/s between 22 and 23 UT. The CCM method is more precise and indicates that the phase velocity varies between 84 and 130 m/s during the same interval.

We have also applied the CCM method to TECP values collected by other GPS receivers that belong to the LISN network and are operating at Piura, Cuzco, and Iquitos, all of them in Peru. We found very good correlation, in excess of 0.9, between the TECP traces corresponding to stations placed hundreds of kms away. We were able to calculate that the angle of propagation $\alpha(t)$ was close to 0° (northward propagation) and the horizontal phase velocity $V_h(t)$ was between 130 and 160 m/s. The latter values are larger than the numbers obtained using the small interferometer array of 3 GPS receivers. However, several geometrical considerations need to be accounted with more precision. One of them is the fact that the projection of the apparent satellite motion over the propagation direction of the TID may vary during the 45 min travel time between the LISN stations. It is also possible that the TID may not be a plane wave, and the wave front may have a circular shape. Figure 11 shows that the region populated by TIDs seems to extend over a large region. TECP images obtained prior to Figure 11 indicate that the TIDs originated near latitude =20°S. We do not know if they were produced by deep convection cells or were a response to orographic features in northern Chile. We believe that the success in getting good agreements is partially due to the scale size of the TIDs that was over 400 kms. This distance is larger than the separation between GPS sites. Further studies will determine the preferred scale sizes, the phase velocity, and the direction of propagation of the TIDs using all the GPS receivers that are operating in South America.

The GPS radio-interferometry (GPS-RI) method is one of the most effective techniques with which to observe GW/TID activity at low latitudes. TIDs are notoriously difficult to characterize due to their large size. However, GPS-RI enables characterization of their large-scale features with ease, including their horizontal wavelength, phase speed, and propagation azimuths. Since GPS-RI is based on radio-frequency measurements, the technique is not limited by weather considerations; data may be taken at all times, unlike with optical observations. GPS receivers are much less expensive to deploy than other instruments of similar or lesser sensitivity to TIDs, including phased-array radars, steerable-beam radar systems, and ionosondes. Moreover, an extensive global array of GPS receivers already exists, allowing for widespread, continuous monitoring of TID activity. Never before has this been possible.

Additional observations, when combined with GPS-RI observations, yield an even more complete TID characterization. In particular, measurements of the thermospheric neutral wind may be used to calculate TID intrinsic frequency by the relation: $\omega = K_H(V_h(t), -U_n)$, where ω is the intrinsic frequency, K_H is the TID horizontal wavenumber of the TID, and U_n is the neutral mean wind in the direction of TID propagation. Further, if the intrinsic frequency is known, TID vertical wavelength may be calculated by using one of the dispersion relations [35, 36]. Thus, a very complete picture of TIDs results with only GPS-RI and thermospheric wind observations: TID horizontal wavelength, vertical wavelength, intrinsic frequency, amplitude, and propagation direction are all specified. When GPS-RI receivers are co-located with other instrumentation, observed and calculated parameters may be cross-validated. Most importantly, TID horizontal wavelengths and propagation azimuths may be compared to GPS-RI results when favorable seeing conditions pertain, providing a crucial validation of this technique. Calculated values of intrinsic frequency and vertical wavelength may be compared with ionosonde-derived values.

Low-latitude TIDs are important to study because they are a key mechanism by which energy and momentum are deposited into the thermosphere. Observation of TIDs at altitudes near where they break would enable a direct investigation of this process for the first time. Additionally, TIDs are thought to be a key seed mechanism of plasma depletions. While evidence of TID seeding was found during the Spread-FEx campaigns [37], never before has it been possible to continuously monitor TID activity on the F-region bottomside during the crucial early-evening hours, when seeding occurs. An investigation of the presence/absence of TID activity on the bottomside will help determine the importance of TIDs relative to other mechanisms toward seeding equatorial plasma depletions.

Characterization of TID large-scale features should help establish which types of TIDs are most effective seeds. For example, is a minimum perturbation amplitude needed to seed depletions? TID wavelength and propagation azimuth, also characterized by GPS-RI arrays, are also thought to influence seeding effectiveness. Observations [38] suggest that TIDs must have a minimum horizontal wavelength to effectively seed depletions. Model results and seeding theory indicate that a TID must propagate nearly parallel to the geomagnetic equator to be maximally effective at seeding plasma depletions. GPS-RI makes it possible to directly investigate the impact of the variation of horizontal wavelength and propagation azimuth on plasma depletion formation.

7. Conclusions

We have presented calculations of the phase velocity, scale size, and angle of propagation of TIDs that are likely originated by gravity waves that are propagating in the northwestern side of the South American continent. The values that were obtained using the SADM-GPS and the CCM methods are consistent with values of medium-scale TIDs observed by other authors. We calculated that on July 20, 2008, the propagation velocity and the azimuth direction of the TIDs were 140 m/s and parallel to geographic north.

We also applied the CCM method to stations separated by hundreds of km, like Huancayo, Cuzco, and Piura. It was unexpected to find a large coherence between the TECP traces from all the adjacent stations, allowing us to determine the morphological characteristics of the TIDs. This result opens the possibility of using measurements of the TEC perturbations by the GPS receivers already deployed in South America to monitor TID activity. We have conducted two more campaigns using a small network of three receivers that were placed near a new ionosonde that operated at the Jicamarca radar site between 2009 and 2010. We expect to publish these results in the near future.

Acknowledgments

The authors would like to thank International GNSS service (IGS), Geocentric Reference System for the Americas (SIRGAS), and Dr. Michael Bevis from Ohio State University-Central and Southern Andes GPS Project (OSU-CAP) for providing GPS data. One of the authors, C. E. Valladares, was partially supported by Air Force Research Laboratory contract FA8718-09-C-0041, NSF Grant ATM-0521487, and NASA LWS Grant NNX11AP02G. The Low-Latitude Ionospheric Sensor Network (LISN) is a project led by Boston College in collaboration with the Geophysical Institute of Peru, and other institutions that provide information in benefit of the scientific community. They thank all organizations and persons that are supporting and operating receivers in LISN. They thank Robert Sheehan for his helpful comments and suggestions on the paper.

References

[1] W. H. Hooke, "Ionospheric irregularities produced by internal atmospheric gravity waves," *Journal of Atmospheric and Terrestrial Physics*, vol. 30, no. 5, pp. 795–823, 1968.

[2] T. M. Georges and W. H. Hooke, "Wave- induced fluctuations in ionospheric electron content. A model indicating some observational biases," *Journal of Geophysical Research*, vol. 75, no. 31, pp. 6295–6308, 1970.

[3] M. J. Davis, "The integrated ionospheric response to internal atmospheric gravity waves," *Journal of Atmospheric and Terrestrial Physics*, vol. 35, no. 5, pp. 929–959, 1973.

[4] M. C. Kelley and S. Fukao, "Turbulent upwelling of the mid-latitude ionosphere. II—theoretical framework," *Journal of Geophysical Research*, vol. 96, pp. 3747–3753, 1991.

[5] T. L. Beach, "Total electron content variations due to non-classical traveling ionospheric disturbances: theory and Global Positioning System observations," *Journal of Geophysical Research A*, vol. 102, no. 4, pp. 7279–7292, 1997.

[6] C. A. Miller, "Electrodynamics of midlatitude spread F 1. Observations of unstable, gravity wave-induced ionospheric electric fields at tropical latitudes," *Journal of Geophysical Research A*, vol. 102, no. 6, pp. 11521–11532, 1997.

[7] A. R. Jacobson, R. Carlos, R. Massey, and G. Wu, "Observations of traveling ionospheric disturbances with a satellite-beacon radio interferometer: seasonal and local time behavior," *Journal of Geophysical Research*, vol. 100, pp. 1653–1665, 1995.

[8] P. J. Sultan, "Linear theory and modeling of the Rayleigh-Taylor instability leading to occurrence of equatorial spread F," *Journal of Geophysical Research*, vol. 101, pp. 26819–26827, 1996.

[9] R. F. Woodman and C. La Hoz, "Radar observations of F region equatorial irregularities," *Journal of Geophysical Research*, vol. 81, no. 31, pp. 5447–5466, 1976.

[10] A. J. Scannapieco and S. L. Ossakow, "Nonlinear equatorial spread F evolution," *Geophysical Research Letters*, vol. 3, pp. 451–454, 1976.

[11] M. C. Kelley, M. F. Larsen, C. LaHoz, and J. P. McClure, "Gravity wave initiation of equatorial spread F: a case study," *Journal of Geophysical Research*, vol. 86, pp. 9087–9100, 1981.

[12] E. Kudeki, A. Akgiray, M. Milla, J. L. Chau, and D. L. Hysell, "Equatorial spread-F initiation: post-sunset vortex, thermospheric winds, gravity waves," *Journal of Atmospheric and Solar-Terrestrial Physics*, vol. 69, no. 17-18, pp. 2416–2427, 2007.

[13] R. T. Tsunoda, "Seeding of equatorial plasma bubbles with electric fields from an E_s-layer instability," *Journal of Geophysical Research A*, vol. 112, no. 6, Article ID A06304, 13 pages, 2007.

[14] S. Fukao, M. Kelley, and T. Shirakawa, "Turbulent upwelling of the mid-latitude ionosphere. I- Observational results by the MU radar," *Journal of Geophysical Research*, vol. 96, pp. 3725–3746, 1991.

[15] W. L. Oliver, S. Fukao, M. Sato, Y. Otsuka, T. Takami, and T. Tsuda, "Middle and upper atmosphere radar observations of the dispersion relation for ionospheric gravity waves," *Journal of Geophysical Research*, vol. 100, no. 12, pp. 23763–23768, 1995.

[16] M. K. Ejiri, K. Shiokawa, T. Ogawa, K. Igarashi, T. Nakamura, and T. Tsuda, "Statistical study of short-period gravity waves on OH and OI nightglow images at two separated sites," *Journal of Geophysical Research D*, vol. 108, no. 21, pp. 1–12, 2003.

[17] S. Suzuki, K. Shiokawa, Y. Otsuka, T. Ogawa, and P. Wilkinson, "Statistical characteristics of gravity waves observed by an all-sky imager at Darwin, Australia," *Journal of Geophysical Research D*, vol. 109, no. 20, pp. D20–DS07, 2004.

[18] A. A. Pimenta, M. C. Kelley, Y. Sahai, J. A. Bittencourt, and P. R. Fagundes, "Thermospheric dark band structures observed in all-sky OI 630 nm emission images over the Brazilian low-latitude sector," *Journal of Geophysical Research A*, vol. 113, no. 1, Article ID A01307, 9 pages, 2008.

[19] D. C. M. Amorim, A. A. Pimenta, J. A. Bittencourt, and P. R. Fagundes, "Long-term study of medium-scale traveling ionospheric disturbances using O i 630 nm all-sky imaging and ionosonde over Brazilian low latitudes," *Journal of Geophysical Research A*, vol. 116, no. 6, Article ID A06312, 7 pages, 2011.

[20] K. Shiokawa, C. Ihara, Y. Otsuka, and T. Ogawa, "Statistical study of nighttime medium-scale traveling ionospheric disturbances using midlatitude airglow images," *Journal of Geophysical Research A*, vol. 108, no. 1, article 1052, 2003.

[21] J. L. Bougeret, "Some effects produced by the ionosphere on radio interferometry—fluctuations in apparent source position and image distortion," *Astronomy & Astrophysics*, vol. 96, pp. 259–266, 1981.

[22] C. Mercier, "Observations of atmospheric gravity waves by radiointerferometry," *Journal of Atmospheric and Terrestrial Physics*, vol. 48, no. 7, pp. 605–624, 1986.

[23] A. R. Jacobson and W. C. Erickson, "A method for characterizing transient ionospheric disturbances using a large radio telescope array," *Astronomy & Astrophysics*, vol. 257, pp. 401–409, 1992.

[24] C. E. Valladares, J. Villalobos, M. A. Hei et al., "Simultaneous observation of traveling ionospheric disturbances in the Northern and Southern Hemispheres," *Annales Geophysicae*, vol. 27, no. 4, pp. 1501–1508, 2009.

[25] M. C. Kelley, *The Earth's Ionosphere*, Academic Press, London, UK, 1989.

[26] M. W. Kirkland and A. R. Jacobson, "Drift-parallax determination of the altitude of traveling ionospheric disturbances observed with the Los Alamos radio-beacon interferometer," *Radio Science*, vol. 33, no. 6, pp. 1807–1825, 1998.

[27] S. L. Vadas, "Horizontal and vertical propagation and dissipation of gravity waves in the thermosphere from lower atmospheric and thermospheric sources," *Journal of Geophysical Research A*, vol. 112, no. 6, Article ID A06305, 23 pages, 2007.

[28] E. L. Afraimovich, K. S. Palamartchouk, and N. P. Perevalova, "GPS radio interferometry of travelling ionospheric disturbances," *Journal of Atmospheric and Solar-Terrestrial Physics*, vol. 60, no. 12, pp. 1205–1223, 1998.
[29] E. L. Afraimovich, E. A. Kosogorov, L. A. Leonovich, K. S. Palamartchouk, N. P. Perevalova, and O. M. Pirog, "Determining parameters of large-scale traveling ionospheric disturbances of auroral origin using GPS-arrays," *Journal of Atmospheric and Solar-Terrestrial Physics*, vol. 62, no. 7, pp. 553–565, 2000.
[30] E. L. Afraimovich, N. P. Perevalova, and S. V. Voyeikov, "Traveling wave packets of total electron content disturbances as deduced from global GPS network data," *Journal of Atmospheric and Solar-Terrestrial Physics*, vol. 65, no. 11–13, pp. 1245–1262, 2003.
[31] D. C. Fritts, M. A. Abdu, B. R. Batista et al., "Overview and summary of the Spread F Experiment (SpreadFEx)," *Annales Geophysicae*, vol. 27, pp. 2141–2155, 2009.
[32] E. Sardón and N. Zarraoa, "Estimation of total electron content using GPS data: how stable are the differential satellite and receiver instrumental biases?" *Radio Science*, vol. 32, no. 5, pp. 1899–1910, 1997.
[33] C. Mercier and A. R. Jacobson, "Observations of atmospheric gravity waves by radio interferometry: are results biased by the observational technique?" *Annales Geophysicae*, vol. 15, no. 4, pp. 430–442, 1997.
[34] K. Hocke and K. Schlegel, "A review of atmospheric gravity waves and travelling ionospheric disturbances: 1982–1995," *Annales Geophysicae*, vol. 14, no. 9, pp. 917–940, 1996.
[35] J. K. Hargreaves, *The Solar-Terrestrial Environment: An Introduction to Geospace*, Cambridge University Press, Cambridge, UK, 1992.
[36] E. A. Kherani, M. A. Abdu, E. R. D. De Paula, D. C. Fritts, J. H. A. Sobral, and F. C. de Meneses Jr., "The impact of gravity waves rising from convection in the lower atmosphere on the generation and nonlinear evolution of equatorial bubble," *Annales Geophysicae*, vol. 27, no. 4, pp. 1657–1668, 2009.
[37] D. C. Fritts, S. L. Vadas, D. M. Riggin et al., "Gravity wave and tidal influences on equatorial spread F based on observations during the Spread F Experiment (SpreadFEx)," *Annales Geophysicae*, vol. 26, no. 11, pp. 3235–3252, 2008.
[38] M. A. Hei, R. A. Heelis, and J. P. McClure, "Seasonal and longitudinal variation of large-scale topside equatorial plasma depletions," *Journal of Geophysical Research*, vol. 110, Article ID A12315, 13 pages, 2005.

2

Precursor-Like Anomalies prior to the 2008 Wenchuan Earthquake: A Critical-but-Constructive Review

Tengfei Ma and Zhongliang Wu

Institute of Geophysics, China Earthquake Administration, Beijing 100081, China

Correspondence should be addressed to Zhongliang Wu, wuzl@cea-igp.ac.cn

Academic Editor: Rodolfo Console

Results published since the last three years on the observations of the precursor-like anomalies before the May 12, 2008, Wenchuan, $M_S8.0$ earthquake are collected and analyzed. These retrospective case studies would have provided heuristic clues about the preparation process of this inland great earthquake and the predictability of this destructive event if the standards for the rigorous test of earthquake forecast schemes were strictly observed. At least in some of these studies, however, several issues still need to be further examined to confirm or falsify the connection of the reported observations with the Wenchuan earthquake. Some of the problems are due to the inevitable limitation of observational infrastructure at the recent time, but some of the problems are due to the lack of communication about the test of earthquake forecast schemes. For the interdisciplinary studies on earthquake forecast, reminding of the latter issue seems of special importance for promoting the works and cooperation in this field.

1. Introduction

At least partly due to the tremendous loss of life and property and the intense social impact [1], the Wenchuan earthquake that occurred on May 12, 2008, in Sichuan Province of southwest China, has attracted widespread attention not only in seismological communities but also in other scientific communities. Since the occurrence of the Wenchuan earthquake, there have been some 300 papers published related to the precursor-like anomaly observations prior to this great earthquake, making it necessary for a systematic collection and comprehensive analysis of these materials. The necessity of such collection and analysis is further highlighted by the fact that over 3/4 of these publications are in Chinese with/without English abstract (with some of them being similar to, or just simply a repetition of the English publications) and quite a few of these publications are actually not known to international seismological communities. This is, similar to the situation of other developing countries [2], a characteristic of the scientific publications in China.

In the study of earthquake forecast/prediction, China seems of special features in that earthquake forecast/prediction has been kept for a long time as a nationwide scientific goal, even if there were intense debates on the predictability of earthquakes in the international seismological communities [3]. From merely the number of papers published, the Wenchuan earthquake might be one of the special, or even unique, events with so many studies on its forecast or pre-shock anomalies. If these anomalies could be confirmed, it would be an important event in the study on earthquake predictability; otherwise if these anomalies could not be confirmed, then it would be a useful sample for reminding of how the studies on earthquake forecast/prediction should be conducted in an efficient way. After the Wenchuan earthquake, there have been several studies trying to collect, compare, and analyze the (published and/or internal) data, including the data from the authors themselves (e.g., [4–11]). There are also some papers on the reflection of the earthquake forecast/prediction approaches based on the lessons of this earthquake [3, 12–15], with a diversity of ideas and data. The present paper is, to much extent, a continuation and extension of such collection and analysis works, but with different emphases in the analysis, highlighting the test of earthquake forecast/prediction schemes, and with a wider range of collection stressing interdisciplinary studies.

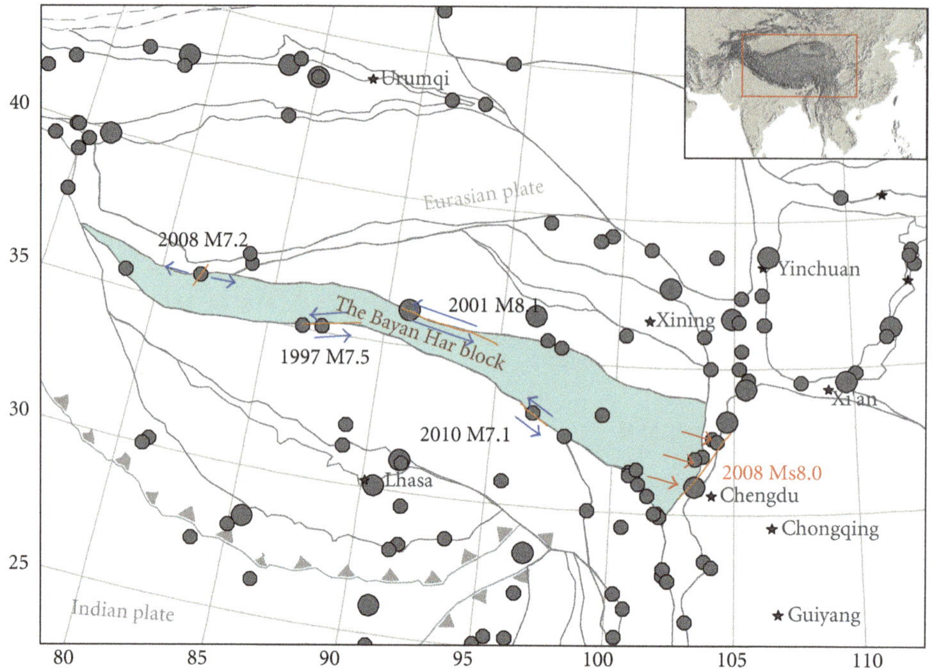

Figure 1: Distribution of major-to-great earthquakes around the Bayan Har block since 1997, with the 2008 Wenchuan earthquake located to the east, as shown by the text and arrows in red. To the top right is the indexing figure showing the position of the map. In the figure, orange lines show the earthquake ruptures, and blue arrows show the focal mechanism types of the earthquakes. Gray dots show the epicenters of historical major to great earthquakes. Gray solid lines indicate the boundaries of the tectonic blocks (according to [19]), with the Bayan Har block highlighted in the figure.

2. Materials Collected and Used for the Analysis

Although it has been only three years since the 2008 Wenchuan earthquake, because publications related to the precursor-like anomalies are within different research fields, the collection of such publications is still difficult to be completed. The collecting process was in two steps. For the materials published in English, searching terms "(ti = Wenchuan or ts = Wenchuan) and (ti = precursor or ts = precursor or ti = before or ts = before or ti = prior or ts = prior)" were used to the SCI-E database via the Web of Science (http://www.isiknowledge.com/). From 2008 to mid-2011, a total of 151 results hit the search. Removing the terms not directly related to the main topics, which are basically on the subjects of earthquake engineering and geological disasters, 61 articles were selected. For the materials published in Chinese, the China National Knowledge Infrastructure (CNKI) database (http://www.cnki.net/) was used with ("topic/title/keywords" = "(wenchuan) and (precursor or before or prior to)", in Chinese) as the search terms. Another two databases, the VIP database (http://www.cqvip.com/) and the Wanfang database (http://www.wanfangdata.com.cn/), were used as complementary sources. Using the screening criteria similar to that of the English publications, 261 articles in Chinese were collected. Abstracts of symposia [16–18] are not included, since most of the results were published after these meetings. Due to the same reason, diplomatic theses are not included in the References. We eliminated the "repeated publications" as much as we could. If there are two papers with exactly the same contents in English and in Chinese, respectively, we just keep the English one in the reference list. If there were two papers, with extremely similar contents and similar author teams, but published in different journals, then we just leave one of them (in the journal with higher impact factors) in the reference list. To keep the paper as concise as it could be, we also screened out the publications which are concentrating on the mechanisms of the phenomenology based on other studies rather than providing the ("fresh") phenomenological report.

Ranking by the number of papers published, these papers appear mainly in Journal of Geodesy and Geophysics (Wuhan, with the English edition Geodesy and Geophysics), Earthquake (Beijing), Acta Seismologica Sinica (the Bulletin of the Seismological Society of China, Beijing, with the English edition Acta Seismologica Sinica, changing to Earthquake Science since 2009), and Chinese Journal of Geophysics (the Bulletin of the Chinese Geophysical Society, Beijing, with the English edition Acta Geophysica Sinica or Chinese Journal of Geophysics), publishing no less than 15 papers in each of them. Journals publishing the related papers numbers up to 53, reflecting the diversity of the related results.

3. Characteristic Anomalies Reported and Characteristic Time Scales

As a background of the geology and seismicity related to the Wenchuan earthquake, Figure 1 shows the distribution and size of the major-to-great earthquakes around the Bayan Har block since 1997 (using the earthquake catalogue from the China Earthquake Networks Center (CENC)). During this time period, the Bayan Har block is the unique contributor to the major-to-great earthquake activity in continental China. Before this period, for more than 2 decades, there was no major-to-great earthquake occurring around the Bayan Har block. This provides the concepts of the geodynamic origin of the Wenchuan earthquake, and its preparation process.

Precursor-like anomalies, observed at different time scales prior to the Wenchuan earthquake, are in a wide range. Following is a brief summary of the main observations.

3.1. Anomalous Seismicity. Decade-scale quiescence along the Longmenshan fault zone [20, 21]; variation of monthly number of earthquakes since 2000 along the Longmenshan fault zone [22]; six-and-half-year-scale gap of seismicity above M_L4.0, disrupted 1 year before the Wenchuan earthquake [23]; five-year-scale PI "hotspots" along the Longmenshan fault zone, and five-year-scale "accelerating seismic release (ASR)" [21, 24, 25]; five-year-scale preshock increasing activity of intermediate-depth earthquakes [26]; three-year scale quiescence of seismicity above M_L6.0 [10, 27]; three-year-scale large-range seismicity pattern [28]; one-to-three-year-scale variation of "load-unload response ratio (LURR)" presented by seismicity [29, 30]; variation of "modulated earthquakes" 2 years before the Wenchuan earthquake [31]; two-year-scale variation of the homogeneity in seismicity [32]; variation of several statistical parameters of seismicity since the beginning of 2008 [33, 34]; half-year-scale seismic activation identified by PI method [35]. Besides, there are also studies on the potential tidal triggering effect which determines the origin time of the earthquake [36], and relation between Earth rotation and microseismicity [37].

3.2. Anomalies in Deformation Measurement. Decade-scale "locking" along the Longmenshan fault [38–40]; tilt variation from 2005 to 2006 [41]; accelerating fault activity since 2006 [42]; three-to-one-year-scale accelerating deformation [43]; "Oscillation anomalies" of GPS time series since 2007 [44]; one-year scale GPS baseline variation [45]; anomalous changes near the epicenter recorded by tiltmeter, since November 2007 [5]; anomalies of deformation (with resolution 3 months) before the Wenchuan earthquake identified retrospectively by wavelet analysis [46]; anomalies of tilt tidal factor in Shaanxi 3 months prior to the earthquake [47]; half-month-scale anomalous tilt [48]; and anomalies of deformation 3 days and 1 hour before the Wenchuan earthquake [40].

3.3. Anomalies in Strain/Stress Measurements. Five-year-scale perturbation of regional stress field before the Wenchuan earthquake by focal mechanism data [49]; increasing compressional strain since 2004 [50]; two-year-scale increase of regional stress [51]; two-year-scale micro-earthquake swarm, with focal mechanisms approaching to homogeneous [52]; months-to-year-scale disturbance in borehole strain measurement [53] and change of predominant focal-mechanisms of small earthquakes [54]; strain anomalies 3 months before the earthquake, with dominant frequencies depending on the epicentral distance [55]; changes of crustal stress since the end of April, 2008 [11]; week-scale variation of in-situ stress [56]; anomalous variation in in-situ stress measurement 48, 30, 8 hours and 37 minutes before the Wenchuan earthquake [57]; half-an-hour-scale abrupt anomaly recorded by strainmeter near to the epicenter [5, 58].

3.4. Possible Structure Variation. Four-year-scale preshock variation of seismic wave velocity [59]; three-year-scale [10, 60], one-year-scale [61], and two-to-one-month-scale variation of Earth resistivity [60, 61]; two-month-scale step-like resistivity anomalies [62]; one-month-scale increase of gas well pressures in a gas-field in Sichuan [63]; variation of noise correlation function (NCF) five days before the Wenchuan earthquake near the Longmenshan fault zone [64].

3.5. Anomalous Signals Observed in Broadband Seismic Recordings and Gravity Recordings. Decade-scale variation of gravity before the Wenchuan earthquake [65–68]; anomalous signals (tremors?) in broadband seismic recordings and gravity recordings, starting from about May 9~10, 2008 [69–75].

3.6. Geomagnetic Anomalies. Geomagnetic anomalies revealed by fractional Brownian motion (fBM) analysis 2 to 3 months before the earthquake [76] and electromagnetic anomalies 2~1 months before the earthquake [77]; and anomalies 3 days before the Wenchuan earthquake, within a large range surrounding the epicenter [78, 79].

3.7. Ionospheric Anomalies. Ionospheric anomalies 13, 6, 5 days [80], 6 days [11, 81–83], 6~7 days [84–86], 5 days [87–89], 8~4 days [90–93], 4 days [94], and 2~3 days [81–83, 88, 94–118] before the Wenchuan earthquake.

3.8. Geothermal and Atmospheric Anomalies. Extreme meteorological condition 21 months and 10 months before the Wenchuan earthquake [119]; temperature variation near the epicenter since November 2007 [5]; half-year-scale decrease of precipitation [120]; temperature variation since January 2008 [121]; large-scale satellite infrared thermal anomaly, appeared since March 2008 [122]; infrared radiation anomalies about 2 months before the Wenchuan earthquake [123–125]; anomalies of outgoing long-wave radiation 40 days before the Wenchuan earthquake [126]; one-month-scale decrease of the SNR of the VLF radio signal detected by satellite [127]; higher temperature in Sichuan in May, 2008, comparing to the last 30 years [128]; abnormal infrasonic waves received 10 days before the Wenchuan earthquake [11]; abnormal increase of temperature since May 5 [129]; abnormal surface latent heat flux 7 days before the earthquake [130]; abnormal variation in thermosphere 3 days before the Wenchuan earthquake [131]; "earthquake cloud" 5 hours before the earthquake [132].

Additionally, preshock macroanomalies [133–137], including animal behaviors [138, 139] and vegetation degeneration [140], are also collected and analyzed. This can act as

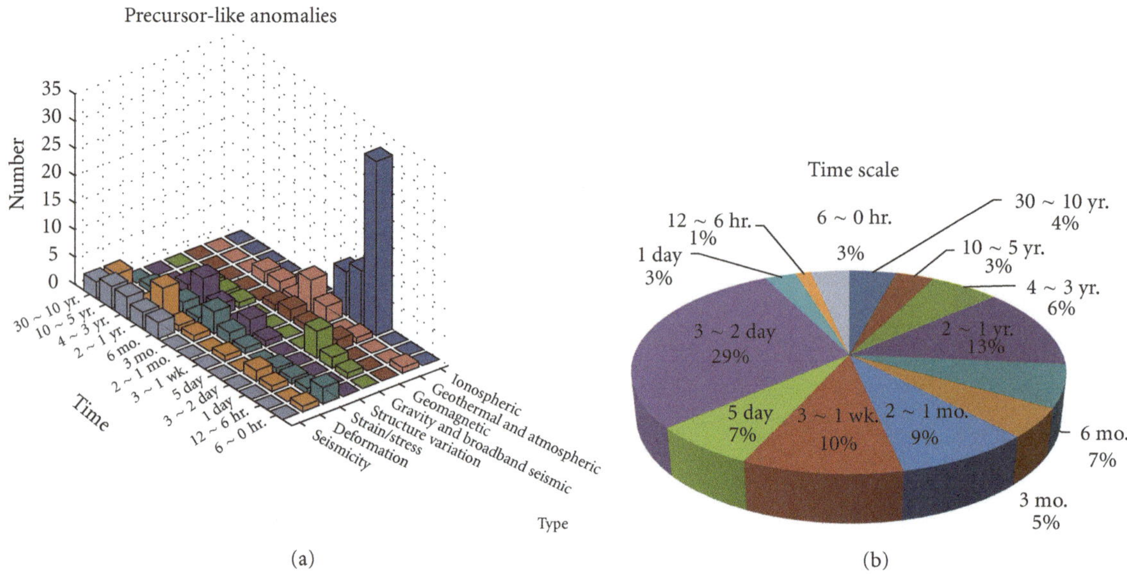

Figure 2: (a) Temporal distribution of the appearance of different types of reported anomalies, as summarized in Section 3. (b) Temporal distribution of the appearance of all the reported anomalies, as summarized in Section 3.

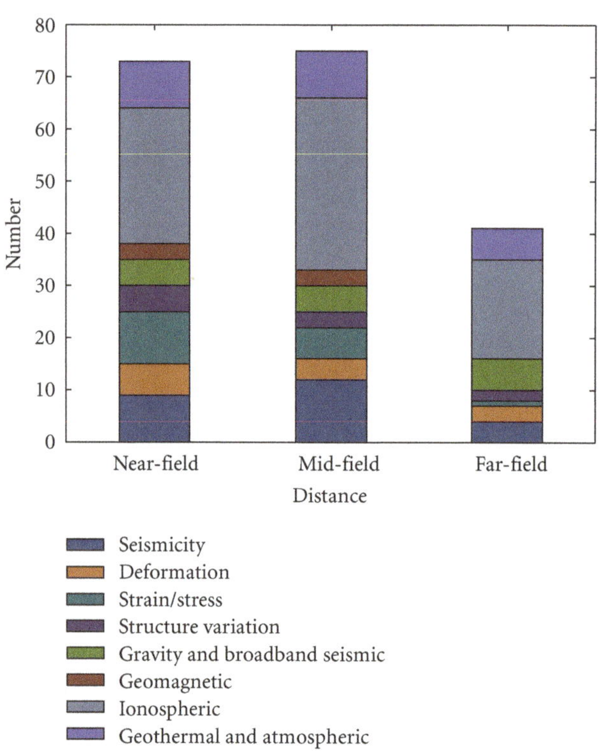

Figure 3: Distance ranges of the reported anomalies, as summarized in Section 3.

a useful reference, but how to confirm or falsify these (almost unrepeatable) observations needs further careful consideration.

At present, it is still hard to draw any definite conclusion on whether the above-mentioned observations, some are independent of others, while some are intercorrelated, may lead to a "unified description" of the earthquake preparation process. Some of the works are not understandable to us at present time regarding their methodology or observational phenomenology (e.g., [141–143]). However, it turns out that some characteristic times seem relatively important. Figure 2 shows, based on the analysis in this section, the "anomaly times" reported. It can be seen from the figure that anomalies detected by seismic, deformation, strain/stress, structure variation, gravity and broadband seismic recordings, geomagnetic, geothermal and atmospheric, and ionosphere observations appeared successively approaching the earthquake. The presently accessible data prevents from a detailed analysis of anomaly-distance dependence. As a conceptual picture, Figure 3 shows the statistics of anomalies with different distance ranges, in which "near-field" means the locations near to the earthquake fault, "mid-field" means the region within the circle centered at the epicenter and radius 1,000 km, and "far-field" means the regions beyond this circle. For those cases with distributed anomalies (an "anomaly field", as detected by satellite TEC), we take the locations within the above three ranges, respectively, in the counting of observation reports. It can be seen that the majority of the reports are within the "near-field" and the "mid-field" regions, with cautions necessary that the sampling of observational sites is by no means homogeneous.

4. Problems in Need of Further Considerations in Future

Except a few studies [10, 24, 25, 35, 79, 144, 145], there are few discussions on the statistical significance of the correlation between the anomalies observed and the earthquake. This is to much extent a problem which needs to be considered seriously in future, because in the test of earthquake forecast/prediction schemes, statistical significance is one of the key factors in need of consideration. Without a rigorous statistical test, some of the arguments, such as the year-to-month-scale long-range low-temperature before strong earthquakes [146], seems questionable.

The role of a single station analysis is, to some extent, double sided: on one hand, there might be some "special" stations which are especially "sensitive" to some precursory anomalies, even if the anomalies are associated with remote earthquakes; on the other hand, it is hard to draw any definite conclusion only by the records of a single station, while other stations have almost no "reflections." Result of Cheng et al. [8] indicates that, despite the anomalies registered at some individual sites, the routine precursor monitoring networks had no significant anomalies recorded. In the publications, anomalies retrospectively reported at individual stations/sites, ranging by distances from the epicenter, include abnormal strain changes observed at Guza, Sichuan, about one year before the Wenchuan earthquake [147], with epicentral distance ~1.6°; increasing compressional stress in Wudu, Gansu, 7 months before the Wenchuan earthquake [148], and electronic anomalies in Longnan, Gansu [149]; water level variation in Guanzhong, Shaanxi [150]; deformation anomalies in Liujiaxia, Gansu [151]; water level and ground gas Hg anomalies in Zhouzhi, Shaanxi [152, 153]; ground fluid anomalies and electronic anomalies in Yunnan [154–157], with epicentral distance ~5.5°; abnormal variation of fluid temperature in Qinghai [158]; abnormal ground fluid variation at sites as far as Huangyuan, Ningxia [159]; anomalous pre-Wenchuan-earthquake deformation/strain/stress recorded in Shanxi [160–164], with epicentral distance ~10°; electromagnetic radiation anomaly in Gaobeidian and Ningjin, Hebei [165]; anomalous tilt in Tai'an, Shandong [166, 167]; anomalous deformation in Yixian, Hebei [168]; anomalous Earth resistivity from the end of April, 2008, in Qingdao, Shandong [169]; anomalous water level and water temperature variation 6 days before the Wenchuan earthquake, observed at Changli, Hebei [170]; anomalies of water temperature and Radon content in Ningbo, Zhejiang [171], with epicentral distance ~15.6°. Variation of cosmic rays deals with the stations from Yangbajing, Tibet [172] to Guangzhou, Beijing, Irkutsk, Nagoya, and Moscow [173], with epicentral distance up to ~51°. Because the Wenchuan earthquake is a great one, its preparation process may have an extremely large spatial scale. Therefore, we try not to be too skeptical about the reliability of these "remote" anomalies. However, when dealing with such a large spatial scale, and dealing with the situation that only a few sites have the anomalies, some statistical test is important.

For some of the observations, the anomalous/alarmed regions are so large that intuitive visual inspection is hard to provide correct judgments. In this case, statistical test is of special importance. An example is the ionospheric anomalies and the satellite-detected thermal anomalies as mentioned in the last section. Several days before the Wenchuan earthquake, abnormal TEC of ionosphere could be observed even in south China [174, 175], with distance about 11°, giving the idea of the size of the "warning region."

Length of data for "baseline comparison" is another concern when reading the related reports. In quite a few studies, the data for checking the "background variations" are only since 2007. This is to much extent an inevitable problem because several observational facilities are just at their beginning stage. However, this limitation prevents from getting concrete conclusions about the anomalies in the case that there is a lack of sufficient knowledge about the *normal* state. Evidently the continuous accumulation of observations is needed. Some of the papers mentioned objectively that there is still lack of the experiences of a great earthquake (e.g., [176]).

Coseismic changes, or changes before and after the earthquake, are presented by a few analysis (e.g., [39, 57, 81, 91, 127, 147, 177]). But generally, lack of analysis on the coseismic variation seems to be one of the problems for some of the investigations. This is also a problem in need of serious consideration, since in the study of the candidate precursors, coseismic variation may provide useful constraints on the mechanism of such precursors.

Based on the above discussions, we suggest that in future works, the following issues should be paid special attention to (1) Statistical evaluation of the correlation between the anomalies reported and the earthquake needs to be considered, semiquantitatively or quantitatively if possible; (2) distance from the observation station and the "target" earthquake has to be taken into serious consideration, especially, if the distance is too large, then theoretical concepts as per the size of earthquake preparation (e.g., [178]) have to be accounted for, and statistical consideration is needed for the large-scale anomalies; (3) information about the "normal" state, or the "baseline" variations, has to be accounted for in identifying the potential anomalies; (4) comparison of preseismic, coseismic, and postseismic changes would be of help to understand the earthquake preparation process as well as the characteristics of the anomalies.

The above-mentioned "problems," however, do not imply that the publications introduced in this paper are not acceptable. As a matter of fact, all these observations are contributions to the study of the predictability of this earthquake. Especially valuable, among the publications, there are papers debating on the causes of the observed variations [179–182]. Some papers provide "negative" results [144, 183–190] which are useful in excluding the misleading information. Some papers objectively report that some of the observational systems did not show significant anomalies [8, 191], or that some of the observation systems are shown to be unable to capture the precursors [192, 193]. There are reports stating that among the whole set of monitoring stations, the stations with anomalies only occupy a small portion (e.g., [194]). Some works also try to exclude the effect of other factors in identifying the anomalies [195, 196]. Some of the reports (e.g., [197–206]) just provided the observation (before, or before and after the earthquake) but were too prudent to reach any direct conclusion related to earthquake precursors. Analysis tools such as ROC test [25], RTL analysis [10], and RTP [207] were used to the forecast test. If applied to geomagnetic data (e.g., [155]), then more objective conclusions about the correlation between the variation or fluctuation of geomagnetic field and the earthquake could be obtained. The same need exists for the tidal data (e.g., [208]), fluid data [209], electromagnetic radiation data [210], or fault deformation data [211]. Even if for “traditional” seismicity analysis (e.g., [212, 213]), such statistical test would be of help. But generally, however, it is somehow “abnormal” that there have been not so many “alternative” explanations in such a field with so many complexities and controversies. Maybe time is a remedy to this problem.

Last but not least, very few discussions (e.g., [214, 215]) are concerning how to apply the knowledge from these retrospective case studies, such as the observed patterns of seismicity, practically to the decision-making approaches to “operational earthquake forecast” [216]. Complexity of the deformation-related precursors and fluid-related precursors has caused some attentions [217–220]. Considering the observation that earthquakes occur after the restore of some of the anomalies, such as LURR [29], the anomaly-based alarm-oriented forecast problem is shown to be more complicated. Based on gravity measurement, a forward intermediate-term forecast was made [221], but the forecast did not contribute to the reduction of earthquake disasters.

5. Discussion and Conclusions

Systematic collection and comprehensive analysis of the cases of earthquakes regarding the precursor-like anomalies have been an academic tradition in China. In China, *Earthquake Cases* series have been published (in Chinese with English abstract, by the Seismological Press in Beijing) regularly since the 1970s. Contemporary level of informatics allows search and analysis of different data flow including scientific publications themselves. In the case of Wenchuan, what can be seen is that different observations may have some intrinsic consistency to each other, providing heuristic clues to the preparation process of this great earthquake. Remarkably, several characteristic times, such as 2~4 days and 1~2 years, may reveal the preparation and approaching process of this inland great earthquake, which is in need of further investigation.

In the study on earthquake forecast/prediction, it is always much easier and much simpler to be critical or skeptical than to conduct concrete observations. Keeping this in mind, the objective of this paper is firstly to summarize what have been done either in China or in other places of the world; secondly to introduce these works, especially to non-Chinese-speaking communities; thirdly to avoid being too demanding or too skeptical in commenting on these works; and at last to propose, in a constructive way, several problems in need of consideration in future—some of them are not complicated but important. We believe that, if these problems were paid special attention to, then to much extent, more useful conclusions would be obtained, and the study on earthquake forecast/prediction would be “accelerated”. And this hint is important not only for China but also for other places all over the world.

Being only 3 years after the Wenchuan earthquake, at the present time this work is still far from the stage of systematic evaluation [222–224] and/or empirical semiquantitative analysis [225–228]. More concrete conclusions need more time, although we have had an apparently good start with pretty rich (but complicated) materials.

Acknowledgments

Thanks to Professor R. Console for invitation to this special issue and patience to tolerate the delay in submission. Discussion with Harsh Gupta, V. Kossobokov, Jie Liu, Xiangchu Yin, and Yongxian Zhang stimulated and improved this investigation. Changsheng Jiang, Hanshu Peng, and Jiancang Zhuang helped in the work. Thanks are due to the staffs of the National Library of China, Beijing, for valuable advice and assistance in accessing to the databases. This work is supported by the WFSD project.

References

[1] Y. Chen and D. C. Booth, *The Wenchuan Earthquake of 2008*, Science Press, Springer, Beijing, China, 2011.

[2] A. M. Cetto, *Scientific journal publishing in the developing world?* ICSU-COSTED Occasional Paper No.3, 1998.

[3] Q. F. Chen and K. L. Wang, “The 2008 Wenchuan earthquake and earthquake prediction in China,” *Bulletin of the Seismological Society of America*, vol. 100, no. 5, pp. 2840–2857, 2010.

[4] Y.-T. Che, C.-L. Liu, J.-Z. Yu, Z.-J. Guan, and J. Li, “Underground fluid anomaly and macro anomaly of Ms 8.0 Wenchuan earthquake and opinions about earthquake prediction,” *Seismology and Geology*, vol. 30, no. 4, pp. 828–838, 2008 (Chinese).

[5] Z. X. Ouyang, H. X. Zhang, Z. Z. Fu, B. Gou, and W. L. Jiang, “Abnormal phenomena recorded by several earthquake precursor observation instruments before the Ms8.0 Wenchuan, Sichuan earthquake,” *Acta Geologica Sinica*, vol. 83, no. 4, pp. 834–844, 2009.

[6] R. Yan, “Analysis of earthquake precursor monitoring data,” in *Research Report on the Wenchuan M8.0 Earthquake*, Monitoring and Forecasting Department of China Earthquake Administration, Ed., chapter 6, pp. 170–193, Seismological Press, Beijing, China, 2009 (Chinese).

[7] X.-M. Zhang, J.-H. Ding, X.-H. Shen et al., “Electromagnetic perturbations before Wenchuan M8 earthquake and stereo electromagnetic observation system,” *Chinese Journal of Radio Science*, vol. 24, no. 1, pp. 1–8, 2009 (Chinese).

[8] W. Z. Cheng, Z. J. Guan, Q. Su, X. Ruan, and Z. W. Zhang, “Precursory anomalies in Sichuan region before 2008 Wenchuan Ms8.0 earthquake and their statistical analysis,” *Acta Seismologica Sinica*, vol. 33, no. 3, pp. 304–318, 2011 (Chinese).

[9] Q. Huang, “Seismicity changes prior to the Ms8.0 Wenchuan earthquake in Sichuan, China,” *Geophysical Research Letters*, vol. 35, no. 23, Article ID L23308, 2008.

[10] Q. H. Huang, “Retrospective investigation of geophysical data possibly associated with the Ms8.0 Wenchuan earthquake in Sichuan, China,” *Journal of Asian Earth Sciences*, vol. 41, no. 4-5, pp. 421–427, 2010.

[11] C. Y. Liu, J. Y. Liu, W. S. Chen, J. Z. Li, Y. Q. Xia, and X. Y. Cui, “An integrated study of anomalies observed before four major earthquakes: 2004 Sumatra M9.3, 2006 Pingtung M7.0, 2007 Chuetsu Oki M6.8, and 2008 Wenchuan M8.0,” *Journal of Asian Earth Sciences*, vol. 41, no. 4-5, pp. 401–409, 2010.

[12] L. Wu, "Misconception analysis and suggestions for earthquake monitoring and predication," *Science & Technology Review*, vol. 26, no. 10, pp. 28–29, 2008 (Chinese).

[13] J. Liu, T. Guo, L. Yang, Y. Su, and G. Li, "Retrospection on the conclusions of earthquake tendency forecast before the Wenchuan Ms8.0 earthquake," *Earthquake Research in China*, vol. 23, no. 2, pp. 119–133, 2009.

[14] Z. C. Zhang and W. Zhang, "Some reflections on the 12 May, 2008 Wenchuan earthquake," *Earthquake*, vol. 29, no. 1, pp. 193–202, 2009 (Chinese).

[15] W. Zhao, "Pondering over the scientific thinking of earthquake prediction from the miss report of Wenchuan earthquake—re-discussing Li Siguang's earthquake prediction thought," *Engineering Science*, vol. 11, no. 6, pp. 4–15, 2009 (Chinese).

[16] The Seismological Society of China, Ed., "Proceedings of the 12th Assembly of the Seismological Society of China," *Recent Development in World Seismology*, no. 11, pp. 1–172, 2008 (Chinese).

[17] The Seismological Society of China, Ed., "Earthquake Prediction Development Forum Paper Abstract Special Collection," *Recent Development in World Seismology*, no. 4, pp. 1–103, 2009 (Chinese).

[18] The Seismological Society of China, Ed., "Abstracts from Cross-Strait Seminar on Wenchuan Earthquake," *Recent Development in World Seismology*, no. 6, pp. 1–24, 2010 (Chinese).

[19] P. Zhang, Q. Deng, G. Zhang et al., "Active tectonic blocks and strong earthquakes in the continent of China," *Science in China D*, vol. 46, supplement 2, pp. 13–24, 2003.

[20] L.-X. Gao, J.-L. Sun, and H. Zhang, "Moderate-to-strong earthquake quiescence is the most significant seismic anomaly before the Wenchuan 8.0 earthquake," *Earthquake*, vol. 30, no. 1, pp. 90–97, 2010 (Chinese).

[21] C. S. Jiang and Z. L. Wu, "Seismic moment release before the May 12, 2008, Wenchuan earthquake in Sichuan of southwest China," *Concurrency Computation Practice and Experience*, vol. 22, no. 12, pp. 1784–1795, 2010.

[22] X.-Z. Chen, X.-Y. Guo, and Y.-E. Li, "Distribution characteristics of the month-scale number for earthquakes occurred on the Longmenshan fault," *Earthquake*, vol. 30, no. 2, pp. 20–28, 2010 (Chinese).

[23] Y. Xue, J. Liu, S. Mei, and Z. Song, "Characteristics of seismic activity before the Ms8.0 Wenchuan earthquake," *Earthquake Science*, vol. 22, no. 5, pp. 519–529, 2009.

[24] C. Jiang, Z. Wu, H. Ma, and L. Zhou, "Sichuan-Yunnan versus Andaman-Sumatra: PI approach and retrospective forecast test," *Acta Seismologica Sinica*, vol. 31, no. 3, pp. 307–318, 2009 (Chinese).

[25] C. S. Jiang and Z. L. Wu, "PI forecast for the Sichuan-Yunnan region: retrospective test after the May 12, 2008, Wenchuan earthquake," *Pure and Applied Geophysics*, vol. 167, no. 6-7, pp. 751–761, 2010.

[26] L. Zou and X. Wang, "Study on the characteristics of the medium-depth seismicity before the Kunlun Mountain earthquake and the Wenchuan earthquake," *Seismological and Geomagnetic Observation and Research*, vol. 31, no. 5, pp. 37–44, 2010 (Chinese).

[27] Y. Su and J. Liu, "Quiescence anomalies of $M \geq 6.0$ earthquakes before the 2008 Wenchuan (M8.0) earthquake in Sichuan-Yunnan region," *Journal of Seismological Research*, vol. 33, no. 2, pp. 119–124, 2010 (Chinese).

[28] H.-B. Zhu, "The similar evolution of seismicity patterns of strong-moderate earthquakes before about Ms8 earthquakes of Qinghai-Tibet block," *Chinese Journal of Geophysics*, vol. 53, no. 7, pp. 1611–1621, 2010 (Chinese).

[29] X.-C. Yin, L.-P. Zhang, Y.-X. Zhang et al., "Large scale LURR anomaly before Wenchuan earthquake," *Earthquake*, vol. 29, no. 1, pp. 53–59, 2009 (Chinese).

[30] H. Yu, J. Cheng, and Y. Wan, "Load/unload response ratio and stress accumulation model before large earthquakes," *Acta Seismologica Sinica*, vol. 32, no. 5, pp. 517–528, 2010 (Chinese).

[31] C. Wang, J. Cao, H. Guo, L. Zhang, and N. Xue, "Short-term earthquake prediction in the region of Sichuan-Yunnan using method of modulated earthquake," *Earthquake Research in China*, vol. 26, no. 2, pp. 210–217, 2010 (Chinese).

[32] P. Lu, X. Zhao, Z. Cui, X. Liu, and Y. Duan, "Space-time evolution characteristics of seismic inhomogeneous degree before Wenchuan earthquake," *Journal of Institute of Disaster-Prevention Science and Technology*, vol. 10, no. 3, pp. 126–130, 2008 (Chinese).

[33] Z. Xie, Y. Zhu, X. Lei, and Z. Song, "Variation of multiple seismicity parameters used to investigate critical behavior of the Wenchuan earthquake preparation," *Acta Seismologica Sinica*, vol. 32, no. 6, pp. 659–669, 2010 (Chinese).

[34] G. Luo, M. Yang, H. Ma, and X. Xu, "Intermediate and short-trem anomalies of seismic activity energy field before the Wenchuan M8.0 earthquake," *Earthquake*, vol. 31, no. 3, pp. 135–142, 2011 (Chinese).

[35] H. C. Li and C. C. Chen, "Characteristics of long-term regional seismicity before the 2008 Wen-Chuan, China, earthquake using pattern informatics and genetic algorithms," *Natural Hazards and Earth System Sciences*, vol. 11, no. 3, pp. 1003–1009, 2011.

[36] S. Zhao, S. Xu, P. Wu, and L. Ma, "Earthquake occurrence correlated with the movement of the Sun and the Moon," *Science & Technology Review*, vol. 29, no. 13, pp. 18–23, 2011 (Chinese).

[37] H. Wang, X. Zhao, Y. Li, and X. Chen, "A study on the relationships between Earth rotation and the occurrences of several strong earthquakes," *Earthquake*, vol. 31, no. 2, pp. 33–41, 2011 (Chinese).

[38] F. Du, X.-Z. Wen, P.-Z. Zhang, and Q.-L. Wang, "Interseismic deformation across the Longmenshan fault zone before the 2008 M8.0 Wenchuan earthquake," *Chinese Journal of Geophysics*, vol. 52, no. 11, pp. 2729–2738, 2009 (Chinese).

[39] G. Gu, W. Wang, G. Meng, and Y. Xu, "Crustal movements before and after the Wenchuan earthquake as detected by GPS observations," *Geomatics and Information Science of Wuhan University*, vol. 34, no. 11, pp. 1336–1339, 2009 (Chinese).

[40] G. Gu, G. Meng, and Y. Fang, "Crustal movement in the earthquake area before and after 2008 Wenchuan earthquake as detected by precise single epoch positioning of GPS observations," *Acta Seismologica Sinica*, vol. 33, no. 3, pp. 319–326, 2011 (Chinese).

[41] C. Zhang, H. Shao, C. Shi, and J. Chen, "Research on surface variation characteristics of ground tilt in Longmenshan fault zone before Wenchuan Ms8.0 earthquake," *Journal of Geodesy and Geodynamics*, vol. 31, supplement, pp. 6–9, 2011 (Chinese).

[42] Q. Jiao, X. Yang, L. Xu, and B. Wang, "Preliminary study on motion characteristics of Longmenshan fault before and after Ms8.0 Wenchuan earthquake," *Journal of Geodesy and Geodynamics*, vol. 28, no. 4, pp. 7–11, 2008 (Chinese).

[43] L. Guo, L. Ta, D. Sun, and X. Du, "Vertical deformation characteristic of Longmenshan fault zone before Wenchuan

Ms8.0 earthquake," *South China Journal of Seismology*, vol. 29, no. 3, pp. 1–8, 2009 (Chinese).
[44] Y. Fang, Z. Jiang, and G. Gu, "Oscillation analysis of GPS horizontal time series before the Wenchuan earthquake," *Journal of Seismological Research*, vol. 33, no. 2, pp. 125–130, 2010 (Chinese).
[45] L. Guo, X. Hu, F. Zhang, C. Chen, and D. Sun, "Baseline variation of GPS continuous sites and current activity of Qinghai-Tibet block," *Journal of Geodesy and Geodynamics*, vol. 29, no. 4, pp. 10–14, 2009 (Chinese).
[46] Y. Zhang and Y. Wu, "Anomaly of fixed deformation data and explain before the 2008 Wenchuan earthquake," *Geomatics and Information Science of Wuhan University*, vol. 35, no. 1, pp. 25–29, 2010 (Chinese).
[47] C. Ke, Z. Li, and M. Dou, "Anomalous response of crust tilt tidal factor γ value in Shanxi before Wenchuan Ms8.0 earthquake," *Journal of Geodesy and Geodynamics*, vol. 28, no. 6, pp. 56–60, 2008 (Chinese).
[48] Y. Jing, H. Zhang, Y. Sun, H. Li, L. Fan, and Y. Xiong, "Abnormal changes of crustal deformation before Wenchuan 8.0 earthquake observed by BSQ model digital tilt meter," *Geological Journal of China Universities*, vol. 15, no. 3, pp. 358–364, 2009 (Chinese).
[49] W. Cheng, X. Ruan, H. Qiao, Z. Zhang, and J. Yong, "Research on the dynamic change of regional stress fields before the Ms8.0 Wenchuan earthquake," *Earthquake Research in China*, vol. 23, no. 3, pp. 244–256, 2009.
[50] Z. S. Jiang, Y. Wu, Y. Fang, P. Li, and W. Wang, "The dynamic characteristics of strain fields and crustal movement before the Wenchuan earthquake (Ms=8.0)," *Earthquake*, vol. 23, no. 3, pp. 257–265, 2009.
[51] G. Fan and Q. Jiao, "Analysis of fault activity characteristics in Sichuan-Yunnan area before Wenchuan Ms8.0 earthquake," *Journal of Geodesy and Geodynamics*, vol. 28, no. 6, pp. 27–30, 2008 (Chinese).
[52] T. Chen, "Seismic rate and small seismic swarm changes before the 2008 Wenchuan M8.0 earthquake," *Earthquake Research in Sichuan*, no. 3, pp. 23–29, 2010 (Chinese).
[53] J. Zhang and Q. Liu, "Processing and analysis of four-component borehole strain observations," *Journal of Geodesy and Geodynamics*, vol. 30, no. 6, pp. 6–9, 2010 (Chinese).
[54] G.-L. Diao, X.-W. Xu, Y. Chen et al., "The precursory significance of tectonic stress field transformation before the Wenchuan Mw7.9 earthquake and the Chi-Chi Mw7.6 earthquake," *Chinese Journal of Geophysics*, vol. 54, no. 1, pp. 128–136, 2011 (Chinese).
[55] Y. Zhang, Y. Wu, and P. Lv, "Characteristics of deformation anomaly obtained from tilt and strain observations before the Wenchuan Ms8.0 earthquake," *Acta Seismologica Sinica*, vol. 31, no. 2, pp. 152–159, 2009 (Chinese).
[56] Q.-L. Guo, C.-H. Wang, H.-S. Ma, and C.-G. Wang, "In-situ hydro-fracture stress measurement before and after the Wenchuan Ms8.0 earthquake of China," *Chinese Journal of Geophysics*, vol. 52, no. 5, pp. 1395–1401, 2009 (Chinese).
[57] H. Peng, X. M. Ma, and J. J. Jiang, "Process analysis of in-situ strain during the Ms8.0 Wenchuan earthquake —data from the stress monitoring station at Shandan," *Acta Geologica Sinica*, vol. 83, no. 4, pp. 754–766, 2009.
[58] A. Niu, L. Zhan, W. Yan, and P. Ji, "Borehole strain measurement and application to earthquake prediction in China," *Journal of Geodesy and Geodynamics*, vol. 31, no. 2, pp. 48–52, 2011 (Chinese).
[59] L. Wang, Y. Li, F. Li et al., "Temporal variation of Vp/Vs, Vp and Vs before and after 2008 Wenchuan earthquake," *Acta Seismologica Sinica*, vol. 33, no. 1, pp. 1–14, 2011 (Chinese).
[60] X. B. Du, "Two types of changes in apparent resistivity in earthquake prediction," *Science China Earth Sciences*, vol. 54, no. 1, pp. 145–156, 2011.
[61] W. Xiao and H. Guan, "Anomalous change features of resistivity before the Wenchuan M8.0 earthquake and other large earthquake in China," *Northwestern Seismological Journal*, vol. 31, no. 4, pp. 349–354, 2009 (Chinese).
[62] X.-M. Zhang, M. Li, and H.-P. Guan, "Anomaly analysis of earth resistivity observations before the Wenchuan earthquake," *Earthquake*, vol. 29, no. 1, pp. 108–115, 2009 (Chinese).
[63] M. Yang, D. Chen, Y. Huang, and H. Gong, "Abnormal gas well pressure of Zhongba gas field, Sichuan, before and after the Wenchuan earthquake," *Acta Seismologica Sinica*, vol. 33, no. 4, pp. 505–514, 2011 (Chinese).
[64] H. S. Peng, Z. L. Wu, and C. S. Jiang, "Pre-seismic changes of noise correlation function (NCF) before the Wenchuan earthquake?" *Concurrency Computation Practice and Experience*, vol. 22, no. 12, pp. 1774–1783, 2010.
[65] Y. Zhu, W. Liang, Y. Xu, S. Guo, and F. Liu, "Dynamic variation of gravity field before and after Wenchuan Ms8.0 earthquake," *Acta Seismologica Sinica*, vol. 32, no. 6, pp. 633–640, 2010 (Chinese).
[66] Y. Q. Zhu, F. B. Zhan, J. C. Zhou, W. F. Liang, and Y. M. Xu, "Gravity measurements and their variations before the 2008 Wenchuan earthquake," *Bulletin of the Seismological Society of America B*, vol. 100, no. 5, pp. 2815–2824, 2010.
[67] Y. Zhu, F. Liu, and S. Guo, "Temporal variation of gravity field before and after Wenchuan Ms8.0 earthquake," *Geodesy and Geodynamics*, vol. 2, no. 2, pp. 33–38, 2011.
[68] Y. Zhu, F. Liu, W. Liang, and Y. Xu, "Gravity variation associated with Wenchuan earthquake in western Sichuan," *Geodesy and Geodynamics*, vol. 2, no. 1, pp. 55–60, 2011.
[69] X. Hao, X. Hu, H. Xu et al., "Gravity disturbance before Wenchuan Ms8.0 earthquake," *Journal of Geodesy and Geodynamics*, vol. 28, no. 3, pp. 129–131, 2008 (Chinese).
[70] X. Hao and X. Hu, "Disturbance before the Wenchuan earthquake detected by broadband seismometer," *Progress in Geophysics*, vol. 23, no. 4, pp. 1332–1335, 2008 (Chinese).
[71] X. Hao and X. Hu, "Are "third microseisms" in anomalous tremor before the great Wenchuan earthquake?" *Progress in Geophysics*, vol. 24, no. 4, pp. 1213–1215, 2009 (Chinese).
[72] L. Yang, "Preliminary study on the tremors with special frequency recorded by seismograph before Wenchuan earthquake and its charaters," *Recent Developments in World Seismology*, no. 1, pp. 14–19, 2009 (Chinese).
[73] L. Yang, J. Wang, J. Feng, Y. Hu, J. Chen, and J. Yao, "Preliminary study and application on the tremors with lower frequency recorded by seismograph before the Wenchuan earthquake," *Earthquake Research in China*, vol. 25, no. 4, pp. 356–366, 2009 (Chinese).
[74] L. Yin and L. Yang, "Research on low-frequency wave of the broadband digital data and the information of short-imminent precursor before strong earthquakes," *Northwestern Seismological Journal*, vol. 32, no. 1, pp. 82–87, 2010 (Chinese).
[75] Y. Zhang, Y. Wu, Y. Lv, J. Hu, and P. Lu, "Anomalies observed by ultra broadband seismometer before Wenchuan earthquake," *Journal of Geodesy and Geodynamics*, vol. 31, no. 1, pp. 15–18, 2011 (Chinese).
[76] M. Li, J. Lu, X. Su, and Z. Feng, "A study on fractal Brownian motion of geomagnetic observations before large earthquakes," *Acta Seismologica Sinica*, vol. 31, no. 6, pp. 650–659, 2009 (Chinese).

[77] Y.-Y. Fan, X.-B. Du, J. Zlotnicki et al., "The electromagnetic phenomena before the Ms8.0 Wenchuan earthquake," *Chinese Journal of Geophysics*, vol. 53, no. 12, pp. 2887–2898, 2010 (Chinese).

[78] J. Hu, W. Liu, M. Guo, and H. Zheng, ""Double low-points" anomaly in daily variation of vertical component of geomagnetic field before the Ms8.0 Wenchuan earthquake," *Earthquake Science*, vol. 22, no. 5, pp. 539–543, 2009.

[79] W. Wang, J. Ding, S. Yu, and Y. Zhang, "Short-term and imminent geomagnetic anomalies of the Wenchuan Ms8.0 earthquake and exploration on earthquake forecast," *Earthquake Science*, vol. 22, no. 2, pp. 135–141, 2009.

[80] H. K. Jhuang, Y. Y. Ho, Y. Kakinami et al., "Seismo-ionospheric anomalies of the GPS-TEC appear before the 12 May 2008 magnitude 8.0 Wenchuan earthquake," *International Journal of Remote Sensing*, vol. 31, no. 13, pp. 3579–3587, 2010.

[81] X. Zhang, X. Hu, and C. Zhang, "Ionospheric response with Wenchuan big earthquake by occulted data," *GNSS World of China*, no. 5, pp. 1–5, 2008 (Chinese).

[82] J. Blecki, M. Parrot, and R. Wronowski, "Studies of the electromagnetic field variations in ELF frequency range registered by DEMETER over the Sichuan region prior to the 12 May 2008 earthquake," *International Journal of Remote Sensing*, vol. 31, no. 13, pp. 3615–3629, 2010.

[83] Y. Y. Zhou, Y. Wu, X. J. Qiao, F. Y. Zhu, and J. Yang, "Anomalous variations of ionospheric VTEC before Ms8.0 Wenchuan earthquake," *Chinese Journal of Geophysics*, vol. 53, no. 3, pp. 556–566, 2010, (with Chinese abstract).

[84] X.-M. Zhang, X.-H. Shen, X.-Y. Ouyang et al., "Ionosphere VLF electric field anomalies before Wenchuan M8 earthquake," *Chinese Journal of Radio Science*, vol. 24, no. 6, pp. 1024–1032, 2009 (Chinese).

[85] H. Yu, H. Zhou, and X. Qiao, "Study on wave propagation of ELF emission anomaly before Ms8.0 Wenchuan earthquake," *Acta Seismologica Sinica*, vol. 32, no. 6, pp. 641–648, 2010 (Chinese).

[86] Z. An, Y. Fan, J. Liu et al., "Analysis on ion temperature variation detected by DEMETER before 2008 Wenchuan Ms8.0 earthquake," *Acta Seismologica Sinica*, vol. 32, no. 6, pp. 754–759, 2010 (Chinese).

[87] C. C. Hsiao, J. Y. Liu, K. I. Oyama et al., "Seismo-ionospheric precursor of the 2008 Mw7.9 Wenchuan earthquake observed by FORMOSAT-3/COSMIC," *GPS Solutions*, vol. 14, no. 1, pp. 83–89, 2009.

[88] Z. Nie, F. Zhu, and N. Fu, "Application of Kalman filtering in detecting ionospheric TEC anomaly prior to earthquake," *Journal of Geodesy and Geodynamics*, vol. 31, no. 3, pp. 47–50, 2011 (Chinese).

[89] J. Xiong, Y. Wu, Y. Zhou, and J. Lin, "Ionospheric anomalies detected before 2008 Wenchuan earthquake," *Journal of Geodesy and Geodynamics*, vol. 31, no. 2, pp. 28–31, 2011 (Chinese).

[90] J. Lin, Y. Wu, and F. Zhu, "Ionosphere TEC anomalous disturbance of pre-seism," *Geomatics and Information Science of Wuhan University*, vol. 34, no. 8, pp. 975–978, 2009 (Chinese).

[91] S. Sarkar and A. K. Gwal, "Satellite monitoring of anomalous effects in the ionosphere related to the great Wenchuan earthquake of May 12, 2008," *Natural Hazards*, vol. 55, no. 2, pp. 321–332, 2010.

[92] J. Xiong, Y. Wu, F. Zhu, J. Lin, Y. Zhou, and J. Yang, "Anomalous disturbance of ionospheric NmF2 during Wenchuan earthquake," *Journal of Geodesy and Geodynamics*, vol. 28, no. 6, pp. 22–26, 2008 (Chinese).

[93] F. Zhu, Y. Wu, J. Lin, Y. Zhou, J. Xiong, and J. Yang, "Study on ionospheric TEC anomaly prior to Wenchuan Ms8.0 earthquake," *Journal of Geodesy and Geodynamics*, vol. 28, no. 6, pp. 16–21, 2008 (Chinese).

[94] J. W. Lin, "Two-dimensional ionospheric total electron content map (TEC) seismo-ionospheric anomalies through image processing using principal component analysis," *Advances in Space Research*, vol. 45, no. 11, pp. 1301–1310, 2010.

[95] B. Zhao, W. Wan, M. Wang, L. Liu, and B. Ning, "Recent advances on the ionospheric pre-cursors of earthquakes and ionospheric variations prior to Wenchuan earthquake," *Science & Technology Review*, vol. 26, no. 11, pp. 30–34, 2008 (Chinese).

[96] J. Li, G. Meng, M. Wang, H. Liao, and X. Shen, "Investigation of ionospheric TEC changes related to the 2008 Wenchuan earthquake based on statistic analysis and signal detection," *Earthquake Science*, vol. 22, no. 5, pp. 545–553, 2009.

[97] J. Y. Liu, Y. I. Chen, C. H. Chen et al., "Seismoionospheric GPS total electron content anomalies observed before the 12 May 2008 Mw7.9 Wenchuan earthquake," *Journal of Geophysical Research A*, vol. 114, no. 4, Article ID A04320, 2009.

[98] T. G. Yu, T. Mao, Y. G. Wang, and J. S. Wang, "Study of the ionospheric anomaly before the Wenchuan earthquake," *Chinese Science Bulletin*, vol. 54, no. 6, pp. 1080–1086, 2009.

[99] Z.-C. Zeng, B. Zhang, G.-Y. Fang, D.-F. Wang, and H.-J. Yin, "The analysis of ionospheric variations before Wenchuan earthquake with DEMETER data," *Chinese Journal of Geophysics*, vol. 52, no. 1, pp. 11–19, 2009 (Chinese).

[100] X. Zhang, X. Shen, J. Liu, X. Ouyang, J. Qian, and S. Zhao, "Analysis of ionospheric plasma perturbations before Wenchuan earthquake," *Natural Hazards and Earth System Science*, vol. 9, no. 4, pp. 1259–1266, 2009.

[101] Y. Y. Zhou, Y. Wu, X. J. Qiao, and X. X. Zhang, "Ionospheric anomalies detected by ground-based GPS before the Mw7.9 Wenchuan earthquake of May 12, 2008, China," *Journal of Atmospheric and Solar-Terrestrial Physics*, vol. 71, no. 8-9, pp. 959–966, 2009.

[102] F. Zhu, Y. Wu, J. Lin, Y. Zhou, J. Xiong, and J. Yang, "Study on method of detecting ionospheric TEC anomaly before earthquake," *Journal of Geodesy and Geodynamics*, vol. 29, no. 3, pp. 50–54, 2009 (Chinese).

[103] F. Zhu, Y. Wu, J. Lin, Y. Zhou, J. Xiong, and J. Yang, "Anomalous response of ionospheric VTEC before the Wenchuan earthquake," *Acta Seismologica Sinica*, vol. 31, no. 2, pp. 180–187, 2009 (Chinese).

[104] Z.-H. Ding, J. Wu, S.-J. Sun, J.-S. Chen, and P.-P. Ban, "The variation of ionosphere on some days before the Wenchuan earthquake," *Chinese Journal of Geophysics*, vol. 53, no. 1, pp. 30–38, 2010 (Chinese).

[105] Y. Kakinami, J. Y. Liu, L. C. Tsai, and K. I. Oyama, "Ionospheric electron content anomalies detected by a FORMOSAT-3/COSMIC empirical model before and after the Wenchuan earthquake," *International Journal of Remote Sensing*, vol. 31, no. 13, pp. 3571–3578, 2010.

[106] S. A. Pulinets, V. G. Bondur, M. N. Tsidilina, and M. V. Gaponova, "Verification of the concept of seismoionospheric coupling under quiet heliogeomagnetic conditions, using the Wenchuan (China) earthquake of May 12, 2008, as an example," *Geomagnetism and Aeronomy*, vol. 50, no. 2, pp. 231–242, 2010.

[107] T. Xu, Y. Hu, J. Wu, Z. Wu, Y. Suo, and J. Feng, "Giant disturbance in the ionospheric F2 region prior to the M8.0 Wenchuan earthquake on 12 May 2008," *Annales Geophysicae,*, vol. 28, no. 8, pp. 1533–1538, 2010.

[108] X. M. Zhang, X. H. Shen, J. Liu, X. Y. Ouyang, J. D. Qian, and S. F. Zhao, "Ionospheric perturbations of electron density before the Wenchuan earthquake," *International Journal of Remote Sensing*, vol. 31, no. 13, pp. 3559–3569, 2010.
[109] B. Q. Zhao, M. Wang, T. Yu, G. R. Xu, W. X. Wan, and L. B. Liu, "Ionospheric total electron content variations prior to the 2008 Wenchuan earthquake," *International Journal of Remote Sensing*, vol. 31, no. 13, pp. 3545–3557, 2010.
[110] Y. Zhao, X. Zhang, and J. Liu, "Perturbation analysis of the ionospheric TEC before and after the Wenchuan earthquake," *Progress in Geophysics*, vol. 25, no. 2, pp. 447–453, 2010 (Chinese).
[111] F. Zhu, Y. Wu, J. Lin, and Y. Zhou, "Temporal and spatial characteristics of VTEC anomalies before Wenchuan Ms8.0 earthquake," *Geodesy and Geodynamics*, vol. 1, no. 1, pp. 23–28, 2010.
[112] F. Zhu, Y. Wu, and N. Fu, "Application of Kalman filter in detecting pre-earthquake ionospheric TEC anomaly," *Geodesy and Geodynamics*, vol. 2, no. 2, pp. 43–47, 2011.
[113] M. V. Klimenko, V. V. Klimenko, I. E. Zakharenkova, S. A. Pulinets, B. Zhao, and M. N. Tsidilina, "Formation mechanism of great positive TEC disturbances prior to Wenchuan earthquake on May 12, 2008," *Advances in Space Research*, vol. 48, no. 3, pp. 488–499, 2011.
[114] J. W. Lin, "Use of principal component analysis in the identification of the spatial pattern of an ionospheric total electron content anomalies after China's May 12, 2008, M = 7.9 Wenchuan earthquake," *Advances in Space Research*, vol. 47, no. 11, pp. 1983–1989, 2011.
[115] Y. Wu, N. Fu, J. Lin et al., "Research on TEC anomalies before Ms8.0 Wenchuan earthquake by using Kalman filtering," *Journal of Geodesy and Geodynamics*, vol. 31, no. 2, pp. 23–27, 2011 (Chinese).
[116] T. Xu, Y. L. Hu, J. A. Wu et al., "Anomalous enhancement of electric field derived from ionosonde data before the great Wenchuan earthquake," *Advances in Space Research*, vol. 47, no. 6, pp. 1001–1005, 2011.
[117] T. Xu, J. Wu, Z. Zhao et al., "Monitoring ionospheric variations before earthquakes using the vertical and oblique sounding network over China," *Natural Hazards and Earth System Sciences*, vol. 11, no. 4, pp. 1083–1089, 2011.
[118] J. Yang, Y. Wu, and Y. Zhou, "Probe into seismo-ionospheric anomaly of Wenchuan Ms8.0 earthquake based on computerized ionospheric tomography," *Journal of Geodesy and Geodynamics*, vol. 31, no. 1, pp. 9–14, 2011 (Chinese).
[119] L. Li, J. Yang, J. Huang, and Q. Wu, "Analysis about the specific background of rare drought and intense heat and great floodwater which should not be ignored before Wenchuan earthquake," *Plateau Earthquake Research*, vol. 20, no. 4, pp. 69–75, 2008 (Chinese).
[120] X. Jiang and Y. Li, "The statistical analysis of earthquake and precipitation in Sichuan Province," *Plateau and Mountain Meteorology Research*, vol. 28, no. 2, pp. 33–36, 2008 (Chinese).
[121] J. Li and S. Liu, "On the anomalies of crustal deformation & surface temperature and its correlation: a case study of Wenchuan earthquake," *Geography and Geo-Information Science*, vol. 25, no. 1, pp. 79–83, 2009 (Chinese).
[122] L. J. Wei, J. F. Guo, J. H. Lu, Z. Q. Lu, H. B. Li, and H. Cai, "Satellite thermal infrared earthquake precursor to the Wenchuan Ms8.0 earthquake in Sichuan, China, and its analysis on geo-dynamics," *Acta Geologica Sinica*, vol. 83, no. 4, pp. 767–775, 2009.
[123] L. Wei, J. Guo, H. Cai, H. Li, and Z. J. John, "Satellite thermal infrared anomaly: a short-term and impending earthquake precursor before the Wenchuan Ms8.0 earthquake in Sichuan, China," *Acta Geoscientica Sinica*, vol. 29, no. 5, pp. 583–591, 2008 (Chinese).
[124] X. Guo, Y. Zhang, M. Zhong, W. Shen, and C. Wei, "Variation characteristics of OLR for the Wenchuan earthquake," *Chinese Journal Geophysics*, vol. 53, no. 6, pp. 980–988, 2010.
[125] Y. S. Zhang, X. A. Guo, M. J. Zhong, W. R. Shen, W. Li, and B. He, "Wenchuan earthquake: brightness temperature changes from satellite infrared information," *Chinese Science Bulletin*, vol. 55, no. 18, pp. 1917–1924, 2010.
[126] F. Jing, X.-H. Shen, C.-L. Kang, Q.-Y. Meng, and P. Xiong, "Anomalies of outgoing longwave radiation before some medium to large earthquakes," *Earthquake*, vol. 29, no. 4, pp. 117–122, 2009 (Chinese).
[127] Y. F. He, D. M. Yang, H. R. Chen, J. D. Qian, R. Zhu, and M. Parrot, "SNR changes of VLF radio signals detected onboard the DEMETER satellite and their possible relationship to the Wenchuan earthquake," *Science in China D*, vol. 52, no. 6, pp. 754–763, 2009.
[128] W. Min and X. Xu, "Primary analysis of the variation of surface air temperature before and after Wenchuan earthquake," *Plateau and Mountain Meteorology Research*, vol. 28, no. 2, pp. 42–46, 2008 (Chinese).
[129] W. Ma, "Abnormal phenomenon of NCEP before Wenchuan earthquake," *Science & Technology Review*, vol. 26, no. 10, pp. 37–39, 2008 (Chinese).
[130] M. Li, C. Kang, Z. Li, F. Jing, Y. Xue, and W. Yan, "Abnormal surface latent heat flux prior to the Wenchuan Ms8.0 earthquake," *Earthquake*, vol. 30, no. 3, pp. 64–71, 2010 (Chinese).
[131] H. Zeng, G. Zhu, G. Qin, H. Chen, and Y. Li, "Abnormal variation of thermosphere atmospheric density during the period of big earthquake," *Chinese Journal of Space Science*, vol. 31, no. 3, pp. 318–322, 2011 (Chinese).
[132] L. Wu, S. Liu, Y. Chen, B. Ma, and L. Li, "Satellite thermal infrared and cloud abnormities before Wenchuan earthquake," *Science & Technology Review*, vol. 26, no. 10, pp. 32–36, 2008 (Chinese).
[133] X.-T. Zhang, Y.-X. Zhang, and D.-H. Xu, "Investigation of the macroscopic anomalies before and after the Wenchuan M8.0 earthquake," *Earthquake*, vol. 29, no. 2, pp. 104–117, 2009 (Chinese).
[134] W. Cheng, X. Wu, Z. Guan, H. Zhu, and Z. Lu, "Discussion on the macroscopic anomaly phenomenae reported before the Wenchuan M8.0 earthquake," *Earthquake Research in Sichuan*, no. 1, pp. 1–7, 2010 (Chinese).
[135] D. Xu, "Field investigation of macro-anomalies before the Wenchuan Ms8.0 earthquake, Sichuan, China," *Earthquake*, vol. 30, no. 2, pp. 121–133, 2010 (Chinese).
[136] X. T. Zhang, Y. X. Zhang, and D. H. Xu, "Investigation of the macroscopic anomalies before and after the M8.0 Wenchuan earthquake," *Earthquake Research in China*, vol. 24, no. 1, pp. 48–69, 2010.
[137] W. Yan, C. Zhu, X. Li, and H. Zhang, "Investingation and analysis on macroscopic anomalies of earthquake before and after Wenchuan Ms8.0 earthquake in Longnan area of Gansu Province," *Northwestern Seismological Journal*, vol. 33, no. 1, pp. 67–70, 2011 (Chinese).
[138] Y. Li, Y. Liu, Z. Jiang et al., "Behavioral change related to Wenchuan devastating earthquake in mice," *Bioelectromagnetics*, vol. 30, no. 8, pp. 613–620, 2009.
[139] L. L. Chen, X. A. Hu, J. A. Zheng et al., "Increases in energy intake, insulin resistance and stress in rats before Wenchuan

earthquake far from the epicenter," *Experimental Biology and Medicine*, vol. 235, no. 10, pp. 1216–1223, 2010.
[140] B. Ma, L. Wu, and S. Liu, "NDVI variation features before Wenchuan Ms8.0 earthquake," *Science & Technology Review*, vol. 28, no. 13, pp. 52–57, 2010 (Chinese).
[141] Y. Li, "Possibility to detect Wenchuan M8 earthguake by the earthquake warning system," *Progress in Geophysics*, vol. 23, no. 7, pp. 969–971, 2008 (Chinese).
[142] F. Jiang and Q. Gao, "Application of seismo-geochemical methods for Sichuan Province in Wenchuan 8.0 magnitude earthquake of middle-term prediction succeed at according analysis," *Frontier Science*, vol. 3, no. 2, pp. 13–21, 2009 (Chinese).
[143] F. Y. Qian, B. R. Zhao, W. Qian et al., "Impending HRT wave precursors to the Wenchuan Ms8.0 earthquake and methods of earthquake impending prediction by using HRT wave," *Science in China D*, vol. 52, no. 10, pp. 1572–1584, 2009.
[144] L. S. Zhou, Z. H. Qiu, and L. Tang, "Statistical test of precursory waves before Wenchuan Ms8.0 earthquake," *Journal of Geodesy and Geodynamics*, vol. 29, no. 2, pp. 24–28, 2009 (Chinese).
[145] D. Rong and Y. Li, "The changing characteristics of seismic correlation length before Wenchuan M8.0 earthquake," *Northwestern Seismological Journal*, vol. 32, no. 1, pp. 54–58, 2010 (Chinese).
[146] L. Ming, "An explanation on short term low-temperature climate before strong earthquakes," *Journal of Institute of Disaster-Prevention Science and Technology*, vol. 12, no. 3, pp. 133–137, 2010 (Chinese).
[147] Z. Qiu, B. Zhang, S. Chi, L. Tang, and M. Song, "Abnormal strain changes observed at Guza before the Wenchuan earthquake," *Science China Earth Sciences D*, vol. 54, no. 2, pp. 233–240, 2010.
[148] S. Gao, X. Du, Y. Su et al., "The variation characteristics of precursor data in Wudu Seismic Station and the earthquake activity in the boundary area between Gansu and Sichuan Province in recent years," *Northwestern Seismological Journal*, vol. 31, no. 2, pp. 167–173, 2009 (Chinese).
[149] S. D. Gao, J. Tang, X. B. Du et al., "The change characteristics of electromagnetic field before to after Wenchuan Ms8.0 earthquake," *Chinese Journal of Geophysics*, vol. 53, no. 3, pp. 512–525, 2010 (Chinese).
[150] X. Xing, J. Mao, H. Shao, and S. Zhang, "Converted wavelet applied in analysis of well level data in Guanzhong area," *Plateau Earthquake Research*, vol. 21, no. 4, pp. 1–8, 2009 (Chinese).
[151] F. An, "Analysis on short-imminent anomalies of Liujiaxia deformation before Wenchuan earthquake," *Plateau Earthquake Research*, vol. 21, no. 2, pp. 28–31, 2009 (Chinese).
[152] S. Zhang, W. Fang, Y. Shu, J. Li, and Y. Peng, "Study on the anomaly characteristics of observation data in Zhouzhi deep borehole before Wenchuan earthquake," *Journal of Institute of Disaster-Prevention Science and Technology*, vol. 11, no. 1, pp. 20–27, 2009 (Chinese).
[153] X. Zhao, F. Huang, X. Wang, S. Zhang, D. Chen, and D. Shi, "Anomalous characteristics of gas Hg in Zhouzhi well, Shaanxi Province during Wenchuan Ms8.0 earthquake and its aftershocks," *Inland Earthquake*, vol. 24, no. 3, pp. 236–240, 2010 (Chinese).
[154] J. Zhang, X. Liu, T. Tang, and J. Ma, "Research on dynamic evolution characteristics of geomagnetic field in Chinese mainland before and after Wenchuan Ms8.0 earthquake," *Journal of Seismological Research*, vol. 32, no. 3, pp. 231–234, 2009 (Chinese).
[155] J. Zhang, X. Liu, L. Yao, X. Ma, Y. Yuan, and X. Yin, "Study on anomalous change characters of electromagnetic disturbance before the Wenchuan Ms8.0 earthquake," *Seismological and Geomagnetic Observation and Research*, vol. 31, no. 5, pp. 56–60, 2010 (Chinese).
[156] L. Zhang, "The typical precursor anomaly of ground fluid before Wenchuan earthquake in Yunnan region," *Seismological and Geomagnetic Observation and Research*, vol. 31, no. 6, pp. 40–44, 2010 (Chinese).
[157] L. Zhang, D. He, B. Shen, and Y. Su, "Probing the long-term anomaly of groundwater level of Shuifu in Yunnan area before the Wenchuan Ms8.0 earthquake," *Journal of Seismological Research*, vol. 33, no. 2, pp. 159–163, 2010 (Chinese).
[158] P. Qiu, Y. Wang, and G. Yang, "Analysis on underground water temperature anomaly in Qinghai before Wenchuan Ms8.0 earthquake," *Journal of Institute of Disaster-Prevention Science and Technology*, vol. 12, no. 1, pp. 73–78, 2010 (Chinese).
[159] Y. Zhang, X. Liu, Q. Chang, Q. Kang, and Y. Chen, "Analysis on the anomalies of Wenchuan earthquake and characteristics of postseismic effect," *Plateau Earthquake Research*, vol. 21, no. 3, pp. 22–27, 2009 (Chinese).
[160] P. Zhang, D. Zhang, W. Zhu, L. Fan, R. Chen, and J. Yan, "Anomalous Earth stress of the Ms8 Wenchuan earthquake in Sichuan, China—recording from piezomagnetic frequency measurement to the Earth stress," *Acta Geologica Sinica*, vol. 82, no. 12, pp. 1788–1799, 2008 (Chinese).
[161] Z. Song, Z. Wang, Y. Lv, S. Li, and J. Li, "Study on body strain characteristics of Xiyang Seismological Station, Shanxi, before Wenchuan, Sichuan, 8.0 earthquake and coseismic effect," *Earthquake Research in Shanxi*, vol. 2, pp. 19–20, 2009 (Chinese).
[162] S. Zhang, R. Liu, Y. Ning, L. Tang, and B. Li, "Analysis of characteristics of precursor wave of low frequency in Shanxi before Wenchuan Ms8.0 earthquake," *Journal of Geodesy and Geodynamics*, vol. 29, no. 6, pp. 35–39, 2009 (Chinese).
[163] Z. Liu, X. Yang, S. Xue, B. Li, and M. Song, "Analysis on anomalous variations in observation data of Shanxi Crustal Deformation Network before several earthquake," *Earthquake Research in Shanxi*, no. 2, pp. 31–33, 2010 (Chinese).
[164] J. Yang and C. Zhang, "Discussion on the correlation between metal pendulum ground tiltmeters data at Hancheng, Shannxi Seismological Station and its surrounding seismic activities," *Earthquake Research in Shanxi*, vol. 4, pp. 27–29, 2010 (Chinese).
[165] M. Li, J. Lu, Y. Chang, and X. Liu, "ULF electromagnetic anomaly observed at Gaobeidian and Ningjin seismic stations before Wenchuan 8.0 earthquake," *Recent Developments in World Seismology*, vol. 7, pp. 76–82, 2009 (Chinese).
[166] H. Yin, J. Li, X. Dong, and P. Xu, "On characteristics of crustal movement of Shandong and surrounding areas before and after Wenchuan Ms8.0 earthquake," *Journal of Geodesy and Geodynamics*, vol. 29, no. 3, pp. 23–27, 2009 (Chinese).
[167] S. Lu, Q. Wang, X. Lin et al., "Analysis of anomalies about pendulum tiltmeter at Taian seismostation before Wenchuan earthquake," *Journal of Geodesy and Geodynamics*, vol. 31, supplement, pp. 19–23, 2011 (Chinese).
[168] S. Su, J. Li, Q. Zhang, S. Yang, and Z. Song, "The response of crustal deformation data in Yixian Seismic Station to Wenchuan Ms8.0 earthquake," *North China Earthquake Sciences*, vol. 27, no. 2, pp. 54–58, 2009 (Chinese).
[169] J. Zhang, G. Zhao, J. Wang, T. Tang, and L. Qi, "Preliminary study on the relationship between the Ms8.0 Wenchuan earthquake of 2008 and the anomalies of earth resistivity

observed at Qingdao seismic station," *Seismology and Geology*, vol. 32, no. 3, pp. 409–416, 2010 (Chinese).

[170] H. Zhang, J. Guo, Z. Xue, C. Yu, F. Li, and J. Zhou, "Analysis on geothermal and water level response to Wenchuan earthquake in Changli Well," *North China Earthquake Sciences*, vol. 27, no. 4, pp. 46–49, 2009 (Chinese).

[171] D. Liu, "Discussion on relation between anomalies of water temperature and Radon in Ningbo station and Wenchuan Ms8.0 earthquake," *Journal of Geodesy and Geodynamics*, vol. 28, no. 6, pp. 53–55, 2008 (Chinese).

[172] J. Zhang, J. Ding, X. Shen et al., "Abnormal count rate of Yangbajing neutron-muon telescope before Wenchuan earthquake," *Chinese Journal of Radio Science*, vol. 25, no. 2, pp. 227–233, 2010 (Chinese).

[173] J. Wang, Y. Chen, Z. Ye, and S. Liu, "Property analysis of cosmic rays before and during the great Wenchuan earthquake," *Progress in Geophysics*, vol. 25, no. 1, pp. 137–142, 2010 (Chinese).

[174] B. Deng, M. Yang, J. Huang, and W. Zhao, "Study on VTEC anomaly of ionosphere in south China before and after Wenchuan earthquake," *South China Journal of Seismology*, vol. 29, no. 4, pp. 8–15, 2009 (Chinese).

[175] M. Yang, B. Deng, W. Zhao, and J. Huang, "Study on abnormal characteristics of ionosphere TEC above South-China responding to two strong earthquakes," *Journal of Geodesy and Geodynamics*, vol. 29, no. 6, pp. 13–17, 2009 (Chinese).

[176] W. Bo and G. Yang, "Crustal deformation measured by mobile observation before Wenchuan Ms8.0 earthquake," *Journal of Geodesy and Geodynamics*, vol. 28, no. 6, pp. 11–15, 2008 (Chinese).

[177] G. Zhao, W. Ma, J. Wang, and A. He, "Geothermal earthquake precursor network and its response to Ms8.0 Wenchuan earthquake," *Journal of Seismological Research*, vol. 32, no. 3, pp. 248–252, 2009 (Chinese).

[178] I. P. Dobrovolsky, S. I. Zubkov, and V. I. Miachkin, "Estimation of the size of earthquake preparation zones," *Pure and Applied Geophysics*, vol. 117, no. 5, pp. 1025–1044, 1979.

[179] X. G. Hu and X. G. Hao, "The short-term anomalies detected by broadband seismographs before the May 12 Wenchuan earthquake, Sichuan, China," *Chinese Journal of Geophysics*, vol. 51, no. 6, pp. 1726–1734, 2008 (Chinese).

[180] X. G. Hu and X. G. Hao, "An analysis of the influences of typhoon on anomalous tremors before the great Wenchuan and Kunlunshan earthquakes," *Chinese Journal of Geophysics*, vol. 52, no. 5, pp. 1363–1375, 2009 (Chinese).

[181] R. S. Fu, K. S. Wan, J. J. Chong, and T. X. Xue, "Earthquake auspice or other factor? Discuss with authors of the paper "The short-term anomalies detected by broadband seismographs before the May 12 Wenchuan earthquake, Sichuan, China"," *Chinese Journal of Geophysics*, vol. 52, no. 2, pp. 584–589, 2009 (Chinese).

[182] X.-G. Hu, X.-G. Hao, and X.-X. Xue, "The analysis of the non-typhoon-induced microseisms before the 2008 Wenchuan earthquake," *Chinese Journal of Geophysics*, vol. 53, no. 12, pp. 2875–2886, 2010 (Chinese).

[183] Z. H. Qiu, L. S. Zhou, and S. L. Chi, "Study on precursory strain changes of Wenchuan earthquake with ORA method," *Journal of Geodesy and Geodynamics*, vol. 29, no. 4, pp. 1–4, 2009 (Chinese).

[184] Z. H. Qiu, L. Tang, L. S. Zhou, and B. X. Kan, "Observed strain changes from 4-component borehole strainmeter network before 2008 Wenchuan earthquake," *Journal of Geodesy and Geodynamics*, vol. 29, no. 1, pp. 1–5, 2009 (Chinese).

[185] Q. Su, H. Zhu, and Y. Yang, "Short-line leveling anomaly at Gengda and Wenchuan Ms8.0 earthquake," *Journal of Geodesy and Geodynamics*, vol. 29, supplement, pp. 103–105, 2009 (Chinese).

[186] J. Wei, H. Hao, K. Kang et al., "Gravity disturbance of high frequcncy at Chengdu seismostation before Wenchuan Ms8.0 earthquake," *Journal of Geodesy and Geodynamics*, vol. 29, supplement, pp. 15–19, 2009 (Chinese).

[187] W. Liu, H. Chen, F. Zhao, X. Liu, B. Yuan, and J. Zhang, "Characteristics of spatio-temporal relation of strong earthquakes in China before and after the Sichuan Wenchuan Ms8.0 earthquake," *Earthquake*, vol. 30, no. 3, pp. 140–146, 2010 (Chinese).

[188] H. Zhu, J. P. Deng, and G. Yang, "Discussion on the causes of 2 anomalies before the Wenchuan M8.0 earthquake," *Earthquake Research in Sichuan*, no. 1, pp. 13–17, 2010 (Chinese).

[189] H. Zhu, Q. Su, T. Yang, X. Z. Wen, and S. X. Wang, "Identifying anomalous change of short-line leveling on Gengda site pre- and after Ms8.0 Wenchuan earthquake," *Acta Seismologica Sinica*, vol. 32, no. 6, pp. 649–658, 2010 (Chinese).

[190] W. Cheng, Q. Su, and Y. Sun, "Study on the datum anomalies of crust deformation in Sichuan region before the 2008 M8.0 Wenchuan earthquake," *Earthquake Research in Sichuan*, no. 1, pp. 1–10, 2011 (Chinese).

[191] W. Cheng, Y. Ren, and X. Wu, "Analyzing the anomalies of geo-electrical and geo-magnetic observation data in Sichuan Province before the 2008 Wenchuan M8.0 earthquake," *Earthquake Research in Sichuan*, no. 2, pp. 2–9, 2010 (Chinese).

[192] P. Cheng, J. Mi, H. Wang, Y. Cai, and Y. Huang, "Time-variations of some domestic IGS sites during the Wenchuan earthquake," *GNSS World of China*, vol. 33, no. 5, pp. 11–15, 2008 (Chinese).

[193] J. Liu, "On geothermal earthquake observation network and Wenchuan 8.0 earthquake," *Journal of Institute of Disaster-Prevention Science and Technology*, vol. 12, no. 3, pp. 48–54, 2010 (Chinese).

[194] P. Qiu, Y. Wang, G. Yang, X. Bai, and S. Zhao, "Research on the anomalous changes of water surface temperature before and after Wenchuan Ms8.0 earthquake," *Northwestern Seismological Journal*, vol. 32, no. 4, pp. 367–375, 2010 (Chinese).

[195] J. Dong, R. Yan, J. Zhang, X. Zhang, and J. Liu, "Analysis of earthquake-related information in ionosphere based on DEMETER satellite data-applicated to Wenchuan and Donghai earthquakes," *Earthquake*, vol. 29, supplement, pp. 67–75, 2009 (Chinese).

[196] G. Gu and W. Wang, "Vertical crustal movement before and after the great Wenchuan earthquake obtained from GPS observations in the regional network," *Earthquake*, vol. 31, no. 3, pp. 1–8, 2011 (Chinese).

[197] W. Bo and L. Zhang, "The Ms8.0 Wenchuan earthquake: precursor analysis of fault deformation," *Journal of Seismological Research*, vol. 31, supplement, pp. 419–423, 2008 (Chinese).

[198] S. Zhang, X. Zhang, S. Wang, F. Xue, and L. Liu, "Analysis of crustal vertical deformation before and after Wenchuan Ms8.0 earthquake," *Journal of Geodesy and Geodynamics*, vol. 28, no. 6, pp. 43–46, 2008 (Chinese).

[199] Z.-S. Jiang, Y.-Q. Wu, Y. Fang, P. Li, and W.-X. Wang, "Dynamic characteristics of the strain fields and crustal movements before the Wenchuan Ms8.0 earthquake," *Earthquake*, vol. 29, no. 1, pp. 68–76, 2009 (Chinese).

[200] Z.-S. Jiang, Y. Fang, Y.-Q. Wu, M. Wang, F. Du, and J.-J. Ping, "The dynamic process of regional crustal movement and

deformation before Wenchuan Ms8.0 earthquake," *Chinese Journal of Geophysics*, vol. 52, no. 2, pp. 505–518, 2009 (Chinese).
[201] X. Shan, X. Song, Y. Han, C. Qu, G. Zhang, and G. Zhang, "The characteristics of surface vertical deformation before the Wenchuan Ms8.0 earthquake from InSAR," *Chinese Journal of Geophysics*, vol. 52, no. 11, pp. 2739–2745, 2009 (Chinese).
[202] R. Xie, Q. Li, J. Qin, R. Zhang, S. Xu, and J. Feng, "Study on gravity change around Wenchuan Ms8.0 earthquake recorded in Ji-Lu-Yu junction area," *Journal of Geodesy and Geodynamics*, vol. 29, no. 1, pp. 27–30, 2009 (Chinese).
[203] G. A. Yang and Y. Mi, "Thermal anomalies and earthquakes: evidence from Wenchuan, China," *Earthquake Research in China*, vol. 23, no. 1, pp. 48–55, 2009.
[204] Q. Cui, Y. Shu, Z. Wang, J. Fan, and W. Liu, "Suspected correlation between anomalies of body strain observation data at Xi'an Seismological Station and Wenchuan M8.0 earthquake," *Earthquake Research in Shanxi*, no. 4, pp. 30–33, 2010 (Chinese).
[205] J. Shi, W. Wang, and J. Wen, "The meteorological anomalies before and after the Wenchuan earthquake," *Astronomical Research & Technology*, vol. 7, no. 1, pp. 78–84, 2010 (Chinese).
[206] L. Wu, L. Zhang, G. Li, and H. Guo, "The relative calibration and its application of 4-component borehole strain observation in HaiYuan Station," *Journal of Seismological Research*, vol. 33, no. 4, pp. 318–322, 2010 (Chinese).
[207] W.-B. Liu, "Reverse tracing of the seismicity before strong earthquakes in Mainland China," *Earthquake*, vol. 30, no. 2, pp. 46–53, 2010 (Chinese).
[208] Z. Chen, P. Lu, Z. Li, B. Zhao, and S. Lin, "Analysis of tidal data before Wenchuan Ms8.0 earthquake," *Journal of Geodesy and Geodynamics*, vol. 29, no. 4, pp. 48–50, 2009 (Chinese).
[209] X.-F. Fan, J.-Y. Wang, and M.-Y. Lu, "Preliminary study on some typical medium-term subsurface fluid anomalies before the Wenchuan 8.0 earthquake," *Earthquake*, vol. 29, no. 1, pp. 132–140, 2009 (Chinese).
[210] Y. Ding, H. Chen, and J. Zhang, "The preliminary research on anomalous variation character of electromagnetic radiation before the Wenchuan Ms8.0 earthquake," *Seismological and Geomagnetic Observation and Research*, vol. 30, no. 3, pp. 11–15, 2009 (Chinese).
[211] S. Zhou, Y. Wu, S. Shi, and Z. Chen, "Anomalous change of fault deformation before Wenchuan Ms8.0 earthquake and its comparison with response of other large earthquakes," *Acta Seismologica Sinica*, vol. 31, no. 2, pp. 140–151, 2009 (Chinese).
[212] S.-R. Mei, Y. Xue, and Z.-P. Song, "Anomalous seismic characteristics before Wenchuan M8.0 and Kunlunshan M8.1 earthquakes and their implications," *Earthquake*, vol. 29, no. 1, pp. 1–14, 2009 (Chinese).
[213] L. Chen, H. Fu, and X. Zhao, "Comparative study of M≥5 seismic activities before some large earthquakes," *Earthquake*, vol. 30, no. 4, pp. 22–31, 2010 (Chinese).
[214] W. Cheng and Y. Zhang, "Discussion on the seismicity and seismic tendency of Sichuan Province before the 2008 Wenchuan M8.0 earthquake," *Earthquake Research in Sichuan*, no. 3, pp. 1–7, 2009 (Chinese).
[215] L. Chen and H. Fu, "The contrast study of Wenchuan M8.0 earthquake precursor anomalies," *Inland Earthquake*, vol. 24, no. 4, pp. 289–297, 2010 (Chinese).
[216] T. H. Jordan, Y. T. Chen, P. Gasparini et al., "Operational earthquake forecasting: state of knowledge and guidelines for utilization," *Annals of Geophysics*, vol. 54, no. 4, pp. 315–391, 2011.
[217] A.-F. Niu, L.-K. Zhang, W. Yan, X.-D. Jia, and X.-F. Li, "On the characteristics of ground deformation anomalies in the middle and northern part of the south-north seismic belt prior to the Wenchuan earthquake," *Earthquake*, vol. 29, no. 1, pp. 100–107, 2009 (Chinese).
[218] L. Zhang, M. Yu, and D. Sun, "Analysis on fault deformation anomaly of Longmenshan fault zone before Ms8.0 Wenchuan strong earthquake," *North China Earthquake Sciences*, vol. 27, no. 1, pp. 34–38, 2009 (Chinese).
[219] M.-Y. Lu, Z.-F. Fang, and L.-K. Zhao, "Discussion on the long-time trend change characteristics of subsurface fluid before the Wenchuan Ms8.0 earthquake," *Earthquake*, vol. 30, no. 1, pp. 61–72, 2010 (Chinese).
[220] H. Zhang, Z. Ouyang, Z. Fu, B. Gou, and W. Jiang, "Abnormal phenomena recorded by several earthquake precursor observation instruments before the Ms8.0 Wenchuan, Sichuan earthquake," *Progress in Geophysics*, vol. 26, no. 1, pp. 99–109, 2011 (Chinese).
[221] Y. Zhu, W. Liang, and Y. Xu, "Medium-term prediction of Ms8.0 earthquake in Wenchuan, Sichuan by mobile gravity," *Recent Developments in World Seismology*, no. 7, pp. 36–39, 2008 (Chinese).
[222] M. Wyss, Ed., *Evaluation of Proposed Earthquake Precursors*, American Geophysical Union, Washington, DC, USA, 1991.
[223] M. Wyss and R. Dmowska, Eds., "Earthquake Prediction, State-of-the-Art," *Pure and Applied Geophysics*, vol. 149, no. 1, pp. 1–264, 1997.
[224] R. D. Cicerone, J. E. Ebel, and J. Britton, "A systematic compilation of earthquake precursors," *Tectonophysics*, vol. 476, no. 3-4, pp. 371–396, 2009.
[225] T. Rikitake, *Earthquake Prediction*, Elsevier, New York, NY, USA, 1976.
[226] T. Rikitake, "Classification of earthquake precursors," *Tectonophysics*, vol. 54, no. 3-4, pp. 293–309, 1979.
[227] T. Rikitake, "Earthquake prediction: an empirical approach," *Tectonophysics*, vol. 148, no. 3-4, pp. 195–210, 1988.
[228] K. Mogi, *Earthquake Prediction*, Academic Press, Orlando, Fla, USA, 1985.

3

Geoelectrical Tomography as an Operative Tool for Emergency Management of Landslide: An Application in Basilicata Region, Italy

G. Colangelo[1] and A. Perrone[2]

[1] *Department of Infrastructure and Civil Protection, Basilicata Region, C.so Garibaldi 139, 85100 Potenza, Italy*
[2] *Institute of Methodologies for Environmental Analysis, CNR, C.da S. Loja, 85050 Tito Scalo, Italy*

Correspondence should be addressed to G. Colangelo, gerardo.colangelo@regione.basilicata.it

Academic Editor: Sabatino Piscitelli

During the landslide emergency many kinds of data, even if preliminary, can help to better understand the complexity of the investigated phenomenon and to give a valid contribution to the successive damage valuation. The electrical resistivity tomography (ERT) method was applied for investigating the deep characteristics of a landslide body that occurred in March 2006 close to Potenza town in Basilicata region (southern Italy): the landslide slid on a road near some farmers' houses which had to be evacuated. The information obtained by the application of this indirect technique appeared to be particularly useful for end users involved in the risk management. The high resolution of the 2D ERT technique allowed the detection of possible sliding surfaces and the characterization of high water content areas in which the increase of the saturation degree and of pore pressures could cause a weakening of the slopes and a reactivation of the movement. Due to the comparison between ERT results and stratigraphical data from boreholes carried out in the area it was possible to decide on the adoption of other evacuation decrees.

1. Introduction

The investigated area is located in Basilicata region, one of the southern Italian areas more involved in heavy meteorological conditions [1]. On March 2006, the intense precipitations have increased the saturation degree and the pore pressures of the terrains. The snow blanket has made heavy the slope changing the equilibrium of the strengths involved in the stability of a slope. These climatic conditions have deteriorated the physical and mechanic characteristics of the terrains outcropping in the region. As consequence of all these alterations, the reactivation of many dormant landslides, which affected the slopes of the region in the past, occurred. The main typologies of reactivation have been earth-flow, translational, or rotational slides.

The new slides have involved buildings and infrastructures on the slopes. The risk for people and assets needed the intervention of the end users involved in the risk management and, in particular, the inspection of Regional Department of Infrastructure and Civil Protection (RDICP). In many involved areas and for many families evacuation decrees have been issued in order to ensure the safety of the people and allow the damage valuation. The study of such complex phenomena required a multidisciplinary approach based on the integration of all the direct and indirect data acquired in the area. An important contribution has been provided by the geophysical data and, in particular, by the 2D ERT that have been carried out in the areas affected by the reactivations some days after the landslide event.

The use of 2D ERT method for investigating landslides is now well tested. Many examples of the ERT application are reported in literature. In many cases the results of its application allowed to reconstruct the geometry of landslide body, to outline the sliding surface, and to locate areas characterized by high water content [2–7].

By using the Mobile Laboratory for chemical-physical and geophysical measurements of the Institute of Methodologies for Environmental Analysis (IMAA) of CNR, some ERTs have been performed in the more damaged areas of the Basilicata region (southern Italy). In particular, the present

FIGURE 1: Location of the study area in Basilicata region.

paper describes the test case of a complex rototranslational slide occurred close to Potenza town in the Picerno village territory during the winter of 2006.

Immediately after the event occurred two ERT, one with transversal direction and the other one with longitudinal direction to the landslide body, were performed in order to try to give an answer to some important questions for the RDICP like what are the geometrical characteristics of the landslide? How many houses and infrastructures could be involved in the evolution of the phenomena? Is it necessary to issue an evacuation decree for others families?

Active landslide boundary
Main scarp
Minor scarp
ERT profile
S1, S2 core drill

FIGURE 2: Geomorphological map of the area (aerial photo by Google Earth) with location of geophysical and geotechnical soundings.

2. Geological Setting of the Landslide Area

The investigated area is characterized by the presence of a diffuse slope instability. On March 2006, due to the meteorological conditions, many reactivations occurred in the area like the one considered in this paper. In order to define the geological and geomorphological features of the area, aerial photogrammetric analysis and in-field observations have been performed.

From a geological point of view, Basilicata region is located along the axial zone of the southern Apennine chain that is mainly composed of sedimentary cover of platform and deep water environments, scraped off from the former Mesozoic Ligurian ocean, the western passive margin of the Adriatic plate, and the Neogene-Pleistocene foredeep deposits of the active margin. From west to east, the main Mesozoic domains are as follows: (1) the internal oceanic to transitional Liguride-Sicilide basinal domains (internal nappes), (2) the Apennine carbonate platform, (3) the Lagonegro-Molise basins, and (4) the Apulian carbonate platform [8].

The study area (Figure 1) is located in the west of the Basilicata region and on the southeastern slope of Li Foi Mountain (1355 m a.s.l.) near the S. Loja Basin along the axial zone of the Lucanian Apennine. The area is characterized by the outcropping of terrains belonging to the Pignola-Abriola facies (calcareous-silica-marly series) of the Lagonegro Unit II [9]. This facies is composed of the Siliceous Schist (Upper Triassic–Jurrasic), the Galestrino Flysch (Lower Cretaceous), the Red Flysch (Upper Cretaceous–Lower Miocene), and the Corleto Perticara Formation (Upper Eocene–Lower Miocene) [10, 11]. The landslide occurred in 2006 involved the terrains belonging to the Corleto Perticara Formation mainly constituted by calcarenites, calcilutites, and whitish marly limestone.

From a geomorphological point of view the area is characterized by a moderate slope angle of 13–16% with an altimetry range varying between 1072 m a.s.l. at the main crown and 970 m a.s.l. at the toe. The landslide can be classified as a complex retrogressive rototranslational slide and is 600 m long and 230 m wide (Figure 2). Due to the movement, counter slope terraces filled with stagnant water formed. Transversal and radial cracks are still evident in the accumulation zone of the landslide that involved infrastructures and buildings some of whose had to be evacuated [12].

3. ERT and Analysis of the Results

ERT is a geoelectrical method widely applied to obtain 2D and 3D high-resolution images of the resistivity subsurface patterns in areas of complex geology [13]. During the field survey, ERT can be carried out by using different electrode configurations (dipole-dipole, Wenner, etc.) placed at the surface to send the electric currents into the ground and to measure the generated voltage signals. Technically, during an electrical resistivity measurement, the electric current is injected into the ground via two electrodes and the potential drop is measured between two other electrodes in line with current ones. The values of the apparent resistivity acquired along a horizontal axis are assigned at a defined depth and position. In a second step, it is necessary to transform

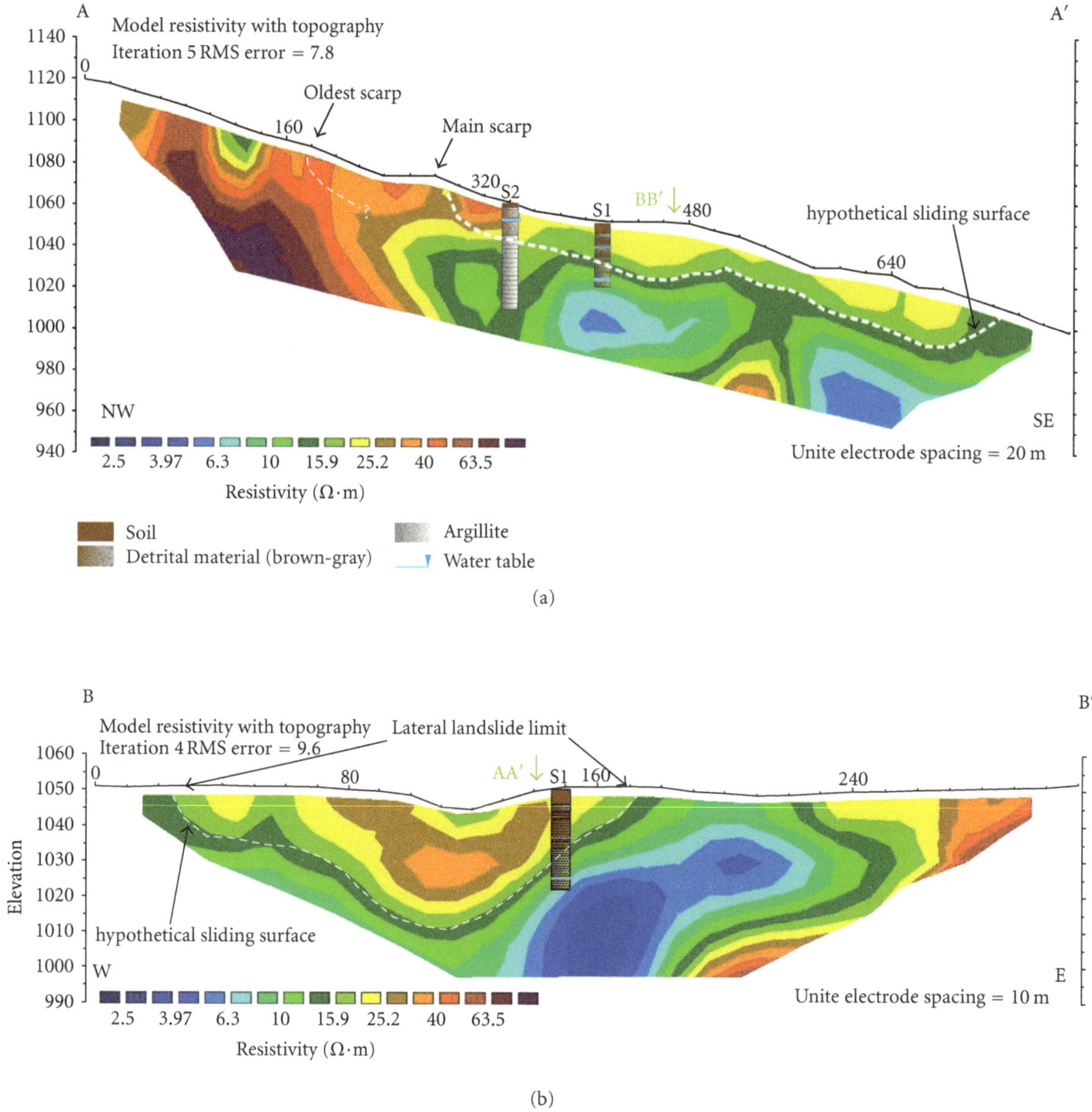

Figure 3: (a) Comparison between AA′ ERT, carried out with longitudinal direction to the landslide body, and S1 and S2 stratigraphical data; (b) comparison between BB′ ERT and S1 stratigraphical data. The horizontal resistivity contrast, highlighted with white dashed line in both the ERT, corresponds to the contact between detrital material (slide material) and argillite material.

the apparent resistivity values obtained during the field survey into real resistivities of the subsoil and the pseudodepths into true depths.

In this work, the algorithm proposed by Loke and Barker [14] for the automatic 2D inversion of apparent resistivity data was used. The inversion routine is based on the smoothness constrained least-squares inversion [15] implemented by a quasi-Newton optimisation technique. The subsurface is divided in rectangular blocks, whose number corresponds to the number of measurement points. The optimisation method adjusts the 2D resistivity model trying to reduce iteratively the difference between the calculated and measured apparent resistivity values. The root mean squared (RMS) error gives a measure of this difference.

The knowledge of local geology associated with the high spatial resolution of the measurements gives us an interpretative tool to explain the ERT obtained for the case study of this work.

ERTs have been carried out with direction longitudinal (AA′) and transversal (BB′) to the axis of the landslide (Figure 2) by using a multielectrode system with 32 electrodes and the dipole-dipole array. Both the ERTs were topographically correct in order to reduce possible mistakes and improve the interpretation of the model.

The longitudinal profile (length 780 m) was carried out by using an electrode spacing of 20 m, reaching an investigation depth of about 55–60 m. The AA′ ERT (Figure 3(a)), obtained along this profile, is characterized by a semivertical discontinuity, located at 280 m from the origin of the profile, which separates relatively high resistivity material (20–80 Ω·m) from conductive one (4–20 Ω·m). This discontinuity represents the main scarp of the landslide here investigated. The remainder of the ERT is characterized by a vertical resistivity variation with a first resistivity ($20 < \rho < 50\,\Omega\cdot$m) layer (about 25 m thick) which covered a more conductivity material ($0 < \rho < 20\,\Omega\cdot$m). The relatively high resistive layer could be associated with the slide material involved in the reactivation.

The high resistivity material characterizing the north-western side of the ERT highlights the presence of the bedrock. At a distance of 180 m from the origin of the profile it is also possible to see the presence of another vertical discontinuity that could be associated with the old scarp of a deep ancient landslide now totally eroded and not visible at the surface.

The transversal profile (length 320 m) has been carried out by using an electrode spacing of 10 m, reaching an investigation depth of about 45–50 m.

The BB′ ERT (Figure 3(b)), obtained along this profile shows a first resistivity ($20 < \rho < 50\,\Omega\cdot$m) layer (about 30 thick), which covered a more conductivity material ($0 < \rho < 20\,\Omega\cdot$m). The contrast between conductive and resistive material could be associated with the presence of a sliding surface. The higher conductive nucleus ($\rho < 10\,\Omega\cdot$m), located at a distance ranging from 160 to 240 m from the origin of the profile, could be associated with an area characterized by high water content.

The ERTs were compared with stratigraphical data from direct boreholes (S1 and S2) carried out in the area by the RDICP (see Figure 2 for the location). In particular, the first 15–16 m of the S1 borehole (a 27 m deep core drilling) can be considered landslide material composed of detrital clay marl deposits with pebbles, interbedded with marly limestone. Below 16 m more consolidated clay can be found down to the bottom of the borehole. Two water tables are found in the S1 borehole at about 14 and 25 m, respectively [12].

The comparison between direct and indirect data (Figures 3(a) and 3(b)) allowed the better definition of the 2D landslide geometry and the correlation between resistivity values, reported in the ERT, and specific lithological characteristics of the slope. The comparison also made possible the detection of high water content areas for which the technicians of RDICP have planned and realized the first drainage works.

4. Discussion and Conclusion

ERT technique was used for emergency management of a landslide that occurred in Basilicata Region close to Potenza town on March 2006. ERT has been applied with the aim to obtain information about the deep characteristics of the landslide body. In particular, two ERTs were carried out with longitudinal and transversal direction to the landslide body, respectively. The results, also compared with stratigraphical data, highlighted the presence of two layers with different resistivity. The sliding surface has been associated with the "layer" corresponding to resistivity contrast and area with high water content has been located. The information obtained by the application of this indirect technique appeared to be particularly useful for the end users involved in the risks management. Indeed, during the first phase of the emergency the main problem was the safety of the people and the relative evacuation of the area. In this case, the houses were principally concentrated on middle and upper part of the slope. RDICP technicians in collaboration with local administration had evacuated only the houses located in the middle part of the slope, in correspondence of the landslide body individuated on longitudinal and transversal ERT. A temporary accommodations for these families was found in a container located in a external part of the landslide (Figure 4).

Figure 4: Aerial photo (modified by Google Earth) reporting the location of evacuated families (circles), temporary accommodation (triangle), warning area (square), and geotechnical instrumentations (stars).

Thanks to the longitudinal ERT, in which the bedrock was detected, it was possible to exclude the retrogressive evolution of the investigated phenomenon so, thanks to this information, to decide that it was not necessary to issue an evacuation decree for the families living in the upper part of the slope. In any case, geotechnical instrumentations (piezometers and inclinometers) and GPS markers were installed inside and outside of the landslide area in order to obtain additional information about the dynamic evolution of the gravitative movement affected the slope (Figure 4).

Moreover, the thickness of the landslide valuated on both longitudinal and transversal ERTs at a depth of about 20–30 m, helped the RDICP to define the slope stabilization plan to adopt during the second phase of the emergency. Indeed, on landslide body only superficial diffuse drainage systems (Figure 5(a)) for the total length of the landslide and naturalistic engineering works to restore the local route (Figure 5(b)) were carried out and any deep works (networks of micropiles, piles, etc.) were realized.

(a)

(b)

Figure 5: (a) Drainage systems effectuated during the second phase of the emergency with 4.0 m depth, and 1.2 km length; (b) Naturalistic engineering works using light and native material (chestnut tree and broom).

Acknowledgment

Thanks are due to G. Calvello for the support during the project and Impresa Curcio of Picerno, Italy for the availability during the different phases of the work.

References

[1] Gli indicatori del clima in Italia nel 2005, I Anno, APAT, http://www.sinanet.isprambiente.it/.

[2] A. Demoulin, A. Pissart, and C. Schroeder, "On the origin of late Quaternary paleolandslides in the Liege (E Belgium) area," *International Journal of Earth Sciences*, vol. 92, pp. 795–805, 2003.

[3] A. Bichler, P. Bobrowsky, P. Best et al., "Three-dimensional mapping of a landslide using a multi-geophysical approach: the Quesnel Forks landslide," *Landslide*, vol. 1, pp. 29–40, 2004.

[4] A. Perrone, A. Iannuzzi, V. Lapenna et al., "High-resolution electrical imaging of the Varco d'Izzo earthflow (southern Italy)," *Journal of Applied Geophysics*, vol. 56, no. 1, pp. 17–29, 2004.

[5] V. Lapenna, P. Lorenzo, A. Perrone, S. Piscitelli, E. Rizzo, and F. Sdao, "2D electrical resistivity imaging of some complex landslides in the Lucanian Apennine chain, southern Italy," *Geophysics*, vol. 70, no. 3, pp. B11–B18, 2005.

[6] O. Meric, S. Garambois, D. Jongmans, M. Wathelet, J. L. Chatelain, and J. M. Vengeon, "Application of geophysical methods for the investigation of the large gravitational mass movement of Séchilienne, France," *Canadian Geotechnical Journal*, vol. 42, no. 4, pp. 1105–1115, 2005.

[7] A. Godio, C. Strobbia, and G. De Bacco, "Geophysical characterisation of a rockslide in an alpine region," *Engineering Geology*, vol. 83, no. 1-3, pp. 273–286, 2006.

[8] D. Scrocca, E. Carminati, and C. Doglioni, "Deep structure of the southern Apennines, Italy: thin-skinned or thick-skinned?" *Tectonics*, vol. 24, no. 3, pp. 1–20, 2005.

[9] P. Scandone, "Studi di geologia lucana: Carta dei terreni della serie calcareo-silicomarnosa e note illustrative," *Bollettino Società dei Naturalisti in Napoli*, vol. 81, pp. 225–300, 1972.

[10] T. Pescatore, P. Renda, and M. Tramutoli, "Rapporti tra le unità lagonegresi e le unità Sicilidi nella media valle del Basento (Appennino lucano)," *Memoriali della Società Geologica Italiana*, vol. 41, pp. 353–361, 1988.

[11] S. Gallicchio, M. Marcucci, P. Pieri, I. Premoli Silva, L. Sabato, and G. Salvini, "Stratigraphical data from a Cretaceus claystones sequence of the "Argille Varicolori" in the Southern Apennines (Basilicata, Italy)," *Palaleopelagos*, vol. 6, pp. 261–272, 1996.

[12] C. de Bari, V. Lapenna, A. Perrone, C. Puglisi, and F. Sdao, "Digital photogrammetric analysis and electrical resistivity tomography for investigating the Picerno landslide (Basilicata region, southern Italy)," *Geomorphology*, vol. 133, no. 1-2, pp. 34–46, 2011.

[13] D. H. Griffiths and R. D. Barker, "Two-dimensional resistivity imaging and modelling in areas of complex geology," *Journal of Applied Geophysics*, vol. 29, no. 3-4, pp. 211–226, 1993.

[14] M. H. Loke and R. D. Barker, "Rapid least-squares inversion of apparent resistivity pseudosections by a quasi-Newton method," *Geophysical Prospecting*, vol. 44, no. 1, pp. 131–152, 1996.

[15] Y. Sasaki, "Resolution of resistivity tomography inferred from numerical simulation," *Geophysical Prospecting*, vol. 40, no. 4, pp. 453–463, 1992.

4

Changes in Precipitation Seasonality in West Africa Predicted by RegCM3 and the Impact of Dynamic Vegetation Feedback

Guiling Wang and Clement A. Alo

Department of Civil and Environmental Engineering, University of Connecticut, 261 Glenbrook Road, Storrs, CT 06269, USA

Correspondence should be addressed to Guiling Wang, gwang@engr.uconn.edu

Academic Editor: Andrea Sealy

Using a regional climate model asynchronously coupled to a dynamic vegetation model, this study examines future climate predictions for the West Africa region and how dynamic vegetation feedback may influence such predictions. Without accounting for the impact of vegetation dynamics, the model predicts a future decrease of annual rainfall over Sahel. Dynamic vegetation feedback reverses this trend, leading to a substantial increase of annual rainfall. Regardless of how vegetation is treated, the predicted future trend of precipitation in the Sahel region follows a specific seasonal pattern, with a decrease during the pre- and early-monsoon season (May-June and early July) due to the warming-induced enhancement of spring convective barrier and an increase after the monsoon is fully established (typically in July-August-September) due to enhanced moisture import from a warmer ocean. Dynamic vegetation feedback reduces the magnitude of the predicted rainfall reduction in the early season and increases the magnitude of the predicted rainfall increase later in the rainy season. The future decrease of early-season rainfall has significant agronomic implications.

1. Introduction

Climate variability in West Africa is characterized by a megadrought in the second half of the 20th century and the alternating occurrence of multidecadal dry and wet spells in recent history (e.g., the fluctuation of lake level in [1]). While the role of large-scale oceanic forcing (e.g., warming of the tropical oceans and changes in the interhemispheric temperature gradient) cannot be neglected [2, 3], land surface dynamics may have played a critical role in triggering, enhancing, and/or sustaining the megadrought and in regulating the multidecadal climate variability of this regional climate system. In particular, past modeling studies have collectively documented that the climate system in West Africa is highly sensitive to both anthropogenic land cover changes and natural vegetation dynamics, in the past, present, and future [4–8].

Studies on paleoclimate indicated that vegetation feedback plays an important role in sustaining the extended grassland cover during the mid-Holocene and in the subsequent abrupt transition from a green Sahara to a desert later during the mid-Holocene [6, 9–12]. Specifically, vegetation dynamics enhances the response of the West African climate system to changes in solar insolation, leading to the drastic differences in climate and vegetation distribution between mid-Holocene and the present day [10]. For the 20th century climate, Xue and Shukla [4] found that desertification in the Sahel region (in the form of land cover change from natural vegetation to desert) causes a decrease of precipitation that resembles the megadrought observed in the Sahel region from the late 1960s on; Wang and Eltahir [5] found that feedback from natural vegetation dynamics enhances (and sustains) the impact of both oceanic forcing and man-made desertification in causing the 20th century Sahel drought. Feedback due to natural vegetation dynamics also acts as a damper to the climate system, enhancing the multidecadal variability of precipitation in the Sahel region [1, 13–16]. Thus, contributions of vegetation dynamics warrant consideration in simulations of future climate. Paeth et al. [7] carried out regional climate predictions up to the year 2050 forced with future CO_2 concentration and land use changes for A1B and B1 emission scenarios and found that the simulated climate response in West and Central Africa is primarily dominated by a drought caused by land use changes. Although the land use change signal may become less dominant if the prediction was done for a higher CO_2 concentration (e.g., for the A2 scenario or for a later stage of the century), their study clearly demonstrated the important role of land use as a climate change forcing. Incorporating natural vegetation dynamics into a regional climate model used for future prediction, Alo and Wang [8] found that natural vegetation feedback leads to a significant increase of future precipitation in most of West Africa.

The strong water dependency of the socioecological system in West Africa and the fragility of food security in the region make it extremely vulnerable to climate changes. However, predictions for precipitation changes in the 2nd half of the 21st century (mostly from atmosphere-ocean general circulation models (AOGCMs)) in the West and North Africa are highly model dependent and lack any general consensus [17]. To specifically address this uncertainty in future predictions, Patricola and Cook [18] generated an ensemble of future predictions for this region using a regional climate model driven with seas surface temperature and lateral boundary conditions from nine AOGCMs and found a relatively good agreement among different ensemble members despite the lack of consensus among the driving AOGCMs. For West Africa, their regional model predicts a generally wetter condition with the exception of a mid-season drought in June and July.

Crop productivity in the Sahel is limited not only by the amount of precipitation during the rainy season but more importantly by the length of the growing/rainy season. In most of the Sahel belt, the length of the rainy season is less than three months. Although the preonset rainfall can amount to a few millimeters on some days, it is scattered and cannot provide reliable source of water for agriculture. The optimal planting time for crops is found to occur within a few days of the large-scale monsoon onset time [19], which is defined as the sudden shift of the ITCZ maximum from around 5 N to around 10 N [20]. Based on projections from multiple general circulation models (GCMs), it was found that most monsoon systems in the 21st century (West Africa monsoon included) will experience a change in precipitation seasonality, with a precipitation decrease in spring/early summer and an increase in late summer/early fall [21, 22]. However, due to the coarse spatial resolution used, most GCMs cannot simulate the sudden onset of the West African monsoon, an important feature of the seasonal variability of this regional climate [23]. Regional climate models (RCMs) have a much better performance in capturing the sudden monsoon onset [24]. Given the strong latitudinal gradient of precipitation in West Africa and the importance of monsoon onset for West African agriculture, it is necessary to assess future changes of precipitation seasonality using RCMs that have a finer spatial resolution than most GCMs.

While still uncommon among the IPCC projections, some GCMs have included dynamic vegetation in future prediction experiments [25, 26]. The inclusion of dynamic vegetation in RCMs however is still at a very early stage [27] and has not been applied to climate predictions. In a preliminary study asynchronously linking a regional climate model with a dynamic vegetation model, Alo and Wang [8] demonstrated that future vegetation in West Africa will become denser, but the vegetation composition will shift towards more drought deciduous trees and grass at the expense of evergreen trees. Compared with the prediction assuming static vegetation, vegetation dynamics acts to increase precipitation amount over most of West Africa except for the narrow band along the coast. This current study is based on the same set of experiments used in Alo and Wang [8], but focuses on changes in precipitation seasonality and monsoon onset in West Africa and how vegetation dynamics may influence them.

2. Model and Experimental Design

The model used is the International Centre for Theoretical Physics (ICTP) regional climate model (RegCM3) asynchronously coupled to the National Center for Atmospheric Research (NCAR) Community Land Model's Dynamic Global Vegetation Model (CLM-DGVM). CLM-DGVM [26] simulates biogeophysical, biogeochemical, and ecosystem dynamics processes at the land surface. Its biogeophysical submodel is CLM3.0 [28, 29] and the biogeochemical and vegetation dynamics submodels are based on the Lund-Potsdam-Jena Dynamic Global Vegetation Model (LPJ-DGVM) [30], with a plant phenology scheme modified based on the Integrated Biosphere Simulator (IBIS) [31]. RegCM3 [32] is a limited-area model with a terrain-following σ-pressure vertical coordinate system and uses the Biosphere-Atmosphere-Transfer Scheme (BATS1E) [33] as its land surface biogeophysical submodel. We configure RegCM3 at 50 km horizontal resolution over a domain spanning approximately 12°S-24°N and 24°W-30°E, with 18 levels in the vertical direction. Among the three different convection parameterizations in RegCM3, the MIT-Emanuel scheme [34] option is chosen based on performance over the study domain. RegCM3 has been shown to perform well in its simulation of precipitation over West Africa [35–37].

In the asynchronous coupling approach, CLM3-DGVM is run at the same spatial and temporal resolutions and over the same domain as RegCM3, but the two models alternate. Specifically, RegCM3 is run for one year with an initial state of vegetation, and the resulting atmospheric forcing for each time step is saved; this saved atmospheric forcing is then read into CLM3-DGVM to drive a transient vegetation simulation for one year, producing an updated vegetation state for RegCM3. This iteration continues through the duration of the simulation. Note that this approach allows the simulation of transient vegetation dynamics, which is different from most other asynchronously coupled models in which an *equilibrium* vegetation state is predicted for each iteration based on the climate state simulated during that iteration [38, 39].

The asynchronously coupled RegCM3/CLM3-DGVM model is used to conduct two simulations, a present-day climate-vegetation run "Present" and a future climate-vegetation run "Future". In addition, a future climate simulation "Future_nv" is conducted using RegCM3 but with vegetation fixed at the present-day state (i.e., prescribed based on output from "Present"), to facilitate quantifying the contribution from dynamic vegetation feedback. Based on this design, climate and hydrological differences between "Future_nv" and "Present" result from the radiative and physiological effects of elevated CO_2, differences between "Future" and "Future_nv" represent the contribution from structural vegetation feedback, and differences between "Future" and "Present" result from the combination of CO_2 radiative and physiological effects and vegetation structural feedback. Each of the simulations is 15-year long, and the first five years of data is discarded as model spinup. The initial and lateral boundary conditions as well as sea surface temperature for RegCM3 are provided by the NCAR global climate system model CCSM3's IPCC AR4 simulations. Although regional climate models are known to be sensitive to lateral boundary conditions, Patricola and Cook [18] found a relatively high level of consensus among predictions by a regional climate model driven with nine different GCMs despite the lack of consensus among these nine driving GCMs. The use of lateral boundary conditions from a single GCM in this study is therefore not expected to a cause too high level of uncertainty. The period 1979–1993 from the CCSM3 20C3M simulation is used for the present-day climate, and 2079–2093 from the CCSM3 SRESA1B experiment is used for the future climate. CO_2 concentration is set to 356 ppmv for the present-day climate and 720 ppmv for future. Initial vegetation conditions for the asynchronously coupled RegCM3/CLM-DGVM runs are derived from CLM-DGVM long-term simulations forced with the 1979–1993 (2079–2093) climate from the CCSM3 20C3M (SRESAIB) output under corresponding CO_2 concentration, interpolated to the RegCM3 spatial resolution. Based on the results of Alo and Wang [8] who examined the vegetation distribution and future changes simulated by CLM-DGVM driven with climate forcing from eight GCMs, the vegetation distribution corresponding to the CCSM3 climate is the most realistic among the eight.

3. Results

When driven by initial and lateral boundary conditions from CCSM3, the asynchronously coupled RegCM3/CLM3-DGVM simulates the present-day precipitation pattern and vegetation distribution reasonably well in comparison with observations. Details about model performance in simulating the regional climate and vegetation are provided in Alo and Wang [8]. The focus of the following analysis is on future changes of precipitation, its seasonal cycle, the West African monsoon onset, and on the impact of vegetation dynamics in these predictions.

In response to the radiative and physiological effects of elevated atmospheric CO_2, RegCM3 without accounting for vegetation dynamics predicts a decrease of annual rainfall in most of West Africa, with an increase over a narrow coastal band and the eastern Sahel region (Figure 1(a)). This decrease may be partly related to the lower stomatal conductance at higher CO_2 concentration, which potentially suppresses transpiration therefore local moisture supply. When the dynamic vegetation feedback is included, the predicted future drought signal is limited to part of the Guinea Coast band only, and annual rainfall is predicted to increase or remain about the same over most of the Sahel (Figure 1(b)). The feedback from dynamic vegetation leads to a substantial increase of precipitation over most of the model domain except for the narrow band along the coast (Figure 1(c)). These changes of precipitation are accompanied by a northward expansion of grassland in the north and increase of broadleaf drought-deciduous tree coverage in the south at the expense of evergreen trees (Figure 2). Compared with the future vegetation cover changes predicted by CLM3-DGVM driven directly with the CCSM3 A1B climate projection [40], the changes predicted by the RegCM3/CLM3-DGVM system as shown in Figure 2 are significantly larger in magnitude (and therefore faster in rate). Across most of West Africa, vegetation is predicted to become denser (regardless of how the vegetation distribution shifts) as a result of the CO_2 fertilization effects dominating over other factors (results not shown; see Alo and Wang [8] for details).

Similar to present-day observations (e.g., [41]), both the present-day and future climates in the model (black and red contour lines in Figure 3) are characterized by substantial amounts of rain during the preonset season followed by the abrupt monsoon onset in late June. Following Sultan and Janicot [20], here we define the monsoon onset as the timing of the sudden shift of rain maximum from 5-6°N to 8–10°N. The monsoon onset is characterized by an abrupt increase of rainfall at about 10°N in late June, which is very apparent from the contour lines in Figure 3. The monsoon rainfall peaks in August and gradually retreats afterwards, which are also consistent with present-day observations of the monsoon characteristics [20].

The predicted future precipitation changes follow a specific seasonal pattern. Over the Sahel region where vegetation growth and crop production are limited by water, future precipitation will decrease early in the rainy season (during late spring/early summer) and increase later (during late summer and early fall) (shading in Figure 3). Despite the rainfall decrease during the preonset season, future rainfall will become more abundant and penetrates further inland after the monsoon is fully developed. This seasonal pattern is similar regardless of whether dynamic vegetation feedback is accounted for. However, vegetation dynamics does make a difference in the magnitude of such changes. Without accounting for vegetation feedback (Figure 3(a)), rainfall is predicted to decrease substantially (by more than 60% in some areas) during spring and early summer across the region, with only minor increase later on. The decrease of rainfall occurs to both the preonset rain and during the peak monsoon. When vegetation dynamics is included (Figure 3(b)), the predicted future decrease of rainfall is much smaller and limited to primarily the preonset rain, with a much larger increase during the peak monsoon and the retreating phase.

These predicted changes in precipitation seasonal cycle in the Sahel region are consistent with previous findings concerning most monsoon systems based on GCM ensembles

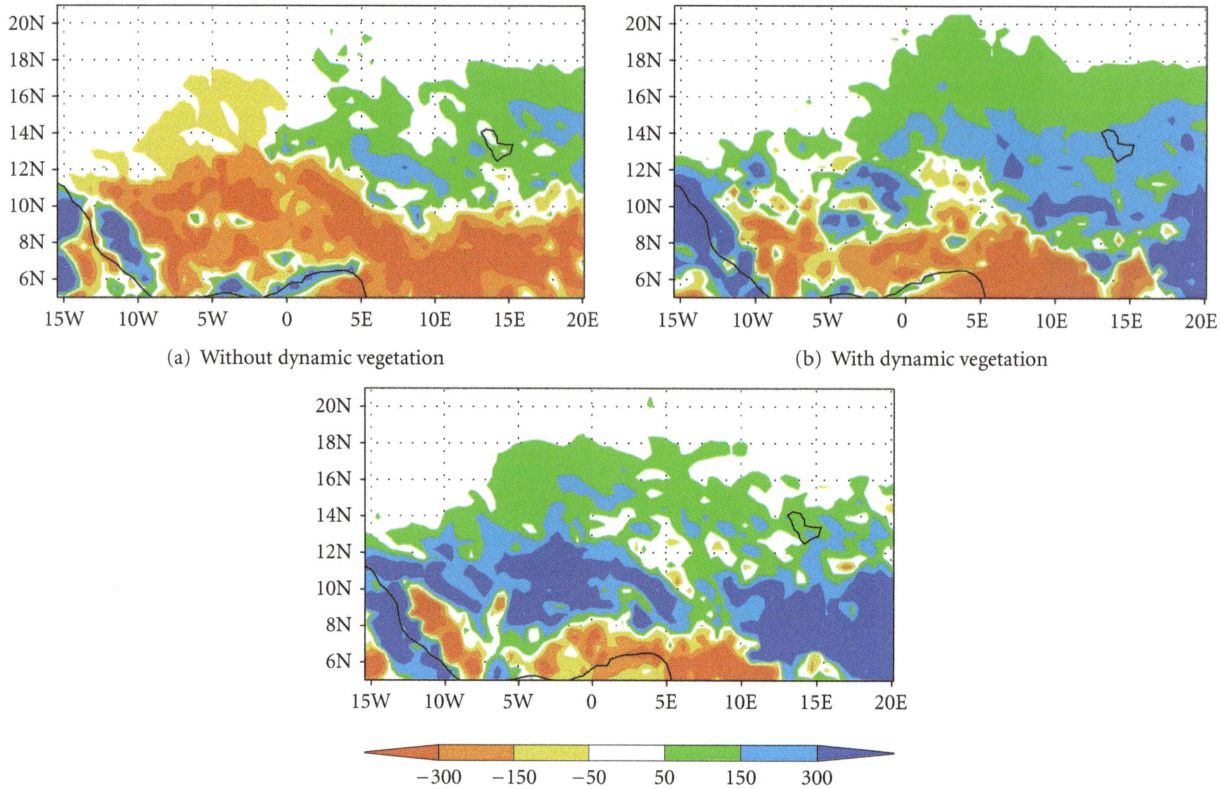

(a) Without dynamic vegetation

(b) With dynamic vegetation

(c) Impact of dynamic vegetation

FIGURE 1: Changes in annual precipitation (in mm) simulated without dynamic vegetation (i.e., "Future_nv"—"Present"), with dynamic vegetation (i.e., "Future"—"Present"), and the contribution due to dynamic vegetation feedback (i.e., "Future"—"Future—nv").

[21, 22]. Seth et al. [21] suggested that such changes might be related to the enhancement of spring convective barrier due to warming-induced drought. Figure 4 presents the predicted future changes in troposphere stability, where the troposphere stability is represented by the moist static energy difference between 200 mb and 850 mb following Seth et al. [21]. A strong increase of troposphere stability is evident in the Sahel region during all seasons except for the active monsoon season when the troposphere stability is weakened. It is apparent that future climate features a stronger contrast in troposphere stability between the wet and dry seasons. Feedback from vegetation dynamics is found to reduce the future troposphere stability and favor more precipitation across all seasons, probably due to the increased specific humidity caused by enhancement of evapotranspiration by denser vegetation.

The changes in precipitation amount and seasonal cycle do not seem to be related to changes in the timing of the monsoon circulation. Using 12°N and 15°N as examples, Figure 5 shows the daily time series of zonally averaged wind and precipitation. Despite the strong decrease in the preonset precipitation, there is no clear evidence for a delay in the timing of zonal or meridional wind reversal. The timing of the wind reversal in the model is remarkably robust against future climate changes and against how vegetation is treated. To define the monsoon onset following the established approach of Sultan and Janicot [20], that is, the timing of a sudden change in the slope of precipitation increase, Figure 6 also shows the 11-day running mean of precipitation at 10°N. The crosses in Figure 6 indicate the approximate timing for monsoon onset. The model predicts a five-day delay in future monsoon onset without accounting for the impact of vegetation dynamics, and dynamic vegetation feedback reduces this delay to two days. As evident in Figure 6, a slight delay of monsoon retreat might be expected, but the magnitude of such delay is very small.

Despite the decrease of preonset rainfall and the slightly delayed monsoon onset, atmospheric specific humidity increases over land after the monsoon onset, as a result of moisture import from over the ocean where warming-induced increase of evaporation leads to a significant increase of atmospheric moisture (Figure 7). This increase of moisture partly contributes to the enhanced precipitation and the weakening of troposphere stability after the monsoon is fully established (Figures 3 and 4). During the peak monsoon season (e.g., August), the ocean-to-land gradient of moist static energy in the lower troposphere is predicted to become much stronger in the future, and the peak of the moist static energy is located further inland in the case with dynamic vegetation feedback (Figure 8). This, according to Eltahir and Gong [42], suggests a stronger monsoon circulation and a more northward penetration of the monsoon rainfall, consistent with results in Figures 3 and 5.

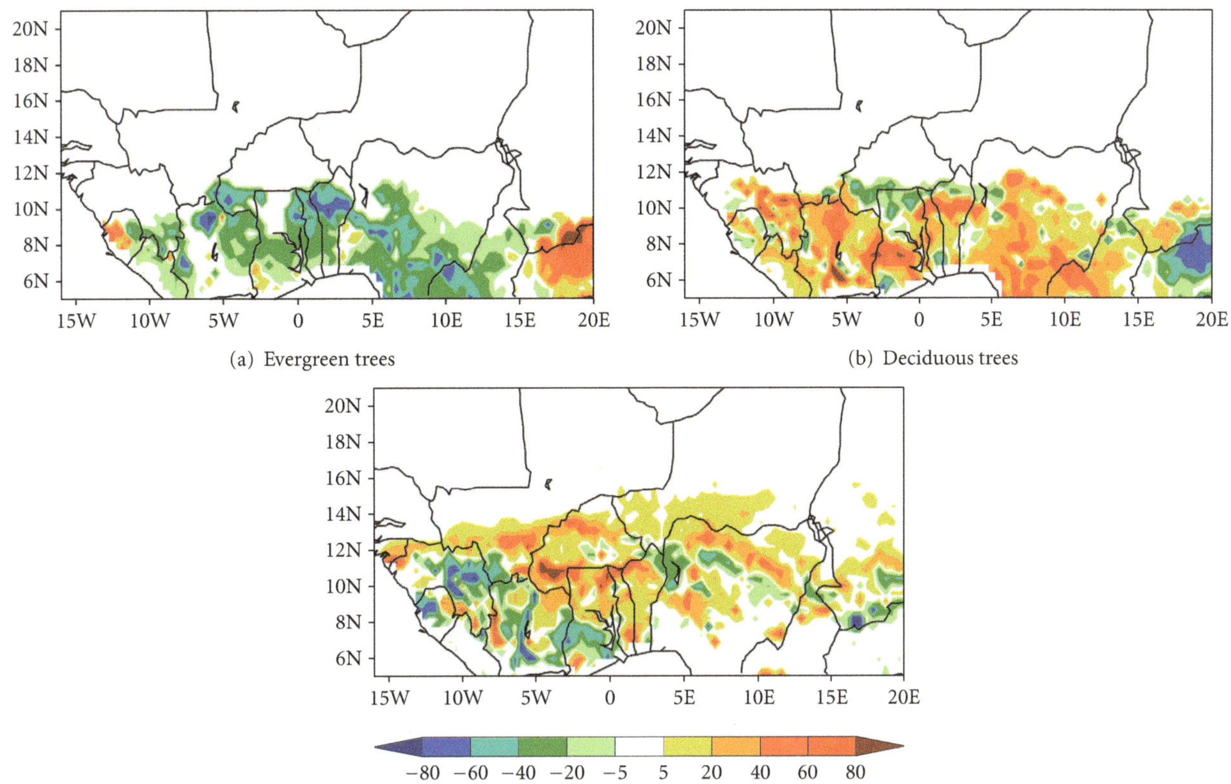

(a) Evergreen trees

(b) Deciduous trees

(c) Grasses

Figure 2: Changes in fractional coverage of vegetation types ("Future"—"Present") as percent of vegetated portion of grid cell.

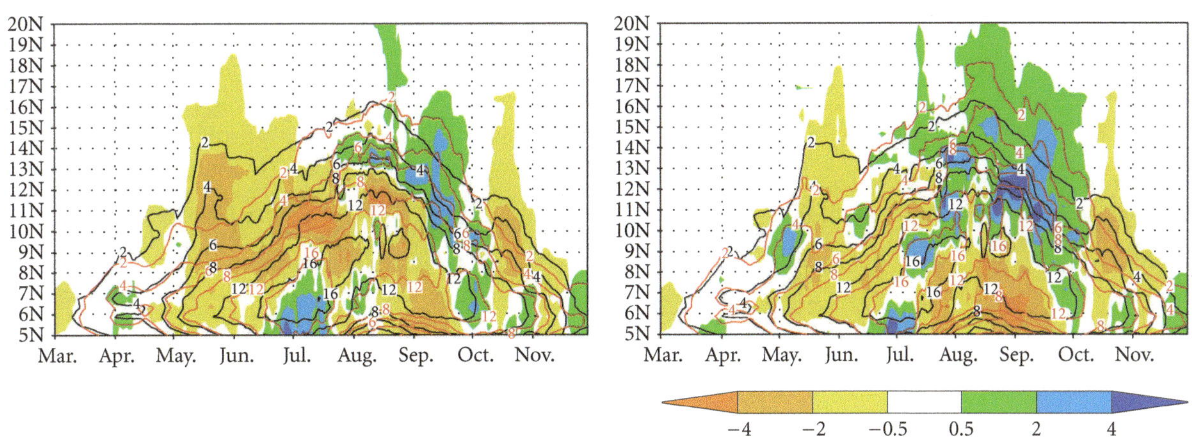

(a) Without dynamic vegetation

(b) With dynamic vegetation

Figure 3: Predicted daily precipitation changes (color shading, in mm/day) during March–November, with variability more frequent than 11 days filtered out. The contour lines represent the 11-day filtered daily precipitation during 1984–1993 (black) and during 2084–2093 (red).

4. Summary and Discussion

Using the regional climate model RegCM3 driven with lateral boundary conditions from the NCAR CCSM3 based on the SRES A1b scenario, this study examines future climate predictions for the West Africa region and how dynamic vegetation feedback may influence such predictions. Major findings are summarized in the following.

(1) Vegetation dynamics reverses the predicted future trend of annual precipitation in the Sahel region, from a drying trend to a wet trend. Over the Guinea Coast region, vegetation dynamics has little impact on future precipitation predictions.

(2) The predicted future trend of precipitation in the Sahel region follows a specific seasonal pattern, with

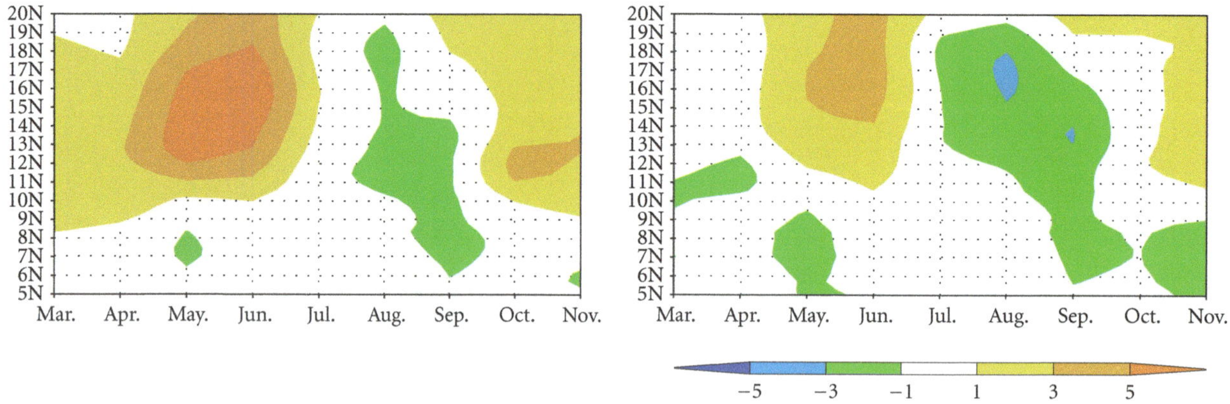

(a) Without dynamic vegetation

(b) With dynamic vegetation

FIGURE 4: Changes in the troposphere stability between 1984–1993 and 2084–2093, averaged over 5°W-5°E, simulated without dynamic vegetation (a) and with dynamic vegetation (b). The troposphere stability is represented by the moist static energy difference between 200 mb and 850 mb (MSE_{200}–MSE_{850}).

a decrease during the preonset and early monsoon seasons and an increase after the monsoon is fully developed. Dynamic vegetation feedback reduces the magnitude of the predicted rainfall reduction early in the season and increases the magnitude of the predicted rainfall increase later in the season.

(3) The decrease of preonset rainfall is related to the warming-induced enhancement of convective barrier during spring, while the increase of precipitation during the peak monsoon results from the increased specific humidity as the monsoon circulation brings more moisture from a warmer ocean.

(4) Despite the substantial decrease of preonset rainfall in most of West Africa, the model predicts no detectable delay in the timing of the monsoon circulation (e.g., timing of wind reversal) and only a slight (if any) delay in the onset of the monsoon rainfall. This statement holds regardless of how vegetation is treated.

Druyan [17] reviewed ten major papers on future precipitation prediction in this region and found no consensus among different models regarding future precipitation trend. There is some similarity between our results and the study of Patricola and Cook [18] in that future precipitation trend differs between different stages of the monsoon development, but the two differ significantly for the pre- and early-monsoon stages. The simulated robust increase of precipitation in this study during the fully developed monsoon phase is consistent with the finding of Fontaine et al. [43]. Fontaine et al. [43] focused on the July-August-September (JAS) precipitation for the Sahel region and found that a vast majority of the 12 GCMs examined predict an increase of JAS precipitation for the Sahel region. The increase of precipitation during JAS however does not necessarily mean rainfall excess for plant growth. As our results indicate, the future trend of precipitation during the preonset season can be very different from the JAS season. The decrease of rainfall in the preonset season found in our study can lead to a delayed and in some areas (e.g., south of 12°N) shortened growing season. North of 12°N, however, the increase of future precipitation is dominant and the growing season may become longer.

The strong increase of precipitation across the rainy season north of 12°N is consistent with the expansion of grass cover in Figure 2. While the rainfall excess supports more growth, the increased grass cover lowers the surface albedo and enhances evapotranspiration, both of which favor more precipitation over this highly sensitive region. The vegetation feedback can take place both at the intraseasonal time scale and over interannual time scale. At the glassland region, it is more straightforward at the intraseasonal time scale through the above-ground vegetative response. At the interannual time scale, this feedback can be facilitated by the root system of perennial grasses and/or seed bank of annual grasses, with the first flush of vegetation growth at the beginning of the rainy season supported by carbon storage in the roots resulting from previous year's carbon production or seeds produced in the previous year. In the real world, both mechanisms can support inter-annual land memory in the Sahel region found in the Normalized Difference Vegetation Index (NDVI) data [44]. In the model used here, vegetation feedback at the inter-annual time sale over grassland region relies on roots of perennial grass, as annual grass is not simulated in the model.

The use of different land surface biogeophysical schemes (BATS versus CLM3) creates a certain level of inconsistency between RegCM3 and CLM3-DGVM, and such inconsistency is intrinsic to all asynchronous coupling approaches (e.g., [38, 39, 45, 46]). This also creates the need for mapping between the BATS vegetation categorization and

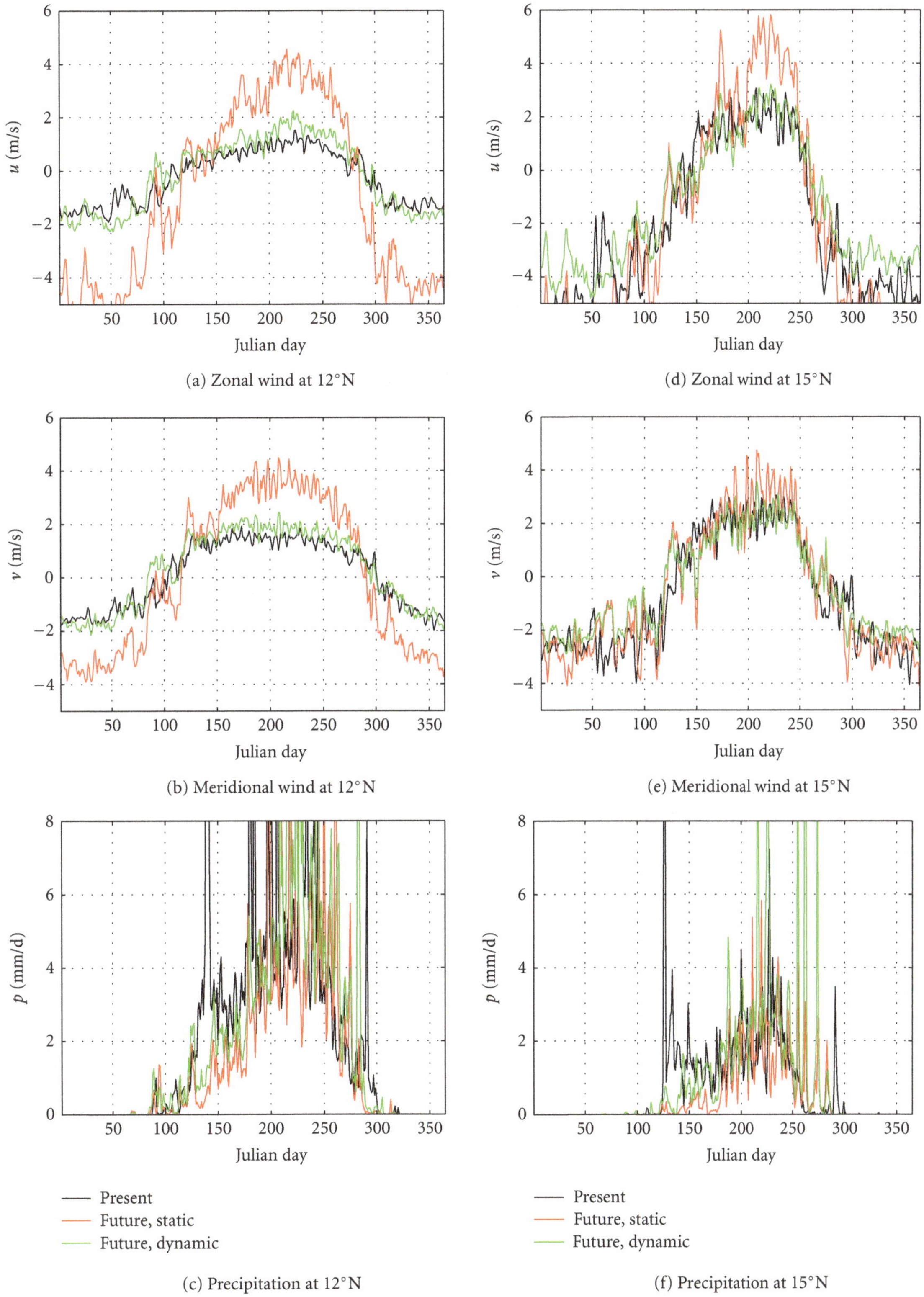

(a) Zonal wind at 12°N

(d) Zonal wind at 15°N

(b) Meridional wind at 12°N

(e) Meridional wind at 15°N

(c) Precipitation at 12°N

(f) Precipitation at 15°N

FIGURE 5: Climatology of daily wind and daily rainfall at 12°N and 15°N, averaged over 5°W-5°E.

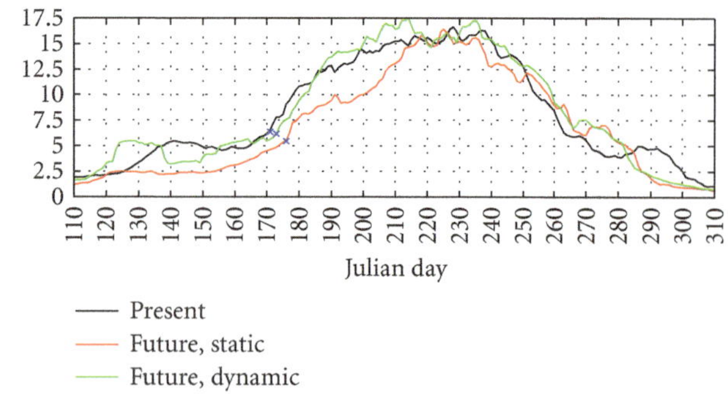

FIGURE 6: 11-day running mean of the zonally averaged daily precipitation (mm) at 10°N, with the monsoon onset approximately marked by "x".

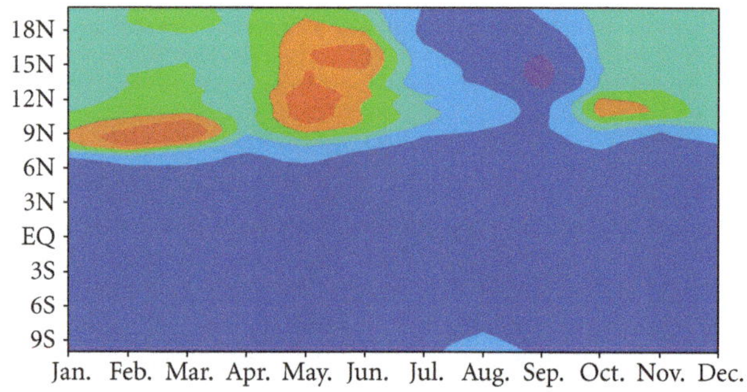

(a) Without dynamic vegetation

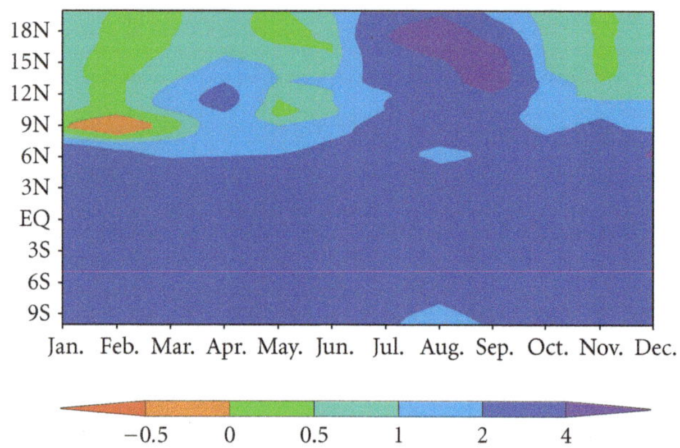

(b) With dynamic vegetation

FIGURE 7: Monthly mean changes in surface specific humidity (g/kg), averaged over 5°W-5°E.

the vegetation plant functional types simulated by CLM-DGVM. It is not clear how this inconsistency may have influenced the results, especially results on the modifications of future prediction by vegetation dynamics. To address this uncertainty, we are currently developing a synchronously coupled version of the model, which involves incorporating vegetation dynamics into the latest version of the regional climate model RegCM4 [47] that uses CLM3.5 (as opposed to BATS in RegCM3) as its land surface model.

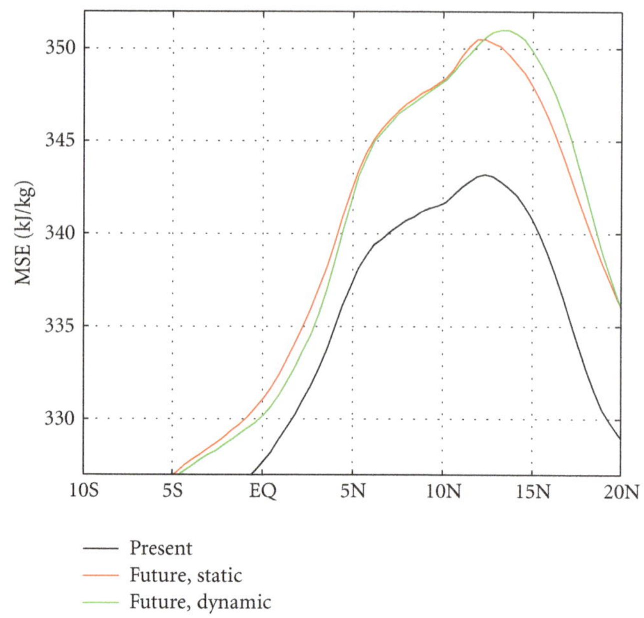

FIGURE 8: Low-level moist static energy averaged during the peak monsoon month August and over 5°W-5°E.

Acknowledgment

This work is supported by funding from the National Science Foundation (NSF) under Agreements ATM-0531485 and AGS-1063986.

References

[1] G. Wang and E. A. B. Eltahir, "Role of vegetation dynamics in enhancing the low-frequency variability of the Sahel rainfall," *Water Resources Research*, vol. 36, no. 4, pp. 1013–1021, 2000.

[2] A. Giannini, R. Saravanan, and P. Chang, "Oceanic forcing of sahel rainfall on interannual to interdecadal time scales," *Science*, vol. 302, no. 5647, pp. 1027–1030, 2003.

[3] M. Biasutti, I. M. Held, A. H. Sobel, and A. Giannini, "SST forcings and Sahel rainfall variability in simulations of the twentieth and twenty-first centuries," *Journal of Climate*, vol. 21, no. 14, pp. 3471–3486, 2008.

[4] Y. Xue and J. Shukla, "The influence of land surface properties on Sahel climate. Part I: desertification," *Journal of Climate*, vol. 6, no. 12, pp. 2232–2245, 1993.

[5] G. Wang and E. A. B. Eltahir, "Ecosystem dynamics and the Sahel drought," *Geophysical Research Letters*, vol. 27, no. 6, pp. 795–798, 2000.

[6] C. M. Patricola and K. H. Cook, "Atmosphere/vegetation feedbacks: a mechanism for abrupt climate change over northern Africa," *Journal of Geophysical Research D*, vol. 113, no. 18, Article ID D18102, 2008.

[7] H. Paeth, K. Born, R. Girmes, R. Podzun, and D. Jacob, "Regional climate change in tropical and Northern Africa due to greenhouse forcing and land use changes," *Journal of Climate*, vol. 22, no. 1, pp. 114–132, 2009.

[8] C. A. Alo and G. Wang, "Role of dynamic vegetation in regional climate predictions over Western Africa," *Climate Dynamics*, vol. 35, no. 5, pp. 907–922, 2010.

[9] M. Claussen, C. Kubatzki, V. Brovkin, A. Ganopolski, P. Hoelzmann, and H. J. Pachur, "Simulation of an abrupt change in Saharan vegetation in the mid-Holocene," *Geophysical Research Letters*, vol. 26, no. 14, pp. 2037–2040, 1999.

[10] M. M. Irizarry-Ortiz, G. Wang, and E. A. B. Eltahir, "Role of the biosphere in the mid-Holocene climate of West Africa," *Journal of Geophysical Research D*, vol. 108, no. 2, pp. 5–15, 2003.

[11] Z. Liu, Y. Wang, R. Gallimore, M. Notaro, and I. C. Prentice, "On the cause of abrupt vegetation collapse in North Africa during the Holocene: climate variability vs. vegetation feedback," *Geophysical Research Letters*, vol. 33, no. 22, Article ID L22709, 2006.

[12] M. Notaro, Y. Wang, Z. Liu, R. Gallimore, and S. Levis, "Combined statistical and dynamical assessment of simulated vegetation-rainfall interactions in North Africa during the mid-Holocene," *Global Change Biology*, vol. 14, no. 2, pp. 347–368, 2008.

[13] N. Zeng, J. D. Neelin, K. M. Lau, and C. J. Tucker, "Enhancement of interdecadal climate variability in the Sahel by vegetation interaction," *Science*, vol. 286, no. 5444, pp. 1537–1540, 1999.

[14] G. Wang, E. A. B. Eltahir, J. A. Foley, D. Pollard, and S. Levis, "Decadal variability of rainfall in the Sahel: results from the coupled GENESIS-IBIS atmosphere-biosphere model," *Climate Dynamics*, vol. 22, no. 6-7, pp. 625–637, 2004.

[15] C. Delire, J. A. Foley, and S. Thompson, "Long-tern variability in a coupled atmosphere-biosphere model," *Journal of Climate*, vol. 17, no. 20, pp. 3947–3959, 2004.

[16] C. Delire, N. de Noblet-Ducoudré, A. Sima, and I. Gouirand, "Vegetation dynamics enhancing long-term climate variability confirmed by two models," *Journal of Climate*, vol. 24, no. 9, pp. 2238–2257, 2011.

[17] L. M. Druyan, "Studies of 21st-century precipitation trends over West Africa," *International Journal of Climatology*, vol. 31, no. 10, pp. 1415–1424, 2011.

[18] C. M. Patricola and K. H. Cook, "Northern African climate at the end of the twenty-first century: an integrated application of regional and global climate models," *Climate Dynamics*, vol. 35, no. 1, pp. 193–212, 2010.

[19] B. Sultan, C. Baron, M. Dingkuhn, B. Sarr, and S. Janicot, "Agricultural impacts of large-scale variability of the West African monsoon," *Agricultural and Forest Meteorology*, vol. 128, no. 1-2, pp. 93–110, 2005.

[20] B. Sultan and S. Janicot, "The West African monsoon dynamics. Part II: the "preonset" and "onset" of the summer monsoon," *Journal of Climate*, vol. 16, no. 21, pp. 3407–3427, 2003.

[21] A. Seth, S. A. Rauscher, M. Rojas, A. Giannini, and S. J. Camargo, "Enhanced spring convective barrier for monsoons in a warmer world? A letter," *Climatic Change*, vol. 104, pp. 403–414, 2011.

[22] M. Biasutti and A. H. Sobel, "Delayed Sahel rainfall and global seasonal cycle in a warmer climate," *Geophysical Research Letters*, vol. 36, no. 23, Article ID L23707, 2009.

[23] Y. Xue, F. de Sales, W. K. M. Lau et al., "Intercomparison and analyses of the climatology of the West African Monsoon in the West African Monsoon Modeling and Evaluation project (WAMME) first model intercomparison experiment," *Climate Dynamics*, vol. 35, no. 1, pp. 3–27, 2010.

[24] L. M. Druyan, J. Feng, K. H. Cook et al., "The WAMME regional model intercomparison study," *Climate Dynamics*, vol. 35, no. 1, pp. 175–192, 2010.

[25] P. M. Cox, R. A. Betts, C. D. Jones, S. A. Spall, and I. J. Totterdell, "Acceleration of global warming due to carbon-cycle feedbacks in a coupled climate model," *Nature*, vol. 408, no. 6809, pp. 184–187, 2000.

[26] G. Bala, K. Caldeira, A. Mirin, M. Wickett, C. Delire, and T. J. Phillips, "Biogeophysical effects of CO_2 fertilization on global climate," *Tellus, Series B*, vol. 58, no. 5, pp. 620–627, 2006.

[27] J. M. Winter, J. S. Pal, and E. A. B. Eltahir, "Coupling of integrated biosphere simulator to regional climate model version 3," *Journal of Climate*, vol. 22, no. 10, pp. 2743–2757, 2009.

[28] Y. Dai, X. Zeng, R. E. Dickinson et al., "The common land model," *Bulletin of the American Meteorological Society*, vol. 84, no. 8, pp. 1013–1023, 2003.

[29] K. Oleson, Y. Dai, G. B. Bonan et al., "Technical description of the community land model (CLM)," NCAR Technical Note TN-461+STR, 2004.

[30] S. Sitch, B. Smith, I. C. Prentice et al., "Evaluation of ecosystem dynamics, plant geography and terrestrial carbon cycling in the LPJ dynamic global vegetation model," *Global Change Biology*, vol. 9, no. 2, pp. 161–185, 2003.

[31] C. J. Kucharik, J. A. Foley, C. Delire et al., "Testing the performance of a dynamic global ecosystem model: water balance, carbon balance, and vegetation structure," *Global Biogeochemical Cycles*, vol. 14, no. 3, pp. 795–825, 2000.

[32] J. S. Pal, F. Giorgi, X. Bi et al., "Regional climate modeling for the developing world: the ICTP RegCM3 and RegCNET," *Bulletin of the American Meteorological Society*, vol. 88, no. 9, pp. 1395–1409, 2007.

[33] R. E. Dickinson, A. Henderson-Sellers, and P. J. Kennedy, "Biosphere-atmosphere transfer scheme (BATS) version 1e as coupled to the NCAR community climate model," NCAR Tech. Note/TN-387+STR, 1993.

[34] K. A. Emanuel, "A scheme for representing cumulus convection in large-scale models," *Journal of the Atmospheric Sciences*, vol. 48, no. 21, pp. 2313–2335, 1991.

[35] G. S. Jenkins, A. T. Gaye, and B. Sylla, "Late 20th century attribution of drying trends in the Sahel from the regional climate model (RegCM3)," *Geophysical Research Letters*, vol. 32, no. 22, Article ID L22705, pp. 1–4, 2005.

[36] E. A. Afiesimama, J. S. Pal, B. J. Abiodun, W. J. Gutowski, and A. Adedoyin, "Simulation of West African monsoon using the RegCM3. Part I: model validation and interannual variability," *Theoretical and Applied Climatology*, vol. 86, no. 1–4, pp. 23–37, 2006.

[37] J. B. Omotosho and B. J. Abiodun, "A numerical study of moisture build-up and rainfall over West Africa," *Meteorological Applications*, vol. 14, no. 3, pp. 209–225, 2007.

[38] N. S. Diffenbaugh, "Atmosphere-land cover feedbacks alter the response of surface temperature to CO_2 forcing in the western United States," *Climate Dynamics*, vol. 24, no. 2-3, pp. 237–251, 2005.

[39] K. H. Cook and E. K. Vizy, "Effects of twenty-first-century climate change on the Amazon rain forest," *Journal of Climate*, vol. 21, no. 3, pp. 542–560, 2008.

[40] C. A. Alo and G. Wang, "Potential future changes of the terrestrial ecosystem based on climate projections by eight general circulation models," *Journal of Geophysical Research G*, vol. 113, no. 1, Article ID G01004, 2008.

[41] S. Janicot, G. Caniaux, F. Chauvin et al., "Intraseasonal variability of the West African monsoon," *Atmospheric Science Letters*, vol. 12, pp. 58–66, 2011.

[42] E. A. B. Eltahir and C. L. Gong, "Dynamics of wet and dry years in West Africa," *Journal of Climate*, vol. 9, no. 5, pp. 1030–1042, 1996.

[43] B. Fontaine, P. Roucou, and P.-A. Monerie, "Changes in the African monsoon region at medium-term time horizon using 12 AR4 coupled models under the A1b emissions scenario," *Atmospheric Science Letters*, vol. 12, no. 1, pp. 83–88, 2011.

[44] N. Philippon, L. Jarlan, N. Martiny, P. Camberlin, and E. Mougin, "Characterization of the interannual and intraseasonal variability of West African vegetation between 1982 and 2002 by means of NOAA AVHRR NDVI data," *Journal of Climate*, vol. 20, no. 7, pp. 1202–1218, 2007.

[45] M. Claussen, "On multiple solutions of the atmosphere-vegetation system in present-day climate," *Global Change Biology*, vol. 4, no. 5, pp. 549–559, 1998.

[46] R. A. Betts, P. M. Cox, and F. I. Woodward, "Simulated responses of potential vegetation to doubled-CO_2 climate change and feedbacks on near-surface temperature," *Global Ecology and Biogeography*, vol. 9, no. 2, pp. 171–180, 2000.

[47] A. L. Steiner, J. S. Pal, S. A. Rauscher et al., "Land surface coupling in regional climate simulations of the West African monsoon," *Climate Dynamics*, vol. 33, no. 6, pp. 869–892, 2009.

5

Comparative Study of the West African Continental, Coastal, and Marine Atmospheric Profiles during the Summer of 2006

Ibrahima Kalil Kante,[1,2] Daouda Badiane,[1] Saïdou Moustapha Sall,[1] Abdoulaye Deme,[1] and Arona Diedhiou[3]

[1] *Laboratoire de Physique de l'Atmosphère et de l'Océan Siméon Fongang, Ecole Supérieure Polytechnique, Université Cheikh Anta Diop Dakar-Sénégal, BP: 5085, Dakar, Senegal*
[2] *Direction Nationale de la Météorologie de Guinée-Conakry, BP: 566 Rep., Guinea*
[3] *IRD/LTHE, BP 53, 38041 Grenoble Cedex 9, France*

Correspondence should be addressed to Daouda Badiane, daouda.badiane@ucad.edu.sn

Academic Editor: Gregory S. Jenkins

We used sounding data of the Multidisciplinary Analysis of the African Monsoon experience in summer 2006 at continental and coastal sites of West Africa, respectively, to analyze the vertical profiles of relative humidity, temperature, dew point, and speed and wind direction for the JJAS rainy period. The vertical gradient method is applied to the profiles of some thermodynamic parameters estimated from sounding data to do a comparative study of the structure and thermal properties, moisture, and static stability of the atmospheric boundary layer of inland, coastal, and marine sites to show consistent differences related to geographic factors. In vertical profiles of relative humidity, the intensity is higher in Dakar than in Niamey particularly in the core of the season. There are dry intrusions in the low levels at the beginning and end of the season in Dakar, which do not exist in Niamey. The mixing layer on the continent during the day can reach a height greater than 1100 m, and the inversion layer height can exceed 1700 m. Therefore, the maximum thickness of the boundary layer is observed on the continent during the day, while at night the marine boundary layer is the thickest. The diurnal evolution shows that the mixing layer thickness decreases during the night over the continent but increases at the coast and at sea. In the night at the continental site there is a division of the mixing layer with a consistent residual mixing layer. Continental boundary layer is more unstable during the day, while at night it is the marine boundary layer that is more unstable than the coastal and inland ones.

1. Introduction

The term boundary layer was first introduced in the literature by Prandtl and Lustig [1]. Since then, many authors such as Turner [2], Monin [3], Zeman [4] and more, recently, Cushman-Roisin and Beckers [5] have studied the detailed description. The atmospheric boundary layer (ABL) has been variously defined, but it is commonly recognized as the lower part of the atmosphere, which is strongly influenced by the presence of the earth's surface and responds to surface forcing with a timescale of about one hour or less. The ABL is the place where many processes, such as turbulence, friction, dispersion, energy dissipation, and wind shear occur. These processes are poorly parameterized in atmospheric models. Several ABL studies have focused on the turbulence but also on the mixing process triggered by significant warming or cooling [6]. Within the ABL, the transport of various quantities (heat, pollution, moisture, momentum, etc.) is mainly influenced horizontally by the wind and vertically by the turbulence. The importance of the entrainment process of the air at the top of the boundary layer over the continent has been studied by Betts et al. [7] by analyzing the variations of potential temperature (θ_e).

However, many gaps exist in understanding the ABL, particularly in West Africa where the ABL is poorly studied due to lack of data. Indeed Johnson [8] studied the GATE (GARP Atlantic Tropical Experiment) data to find a link between the thickness of the mixing layer and deep convection and inferred a simple model of the boundary layer associated with the passage of mesoscale precipitation systems with the help GATE experimental data. Recently, data obtained during the African Monsoon Multidisciplinary Analysis (AMMA) experience allowed more specific studies of ABL in West Africa [9–11].

In this paper, we analyze and compare relative humidity, dew point temperature, speed and wind direction vertical profiles, and atmospheric boundary layers of continental (Niamey), coastal (Dakar), and marine (Sal) sites located in West Africa. We use data from the special observations phase called SOP3 (Special Observing Period 3) of summer 2006 during the AMMA experience, when there were many sounding observations at various sites in West Africa [12, 13]. The objective is to analyze the seasonal evolution over the period June, July, August, and September (JJAS) and the profiles of some characteristic parameters of the boundary layer such as relative humidity, dew point temperature, wind speed, and wind direction in order to make a finer characterization and understand the variability. It is also an issue if there are consistent differences in the ABL related to several factors such as the continentality, the proximity of the ocean, the state of static stability, and mesoscale weather systems. In addition, we attempt to identify the characteristic parameters that determine the ABL structure in our study area and highlight any differences between the structures and thermodynamic properties of the ABL of, respectively, continental, coastal and coastal sites. Finally, we seek to determine the diurnal evolution of the structure and thermodynamic properties of these different types of ABL.

2. Data and Methodology

2.1. Data. We used sounding data from the AMMA SOP3 experience in summer 2006 in West Africa. We have made a first step for a diagnosis of sounding data to select those that seem appropriate for our study. The data availability is illustrated by Figures 1(a), 1(b), 1(c), and 1(d) for the continental, coastal, and marine sites.

2.1.1. Availability and Consistency of Data by Category of Site and by Month. Figure 1(a) shows that continental (Bamako and Niamey) and coastal (Dakar) sites have more than 80% of data available at 0000 UTC and 1200 UTC in June, while those marine (Sal, Praia) have no information for that month. Figure 1(b) shows that continental (Abidjan, Agadez, Bamako, Niamey) and coastal (Dakar) sites exceed 80% of data available for 0000 UTC and 1200 UTC in July. While among the marine sites, only Sal has few data at 1200 UTC. We dismissed the month of July. Figure 1(c) shows that for August, the continental (Agadez, Niamey, and Bamako) and coastal (Abidjan and Dakar) sites exceed 80% of data availability at 0000 UTC and 1200 UTC.

Sounding data are quite not available for the marine sites (Sal, Praia) at 0000 UTC and 1200 UTC (about 19%). Because Sal has limited data at 1200 UTC, we did not use the month of July. Finally in Figure 1(d), we find that continental (Bamako and Niamey) and coastal (Abidjan and Dakar) sites have more than 80% data available at 0000 UTC and at 1200 UTC.

Following the diagnosis of availability and data consistency, we use the Niamey (inland) and Dakar (coastal) sites for the first part concerning the study of vertical profiles of the troposphere on the entire seasonal period. For the second part of our study focusing on the boundary layer based on sounding data case studies, we consider in addition to these two first sites and Sal (marine).

2.1.2. Case Studies. Three case studies were selected according to the criterion that the 1200 UTC sounding data have the maximum values of CAPE during the month of August at each of all three sites (Table 1). To study the diurnal evolution of the ABL, we take the sounding at 0000 UTC of the night when the next or previous sounding data exist. This is the case for Niamey and Dakar sites. For Sal, we take the first sounding data available at 0000 UTC four days later. Thus the three case studies consist of Niamey, Dakar, and Sal sounding data at 1200 UTC according to the instability criterion on, respectively, 26, 17, and 19 August 2006. The sounding data at 000 UTC used are those from 25, 18, and 23 August 2006, respectively, in Niamey, Dakar, and Sal sites. August is the time of year when the convective activities are more intensive in our study area [14]. Figure 2 shows Meteosat-8 water vapor channel images corresponding to the case study dates. Figures 2(a) and 2(b) show convective systems on or near Niamey site. While Figures 2(c) and 2(d) indicate no deep convection systems over or near Dakar site. Finally, Figures 2(e) and 2(f) show the presence at the Sal site of the Saharan Air Layer (SAL), which is a mass of dry air.

2.2. Methodology. We present the different methods used to process the sounding data and deduce the vertical thermodynamic profiles, structures, thermal, and moisture properties of boundary layers in order to analyze and inter-compare them.

2.2.1. Hovmoêller Diagrams. We plotted relative humidity, dew point temperature, speed and wind direction hovmoêller diagrams (altitude-time) of 1200 UTC and 0000 UTC sounding data. The time period in hovmoêller diagrams is June, July, August, and September, denoted in the following JJAS. This diagram is used to analyze the seasonal thermodynamic vertical profiles of the troposphere of continental (Niamey) and coastal (Dakar) sites.

2.2.2. Determination of the ABL Structure and Temperature and Moisture Properties. Many authors have proposed methods for determining the structure of the atmospheric boundary layer [15–17]. These methods allow identifying the ABL structural elements as sublayers that are the surface layer, mixing layer, and the capping inversion generally above the mixing layer. These ABL structural elements are described extensively in the literature [17]. Each of these elements is associated with characteristic profiles of key parameters and physical variables. The key parameters to be chosen depend on the nature of the data and applications

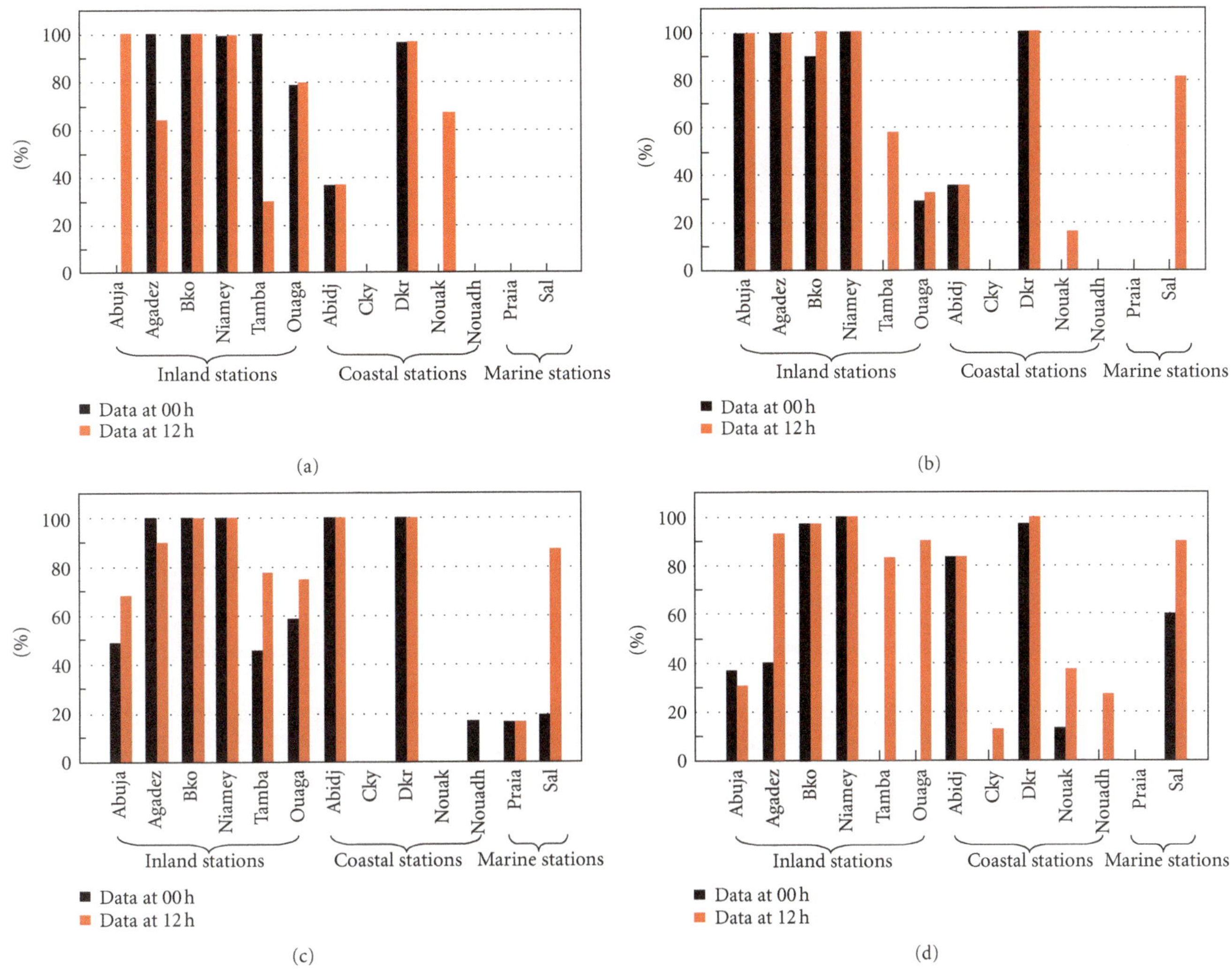

FIGURE 1: Data availability for the different categories of sites for the months of June (a), July (b), August (c), September (d) of 2006.

TABLE 1: CAPE maximum values for the selected case studies.

Stations	CAPE (JKg^{-1})			
	Maximum at 12 h		Maximum at 00 h	
Niamey	08/26/2006	3789	08/25/2006	3721
Dakar	08/17/2006	4473	08/18/2006	3589
Sal	08/19/2006	2196	08/23/2006	2394

used in the study of the boundary layer [16, 18]. In this study, we used sounding data and apply the vertical gradient methods of some thermodynamic parameters such as potential temperature θ, virtual potential temperature θ_v, equivalent potential temperature θ_e, specific humidity q, mixing ratio r, and saturation mixing ratio r_w. Figure 3 shows virtual temperature θ_v, potential temperature θ, equivalent potential temperature θ_e, and mixing ratio saturating r_w profiles deducted from sounding data at 1200 UTC case studies of the three sites. Examination of these profiles reveals that those of virtual potential temperature θ_v and saturating mixing ration r_w are quite similar. With the profile of potential temperature θ, we can easily identify the structural elements of the boundary layer. In the following part, we explain how we apply the gradients method on the profiles of the various parameters used to determine the ABL structure and temperature and moisture properties.

(1) ABL Characteristic Parameters. Simplified diagrams of the vertical profile of potential temperature θ have been proposed to model the structure of the boundary layer [15, 19, 20].

It is an ABL conceptual representation and shows the structural elements and their descriptive parameters. The mixing layer has been variously defined [16, 21], but we

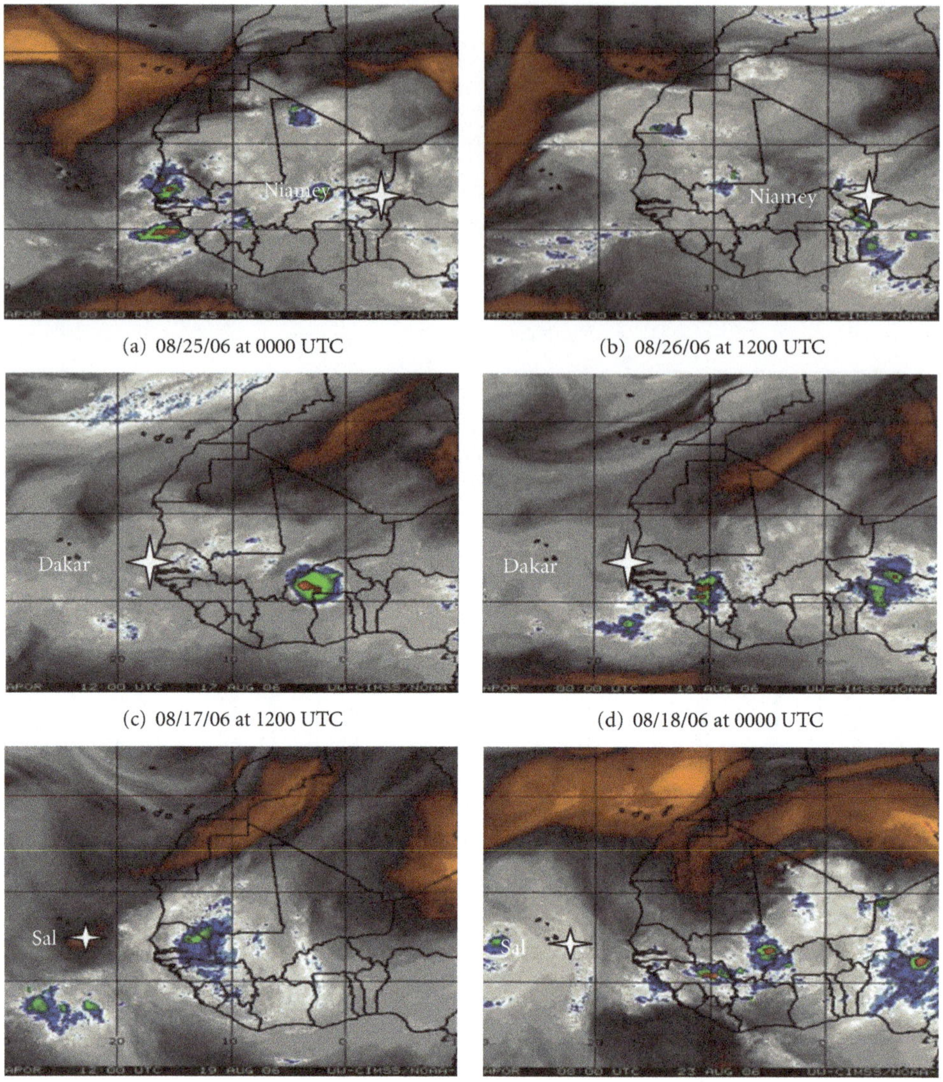

(a) 08/25/06 at 0000 UTC

(b) 08/26/06 at 1200 UTC

(c) 08/17/06 at 1200 UTC

(d) 08/18/06 at 0000 UTC

(e) 08/19/06 at 1200 UTC

(f) 08/19/06 at 1200 UTC

FIGURE 2: Images Meteosat-8-Upper Level Water Vapor in Niamey (a) and (b), Dakar (c) and (d), and Sal (e) and(f) 00 UTC and 12 UTC. The orange color represents the very dry air associated with airborne dust. The colors green, blue, and red represent areas of deep convection. (source: uw-CIMSS/NOAA-HR).

consider it here as a part of the boundary layer in which the vertical gradient of potential temperature θ is very low or even zero. In other words, in the mixing layer, the value of θ remains almost constant with height. The part surmounting the mixing layer where potential temperature θ increases rapidly and presents a sharp discontinuity is the capping inversion layer. Above the inversion layer, we find the free atmosphere that is the upper limit of the boundary layer. The third structural element of the boundary layer is the surface layer, which is located in the immediate vicinity of the ground.

The parameters used to characterize the structural elements of the boundary layer are heights h_s, h_0, h_1, the thickness Δh, and the thermal quantities such as θ_m, θ_{00}, $\Delta\theta$, γ (Figure 4). The heights h, h_0, h_1 represent, respectively, those of the surface layer, of the mixing layer, and of the inversion layer. Δh is the capping inversion layer thickness. The thermal parameters θ_m, θ_{00}, $\Delta\theta$, γ correspond, respectively, to the potential temperature of the mixed layer, the potential temperature in the free atmosphere interpolated at altitude $z = 0$, the amplitude of the potential temperature in the capping inversion layer, and the asymptotic gradient of potential temperature in the free atmosphere [18].

In a real sounding, in general the profile of potential temperature θ may deviate from that of the conceptual model in Figure 4 as a function, among others, of the degree of convective activity, the amount of moisture in the lower atmosphere [17], and of large and medium scales weather conditions [8]. Here, we apply the gradients method to potential temperature θ, which seems more efficient for our individual sounding data to determine the structure of the boundary layer (Figure 3).

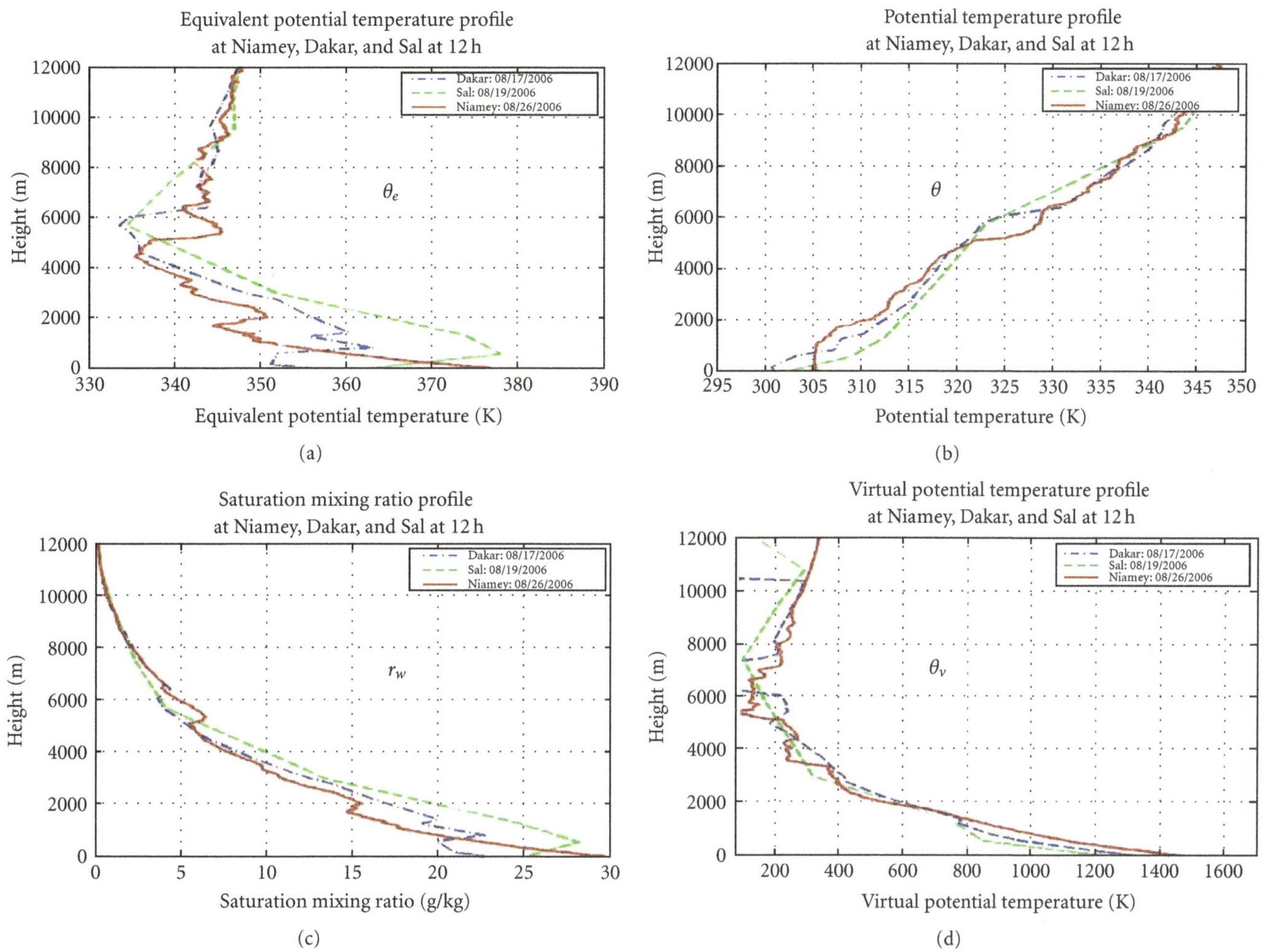

FIGURE 3: Profiles of virtual potential temperature θ_v, potential temperature θ, equivalent potential temperature θ_e, and saturating mixing ratio r_w of the three case studies selected.

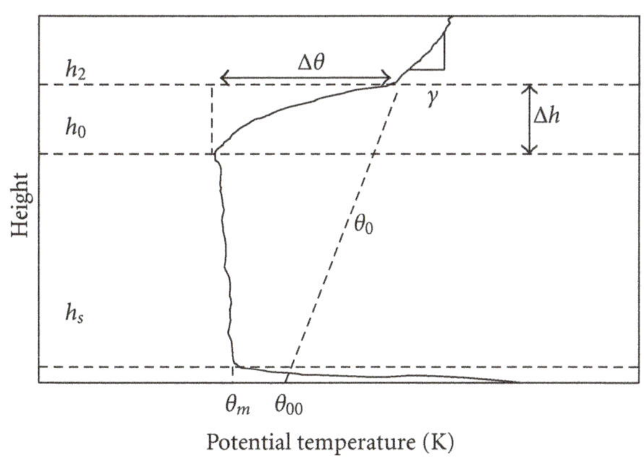

FIGURE 4: ABL Conceptual representation of the structure with the profile of potential temperature θ (Adapted from [18]).

2.2.3. Static Stability. Unlike some authors who have used it to determine the height of the boundary layer [17, 21], we use the gradient of equivalent potential temperature θ_e here to determine the static stability of each layer. Thus, when within an atmospheric layer, the variation of the parameter θ_e is monotone, we determine the static stability as follows.

(i) If θ_e decreases with height in a layer, the later is unstable.

(ii) If θ_e remains almost constant with height in a layer, the later is neutral.

(iii) If θ_e increases with height in a layer, the later is stable.

2.2.4. ABL Thermal and Moisture Properties. It is also the profile of equivalent potential temperature θ_e that we used to determine the thermal properties of the ABL different structural elements. The profile of the mixing ratio r is used for determining the ABL moisture properties.

3. Results and Discussion

The results presented concern the continental (Niamey), coastal (Dakar), and marine (Sal) sites. This is due to the quality and availability of data from these sites during

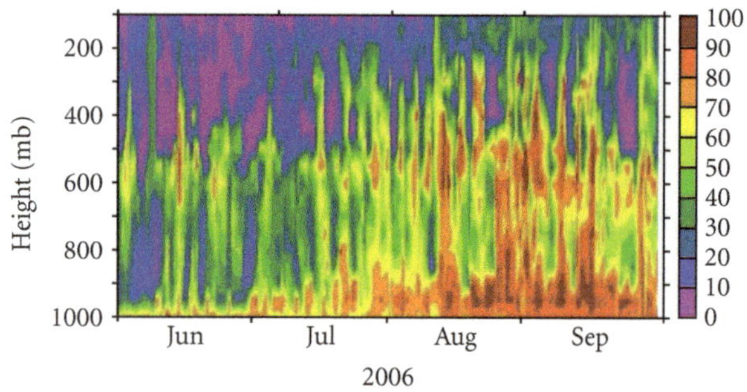

Time evolution of the vertical profile: the mean daily relative humidity during JJAS at Dakar in 2006

(a) Dakar

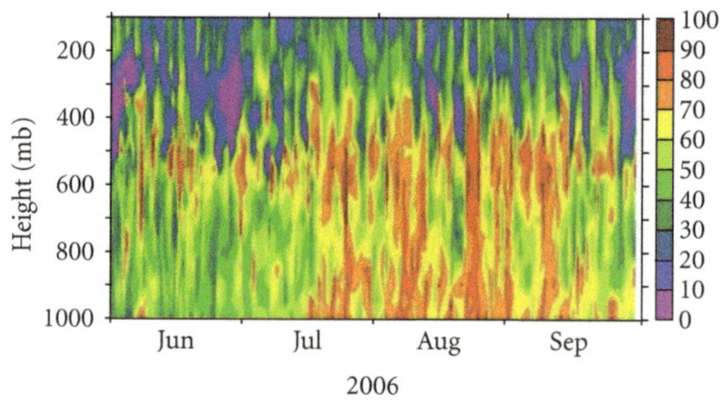

Time evolution of the vertical profile: the mean daily relative humidity during JJAS at Niamey in 2006

(b) Niamey

FIGURE 5: Daily humidity profiles in Dakar and Niamey at 0000 UTC and 1200 UTC of 2006 JJAS.

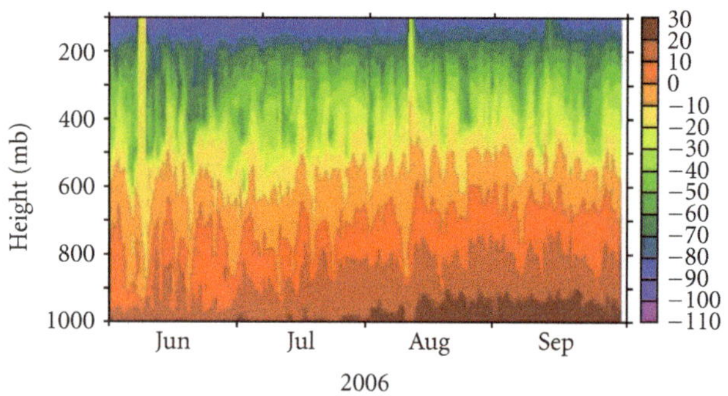

Time evolution of the vertical profile: the mean daily dew point temperature during JJAS at Dakar in 2006

(a) Dakar

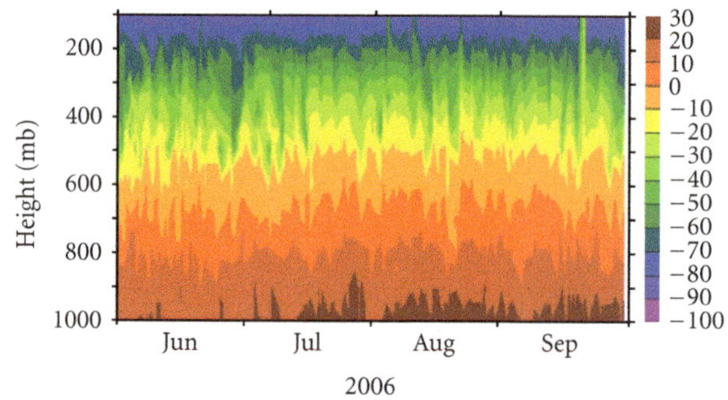

Time evolution of the vertical profile: the mean daily dew point temperature during JJAS at Niamey in 2006

(b) Niamey

FIGURE 6: Dew point temperature profiles in Dakar and Niamey at 0000 UTC and 1200 UTC of JJAS 2006.

AMMA SOP3 experience. The hovmoêller outputs (Altitude-time) of the relative humidity, dew point temperature, and wind speed and wind direction parameters for the period JJAS are plotted with data from Dakar and Niamey sites. The profiles of potential temperature θ, of equivalent potential temperature θ_e, and of mixing ratio r were analyzed for the case studies. The case studies are the most unstable atmospheres in August in Dakar, Niamey and Sal after a systematic analysis of the CAPE and CIN values during JJAS period. The profiles allow us to compare the structures of continental, coastal, and marine boundary layers.

3.1. Profiles of Moisture at Coastal (Dakar) and Continental (Niamey) Sites in JJAS 2006. There is increased moisture at Dakar site ranging from 70% to 80% from June, at the low levels (Figure 5). In early summer (June), dry air intrusions are observed both in the low and mid levels, which limit the development of deep convection. In the core of the rainy season (August and September), the humidity is intense even at the mid levels (400 hPa). However, there are dry air intrusions during the last week of September at the mid levels.

Figure 5(b) indicates that Niamey site compared to Dakar site has a low relative humidity (50% and 70%) at low levels and comparable or greater humidity intensity than Dakar site at the mid levels. In addition, there is no dry air intrusion in early summer at the low levels; the dry air intrusions are confined only to mid levels.

The low levels at Dakar site are relatively wetter than those to Niamey, probably because of the proximity of the ocean. Dry air intrusions are observed in the low levels in the early season in Dakar in contrast to Niamey. From July, the moisture content is more homogeneous with height at Niamey site than at Dakar site, that is to say that ABL is more developed in Niamey than in Dakar (Figure 5).

3.2. Profiles of Dew Point Temperature Td at Coastal (Dakar) and Continental (Niamey) Sites in JJAS Period. Figure 6(a) shows that at the low levels, the dew point temperature (Td) is low at the start of the season JJAS, between 10 and 20°C, and it increases during the season. It reaches values between 20°C and 30°C during the core of the wet season (August and September) at the low levels. This corresponds to a strong moistening of the ABL at the coast. However, there was in early season (June), low dew point temperature at the low levels, indicating dry air intrusions observed in the moisture profiles (Figures 5(a) and 5(b)).

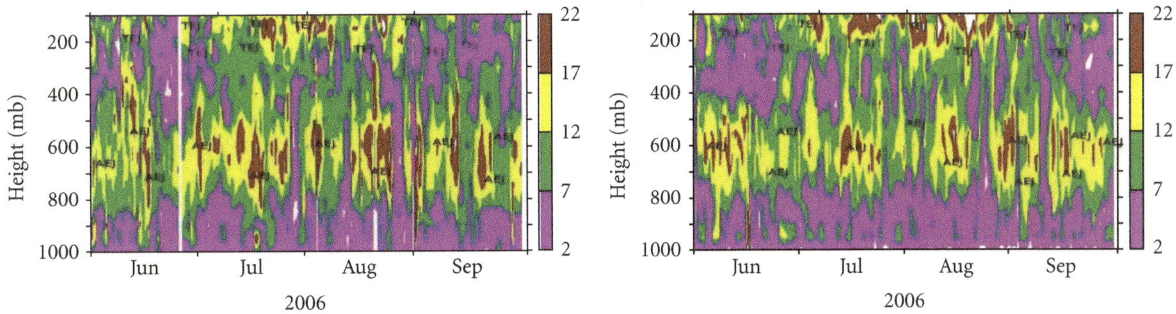

Time evolution of the vertical profile: the mean daily wind speed during JJAS at Dakar in 2006

(a) Dakar

Time evolution of the vertical profile: the mean daily wind speed during JJAS at Niamey in 2006

(b) Niamey

FIGURE 7: Wind speed profiles in Dakar and Niamey at 0000 UTC and 1200 UTC during JJAS 2006.

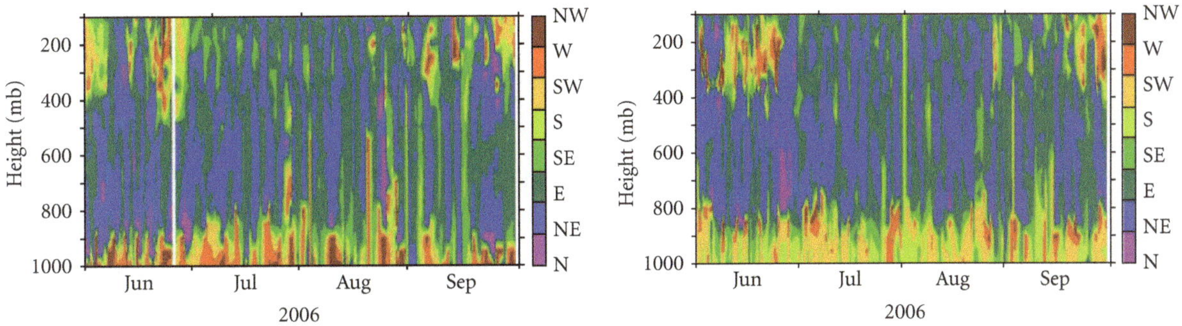

Time evolution of the vertical profile: the mean daily wind direction during JJAS at Dakar in 2006

(a) Dakar

Time evolution of the vertical profile: the mean daily wind direction during JJAS at Niamey in 2006

(b) Niamey

FIGURE 8: Daily wind direction profiles in Dakar and Niamey at 0000 UTC and 1200 UTC of JJAS 2006.

Figure 6(b) shows that at the low levels, the dew point temperatures Td are higher in Niamey, between 20°C and 30°C, with frequent temporal discontinuities. However, the values of Td between 10°C and 20°C are observed consistently in the low levels throughout the JJAS period; low values of Td occur only in the mid levels. This situation suggests a dry air intrusions at the low levels at Niamey (Figure 6(b)).

The profiles of the dew point temperature patterns at the low levels in Dakar and Niamey are quite similar to those of moisture. The presence at the low levels of dry air at Dakar site in early season can be particularly noted.

3.3. Profiles of the Wind Speed at Coastal (Dakar) and Continental (Niamey) Sites in JJAS 2006. Figure 7(a) shows that early in the season (June), the African Easterly Jet (AEJ) is weaker at the mid levels at Dakar site. The AEJ starts increasing from mid-June with intensity between 15 and 25 $m\cdot s^{-1}$ until the end of the season; its maximum intensity is observed in August. Note also that the AEJ can weaken or disappear sometimes during the season. The Tropical Easterly Jet (TEJ) appears at the beginning of July and continues until August with an intensity ranging from 10 to 25 $m\cdot s^{-1}$.

Figure 7(b) shows, in contrast to Dakar, that the AEJ at Niamey is intense with speeds up to 25 $m\cdot s^{-1}$ in June. It presents discontinuities until the end of the season. The TEJ, with an intensity ranging from 10 $m\cdot s^{-1}$ and 25 $m\cdot s^{-1}$, is visible only from July and disappears in late August.

The AEJ is more regular and stronger in Dakar than in Niamey, and the TEJ is stronger in Niamey than in Dakar. The AEJ is established earlier in Niamey than in Dakar. The TEJ is much more intense in Niamey.

3.4. Profiles of Wind Direction at Coast (Dakar) and Continental (Niamey) Sites in JJAS 2006. Figure 8(a) shows that at low levels, wind direction varies at Dakar site from Northwest (Harmattan), Southeast and Southwest during JJAS period. But the prevailing winds are South-West (monsoon) and begin early in the season (June). We observe Southwest winds at up levels (200 hPa) at the beginning and the end of the season.

Figure 8(b) shows that in Niamey at the low levels, wind direction varies from southwest (Monsoon), south-southwest. The prevailing wind direction is Southwest (monsoon) with fluctuations in other directions. At up levels in Niamey, at the beginning and end of the season, we observe sometimes southwest and south winds. The dominant wind

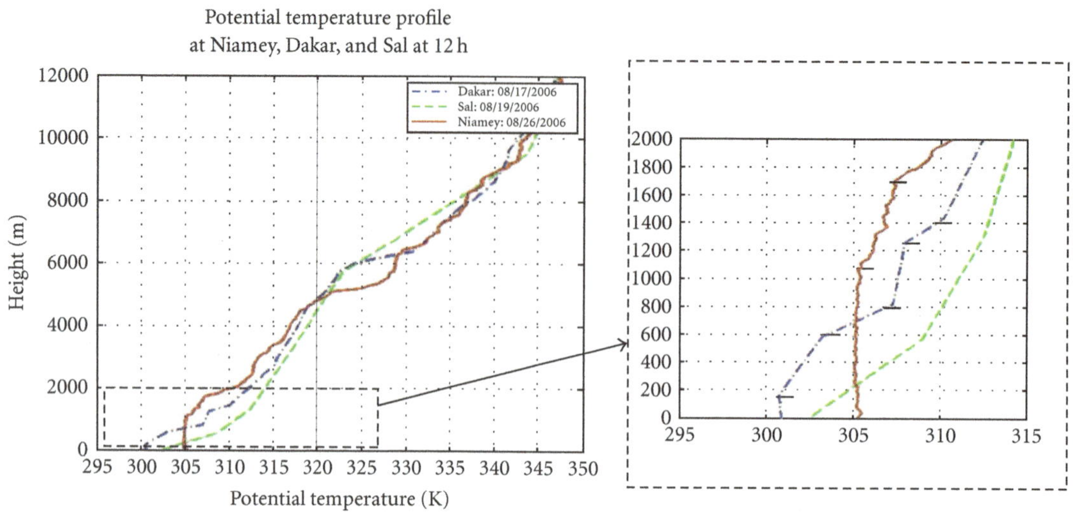

FIGURE 9: Profiles of potential temperature θ at 1200 UTC of 08/17/06 in Dakar (blue), of 08/26/06 in Niamey (red), and of 08/19/06 in Sal (green).

directions in Dakar and Niamey are Southwest at low levels and indicate the presence of the monsoon. At the up levels (200 hPa) at the beginning and the end of the season, we have the same dominant directions from east in Dakar and in Niamey.

3.5. Comparison of Atmospheric Boundary Layer in Niamey, Dakar, and Sal. We compare the structures of inland, coastal, and marine boundary layers in the lower troposphere in the range 0 to 2000 meters using their characteristic parameters, their static stability stratification and thermal properties, and moisture content. To do this, we use the sounding data at 1200 UTC and 0000 UTC of the case studies presented in Section 2.

3.5.1. Comparison of Boundary Layers at 1200 UTC

(1) ABL Structure. Table 2 gives the values of characteristic parameters of the boundary layer obtained from the profiles of potential temperature θ deduced from 1200 UTC case studies soundings (Figure 9). We observe the deepest mixing layer of 1000 m thick at inland (Niamey), while at the coast (Dakar), the mixing layer is shallow, and at the sea (Sal), it does not exist. The capping inversion is thick (600 m) in Niamey (red), but its temperature magnitude is very small (1 K). The ground-based inversion in Sal (green) is thick enough (550 m) with a strong temperature magnitude (6 K). We observe that the inversion amplitude is greatest in Sal and lowest in Niamey.

(2) Static Stability. We indicate in Table 3 the static stability stratification of the boundary layers for inland, coastal, and marine profiles deduced from equivalent potential temperature θ_e (Figure 10). In Niamey, there is a thick unstable layer from 0 to 1100 m topped by a neutral layer up to 2000 m. This unstable layer thickness coincides with the mixing layer (Table 2).

TABLE 2: Characteristic parameters of continental (Niamey), coastal (Dakar), and marine (Sal) boundary layers at 1200 UTC for respective case studies.

Variable	Niamey 1200 UTC	Dakar 1200 UTC	Sal 1200 UTC
h_0 (m)	1100	200	0
h_s (m)	0	0	0
h_2 (m)	1700	600	550
Δh (m)	600	200	550
$\Delta\theta$ (K)	1	3	6
θ_m (K)	306	301	—
θ_{00} (K)	306,5	302	307,5
γ (K·km^{-1})	8,5	22,5	8,1

In Dakar, there is a complex stratification with an unstable layer surmounted by a neutral layer and a double layer alternating unstable and stable. The unstable first layer overlaps the mixing layer thickness.

At Sal, the boundary layer has a stratification characterized by a thick layer of ground-based stability, which explains the absence of the mixing layer. This stable layer is surmounted by a thick layer of instability from 600 to 2000 m.

(3) Thermal Properties. Figure 10 also indicates that the air in Sal (green) is much warmer than in Dakar (blue) and Niamey (red) from 200 m. Between 0 and 200 m, the air in Niamey is warmer than in other sites. It is the coolest above 700 m. Below 700 m, the air of Dakar is the coolest of all. The warmer air in Sal in the layer 200–2000 m is due to the existence of the Saharan Air Layer (SAL). Warm air in Niamey in the range 0–200 m is an expected result due to the stronger daytime heating over a land surface. The existence of a warmer air in Dakar above 700 m may be explained by the influence of Saharan Air Layer in Dakar.

TABLE 3: Static stability stratification of ABL determined from equivalent potential temperature θ_e at 1200 UTC for case studies.

Stations	Study case	Inversion of θ_e	Characterization
Niamey (red)	08/26/06 at 1200 UTC	0–1000 m	Unstable layer
		1000–1700 m	Neutral layer
		1700–2000 m	Stable layer
Dakar (blue)	08/17/06 at 1200 UTC	0–200 m	Unstable Layer
		200–600 m	Neutral layer
		600–800 m	Stable layer
		800–1300 m	Unstable layer
		1300–1400 m	Unstable layer
		1400–2000 m	Unstable layer
Sal (green)	08/19/06 at 1200 UTC	0–600 m	Stable layer
		600–2000 m	Unstable layer

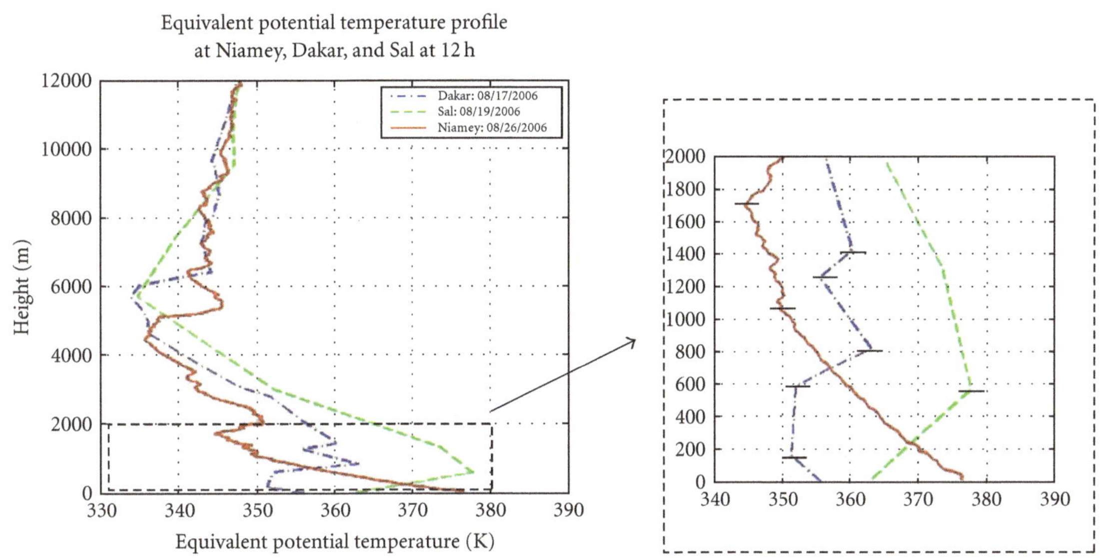

FIGURE 10: Equivalent potential temperature (θ_e) profiles at 1200 UTC of 08/17/06 in Dakar (blue), 08/26/06 in Niamey (red), and 08/19/06 in Sal (green).

(4) Moisture Properties. Figure 11 shows profiles of mixing ratio. The values of the mixing ratio r of continental (Niamey), coastal (Dakar), and marine (Sal) sites are, respectively, 12, 16.5, and 15.5 g/kg at the ground level and decreases linearly with height. Note that the boundary layer in Niamey (red) is significantly wetter than in Dakar (blue) and Sal (green) between 500–1800 m, but it is drier between 0–500 m. The boundary layer of Dakar (blue) is more humid than in Sal (green) and Niamey (red) significantly between 0–500 m. Then, it is drier between 600–2000 m in Dakar. The boundary layer of Sal (green) is more humid than in Dakar (blue) and Niamey (red) between 600–1800 m. These moisture properties differences can be explained by the fact that during the JJAS period, the Dakar and Sal sites were influenced by the Saharan Air Layer (SAL), which is characterized by drier air than in Niamey.

3.5.2. Comparison of Boundary Layers at 0000 UTC

(1) ABL Structure. Table 4 gives the values of characteristic parameters obtained from the profiles of potential temperature θ (figure not shown) deduced from 0000 UTC case studies soundings.

Here the thickest mixing layer (400 m) is observed at the coast (Dakar), while on the continent (Niamey), the mixing layer is shallow and does not exist at all on the sea (Sal). However, there is a residual mixing layer between 300–1000 m in Niamey. The capping inversion is very thick (800 m) in Dakar, and its thermal amplitude is strong enough (4 K). The ground-based inversion in Sal is thick enough (500 m) with a low temperature range (2 K). It is found that the amplitude of the inversion is lower in Sal.

(2) Static Stability. Table 5 shows the stratification of the static stability of boundary layers for inland, coastal, and

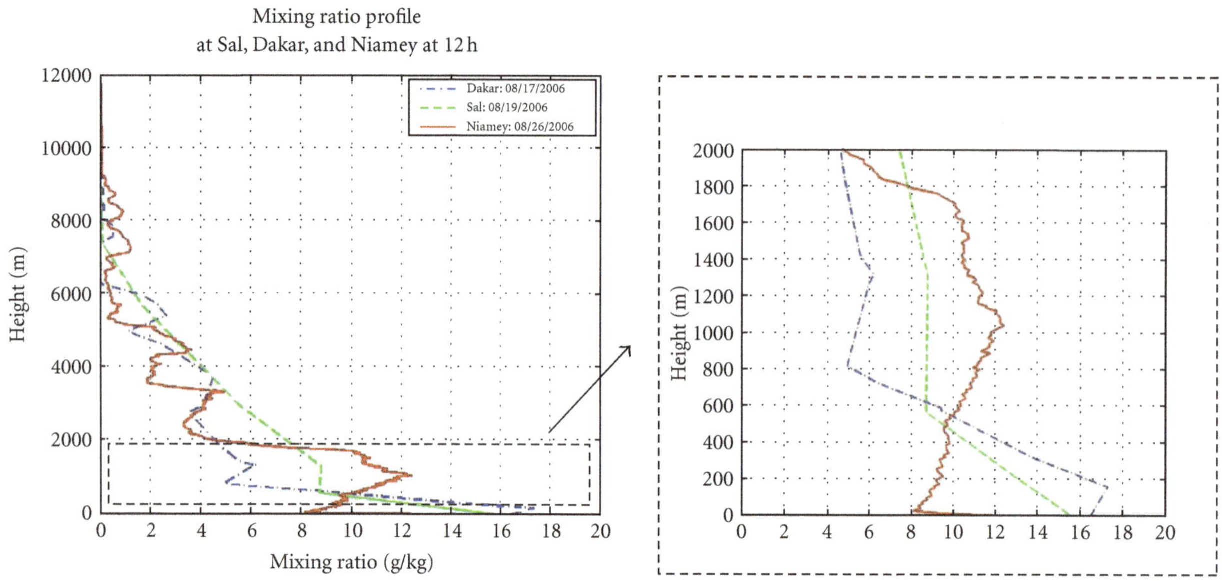

FIGURE 11: Mixing ratio (r) profiles at 1200 UTC of 08/26/06 in Niamey (red), at 08/17/06 in Dakar (blue), and at 08/19/06 in Sal (green).

TABLE 4: Characteristic parameters of the continental boundary layer (Niamey), coastal (Dakar), and marine (Sal) at 0000 UTC for case study.

Parameters	Niamey 0000 UTC	Dakar 0000 UTC	Sal 0000 UTC
h_0 (m)	180	400	0
h_s (m)	0	0	0
h_2 (m)	310	800	500
Δh (m)	120	200	500
$\Delta\theta$ (K)	4	4	2
θ_m (K)	302	302	—
θ_{00} (K)	303	307	299
γ (K·km^{-1})	8,7	47,8	8
h'_0 (m)	1000		
h'_s (m)	300		
h'_2 (m)	1050		
$\Delta h'$ (m)	50		
$\Delta\theta'$ (K)	1		
θ'_m (K)	303		

The "prime" denotes the same parameter as in the first list when there is a residual mixed layer.

marine sites derived from profiles of equivalent potential temperature θ_e (figure not shown).

In Niamey, there is a very thin unstable layer from 0 to 180 m surmounted by a stable layer up to 300 m. Then there is a thick unstable layer from 300 to 1000 m. Then there is a neutral layer from 1000 to 1800 m. Beyond and up to 2000 m, there is a stable layer. The unstable layer 300–1000 m coincides in thickness with the residual mixing layer (Figure 12(b)).

In Dakar, there is a complex stratification with a thin unstable layer surmounted by double alternating unstable and stable layers. The unstable first layer in Dakar is the mixing layer.

In Sal, the boundary layer is characterized by a layer of instability 0–500 m capped by a thick layer of stability from 500 to 1200 m. Then there is an alternation of a thin unstable layer (1200–1400 m) and a stable layer (1400–2000 m). The ground-based thick layer of instability at Sal is a mixing layer not detectable on the profile of potential temperature θ.

(3) Thermal and Moisture Properties. At night, the warmest air in Dakar is aloft from 400 to 2000 m. Below 400 m, the warmest air is in Niamey. From the surface to about 800 m, the coolest air is found in Sal. Above 800 m, the coolest air is observed in Niamey (figure not shown).

This configuration of the temperature profiles during the night is a bit unexpected. The warmer air in Dakar could be explained by the influence of the Saharan air layer, which would not have influenced the site of Sal during the night (Figure 2).

Note that the boundary layer in Niamey is significantly moister than in Dakar and Sal from the surface to 2000 m (figure not shown). The boundary layer of Dakar is more humid than Sal up to 400 m, and moisture in Dakar and Sal decreases above 400 m. In Dakar, there was a significant dryness above 400 m, confirming the influence of SAL (Sahara Air Layer).

3.6. Evolution of the Daytime Boundary Layer. Let us compare, between 1200 UTC and 0000 UTC, the structures and thermodynamic properties of boundary layers for case studies. Remember that 0000 UTC sounding data in question is that of the previous day or the next day at Niamey and Dakar site. But at Sal site, because of the nonexisting data, we took the night sounding data closest in time, which happens 4 days after. This causes that the evolution of the boundary layer is not of diurnal scale at Sal but the comparison is made

Table 5: Static stability stratification of ABL determined from the equivalent potential temperature θ_e at 0000 UTC for case studies.

Stations	Study case	Stratification of θ_e	characterization
Continental (Niamey)	08/25/06 at 0000 UTC	0–180 m	unstable layer
		180–300 m	stable layer
		300–1000 m	unstable layer
		1000–1800 m	neutral layer
Coastal (Dakar)	08/18/06 at 0000 UTC	0–380 m	unstable layer
		380–580 m	stable layer
		580–840 m	unstable layer
		840–1400 m	stable layer
		1400–2000 m	unstable layer
Marine (Sal)	08/23/06 at 0000 UTC	0–500 m	unstable layer
		500–1200 m	stable layer
		1200–1400 m	unstable layer
		1400–2000 m	stable layer

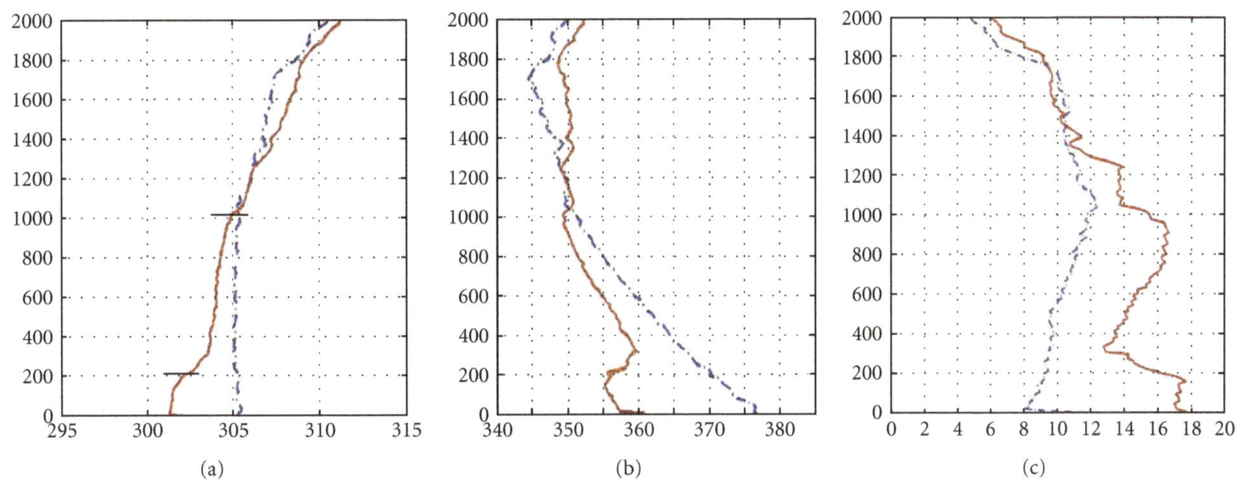

Figure 12: Potential temperature θ (a), equivalent potential temperature θ_e (b), and mixing ratio r (c) profiles 08/26/06 at 0000 UTC (red) and 1200 UTC (blue) in Niamey.

between a daytime boundary layer and nocturnal boundary layer.

3.6.1. Change in the ABL Structure. A joint reading of Tables 2 and 4 shows that the height of the Niamey mixing layer decreased significantly from 1600 to 120 m from day to night, and also this is shown in Figure 8(a). Meanwhile, the height of the capping inversion height fell from 1700 to 300 m. The thickness of the inversion layer has also decreased from 600 to 120 m, but its thermal amplitude is increased from 1 to 4 K. Furthermore, it has established a residual mixing layer whose thickness is considerable. In fact, the daytime mixing layer ground-based, was divided into two mixing layers with a mixing layer at the bottom very shallow 0–180 m and a residual mixing layer depth of 300–1000 m. Between the two layers is a layer corresponding to the stable nocturnal inversion.

In Dakar, the height of the mixing layer increased from day to night; from 200 to 400 m along the height of the inversion layer is increased from 600 to 800 m at night (Figure 10(a)). The thickness of the inversion layer remained unchanged but its thermal amplitude was increased by 3 to 4 K from daytime to nighttime. In short, the diurnal evolution has resulted in an increase in thickness of the mixing layer due to turbulence associated to nonthermal phenomena surely mesoscale or large-scale one.

In Sal, the nocturnal boundary layer structure is similar to that of the daytime: with the absence of a mixing layer and the existence of a thick layer of ground-based inversion, the height has increased slightly from 500 to 550 m from night to day (Figure 11(a)).

3.6.2. Diurnal Evolution of Static Stability. In Niamey, there is the establishment of a stable layer at night between 180 and 300 m and a neutral layer between 1000 and 1800 (Figure 12(b)). Thus the nocturnal boundary layer comprises two unstable layers separated by a stable layer corresponding to the component called the stable boundary layer (SBL) and a neutral layer between 1000–1800 m. The diurnal evolution of the static stability of the boundary layer

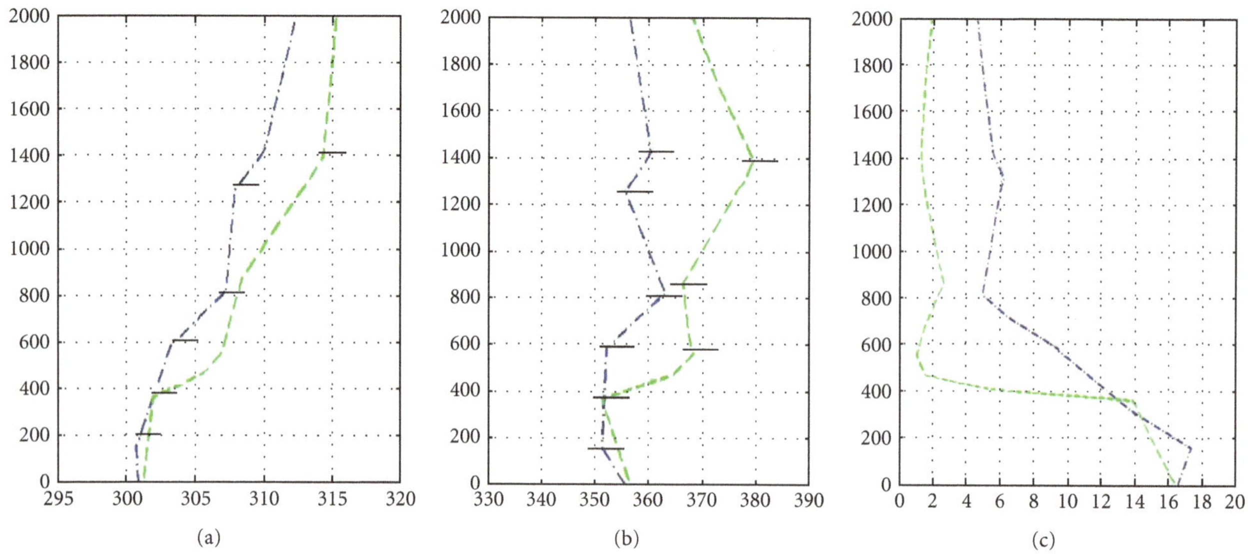

FIGURE 13: Potential temperature θ (a), equivalent potential temperature θ_e (b), and mixing ratio r profiles (c) of 08/17/06 at 1200 UTC (blue) and 08/18/06 at 0000 UTC (green) in Dakar.

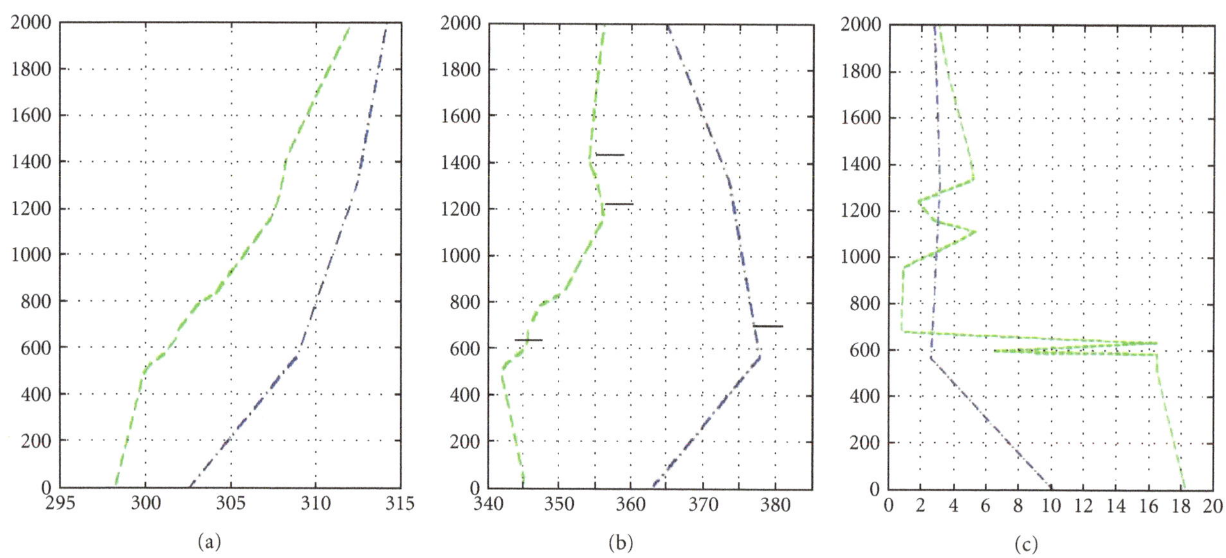

FIGURE 14: Potential temperature θ (a), equivalent potential temperature θ_e (b), and mixing ratio r (c) profiles of 08/19/06 at 1200 UTC (blue) and 08/23/06 at 0000 UTC (green) in Sal.

in Niamey has led to a stabilization of a portion of the lower half of the daytime unstable layer and a moving to the neutral state of the upper half. Overall, the instability of the daytime boundary layer was maintained during the whole night but was somewhat weakened in its top and bottom.

In Dakar, the diurnal evolution of the static stratification is characterized by the destruction during the night of the neutral layer whose lower part became unstable and whose upper part was stabilized. Also the stable layer 600–800 m during daytime became unstable (Figure 13(b)). And conversely, the unstable layer 800–1300 m stabilized during the nighttime between 850–1400 m. The unstable layer 1400–2000 m kept the same static state during nighttime. The thickness of the unstable layer corresponding to ground-based mixing layer increased from 200 m to 400 m from daytime to nighttime (Figure 13(b)).

In Sal, the mixing ratio profiles indicate that low levels of humidity are almost between 0–300 m and a slight moistening during nighttime 300 to 400 m. And a nocturnal drying between 400 and 2000 m is observed (Figure 13(c)).

In Sal, the instability of the layer 0–500 m at nighttime has replaced the stable layer from 0 to 600 m observed in daytime (Figure 14(a)). Thick and unstable 600–2000 m layer, which is stabilized during daytime, except for a narrow unstable layer 1200–1400 m, was stable at the nighttime (Figure 14(b)). Overall, the static stratification is modified at night with the establishment of instability in low levels between 0–500 m and a stabilization of the upper troposphere.

3.6.3. Diurnal Evolution of Thermal and Moisture Properties. In Niamey (Figure 12(c)), there is a nighttime cooling of the boundary layer between 0–1500 m, and it is greater in the mixing layer (1500–1700 m). Then, during the daytime, air in the mixing layer is dried between 0 and 1500 m and 1800–2000 m (Figure 12(c)).

The humidity remained unchanged in the boundary layer between 0–2000 m in Niamey, but there is a slight increase between 400–2000 m. during nighttime. A nocturnal slight drying is confined in the layer 0–200 m corresponding to the lower mixing layer (Figure 12(a)). Due to the lack of vertical turbulent moving, the air in the stable layer 200–400 m retained the same level of humidity during daytime as well as nighttime.

In Dakar, a nighttime warming occurs over the entire troposphere (Figure 13(b)) and is low between 0 to 400 m and becomes more considerable with height and has a maximum between 800 and 2000 m. Humidity profiles are identical in between 0–400 m. Between 400–2000 m, the nocturnal air became drier with a maximum drying at about 600 m.

In Sal, the lower troposphere has cooled a lot during the nighttime between 0–2000 m with a maximum difference at about 600 m (Figure 14(b)). Also, the mixing ratio profiles indicate that there are nighttime drying between 0–700 m and a slight moistening between 700–1000 m and 1300–2000 m (Figure 14(c)).

4. Conclusion

The profiles of relative humidity during JJAS 2006 indicate a higher humidification of Dakar in the month of June and later in Niamey. There is a boundary layer with higher moisture levels in Dakar and Niamey especially in the season core (July and August). However, there are dry intrusions in the low levels at the beginning and end of the season in Dakar that do not exist in Niamey.

The profiles of wind speed show that the African Easterly Jet (AEJ) and the Tropical Easterly Jet (TEJ) are observed at both sites, but the AEJ is stronger and more regular in Dakar than in Niamey. TEJ is stronger in Niamey than in Dakar.

The profile analysis of some thermodynamic parameters deduced from sounding data of our case studies selected by the criterion of the CAPE monthly maximum shows the following main findings.

The profile of equivalent potential temperature θ_e also determines the static stability of each of the different sub-layers of the boundary layer and defines their respective thicknesses. The mixing layer always corresponds to an unstable layer and the inversion layer to a stable layer.

The boundary layer comparison showed that the mixing layer on the continent during daytime can reach a height h_1 greater than 1100 m, and the inversion layer height h_2 can exceed 1700 m. Therefore, the maximum thickness of the boundary layer is observed on the continent during daytime, while at nighttime the marine boundary layer is the thickest.

The diurnal evolution determined from our case study shows that the thickness of the mixing layer decreases during nighttime over the continent but increases at the coast and at sea. During nighttime, at the continental site of Niamey, we observe a division of the mixing layer into two parts; the upper mixing layer is the residual layer.

The continental boundary layer is more unstable at daytime while at nighttime it is the marine boundary layer that is more unstable than the coastal and continental ones.

The coastal mixing layer (Dakar) is significantly wetter than the continental (Niamey) and marine (Sal) ones. The marine mixing layer (Sal) is significantly wetter than the continental one (Niamey). The nighttime cooling is greater at sea than at the coast and at the continent.

Our future work will focus on the seasonal evolution of the structure and properties of temperature and humidity of the atmospheric boundary layer, respectively, at the inland, coastal, and marine sites to validate the preliminary findings of the present study which focused on a few case studies.

Acknowledgments

The authors present their sincere thanks to the RIPIESCA project (Interdisciplinary and Participatory Research on the Interactions between Ecosystems, Climate and Societies in West Africa), the Institute of Development Research (IDR), and the University Agency of the Francophony (UAF), which has enabled this work to be carried out.

References

[1] W. Prandtl and F. Lustig, "Ueber complexe verbindungen des fünfwerthigen vanadins mit vierwerthigen elementen," *Berichte der Deutschen Chemischen Gesellschaft*, vol. 38, pp. 1305–1310, 1905.

[2] J. Turner, *Buoyancy Effects and Fluids*, Cambridge University, 1973.

[3] A. S. Monin, "The atmospheric boundary layer," *Annual Review of Fluid Mechanics*, vol. 2, pp. 225–250, 1970.

[4] O. Zeman, "Progress in the modeling of planetary boundary layers," *Annual Review of Fluid Mechanics*, vol. 13, pp. 253–272, 1981.

[5] B. Cushman-Roisin and J. M. Beckers, *Introduction to Geophysical Fluid Dynamics: Physical and Numerical Aspects*, Accademic Press, 2009.

[6] C. Renaudie, *Etude et validation des couches limites atmosphériques et océaniques locales*, Ph.D. thesis, université de Toulouse III-Paul Sabatier, 2009.

[7] A. K. Betts, J. H. Ball, A. C. M. Beljaars, M. J. Miller, and P. A. Viterbo, "The land surface-atmosphere interaction: a review based on observational and global modeling perspectives," *Journal of Geophysical Research D*, vol. 101, no. 3, pp. 7209–7225, 1996.

[8] R. H. Johnson, "Large-scale effects of deep convection on the GATE tropical boundary layer," *Journal of the Atmospheric Sciences*, vol. 38, no. 11, pp. 2399–2413, 1981.

[9] M. Lothon, F. Couvreux, S. Donier et al., "Impact of coherent eddies on airborne measurements of vertical turbulent fluxes," *Boundary-Layer Meteorology*, vol. 124, no. 3, pp. 425–447, 2007.

[10] F. Saïd, G. Canut, P. Durand, F. Lohou, and M. Lothon, "Seasonal evolution of boundary-layer turbulence measured by aircraft during the AMMA 2006 special observation period," *Quarterly Journal of the Royal Meteorological Society*, vol. 136, no. 1, pp. 47–65, 2010.

[11] G. Canut, M. Lothon, F. Saïd, and F. Lohou, "Observation of entrainment at the interface between monsoon flow and the Saharan air layer," *Quarterly Journal of the Royal Meteorological Society*, vol. 136, no. 1, pp. 34–46, 2010.

[12] J. Redelsperger, C. Thorncroft, A. Diedhiou, T. Lebel, D. J. Parker, and J. Polcher, "African monsoon multidisciplinary analysis: an international research project and field campaign," *Bulletin of the American Meteorological Society*, vol. 87, no. 12, pp. 1739–1746, 2006.

[13] P. Drobinski, "Dynamique de la couche limite atmosphérique: de la turbulence aux systems de méso-échelle," Synthèse d'activité scientifique proposée pour une thèse d'habilitation à diriger des recherches de l'Université Paris VI, p. 150, 2009.

[14] J. M. Prospero and T. N. Carlson, "Vertical and areal distribution of. Saharan dust over the westem equatorial north atlantic ocean," *Journal of Geophysical Research*, vol. 77, no. 27, pp. 5255–5265, 1972.

[15] J. W. Deardoff, "Prediction of convective mixed-layer entrainment for realistic capping inversion structure," *Journal of Atmospheric Sciences*, pp. 558–567, 1978.

[16] P. Seibert, F. Beyrich, S. E. Gryning, S. Joffre, A. Rasmussen, and P. Tercier, "Review and intercomparison of operational methods for the determination of the mixing height," *Atmospheric Environment*, vol. 34, no. 7, pp. 1001–1027, 2000.

[17] S. K. Nair, T. J. Anurose, D. Bala Subrahamanyam et al., "Characterization of the vertical structure of coastal atmospheric boundary layer over thumba (8.5°N, 76.9°E) during different seasons," *Advances in Meteorology*, vol. 2011, Article ID ID 390826, 9 pages, 2011.

[18] G. Rampanelli and D. Zardi, "A method to determine the capping inversion of the convective boundary layer," *Journal of Applied Meteorology*, vol. 43, no. 6, pp. 925–933, 2004.

[19] H. Tennekes, "A model for the dynamics of the inversion above a convective boundary," *Journal of Atmospheric Science*, vol. 30, no. 4, pp. 558–567, 1973.

[20] A. G. M. Driedonks, "Models and observations of the growth of the atmospheric boundary layer," *Boundary-Layer Meteorology*, vol. 23, no. 3, pp. 283–306, 1982.

[21] X. Zeng, M. A. Brunke, M. Zhou, C. Fairall, N. A. Bond, and D. H. Lenschow, "Marine atmospheric boundary layer height over the Eastern pacific: data analysis and model evaluation," *Journal of Climate*, vol. 17, no. 21, pp. 4159–4170, 2004.

6

Ground-Penetrating Radar Investigations along Hajipur Fault: Himalayan Frontal Thrust—Attempt to Identify Near Subsurface Displacement, NW Himalaya, India

Javed N. Malik, Ashutosh Kumar, Sravanthi Satuluri, Bishuddhakshya Puhan, and Asmita Mohanty

Department of Civil Engineering, Indian Institute of Technology Kanpur, Kanpur 208016, India

Correspondence should be addressed to Javed N. Malik, javed@iitk.ac.in

Academic Editor: Raffaele Solimene

The study area falls in the mesoseismal zone of 1905 Kangra earthquake (Mw 7.8). To identify appropriate trenching site for paleoseismic investigation and to understand the faulting geometry, ground-penetrating radar (GPR) survey was conducted across a Hajipur Fault (HF2) scarp, a branching out fault of Himalayan Frontal Thrust (HFT) in a foot hill zone of NW Himalaya. Several 2D and 3D profiles were collected using 200 MHz antenna with SIR 3000 unit. A 2D GPR profile collected across the HF2 scarp revealed prominent hyperbolas and discontinuous-warped reflections, suggesting a metal pipe and a zone of deformation along a low-angle thrust fault, respectively. The 3D profile revealed remarkable variation in dip of the fault plane and pattern of deformation along the strike of the fault.

1. Introduction

Paleoseismology defines the study of prehistoric earthquakes, especially their location, timing, and size [1–8]. The paleoseismic studies have attained a great pace in seismically active regions of the world [7, 9]. The investigation is mainly emphasized towards identification of surface manifestation of deformation and near subsurface displacements on active fault related to past earthquakes. It has usually been experienced that the preservation of active tectonic landforms along the mountain fronts is difficult because of ongoing dynamic fluvial processes causing high degree of erosion and deposition [10, 11]. Due to the removal of such feature, it becomes difficult to identify appropriate site for trenching to undertake paleoseismic studies. Several geophysical techniques to map the subsurface deformation, for example, seismic reflection and refraction, resistivity survey, ground-penetrating radar (GPR) have been used. The seismic reflection-refraction method has been used to get good resolution subsurface profile up to ≥100 m, but submeter resolution is not possible. Since the Ground-penetrating Radar (GPR) technique gives a high-resolution profile of shallow subsurface succession up to submeter resolution, it has been extensively used for geological studies since the 1990s [12].

Since a decade, the GPR technique has been widely used for different studies for mapping near subsurface displacements in the young Quaternary sediments related to active faulting, for identifying neotectonic faulting activity and liquefaction features like sand-blows (e.g., [10–20]). It has also proved as an important tool in locating archeological sites (e.g., [21–24]). Along with this GPR has been quite useful towards identifying appropriate site to undertake trenching for detailed paleoseismic studies along active faults in an area, where surface manifestation of active faulting is sparsely preserved [10, 11, 16].

The convergence and ongoing collision between Indian and Eurasian plates has kept the Himalayan arc to be one of the most seismically active regions of the world; therefore, seismic hazard assessment is one of the major concern. The built-up strain is released episodically along the active fault systems in Himalaya [20, 25]. This is well justified by the occurrence of large-magnitude earthquakes, namely,

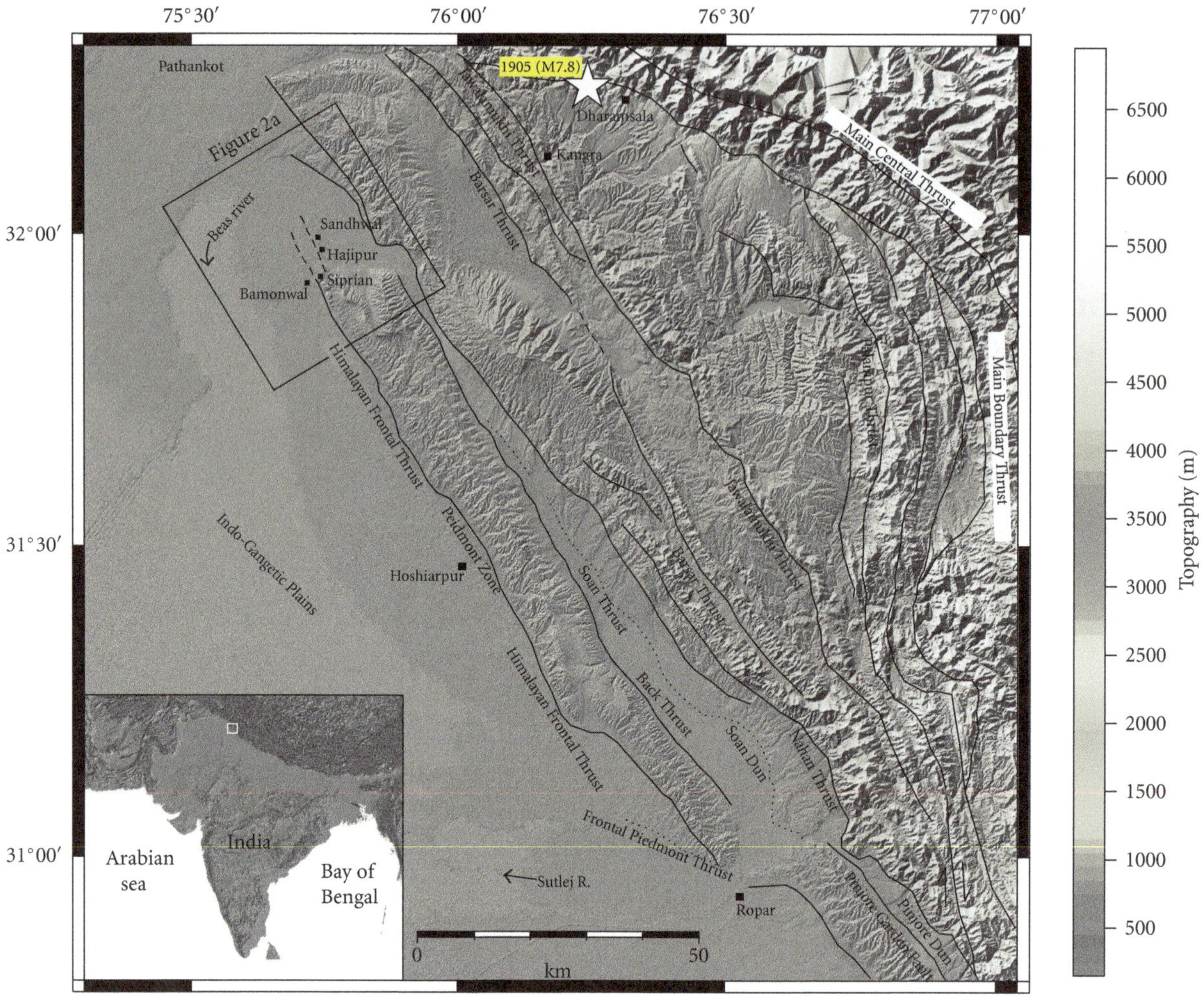

Figure 1: SRTM data of study area illustrating major faults marked by continuous lines and major geomorphic divisions around north west Himalaya. Box on top left marks the area of study which was shown in Figure 2. Inset shows DEM of India with location of study area.

1905 Kangra (Mw 7.8), 1934 Bihar (Mw 8.1), 1950 Upper Assam (Mw 8.4), and recent 2005 Muzaffarabad (Mw 7.6) earthquakes [8, 20, 26–28] (Figure 1). A very few studies have been taken up so far in Himalayan region attempting to map the surface manifestation of the past earthquake and to undertake paleoseismic investigation by excavating trenches across active fault scarps. Identification and mapping of such features, understanding the pattern of deformation along with known paleoseismic event are extremely important for proper seismic hazard assessment. In this paper, we discuss GPR studies carried along Hajipur Fault (HF2) a branching out fault of Himalayan Frontal Thrust (HFT) in the foot hill zone of NW Himalaya (Figure 1). To identify the appropriate site for trenching and to understand the geometry of faulting, we collected 2D and 3D GPR profiles.

2. Methodology

Ground-penetrating Radar (GPR) is a noninvasive geophysical technique, where the high-frequency (MHz) electromagnetic wave is transmitted into the ground that detects the electrical discontinuities in shallow subsurface earth material [12]. The propagation of the georadar waves is dependent on electrical properties of the material such as electric conductivity (σ), magnetic permeability (μ), and dielectric permittivity (ε) [12]. When the georadar waves propagate down into the earth surface, they are refracted-reflected back to the surface depending upon the properties of the lithology [29, 30]. The unconsolidated sediments, construction material, and weathered or fractured rocks are porous in nature, hence if filled with water, affect dielectric permittivity or dielectric constant [30]. As the contrast in the subsurface layers is greater, the amount of reflected energy will be greater which is given by the reflection coefficient, R, and can be determined by the contrast in relative velocities and also by the contrast in relative dielectric constants [12]:

$$R = \frac{V1 - V2}{V1 + V2} \quad \text{or} \quad \frac{\sqrt{\varepsilon 2} - \sqrt{\varepsilon 1}}{\sqrt{\varepsilon 2} + \sqrt{\varepsilon 1}}, \tag{1}$$

where, $V1$ and $V2$ are relative velocities, and $\varepsilon 1$ and $\varepsilon 2$ are relative dielectric constants of the adjacent layers. Velocity of the medium (V) is dependent upon the velocity of light in air (0.3 m/ns) and is given by a simplified equation:

$$V = \frac{c}{\sqrt{\varepsilon}}, \quad (2)$$

where c is velocity of light in air. Further, depth of penetration (D) can be determined using two-way travel time (T) and the velocity of the medium, which can be obtained from the graphical interpretation of the GPR signals:

$$D = \frac{T * V}{2}. \quad (3)$$

2.1. Data Acquisition along Hajipur Fault. GPR profiles were collected with SIR-3000 system using 200 MHz-shielded antenna in common-offset continuous mode. Variable settings available in the portable unit were used to determine the best acquisition parameters. For a better vertical resolution, the data was collected with 16-bit format and 512 Sample/scan whereas, 64 Scan/unit (meter) was set for optimum horizontal resolution. Since the data was collected in a distance mode using survey wheel with 512 Sample/scan the rate of 100 Scan/sec was preferred. The floodplain deposits of Beas River is mainly comprised of sand + gravel deposits, with the top surface made up of medium-to-coarse sand with scattered gravels. Keeping in mind the usual dielectric constant of sand ranging from 3–6, higher value was attributed since the area was slightly moist. Finally, after having these parameters GPR profiles were collected, which gave a good resolution with penetration up to 6 m (Table 1). A calibrated odometer fixed on survey wheel was attached to the antenna to get the horizontal distance. Along with this, appropriate gain and IIR filter was applied to the field data to get better resolution and to remove noise with lower and higher frequencies, respectively.

At Sandhwal village, both 2D and 3D GPR profiling was carried out across the active fault scarp along Hajipur Fault (HF2), (Figure 2(a)). The 2D profile was 28 m long and was collected with the acquisition parameters given in Table 1. Out of several profiles taken along this line, the profile shown in Figure 3 was the best one that exhibits ample information about the fault geometry. However, to have further understanding of the fault geometry, a 3D profile was collected for a grid-ABCD with dimensions of 20 m × 6 m (Figures 2(b) and 4). For 3D profiling, the grid was prepared by taking 20 m along the X-axis and 6 m along the Y-axis and was further divided into smaller grids of 1 m spacing with which we have collected 21 grid lines along X-axis and 07 gridlines along Y-axis. The western corner (A) of the grid was considered as origin (Figure 2(b)). The data acquisition parameters for the 3D grid profile were the same as that of the 2D profile (Table 1). Since the surface across which the profile was collected was not horizontal, topographic profile was collected.

Table 1: Acquisition parameters used for collection of 2D and 3D profiles across HF2 scarp at Sandhwal village.

Data acquisition parameters	Corresponding values
Sample/scan	512
Bits/sample	16
Scan/sec	100
Scans/m	64
Dielectric constant	6

Table 2: Data processing parameters used for 2D and 3D GPR profiles across Hajipur Fault (HF2).

Parameters		2D profile	3D grid profile
Time-zero correction		5.86 ns	5.66 ns
FIR filters	Low pass	720 MHz	690 MHz
	High pass	38 MHz	40 MHz
3D grid	x-(start–end)	—	0 m–20 m
	y-(start–end)	—	0 m–6 m

3. Data Processing and Interpretation

The radar profiles collected in the field were processed with RADAN 6.5 software manufactured by Geophysical Survey Systems, Inc. The 2D profile was processed in the sequence starting from time-zero correction, application of filters, surface normalization, and autogain restoration. The processing parameters for the 2D profiles and 3D grid are given in Table 2. The time-zero correction was applied to remove the uppermost high amplitude reflections obtained from the direct air wave and ground wave (Figure 3(a)) and for the successful realignment of all reflections beneath. The low- and high-pass filters along with background removal were applied to remove noise (interference). Topographic data collected along the same transect was incorporated in the 2D profile, and surface normalization was performed (Figure 3(a)). Automatic gain control (AGC) was applied to enhance the visibility of low-frequency features. The processing of the 3D grid profile was also done following the same steps as for the 2D GPR profiles. To have better view of the variation in fault geometry along the strike, the 3D profiles were vertically sliced.

3.1. Interpretation. Previous studies suggest that for the interpretation of GPR data and for locating near subsurface displacements in young Quaternary deposits, it is necessary to look for the offset of radar reflections on both sides of the fault strand and for the warping of the GPR reflectors or discontinuity in the reflections (e.g., [10, 11, 30, 31]). Keeping the above points into consideration and with the background of the earlier paleoseismic and GPR surveys carried out along the Himalayan Frontal Thrust by Malik et al. [20, 23], attempts were made to interpret the GPR profiles taken for this study.

The 2D profile revealed prominent reflections in the form of hyperbolas between 4 m and 5 m horizontal markers and inclined discontinuities georadar reflections between 10 m and 28 m (Figure 3(a)). The hyperbolic reflections extending up to the depth of 2.5 m are interpreted as a metal pipe. In field, the local villagers informed that this was a main

(a)

(b)

FIGURE 2: (a) CORONA satellite photo showing distribution of active fault traces named as "Hajipur faults"—HF1 and HF2 on the left bank of Beas River near Hajipur village [25]. Box shows the location of trench and GPR profile line across the HF2 fault scarp at Sandhwal village. Empty arrow marks the direction of fault propagation towards NW. Young tectonic movements along HF1 and HF2 faults resulted into uplift of floodplain deposits, responsible for the shifting/deflection of the Beas River channel along it present course. (b) Panoramic view of HF2—Hajipur fault scarp. PQ indicates the 2D scan, and ABCD indicates the 3D grid collected, respectively across the fault which is marked in red arrows.

pipeline that supplies water to a few surrounding villages. Looking the typical characteristic of the multiple inclined radar reflectors marked by discontinuous nature, with warping close to the dipping reflections, it is suggested that such a pattern was due to displacement of near subsurface young sediment succession caused by faulting (Figures 3(a) and 3(b)). The strong warping observed between 24 m and 28 m horizontal markers extending from 1 m to 5.5 m in

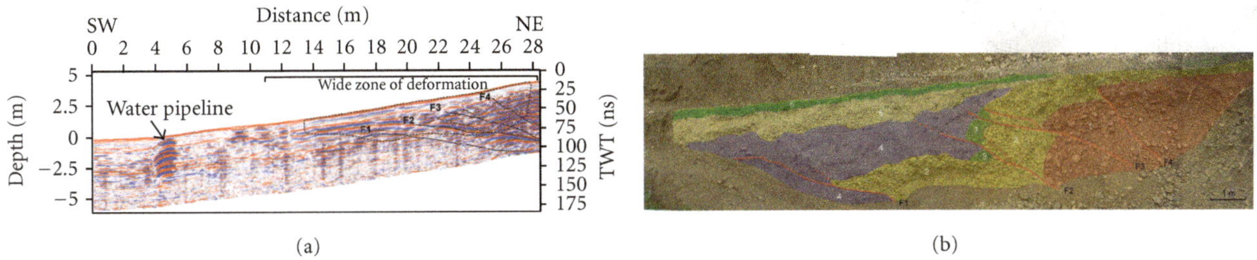

FIGURE 3: (a) Processed 2D GPR profile collected using 200 MHz antenna across the HF2 scarp at Sandhwal Village (refer to Figures 2(a) and 2(b) for location). The profile was about 29 m long showing the processed profile after topographic correction. Reflections at 4-5 m indicate water pipeline, whereas the reflections from 14–28 m show the fault traces. Box with dotted line shows approximate area of a trench. (b) Interpretation with fault traces and stratigraphic units on north wall of trench excavated across HF2 fault scarp. F1, F2, F3, and F4 are the thrust faults dipping in NE direction. Latest event has been observed along F1, F2, and F3 displacing the units 1, 2, 3, and 4.

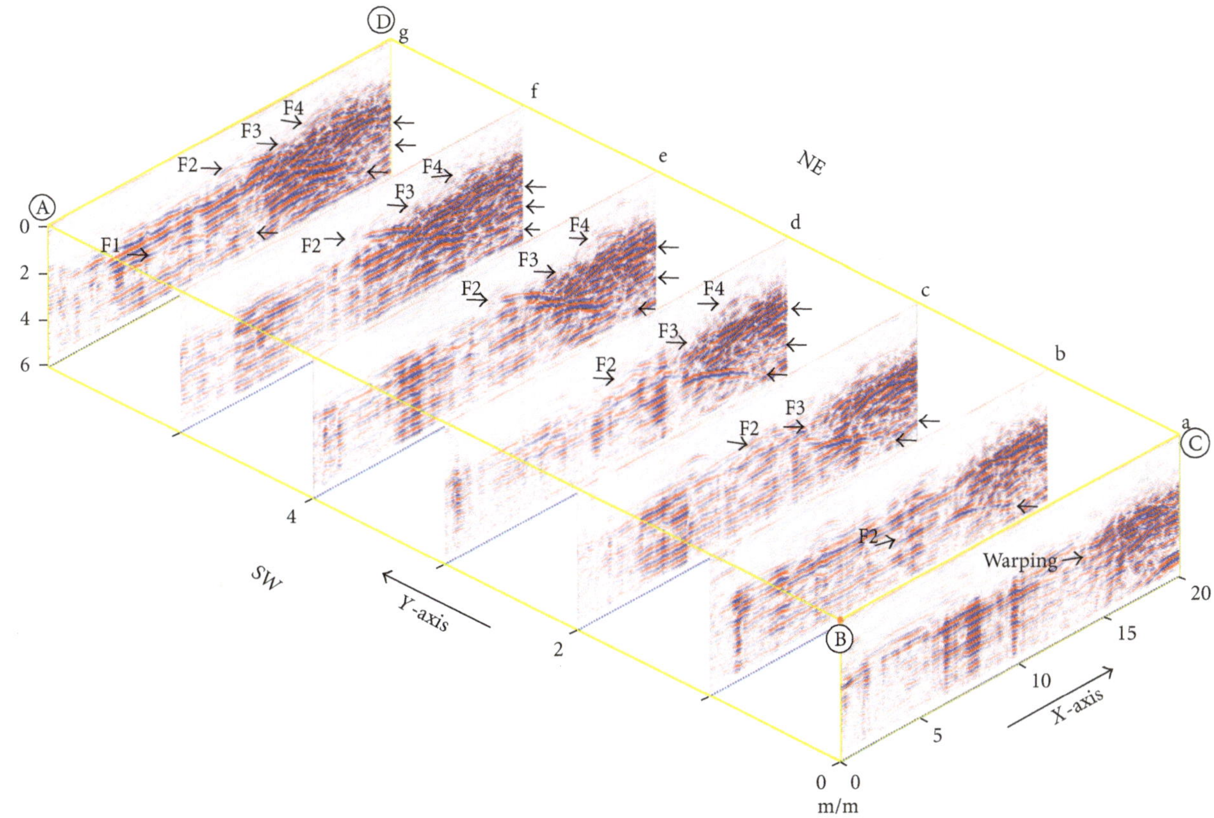

FIGURE 4: Sliced 3D grid profile across the HF2 scarp at Sandhwal Village collected with 200 MHz antenna (refer to Figures 2(a) and 2(b) for location). The profile was collected for a grid-ABCD with dimensions of 20 m × 6 m. The dashed lines and arrowheads represent the faults. The grid was sliced along the Y-axis along the strike of the scarp at 1 m interval (seven slices from a–g). Changes in the pattern of deformation along the strike can be noticed from the slices. For location of the grid, refer to Figure 2.

depth is related to the deformation on hanging wall side (Figures 3(a) and 3(b)). Considering the angle of dip in the 2D profile, it can be suggested that the deformation is due to movement from NE to SW along a major low-angle thrust fault (Figures 3(a) and 3(b)). Discontinuity and warping of the reflectors along more than one inclined reflection suggest splays of branching faults from the main fault.

To have proper understanding of the fault geometry along the strike, 3D profile collected was vertically sliced at an interval of 1 m (Figure 4). The grid was sliced along the Y-axis (i.e., along the strike of the fault scarp) into seven layers with 1 m spacing, namely a, b, c, d, e, f, and g (Figure 4). Each layer represents the line scan profile at 1 m interval. After close analysis, four low-angle fault strands were identified in slices (a–g) which were named as F1, F2, F3, and F4. The F2 strand was identified in all slices, where in slices a, b, and c, it is marked by low-angle dipping towards NE between 10 m and 20 m horizontal markers and clearly traceable from surface up to 6 m of depth (Figure 4). Prominent warping of georadar reflections was observed on the northeast side of the fault plane, which represents the hanging wall. A change in dip of F2 strand was observed in slice *b* (Figure 4). In addition to this other fault, strands F3 and F4 were seen in slices *d*–*g*. These strands are also dipping in NE direction with slightly higher angle. Strand F1 was observed between 3 m and 10 m horizontal markers in slice *g* only. It is marked by slightly steeper angle in basal part and becomes almost subhorizontal towards the surface (Figure 4).

Based on the multiple fault strands identified in 2D and 3D profiles, it is suggested that the F1 is probably the main fault propagating towards foreland and F2, F3, and F4 are the branching out faults. The 3D slice *g* and 2D profile collected gave similar features; hence, with this information for detailed paleoseismic investigation a trench with dimension—16 m long, 4-5 m wide 1–3 m deep was excavated across the HF2 fault scarp in NE-SW direction [20] (Figure 3(b)). The trench wall matched the fault strands identified in GPR profiles (Figures 3(a) and 3(b)).

The splays, namely, F1, F2, F3, and F4 identified in GPR profiles were comparable with the exposed trench section (Figures 3(a) and 3(b)). Trench section revealed five sedimentary units 1–6. Units 1 and 2 are rounded to sub-rounded gravel-poorly sorted cobble-pebble with sandy matrix. These units are marked by high-prominent geo-radar reflections observed on the NE part of the profiles. The warping reflection can well be justified by the preferred orientation of gravel clasts, suggestive of folding related to faulting along the F1–F4 faults. The southern part of the trench is comprised of finer deposits as compare to the northern. These deposits are comprised of medium-to-fine sand (units 3-4). Unit 5 is made up of medium-to-coarse sand with scattered gravel finally caps the sequence. It is suggested that F1, F2, F3, and F4 are the thrust faults dipping in NE direction, and the latest event occurred along F1, F2, and F3 displacing the units 1, 2, 3, and 4. Because of the instability of trench wall due to coarser gravel clasts, we were not able to excavate the trench to see deeper section. However, the GPR profiles (2D and 3D) helped us tracing the fault at greater depth.

4. Conclusion

The GPR survey, 2D and 3D helped us in identifying the suitable location for trenching. The 3D profiling with slicing at 1 m interval gave a complete picture of variation in deformation in terms of the geometry of the fault and related deformation along the strike of the active fault (HF2). The 2D profile and slice *g* of 3D profile were considered to open a trench for paleoseismic investigations. The prominent georadar reflections marked with discontinuities on either side of the inclined planes, warping suggested faulting along thrust faults. Also with the limitation to excavate trench up to greater depth, GPR profiles provide information on deformation and structure of greater depth. It is, therefore, suggested that GPR is one of the important tools which can be used for mapping the near subsurface displacements and variation in deformation along the strike of the active fault.

Acknowledgments

Financial support provided to the research project sponsored by DST, New Delhi, (vide project no. SR/S4/ES-417/2009) is duly acknowledged. We are grateful to two anonymous referees for providing valuable suggestions and comments, this helped us in improving the paper and brought more clarity to our findings. We are also thankful to our institute IIT Kanpur for providing support to our students.

References

[1] V. P. Solonenko, "Paleoseismogeology," *Izvestiya, Academy of Sciences, USSR, Physics of the Solid Earth*, vol. 9, pp. 3–16, 1973.

[2] K. Sieh, "Pre-historic large earthquakes produced by slip on the San Andreas Fault at Pallett Creek," *Journal of Geophysical Research*, vol. 83, pp. 3907–3939, 1978.

[3] R. E. Wallace, "Active faults, Palaeoseismology, and earthquake hazards in the Western United States," in *Earthquake Prediction. An International Review*, D. W. Simpson and P. G. Richards, Eds., pp. 209–216, American Geophysical Union, Washington, DC, USA, 1981.

[4] R. E. Wallace, "Variations in slip rates, migration, and grouping of slip events on faults in the Great Basin Province," *Bulletin of Seismological Society of America*, vol. 77, pp. 868–876, 1987.

[5] D. P. Schwartz and K. J. Coppersmith, "Fault behavior and characteristic earthquakes: examples from the Wasatch and San Andreas fault zones (USA)," *Journal of Geophysical Research*, vol. 89, no. 7, pp. 5681–5698, 1984.

[6] C. R. Allen, "Seismological and paleoseismological techniques of research in active tectonics," in *Active Tectonics*, R. E. Wallace, Ed., pp. 148–154, National Academy Press, Washington, DC, USA, 1986.

[7] J. P. Mc Calpin, Ed., *Paleoseismology*, Academic Press, San Diego, Calif, USA, 1996.

[8] R. S. Yeats, K. Sieh, and C. R. Allen, *Geology of Earthquakes*, Oxford University Press, 1997.

[9] R. S. Yeats and V. C. Thakur, "Reassessment of earthquake hazard based on a fault-bend fold model of the Himalayan plate-boundary fault," *Current Science*, vol. 74, no. 3, pp. 230–233, 1998.

[10] K. B. Anderson, J. A. Spotila, and J. A. Hole, "Application of geomorphic analysis and ground-penetrating radar to characterization of paleoseismic sites in dynamic alluvial environments: an example from southern California," *Tectonophysics*, vol. 368, no. 1–4, pp. 25–32, 2003.

[11] J. N. Malik and C. Mohanty, "Active tectonic influence on the evolution of drainage and landscape: geomorphic signatures from frontal and hinterland areas along the Northwestern Himalaya, India," *Journal of Asian Earth Sciences*, vol. 29, no. 5-6, pp. 604–618, 2007.

[12] A. Neal, "Ground-penetrating radar and its use in sedimentology: principles, problems and progress," *Earth-Science Reviews*, vol. 66, no. 3-4, pp. 261–330, 2004.

[13] J. P. Busby and J. W. Merritt, "Quaternary deformation mapping with ground penetrating radar," *Journal of Applied Geophysics*, vol. 41, no. 1, pp. 75–91, 1999.

[14] L. Liu and Y. Li, "Identification of liquefaction and deformation features using ground penetrating radar in the New Madrid seismic zone, USA," *Journal of Applied Geophysics*, vol. 47, no. 3-4, pp. 199–215, 2001.

[15] J. C. Audru, M. Bano, J. Begg, K. Berryman, S. Henryys, and B. Niviere, "GPR investigations on active faults in urban areas: the Georisc-NZ project in Wellington, New Zealand," *Comptes Rendus de l'Académie des Sciences-Series IIA- Earth and Planetary Science*, vol. 333, no. 8, pp. 447–454, 2001.

[16] R. Gross, A. Green, K. Holliger, H. Horstmeyer, and J. Baldwin, "Shallow geometry and displacements on the San Andreas fault near Point Arena based on trenching and 3-D georadar surveying," *Geophysical Research Letters*, vol. 29, no. 20, pp. 34–1, 2002.

[17] A. Green, R. Gross, K. Holliger, H. Horstmeyer, and J. Baldwin, "Results of 3-D georadar surveying and trenching the San Andreas fault near its northern landward limit," *Tectonophysics*, vol. 368, no. 1–4, pp. 7–23, 2003.

[18] L. Slater and T. M. Niemi, "Ground-penetrating radar investigation of active faults along the Dead Sea Transform and implications for seismic hazards within the city of Aqaba, Jordan," *Tectonophysics*, vol. 368, no. 1–4, pp. 33–50, 2003.

[19] P. J. Alasset and M. Meghraoui, "Active faulting in the western Pyrénées (France): paleoseismic evidence for late Holocene ruptures," *Tectonophysics*, vol. 409, no. 1–4, pp. 39–54, 2005.

[20] J. N. Malik, A. A. Shah, A. K. Sahoo et al., "Active fault, fault growth and segment linkage along the Janauri anticline (frontal foreland fold), NW Himalaya, India," *Tectonophysics*, vol. 483, no. 3-4, pp. 327–343, 2010.

[21] B. K. Sternberg and J. W. McGill, "Archaeology studies in southern Arizona using ground penetrating radar," *Journal of Applied Geophysics*, vol. 33, no. 1–3, pp. 209–225, 1995.

[22] B. M. Whiting, D. P. McFarland, and S. Hackenberger, "Three-dimensional GPR study of a prehistoric site in Barbados, West Indies," *Journal of Applied Geophysics*, vol. 47, no. 3-4, pp. 217–226, 2001.

[23] J. N. Malik, A. K. Sahoo, and A. A. Shah, "Ground-penetrating radar investigation along Pinjore Garden Fault: implication toward identification of shallow subsurface deformation along active fault, NW Himalaya," *Current Science*, vol. 93, no. 10, pp. 1422–1427, 2007.

[24] S. Satuluri, "Ground penetrating radar application in archaeological investigations at ahichhatra, bareilly (Distt)," M.Tech Dissertation Report, Department of Civil Engineering, Indian Institute of Technology, Kanpur, India, 2009.

[25] J. N. Malik, A. K. Sahoo, A. A. Shah, D. P. Shinde, N. Juyal, and A. K. Singhvi, "Paleoseismic evidence from trench investigation along Hajipur fault, Himalayan Frontal Thrust, NW Himalaya: implications of the faulting pattern on landscape evolution and seismic hazard," *Journal of Structural Geology*, vol. 32, no. 3, pp. 350–361, 2010.

[26] L. Seeber and J. G. Armbruster, "Great detachment earthquakes along the Himalayan arc and long-term forecasting," in *Earthquake Prediction. An International Review*, D. W. Simpson and P. G. Richards, Eds., vol. 4 of *Maurice Ewing Series*, pp. 259–279, AGU, 1981.

[27] N. Ambraseys and R. Bilham, "A note on the Kangra $M_s = 7.8$ earthquake of 4 April 1905," *Current Science*, vol. 79, no. 1, pp. 45–50, 2000.

[28] N. N. Ambraseys and J. Douglas, "Magnitude calibration of north Indian earthquakes," *Geophysical Journal International*, vol. 159, no. 1, pp. 165–206, 2004.

[29] C. S. Bristow and H. M. Jol, *Ground Penetrating Radar in Sediments, NO. 211*, Geological Society, London, UK, 2003.

[30] Jol H. M., *Ground Penetrating Radar: Theory and Applications*, Elsevier Scientific, Oxford, UK, 2009.

[31] J. Chow, J. Angelier, J. J. Hua, J. C. Lee, and R. Sun, "Paleoseismic event and active faulting: from ground penetrating radar and high-resolution seismic reflection profiles across the Chihshang Fault, eastern Taiwan," *Tectonophysics*, vol. 333, no. 1-2, pp. 241–259, 2001.

7

Spatiotemporal Relationship between Geodetic and Seismic Quantities: A Possible Clue to Preparatory Processes of M ≥ 6.0 Inland Earthquakes in Japan

Masashi Kawamura,[1] Takeshi Kudo,[2] and Koshun Yamaoka[3]

[1] *Department of Earth Sciences and Graduate Institute of Geophysics, National Central University, Jhongda Road no. 300, Taoyuan, Jhongli 32001, Taiwan*
[2] *General Education Division, College of Engineering, Chubu University, Matsumoto-cho 1200, Aichi, Kasugai 487-8501, Japan*
[3] *Research Center for Seismology, Volcanology and Disaster Mitigation, Graduate School of Environmental Studies, Nagoya University, Chikusa-ku, Nagoya 464-8601, Japan*

Correspondence should be addressed to Masashi Kawamura, mkawamu@ncu.edu.tw

Academic Editor: Rodolfo Console

Constructing a statistical model to characterize the physical conditions associated with large earthquake occurrence is crucial for disaster mitigation. With the aim to formulate such a model, we previously developed a statistical evaluation system which assesses the correlation of the spatiotemporal relationship between different kinds of physical quantities with the occurrence times of large inland earthquakes. In this study, we focused on assessing the relationship between geodetic and seismic quantities and attempted to find the pair of related quantities that most likely indicates preparatory processes of large earthquakes in Japan. We assessed the quantities prior to M ≥ 6.0 inland earthquakes for the period of 2001–2007 in terms of probability gains and error diagram. Our system revealed that the pair of absolute value of dilatation rate and seismic energy showed the highest statistical performance. Further validation of this result is required by updating the database of physical quantities.

1. Introduction

It is an urgent issue for disaster mitigation to develop a statistical model well reflecting crustal activities, which are expected to include the preparatory processes of large inland earthquakes. Resolution of this issue requires a comprehensive understanding and monitoring of crustal activities through more than one kind of physical observation. Therefore, keeping this in mind, in this study, we focused on examining the spatiotemporal relationships between different physical quantities, at least one of which is time-variable. By determining informative combinations of physical quantities, that is, those having some correlation, we aim to develop a monitoring index/indices in crustal activities which feature the preparatory processes of large inland earthquakes for Japan. Imoto [1, 2] insisted that their seismicity model shows higher performance by taking into consideration the correlation between different kinds of geophysical parameters. This implies that there are some correlations between different kinds of observations that may represent different aspects of crustal phenomena or show causality between apparently different crustal phenomena.

As the first step toward formulating a model to capture statistical and physical conditions for the occurrence of large inland earthquakes, we created a database of physical observations with spatially and temporally gridded formats (Figure 1). Furthermore, we constructed a statistical evaluation system to assess the relationship between the occurrence times of target events and interquantity relationship in the physical observation data, at least one of which is time variable [4]. This system comprises some calculation modules corresponding to respective steps in Section 2 and their subordinate functions (not shown in Section 2) to output several parameters which are required for next steps

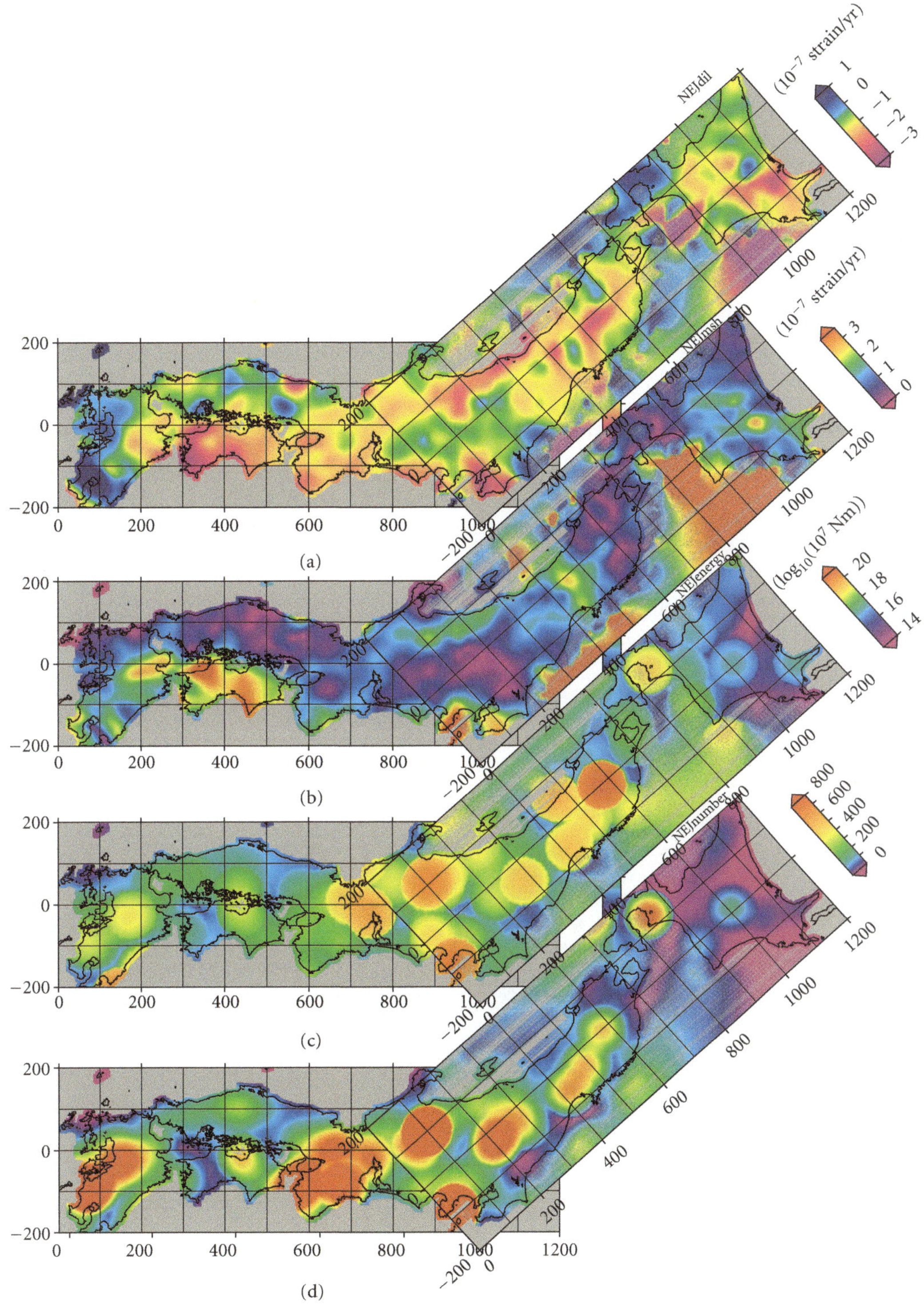

FIGURE 1: Gridded maps for (a) dilatation rate, (b) maximum shear strain rate, (c) seismic energy [3] within 30 km search radius, and (d) the number of earthquakes within 30 km search radius, respectively, over the Japanese islands.

or to display the final result of statistical evaluation at step 4 (Table 1, Figure 2).

Here, we actually operated the system to examine the relationship between two physical quantities and the occurrence times of M ≥ 6.0 mainshocks in Japan. We adopted the spatiotemporal relationship between geodetic quantities (dilatation rate and maximum shear strain rate) and seismic quantities (seismic energy and number of earthquakes) for representative physical observation. We look for the relationship for the earthquake from 2001 through 2007

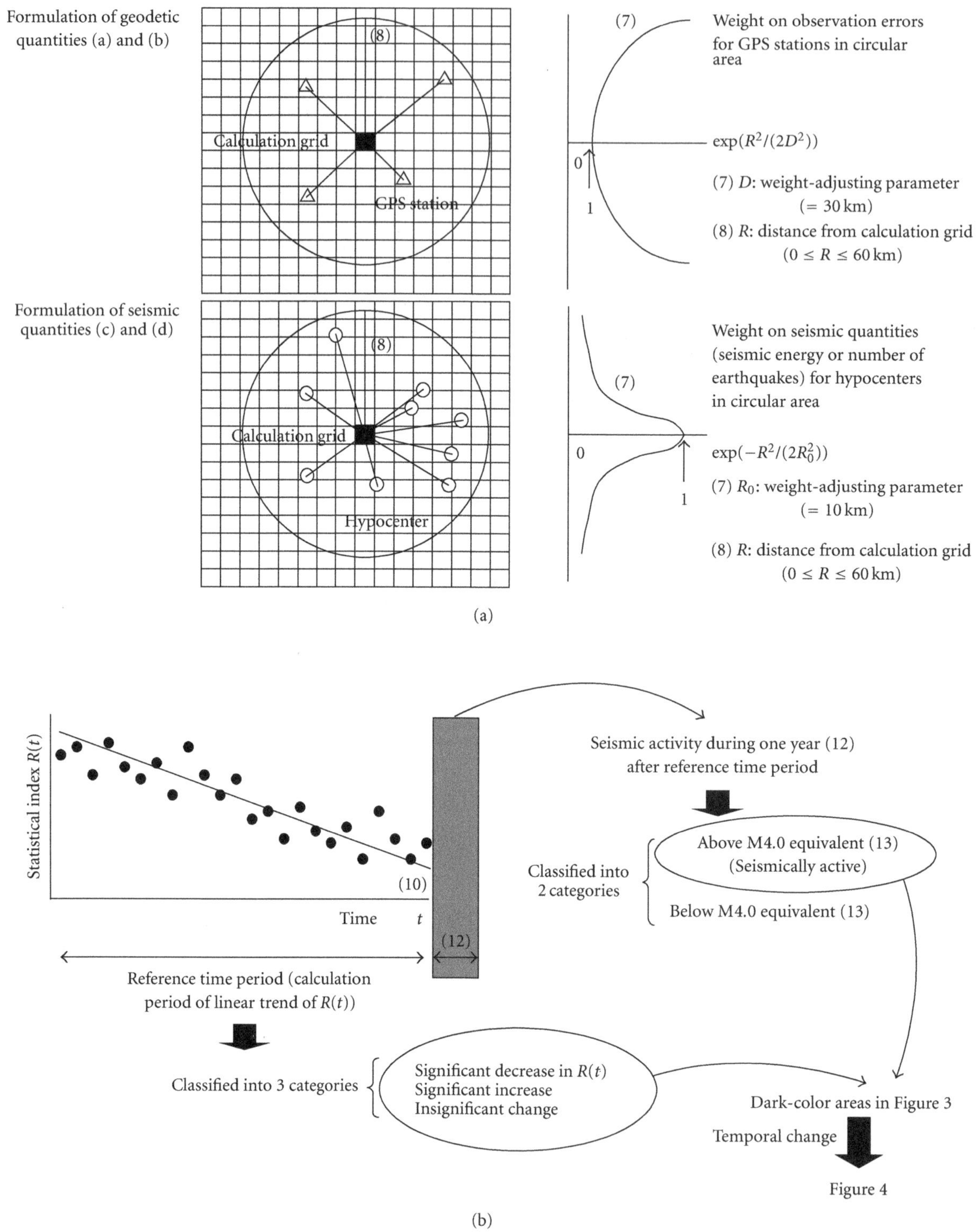

FIGURE 2: (a) Schematic explanation on the method of obtaining geodetic quantities (dilatation rate and maximum shear strain rate) and seismic quantities (seismic energy and number of earthquakes) for each grid, which were calculated by relating them to the displacement rates for GPS stations (open triangle) and the magnitudes for hypocenters (open circle) in circular area, respectively. We weighted the observation errors for GPS stations and seismic quantities for hypocenters depending on the distances between the calculation grid (black square) and the other grids in circular area, respectively. The vertical axis of each right figure in which weight curve is delineated shows the distances relative to the calculation grid. (b) Schematic explanation on the approach of obtaining Figures 3 and 4. See also text for detail.

Table 1: Tabulated information on input parameters required at each step of statistical evaluation system. The related schematic explanation should be referred to Figure 2.

No.	Parameter
(1)	Range of region for analysis (inland Japan)
(2)	Interval between spatial grids (0.05°)
(3)	Interval between time grids (three month)
(4)	Time period referred to for calculating physical quantities such as dilatation rate for each time grid (two years)
(5)	Range of depth ($0 \le \text{depth} \le 30$ km)
(6)	Range of magnitude ($M \ge 1.0$)
(7)	Smoothing (weight-adjusting) parameters for obtaining spatially gridded physical quantities (refer to Figure 2(a))
(8)	Search radius for applying smoothing parameters to area centered at a calculation grid (Figure 2(a))
(9)	Search radius for calculating statistical index for a calculation grid (60 km)
(10)	Approval method for classifying the temporal change in statistical index $R(t)$ into three categories (Figure 2(b))
(11)	Confidence level for statistical approval (95%)
(12)	Calculation period of total seismic energy released following calculation period of temporal change in statistical index $R(t)$ (Figure 2(b))
(13)	Lower limit of total seismic energy released during period (12) which is regarded as seismically active (Figure 2(b))
(14)	Definition of large inland earthquakes (lower limit of magnitudes) ($M \ge 6.0$)

for inland Japan. In evaluating this relationship, the system computes the statistical performance of a proposed monitoring index which would reflect an aspect of the preparatory process of large earthquake occurrence. We searched for the most informative combination between geodetic and seismic quantities based on evaluating the statistical performance of the proposed monitoring index for $M \ge 6.0$ inland mainshocks. Evaluation of the statistical performance revealed that the time variation of the monitoring index created based on the relationship between the absolute value of dilatation rate and seismic energy was most closely associated with the occurrence times of $M \ge 6.0$ inland mainshocks. Our result revealed the relationship between the absolute value of dilatation rate and seismic energy would best reflect the crustal activities involving preparatory processes of the occurrence of $M \ge 6.0$ inland mainshocks. An important future issue is to further validate the highest performance of the proposed monitoring index by updating the database to increase the number of samples, or large inland earthquakes.

2. Statistical Evaluation of the Relationship between Geodetic and Seismic Quantities

The statistical evaluation system developed by Kawamura et al. [4] statistically assesses the relationship between the spatiotemporal connection in any two kinds of physical quantities and the occurrence times of target physical events of interest We actually operated the system as follows to find a monitoring index, which is created based on a pair of items (a) to (d) in Figure 1, or a pair of geodetic and seismic quantities, such that the index can feature crustal activities which are expected to reflect the preparatory processes of large inland earthquakes.

2.1. Step 1: Making a Database with Spatially and Temporally Gridded Formats. We created a database comprising items (a) to (d) in Figure 1. It shows spatially and temporally gridded maps of four physical quantities: (a) dilatation rate, (b) maximum shear strain rate, (c) seismic energy [3] within 30 km search radius, and (d) the number of earthquakes within 30 km search radius, respectively, over inland Japan. Items (a) and (b) were computed from GEONET (GPS Earth Observation NETwork) data (Figure 2(a)). Items (c) and (d) were calculated from unified hypocenter data operated by Japan Meteorological Agency (Figure 2(a)).

We here explain the method of making physical quantities (a) to (d) (Figure 2(a)). To obtain dilatation rate (a) and maximum shear strain rate (b) with grid formats, we first removed abnormal values from the GEONET coordinate data, where the coordinate data with a standard deviation >3 over half a month before and after each time was regarded as abnormal value. We next calculated the displacement rate at each observation point. We assumed that the displacement rate on an arbitrary coordinate grid can be represented by using a linear function of the displacement rate at its nearby observation point. Then, the displacement rate at observation point is related to the position of grid point as follows:

$$U_x = A_{11} + A_{12}(x - x_g) + A_{13}(y - y_g) + E_x,$$
$$U_y = A_{21} + A_{22}(x - x_g) + A_{23}(y - y_g) + Ey, \quad (1)$$

where U_x and U_y are east-west and north-south components of displacement rate at each observation point; $x - x_g$ and $y - y_g$ are east-west and north-south components of the position (x, y) of observation point relative to the position (x_g, y_g) of grid point. E_x and E_y are the error term of respective components. Then, dilatation rate Δ and maximum shear strain rate $\sum$ can be obtained as follows:

$$\left(\Delta, \sum\right) = \left(A_{12} + A_{23}, \ \left((A_{12} - A_{23})^2 + (A_{13} + A_{22})^2\right)^{1/2}\right). \quad (2)$$

Furthermore, the error term E_x, E_y was weighted depending on the distance between observation and grid points as follows:

$$E_i = \sigma_i \exp\left(\frac{R^2}{2D^2}\right) \quad (i = x, y), \tag{3}$$

where σ_i is each component of the estimation error of the displacement rate at observation point. R is the distance between observation and grid points. D is weight-adjusting parameter. In the following sections, we show the system operation result obtained by setting $D = 30$ km and $0 \leq R \leq 60$ km.

To obtain seismic energy (c) and the number of earthquakes (d) for each grid point, the earthquakes within a radius of 60 km of grid point with depths shallower than 30 km and magnitudes ≥ 1.0 were collected (Figure 3(b)). Then, the total seismic energy and the total number of earthquakes were calculated, where the seismic energy and the number of each earthquake were weighted depending on the distance between observation and grid points as follows:

$$P = P_i \exp\left(-\frac{R^2}{2R_0^2}\right) \quad (i = 1, n), \tag{4}$$

where P_i is physical quantity related to the ith earthquake, to which 1 is assigned in the case physical quantity is defined as the number of earthquakes, n is the total collected number of earthquakes for grid point, and R is the distance between observation and grid points. R_0 is weight-adjusting parameter. In this way, by taking the distance-depending weight into consideration, the essential event collection radius becomes 30 km. In the following sections, we show the system operation result obtained by setting $R_0 = 10$ km and $0 \leq R \leq 60$ km.

2.2. Step 2: Comparison between Geodetic and Seismic Quantities. After the selection of one of the four possible pairs (here, the absolute value of dilatation rate and seismic energy (per two years), maximum shear strain rate and seismic energy, absolute value of dilatation rate and the number of earthquakes (per two years), and maximum shear strain rate and the number of earthquakes), its relationship was examined using scatter diagrams for circular regions, each having 60-km radius centered in 0.05°-by-0.05° spatial grids, over inland Japan; each pair was compared for two-year time windows centered in a three-month-interval time grid.

2.3. Step 3: Quantification of the Relationship Between Geodetic and Seismic Quantities. Quantification of the relationship between geodetic and seismic quantities requires us to define an index well characterizing the relationship. We here define the index as a statistical index $R(t)$, where t represents time. One of the simplest indices is correlation coefficient. Scatter diagrams for respective time windows, however, implied that the relationship between geodetic and seismic quantities showed relatively complicated patterns, which forced us to define another index as follows.

(1) For a particular circular region centered in a spatial grid and for a two-year time window centered in a time grid, the mean of the absolute value of dilatation rate or the maximum shear strain rate was calculated from the grids ranking in the top 20% in terms of seismic energy or number of earthquakes. We defined this parameter as statistical index $R(t)$, where t represents the midpoint of each two-year time window.

(2) Process (1) was carried out for each time window which is moved by three months. The linear trend of $R(t)$ was then calculated for each time period from 1 January 2001 through 1 October 2007 with reference to the fixed time grid of 1 April 1998. The linear trend of $R(t)$ thus obtained was classified into three categories on the basis of trend significance with a 95% confidence level: significant decrease, significant increase, and insignificant change.

(3) Processes (1) and (2) are carried out for every spatial grid with an interval of 0.05°.

Table 2: Model contingency table for explaining the relationship between potential conditions for/against the occurrence of physical events such as large inland earthquakes and their actual occurrence. The relation can be classified into the following four elements: *a*: the number of "yes" warnings accompanied by the occurrence of events, *b*: the number of "yes" warnings not accompanied by the occurrence of events, *c*: the number of "no" warnings accompanied by the occurrence of events, *d*: the number of "no" warnings not accompanied by the occurrence of events.

Potential condition for event occurrence	Event occurrence	
	Yes	No
Positive	*a*	*b*
Negative	*c*	*d*

2.4. Step 4: Statistical Evaluation of the Relationship between Geodetic-Seismic Quantity Connection and Large Event Occurrence Times. The relationship between the occurrence times of large inland mainshocks and different types of statistical index $R(t)$ was evaluated by calculating the probability gains of alarm rate (AR) and success rate (SR) for M $\geq$ 6.0 inland mainshocks and drawing an error diagram based on a contingency table. To calculate AR and SR for each pair of physical quantities, we need to tabulate the relationship between the proposed potential condition, which is proposed for the possible occurrence of physical events of interest, and the actual occurrence of the event (Table 2). The items in the row and column of contingency table correspond to the potential condition for event occurrence and the actual occurrence, respectively. Defining the time period for positive potential condition in which the events will be predicted to occur as the period of A and the period of negative potential condition as the period of B, each element

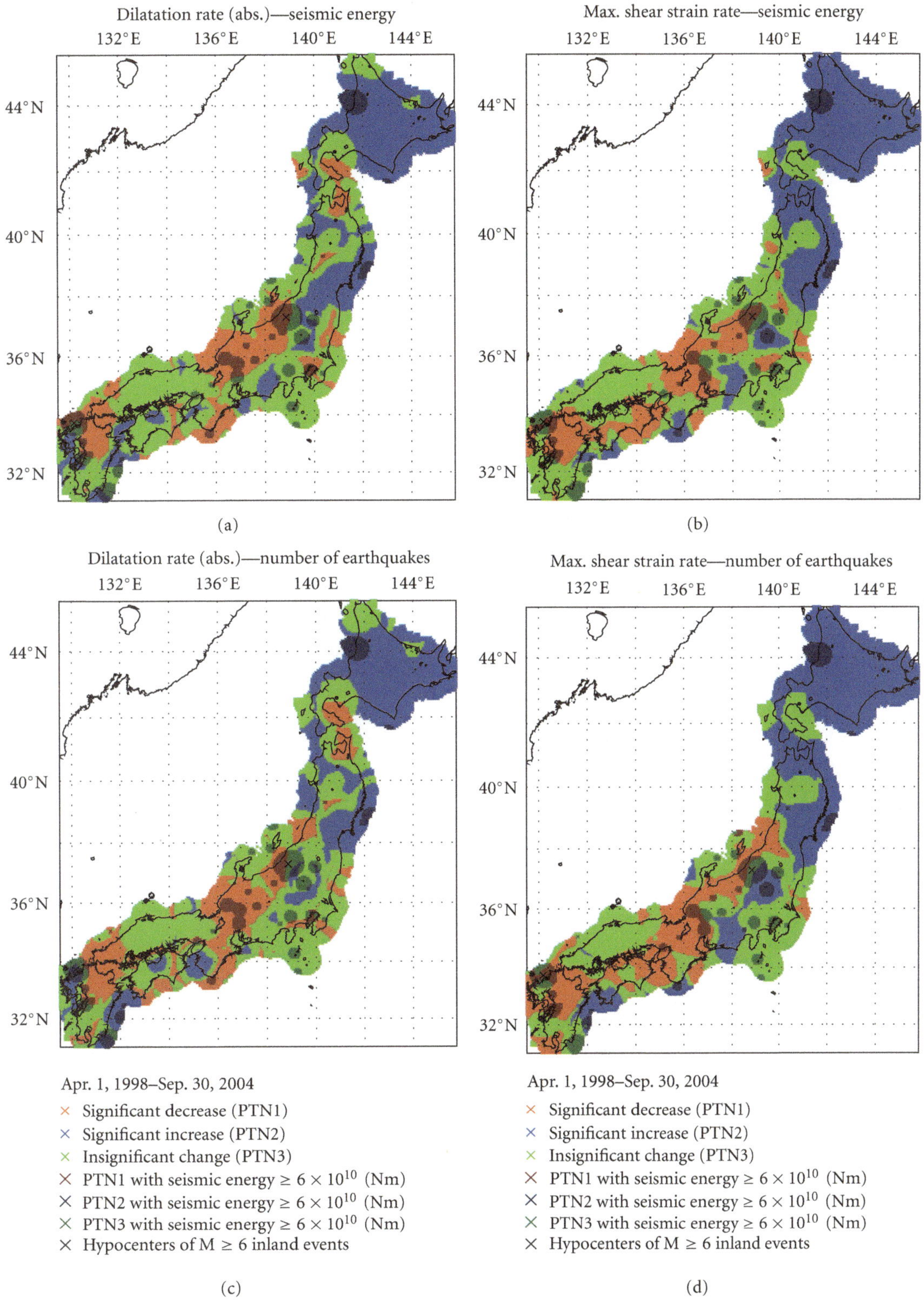

Figure 3: Spatial distributions of linear trends of statistical index $R(t)$ (pure red: significant decrease, pure blue: significant increase, pure green: insignificant change), which is obtained by comparison between (a) the absolute value of dilatation rate and seismic energy per two years, (b) maximum shear strain rate and seismic energy per two years, (c) absolute value of dilatation rate and number of earthquakes per two years, (d) maximum shear strain rate and number of earthquakes per two years. Time period for calculating linear trends of $R(t)$ is from 1 April 1998 through 30 September 2004, which is shown in panel (c) and (d). Superimposed in dark colors are the same distributions of areas in which seismic energy more than or equal to M4.0 earthquake (6×10^{10} Nm) was radiated during one year after the time period shown in panel (c) and (d). Linear trends of $R(t)$, which were obtained by least squares fitting, were classified using T approval with a confidence level of 95%. The numbers of dark-red, dark-blue, and dark-green grids, which are normalized by those of their corresponding pure-color ones, are here defined as no. PTN1, no. PTN2, and no. PTN3, respectively, for convenience in Figures 4 and 5.

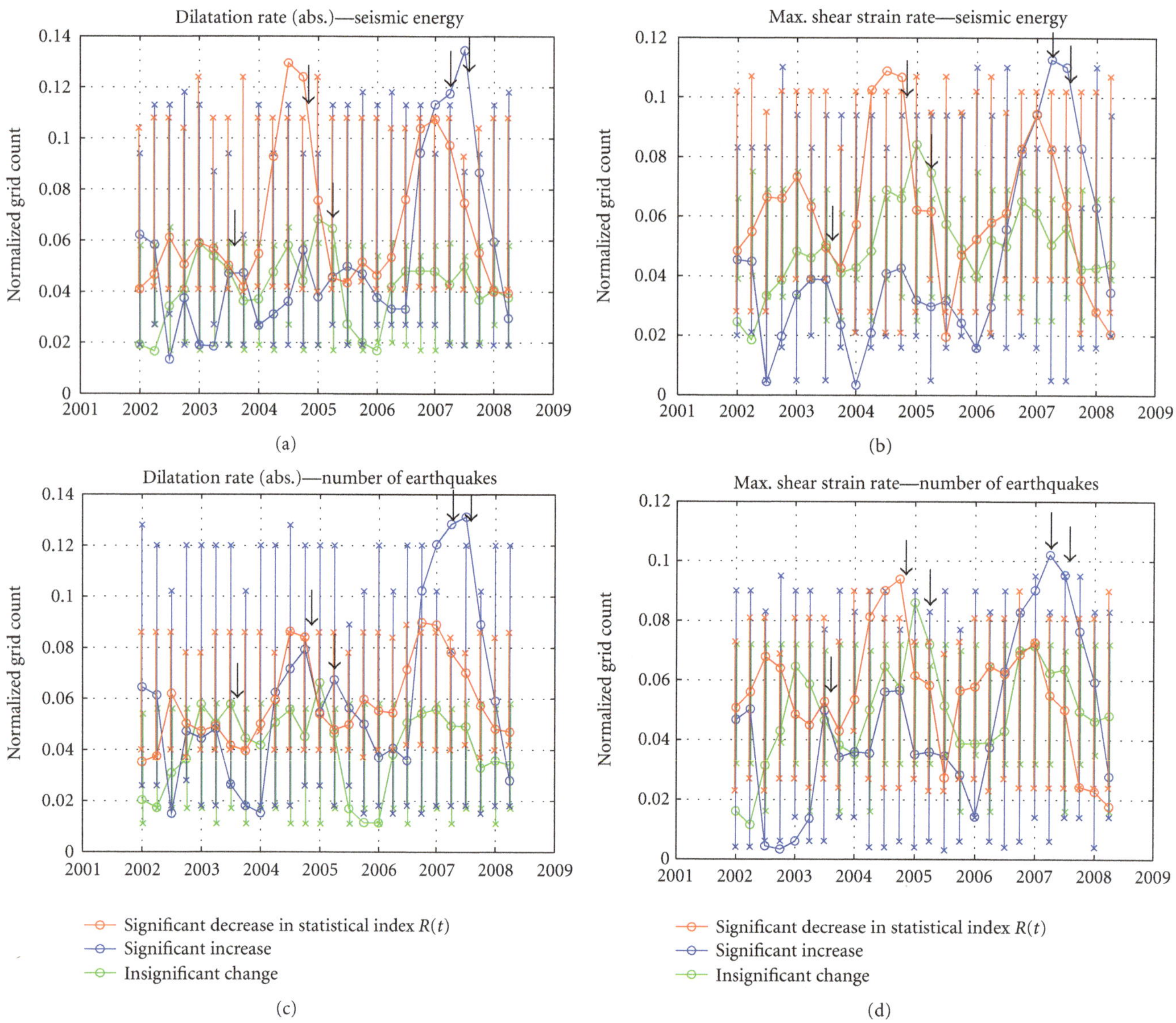

FIGURE 4: Temporal variation in no. PTN1, no. PTN2, and no. PTN3 (ref. Figure 3). Black arrows correspond to the occurrence date of $M \geq 6.0$ inland mainshocks. Also shown are the error bars which were obtained by 100 times of shuffling of no. PTN1, no. PTN2, and no. PTN3 in time, respectively. The horizontal axis denotes the time corresponding to the end of the time period referred to for calculation of linear trend of $R(t)$. The beginning of each time period is fixed to April 1, 1998. Panels (a) to (d) correspond to different four pairs of geodetic and seismic quantities, which are the same as those in Figure 3.

of (a, b, c, d) in Table 2 can be explained as follows:

(a) the number of cases in which physical events of interest occur during the period of A,

(b) the number of cases in which physical events did not occur during the period of A,

(c) the number of cases in which physical events of interest occur during the period of B,

(d) the number of cases in which physical events did not occur during the period of B.

By this definition, $a+b$ and $a+c$ can be interpreted as follows.

($a+b$): Total number of the periods of A.

($a+c$): Total number of periods in which physical events of interest occurred.

If an event occurs during the time period of A, one count is added to element a. Using elements of (a, b, c, d), alarm rate (AR), and success rate (SR) can be calculated as follows [5, 6]:

$$\text{AR} = \frac{a}{a+b}, \qquad \text{SR} = \frac{a}{a+c}. \tag{5}$$

The role of this step is to find a useful monitoring index/indices which feature the preparatory crustal activity

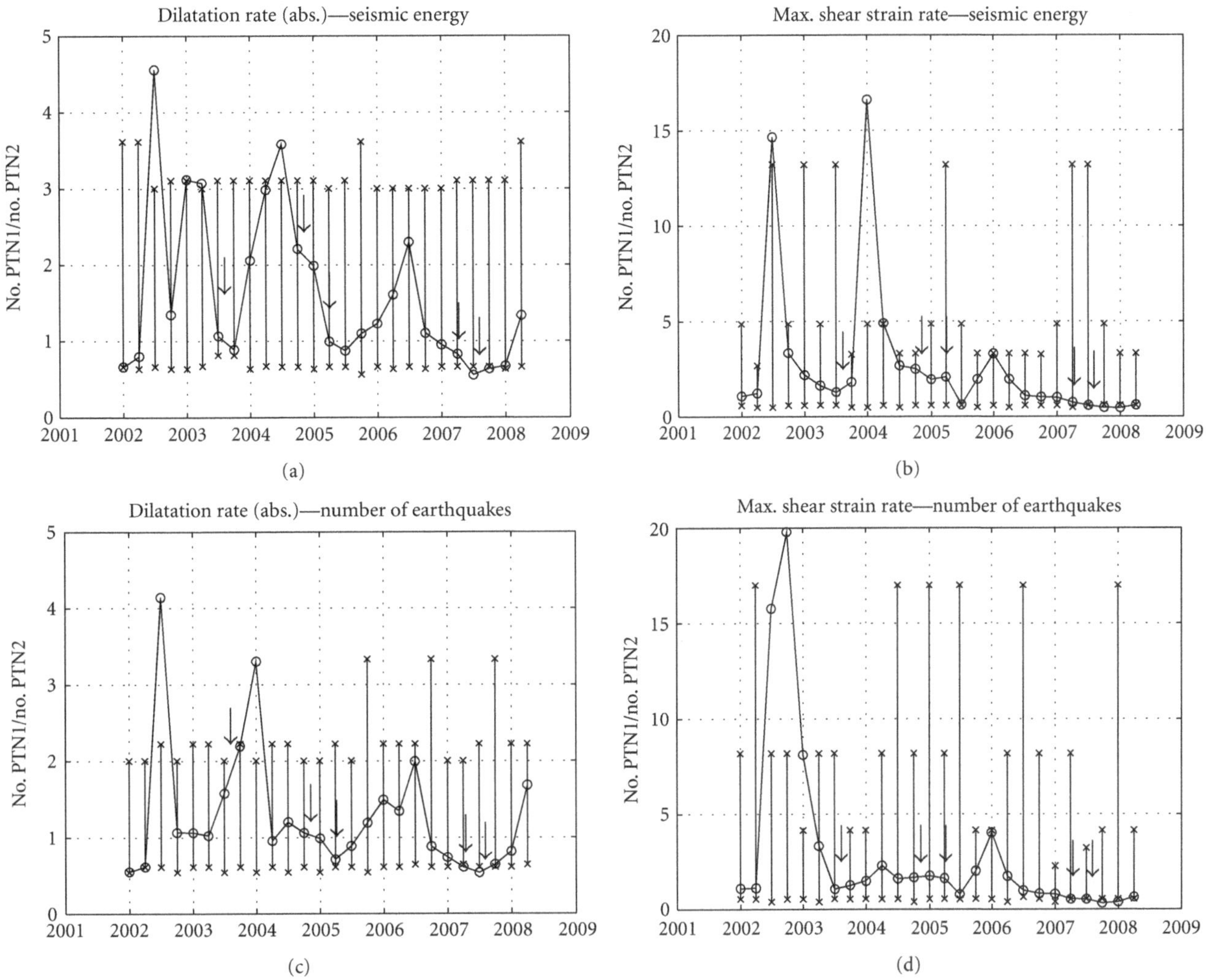

Figure 5: Temporal variation in the ratio of no. PTN1 to no. PTN2 (ref. Figure 3). Black arrows and the horizontal axis denote the same parameters as in Figure 4. Also shown are the error bars which were obtained by calculating the ratio between no. PTN1 and no. PTN2 shuffled in Figure 4. Panels (a) to (d) correspond to different four pairs of geodetic and seismic quantities, which are the same as those in Figures 3 and 4.

of a large inland earthquake through a statistical evaluation process. Such a monitoring index is defined based on the information on grid counts showing a specific (ex. negative significant) trend of $R(t)$ for a time window which are accompanied by high seismic activities. It should be noted that the performance of the monitoring index must be verified and validated in prospective way by updating the database.

We here focused on time variations of spatial grid counts which are accompanied by one-year release of total seismic energy larger than or equal to M4.0 earthquake (6.0×10^{10} Nm); the seismically active grid counts are specifically defined as no. PTN1 for those of significant decrease in $R(t)$, no. PTN2 for those of significant increase in $R(t)$, and no. PTN3 for those of insignificant change in $R(t)$. In addition, the three kinds of grid counts were normalized by their respective total grid counts (pure-color plus dark-color grids in Figure 3) for each time period of calculation. Step 4 tabulates thus calculated spatiotemporal distributions of no. PTNs1–3 and outputs the following figures: Figure 4(a) spatial distributions of no. PTNs1–3 for each time period (Figure 3), Figure 4(b) temporal changes in no. PTN1, no. PTN2, and no. PTN3 (Figure 4), and Figure 4(c) temporal changes in no. PTN1/no. PTN2 (Figure 5); this figure represents superiority or inferiority in the normalized area of PTN1 and PTN2. Assuming the temporal changes in no. PTN1/no. PTN2 reflect the preparatory processes of large inland earthquakes, we attempted to formulate a positive potential condition for the occurrence of large inland earthquakes in terms of a way of temporal changes in no. PTN1/no. PTN2 as shown in the first row of Table 3.

Statistical evaluation process at Step 4 requires the probability gains [7, 8] of alarm rate (AR) and success rate (SR). To calculate the probability gains of AR and SR, we prepared reference AR and SR for each pair of geodetic and seismic quantities in the following way.

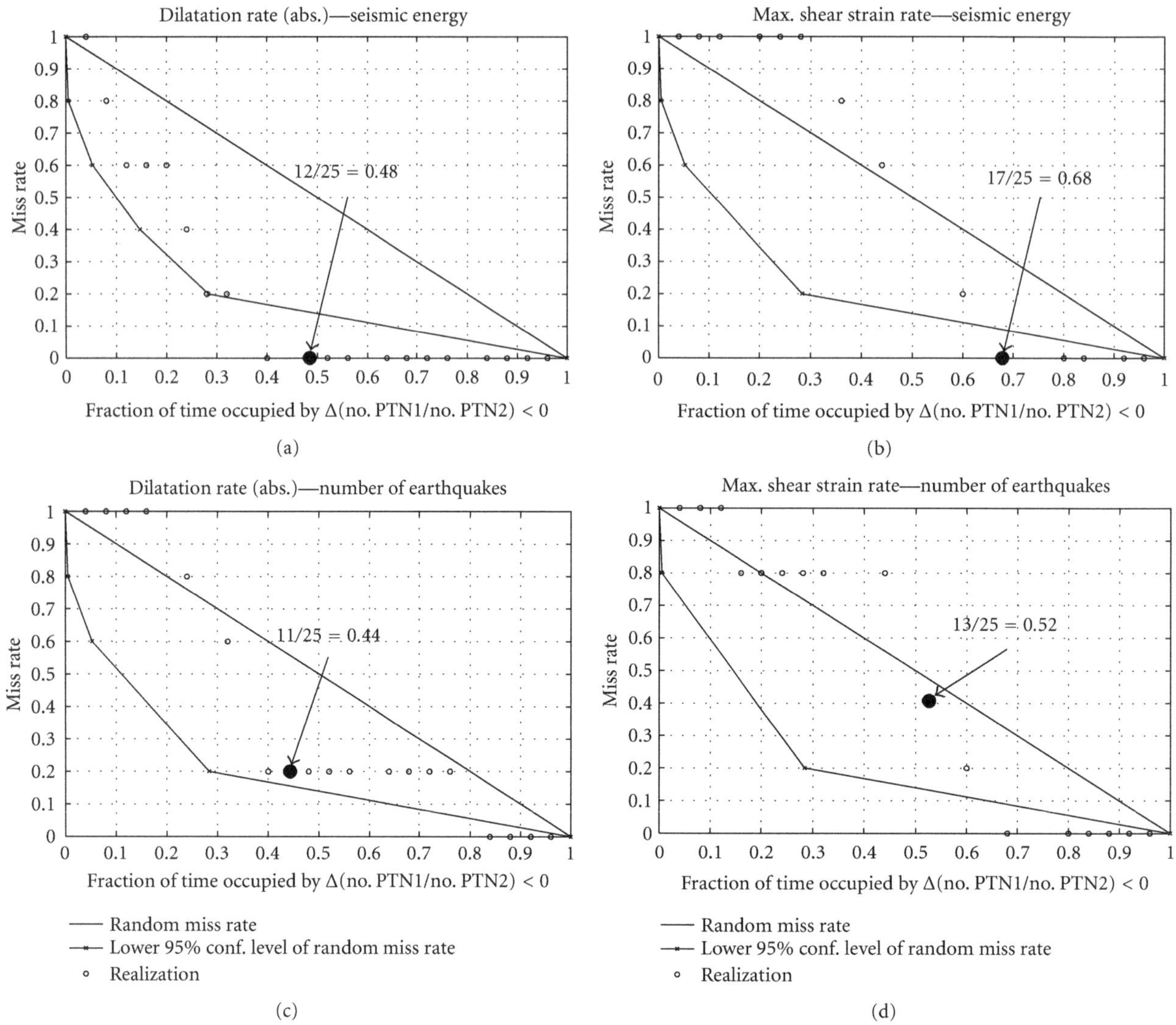

FIGURE 6: Error diagram (Molchan diagram) for evaluating the significance of the performance of a proposed potential condition for the occurrence of large inland earthquakes.

(1) We applied the bootstrap method to shuffle the spatial distributions of no. PTNs1–3 grids and their corresponding seismically inactive grids for a particular time window.

(2) Process (1) was carried out by moving the time window with its beginning time grid fixed. By repeating this, we obtained the temporal variation of the shuffled spatial distributions.

(3) We applied the bootstrap method to shuffle the temporal variation obtained in process (2).

(4) Processes (1) to (3) were repeated 100 times.

(5) The 100 ARs and SRs obtained were averaged.

Thus, calculated references AR and SR were considered as being due to a random phenomenon.

3. Most Informative Pair of Geodetic and Seismic Quantities

Among the four pairs of geodetic and seismic quantities examined, the pair comprising the absolute value of dilatation rate and seismic energy was found to be the most closely related in terms of probability gains of both alarm rate (AR) and success rate (SR) for $M \geq 6.0$ inland mainshocks (3.56 and 3.58, resp.) (Table 3). This pair showed the best performance even from the viewpoint of the error diagram (Figure 6). This is because the actual plot was located closest to position of the origin in Figure 6, which corresponds to the best performance of a proposed potential condition for event occurrence. The condition is represented in the first row of Table 3 or as the horizontal axis of Figure 6. This result was the same regardless of the approval methods and confidence levels. Figures 4(a) and 5(a) and the leftmost figures of respective elements in Table 3, which correspond to

TABLE 3: Contingency table showing the relationship between proposed potential conditions associated with the occurrence of M ≥ 6.0 inland mainshocks for four pairs of geodetic and seismic quantities and their actual occurrence times. The four figures for each element correspond to the following four pairs of geodetic and seismic quantities from left: the absolute value of dilatation rate and seismic energy, maximum shear strain rate and seismic energy, absolute value of dilatation rate and the number of earthquakes, and maximum shear strain rate and the number of earthquakes. A proposed potential condition occurring prior to the occurrence of inland mainshocks, Δ(no. PTN1/no. PTN2)<0, implies that the trend of no. PTN1/no. PTN2 is negative (Figure 5).

	Occurrence of inland earthquakes (M ≥ 6)	Nonoccurrence	Alarm rate (AR)
Δ(no. PTN 1/no. PTN2) < 0	5/5/4/3	7/12/7/10	0.417/0.294/0.364/0.231
Δ(no. PTN 1/no. PTN2) > 0	0/0/1/2	13/8/13/10	
Success rate (SR)	1.00/1.00/0.80/0.60		

Probability gains of AR and SR: 3.56/2.49/3.00/1.97 and 3.58/3.49/2.73/2.13.

the pair comprising the absolute value of dilatation rate and seismic energy, indicate that M ≥ 6.0 inland mainshocks are most likely to occur during the time period of the decrease in no. PTN1 relative to no. PTN2 or the increase in no. PTN2 relative to no. PTN1.

On the assumption that this tendency holds for even a small region influenced by a change in Coulomb failure stress (ΔCFS), a possible physical occurrence mechanism of a M ≥ 6.0 mainshock would be related to its preparatory process over a fault system. A mainshock would be triggered at the stage after microseismicity becomes active in the fault system, where stresses are considered to be accumulated on its asperities. Such a region may be accompanied by smaller strain rates. Because only a few samples of M ≥ 6.0 inland mainshocks were available to us, we need to continue to collect observations and further evaluate and validate the condition for event occurrence and the closest relation between geodetic and seismic quantities in the same manner as the one described in this study. We emphasize that this manner must not be changed in validating the condition for event occurrence. Furthermore, it is important to operate the statistical evaluation system to search for other informative pairs of physical quantities to gain a physical understanding of crustal activities as preparatory processes of large inland earthquakes.

4. Summary

Based on the notion that there are some correlations between different kinds of observations that may represent different aspects of crustal phenomena, we began with creating a database of various physical quantities and further constructed a basic system for statistically evaluating and validating a monitoring index which would reflect crustal activities associated with the occurrence of physical events such as large inland earthquakes. We have indicated that the most informative pair of geodetic and seismic quantities comprises the absolute value of dilatation rate and seismic energy in terms of probability gains of alarm rate (AR) and success rate (SR) and error diagram for M ≥ 6.0 inland mainshocks. The probability gains were calculated on the basis of contingency tables which show the relationship between the temporal changes in linear trends of statistical index $R(t)$ in seismically active areas and the occurrence times of M ≥ 6.0 inland mainshocks. Further evaluation and validation of the best pairs and the search for other informative combinations of physical quantities are necessary to gain a physical understanding of crustal activities, particularly, the preparatory processes of large inland earthquakes.

Acknowledgments

The authors deeply acknowledge valuable comments provided by two anonymous reviewers, which highly contributed to the improvement of the manuscript. They are also grateful to Geospatial Information Authority of Japan, and Japan Meteorological Agency for letting them utilize GEONET (GPS Earth Observation NETwork) data and unified hypocenter catalog, respectively.

References

[1] M. Imoto, "Earthquake probability based on multidisciplinary observations with correlations," *Earth, Planets and Space*, vol. 58, no. 11, pp. 1447–1454, 2006.

[2] M. Imoto, "Information gain of a model based on multidisciplinary observations with correlations," *Journal of Geophysical Research B*, vol. 112, no. 5, Article ID B05306, 2007.

[3] B. Gutenberg and C. F. Richter, "Magnitude and energy of earthquakes," *Annali di Geofisica*, vol. 9, pp. 1–15, 1956.

[4] M. Kawamura, T. Kudo, K. Yamaoka, and M. Furumoto, "Understanding of crustal activity with statistical approaches: I. Construction of spatiotemporal database with a unified data format,," *Journal of the Seismological Society of Japan (Second Series)*, vol. 63, pp. 21–33, 2010 (Japanese).

[5] I. T. Jolliffe and D. B. Stephenson, *Forecast Verification*, John Wiley & Sons, New York, NY, USA, 2003.

[6] S. Uyeda and A. Kumamoto, "Evaluation of the Kushida method of short-term earthquake prediction," *Proceedings of the Japan Academy Series B*, vol. 80, no. 3, pp. 140–147, 2004.

[7] K. Aki, "A probabilistic synthesis of precursory phenomena," in *Earthquake Prediction: An International Review*, D. W. Simpson and P. G. Richards, Eds., Maurice Ewing Seires 4, pp. 566–574, AGU, Washington, DC, USA, 1981.

[8] A. Helmstetter, Y. Y. Kagan, and D. D. Jackson, "High-resolution time-independent grid-based forecast for M ≥ 5 earthquakes in California," *Seismological Research Letters*, vol. 78, no. 1, pp. 78–86, 2007.

8

Evaluation of Vapor Pressure Estimation Methods for Use in Simulating the Dynamic of Atmospheric Organic Aerosols

A. J. Komkoua Mbienda,[1] C. Tchawoua,[2] D. A. Vondou,[1] and F. Mkankam Kamga[1]

[1] *Laboratory of Environmental Modeling and Atmospheric Physics, Department of Physics, Faculty of Science, University of Yaounde 1, P.O. Box 812, Yaounde, Cameroon*

[2] *Laboratory of Mechanics, Department of Physics, Faculty of Science, University of Yaounde 1, P.O. Box 812, Yaounde, Cameroon*

Correspondence should be addressed to A. J. Komkoua Mbienda; kombiend@gmail.com

Academic Editor: Robert Tenzer

The modified Mackay (mM), the Grain-Watson (GW), Myrdal and Yalkovsky (MY), Lee and Kesler (LK), and Ambrose-Walton (AW) methods for estimating vapor pressures (P^{vap}) are tested against experimental data for a set of volatile organic compounds (VOC). P^{vap} required to determine gas-particle partitioning of such organic compounds is used as a parameter for simulating the dynamic of atmospheric aerosols. Here, we use the structure-property relationships of VOC to estimate P^{vap}. The accuracy of each of the aforementioned methods is also assessed for each class of compounds (hydrocarbons, monofunctionalized, difunctionalized, and tri- and more functionalized volatile organic species). It is found that the best method for each VOC depends on its functionality.

1. Introduction

Atmospheric aerosols (AA) have a strong influence on the earth's energy balance [1] and a great importance in the understanding of climate change and human health (respiratory and cardiac diseases, cancer). They are complex mixtures of inorganic and organic compounds, with composition varying over the size range from a few nanometers to several micrometers. Given this complexity and the desire to control AA concentration, models that accurately describe the important processes that affect size distribution are crucial. Therefore, the representation of particle size distribution is of interest in aerosol dynamics modeling. However, in spite of the impressive advances in the recent years, our knowledge of AA and physical and chemical processes in which they participate is still very limited, compared to the gas phase [2]. Several models have been developed that include a very thorough treatment of AA processes such as in Adams and Seinfeld [3], Gons et al. [4], and Whitby and McMurry [5]. Indeed, the evolution of size distribution of AA is made by a mathematical formulation of processes called the general dynamic equation (GDE). It is well known that the first step in developing a numerical aerosol model is to assemble expressions for the relevant physical processes. The second step is to approximate the particle size distribution with a mathematical size distribution function. Thus, the time evolution of the particle size distribution of aerosols undergoing coagulation, deposition, nucleation, and condensation/evaporation phenomena is finally governed by GDE [1]. This latter phenomenon is characterized by the mass flux I_i for volatile species i between gas phase and particle which is computed using the following expression [6]:

$$I_i = \frac{dm_i}{dt} = 2\pi D_i^g d_p f_{\text{FS}}\left(K_{n_i}, \alpha_i\right)\left(c_i^g - c_i^s\right). \quad (1)$$

f_{FS} describes the noncontinuous effects [7]. When $I_i \geq 0$ ($I_i \leq 0$), there is condensation (evaporation). c_i^s is assumed to be at local thermodynamic equilibrium with the particle composition [8] and can be obtained from the vapour pressure (P_i^{vap}) of each volatile compound i with average molar mass of the atmospheric aerosol, mole fraction, and activity coefficient, through the following equation:

$$c_i^s = x\frac{\gamma P_i^{\text{vap}} m 10^6}{RT}, \quad (2)$$

where P_i^{vap} is given by the Clausius-Clapeyron law:

$$P_i^{\text{vap}}(T) = P_i^{\text{vap}}(298)\exp^{[-(H_{\text{vap}}/R)((1/T)-(1/298))]}. \quad (3)$$

Parameters in (1)–(3) are named in Table 1.

In (3), H_{vap} is equal to 156 kJ/mol as stated by Derby et al. [8]. Consequently, the vapour pressures of all aerosol compounds are needed to calculate the mass flux of condensing and evaporating compounds. The knowledge of P_i^{vap} for organic compound i at the atmospheric temperature T is required whenever phase equilibrium between gas phase and particle is of interest. Often, most of the compounds able to condense have experimental vapor pressures unavailable, and because of that, their estimation becomes necessary. To solve this problem of estimation, many methods have been developed. For example, in Tong et al. [9], a method based on atomic simulation is applied only for compounds bearing acid moieties. Quantum-mechanical calculations are making steady effort in vapor pressure prediction (Banerjee et al. [10]; Diedenhofen et al. [11]). Furthermore, current models describing gas-particle partitioning use semiempirical methods for vapor pressure estimation based on molecular structure, often in the form of a group contribution approach. Therefore, these methods require in most cases molecular structures (e.g., boiling point T_b, critical temperature T_c, and critical pressure P_c), which usually have themselves to be estimated. For example, The MY method [12] was used by Griffin et al. [13] and Pun et al. [14] for modeling the formation of secondary organic aerosol. Jenkin [15], in a gas particle partitioning model, used the modified form of the Mackay method [16]. Some methods (Pankow and Asher [17], Capouet and Müller [18]) assume a linear logarithmic dependence ($\ln P_i^{vap}$) on several functional groups, but this consideration fails when multiple hydrogen bonding groups are present. Furthermore, more investigations are needed to clarify which method can give values closer to the experimental data. Camredon and Aumont [19], Compernolle et al. [20], and Barley and McFiggans [21] have made an assessment of different vapor pressure estimation methods with experimental data for compounds of relatively higher volatility. For this latter reason, large differences in the estimated vapor pressure have been reported.

In this paper, our focus will be on the (i) evaluation of a number of vapor pressure estimation methods against experimental data using all volatile organic compounds present in our database and (ii) assessment of the accuracy of each of these methods on the base of each class of compounds.

The rest of the paper is organized as follows. In Section 2, we describe experimental data and methods. The results are presented in Section 3. Finally, we summarize our finding in Section 4.

TABLE 1: Nomenclature.

Parameters	Names
I_i	Mass flux for volatile species i
d_p	The particle wet diameter
m_i	Mass of species i
D_i^g	Molar diffusivity in the air of species i
c_i^g	Gas-phase concentration of species i
f_{FS}	Fuchs-Sutugin function
c_i^s	Concentration at the surface of species i
R	Gas constant
T	Temperature in Kelvin
H_{vap}	Vaporisation enthalpy
x	Mole fraction
γ	Activity coefficient
P_i^{vap}	Vapour pressure of each volatile compound i

2. Data and Vapor Pressure Estimation Methods

2.1. Experimental Data. The molecules selected in this study have been identified during in situ campaigns [22] and during chamber experiments [23]. In fact, they are hydrocarbons, monofunctionalized and multifunctionalized species, and bearing alcohol, aldehyde, ketone, carboxylic acid, ester, ether, and alkyl nitrate functions. The experimental vapor pressures are taken from NIST chemistry website (http://www.nist.gov/chemestry) and from Myrdal and Yalkowsky [12], Asher et al. [24], Lide [25], Yams [26], and Boulik et al. [27], and they have a range from 10^{-8} atm to 1 atm. Molecular properties (boiling point, critical temperature, and critical pressure) are also taken from the NIST chemistry website. All vapor pressure estimation methods used in this study take into account these properties. In most cases, these properties have also to be estimated.

2.2. Estimation of Molecular Properties

2.2.1. Boiling Temperature T_b. Using the boiling temperature of Joback [28] and its extension, the group contribution technique denoted by T_b^{Job} is written as

$$T_b^{Job} = 198 + \sum_i N_i t_{bi}, \tag{4}$$

where t_{bi} is the contribution of group i, and N_i is the occurrence of this group in the molecule. As in Camredon and Aumont [19], the extension is made by adding some other group contribution to take into account molecules bearing hydroperoxide moiety (–OOH), alkyl nitrate moiety (ONO_2), and peroxyacyl nitrate moiety (–C(=O)$OONO_2$). Thus, the first group is divided into the existing Joback groups –O– and –OH. The second group value is provided by the NIST chemistry website, and the last group value is provided by Camredon and Aumont [19] using boiling point value from Bruckmann and Willner [29].

2.2.2. Critical Temperature T_c. We have used Joback [28] and Lydersen's [30] techniques to estimate T_c. Denoting by T_C^{Job} and T_c^{Lyd} the Joback and Lydersen critical temperatures, respectively, we have

$$T_c^{Job} = \frac{T_b}{0,584 + 0,965 \sum_i N_i t_i - \left(\sum_i N_i t_i\right)^2},$$
$$T_c^{Lyd} = \frac{T_b}{0,567 + \sum_i N_i t_i - \left(\sum_i N_i t_i\right)^2}. \tag{5}$$

New group contributions have been added by Camredon and Aumont [19] for hydroperoxide, alkyl nitrate, and peroxyacyl nitrate moieties:

$$t_c^{\text{-OOH}} = t_c^{\text{-O-}} + t_c^{\text{-OH}},$$
$$t_c^{\text{-ONO}_2} = t_c^{\text{-O-}} + t_c^{\text{-NO}_2}, \quad (6)$$
$$t_c^{\text{-C(=O)OONO}_2} = t_c^{\text{-C(=O)O}} + t_c^{\text{-O-}} + t_c^{\text{-NO}_2}.$$

2.2.3. Critical Pressure P_c. The two techniques listed in the previous section have been used to estimate critical pressure P_c. Here, it is also assumed that P_c is the sum of group contributions. Therefore, denoting by P_c^{Job} and P_c^{Lyd} the Joback and Lydersen critical pressures, respectively, we can write

$$P_c^{\text{Job}} = \frac{1}{\left(0,113 + 0,0032n - \sum_i N_i p_i\right)^2},$$
$$P_c^{\text{Lyd}} = \frac{M}{\left(0,34 - \sum_i N_i p_i\right)^2}, \quad (7)$$

where M is the molar mass, n is the number of atoms in the molecule, and p_i is the critical pressure contribution of group i. The new group contributions described in the previous subsection are taken into account here.

2.3. Vapor Pressure Estimation Methods. As said earlier, many methods for vapor pressure estimation have been developed and are based on the Antoine or on the extended form of the Clausius-Clapeyron equation. Let us present in what follows each of the five methods used.

2.3.1. The Myrdal and Yalkowsky (MY) Method. The MY method [12] starts from the extended form of the Clausius-Clapeyron equation obtained by using Euler's cyclic relation. Here, the expression of P^{vap} is given by

$$\ln P^{\text{vap}} = \frac{\Delta S_b \left(T_b - T\right)}{RT} + \frac{\Delta C_{pb}}{R}\left(\frac{T_b - T}{T} - \ln\frac{T_b}{T}\right), \quad (8)$$

where ΔS_b is the vaporization entropy at the boiling temperature, C_{pb} is the gas-liquid heat capacity, and R is the gas constant. ΔS_b used in this method is an empirical expression given by Myrdal et al. [31]:

$$\Delta S_b = 86 + 0.4\tau + 1421\text{HBN}. \quad (9)$$

In (9), the parameters τ and HBN which characterize the molecular structure represent the torsional bond (see Vidal [32]) and the hydrogen bonding number (see [19, Section 3.1.2]), respectively. ΔC_{pb} is a linear dependence of τ [32]:

$$\Delta C_{pb} = -\left(90 + 2.5\tau\right). \quad (10)$$

Thus, in the MY method, vapor pressure is estimated by the relatively simplified formula

$$\ln P^{\text{vap}} = -\left(21,2 + 0,3\tau + 177\text{HBN}\right)\left(\frac{T_b - T}{T}\right) + \left(10,8 + 0,25\tau\right)\ln\frac{T_b}{T}. \quad (11)$$

2.3.2. The Modified Mackay (mM) Method. Often called simplified expression of Baum [33], this method is also based on the extended form of the Clausius-Clapeyron equation. The simplifying assumption here is to consider the ratio $\Delta C_{pb}/\Delta S_b$ to be constant [34]:

$$\frac{\Delta C_{pb}}{\Delta S_b} = -0.8. \quad (12)$$

In (12), the vaporization entropy, ΔS_b, takes into account the van der Waals interactions and is based upon the Trouton's rule. For its calculation, Lyman [35] has suggested the following expression:

$$\Delta S_b\left(T_b\right) = K_f\left(36,6 + 8,31\ln T_b\right), \quad (13)$$

where K_f is a structural factor of Fishtine [36] which corrects many polar interactions. It has different values as follows:

(i) $K_f = 1$ for nonpolar and monopolar compounds;

(ii) $K_f = 1.04$ for compounds with a weak bipolar character;

(iii) $K_f = 1.1$ for primary amines;

(iv) $K_f = 1.3$ for aliphatic alcohols.

Finally, the mM method is reduced to the following equation:

$$\ln P^{\text{vap}} = K_f\left(4,4 + \ln T_b\right)\left(1,8\frac{T_b - T}{T} - 0,8\ln\frac{T_b}{T}\right). \quad (14)$$

2.3.3. The Grain-Watson (GW) Method. The GW method is based on the following equation [37]:

$$\ln P^{\text{vap}} = \frac{\Delta S}{R}\left[1 - \frac{\left(3 - 2T_p\right)^m}{T_p} - 2m\left(3 - 2T_p\right)^{m-1}\cdot\ln T_p\right], \quad (15)$$

where $m = 0.4133 - 0.2575T_p$ and T_p is inversely proportional to T_b ($T_p = T/T_b$). ΔS has the same form like that used in the mM method.

2.3.4. The Lee and Kesler (LK) Method. Like the methods described previously, the LK method required critical temperature, critical pressure, and boiling temperature. Here, the vapor pressure is estimated on the base of Pitzer expansion [38]:

$$\ln P_r^{\text{vap}} = f^0\left(T_r\right) + wf^1\left(T_r\right), \quad (16)$$

where P_r^{vap} is reduced vapor pressure, T_r is reduced temperature, and w is the Pitzer's acentric factor which accounts for the nonsphericity of molecules:

$$w = \frac{-\ln P_c - f^0(\theta)}{f^1(\theta)} \tag{17}$$

with $\theta = T_b/T_c$. In (16) and (17), f^0 and f^1 are the Pitzer's functions which are polynomials in T_r. Lee and Kesler have suggested the following equations [39]:

$$\begin{aligned} f^0(T_r) &= 5,92714 - \frac{6.09648}{T_r} \\ &\quad - 1.28862 \ln T_r + 0.169347 \ln T_r^6, \\ f^1(T_r) &= 15.2518 - \frac{15.6875}{T_r} \\ &\quad - 13.4721 \ln T_r + 0.43577 \ln T_r^6. \end{aligned} \tag{18}$$

2.3.5. The Ambrose-Walton (AW) Method. AW method [40] is also based on the Pitzer expansion. They have reported their analytical expressions of Pitzer's functions in the form of a Wagner type of vapor pressure equation:

$$\begin{aligned} &f^0(T_r) \\ &= \frac{-5.9761\tau + 1.29874\tau^{1.5} - 0.60394\tau^{2.5} - 1.06841\tau^5}{T_r}, \\ &f^1(T_r) \\ &= \frac{-5.03365\tau + 1.11505\tau^{1.5} - 5.41217\tau^{2.5} - 7.46628\tau^5}{T_r}. \end{aligned} \tag{19}$$

In (19), $\tau = 1 - T_r$.

3. Results

3.1. Molecular Properties. We present in this section the results obtained for the Joback and Lydersen techniques described previously. The accuracy of each of the five vapor pressure estimation methods used in this study is assessed taking into account the reliability of pure substance property estimates. The reliability of the two techniques presented in Section 2.2 is therefore crucial.

In Figure 1, where the results of Joback technique are displayed, T_b^{Job} is plotted against experimental values for a set of 253 volatile organic compounds. The scatter tends to be larger for boiling temperature higher than 500 K. The correlation coefficient ($R^2 = 0.97$) shows that estimated values match very well experimental T_b. The root mean square error (RMSE) and the mean absolute error (MAE) are, respectively, 17.60 K and 12.65 K. This MAE agrees with 12.9 K and 12.1 K calculated in Reid et al. [39] and Camredon and Aumont [19] for a set of 252 and 438 volatile organic compounds, respectively. Hence, these results show that T_b^{Job}

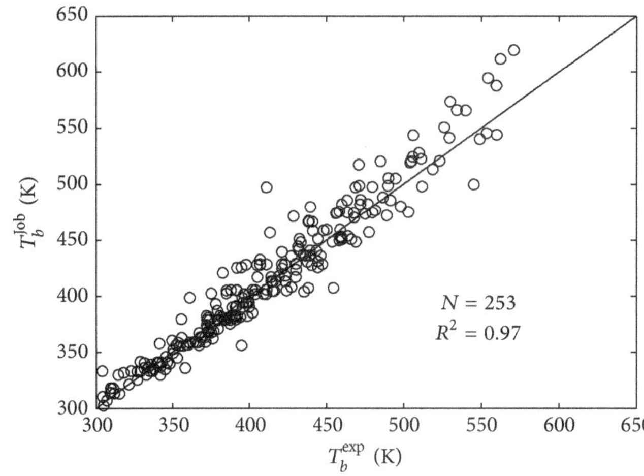

FIGURE 1: Estimated boiling point for a set of 253 species versus experimental values for the Joback technique. The black line is the 1 : 1 diagonal.

can be used in vapor pressure estimation methods. Moreover, Joback reevaluated Lydersen's group contribution scheme. He added several new functional groups and deducted new contribution values.

The two techniques for the estimation of T_c are compared to experimental data of 138 compounds in Figure 2.

Figures 2(a) and 2(b) show that the Joback and Lydersen techniques give similar results for T_c values lower than 700 K. Joback technique shows a negative bias for T_c higher than 700 K. This technique gives for the overall compounds an RMSE of 24.98 K. This value is higher than the 19.81 K provided by the Lydersen technique. The mean bias error (MBE) for T_c^{Job} and T_c^{Lyd} is −4.9 and −1.9, respectively. These results and Figures 2(a) and 2(b) show clearly that Joback technique underpredicts critical temperature, mostly for compounds which can be condensed onto particle phase, with high boiling temperature. The experimental group contributions provided by Lydersen are therefore more accurate than those provided by Joback. Thus, the Lydersen technique is more reliable than the Joback technique to estimate T_c.

Figure 3 shows P_c^{Job} and P_c^{Lyd} versus experimental values for a set of 117 compounds. The RMSE is 4.5 atm and 2.6 atm for Joback and Lydersen, respectively. According to Figures 3(a) and 3(b), P_c estimated by Lydersen matches fairly better ($R^2 = 0.95$) experimental data than P_c estimated by Joback ($R^2 = 0.85$). Joback technique considerably overpredicts critical pressure with MBE of 1,95 K higher than 0,39 K obtained with the Lydersen technique. In fact, besides group contribution, Lydersen technique takes into account molecular weight. Therefore, the Lydersen technique is retained to the critical pressure estimation in this paper. This is in agreement with Poling et al. [38] who found that the Lydersen technique is one of the best techniques for estimating critical properties. For the five vapor pressure estimation methods described previously, we will use estimated T_b, T_c, and P_c because there is in general a lack of experimental data.

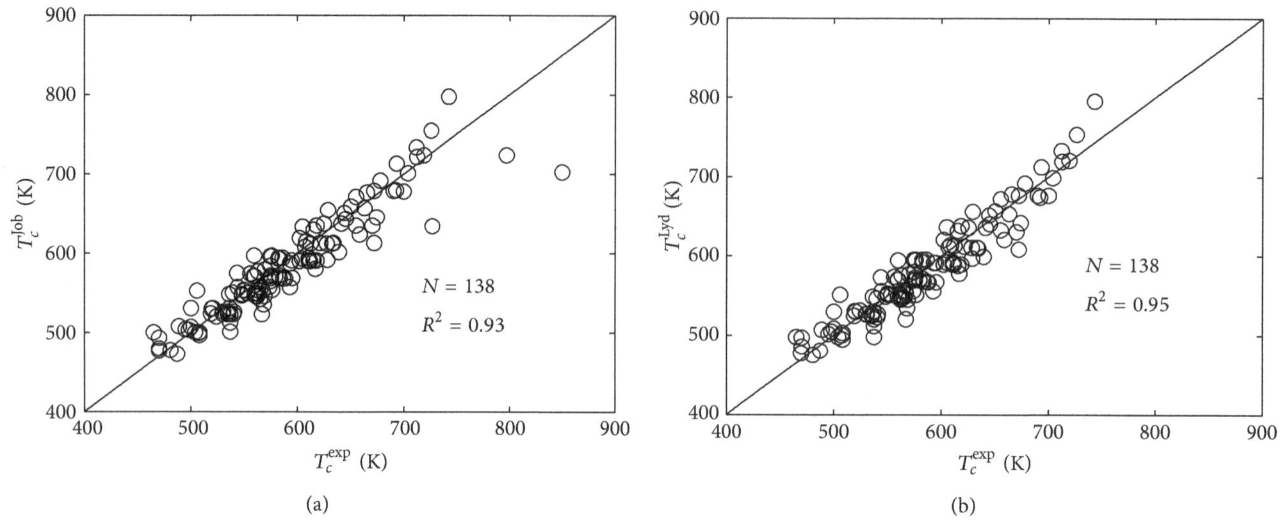

FIGURE 2: Estimated critical temperature for a set of 138 species versus experimental values for the (a) Joback technique and (b) Lydersen technique. The black line is the 1 : 1 diagonal.

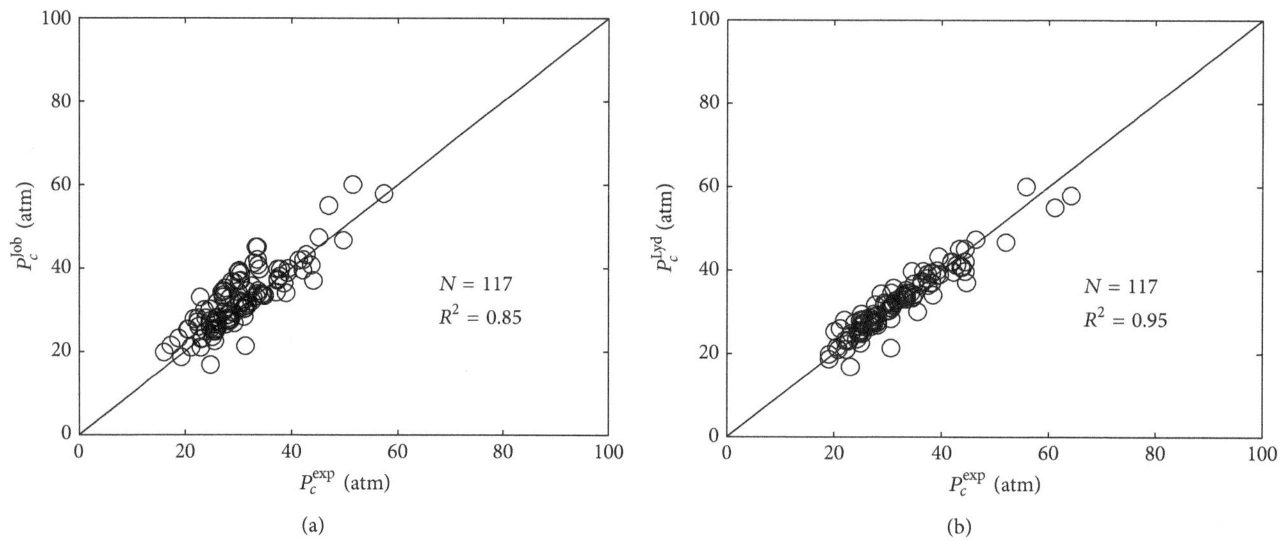

FIGURE 3: Estimated critical pressure for a set of 117 species versus experimental values for the (a) Joback technique and (b) Lydersen technique. The black line is the 1 : 1 diagonal.

Some of the five methods described in Section 2.3 need P_c, T_c, and T_b, while others need only T_c and T_b. This last pure substance property is estimated by the Joback technique, and the two critical properties are estimated by the Lydersen technique.

3.2. Evaluation of Vapor Pressure Estimation Methods. The accuracy of each method is assessed in terms of the mean absolute error (MAE), the main bias error (MBE), and the root mean square error (RMSE) (Table 3). The MBE measures the average difference between the estimated and experimental values, while the MAE measures the average magnitude of the error. The RMSE also measures the error magnitude, but gives some greater weight to the larger errors. Their expressions are given by

$$\mathrm{MBE} = \frac{1}{N}\sum_{i=1}^{N}\left(\log P_{\mathrm{est},i}^{\mathrm{vap}} - \log P_{\mathrm{exp},i}^{\mathrm{vap}}\right),$$

$$\mathrm{MAE} = \frac{1}{N}\sum_{i=1}^{N}\left|\log P_{\mathrm{est},i}^{\mathrm{vap}} - \log P_{\mathrm{exp},i}^{\mathrm{vap}}\right|, \quad (20)$$

$$\mathrm{RMSE} = \sqrt{\frac{1}{N}\sum_{i=1}^{N}\left(\log P_{\mathrm{est},i}^{\mathrm{vap}} - \log P_{\mathrm{exp},i}^{\mathrm{vap}}\right)^2}.$$

Table 2: MAE, MBE, and RMSE of vapor pressure computed based on various methods.

	GW	LK	mM	MY	AW
MAE	0.3235	0.3240	0.3186	0.2652	0.3013
MBE	0.0736	−0.1513	0.0263	0.0270	−0.0980
RMSE	0.4496	0.4909	0.4426	0.3811	0.4525

Table 3: MAE, MBE, and RMSE of vapor pressure computed based on various methods for each class of compounds.

	GW	LK	mM	MY	AW
Hydrocarbons					
MAE	0.1455	0.1303	0.1424	0.1251	0.1384
MBE	0.0547	0.0427	0.0486	0.0054	0.0603
RMSE	0.1983	0.1750	0.1930	0.1711	0.1848
Monofunctionalized species					
MAE	0.3692	0.3168	0.1424	0.2699	0.2900
MBE	0.0355	−0.1712	−0.0189	−0.0402	−0.1141
RMSE	0.4992	0.4471	0.4955	0.3891	0.4130
Difunctionalized species					
MAE	0.4494	0.5641	0.4109	0.4291	0.5063
MBE	0.2875	−0.3322	0.2084	0.2378	−0.2456
RMSE	0.5506	0.6978	0.5120	0.5398	0.6291
Tri- and more functionalized species					
MAE	0.4407	0.5946	0.4596	0.4261	0.5489
MBE	0.0536	−0.3661	−0.0339	0.1632	−0.2739
RMSE	0.5494	0.8386	0.5592	0.5065	0.7704

In (20), $P^{\text{vap}}_{\text{est},i}$ and $P^{\text{vap}}_{\text{exp},i}$ are the estimated and experimental values of VOC i, respectively, and N is the total number of VOC. We have also used linear correlation coefficient which measures the degree of correspondence between the estimated and experimental distributions.

The logarithms of vapor pressures estimated at $T = 298$ K for different methods are compared in the scatter plots shown in Figure 4 for a set of 262 VOC. Corresponding MAE, MBE, and RMSE are given in Table 1. In this figure, it is clear that all the five methods give similar scatter for vapor pressures higher than 10^{-5} atm. The species concerned are hydrocarbons (Figure 5). For the set of 28 tri- and more functionalized species used in this study, vapor pressures are lower than 10^{-2} atm (Figure 8).

Figure 4(d) shows that MY method is well correlated with experimental values. For this method, we have one of the best correlation coefficients $R^2 = 0.968$. This method shows no systematic bias for vapor pressure lower than 10^{-5} atm, while it is not the case for other methods. The MBE found here is 0.027. This value is one of the lowest ones of the total VOC (see Table 1). Thus, the MY method does not show any systematic bias. This method also provides the smallest values of MAE and RMSE (0.265 and 0.381, resp.). These results are in agreement with those found by Camredon and Aumont [19] using Ambrose technique to estimate critical properties. It is also found that this method provides the smallest values of these errors for a set of 74 hydrocarbons (see Table 2). Those are compounds with vapor pressure higher than 10^{-2} atm. This result is not the same for mono- and difunctionalized species whose errors are some of the largest ones. The vapor pressures higher than 10^{-4} atm are fairly well estimated. Using a set of 45 multifunctional compounds, Barley and McFiggans [21] found that MY method tends to overpredict vapor pressure of lower volatility compounds. Furthermore, it is important to note that MY method provides one of the poor results for difunctionalized VOC (see Table 2). Thus, for a set of 32 difunctionalized VOC, estimation values fit the experimental ones with a coefficient $R^2 = 0.957$ (Figure 7) which is the smallest value obtained for this class of species. Vapor pressures are overpredicted with a bias of 0.24, while the RMSE = 0.54 is of the same order of magnitude as those obtained by other methods. In contrast, Figure 8 shows that MY method has the best correlation for tri- and more functionalized species and is therefore the best method to estimate vapor pressure for this class of species. The systematic errors reported in Table 2 allow us to conclude that assumption. Indeed, the peculiarity of the MY method is that it takes into account the molecular structure.

Figure 4(c) displays the results for the mM method. This method gives one of the lowest scatterings with a coefficient $R^2 = 0.957$ and agrees with other methods for the highest vapor pressures. It shows a positive bias (MBE = 0.026) for the total set of 262 VOC. As the GW method, the mM method tends to overpredict vapor pressures lower than 10^{-6} atm. Furthermore, these methods describe vaporisation entropy by taking into account van der Waals interactions.

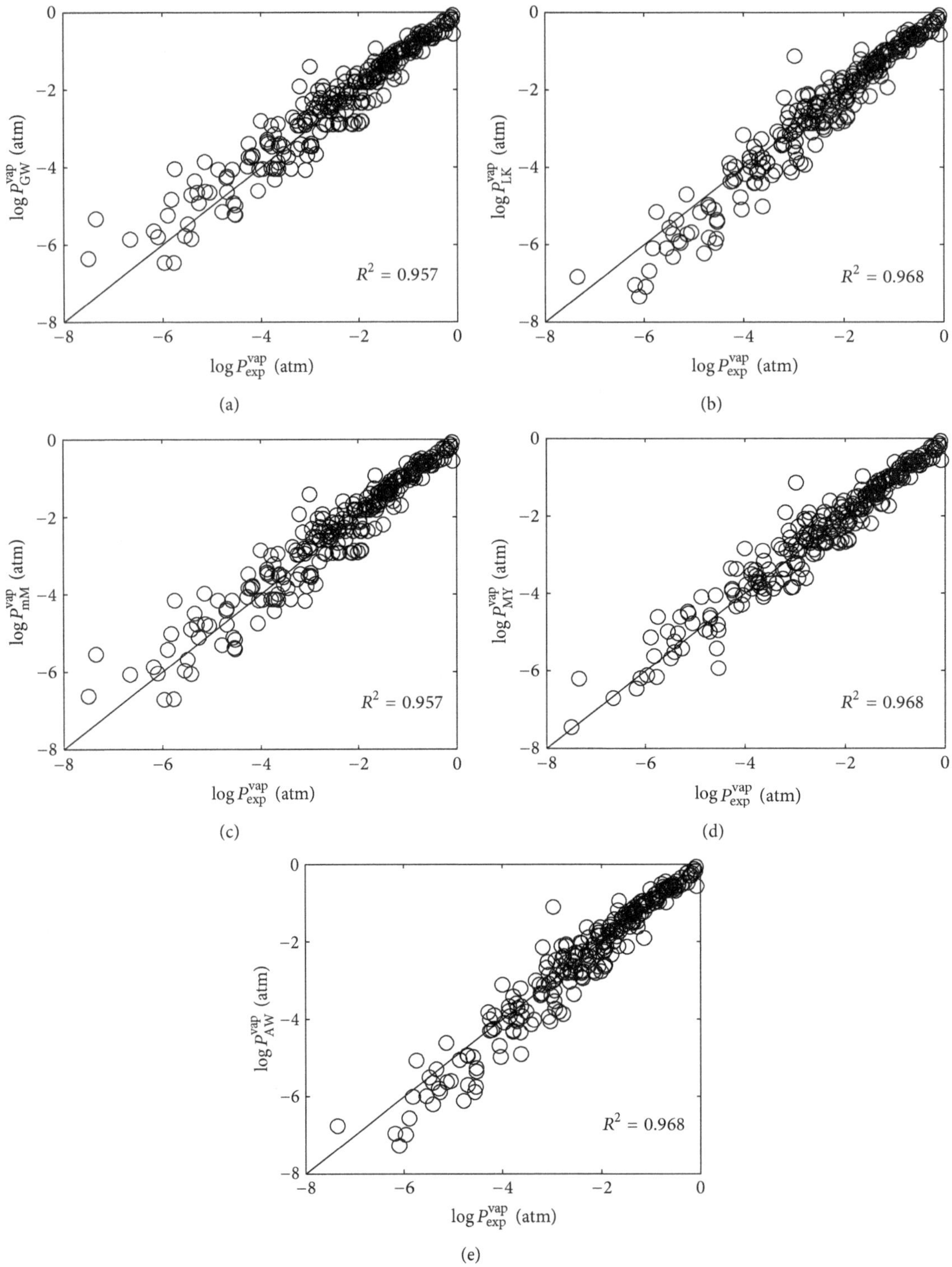

FIGURE 4: Logarithm of estimated vapor pressure of all VOCs used in this study versus experimental values for the (a) GW, (b) LK, (c) mM, (d) MY, and (e) AW methods. The black line is the 1 : 1 diagonal.

The root mean square error (RMSE = 0.442) is close to those provided by GW and AW methods. The mM method is then less appropriate than the four others for all classes of VOC.

Figure 5(c) shows that the predicted values match the experimental values with a coefficient $R^2 = 0.96$ for 32 difunctionalized species. For this class of species, estimates are provided with a positive bias (MBE = 0.208) and an RMSE of 0.512. These are the best values obtained from all the five methods (see Table 2) for difunctionalized species. Thus, mM method is more accurate than others to estimate vapor pressure for difunctionalized species, but does not provide

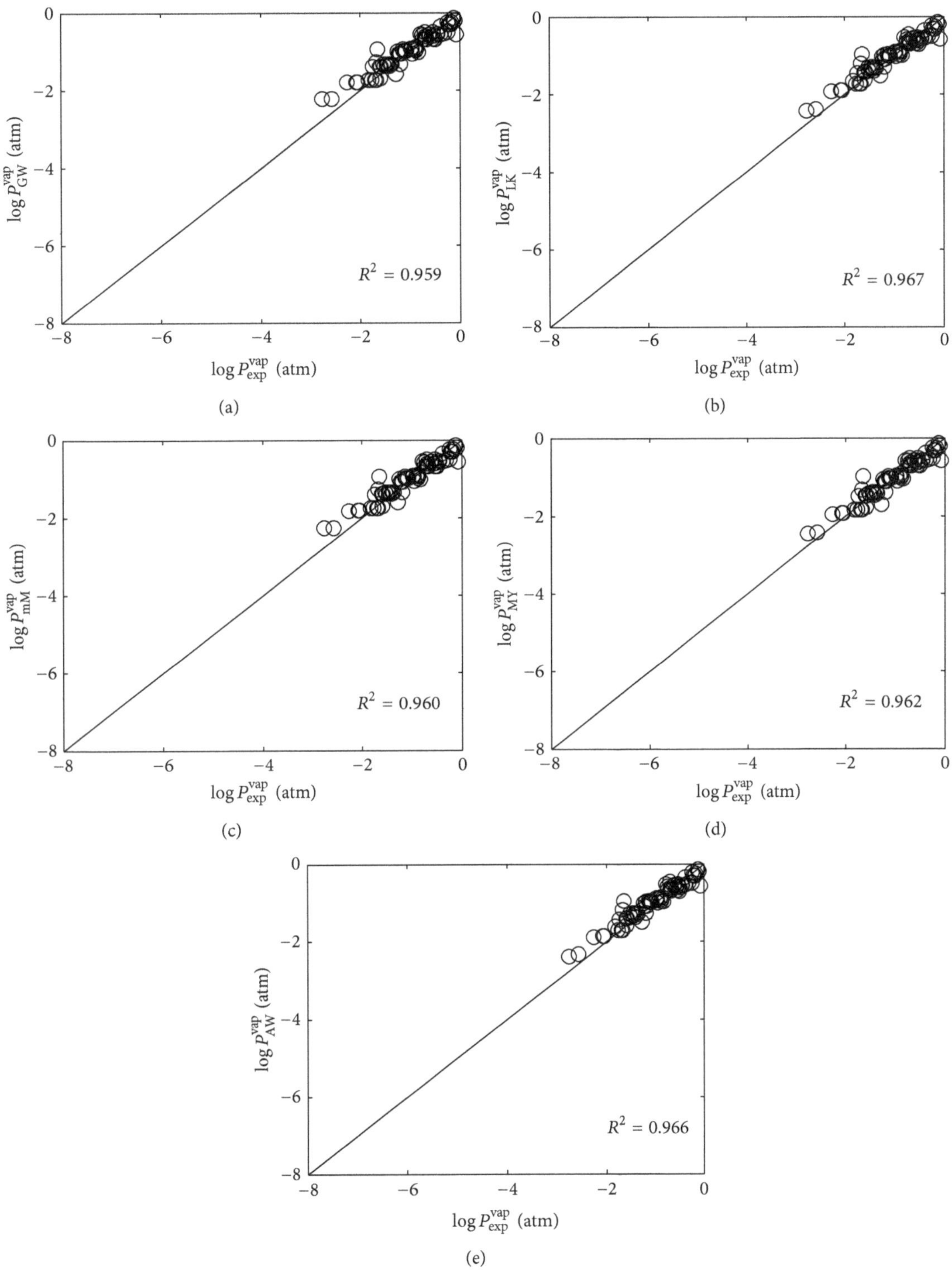

FIGURE 5: Logarithm of estimated vapor pressure for a set of 74 hydrocarbons versus experimental values for the (a) GW, (b) LK, (c) mM, (d) MY, and (e) AW methods. The black line is the 1 : 1 diagonal.

good results for monofunctionalized (Figure 3) and tri- and more functionalized species (Figure 5).

It can be seen in Figure 7 that estimated values provided by GW method are strongly correlated with experimental values ($R^2 = 0.960$) for difunctionalized species. This method tends to overpredict vapor pressure for this class of species (MBE = 0.288), but does not show any bias for other classes of species (Table 2). Figure 8 shows that we have very acceptable results for tri- and more functionalized species with RMSE and MAE equal to 0.549 and 0.440, respectively.

Except for tri- and more functionalized species (see Figure 8), it is clear from Figures 4 to 7 that the LK method

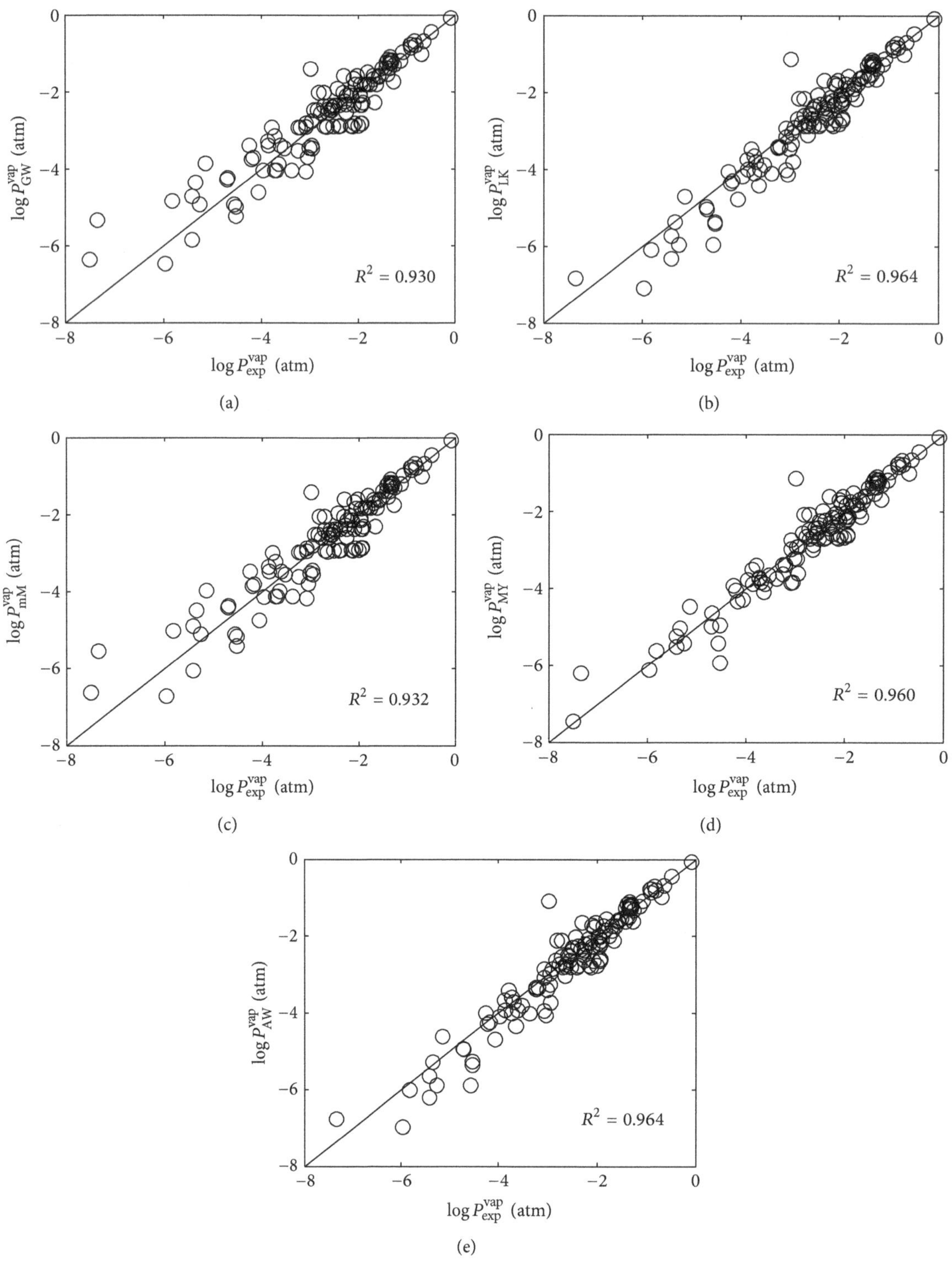

FIGURE 6: Logarithm of estimated vapor pressure for a set of 128 monofunctionalized species versus experimental values for the (a) GW, (b) LK, (c) mM, (d) MY, and (e) AW methods. The black line is the 1 : 1 diagonal.

gives accurate values, based upon best correlation coefficient values. This method is the second best one of the five methods, but it has the greatest systematic bias (RMSE = 0.490, MAE = 0.32) for the total set of VOC and for difunctionalized species (RMSE = 0.70, MAE = 0.56). The MAE of hydrocarbons and monofunctionalized species are 0.142 and 0.36, respectively. For these species, Figures 5 and 6 give the best correlations.

The AW and LK methods are both based on Antoine's equation. According to all figures plotted, it is clear that these two methods give very similar results. The peculiarity of AW is that, for the monofunctionalized compounds,

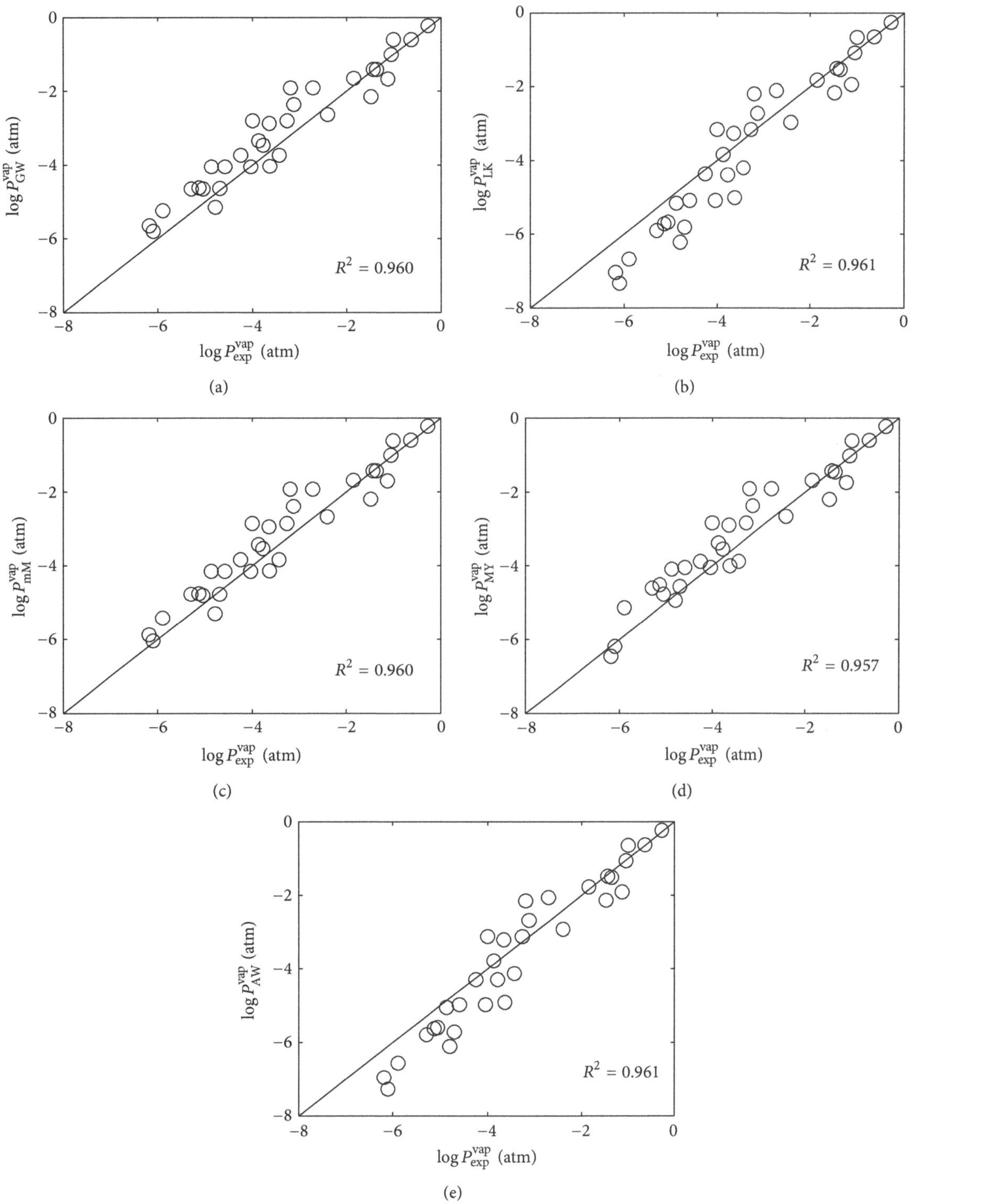

FIGURE 7: Logarithm of estimated vapor pressure for a set of 32 difunctionalized species versus experimental values for the (a) GW, (b) LK, (c) mM, (d) MY, and (e) AW methods. The black line is the 1:1 diagonal.

the predicted and experimental values are strongly correlated with a coefficient $R^2 = 0.9638$.

Vapor pressures for a set of 74 hydrocarbons are higher than 10^{-3} atm. It is found for the five methods that estimated values for this class of species are well correlated (Figure 5). Furthermore, vapor pressures of tri- and more functionalized species are below 10^2 atm (Figure 8). For this class of species, LK method yields a weak correlation and has the largest positive bias. Therefore, this method is the least reliable to estimate vapor pressure for tri- and more functionalized species.

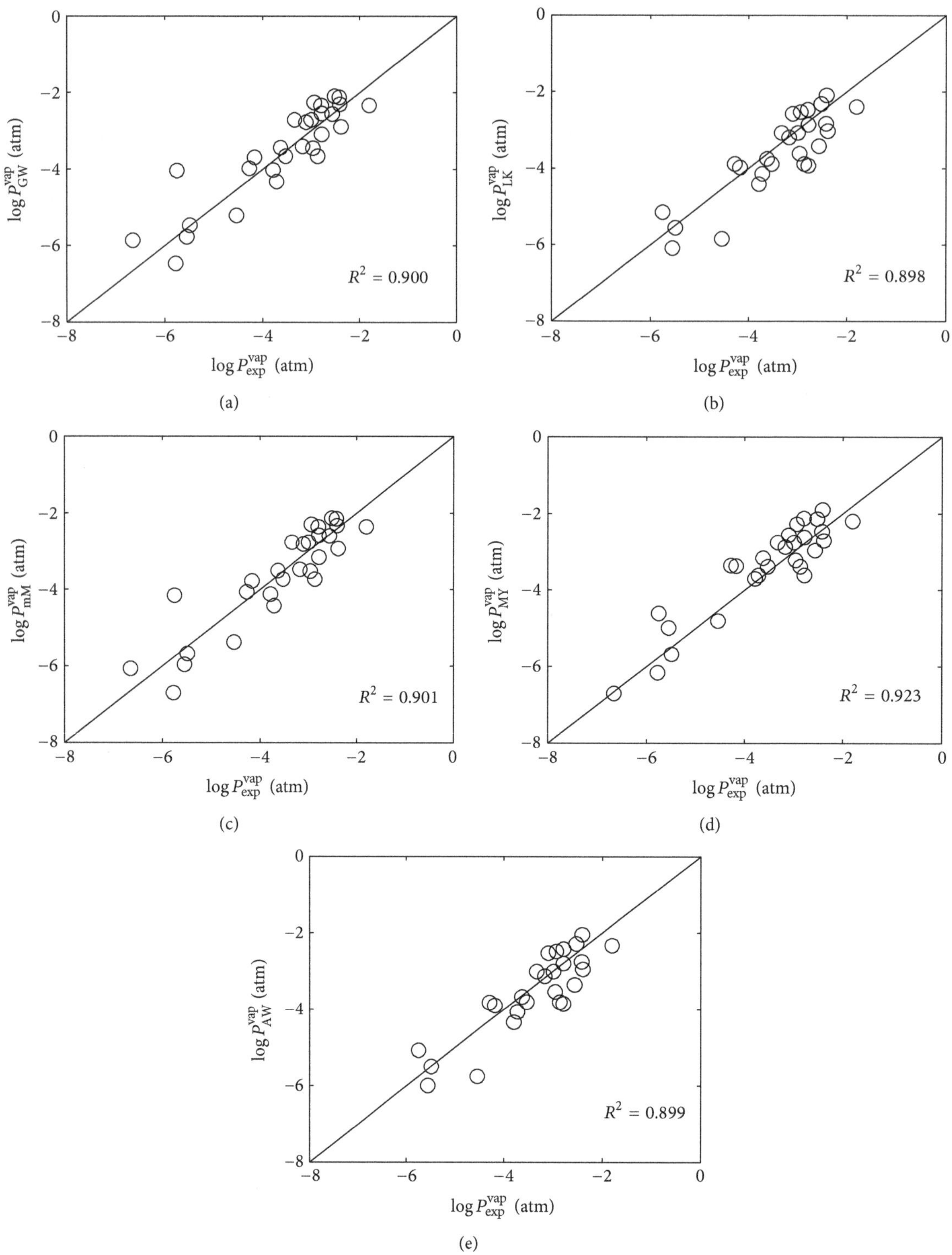

FIGURE 8: Logarithm of estimated vapor pressure for a set of 28 tri- and more functionalized species versus experimental values for the (a) GW, (b) LK, (c) mM, (d) MY, and (e) AW methods. The black line is the 1 : 1 diagonal.

4. Conclusion

We have evaluated in this study five vapor pressure estimation methods useful for simulating the dynamics of atmospheric organic aerosols. These are the Myrdal and Yalkovsky (MY), the Lee and Kesler (LK), the Grain-Watson (GW), the modified Mackay (mM), and the Ambrose-Walton (AW) methods. Some of them are based on the Antoine equation, while others are based on the extended form of the Clausius-Clapeyron equation. But all of them take into account boiling temperature T_b and (or) critical temperature T_c. Therefore, Joback technique has been used to estimate T_b, while the Lydersen technique was found to be better for T_c estimation.

When using Joback to provide the T_b values, LK, AW, and MY are the best three methods for all classes of species. Moreover, for a set of 262 volatile organic compounds and as illustrated in the scatter plots and errors computed, the MY method which appears to be the best one fails for difunctionalized species. For these latter species, the mM method provides good results, according to the correlation coefficient $R^2 = 0.960$ and the least errors reported in Table 2. GW method is the least reliable, which provides the lowest results for all VOC and also for each class of species. Predictions made with the AW method for monofunctionalized species are more reliable than those made with the other four methods employing the Joback technique to provide the T_b. For vapor pressure higher than 10^{-2} atm, all the five methods give similar results.

This work highlights that the choice of a method to predict vapor pressure of volatile organic compounds depends on the number of functional groups existing in the species.

References

[1] J. Seinfeld and S. Pandis, *Amospheric Chemestry and Physics: From a Pollution to Climate Change*, John Wiley & Sons, New York, NY, USA, 1998.

[2] M. Tombette, *Modélisation des aérosols et de leurs propriãltés optiques sur l'Europe et l'île de France: validation, sensibilité et assimilation des données [Thèse]*, Ecole nationale des ponts et chaussées, 2007.

[3] P. Adams and J. Seinfeld, "Predicting global aerosol size distributions in general circulation models," *Journal of Geophysical Research*, vol. 107, no. 19, pp. AAC 4-1–AAC 4-23, 2002.

[4] S. Gons, L. A. Barrie, J. P. Blanchet et al., "Canadian aerosols module: a size-segregated simulation of atmospheric aerosol processes for climate to and air quality models. I Module development," *Journal of Geophysical Research*, vol. 108, no. 1, pp. AAC 3-1–AAC 3-16, 2003.

[5] E. R. Whitby and P. H. McMurry, "Modal aerosol dynamics modeling," *Aerosol Science and Technology*, vol. 27, no. 6, pp. 673–688, 1997.

[6] E. Debry, *Modélisation et simulation de la dynamique des aérosols atmosphériques [Thèse]*, Ecole nationale des ponts et chaussées, 2004.

[7] B. Dahneke, *Theory of Dispersed Multiphase Flow*, Academic press, New York, NY, USA, 1983.

[8] E. Debry, K. Fahey, K. Sartelet, B. Sportisse, and M. Tombette, "Technical note: a new SIze REsolved Aerosol Model (SIREAM)," *Atmospheric Chemistry and Physics*, vol. 7, no. 6, pp. 1537–1547, 2007.

[9] C. Tong, M. Blanco, W. A. Goddard III, and J. H. Seinfeld, "Thermodynamic properties of multifunctional oxygenates in atmospheric aerosols from quantum mechanics and molecular dynamics: dicarboxylic acids," *Environmental Science and Technology*, vol. 38, no. 14, pp. 3941–3949, 2004.

[10] T. Banerjee, M. K. Singh, and A. Khanna, "Prediction of binary VLE for imidazolium based ionic liquid systems using COSMO-RS," *Industrial and Engineering Chemistry Research*, vol. 45, no. 9, pp. 3207–3219, 2006.

[11] M. Diedenhofen, A. Klamt, K. Marsh, and A. Schäfer, "Prediction of the vapor pressure and vaporization enthalpy of 1-n-alkyl-3-methylimidazolium-bis-(trifluoromethanesulfonyl) amide ionic liquids," *Physical Chemistry Chemical Physics*, vol. 9, no. 33, pp. 4653–4656, 2007.

[12] P. B. Myrdal and S. H. Yalkowsky, "Estimating pure component vapor pressures of complex organic molecules," *Industrial and Engineering Chemistry Research*, vol. 36, no. 6, pp. 2494–2499, 1997.

[13] R. J. Griffin, K. Nguyen, D. Dabdub, and J. H. Seinfeld, "A coupled hydrophobic-hydrophilic model for predicting secondary organic aerosol formation," *Journal of Atmospheric Chemistry*, vol. 44, no. 2, pp. 171–190, 2003.

[14] B. K. Pun, R. J. Griffin, C. Seigneur, and J. H. Seinfeld, "Secondary organic aerosol 2. Thermodynamic model for gas/particle partitioning of molecular constituents," *Journal of Geophysical Research D*, vol. 107, no. 17, pp. AAC 4-1–AAC 4-15, 2002.

[15] M. E. Jenkin, "Modelling the formation and composition of secondary organic aerosol from α- and β-pinene ozonolysis using MCM v3," *Atmospheric Chemistry and Physics*, vol. 4, no. 7, pp. 1741–1757, 2004.

[16] D. Mackay, A. Bobra, D. W. Chan, and W. Y. Shiu, "Vapor pressure correlations for low-volatility environmental chemicals," *Environmental Science and Technology*, vol. 16, no. 10, pp. 645–649, 1982.

[17] J. F. Pankow and W. E. Asher, "SIMPOL.1: a simple group contribution method for predicting vapor pressures and enthalpies of vaporization of multifunctional organic compounds," *Atmospheric Chemistry and Physics*, vol. 8, no. 10, pp. 2773–2796, 2008.

[18] M. Capouet and J.-F. Müller, "A group contribution method for estimating the vapour pressures of α-pinene oxidation products," *Atmospheric Chemistry and Physics*, vol. 6, no. 6, pp. 1455–1467, 2006.

[19] M. Camredon and B. Aumont, "Assessment of vapor pressure estimation methods for secondary organic aerosol modeling," *Atmospheric Environment*, vol. 40, no. 12, pp. 2105–2116, 2006.

[20] S. Compernolle, K. Ceulemans, and J.-F. Müller, "Technical Note: vapor pressure estimation methods applied to secondary organic aerosol constituents from α-pinene oxidation: an intercomparison study," *Atmospheric Chemistry and Physics*, vol. 10, no. 13, pp. 6271–6282, 2010.

[21] M. H. Barley and G. McFiggans, "The critical assessment of vapour pressure estimation methods for use in modelling the formation of atmospheric organic aerosol," *Atmospheric Chemistry and Physics*, vol. 10, no. 2, pp. 749–767, 2010.

[22] L. E. Yu, M. L. Shulman, R. Kopperud, and L. M. Hildemann, "Characterization of organic compounds collected during Southeastern Aerosol and Visibility Study: water-soluble organic species," *Environmental Science and Technology*, vol. 39, no. 3, pp. 707–715, 2005.

[23] M. Jang and R. M. Kamens, "Characterization of secondary aerosol from the photooxidation of toluene in the presence of NOx and 1-propene," *Environmental Science and Technology*, vol. 35, no. 18, pp. 3626–3639, 2001.

[24] W. E. Asher, J. F. Pankow, G. B. Erdakos, and J. H. Seinfeld, "Estimating the vapor pressures of multi-functional oxygen-containing organic compounds using group contribution methods," *Atmospheric Environment*, vol. 36, no. 9, pp. 1483–1498, 2002.

[25] D. R. Lide, *Handbook of Chemistry and Physics*, 86th edition, 1997.

[26] C. L. Yams, *Hanbook of Vapour Pressure, Vols. 2 and 3*, Guef Publishing Compagny, Houston, Tex, USA, 1994.

[27] T. Boulik, V. Freid, and E. Hala, *The Vapour Pressure of Pure Substances*, Elsevier, Amsterdam, The Netherlands, 1984.

[28] K. J. Joback and R. C. Reid, "Estimation of pure component properties from group contribution," *Chemical Engineering Communications*, vol. 57, pp. 233–243, 1987.

[29] P. W. Bruckmann and H. Willner, "Infrared spectroscopic study of peroxyacetyl nitrate (PAN) and its decomposition products," *Environmental Science and Technology*, vol. 17, no. 6, pp. 352–357, 1983.

[30] Y. Nannoolal, *Development and critical evaluation of group contribution methods for the estimation of critical properties, liquid vapour pressure and liquid viscosity of organic compounds [Ph.D. thesis]*, University of Kwazulu Natal, 2006.

[31] P. B. Myrdal, J. F. Krzyzaniak, and S. H. Yalkowsky, "Modified Trouton's rule for predicting the entropy of boiling," *Industrial and Engineering Chemistry Research*, vol. 35, no. 5, pp. 1788–1792, 1996.

[32] J. Vidal, *Thermodynamique: Application au Génie Chimique et Ãa l'industrie du Pétrolière*, Technip, Paris, Farnce, 1997.

[33] E. J. Baum, *Chemical Property Estimation-Theory and Application*, section 6.3, Lewis, Boca Raton, Fla, USA, 1998.

[34] R. P. Schwarzenbasch, P. M. Gschwend, and D. M. Imboden, *Environmental Organic Chemistry*, John Wiley & Sons, New York, NY, USA, 2nd edition, 2003.

[35] W. J. Lyman, *Environmental Exposure from Chemicals*, vol. 1, chapter 2, CRC Press, Boca Raton, Fla, USA, 1985.

[36] S. H. Fishtine, "Reliable latent heats of vaporization," *Environmental Science & Technology*, vol. 55, pp. 47–56, 1963.

[37] C. F. Grain, "Vapor pressure," in *Handbook of Chemical Property Estimation Methods*, chapter 14, McGraw-Hill, New York, NY, USA, 1982.

[38] B. E. Poling, J. M. Prausnitz, and J. P. O'Connell, *The Properties of Gases and Iquids*, McGraw-Hill, New York, NY, USA, 5th edition, 2001.

[39] R. C. Reid, J. M. Prausnitz, and B. E. Poling, *The Properties of Gases and Liquids*, McGraw-Hill, New York, NY, USA, 4th edition, 1986.

[40] D. Ambrose and J. Walton, "Vapor pressures up to their critical temperatures of normal alkanes and 1-alkohols," *Pure and Applied Chemistry*, vol. 61, pp. 1395–1403, 1989.

9

Rapid Mapping and Deformation Analysis over Cultural Heritage and Rural Sites Based on Persistent Scatterer Interferometry

D. Tapete[1] and F. Cigna[2]

[1] *Department of Earth Sciences, University of Florence, Via G. La Pira 4, 50121 Florence, Italy*
[2] *British Geological Survey, Natural Environment Research Council, Nicker Hill, Keyworth, Nottingham NG12 5GG, UK*

Correspondence should be addressed to D. Tapete, deodato.tapete@gmail.com

Academic Editor: Francesco Soldovieri

We propose an easy-to-use procedure of "PSI-based rapid mapping and deformation analysis," to effectively exploit Persistent Scatterer Interferometry (PSI) for multispatial/temporal hazard assessment of cultural heritage and rural sites, update the condition report at the scale of entire site and single building, and address the conservation strategies. Advantages and drawbacks of the methodology are critically discussed based on feasibility tests performed over Pitigliano and Bivigliano, respectively, located in Southern and Northern Tuscany, Italy, and representative of hilltop historic towns and countryside settlements chronically affected by natural hazards. We radar-interpreted ERS-1/2 (1992–2000) and ENVISAT (2003–2010) datasets, already processed, respectively with the Permanent Scatterers (PSs) and Persistent Scatterers Pairs (PSPs) techniques, and assigned the levels of conservation criticality for both the sites. The PSI analysis allowed the zoning of the most unstable sectors of Pitigliano and showed a good agreement with the most updated hazard assessment of the cliff. The reconstruction of past/recent deformation patterns over Bivigliano confirmed the criticality for the Church of San Romolo, supporting the hypothesis of a correlation with local landslide phenomena, as also perceived from the annual motions observed over the entire site, where several landslide bodies are mapped.

1. Introduction

One of the main challenges in conducting a technological transfer process consists in communicating developments and achievements of science to the potential stakeholders, to effectively contribute to the building and strengthening of their capabilities in the use of new technologies for the specific fields of application. This issue is currently particularly relevant in the perspective of spreading and encouraging the use of satellite radar data for deformation analyses, not only in geological and environmental applications but also in the management and conservation of built heritage in cultural and rural sites.

In this regard, in the last decades, the standard approach of Synthetic Aperture Radar Interferometry (InSAR) [1, 2] has increasingly found successful implementation in ground motions monitoring and deformation analyses at multiple scales, mainly relating to natural hazard events [3–6], geological processes [7, 8], and human-induced phenomena [9, 10]. But a significant improvement, especially for analyses over urbanized areas, infrastructure, and human settlements [11–14], has been achieved with the multi-interferogram approaches referred to as Persistent Scatterer Interferometry (PSI). The latter are actually performed by means of different algorithms (among which those developed by different authors [15–19]), which share the common feature of providing point wise information regarding displacements of reflective targets recognized as Persistent Scatterer (PS) throughout long data stacks of processed SAR images.

Hence, the PSI approaches are nowadays validated techniques which can allow a quantitative estimation of the degree of (in)stability of an entire region, an enclosed sector and even single clusters of buildings/elements on the ground to be performed, with a temporal reconstruction up to 1992 in case of availability of ERS-1/2 satellite imagery. This actually corresponds to the so-called back monitoring, as firstly defined by Cigna et al. [12] and specifically adapted to structural monitoring and early stage warning of cultural heritage sites by Tapete et al. [20]. The demonstration that PSI data can support the preventive diagnosis of monuments and historical buildings induces to consider such techniques as potential routine tools to monitor and assess the stability and, more generally, the condition over time, to be used by heritage bodies, superintendences, public administrations in charge of cultural heritage management, conservators, and practitioners.

A further chance for that is currently offered by the increasing number of projects and acquisition programmes at regional, national, and, sometimes, transnational level, which are aimed at the creation of databases containing processed PSI data thought to be finally used by public bodies, local authorities, and a wide spectrum of stakeholders and end-users, in their ordinary activities of land and urban environment planning and management. We can here cite, for instance, the ESA GMES project Terrafirma [21] and the Seventh Framework Programme EU project PanGeo [22], as well as the Extraordinary Plan of Environmental Remote Sensing (EPRS-E), that is, the Italian project established between the Italian Ministry of Environment Territory and Sea (METS), Chairperson of the Council of Ministers Department of Civil Protection and the Ministry of Defense in agreement with the Regions and Autonomous Provinces, to acquire, process, and periodically update PSI data over the whole Italian territory for environmental and civil protection activities [23].

Although such initiatives are frequently designed for other uses than application on cultural heritage sites, it cannot exclude that this typology of PSI data can constitute a precious reservoir of information for the preservation of monuments and historic centres, as also recently suggested by Trigila et al. [24] for landslide risk assessment in cultural heritage sites.

The present work is specifically aimed at proposing a simple and easy-to-use methodology to exploit PSI data for rapid mapping and deformation analysis over hilltop historic centres and rural sites, based on the results of feasibility tests performed on two case studies located in Tuscany, Central Italy. The latter were purposely selected as examples of real-world situations and related analytical issues that potential end-users (e.g., superintendence officers, conservators) might have to tackle, especially if they have to radar-interpret PSI data not necessarily processed to be used for cultural heritage applications or to be focused over a particular area of interest at an appropriate scale of analysis. Under this perspective, ERS-1/2 (1992–2000) and ENVISAT (2003–2010) PSI data, already processed in the framework of the EPRS-E project [23] by means of, respectively, the PSInSAR technique [16] and PSP-DIFSAR approach (Persistent Scatterers Pairs-DIFferential InSAR; [15]), were radar-interpreted as they are, without any further processing. The results of these site-specific analyses are here presented to critically discuss the potentials and limits of such PSI-based deformation analysis, in terms of real usefulness for preventive diagnosis of cultural heritage and land management in rural environments.

2. PSI-Based Procedure of Preventive Diagnosis

2.1. PSI Techniques. Superficial deformation affecting the area covered by the processed radar imagery (Area Of Interest (AOI)) acquired by means of space-borne SAR sensors can be detected after a PSI processing, since the employed algorithms allow the identification of natural/artificial reflectors on the ground. The latter keep their dielectric properties quite constant over the monitoring period and, consequently, act as permanent reflectors (Persistent Scatterer—(PS)) with respect to the microwave wavelengths generated and sent by the active satellite sensors. For each identified PS, the multi-interferogram processing of long data stacks of SAR images covering the AOI highlights the occurrence of changes in its spatial position along the Line Of Sight (LOS) of the satellite and provides an estimate, over time, of the LOS component of the real displacement vector, which geometrically describes the deformation of the point wise target.

The final output of such PSI processing consists of a distribution, over the AOI, of clusters and/or isolated PS; for each of them the following parameters and information are retrieved:

(i) longitude and latitude coordinates, associated with the height (h) and related standard deviation (SD_h);

(ii) estimate along the LOS of the yearly deformation rate (V_{LOS}), usually expressed in mm/yr, and related standard deviation (SD_{VLOS});

(iii) deformation time series, where the single values of LOS displacements (expressed in mm) estimated according to the sampling frequency of the satellite are plotted versus the monitoring interval;

(iv) PS coherence, that is, one of the quality parameters, together with the LOS velocity and height deviation standards, usually employed to evaluate the informational content of each PS estimate. Its dimensionless value, ranging between 0 and 1, gives a measure of how much the single time series fits with the deformation model used for the PSI processing.

Different authors [25, 26] have demonstrated that the PS usually corresponds to boulders and outcrops within natural contexts, man-made structures and infrastructure in built-up areas. In this regard, monuments, architectural elements with prominent and multiangular shape (e.g., pillars, towers, spires), archaeological structures (e.g., pinnacles, columns, ruins), and findings scattered on the ground (e.g., column drums, fallen architraves) can generate PS after PSI processing. This evidence has been recently confirmed by the results of different site-specific analyses, such as those discussed elsewhere [13, 20, 27, 28].

The papers [19, 27] also provide an extended theoretical discussion of the basic principles of the PSI processing (particularly the PSInSAR technique [16], used for the PSI processing of the radar data here analyzed) with specific reference to cultural heritage contexts, as well as schematic examples of the deterioration processes which can be monitored (e.g., land subsidence, masonry toppling, crack opening, collapses). The wide spectrum of detectable phenomena is evidently coupled by intrinsic pros and cons. If on one hand the PSI techniques can be successfully exploited for different typologies of instability processes, on the other hand the retrieved estimates only give information about the superficial effects of deterioration and instability phenomena which might be actually attributed to different triggering causes. Hence, solid background knowledge of the AOI and on-site inspections are frequently the key elements to achieve a correct radar interpretation of the PSI data.

Nevertheless, the last decade of PSI applications on environmental, geological, and urban topics confirms the reliability of these techniques, especially in terms of interoperability with conventional field observations and investigations [29, *cum biblio*]. This property is evidently essential for the purposes of a rapid mapping based on PSI data, like that proposed in this paper, which starts from an evaluation, carried out remotely, of the conservation issues affecting the built heritage over the AOI, as preparatory action to the execution of targeted and tailored terrestrial surveys.

2.2. PSI-Based Rapid Mapping. With the term "PSI-based rapid mapping" we mean the methodology and related final product to assess the (in)stability conditions over the AOI and reconstruct their past and/or recent evolution, by means of a multispatial and temporal analysis of PSI data which uses the LOS velocity estimates and the associated directions of movement (towards or away from the satellite) as reliable satellite evidences of "conservation criticalities." In particular, the latter term refers to natural and/or human-induced factors, conditions, and dynamics which can bring to the triggering of deterioration processes and more generally to key areas of concern for the conservation of the AOI.

The main objective of a PSI-based rapid mapping is the production of a map (updated as much as possible, dependently on the last SAR image of the processed data stacks) of the deformation patterns recognized after the PSI processing, which are likely to be related to recent, ongoing, or, even, upcoming deterioration phenomena. This map is expected to preventively highlight the unstable sectors of a monument, archaeological site, historic centre, or rural settlement rather than the stable ones ("preventive diagnosis"), thereby becoming a practical guiding tool for the selective zoning of the critical areas, where the execution of surveys and stabilization interventions is found as a priority.

The proposed procedure (Figure 1) is aimed at the extraction of the maximum amount of information in terms of deformation estimates, especially in cases of already existing PSI data, which were not processed with the main purpose of being used for cultural heritage applications. In doing that, the sequence of the different phases constituting the methodology is thought to avoid both under- and overestimation of the identified PS, as well as possible misinterpretation of those PSs which are actually pertinent to unrelated elements on the ground, located close to the objects of interest.

In this regard, the first step is the "background phase." It is specifically dedicated to the collection and review of all the existing information about the AOI and its heritage, to build a background knowledge which should address the subsequent multispatial and temporal analysis of the PSI data.

Before running through the "rapid mapping phase", which is the core of the PSI-based procedure of preventive diagnosis, SAR imagery to process and the deformation model to employ for the PSI processing are to be carefully selected (a specific focus on this issue can be found in [20]). This intermediate step is not tackled in the present paper, since we here consider the case of rapid mapping and deformation analysis carried out on already available and processed multiplatform PSI data. Nevertheless, it is worth mentioning that the choice of the processing parameters can significantly influence the obtained PSI data and, consequently, the reliability of the analysis. Indeed, they might have nonnegligible implications on the actual capability of monitoring deformation at the scale of single building/monument, as well as very localized small-area phenomena, as properly suggested by Colesanti and Wasowski [30].

During the rapid mapping phase, each of the employed PS datasets is progressively analyzed in terms of velocity, spatial distribution, and temporal frame.

The first evaluation concerns the LOS velocity distribution of the PS identified over the AOI. This initial assessment gives a rough idea of the overall stability of the AOI, although it is to be considered only a preliminary impression of the deformation pattern(s) to be confirmed during the subsequent phases. According to the sensitivity of the PSI techniques, the threshold commonly chosen as "interval of relative stability" is fixed at $V_{\mathrm{LOS}} = \pm 1.5$ mm/yr (where negative values correspond to LOS displacements away from the satellite, while the positive ones towards the satellite).

It means that PSs with value of $-1.5\,\mathrm{mm/yr} \leq V_{\mathrm{LOS}} \leq +1.5\,\mathrm{mm/yr}$ are classified as relatively stable, pending the verification that any displacement trend or acceleration/deceleration(s) are recognizable within their respective time series. An area apparently stable might have been affected by deformation in the past. At the opposite, high values of V_{LOS} can correspond to LOS displacements with acceleration(s) concentrated in temporal intervals limited within the whole monitoring period, which might have consequently influenced the entire PS deformation time series.

Prior to a temporal evaluation of the detected deformation, the PS spatial distribution is analyzed with regard to the location of the site/monuments, to highlight if significant deformation has taken place in correspondence with elements of cultural/civil value, their surroundings, and/or their foundation substratum. In relation to the spatial configuration of the PS clusters, the detected deformation can be distinguished into three main typologies, following the classification proposed in the papers [20, 28]—"macropattern," "micropattern," and "localized deformation" (Figure 2).

A macropattern is frequently found over areas affected by natural and/or anthropogenic phenomena active at large scale, in some cases at regional scale. An example is provided

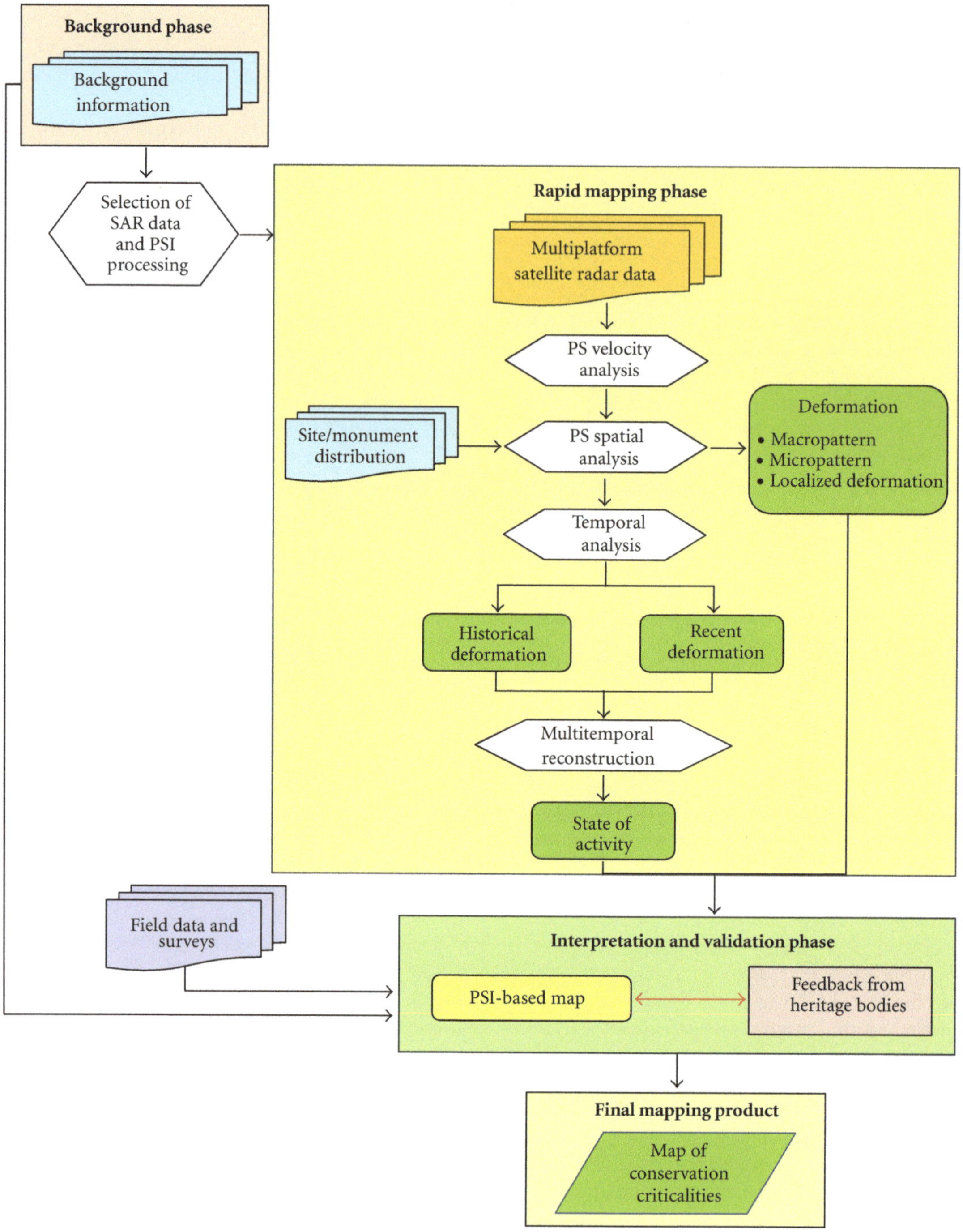

FIGURE 1: Procedure of PSI-based rapid mapping and deformation analysis to retrieve the map of conservation criticalities over cultural heritage and rural sites.

by the subsidence movements observed over recent alluvial deposits of the Tiber river, centre of Rome (Italy), which might have implications for the conservation of urban monuments like the Aurelian Walls (Figure 2(a)). Deformation macropatterns may also affect human settlements, such as in the case of Torrita di Siena, Tuscany (Italy), where a large portion of the town shows LOS displacements away from the satellite (Figure 2(b)). Differently, micropatterns are usually much less extended and typically mark single portions of a site/monument (Figure 2(c), with a zoomed aerial view of Fortezza da Basso, centre of Florence, Italy), likely over buildings located on unstable ground or structural weakness points.

While small clusters of PSs (at least two/three PSs) can identify a micropattern, especially if they show common LOS displacement trends over the same time interval, isolated PS can be referred to localized deformation (Figure 2(c), PS in the yellow circle). The major issue during the interpretation

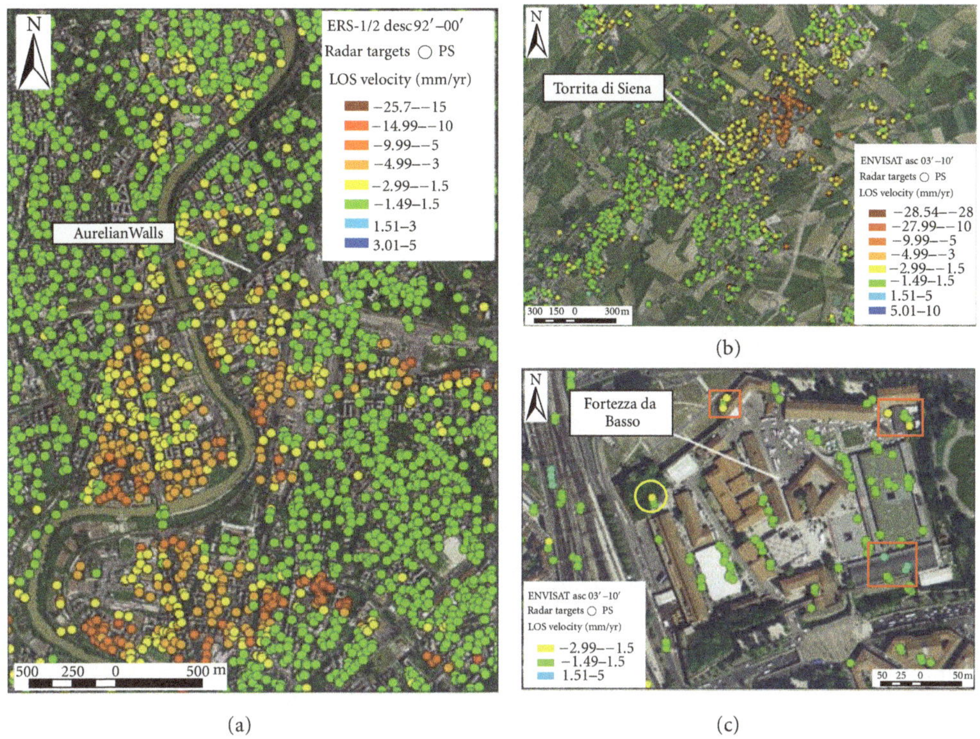

FIGURE 2: PSI-based deterioration patterns classifiable as macropattern ((a)-(b)); micropattern (i.e., the PS clusters within the red squares in (c)); localized deformation (i.e., the single PS within the yellow circle in (c)).

of such localized deformation concerns the reliability of their informational value. The comparison between their coordinates and height value and the spatial distribution of the built heritage, as well as the knowledge of the local AOI topography, significantly facilitates the attribution of single PS to objects of interest rather than neighbouring unrelated elements (e.g., light poles, infrastructure, public bins). Once the PS has been correctly spatially located, its value in terms of detected deformation is to be carefully assessed in the context of a multitemporal PSI analysis, before classifying it as an occasional displacement event or a negligible deformation estimate.

In this regard, the temporal analysis of multiplatform PS datasets covering distinct and subsequent temporal intervals of observation allows the reconstruction of the evolution of the detected deformation in time, although the different acquisition geometries which characterize the employed satellites still remain a technical constraint to achieve a perfect comparison between multi-platform deformation estimates. Nevertheless, the persistence of deformation over a same area, monument, or architectural element, throughout the different analyzed datasets (i.e., from historical data to recent ones), is generally a reliable indicator of ongoing deterioration processes. Frequently, the analysis of more recent PSI data permits the operators to better and more correctly radar-interpret PS belonging to older datasets, and further confirmation can be retrieved by means of field surveys.

As main outcome of the multitemporal analysis, the state of activity of a deformation is assessed. Regarding the historical data, the absence of any previous PSI-based classification implies that the areas found as subjected to deformation can be classified as follows:

(i) absence of PSI data → N.C. (i.e., not classifiable);

(ii) single or clusters of PS with $-1.5\,\mathrm{mm/yr} \leq V_{\mathrm{LOS}} \leq +1.5\,\mathrm{mm/yr}$ (independently of the PS spatial distribution) → stable;

(iii) single or clusters of PS with V_{LOS} exceeding the stability threshold, that is, $V_{\mathrm{LOS}} \leq -1.5\,\mathrm{mm/yr}$ or $V_{\mathrm{LOS}} \geq +1.5\,\mathrm{mm/yr}$ → active.

In the latter case, the definition of the state of activity is coupled with the assessment of the typology of deformation pattern (i.e., macropattern, micropattern or localized deformation).

This type of classification is also employed in those cases where the rapid mapping is performed only based on single-platform PSI data (e.g., PS obtained after multi-interferogram processing of ERS-1/2 radar imagery rather than ENVISAT one). The nonavailability of other platform PS datasets should always be avoided, since it leads to a partial evaluation of the deformational behaviour over the AOI. Nevertheless, it is not rare that, due to economic reasons and/or gaps in the satellite archives, the rapid mapping over the AOI can be carried out with a limited number of PS datasets. Furthermore, even if both the ascending and descending geometries of a same satellite are available, any difference in spatial resolution (e.g., Standard Beam Mode rather than Fine Beam Mode in the case of

Classification based on historical versus recent PSI data			
State of activity	LOS velocity	Spatial distribution	Level of conservation criticality
Active	Intensified	More extended	+
Active	Intensified	Stable	
Active	Intensified	Reduced	
Active	Less intense	More extended	
Active	Less intense	Stable	
Active	Less intense	Reduced	
Stabilized	—	—	
Stable	—	—	−

FIGURE 3: Scale of conservation criticality based on multi-temporal analysis of historical and recent PSI data. Increasing level of conservation criticality is associated to deformation, which worsen in terms of state of activity, intensity (LOS velocity—V_{LOS}) and/or spatial distribution (i.e., typology of deformation pattern).

RADARSAT-1/2 data stacks) or type of processing can affect the quality of the radar interpretation.

Conversely, when comparing recent PSI data with the historical ones, the following aspects should be taken into account:

(i) if new deformation is detected where no PSs were identified in the previous dataset(s);

(ii) if a previously detected phenomenon is still present and, in such a case, if it is still active or not;

(iii) if the deformation has changed its intensity or not;

(iv) if the new PS spatial distribution still belongs or not to the same class of deformation pattern typology.

According to the previously listed elements, a level of conservation criticality can be assigned, following the scale of progressive alert and priority shown in Figure 3.

The maximum level is attributed to deformation which is found as intensified (i.e., $V_r \geq V_h$, where V_h is the LOS velocity estimated in the historical PSI data and V_r in the recent ones) and becomes more spatially extended (e.g., localized deformation changed into micropattern or micropattern changed into macropattern) from the historical data to the recent ones. Similarly, if a new deformation not previously mapped is identified, worse typologies of deformation pattern will be classified with higher alert levels.

On the contrary, if the more updated PSI data record lower yearly deformation rates (i.e., $V_r \leq V_h$), an accurate detailed analysis of the deformation time series is highly suggested, in addition to the mere comparison between the two LOS velocity values. Also, following such a cautionary and conservative approach (similar to that currently employed in PSI-based updating of landslide inventories; [28]), the execution of on-site inspections and feedback from the heritage bodies can provide, during the "interpretation and validation phase," further evidence to classify a sector of the AOI as stabilized (Figure 1).

If no changes occurred over the area, this sector can be classified as stable with a relatively high degree of confidence. Conversely, the absence of PSs over previously moving areas does not necessarily mean that these sectors are now stable. Especially when comparing multiplatform PS datasets, the difference of both the acquisition geometry and temporal period of observation is to be carefully considered. Deformation that was observed and estimated by one satellite at a certain time according to its acquisition geometry might not have been appreciated by another satellite. Lack of PS is evidently an intrinsic weakness of such a type of rapid mapping.

2.3. Map of Conservation Criticalities. The output of the rapid mapping phase is a "PSI-based map" which shows the location, extent, and distribution of the unstable areas of a site and critical sectors of a monument, dependently on the scale and purposes of the analysis. This map is, at the same time, a guide for the execution of targeted on-site checks rather than extensive and expensive surveys and a tool to be validated by the ground truth collected during the inspections themselves.

Indeed, the latter, combined with the background knowledge and the feedback from the heritage bodies (and, more generally, from all the keepers of the cultural and civil elements of interest subjected to deformation during the observation temporal interval) can help to confirm the satellite evidences reported in the PSI-based map. Also, a reliable and strong correlation between the detected deformation and the deterioration phenomena observable on the exposed surfaces can be found, thereby retrieving essential elements to propose a robust interpretation of the analyzed PSI data.

The multilayering analysis of (i) the local natural and/or human-induced factors of deterioration, (ii) the multitemporal deformation estimates detected from satellite, and (iii) the information of the past/recent history of the AOI

Figure 4: (a) Aerial view of the Pitigliano cliff, Southern Tuscany and (b) a detail of the southern side. The built heritage along the cliff edge is exposed to rock mass instability mechanisms (c), while an additional hazard can derive from inner collapses of the cavities cut in the tuff substratum underneath the buildings ((d); photo by M. Capretti).

finally translates into a same mapping product, called map of conservation criticalities. This actually represents the final output of the process of rapid mapping and deformation analysis, and it is expected to provide an inventory map in support of the condition reporting of the site/monument of interest. Its degree of updating is evidently strictly dependent on the temporal coverage of the analyzed PSI data. The map of conservation criticalities can be ideally updated, as soon as new satellite acquisitions and more recent PS datasets are added to the previous analysis.

3. Feasibility Tests

3.1. Sites Selection, PSI Data, and Rationale. The applicability of the proposed methodology was evaluated by means of feasibility tests carried out on two case studies located in Tuscany, Central Italy, which were purposely selected to be representative of the following typologies of cultural heritage and countryside settlements: (i) hilltop historic towns and (ii) rural sites.

The first category allows the discussion of the conservation issues concerning architectural heritage built at the top of unstable rock masses, whose instability mechanisms can progressively cause structural damages to the monuments located along the edge, even their collapse and consequent disappearance. In these terms, the case of the historic centre of Pitigliano, Southern Tuscany (Figure 4) is a quite demonstrative example of sites where the built heritage potentially at risk is distributed along the entire cliff edge and is exposed to rock mass instability mechanisms, as well as cavities collapse.

For the second category of cultural heritage, the countryside settlement of Bivigliano, north of Florence (Figure 5), was selected, since its cultural, landscape, and built heritage and related conservation issues well exemplify situations of land management quite common over the Italian territory.

These feasibility tests were thought not only to provide a multiple validation of the usefulness of the PSI-based methodology and techniques employed on different contexts of application but also to demonstrate the potential range of stakeholders and end-users who can be interested in the use of PSI data, especially if the latter are made available in the framework of acquisition and processing programmes at local, regional, or national scale.

In this perspective, we exploited the deformation estimates of the previously cited EPRS-E database [23], obtained after PSI processing of C-band data of ENVISAT and ERS-1/2 satellites of the European Space Agency. ENVISAT images, acquired with wavelength (λ) of 5.63 cm and nominal revisiting time of 35 days, were processed with the PSP-DIFSAR approach (Persistent Scatterers Pairs-DIFferential InSAR; [15]), while ERS-1/2 images, acquired with λ = 5.66 cm and the same nominal revisiting time as ENVISAT satellite, were processed with the PSInSAR technique [16]. Details about the processing parameters, as well as accuracy and error bars of these PSI estimates, are reported in [31], while the main characteristics of the analyzed PS datasets with specific regard to each test site are summarized in Table 1.

FIGURE 5: Aerial view of the countryside settlement of Bivigliano, Northern Tuscany, with location of the main historic heritage: (a) Church of San Romolo; (b) Convent of Monte Senario; (c) Villa Pozzolini. Further elements of interest for the present feasibility test are (d) the modern cemetery and (e) the area of the local craftsmen (Marroneto).

TABLE 1: Main characteristics of the analyzed datasets; Asc: ascending; Desc: descending. PS density was calculated with regard to the extent of the AOI, which was equal to 0.11 km^2 for the Pitigliano cliff and 1.73 km^2 for the territory of Bivigliano.

Site	Data stack	Orbit	Repeat cycle (days)	Time interval	*N.* images	Processing technique	Total *N.* PS	PS density (PS km^{-2})
Pitigliano	ERS-1/2	Desc	35	10/05/1992 13/12/2000	60	PSInSAR	44	400
	ENVISAT	Asc	35	16/10/2003 27/05/2010	34	PSP-DIFSAR	159	1446
	ENVISAT	Desc	35	07/05/2003 09/06/2010	51	PSP-DIFSAR	193	1775
Bivigliano	ERS-1/2	Desc	35	24/04/1992 27/11/2000	79	PSInSAR	145	84
	ENVISAT	Asc	35	16/10/2003 27/05/2010	35	PSP-DIFSAR	130	75
	ENVISAT	Desc	35	10/02/2003 28/06/2010	35	PSP-DIFSAR	504	291

As previously mentioned, these PSI data were radar-interpreted as they are, without any further processing. The rationale behind this technical choice was the possibility to simulate real-world situations that any public administration officer or stakeholder might have to tackle, when exploiting this type of data for applications of rapid mapping and deformation analysis, dealing with both their advantages and shortcomings.

3.2. Hilltop Historic Sites: Pitigliano, Southern Tuscany (Italy). The historic centre of Pitigliano developed through centuries over a flat slab at the top of 25-metre-thick plateau of Pleistocene tuffs, with a present elevation of about 310 m a.s.l. and approximately oriented along the E-W direction (Figures 4(a)-4(b)). The geological sequence which constitutes the cliff is the result of an intense volcanic activity, mainly related to the volcanic complexes located in the northern Latium [32, 33], and the rock formations originated from the corresponding phases of pyroclastic deposits. Referring to Canuti et al. [34] and Fanti et al. [35] for a detailed geological setting of the Pitigliano cliff, it is here worth mentioning that the welded tuffs belonging to the Sorano/Grotte di Castro Formation represent the substratum underneath the historical buildings. These tuffs were also extensively cut and excavated in depth since Etruscan times, thereby creating a huge network of subterranean caves and corridors, which is actually a sort of "second town" and, at the same time, a source of instability issues due to inner collapses (Figure 4(d)).

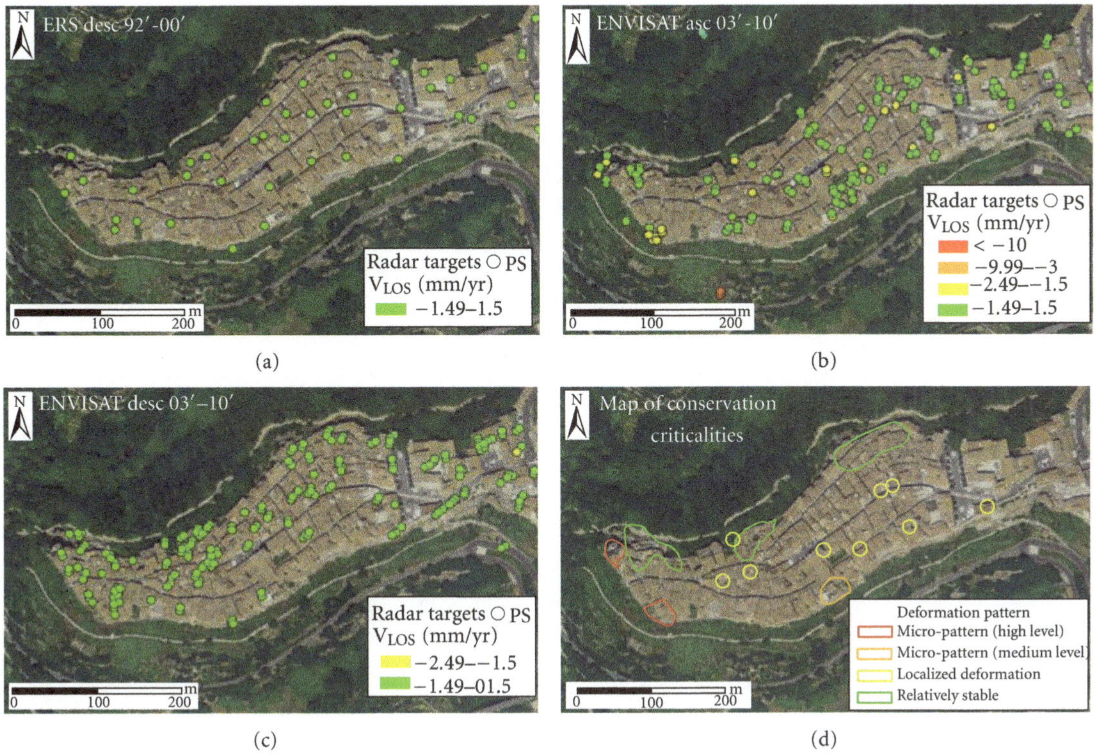

FIGURE 6: PS spatial distribution and related V_{LOS} field over the Pitigliano cliff for the datasets: (a) ERS descending (1992–2000); (b) ENVISAT ascending (2003–2010); (c) ENVISAT descending (2003–2010). The PSI-based rapid mapping led to draw a map of the conservation criticalities (d), which reports the zoning of the unstable sectors and localized deformation, distinguished from the relatively stable areas.

In addition to this factor of criticality, the built heritage of Pitigliano located along the cliff edges is chronically affected by the instability mechanisms of the exposed rock surfaces, due to a combination of several weathering processes, such as chemical alteration, erosion, cryoclastism, and crack patterns opening, which progressively create the conditions for the triggering of rock falls and toppling events (Figure 4(c)). Hence, the continuous geomorphologic evolution of the cliff can be nowadays recognized as one of the main threats for the preservation of the Pitigliano heritage [35].

PSI data were specifically radar-interpreted to perform a zoning of the sectors along the cliff edges classifiable as potentially critical. At the same time, detailed time series analysis was exploited to verify if the unstable PSs identified over the historical buildings located within the town centre could highlight the occurrence of sudden displacements, to be correlated to potential collapses of subterranean cavities or predisposing conditions for near future collapses.

Despite the relevant number of the processed SAR images, the PSInSAR processing of the ERS-1/2 descending (1992–2000) data stack led to the identification of a reduced set of PSs (Table 1), whose spatial distribution does not homogeneously cover the historic centre (Figure 6(a)). Although this result partially makes the deformation analysis less spatially representative over the different sectors of the town, the ERS data are sufficient to confirm the absence, in the period 1992–2000, of deformation macropatterns referable to instability phenomena extended to the entire site. Indeed, all the ERS descending PSs show values of V_{LOS} not exceeding the stability threshold (±1.5 mm/yr). This first impression of relative stability is supported by the detailed analysis of the PS time series, which in few cases highlight a slight tendency of the LOS displacements to movements going towards the satellite, only occasionally with $V_{LOS} >$ +1 mm/yr. In the absence of the corresponding ascending dataset, very low level of conservation criticality can be assigned over the whole historic centre for the period (1992–2000).

Passing to the more recent data, that is, the ENVISAT (2003–2010) ones, a significant increase in the total number of PS is immediately appreciable in both the ascending and descending datasets (Table 1), although their respective PS spatial distribution shows some lack over certain sectors of the town, reflecting the different acquisition geometry of the processed SAR images (Figures 6(b)-6(c)). Particularly, the ascending dataset mainly covers the central-southern parts of the town (Figure 6(b)), while the descending one results in being better distributed over the central-northern sectors (Figure 6(c)).

Being aware that the two datasets are not perfectly temporally overlapped (October 2003–May 2010 for the ascending data, May 2003–June 2010 for the descending ones), the integrated analysis of the deformation estimates highlights the presence of two micropatterns, with downwards LOS

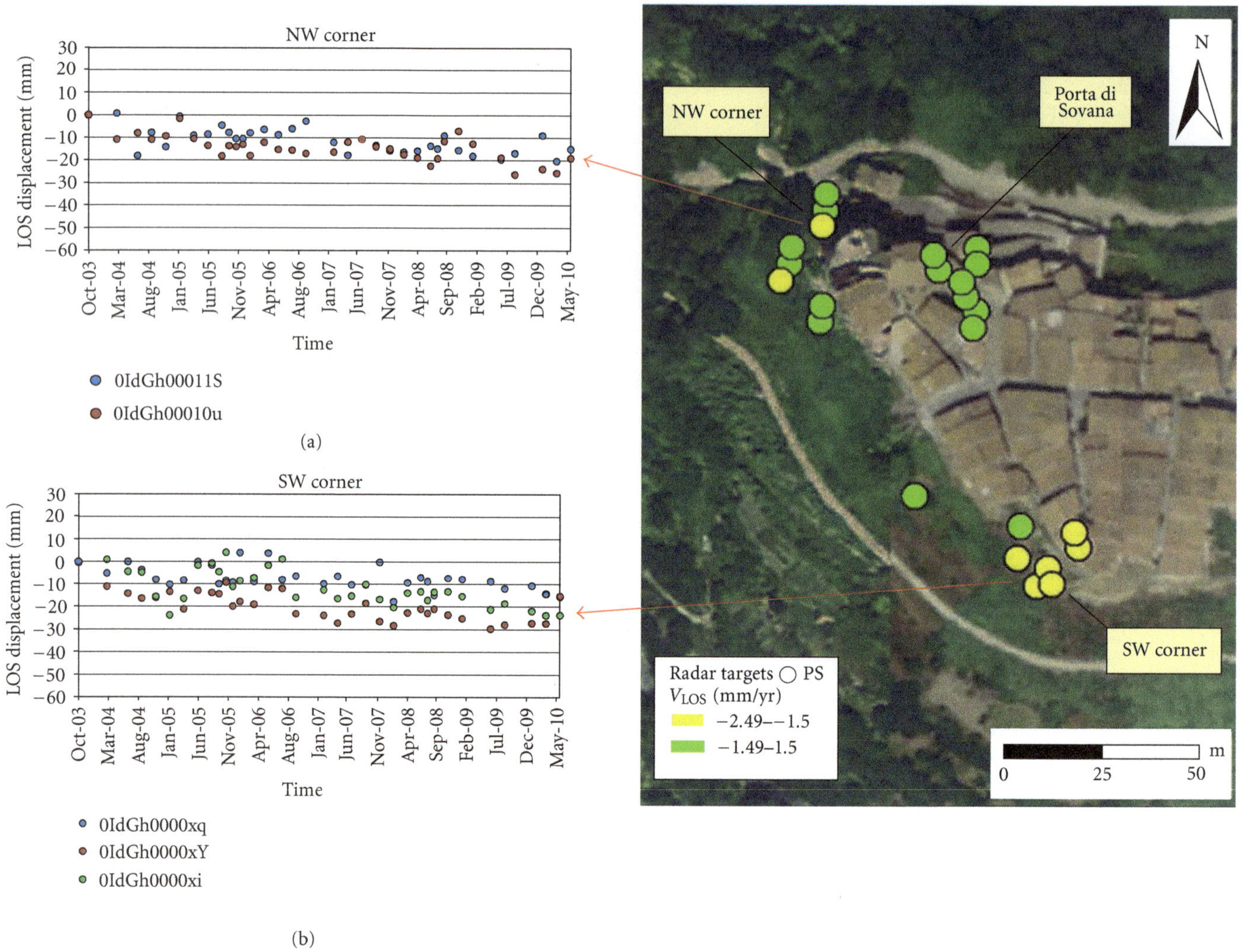

Figure 7: Spatial distribution and related V_{LOS} field for the ENVISAT ascending (2003–2010) dataset over the western side of the Pitigliano cliff. Deformation micropatterns are identified over the north-western and south-western corners of the cliff, with downwards moving PS, whose time series are reported, respectively, in the plots (a) and (b).

displacements in the ascending data, concentrated, respectively, over the north-western and south-western corners of the cliff (Figures 6(b), 6(d), and 7).

The PS cluster over the north-western corner is characterized by two PS ENVISAT ascending (2003–2010) with V_{LOS} up to −2.1 mm/yr (Figure 7), whose time series show displacement trends progressively going away from the satellite, especially since March 2005 (Figure 7(a)). Slight tendency to downwards movements is also recognizable within the time series of neighbouring PSs, which appear stable as their V_{LOS} only occasionally exceed the value of −1 mm/yr. The presence of several deformation estimates which share similar displacement trends induces to assign a medium-high level of conservation criticality to the buildings located on the north-western corner of the cliff.

On the other hand, there is no doubt to classify the built-up area on the south-western corner as highly critical (Figures 6(d) and 7), due to the significant number of downwards moving PS ENVISAT ascending (2003–2010) concentrated over the buildings and along the slope; the intensity of the detected deformation (V_{LOS} up to −2.9 mm/yr); the persistence, throughout the monitoring period, of a displacement trend progressively going away from the satellite (Figure 7(b)); the value of the estimated total LOS displacement (up to −25 mm over approximately seven years). Also, the time series of the PS located, respectively, along the slope, close to the cliff edge and over the buildings show similar LOS displacement trends (Figure 7(b)).

The priority level to assign to these two sectors, as assessed based on the PSI data, is further appreciable if compared with the relative stability that both the ascending and descending ENVISAT datasets show over other sectors of the cliff, such as the area of Porta di Sovana, located just few metres from the north-western corner, along the northern edge (Figures 6(d) and 7). Here, the detailed analysis of PS time series completely excludes the occurrence of clear deformation trends potentially associable to instability processes active in the period 2003–2010.

Figure 8: Detailed view of the south-western corner of the Pitigliano cliff. Close examination of the rock surfaces highlights the boundaries of rock wedges underneath the masonry walls.

As further element to strengthen the relevance of these satellite findings in the perspective of an effective early stage warning of near future instability, it is to be mentioned that both the sectors classified as unstable belong to a portion of the Pitigliano town which was historically damaged by progressive retreat of the cliff likely due to rock falls. As reported by Fanti et al. [35], a recent archive-based research comparing historical maps and old photographs has clarified that the western side of the cliff is nowadays completely modified from the past, and several buildings totally disappeared.

The high level of conservation criticality of both the sectors is also confirmed by the hazard assessment performed by Canuti and Fanti [36] and recently published by Fanti et al. [35] as 3D quantitative kinematic analysis directly performed on high detailed terrestrial laser scanning survey of the cliff. Based on these data and direct observations on the rock surfaces (Figure 8), the mechanisms known as "wedge failure," "flexural block toppling," and, secondarily, "toppling" are the highest probable for the south-western corner, while the flexural toppling is predominant for the north-western corner. In these terms, the persistence, throughout the monitoring period, of LOS displacements going away from the satellite in the ascending geometry leads not to exclude the hypothesis of tilting movements of the exposed elements as a consequence of active rock mass instability.

A second level of priority seems reasonable to be attributed to localized deformation, which was detected over single buildings located along the southern side of the cliff (orange micropattern in Figures 6(d) and 9). The similarity in terms of LOS displacement trend between a downwards moving PS with V_{LOS} up to -1.7 mm/yr and two PSs apparently stable suggests the presence of a more extended deformation pattern (Figure 9(b)). In particular, mutually comparable records of LOS displacements are recognizable within all the respective PS time series, especially since February 2007.

Differently, the ENVISAT ascending (2003–2010) dataset does not show particular evidence of instability processes for the northern cliff side, except for a localized deformation found over an enclosed sector of buildings in the central part of the northern edge (yellow circle in Figure 6(d)). Although the poor coverage of this dataset along the northern side does not allow a complete deformation analysis to be performed, the corresponding descending data show relatively stable PS time series and suggest a quite relative stability in the period 2003–2010 (Figures 6(c)-6(d)).

Interestingly, over several buildings within the town centre, the time series of downwards moving PS ENVISAT ascending (2003–2010) do not reveal the occurrence of significant accelerations. No evidences of sudden displacements due to cavities collapses are found (Figures 6(b) and 9). Indeed, most of the PS time series are characterized by LOS displacement trends progressively going away from the satellite, with V_{LOS} up to -2.2 mm/yr (Figure 9(a)). If on one hand this evidence does not give proof about past/recent instability events, on the other hand it is not to be excluded that further SAR acquisitions after the last updating of the analyzed PSI data (i.e., May 2010) could confirm the evolution of these progressive displacement trends. Hence, on-site checks and terrestrial surveys, performed also within the subterranean cavities underneath the buildings, seem to be a necessary step to validate and correctly interpret the data.

Similar considerations are also valid for the time series of the single downwards moving PS identified in the ENVISAT descending data (2003–2010) in the middle of the town centre (Figure 6(c)).

3.3. Rural Sites: Bivigliano, Northern Tuscany (Italy). Sited 20 km north of Florence in the municipal territory of Vaglia, the countryside settlement of Bivigliano is mainly built along the western slope of the mountain called Monte Senario (Figure 5). Geologically, the site is founded on Holocene colluvial sediments overlying the bedrock constituted by the Monte Senario Sandstones Formation, that is, Middle-Lower Eocene feldspathic-quartz sandstones containing coarse quartz pebbles, mica, phyllite, and calcareous material. In terms of historic and artistic heritage, since medieval times the parish church (Figure 5(a)) based in Bivigliano was a landmark for the social life of all the territory and, starting from the XIII century, it was flanked by the neighbouring Convent of Monte Senario (Figure 5(b)). Despite this ancient origin (the name "Bivigliano" derives from a member of the Medici family, already powerful in the Medieval-Renaissance Florence), the current urban arrangement of Bivigliano was strongly influenced by the recent history. The village was a renowned place for holidays since the end of the XIX century and flourished with a significant urban development phase in the decades 1900–1950s. In this period, for instance, the

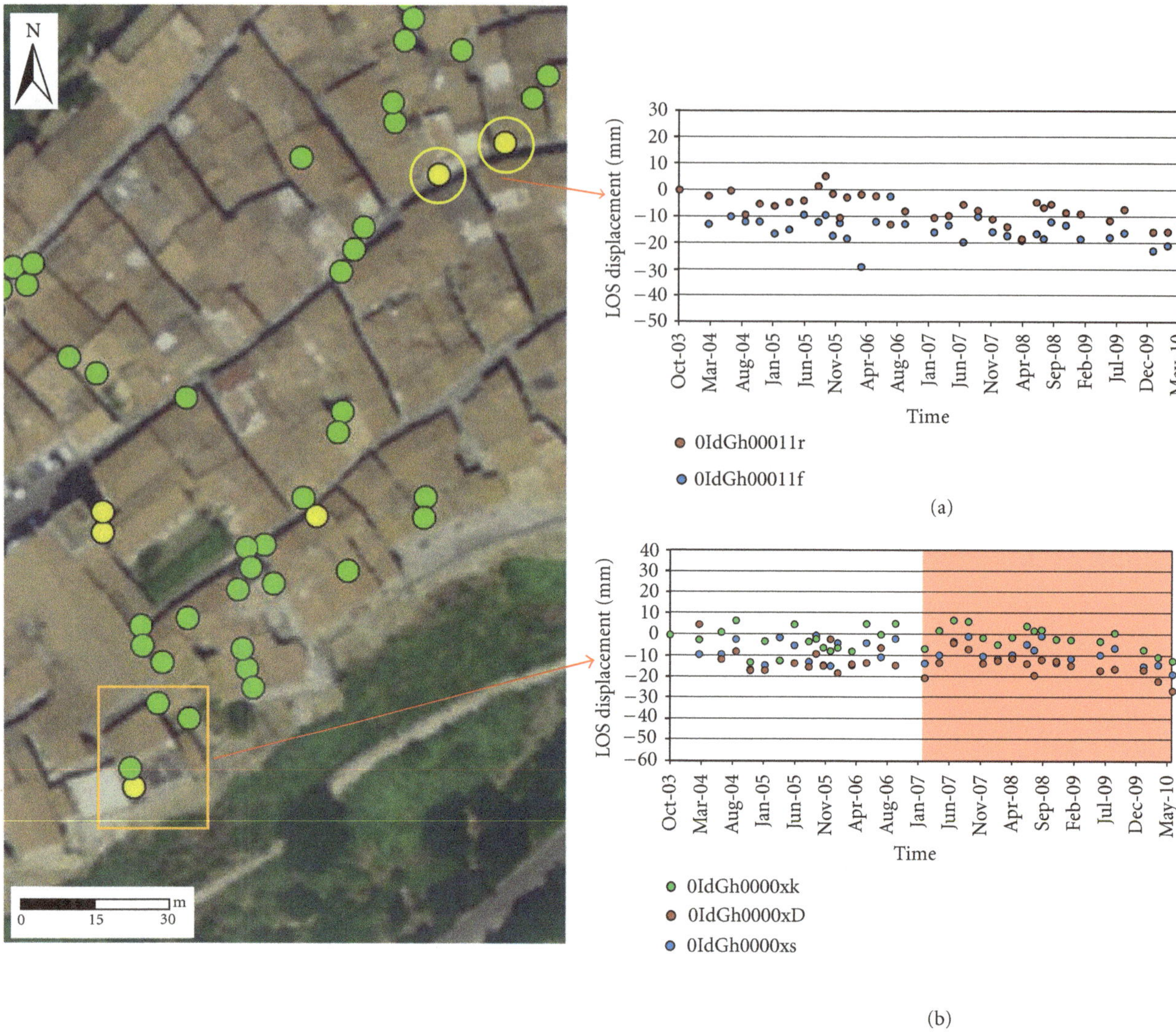

Figure 9: Spatial distribution and related V_{LOS} field for the ENVISAT ascending (2003–2010) dataset over the central-eastern part of the Pitigliano town centre. Localized deformation is identified over some buildings, with downwards moving PS showing a progressive LOS displacement trend away from the satellite (a), while one downwards moving PS and some apparently stable PSs located close to the cliff edge share a common LOS displacement trend, especially since February 2007 (b).

road still connecting the village with the Monte Senario convent was cut along the slope. Among the built heritage of Bivigliano are to be mentioned the Church of San Romolo (Figure 5(a)), the late XVI century noble residency called Villa Pozzolini (or "Villa di Bivigliano"; Figure 5(c)), and the Monte Senario convent itself (Figure 5(b)). All these historical evidences are still inserted within a landscape context, which is nowadays one of the main features to be preserved.

Nevertheless, effective strategies of land management have to tackle the landslide susceptibility of the territory, as testified from different historical records of terrain motions and ground instability, such as the "landslide of Bivigliano" mentioned in a document and which occurred in 1729 [37]. Indeed, several landslide bodies are also reported in the Regional Geological Map (*Carta Geologica Regionale* (CGR)) Sheet 263120 scale 1 : 10,000 (Figure 10), as well as in the Italian Landslide Inventory (*Inventario Fenomeni Franosi in Italia* (IFFI)) produced by the Italian Institute for Environmental Protection and Research in 2007. Due to landslipping, in modern times the old cemetery of the village was moved to the present place (area d in Figure 5).

In light of these conservation issues, the feasibility test was aimed at exploiting the PSI data to assess the evolution of the condition for the main monuments and historical buildings of Bivigliano since 1992, within the broader environmental context and related deformation patterns.

At the scale of entire site, the multitemporal analysis of the PS spatial distribution of both the ERS and ENVISAT datasets highlights the persistence, from the historical period (1992–2000) to the recent one (2003–2010), of an extended deformation macropattern covering most of the residential

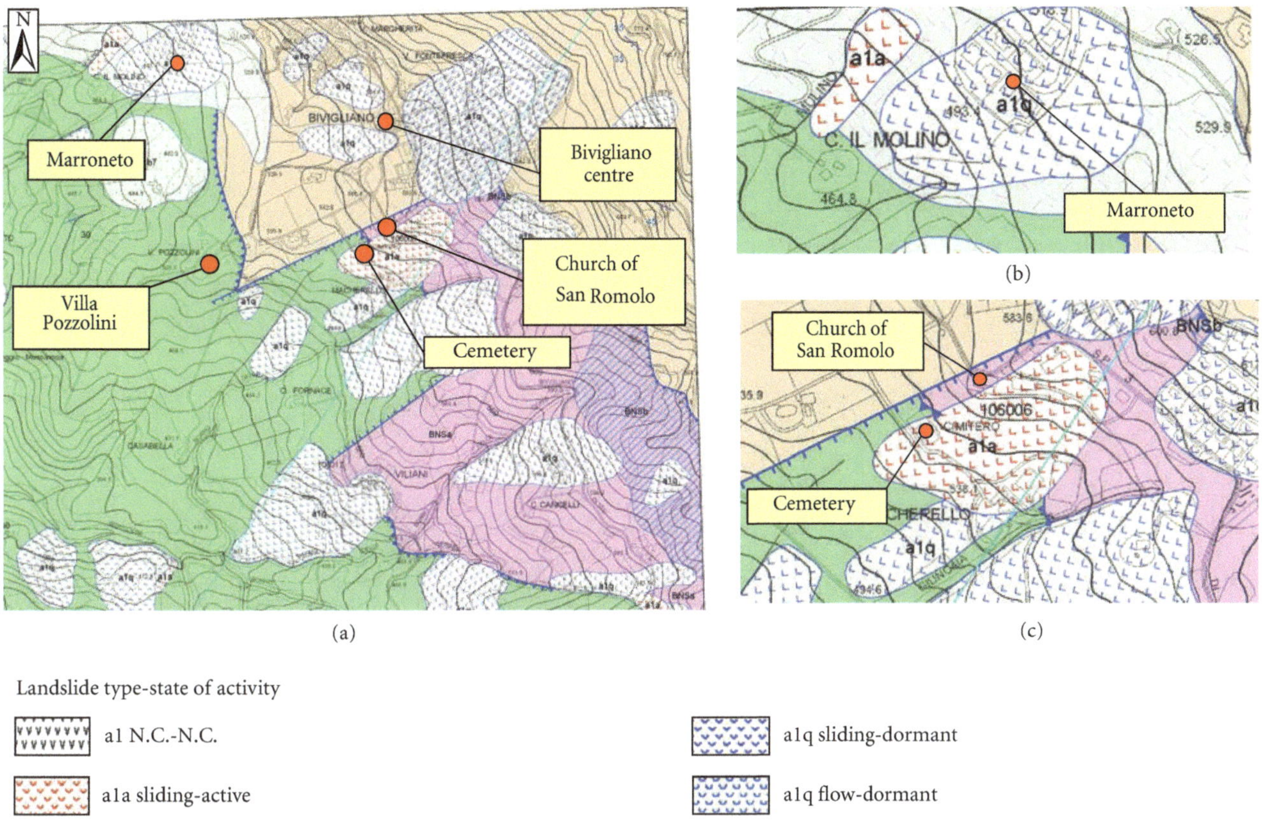

Figure 10: (a) Detail of the Regional Geological Map Sheet 263120 scale 1 : 10,000 over the Bivigliano territory, with the landslide inventory reporting several landslide bodies, among which: (b) dormant slide over Marroneto; (c) active slide close to the church and including the cemetery.

sectors of Bivigliano, with LOS displacement field mainly characterized by movements going away from the satellite (Figure 11). This deformation macropattern is clearly detected in both ERS descending (1992–2000) and ENVISAT descending (2003–2010) datasets (Figures 11(a) and 11(c)), while in the ENVISAT ascending (2003–2010) dataset only few and isolated PSs show V_{LOS} exceeding the stability threshold (± 1.5 mm/yr; Figure 11(b)).

In particular, for the ERS descending (1992–2000) data, the modal value (i.e., the value that occurs most frequently) of coherence is 0.73, with 35% of the identified MP assuming coherence values ≥ 0.80. The associated V_{LOS} field spans from -8.8 to -0.5 mm/yr, with a V_{LOS} modal value of -2.2 mm/yr. On the other hand, the ENVISAT descending (2003–2010) data, here used as recent dataset, shows coherence modal value of 0.76, with 25% of the identified PS assuming coherence values ≥ 0.78. The associated V_{LOS} field ranges from -4.7 to -1.2 mm/yr, with a V_{LOS} modal value of -1.0 mm/yr.

These simple statistics confirm the impression of a less concentrated and intense deformation macropattern covering the settlement during the period 2003–2010, than the corresponding one observed in the historical data (1992–2000). Nevertheless, both the datasets suggest a potential correlation between the spatial distribution of most of the identified PS clusters and the presence of landslide bodies, especially if considering the boundaries and extent of the landslide phenomena mapped in the previously mentioned CGR scale 1 : 10,000 (Figures 10 and 11). At the same time, the PS spatial distribution of the ERS and ENVISAT descending data opens to the need of an accurate updating of the knowledge basis about the current state of activity of the already mapped landslide phenomena. Such analysis will clarify whether the deformation detected from the satellite can be attributed only to natural hazards or also to human-induced impacts (e.g., water pumping).

In these terms, it is quite interesting to note the persistence of a deformation micropattern since 1992 over the peripheral area of Bivigliano called Marroneto (Figures 11 and 12), historically occupied by the local craftsmen and located north-west of the main residential quarter (area (e) in Figure 5). Here, the CGR reports the presence of a huge landslide body classified as "dormant slide" ("*frana di scorrimento quiescente*"; Figure 10(b)).

Over there, the ERS descending (1992–2000) data show a concentrated micropattern consisting of four downwards moving PS (Figure 12(a)), with V_{LOS} up to -5.1 mm/yr and a linear displacement trend progressively going away from the satellite, reaching a total value of LOS displacement up to -50 mm over approximately eight years (Figure 12(d)). Not less severe is the corresponding micropattern observable in the ENVISAT descending (2003–2010) data (Figure 12(c)).

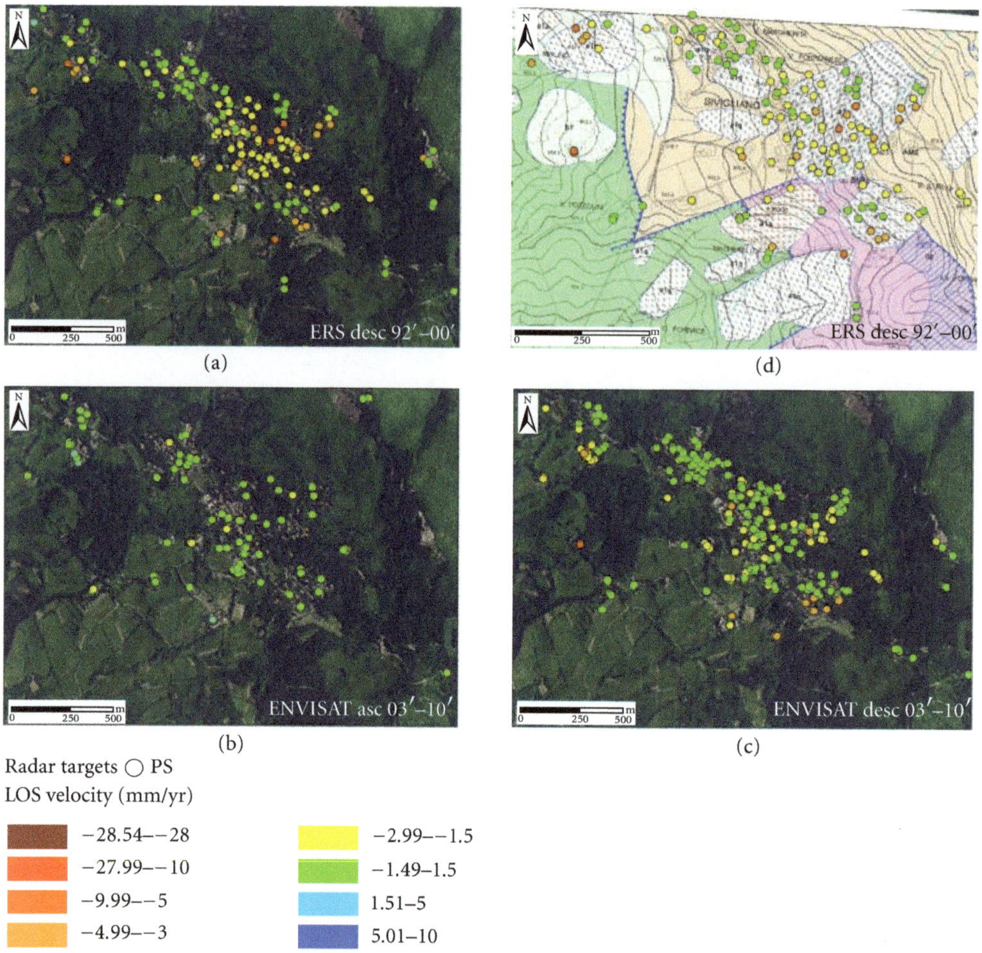

FIGURE 11: PS spatial distribution and related V_{LOS} field over the Bivigliano territory for the datasets: (a) ERS descending (1992–2000); (b) ENVISAT ascending (2003–2010); (c) ENVISAT descending (2003–2010). Superimposition of PS datasets over the landslide inventory (see Figure 10) suggests a potential correlation with the distribution and extent of the mapped landslides, as perceivable observing the ERS data (d).

Nineteen downwards moving PSs were identified over the whole area, with V_{LOS} ranging from −5.3 to −1.5 mm/yr, and modal value of −3.7 mm/yr. All the PS time series share the same LOS displacement trend, which progressively goes away from the satellite throughout the monitoring period (Figure 12(f)), with a progression not as linear as that found in the time series of the historical ERS descending data (Figure 12(d)). The total LOS displacement reaches up to −45 mm over about seven years. Correspondingly, despite the reduced number of the identified PS, the ENVISAT ascending (2003–2010) dataset shows a localized micropattern consisting of two isolated upwards moving PSs (Figure 12(b)), with V_{LOS} of +1.8 mm/yr and LOS displacement trends characterized by an annual seasonal component of about 5 mm (Figure 12(e)).

Passing to the scale of single building, the ancient and modern built heritage of Bivigliano is characterized by differential conditions in terms of PSI-based levels of conservation criticality. The worst situation was found for the Church of San Romolo (Figure 13). Although it is only a single PS not surrounded by other deformation estimates, the downwards moving PS ERS descending (1992–2000) identified over the bell tower of the church (Figure 13(a)) assumes a particular relevance, especially if analyzed in integration with the more recent PSI data (Figure 13(c)). This PS shows a V_{LOS} of −2.2 mm/yr and an approximately linear LOS displacement trend progressively going away from the satellite until March 1999 (Figure 13(b)).

While a lack of PS characterizes the ENVISAT ascending (2003–2010) dataset, a clear deformation micropattern affects the entire complex of the church in the corresponding descending (2003–2010) dataset (despite the shift in the PS georeferencing; Figure 13(c)). All the five identified PSs, including the one which appears stable, share a common LOS displacement trend going progressively away from the satellite, with V_{LOS} reaching the maximum value of −2.7 mm/yr and a total LOS displacement up to −20 mm over approximately seven years (Figure 13(d)). These satellite evidences confirm a high level of conservation criticality for the entire architectural complex.

Indeed, the recent history of the church and on-site inspections provide a strong validation of the analyzed PSI

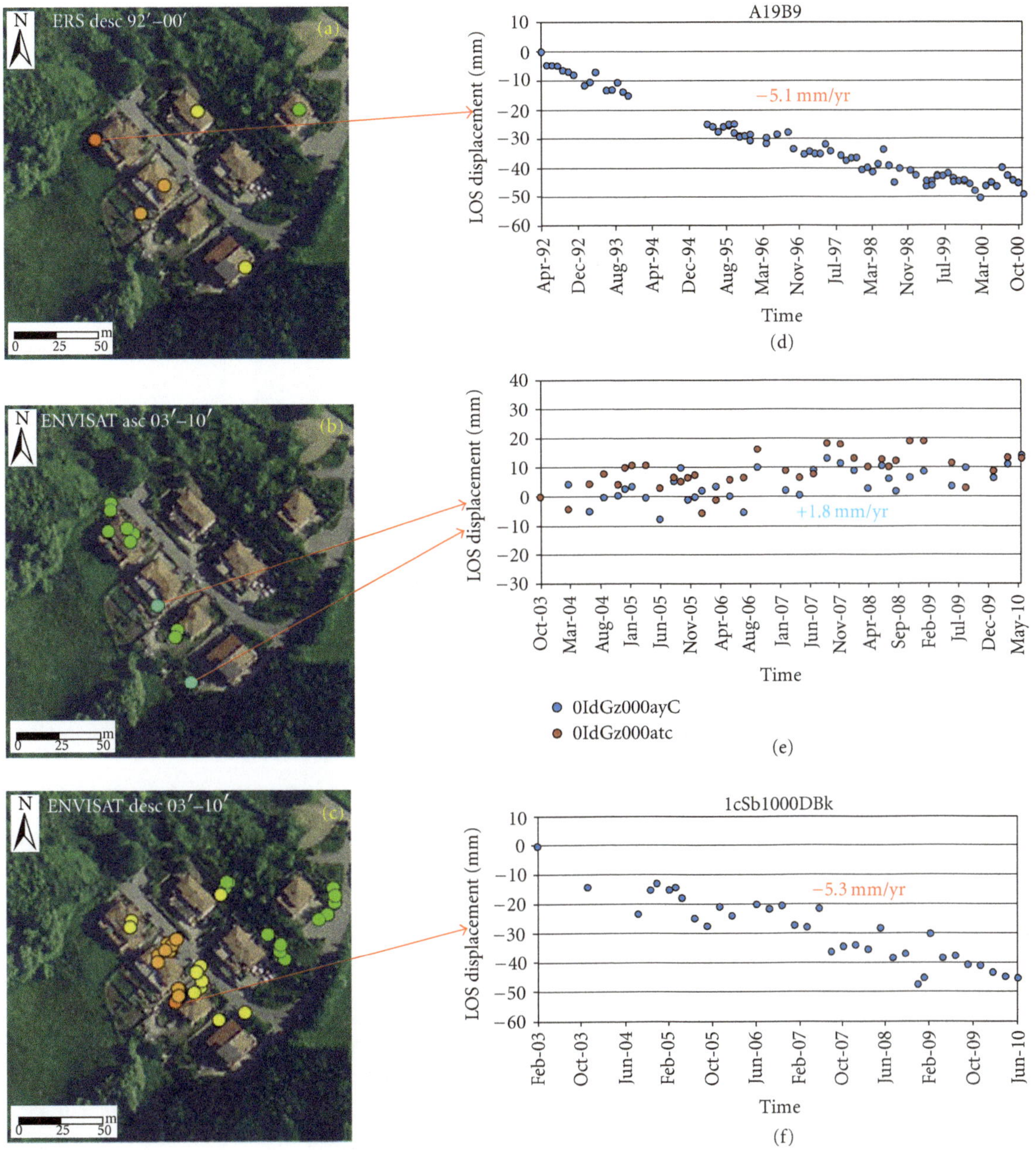

FIGURE 12: PS spatial distribution and related V_{LOS} field over Marroneto for the datasets: (a) ERS descending (1992–2000); (b) ENVISAT ascending (2003–2010); (c) ENVISAT descending (2003–2010), associated to examples of time series representative of the respective LOS displacement trends ((d)–(f)).

data. During the monitoring period (1992–2010), several interventions of stabilization and consolidation were carried out by the local authorities and heritage bodies. Particularly, substantial repairs were carried out between 1990s and 2000s on the containing wall, which supports and protects the substratum on which the church is founded.

Nevertheless, new cracks recently opened and currently cross over the pavement in front of the main entrance of the church. Also, new fissures progressively developed on the brick floor inside the church, starting from the main entrance and reaching the middle of the nave and the southern wall, as well as huge cracks which run on both the exterior and interior façades (Figure 13(f)).

Curiously, looking at the landslide inventory reported in the CGR, the boundary of an "active slide" ("*frana di scorrimento attiva*") runs close to the church but does not include it (Figure 10(c) and red dashed line in Figure 13(a) and 13(c)). Based on the PSI data and ground truth, it seems reasonable to propose a modification of the landslide boundaries, to be extended also to the area of the church. Execution of detailed ground investigations might be preparatory to the design of appropriate consolidation

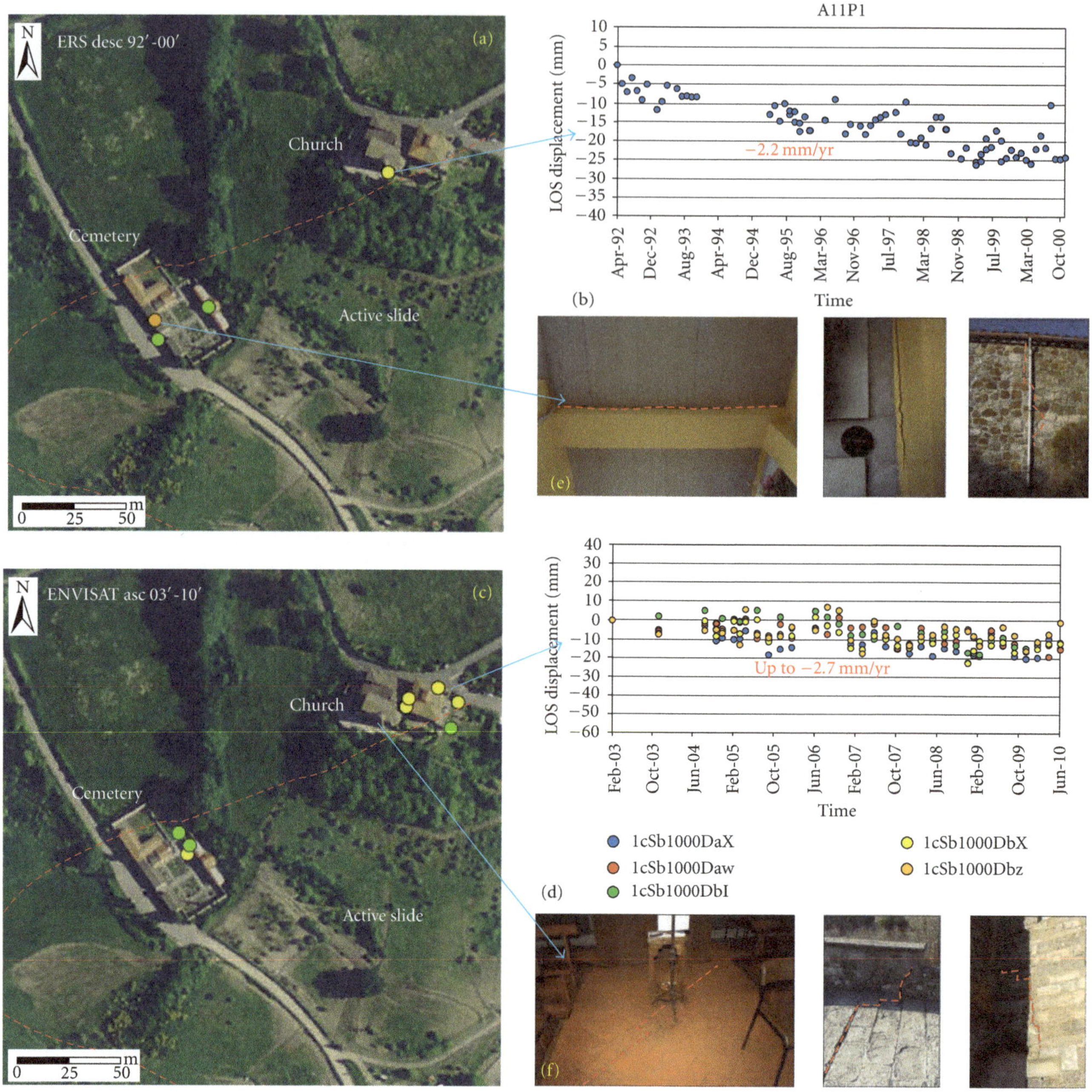

FIGURE 13: PS spatial distribution and related V_{LOS} field over the Church of San Romolo and the cemetery for the datasets: (a) ERS descending (1992–2000) with (b) a single downwards moving PS in correspondence with the bell tower; (c) ENVISAT descending (2003–2010), with several downwards moving PSs sharing a common LOS displacement trend (d). Severe crack patterns affect the exterior and interior wall surfaces and floor of both the cemetery (e) and the church of San Romolo (f). The red dashed line in (a) and (c) marks the boundary of an active slide mapped in the inventory of the CGR (cf. Figure 10).

works, at the level of both the unstable ground and the deteriorated structures.

The same active slide actually includes most of the modern cemetery, located south of the church, along the slope (Figure 10(c)). Again, the ERS descending (1992–2000) and ENVISAT descending (2003–2010) datasets suggest the presence of superficial deformations which might be the effects of landsliding phenomena and consequent structural instability of the masonries (Figure 13(a) and 13(c)). A single PS with a linear LOS displacement trend going away from the satellite shows a V_{LOS} of −3.5 mm/yr, and it is coupled with an apparently stable PS. The latter is characterized by a phase of downwards movements until May 1999, followed by an upwards phase till November 2000. Both the PSs are located over the southern wall of the cemetery (Figure 13(a)).

Differently from the ENVISAT ascending (2003–2010) data where a lack of PS affects most of the area of the cemetery, the corresponding ENVISAT descending data

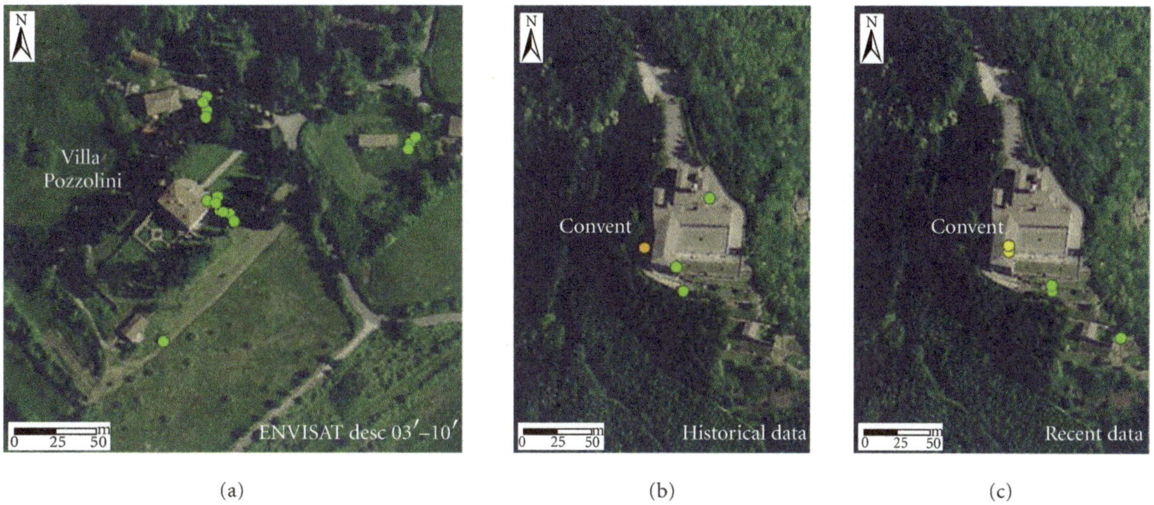

FIGURE 14: (a) Spatial distribution and related V_{LOS} field of the ENVISAT descending (2003–2010) dataset confirm a relative stability for Villa Pozzolini and surroundings, while the comparison between (b) historical data (ERS descending 1992–2000) and (c) recent ones (ENVISAT descending 2003–2010) over the Convent of Monte Senario highlights the persistence of localized deformation over the southern part of the complex.

highlight a slight tendency to downwards LOS displacements throughout the monitoring period within the time series of all the three identified PSs.

Although the reduced number of the identified PS and their spatial configuration lead to consider them only localized deformation, a stronger evidence to validate these satellite findings can be retrieved from on-site checks. The crack pattern survey over the concrete and stone-faced masonries of both the southern and northern walls highlights a good agreement between the spatial distribution of the fissures and the PS positions (Figure 13(e)). The cracks mainly developed in correspondence with the conjunction of masonries related to different phases of construction, that is, the weakest structural points on which active landslide events can have higher impacts.

Conversely, low levels of conservation criticalities are to be assigned to the monumental complexes of Villa Pozzolini and the Convent of Monte Senario (Figure 14). In the first case, all the three analyzed PS datasets show a quite relative stability over the Villa, as well as for the surrounding buildings and the homestead in front of the Villa which belong to the estate. This is particularly observable in the ENVISAT descending (2003–2010) dataset, which is the most completed one in terms of PS spatial distribution (Figure 14(a)). Nevertheless, this dataset clearly exemplifies the issue that sometimes an operator could find, that is, the error in the PS georeferencing and consequent shift of the PS with respect to the orthophoto used as visual reference. In this case, the shift is very easily appreciable by comparing the spatial configuration of the identified PS clusters and the geometry of the buildings to which the PSs are reasonably to be referred.

Similarly, no relevant deformation patterns are detected over the Convent of Monte Senario, except for one single PS ERS descending (1992–2000) with V_{LOS} of -3.4 mm/yr and an almost constant linear trend throughout the monitoring period, reaching a total LOS displacement of -40 mm over eight years (Figure 14(b)). The impression that this deformation is to be recognized as a localized deformation is confirmed by the persistence of downwards LOS displacements also within the time series of two PS ENVISAT descending (2003–2010), located over the southern part of the convent (Figure 14(c)). These PSs share a common LOS displacement trend progressively going away from the satellite and reach a total LOS displacement of about -15 mm over eight years, associated to a V_{LOS} of, respectively, -2.5 and -2.2 mm/yr.

4. Conclusions

The methodology of rapid mapping and deformation analysis proposed here has the potential to become a routine tool to be increasingly employed by the practitioners and heritage bodies, in support of the ordinary activities of noninvasive diagnosis and preservation of cultural heritage, both concentrated within enclosed areas (e.g., historic centres) and spread within urban and rural environments. The implementation of PSI-based analyses is also expected to be more and more complementary to the activities of built heritage management, in the framework of broader strategies of land planning.

The feasibility tests discussed in this paper confirm these potentials of the PSI techniques, especially for the purposes of preventive diagnosis, thanks to the capability of detecting superficial deformation as indicators of ongoing deterioration phenomena, or even evolving processes which might develop into severer instability events. Such diagnostic feature is further enriched by the possibility of back monitoring the displacement field of the AOI up to 1992 with the ERS-1/2 imagery, thereby reconstructing the deformational behaviour of the observed phenomena during a sufficiently long period of observation.

Nevertheless, some drawbacks can arise, such as the lack of PS over the AOI while passing from a dataset to another, especially if the analyzed datasets are acquired from different satellite platforms. The reduced number of identified PSs over the elements of major interest and even the total absence of deformation estimates, respectively, limit and inhibit the execution of a reliable and complete deformation analysis. Frequently, the deformational behaviour is actually assessed based on one of the two acquisition geometries (ascending or descending), and, consequently, the retrieved information can be only referred to the LOS component of the real displacement vector which represents the deformation affecting the radar targets. This issue is only partially solvable by refining the processing setting and better focusing the scale of analysis, but not always significant improvements are achieved. Certainly, the installation of corner reflectors over the AOI and in correspondence with the buildings to monitor is a feasible solution but requires additional costs, is not retroactive before the date of installation, and implies issues in terms of landscape impacts on the monuments.

Although the PSI data radar-interpreted here were not formerly processed for cultural heritage applications, both the feasibility tests of rapid mapping provide interesting results, showing a good agreement between the satellite evidences and the ground truth, as well as the information retrieved from previous research (case of Pitigliano) and existing mapping products (case of Bivigliano). With reference to this second case, the PSI data also demonstrated their potential to provide informational elements to update the background knowledge, suggesting, for instance, a review of the landslide inventory.

In terms of selective analysis of differential conditions, the PSI data over Pitigliano led to zone the most unstable sectors of the built-up areas along the cliff edges, while those covering the territory of Bivigliano allowed localized criticalities to be identified over some historical buildings rather than others, within the broader context of deformation patterns detected at the scale of entire site.

In light of the good results retrieved on ERS-1/2 and ENVISAT data stacks at medium spatial resolution and monthly sampling, further advances in such PSI-based feasibility tests are represented by experiments on PS obtained from the PSI processing of SAR imagery characterized by higher spatial and temporal resolutions.

Undoubtedly, the collection of ground truth and the integration with ancillary data and background information still remain essential elements, without which the radar interpretation of the PSI data might be affected by misinterpretation risks, especially in relation to the distinction between related and unrelated PSs (the latter correspond to PS actually not attributable to radar targets belonging to the element of interest). In this regard, the high level of agreement between the satellite evidences and other source data found during the two feasibility tests encourages the implementation of the experimented PSI techniques, also in the perspective of addressing the execution of terrestrial survey activities and the planning of targeted mitigation works, with consequent benefits in terms of cost effectiveness and sustainability.

Acknowledgments

PSI data were processed in the framework of the Extraordinary Plan of Environmental Remote Sensing (EPRS-E) project and made available through the WMS service of the Italian Ministry of Environment, Territory and Sea. Access to the Regional Geological Map (*Carta Geologica Regionale* (CGR)) Sheet 263120 scale 1 : 10,000 was performed through the Geoportal of Lamma Consorzio (http://geoportale.lamma.rete.toscana.it/).

References

[1] D. Massonnet and K. L. Feigl, "Radar interferometry and its application to changes in the earth's surface," *Reviews of Geophysics*, vol. 36, no. 4, pp. 441–500, 1998.

[2] P. A. Rosen, S. Hensley, I. R. Joughin et al., "Synthetic aperture radar Interferometry," *Proceedings of the IEEE*, vol. 88, no. 3, pp. 333–382, 2000.

[3] D. Massonnet, K. Feigl, M. Rossi, and F. Adragna, "Radar interferometric mapping of deformation in the year after the Landers earthquake," *Nature*, vol. 369, no. 6477, pp. 227–230, 1994.

[4] C. Squarzoni, C. Delacourt, and P. Allemand, "Nine years of spatial and temporal evolution of the La Valette landslide observed by SAR interferometry," *Engineering Geology*, vol. 68, no. 1-2, pp. 53–66, 2003.

[5] A. M. Goudarzi, T. Woldai, and V. A. Tolpekin, "Surface deformation caused by April 6th 2009 earthquake in L'Aquila (Italy): a comparative analysis from ENVISAT ASAR, ALOS PALSAR and ASTER," *International Journal of Applied Earth Observation and Geoinformation*, vol. 13, no. 5, pp. 801–811, 2011.

[6] S. Bianchini, F. Cigna, G. Righini, C. Proietti, and N. Casagli, "Landslide HotSpot mapping by means of persistent scatterer interferometry," *Environmental Earth Sciences*. In press.

[7] H. Rott and T. Nagler, "The contribution of radar interferometry to the assessment of landslide hazards," *Advances in Space Research*, vol. 37, no. 4, pp. 710–719, 2006.

[8] F. Gutiérrez, J. P. Galve, P. Lucha, C. Castañeda, J. Bonachea, and J. Guerrero, "Integrating geomorphological mapping, trenching, InSAR and GPR for the identification and characterization of sinkholes: a review and application in the mantled evaporite karst of the Ebro Valley (NE Spain)," *Geomorphology*, vol. 134, no. 1-2, pp. 144–156, 2011.

[9] S. Stramondo, F. Bozzano, F. Marra et al., "Subsidence induced by urbanisation in the city of Rome detected by advanced InSAR technique and geotechnical investigations," *Remote Sensing of Environment*, vol. 112, no. 6, pp. 3160–3172, 2008.

[10] F. Cigna, B. Osmanoğlu, E. Cabral-Cano et al., "Monitoring land subsidence and its induced geological hazard with synthetic aperture radar interferometry: a case study in Morelia, Mexico," *Remote Sensing of Environment*, vol. 117, pp. 146–161, 2012.

[11] M. Crosetto, O. Monserrat, and G. Herrera, "Urban applications of persistent scatterer interferometry," *Radar Remote Sensing of Urban Areas, Remote Sensing and Digital Image Processing*, vol. 15, pp. 233–248, 2010.

[12] F. Cigna, C. Del Ventisette, V. Liguori, and N. Casagli, "Advanced radar-interpretation of InSAR time series for mapping and characterization of geological processes," *Natural Hazards and Earth System Science*, vol. 11, no. 3, pp. 865–881, 2011.

[13] G. Gigli, W. Frodella, F. Mugnai et al., "Instability mechanisms affecting cultural heritage sites in the Maltese Archipelago," *Natural Hazards and Earth System Sciences*, vol. 12, no. 6, pp. 1883–1903, 2012.

[14] P. Teatini, L. Tosi, T. Strozzi et al., "Resolving land subsidence within the Venice Lagoon by persistent scatterer SAR interferometry," *Physics and Chemistry of the Earth, Parts A/B/C*, vol. 40-41, pp. 72–79, 2012.

[15] M. Costantini, A. Iodice, L. Magnapane, and L. Pietranera, "Monitoring terrain movements by means of sparse SAR differential interferometric measurements," in *Proceedings of the International Geoscience and Remote Sensing Symposium (IGARSS'00)*, pp. 3225–3227, Honolulu, Hawaii, USA, July 2000.

[16] A. Ferretti, C. Prati, and F. Rocca, "Permanent scatterers in SAR interferometry," *IEEE Transactions on Geoscience and Remote Sensing*, vol. 39, no. 1, pp. 8–20, 2001.

[17] A. Ferretti, A. Fumagalli, F. Novali, C. Prati, F. Rocca, and A. Rucci, "A new algorithm for processing interferometric data-stacks: SqueeSAR," *IEEE Transactions on Geoscience and Remote Sensing*, vol. 49, no. 9, pp. 3460–3470, 2011.

[18] C. Werner, U. Wegmüller, T. Strozzi, and A. Wiesmann, "Interferometric point target analysis for deformation mapping," in *Proceedings of the IGARSS: Learning From Earth's Shapes and Colours*, pp. 4362–4364, Toulouse, France, July 2003.

[19] A. Hooper, H. Zebker, P. Segall, and B. Kampes, "A new method for measuring deformation on volcanoes and other natural terrains using InSAR persistent scatterers," *Geophysical Research Letters*, vol. 31, no. 23, Article ID L23611, 5 pages, 2004.

[20] D. Tapete, R. Fanti, R. Cecchi, P. Petrangeli, and N. Casagli, "Satellite radar interferometry for monitoring and early-stage warning of structural instability in archaeological sites," *Journal of Geophysics and Engineering*, vol. 9, pp. S10–S25, 2012.

[21] The Terrafirma Atlas, "Terrain-motion across Europe. A compendium of results produced by the European space agency GMES service element project Terrafirma 2003–2009," 94 pages, June 2009, http://esamultimedia.esa.int/multimedia/publications/TerrafirmaAtlas/pageflip.html.

[22] L. Bateson, M. Cuevas, M. Crosetto, F. Cigna, and M. Schijf, D3. 5: PanGeo Production Manual Version 1, 2012, http://www.pangeoproject.eu/sites/default/files/pangeo_other/D3.5_PanGeo_Production_Manual_v1_of_2nd_April_2012.pdf.

[23] National Geoportal, "Extraordinary plan of environmental remote sensing," 2012, http://www.pcn.minambiente.it/GN//progetto_pst.php?lan=en.

[24] A. Trigila, C. Cacace, C. Iadanza, S. Del Conte, D. Spizzichino, and S. Cespa, "Cultural heritage, landslide risk and remote sensing in Italy," in *The Second World Landslide Forum—Abstract Book*, F. Catani, C. Margottini, A. Trigila, and C. Iadanza, Eds., p. 603, FAO, Rome, Italy, 2011, ISPRA, WLF2-2011-0397.

[25] G. Savio, A. Ferretti, F. Novali, S. Musazzi, C. Prati, and F. Rocca, "PSInSAR validation by means of a blind experiment using dihedral reflectors," in *Proceedings of FRINGE 2005 Workshop*, pp. 1–6, Frascati, Italy, December 2005.

[26] D. Perissin and A. Ferretti, "Urban-target recognition by means of repeated spaceborne SAR images," *IEEE Transactions on Geoscience and Remote Sensing*, vol. 45, no. 12, pp. 4043–4058, 2007.

[27] G. Zeni, M. Bonano, F. Casu et al., "Long-term deformation analysis of historical buildings through the advanced SBAS-DInSAR technique: the case study of the city of Rome, Italy," *Journal of Geophysics and Engineering*, vol. 8, no. 3, pp. S1–S12, 2011.

[28] D. Tapete and F. Cigna, "Site-specific analysis of deformation patterns on archaeological heritage by satellite radar interferometry," in *20th International Materials Research Congress, Symposium 8 Cultural Heritage and Archaeological Issues in Materials Science*, vol. 1374 of *MRS Proceedings*, Cambridge University Press, 2012.

[29] F. Cigna, S. Bianchini, and N. Casagli, "How to assess landslide activity and intensity with persistent scatterer interferometry (PSI): the PSI-based matrix approach," *Landslides*. In press.

[30] C. Colesanti and J. Wasowski, "Investigating landslides with space-borne synthetic aperture radar (SAR) interferometry," *Engineering Geology*, vol. 88, no. 3-4, pp. 173–199, 2006.

[31] National Geoportal, "Progetto persistent scatterers interferometry," 2012, http://www.pcn.minambiente.it/GN/progetto_psi.php?lan=en.

[32] L. Vezzoli, "Stratigraphy of the latera volcanic complex: proposals for a new nomenclature," *Periodico di Mineralogia*, vol. 56, no. 2-3, pp. 89–110, 1987.

[33] S. Conticelli, L. Francalanci, P. Manetti, and A. Peccerillo, "Evolution of Latera Volcano, Vulsinian district (central Italy): stratigraphical and petrological data," *Periodico di Mineralogia*, vol. 56, no. 2-3, pp. 175–199, 1987.

[34] P. Canuti, N. Casagli, R. Fanti, A. Iotti, E. Pecchioni, and A. P. Santo, "Rock weathering and failure of the "Tomba della Sirena" in the Etruscan necropolis of Sovana (Italy)," *Journal of Cultural Heritage*, vol. 5, no. 3, pp. 323–330, 2004.

[35] R. Fanti, G. Gigli, L. Lombardi, D. Tapete, and P. Canuti, "Terrestrial laser scanning for rockfall stability analysis in the cultural heritage site of Pitigliano (Italy)," *Landslides*. In press.

[36] P. Canuti and R. Fanti, "The instability of Pitigliano's cliff: the role of the anthropic cavities, the geotechnical studies and the monitoring system," in *Conservation and Sustainable Development of the Tuff Towns. An Exploration of Pitigliano*, World Monuments Fund Publication, Civita di Bagnoregio, Italy, 2007, http://www.wmf.org/sites/default/files/wmf_publication/tuff_canuti-en.pdf.

[37] Archivio Galilei 18 ins. 1, Memorie per Pratolino Frana di Bivigliano dell'agosto 1729.

10

Regional Climate Model Sensitivity to Domain Size for the Simulation of the West African Summer Monsoon Rainfall

Nana A. K. Browne[1] and Mouhamadou B. Sylla[2]

[1] *Department of Physics, University of Cape Coast, Cape Coast, Ghana*
[2] *Earth System Physics (ESP) Section, International Centre for Theoretical Physics (ICTP), 34151 Trieste, Italy*

Correspondence should be addressed to Nana A. K. Browne, amabrowne@gmail.com

Academic Editor: Alessandra Giannini

We use the International Centre for Theoretical Physics (ICTP) Regional Climate Model (RegCM3) to study the impact of different domain sizes on the simulation of the West African summer monsoon rainfall and circulation features. RegCM3 simulates drier conditions over the default domain (RegCM-D1) and its westward extension (RegCM-D2), much less dryness over the eastward extended domain (RegCM-D3) and excessive wetness in the domain extended northward into the extratropical regions (RegCM-D4). This overestimation is related to the existence of larger source of humidity due to the inclusion of a more significant portion of the Atlantic Ocean and to a weakening of the African Easterly Jet (AEJ), which both favor stronger westerlies advecting moisture towards the land. The best performance is, however, captured in the RegCM-D3 experiment, and this originates from a simulation of moderate westerly moisture fluxes along with a stronger AEJ and occurrences of more frequent African Easterly Waves (AEWs). Therefore, the choice of the domain for regional climate model simulation of the West African summer monsoon rainfall is of critical importance, and caution needs to be taken to account for the main regional forcings including mostly the necessary humidity sources of the tropical Atlantic Ocean and the AEWs genesis region upstream of Sudanese Highlands.

1. Introduction

The West African Monsoon (WAM) system consists of many atmospheric features including the low-level monsoon flow, the midtropospheric African Easterly Jet (AEJ), and the synoptic disturbances along the jet axis so-called African Easterly Waves (AEWs) and the upper level Tropical Easterly Jet (TEJ). The low-level monsoon flow originates from meridional surface pressure gradient established between the land and the ocean during the summer season [1]. The AEJ appears over West Africa during the boreal summer as a result of the strong meridional surface moisture and temperature gradients between the Sahara and equatorial Africa tied to the deep convective heating [2–4]. The AEWs are generated upstream of Sudanese highlands and propagate across West Africa around the midtropospheric AEJ through combined baroclinic and barotropic conversion [4, 5]. The TEJ, which is associated with the upper-level outflow from the Asian monsoon [6], also circulates across West Africa during the boreal summer season. These features interact in a complex way and are responsible for the summer monsoon precipitation over the region. For example, the low-level monsoon flow is the major component for the moisture transport into the West Africa continent from the Atlantic Ocean; the position and strength of the AEJ cause rainfall variability by transporting moisture away from the continent [7] and by interacting with mesoscale convective systems embedded in the AEWs [8] while the strength of the TEJ acts mainly in the lifetime of these systems [9] which are responsible for most of rain over West Africa [10].

Although it is complex, the WAM system is of major economic and social importance to the population of the region whose economy heavily relies on rain-fed agriculture. Therefore, accurate simulation of its features will help the understanding of its dynamics and variability to improve our skill in predicting its onset and evolution. Many studies have used regional climate models to understand the WAM system. For instance, Vizy and Cook [11] studied the response of the WAM to tropical oceanic heating. Jenkins et al. [9] and Sylla et al. [12] investigated the main drivers of the present day and future Sahel droughts. Abiodun et al. [7] documented the impact of deforestation and desertification over West Africa while Moufouma-Okia and Rowell [13] examined the effect of soil moisture initialization on the WAM simulation.

All the studies mentioned above show a substantial progress in understanding the WAM dynamical and physical features using regional climate models. The potential of the RegCM3 as a tool to study WAM circulation have been highlighted in many studies including (but not limited to) Afiesimama et al. [14], Sylla et al. [15], and Sylla et al. [16]. The model has been shown to be capable of reproducing the mean climatology and the variability of the WAM climate. However, since the system consists of atmospheric features that are generated in different locations and sensitive to different topography and surface conditions of the region, selecting the domain size for the regional climate simulation is crucial to be able to get the important sources of forcing features from the boundaries.

Regional climate models are driven by time-dependent large-scale meteorological fields specified at the boundaries of the chosen domain. Anthes et al. [17] and Giorgi and Mearns [18] showed that the choice of domain size and location affects the balance between the boundary and internal model forcings in the simulation. The location of boundaries in relation to the regional sources of forcings in a particular climatic region can also affect the regional climate model solution [19]. As a matter of fact, Seth and Giorgi [20] showed that the lateral boundaries must be placed well outside the region of interest to avoid unrealistic response to internal forcings while Jones et al. [21] showed that the regional domain must be large enough to allow the full development of small-scale features over the area of interest.

These indicate that the domain size can have significant effects on the simulation of regional climate models and therefore a careful choice of the domain is needed for particular studies. In this paper, we investigate how the sizes/extensions of regional model domain affect the WAM circulation and consequently the simulation of rainfall during the peak (JJA) of the rainy season. The model and experiments are briefly described in Section 2 and the results are discussed in Section 3 while final considerations are given in Section 4.

2. Model Description and Experiments

The ICTP Regional Climate Model, RegCM3 [22–24] is used in this study. RegCM3 is a primitive equation, sigma vertical coordinate model based on the hydrostatic dynamical core of the NCAR/PSU's mesoscale meteorological model MM5 [25]. Radiation is represented by the CCM3 parameterization of Kiehl et al. [26] and the planetary boundary layer scheme is by Holtslag et al. [27]. Interactions between the land surface and the atmosphere are described using the Biosphere Atmosphere Transfer Scheme (BATS1E; [28]). The scheme of Zeng et al. [29] is used to represent fluxes from water surfaces. Convective precipitation is calculated with the scheme of Grell et al. [25] applying the Fritsch and Chappell [30] closure assumption. Resolvable precipitation processes are treated with the subgrid explicit moisture scheme (SUBEX) of Pal et al. [31], which is a physically based parameterization including subgrid scale cloud fraction, cloud water accretion, and evaporation of falling raindrops. This model configuration is the same used by Sylla et al. [16] and Sylla et al. [4]. This choice was based on an in-depth analysis of the model performance in multidecadal simulations of present day climate.

Four sets of experiments, covering each 5 years (2001–2005) as in Sylla et al. [4], have been conducted over different domain sizes (see Figure 1(a)) using 50 km of horizontal resolution and 18 verticals level with a top at 5 mb. The first (RegCM-D1) domain covers mostly West Africa and a little part of the Atlantic Ocean (D1: 25E–25W; 10S–25N). The second experiment (RegCM-D2) extends this domain further west with much larger Ocean area up to 45°W (D2: 45E–25W; 10S–25N). The third simulation (RegCM-D3) is carried out in a domain that covers D2 but extended eastward up to East Africa (35°E) including some of the highlands where AEWs are generated (D3: 45E–35W; 10S–25N). The last simulation (RegCM-D4) extends D3 northward up to 45°N to allow RegCM3 to simulate its own circulation around the Azores high (D4: 45E–35W; 10S–45N). Each of these domain comprises a buffer zone is just 12 grid-points (~5°). The topography exhibits some localized highlands (e.g., Figure 1(a)): Guinea Highlands (GH), Jos Plateau (JP), and Cameroun Mountains (CM) while the eastward extended domain comprises also part of the East African complex terrains for instance the Sudanese Highlands (SH). The topography is derived from the United States Geological Survey's (USGS) GTOPO 30 seconds (~1 km of resolution) global elevation data which is interpolated onto the model grid (50 km).

All the experiments are conducted using initial and lateral boundary conditions created from the ERA-Interim 1.5°× 1.5° gridded reanalysis [32], which is the third generation of European Centre of Medium-Range Weather Forecast (ECMWF) reanalysis product. Sea surface temperatures (SSTs) for all experiments is obtained from the National Oceanic and Atmospheric Administration (NOAA) Optimum Interpolation (OI) SST. The OISST used here is produced weekly on a one-degree grid [33]. The boundary conditions and SST are updated four times daily in RegCM3 using the relaxation procedure described by Giorgi et al. [23].

These sets of experiments will help addressing the effect of the western, eastern, and northern boundary on the simulation of the different West African summer monsoon features and rainfall. As a matter of fact, the analysis is carried out by considering the simulated mean rainfall but also the frequency and intensity of daily rainfall events. We should mention that only the core of the West African rainfall season June-August (JJA) is considered in the study because of its large contribution on the region's annual mean precipitation. We evaluate the effect of (1) the westward extension of the domain by comparing RegCM-D2 and RegCM-D1, (2) the eastward extension by comparing RegCM-D3 and

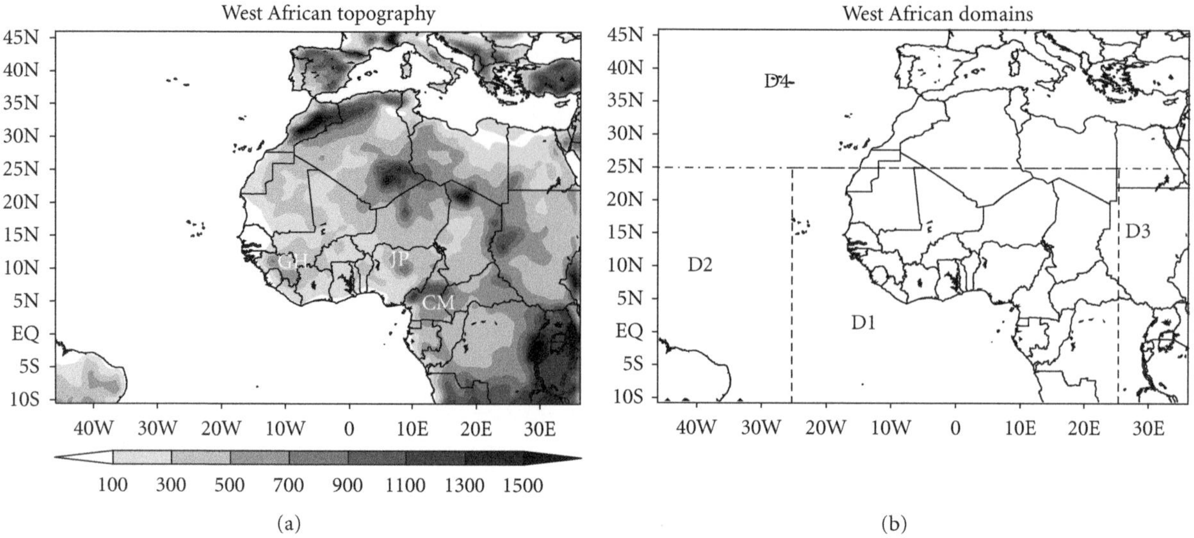

FIGURE 1: (a) The West African domain sizes and (b) topography (in meters). The analyzed domain for quantitative assessment extends from 20°W to 20°E and 0 to 20°N in red. The highlands are highlighted by their respective acronyms: Guinea Highlands (GH), Jos Plateau (JP), Cameroun Mountains (CM), and Sudanese Highlands (SH).

RegCM-D2, and (3) the northward extension by comparing RegCM-D4 and RegCM-D3.

The performance of each experiment is assessed by comparing with the high-resolution (0.25°× 0.25°) product of Tropical Rainfall Measuring Mission (TRMM; [34, 35]). TRMM is satellite-derived daily and monthly rainfall providing data for the entire tropics since November 1997. The GPCP (Global Precipitation Climatology Project, 2.5°× 2.5° resolution; [36], product is also used as additional reference for the baseline validation to account for uncertainties in rainfall observations [37]. Although both of them are mainly based on satellite observations, they have also been compared and merged with ground station rain gauges to create the final products.

For quantitative assessment, we evaluate the spatial patterns of the experiments' bias but also some quantitative measures being mean bias (MB), root mean square difference (RMSD), and pattern correlation coefficient (PCC) with respect to the TRMM rainfall. Although we plot the common region across the simulations, the statistic measures are computed over the analyzed domain (Figure 1(b)), thus excluding the buffer zone of D1. The MB, RMSD, and PCC can be considered as measures of model systematic errors and performance as they provide information at both the regional and the grid point levels. The origins of the different performance of the regional climate model to simulate the seasonal rainfall will therefore be examined through the WAM features for instance the monsoon flow, the subsequent vertically integrated moisture flux, the AEJ, the TEJ, the upward motion between the jets axis, and the AEWs.

3. Results

3.1. The Simulated Rainfall Patterns. The effect of various domain sizes (e.g., Figure 1(b)) is investigated for rainfall and the relevant monsoon circulation features. For instance, we intercompare D1 and D2, D2 and D3, and D3 and D4 as mentioned above. Figures 2(a)–2(f) shows the summer (JJA) mean rainfall from GPCP and TRMM observations along with the RegCM3 simulations over the different domains. Observations (both GPCP and TRMM) display the ITCZ in a zonal and tilted band between 8°N and 13°N, with rainfall decreasing south and north of it. Precipitation maxima are, however, located in orographic regions such as the Guinea highlands, Jos plateau, and Cameroun Mountains (e.g., Figure 1(a)). TRMM shows some fine-scale localized intense rainfall amount, not captured in GPCP, which are related to local topography and are mostly artifact of the higher resolution.

RegCM3 reproduces quite well this spatial distribution, in particular the ITCZ position, the northern and southern gradients of rainfall and the location of maxima over orographic regions in all domains with PCCs of around 0.80. Note that some cores of large rainfall amount are present at the margins of the eastern boundary of both D1 and D2. We should mention that these are just due to the presence of boundary effects in regional climate model simulation [20] because of the proximity of this area to the eastern boundary. Therefore, they are not considered here.

The different domains exhibit different bias patterns as displayed in Figures 3(a)–3(d). In addition, Table 1 summarizes the quantitative measures MB, RMSD, and PCC calculated for each of the simulations with respect to TRMM. RegCM-D1 and RegCM-D2 are drier than TRMM over most of land and oceanic areas resulting, respectively, in an MB of −7.5% and −4.5% and an RMSD of 1.97 mm/day and 1.88 mm/day. This dryness is reduced in both magnitude and spatial extent in RegCM-D3 experiment where the model also simulates some few wet bias along the Gulf of Guinea and the ITCZ regions. The larger and more extended wet

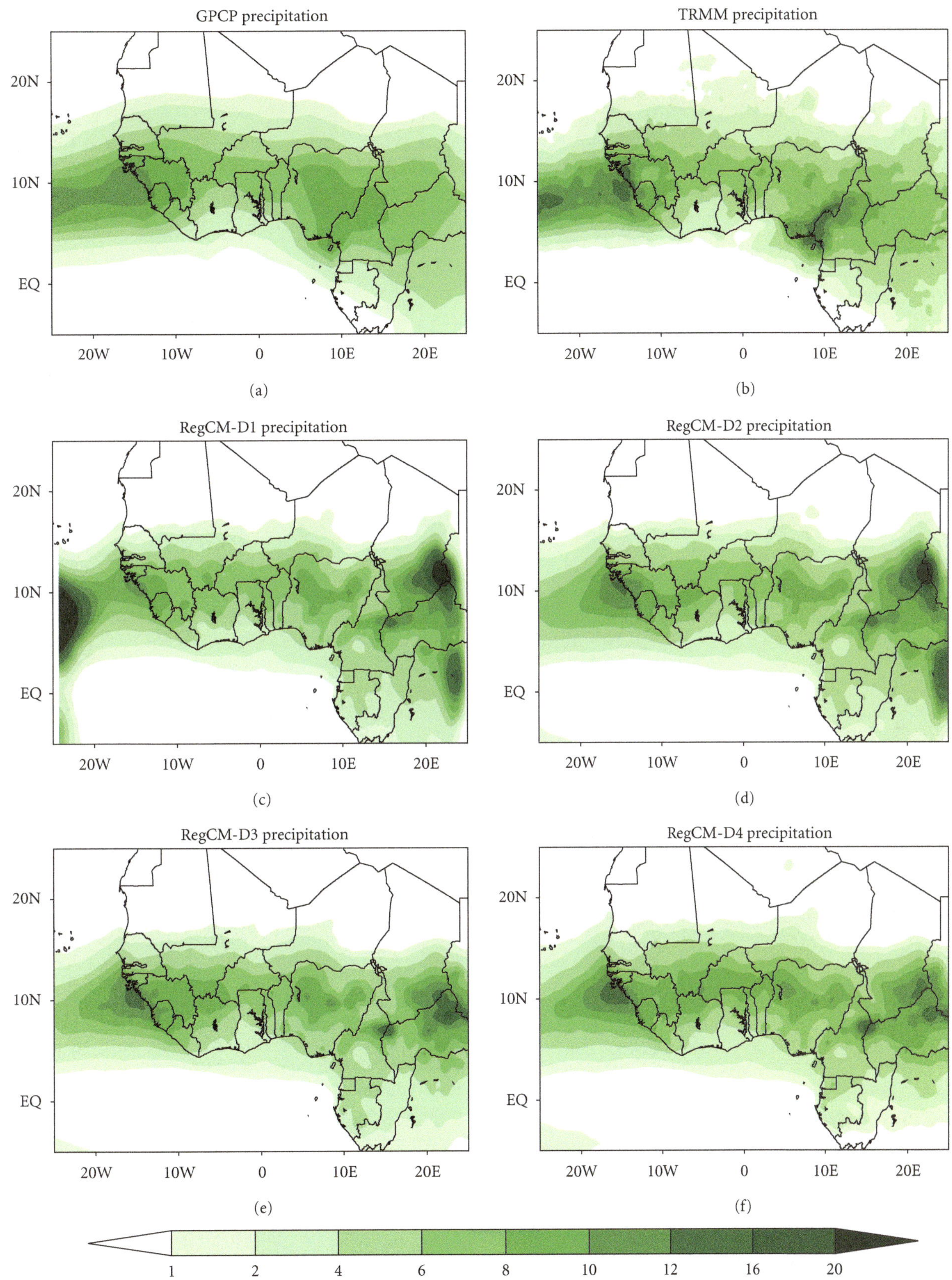

FIGURE 2: Mean JJA rainfall averaged for the period 2001–2005 from (a) GPCP, (b) TRMM, (c) RegCM-D1, (d) RegCM-D2, (e) RegCM-D3, and (f) RegCM-D4. Units are expressed in mm/day.

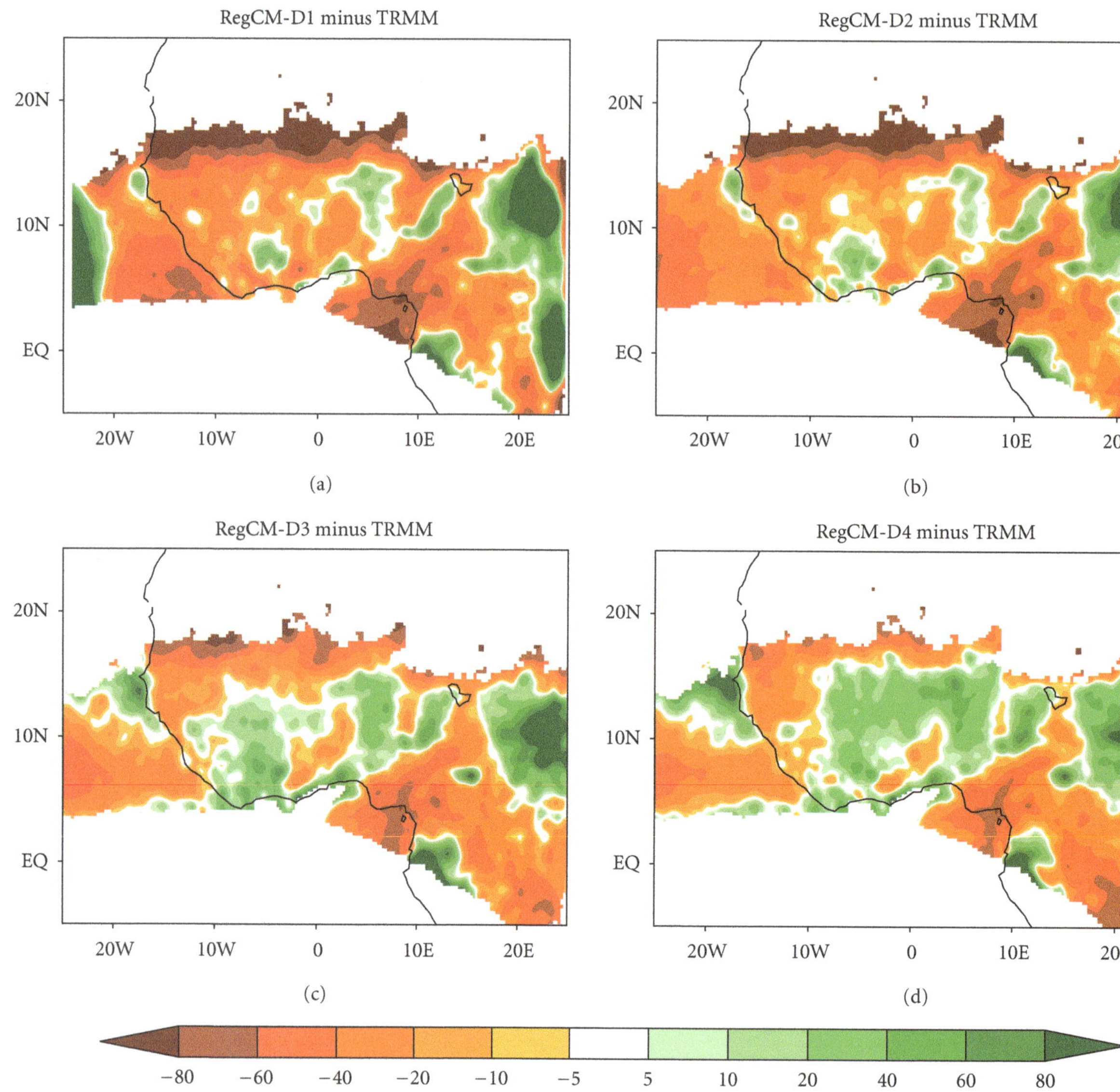

FIGURE 3: Differences in mean JJA rainfall between (a) RegCM-D1 and TRMM, (b) RegCM-D2 and TRMM, (c) RegCM-D3 and TRMM and (d) RegCM-D4, and TRMM. Units are expressed as percentage of TRMM values.

biases are, however, found in RegCM-D4. This is consistent with the lower MB (−2.95%) and RMSD (1.80 mm/day) simulated in RegCM-D3 compared to those of RegCM-D4 (4.07% and 1.93 mm/day, resp.). It is worth noting that RegCM-D3 exhibits also the larger PCC.

It is thus evident that substantial discrepancies exist across the different simulations. This is highlighted in Figures 4(a)–4(c). Differences between RegCM-D2 and RegCM-D1 (e.g., Figure 4(a)), which represents the extension of the western boundary effect, show mostly a noisy pattern in the land where rainfall increases and decreases for about 20% and −20%. The most significant change (up to 40%) is shown at the Guinea coast between 15°W and 5°W and off the west coast. These changes have lead to a larger maximum over the orographic regions and the ocean ITCZ in RegCM-D2 compared to RegCM-D1. The effect of the extension of the eastern boundary, illustrated by the difference between RegCM-D3 and RegCM-D2 (Figure 4(b)) impacts mostly the zonal rainfall band around 12°N with an increase of the intensity of about 20–30%, contributing

TABLE 1: Mean Bias (MB), Root Mean Square Difference (RMSD) and Pattern Correlation Coefficient (PCC) for each of the simulation with respect to TRMM over the analyzed domain (e.g., Figure 1(b)). MB is expressed as percentage of TRMM values while RMSD as mm/day.

	Domain 1	Domain 2	Domain 3	Domain 4
MB	−7.53	−5.42	−2.95	4.07
RMSD	1.97	1.88	1.80	1.93
PCC	0.79	0.80	0.82	0.80

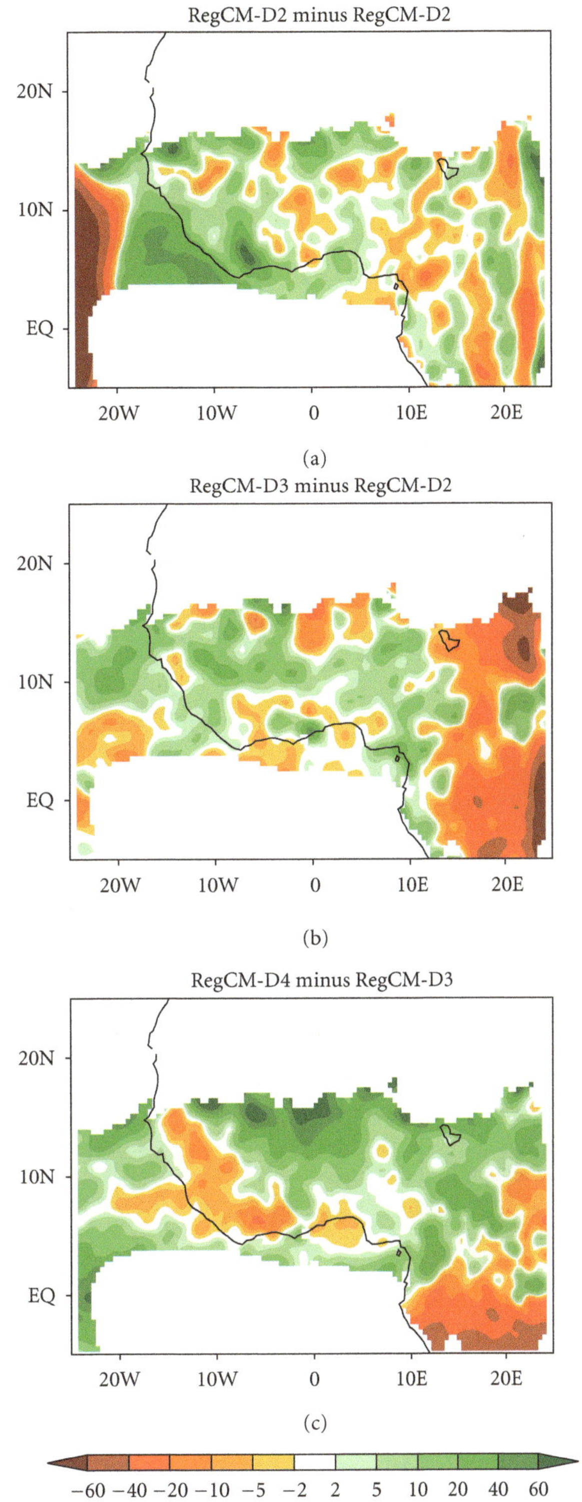

FIGURE 4: Differences among the simulations of mean JJA rainfall between (a) RegCM-D2 and RegCM-D1, (b) RegCM-D3 and RegCM-D2, and (c) RegCM-D4 and RegCM-D3. Units are expressed as percentage of, respectively, RegCM-D1, RegCM-D2, and RegCM-D3 values.

to the inland extension and stronger but more realistic orographic maximum over the Guinea Highlands and ITCZ. The most significant changes are obtained when the northern boundary of D3 is extended further north up to 45°N. In fact, RegCM-D4 shows a sharper and stronger ITCZ, causing an overestimation of rainfall in these convective regions and also over the Sahel of about more than 30–50%. Note that some slight decrease is shown in the western coastal areas.

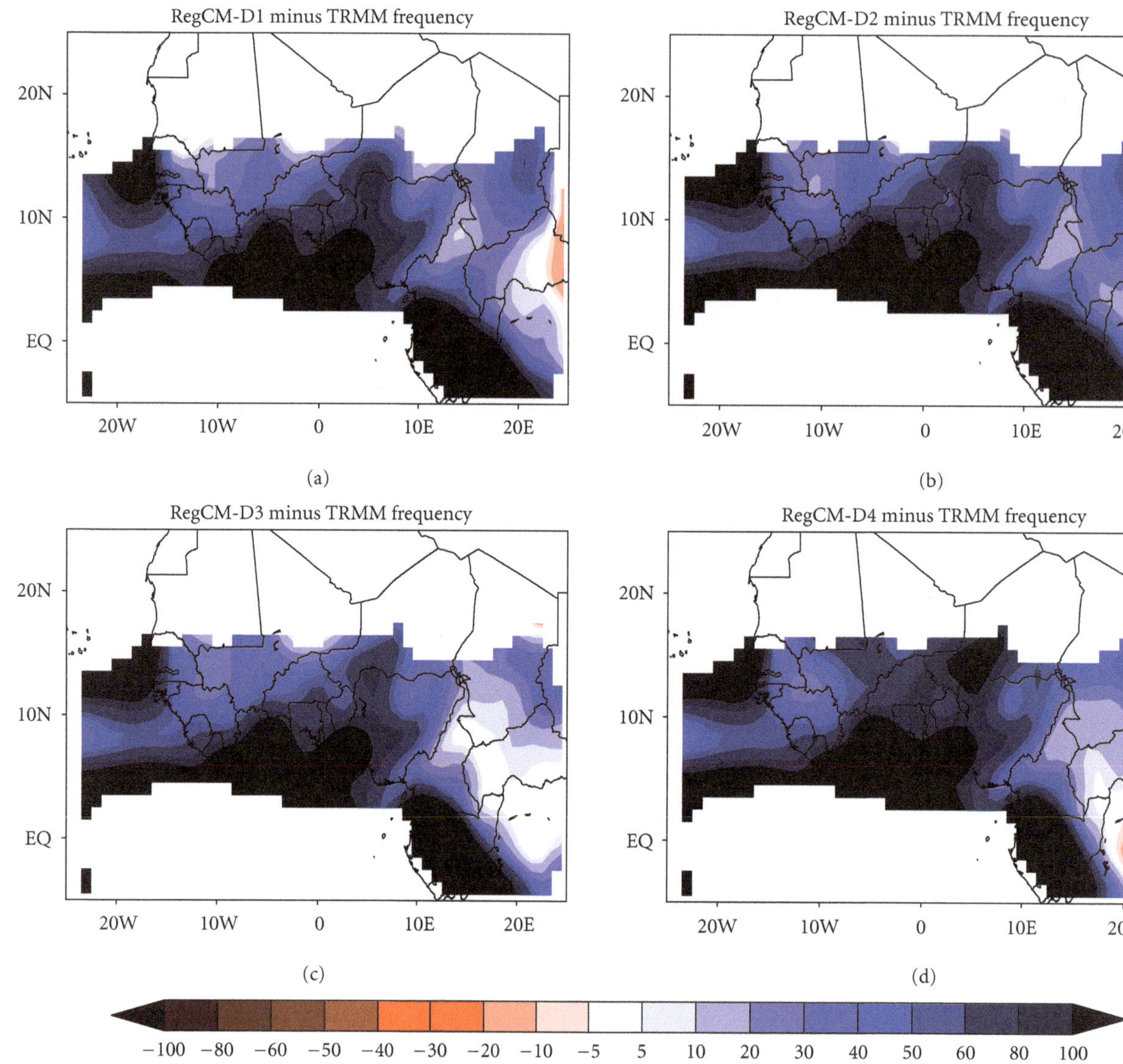

Figure 5: Differences in mean number of rainfall events between (a) RegCM-D1 and TRMM, (b) RegCM-D2 and TRMM, (c) RegCM-D3 and TRMM, and (d) RegCM-D4 and TRMM. Units are expressed as percentage of TRMM values.

Overall, based on the spatial distribution, the mean and the pattern of bias, the RMSD and the PCC analysis, we conclude that the RegCM-D3 is performing better than the other model configurations in simulating the West African summer monsoon rainfall in both magnitude and spatial distribution.

Although changes in the boundary locations trigger internal variability in the regional climate model, that latter does affect only the day-to-day model solution and not the mean climatology [38–40]. This suggests rationales to further analyze how the distribution of daily rainfall, specifically the frequency and intensity of wet days, is affected by the choice of the domain size. The difference in the frequency of wet days between each of the model configurations and the TRMM observation are displayed in Figures 5(a)–5(d). All the RegCM3 simulations consistently and substantially overestimate the number of wet days over the entire analyzed domain, mostly along the Gulf of Guinea, off the Senegal coast, and over central West Africa. In the first three domains (D1, D2, and D3), the difference maps exhibit similar characteristics among the simulations and this indicates that the smaller domain and its westward and eastward extension do not significantly impact the occurrence of wet days. The number of these events is only affected by the northern boundary location (RegCM-D4) which yields to the largest overestimation in the central regions of West Africa. The corresponding differences between the simulations and TRMM observation for the intensity of the rainfall events are presented in Figures 6(a)–6(d). As opposite to the frequency spatial distribution, the regional model generally simulates less intensity of rainfall events with respect to TRMM. Key difference among

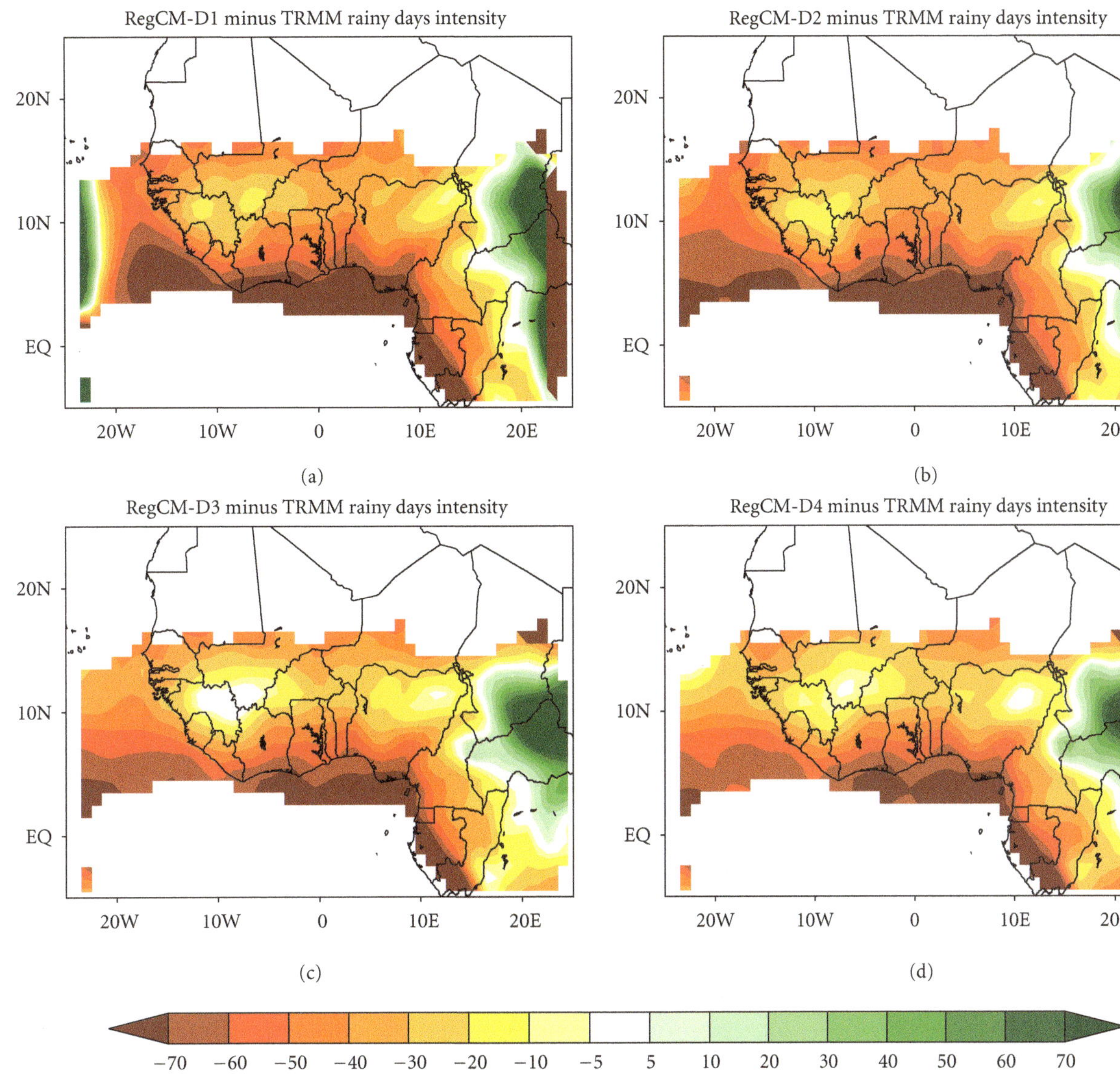

FIGURE 6: Differences in mean intensity of rainfall events between (a) RegCM-D1 and TRMM, (b) RegCM-D2 and TRMM, (c) RegCM-D3 and TRMM, and (d) RegCM-D4 and TRMM. Units are expressed as percentage of TRMM values.

the different model configurations is that RegCM-D1 and RegCM-D2 show larger underestimation along the ITCZ compared to RegCM-D3 and RegCM-D4.

Overall, an intercomparison of the different simulations indicates that both the default domain (RegCM-D1) and its westward extension (RegCM-D2) produce lesser intensity of rainfall events which explain the larger dry bias and that the northward extension (RegCM-D4) favors more frequent and more intensity of rainfall events (compared to other domains' simulation) consistent to the larger wet bias over the ITCZ. We should note that RegCM-D3 is mostly similar to RegCM-D2 for the frequency and to RegCM-D4 for the intensity. Therefore, RegCM-D3 has an intermediate behavior which supports its best performance in capturing the mean summer monsoon rainfall.

From the above analysis, it is clear that the locations of the western, eastern, and northern boundaries have strong effects on the regional climate model simulation of the monsoon precipitation pattern over West Africa. These effects are, respectively, (1) increases of rainfall over the orographic regions adjacent to the west coast and along the ocean ITCZ, (2) inland extension of the Guinea highlands maxima and strengthening of the ITCZ, and (3) sharpening of the ITCZ and large increase of rainfall in the Sudano-Sahel and the Sahel regions. The best performance is, however, shown in RegCM-D3. As the rainfall pattern over West Africa is associated with the migration of the low-pressure system ITCZ to the north [41–43] and the mesoscale convective systems, which are related to the dynamics of the WAM [9, 10, 16, 44], these differences in the simulations due to the

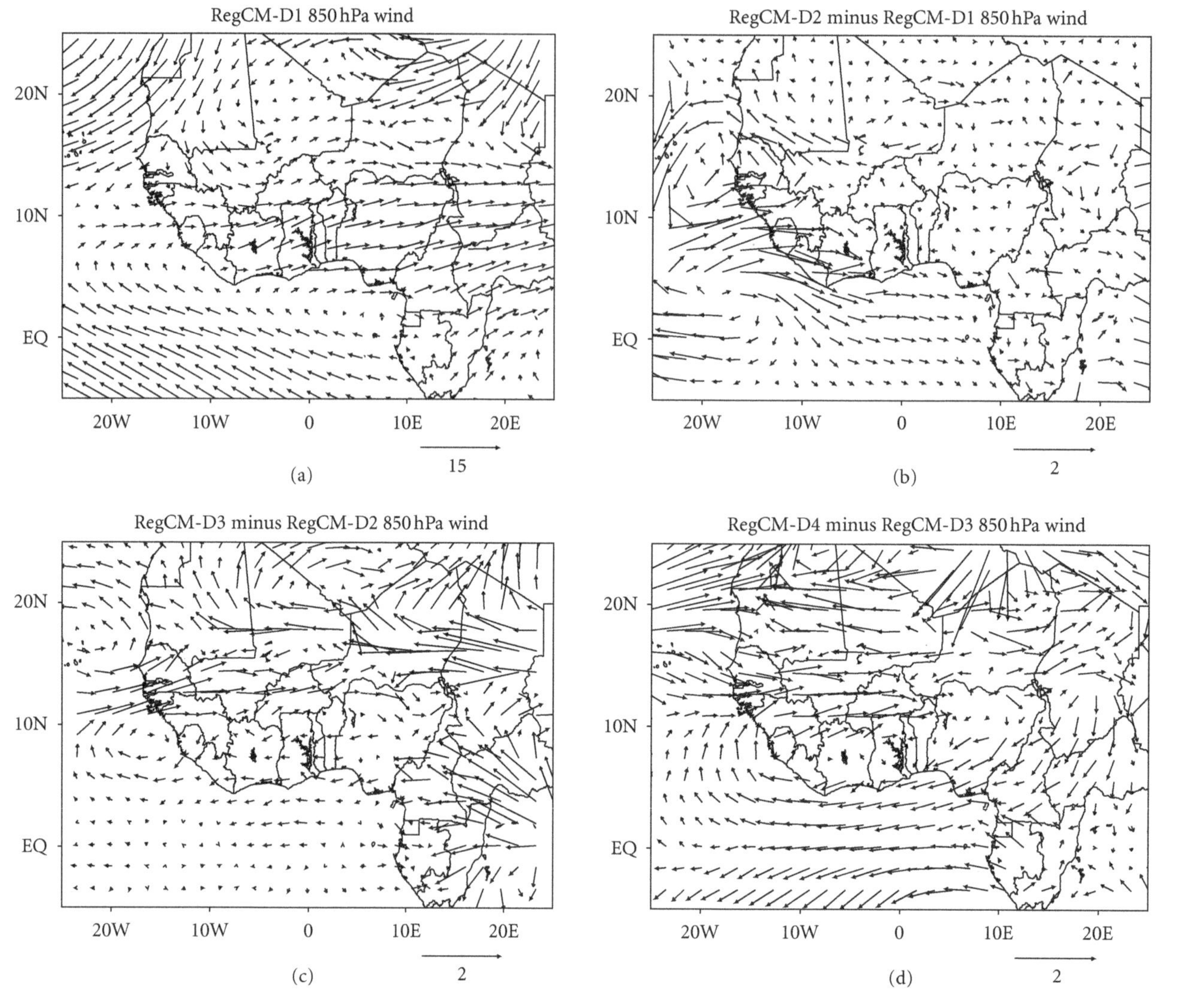

Figure 7: Mean JJA low-level (850 hPa) monsoon flow averaged for the period 2001–2005 for (a) RegCM-D1, (b) the differences between RegCM-D2 and RegCM-D1, (c) the differences between RegCM-D3 and RegCM-D2, and (d) the differences between RegCM-D4 and RegCM-D3. Units are expressed in m/s.

selection of different domain sizes must be originated from the simulated monsoon circulation features.

3.2. The Simulated Monsoon Circulation Features. The low-level (850 hPa) wind field is shown in Figures 7(a)–7(d) for respectively RegCM-D1, RegCM-D2 minus RegCM-D1, RegCM-D3 minus RegCM-D2 and RegCM-D4 minus RegCM-D3. RegCM-D1 shows the monsoon flow, primarily southeasterlies in the Southern Hemisphere becoming southwesterlies and westerlies as they cross the equator. Compared to RegCM-D1, RegCM-D2 simulates stronger monsoon flow off the west coast, in the Guinea highlands, and the Gulf of Guinea. Similarly, differences between RegCM-D3 and RegCM-D2, and on the other hand between RegCM-D4 and RegCM-D3, indicate significant increase of the inflow from the Atlantic Ocean to the land around the 12°N zonal band which lies along the ITCZ. Since the monsoon flow is the major source of humidity for West Africa, the above suggests more humidity transport as the domain extends. To test this hypothesis, we display, respectively, the vertically integrated meridional and zonal moisture fluxes in Figures 8(a)–8(d) and Figures 9(a)–9(d) for each simulation. The simulated meridional moisture flux is mostly southerly over land and off the west coast. It reveals very few differences among the different model configurations mostly located around and north of Guinea highlands. The zonal moisture flux, however, exhibits marked differences mostly in the westerly component. In fact, RegCM-D2 simulates stronger zonal moisture flux (with respect to RegCM-D1) off the west coast, the Gulf of Guinea and along the Sahel band. These fluxes are mainly westerlies and they originate from the availability of larger source of moisture due to the larger portion of the Atlantic Ocean included in the domain. This contributes to enhance the moisture advection inland. Moreover compared to RegCM-D2, the simulation over the Domain D3 shows a widening of the

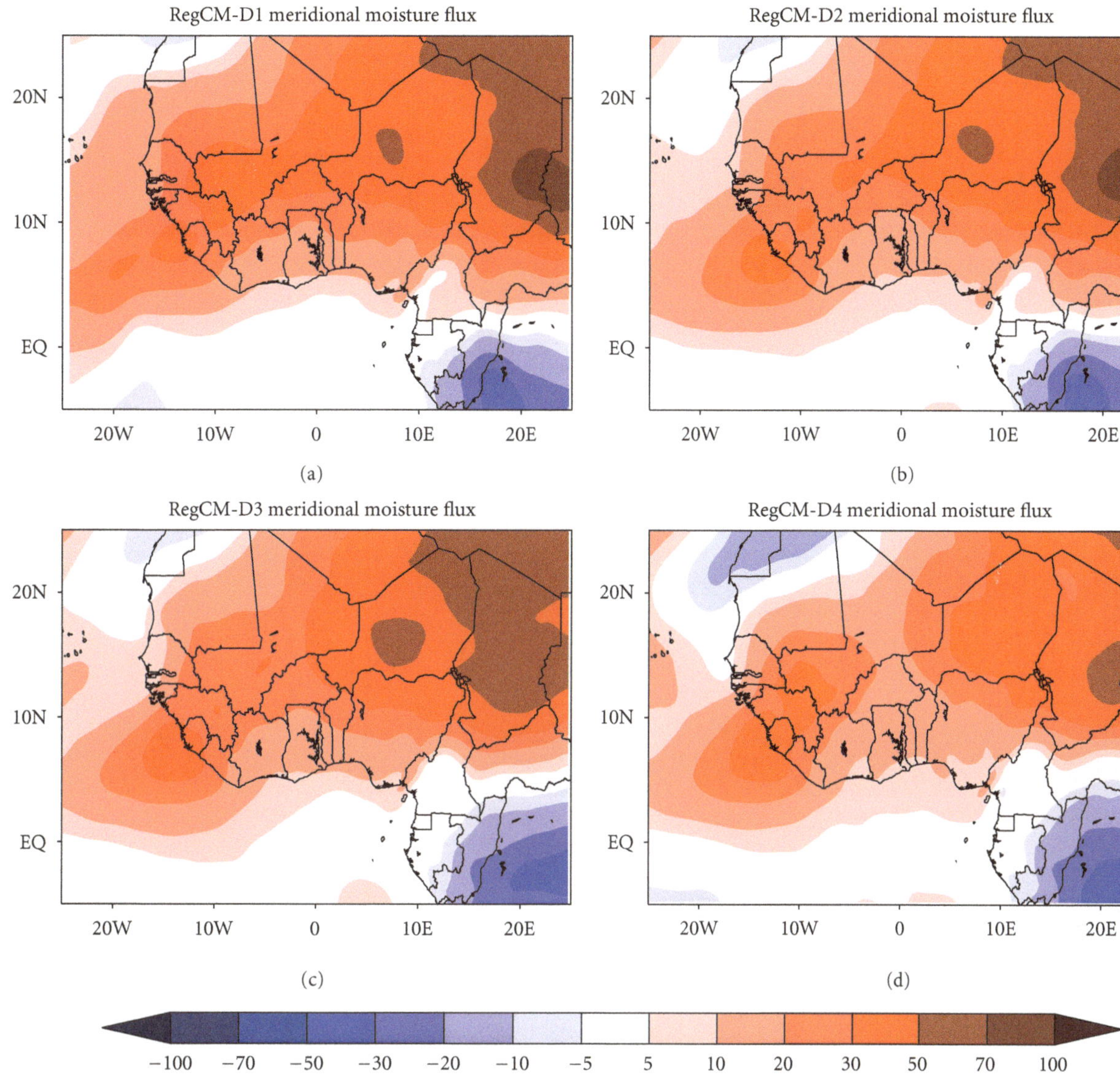

FIGURE 8: Mean JJA vertically integrated meridional moisture flux averaged for the period 2001–2005 for (a) RegCM-D1, (b) RegCM-D2, (c) RegCM-D3, and (d) RegCM-D4. Positive values denote southerly flow. Units are expressed in kg m^{-1} s^{-1}.

maxima off the west coast and also a northward shift and an intensification of the core in the Sahel as a consequence of the weakening of the easterly flow. In RegCM-D4, both the increase of the moisture sources and disappearance of the easterly flow act to further enhance the westerly component of the zonal moisture inflow. Therefore, the zonal moisture flux tends to amplify as the domain extends. The results appear to be consistent to the larger, intermediate, and lower amount of rainfall captured, respectively, in RegCM-D4, RegCM-D3, and RegCM-D2 and RegCM-D1 simulations.

Having examined the low-level monsoon flow and the corresponding integrated moisture fluxes, we now turn our attention to the AEJ. The 650 hPa mean zonal wind depicting the AEJ from the different simulations shows, from Figures 10(a)–10(d), that RegCM-D1 and RegCM-D2 capture similar jet core with speed reaching 14 m/s, getting a bit narrower in RegCM-D3 and much weaker (for about 4 m/s) in RegCM-D4. This suggests that the western and eastern boundaries do not significantly affect the simulation. The reason is that in these domains, the surface conditions (mostly temperature gradient) are not substantially different. In turn, the location of the northern boundary strongly impacts the AEJ strength and spatial pattern. This is originated from the weakening of the lower-level baroclinicity due to decrease in temperature related to the increase in precipitation between 10°N and 15°N [4, 45]. The weaker AEJ is consistent to the disappearance of the easterly component of the integrated zonal moisture flux which led to an increase of the westerly flow and large overestimation of the monsoon rainfall. The simulations of Tropical Easterly Jet (TEJ) across the domains do not show any substantial differences and therefore they are not shown for brevity.

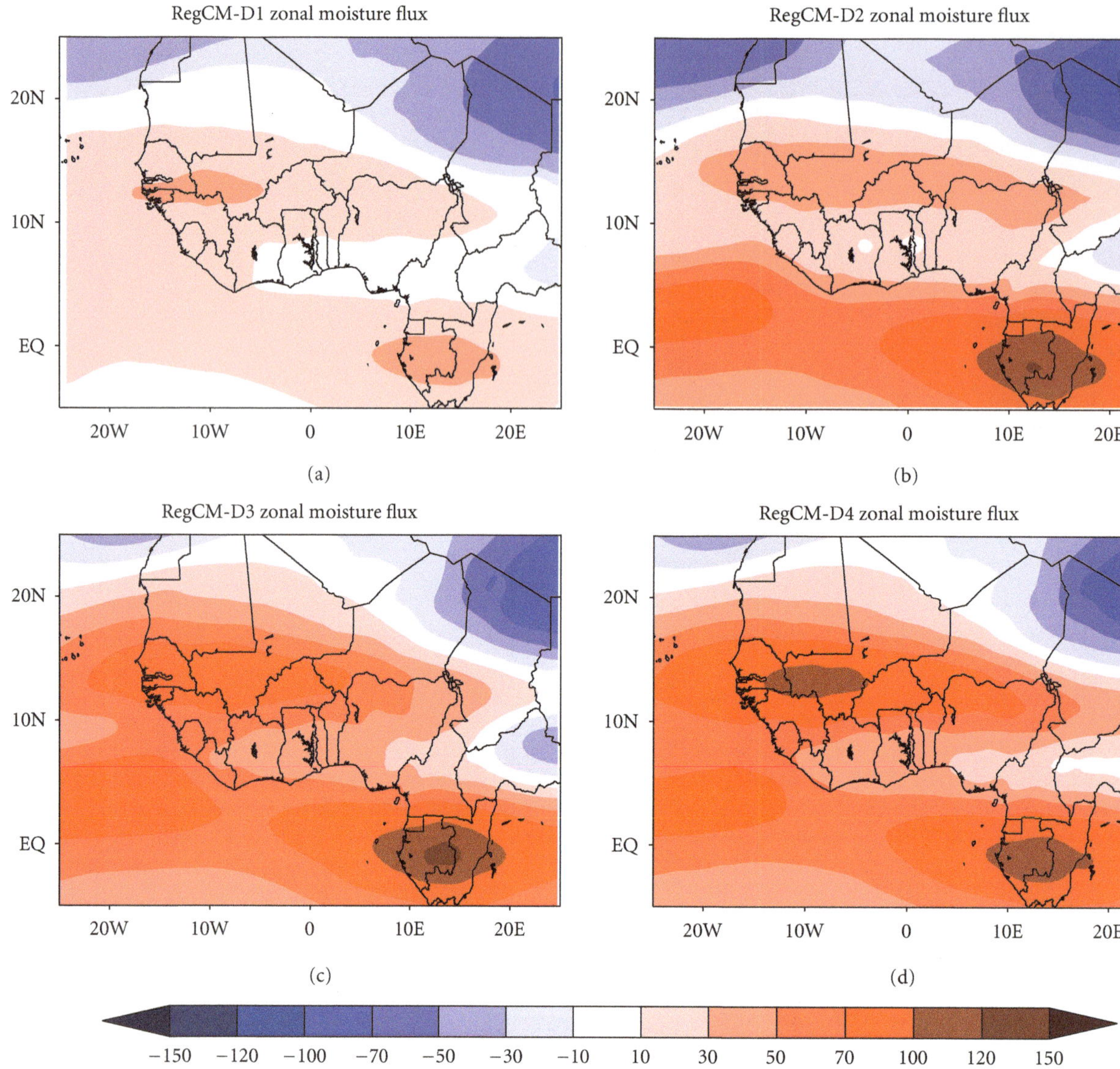

FIGURE 9: Mean JJA vertically integrated zonal moisture flux averaged for the period 2001–2005 for (a) RegCM-D1, (b) RegCM-D2, (c) RegCM-D3, and (d) RegCM-D4. Positive values denote westerly flow. Units are expressed in kg m^{-1} s^{-1}.

Another important feature for the WAM circulation is the strong ascent in the vertical wind bounded by the jet axes and responsible for the location and strength of the ITCZ [46]. This is characterized in a latitude-height cross-section of the vertical wind averaged between 10°E and 10°W and displayed in Figures 11(a)–11(d) for each simulation respectively. As documented in Sylla et al. [16], all the RegCM3 simulations exhibit an upward motion at around 5°N, 10°N and 15–25°N, respectively, connected to friction occurring at the interface between the ocean and land surface in the lower-level [41], to the midtropospheric ITCZ strong ascent and to the Saharan Thermal Low uplift. This deep midtropospheric ascent core is not substantially different in RegCM-D1, RegCM-D2, and RegCM-D3 simulations in both shape and magnitude whereas in RegCM-D4, it is more extended towards the northern latitudes suggesting a wider area of strong convective activity and upward transport of moisture from the monsoon flow level to the upper levels. Furthermore RegCM-D1, RegCM-D2, and RegCM-D3 show some subsidence just below the midtropospheric ascent, which is missing in the RegCM-D4. The wider area of convective activity and the absence of subsidence just below to counterbalance it have led to the excessive rainfall along the ITCZ and the Sahel regions in the RegCM-D4. These larger rainfall amounts and the wider convective activity's region can be assimilated to a poleward extent of the monsoon. This poleward extent of the monsoon rainfall along with the weaker AEJ both found with a northern boundary placed further north around the Azores High region (RegCM-D4) suggest the existence of some

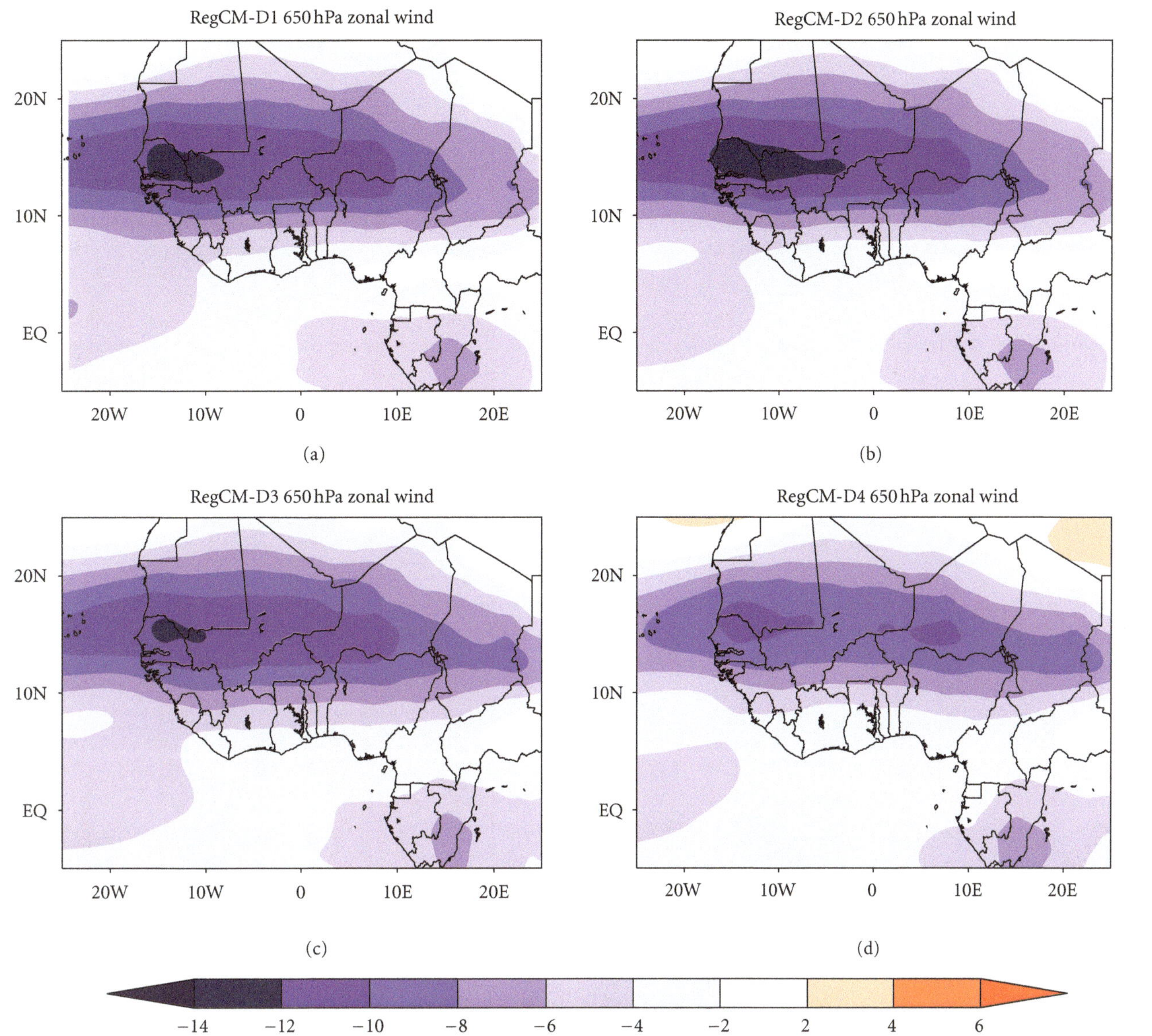

FIGURE 10: Mean JJA zonal wind at 650 hPa averaged for the period 2001–2005 for (a) RegCM-D1, (b) RegCM-D2, (c) RegCM-D3, and (d) RegCM-D4. Units are expressed in m/s.

interactions between the extratropics and the WAM [47, 48]. However, this is out of the scope of the paper and is not analyzed here.

The last WAM feature examined is the AEWs that connect weather and climate and therefore are key drivers of climate variability in the region. Figures 12(a)–12(d) displays the 3–5 days bandpass filtered JJA 650 hPa meridional wind variance averaged over the whole simulation period depicting the AEW activity [4, 49, 50] for each of the model configuration. Their simulations by RegCM3 and how they are related to the atmospheric deep convection are analyzed in details by Sylla et al. [4]. In the smallest (default: D1) domain, the main AEW activity is confined over West Africa while the extension of the western boundary extends it over the Atlantic Ocean causing its weakening in most of the land. The highest changes occur with the extension of the eastern boundary, when the domain (D3) includes regions where AEWs are generated. Substantial increases of the activity are simulated compared to RegCM-D1 and RegCM-D2. This is related to a general overestimation of the frequency of the waves simulated by RegCM3 over the genesis region and to the tendency of this model to produce more instability along the ITZC [4]. Comparison between the RegCM-D4 and RegCM-D3 simulations suggests that the northern boundary location has only a little impact on the waves' activity. The larger wave activity in RegCM-D3 and RegCM-D4 compared to other domains simulation is an indication of more mesoscale convective systems embedded in the AEWs and responsible for most of the rain over West Africa (e.g., [8, 10]). This is thus consistent to the more rainfall amount simulated along the ITCZ in RegCM-D3 and RegCM-D4 compared to RegCM-D1 and RegCM-D2.

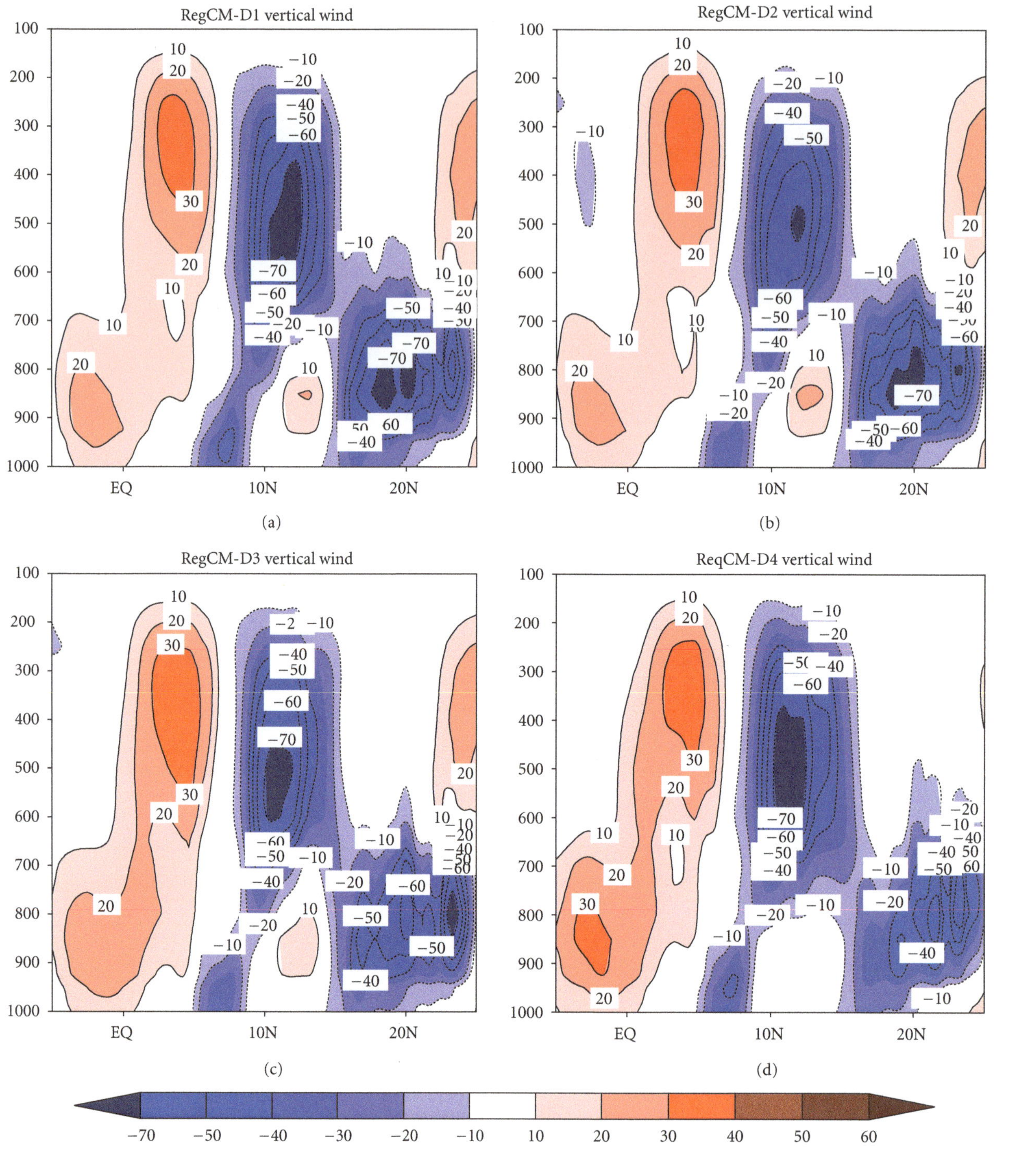

FIGURE 11: Mean JJA latitude-height cross section of vertical velocity averaged between 10° E–10° W for (a) RegCM-D1, (b) RegCM-D2, (c) RegCM-D3, and (d) RegCM-D4. Units are expressed in mm/s.

From this analysis, it is evident that the location of the regional climate model domain boundaries significantly impacts the simulation of the different components of the WAM system along with the summer monsoon rainfall. The monsoon flow and the vertically integrated moisture fluxes are sensitive to any extension of the domain size, the impact on AEJ occurs mostly with the northern boundary while the AEWs activity, exhibiting the largest effects, is more pronounced by an eastward extension of the domain. Given that the simulation of rainfall is more realistic in D3, this overall intercomparison suggests that domain choice for studying the West African summer monsoon climate should be located along the tropical belt and include a large portion of the Atlantic Ocean and regions over the East African

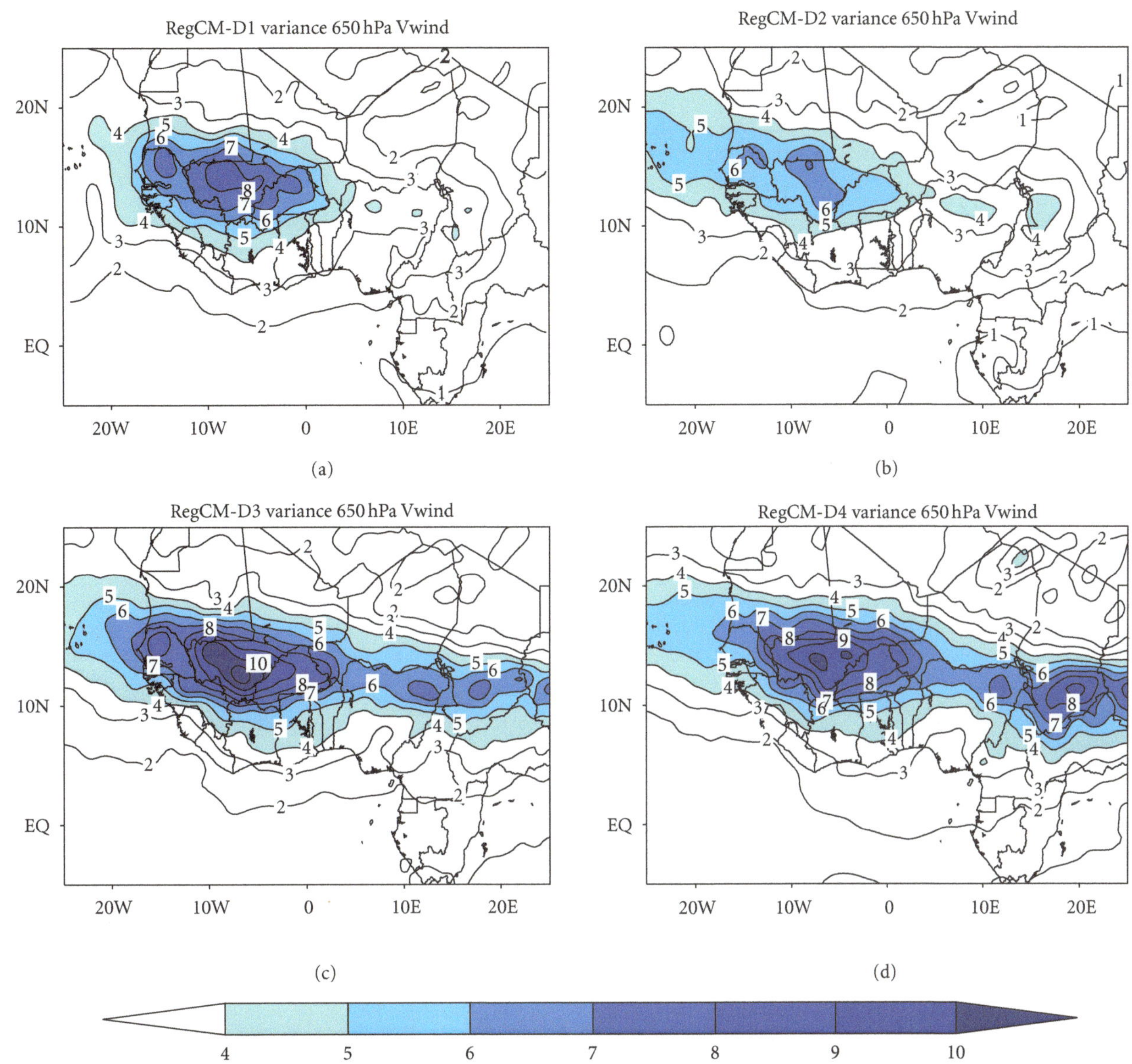

FIGURE 12: Mean variance of the 3–5 days bandpass filtered of the 650 hPa JJA meridional wind for (a) RegCM-D1, (b) RegCM-D2, (c) RegCM-D3, and (d) RegCM-D4. Units are expressed in m/s.

complex terrains (upstream of Sudanese Highlands) to allow the regional climate model to develop a strong monsoon flow, produce large vertically integrated zonal moisture flux, and generate its own AEW features.

4. Summary and Conclusion

In this paper the impact of different domain sizes in the simulation of the West African summer monsoon rainfall is examined using the ICTP-RegCM3. The four different domains in which the simulations are conducted comprise a default one covering mostly West Africa and some part of the Atlantic Ocean; an extended domain toward the west with much larger oceanic region; an eastward extension of that latter up to the East Africa Highlands and finally a last domain covering the regions around the Azores High and extended up to 45°N.

All of the RegCM3 simulations capture well the observed precipitation pattern, particularly the ITCZ position, the northern and southern gradients of rainfall, and the location of maxima over orographic regions. However substantial discrepancies regarding the amount of mean summer monsoon rainfall and the frequency and intensity of rainfall events occur across the different domains. In fact, when the domain is extended westward, the regional climate model simulates higher rainfall amount off the west coast along with an overestimation of rainfall events and much less rainfall intensity. Increases in mean seasonal rainfall over the Guinea Highlands and along the ITCZ arise with an eastward extension of that latter domain. Finally, when the northern boundary is displaced further north up to 45°N, the simulation produces a more intense ITCZ in both land and the Atlantic Ocean, thus overestimating rainfall over much part of central West Africa. It should be noted that the more realistic simulation of the rainfall pattern and intensity is achieved with both a westward and eastward extension of the default domain (e.g., in RegCM-D3). This originates from a more moderate overestimation of the frequency of the rainfall events along with lesser underestimation of their intensity. We thus investigate whether the reasons for these discrepancies in the simulations of the summer monsoon rainfall over West Africa can be found in the behavior of the simulated monsoon circulations features.

Firstly, the monsoon flow and the subsequent vertically integrated zonal moisture flux (mostly westerly) become stronger as the domain extends in both the Atlantic Ocean and towards the East African highlands. In fact, the westward (RegCM-D2) and northward (RegCM-D4) extension of the domain has provided larger source of humidity which favors more moisture advection inland while the eastward (RegCM-D3 and RegCM-D4) extension decreases the easterly flux thereby increases the westerlies. This explains the larger westerly moisture inflow in RegCM-D4, the moderate one in RegCM-D3, and the lower amount in both RegCM-D1 and RegCM-D2 over land.

Secondly, the AEJ and the ascent along the ITCZ are not substantially different in RegCM-D1, RegCM-D2, and RegCM-D3 and in fact, are quasiequally strong, indicating that the eastward and westward locations of the boundary forcings are of minor importance for the simulation of these features because they are mainly driven by surface conditions surface conditions which are similar across these simulations. However, when the northern boundary is placed further north to include the region of Azores High, the AEJ becomes weaker and the width of the deep core of ascent wider due to decreased meridional surface air temperature gradient and suggesting some influence of the extratropics on the WAM climate. This weaker AEJ is also consistent with the larger integrated westerly moisture flux found in this simulation.

Finally, the AEWs activity is lower with a westward extension of the default domain but largest when the domain covers parts of the East African Highlands (in both RegCM-D3 and RegCM-D4). This indicates that the eastward extension tends to increase instability in the simulations and that regional climate model produces excessive numbers of AEWs in the genesis region that would favor more rainfall.

It is thus evident that each of the different domain extensions gives rise to different behavior of the circulation features and that the more accurate simulation of rainfall occurring in Domain D3 is mostly connected to a strong monsoon flow, an intermediate integrated zonal moisture flux, and larger magnitude of the AEWs activity. Therefore domain choice for studying the West African summer monsoon climate must be situated mostly in the tropics and should include a large portion of the Atlantic Ocean and regions upstream of Sudanese Highlands to allow the regional climate model to, respectively, develop enough zonal moisture advection and to generate its own AEW features.

Acknowledgments

N. A. K. Browne is supported by the Abdus Salam International Centre for Theoretical Physics (ICTP) through the Associate Scheme. Therefore, the authors gratefully acknowledge the ICTP grant and the Earth System Physics section. Thanks also to the two anonymous reviewers and the editor whose comments significantly improved the quality of this paper.

References

[1] C. D. Thorncroft, H. Nguyen, and C. Zhang, "Annual cycle of the West African monsoon: regional circulations and associated water vapour transport," *Quarterly Journal of the Royal Meteorological Society*, vol. 137, no. 654, pp. 129–147, 2011.

[2] K. H. Cook, "Generation of the African easterly jet and its role in determining West African precipitation," *Journal of Climate*, vol. 12, no. 5 I, pp. 1165–1184, 1999.

[3] C. D. Thorncroft and M. Blackburn, "Maintenance of the African easterly jet," *Quarterly Journal of the Royal Meteorological Society*, vol. 125, no. 555, pp. 763–786, 1999.

[4] M. B. Sylla, F. Giorgi, P. M. Ruti, S. Calmanti, and A. Dell'Aquila, "The impact of deep convection on the West African summer monsoon climate: a regional climate model sensitivity study," *Quarterly Journal of the Royal Meteorological Society*, vol. 137, no. 659, pp. 1417–1430, 2011.

[5] J. S. Hsieh and K. H. Cook, "A study of the energetics of African easterly waves using regional climate model," *Journal of the Atmospheric Sciences*, vol. 64, no. 2, pp. 421–440, 2007.

[6] V. Sathiyamoorthy, P. K. Pal, and P. C. Joshi, "Intraseasonal variability of the Tropical Easterly Jet," *Meteorology and Atmospheric Physics*, vol. 96, no. 3-4, pp. 305–316, 2007.

[7] B. J. Abiodun, J. S. Pal, E. A. Afiesimama, W. J. Gutowski, and A. Adedoyin, "Simulation of West African monsoon using RegCM3 Part II: impacts of deforestation and desertification," *Theoretical and Applied Climatology*, vol. 93, no. 3-4, pp. 245–261, 2008.

[8] K. I. Mohr and C. D. Thorncroft, "Intense convective systems in West Africa and their relationship to the African easterly jet," *Quarterly Journal of the Royal Meteorological Society*, vol. 132, no. 614, pp. 163–176, 2006.

[9] G. S. Jenkins, A. T. Gaye, and B. Sylla, "Late 20th century attribution of drying trends in the Sahel from the Regional Climate Model (RegCM3)," *Geophysical Research Letters*, vol. 32, no. 22, Article ID L22705, pp. 1–4, 2005.

[10] A. Gaye, A. Viltard, and P. de Félice, "Squall lines and rainfall over Western Africa during summer 1986 and 87," *Meteorology and Atmospheric Physics*, vol. 90, no. 3-4, pp. 215–224, 2005.

[11] E. K. Vizy and K. H. Cook, "Development and application of a mesoscale climate model for the tropics: influence of sea surface temperature anomalies on the West African monsoon," *Journal of Geophysical Research D*, vol. 107, no. 3, Article ID 4023, 2002.

[12] M. B. Sylla, A. T. Gaye, G. S. Jenkins, J. S. Pal, and F. Giorgi, "Consistency of projected drought over the Sahel with changes in the monsoon circulation and extremes in a regional climate model projections," *Journal of Geophysical Research D*, vol. 115, no. 16, Article ID D16108, 2010.

[13] W. Moufouma-Okia and D. P. Rowell, "Impact of soil moisture initialisation and lateral boundary conditions on regional climate model simulations of the West African Monsoon," *Climate Dynamics*, vol. 35, no. 1, pp. 213–229, 2010.

[14] E. A. Afiesimama, J. S. Pal, B. J. Abiodun, W. J. Gutowski, and A. Adedoyin, "Simulation of West African monsoon using the RegCM3. Part I: model validation and interannual variability," *Theoretical and Applied Climatology*, vol. 86, no. 1–4, pp. 23–37, 2006.

[15] M. B. Sylla, A. T. Gaye, J. S. Pal, G. S. Jenkins, and X. Q. Bi, "High-resolution simulations of West African climate using regional climate model (RegCM3) with different lateral boundary conditions," *Theoretical and Applied Climatology*, vol. 98, no. 3-4, pp. 293–314, 2009.

[16] M. B. Sylla, A. Dell'Aquila, P. M. Ruti, and F. Giorgi, "Simulation of the intraseasonal and the interannual variability of rainfall over West Africa with RegCM3 during the monsoon period," *International Journal of Climatology*, vol. 30, no. 12, pp. 1865–1883, 2010.

[17] R. A. Anthes, Ying-Hwa Kuo, Eirh-Yu Hsie, S. Low-Nam, and T. W. Bettge, "Estimation of skill and uncertainty in regional numerical models," *Quarterly Journal of the Royal Meteorological Society*, vol. 115, no. 488, pp. 763–806, 1989.

[18] F. Giorgi and L. O. Mearns, "Introduction to special section: regional climate modeling revisited," *Journal of Geophysical Research*, vol. 104, pp. 6335–6352, 1999.

[19] S. A. Rauscher, A. Seth, J. H. Qian, and S. J. Camargo, "Domain choice in an experimental nested modeling prediction system for South America," *Theoretical and Applied Climatology*, vol. 86, no. 1–4, pp. 229–246, 2006.

[20] A. Seth and F. Giorgi, "The effects of domain choice on summer precipitation simulation and sensitivity in a regional climate model," *Journal of Climate*, vol. 11, no. 10, pp. 2698–2712, 1998.

[21] R. G. Jones, J. M. Murphy, and M. Noguer, "Simulation of climate change over Europe using a nested regional-climate model. I: assessment of control climate, including sensitivity to location of lateral boundaries," *Quarterly Journal of the Royal Meteorological Society*, vol. 121, no. 526 B, pp. 1413–1449, 1995.

[22] F. Giorgi, M. R. Marinucci, and G. T. Bates, "Development of a second-generation regional climate model (RegCM2). Part I: boundary-layer and radiative transfer processes," *Monthly Weather Review*, vol. 121, no. 10, pp. 2794–2813, 1993.

[23] F. Giorgi, M. R. Marinucci, G. T. Bates, and G. De Canio, "Development of a second-generation regional climate model (RegCM2). Part II: convective processes and assimilation of lateral boundary conditions," *Monthly Weather Review*, vol. 121, no. 10, pp. 2814–2832, 1993.

[24] J. S. Pal, F. Giorgi, X. Bi et al., "The ICTP RegCM3 and RegCNET: regional climate modeling for the developing world," *Bulletin of American Meteorological Society*, vol. 88, pp. 1395–1409, 2007.

[25] G. A. Grell, J. Dudhia, and D. R. Stauffer, "Description of the fifth generation Penn State/NCAR mesoscale model (MM5)," Technical Note NCAR/TN-398+STR, 1994.

[26] J. T. Kiehl, J. J. Hack, G. B. Bonan et al., "Description of the NCAR Community Climate Model (CCM3)," Tech. Rep. TN-420+STR, NCAR, Boulder, Colo, USA, 1996.

[27] A. A. M. Holtslag, E. I. F. De Bruijn, and H. L. Pan, "A high resolution air mass transformation model for short-range weather forecasting," *Monthly Weather Review*, vol. 118, no. 8, pp. 1561–1575, 1990.

[28] R. E. Dickinson, A. Henderson-Sellers, and P. J. Kennedy, "Biosphere–atmosphere transfer scheme (BATS) version 1E as coupled to the NCAR Community Climate Model," Technical Note NCAR/TN-387+STR, 1993.

[29] X. Zeng, M. Zhao, and R. E. Dickinson, "Intercomparison of bulk aerodynamic algorithms for the computation of sea surface fluxes using TOGA COARE and TAO data," *Journal of Climate*, vol. 11, no. 10, pp. 2628–2644, 1998.

[30] J. M. Fritsch and C. F. Chappell, "Numerical prediction of convectively driven mesoscale pressure systems. Part I: convective parameterization," *Journal of the Atmospheric Sciences*, vol. 37, no. 8, pp. 1722–1733, 1980.

[31] J. S. Pal, E. E. Small, and E. A. B. Eltahir, "Simulation of regional-scale water and energy budgets: representation of subgrid cloud and precipitation processes within RegCM," *Journal of Geophysical Research D*, vol. 105, no. 24, pp. 29579–29594, 2000.

[32] S. Uppala, D. Dee, S. Kobayashi, P. Berrisford, and A. Simmons, "Towards a climate data assimilation system: status update of ERA-Interim," *ECMWF Newsletter*, vol. 115, pp. 12–18, 2008.

[33] R. W. Reynolds, N. A. Rayner, T. M. Smith, D. C. Stokes, and W. Wang, "An improved in situ and satellite SST analysis for climate," *Journal of Climate*, vol. 15, no. 13, pp. 1609–1625, 2002.

[34] C. Kummerow, Y. Hong, W. S. Olson et al., "The evolution of the Goddard profiling algorithm (GPROF) for rainfall estimation from passive microwave sensors," *Journal of Applied Meteorology*, vol. 40, no. 11, pp. 1801–1820, 2001.

[35] G. J. Huffman, R. F. Adler, D. T. Bolvin et al., "The TRMM multisatellite precipitation analysis (TMPA): quasi-global, multiyear, combined-sensor precipitation estimates at fine scales," *Journal of Hydrometeorology*, vol. 8, no. 1, pp. 38–55, 2007.

[36] R. F. Adler, G. J. Huffman, A. Chang et al., "The version-2 global precipitation climatology project (GPCP) monthly precipitation analysis (1979-present)," *Journal of Hydrometeorology*, vol. 4, no. 6, pp. 1147–1167, 2003.

[37] M. B. Sylla, F. Giorgi, E. Coppola, and L. Mariotti, "Uncertainties in daily rainfall over Africa: assessment of observation products and evaluation of a regional climate model simulation," submitted to *International Journal of Climatology*.

[38] F. Giorgi and X. Bi, "A study of internal variability of a regional climate model," *Journal of Geophysical Research D*, vol. 105, no. 24, pp. 29503–29521, 2000.

[39] M. M. Kgatuke, W. A. Landman, A. Beraki, and M. P. Mbedzi, "The internal variability of the RegCM3 over South Africa," *International Journal of Climatology*, vol. 28, no. 4, pp. 505–520, 2008.

[40] E. Vanvyve, N. Hall, C. Messager, S. Leroux, and J. P. van Ypersele, "Internal variability in a regional climate model over West Africa," *Climate Dynamics*, vol. 30, no. 2-3, pp. 191–202, 2008.

[41] S. E. Nicholson, B. Some, and B. Kone, "An analysis of recent rainfall conditions in West Africa, including the rainy seasons of the 1997 El Nino and the 1998 La Nina years," *Journal of Climate*, vol. 13, no. 14, pp. 2628–2640, 2000.

[42] V. Mathon and H. Laurent, "Life cycle of Sahelian mesoscale convective cloud systems," *Quarterly Journal of the Royal Meteorological Society*, vol. 127, no. 572, pp. 377–406, 2001.

[43] M. B. Sylla, E. Coppola, L. Mariotti et al., "Multiyear simulation of the African climate using a regional climate model (RegCM3) with the high resolution ERA-interim reanalysis," *Climate Dynamics*, vol. 35, no. 1, pp. 231–247, 2010.

[44] B. Sultan, S. Janicot, and A. Diedhiou, "The West African monsoon dynamics. Part I: documentation of intraseasonal variability," *Journal of Climate*, vol. 16, no. 21, pp. 3389–3406, 2003.

[45] M. B. Sylla, F. Giorgi, and F. Stordal, "The large-scale origins of rainfall and temperature bias in high resolution simulations over southern Africa," *Climate Research*. In press.
[46] S. E. Nicholson, "The intensity, location and structure of the tropical rainbelt over west Africa as factors in interannual variability," *International Journal of Climatology*, vol. 28, no. 13, pp. 1775–1785, 2008.
[47] M. J. Rodwell and B. J. Hoskins, "Subtropical anticyclones and summer monsoons," *Journal of Climate*, vol. 14, no. 15, pp. 3192–3211, 2001.
[48] C. Chou and J. D. Neelin, "Mechanisms limiting the northward extent of the northern summer monsoons over North America, Asia, and Africa," *Journal of Climate*, vol. 16, no. 3, pp. 406–425, 2003.
[49] A. Diedhiou, S. Janicot, A. Viltard, and P. De Felice, "Evidence of two regimes of easterly waves over West Africa and the tropical Atlantic," *Geophysical Research Letters*, vol. 25, no. 15, pp. 2805–2808, 1998.
[50] A. Mekonnen, C. D. Thorncroft, and A. R. Aiyyer, "Analysis of convection and its association with African easterly waves," *Journal of Climate*, vol. 19, no. 20, pp. 5405–5421, 2006.

11

Implications of Self-Potential Distribution for Groundwater Flow System in a Nonvolcanic Mountain Slope

Tada-nori Goto,[1] Kazuya Kondo,[2] Rina Ito,[2] Keisuke Esaki,[2] Yasuo Oouchi,[2] Yutaka Abe,[3] and Maki Tsujimura[3]

[1] *Graduate School of Engineering, Kyoto University, C1-1-113, Kyodai-Katsura, Nishikyo, Kyoto 615-8540, Japan*
[2] *Department of Environmental Earth Sciences, Aichi University of Education, 1 Hirosawa, Igaya-cho, Kariya, Aichi 448-8542, Japan*
[3] *Graduate School of Life and Environmental Sciences, University of Tsukuba, 1-1-1 Tennodai, Tsukuba 305-8572, Japan*

Correspondence should be addressed to Tada-nori Goto, goto.tadanori.8a@kyoto-u.ac.jp

Academic Editor: Tsuneo Ishido

Self-potential (SP) measurements were conducted at Mt. Tsukuba, Japan, which is a nonvolcanic mountain, to infer groundwater flow system in the mountain. Survey routes were set around the northern slope, and the reliability of observed SP anomaly was checked by using SP values along parallel survey routes; the error was almost within 10 mV. The FFT analysis of the spatial SP distribution allows us a separation of raw data into two components with shorter and longer wavelength. In the shorter (altitudinal) wavelength than ~200 meters, several positive SP peaks of more than 100 mV in magnitude are present, which indicate shallow perched water discharges along the slope. In the regional SP pattern of longer wavelength, there are two major perturbations from the general trend reflecting the topographic effect. By comparing the SP and hydrological data, the perturbation around the foothill is interpreted to be caused by heterogeneous infiltration at the ground surface. The perturbation around the summit is also interpreted to be caused by heterogeneous infiltration process, based on a simplified numerical modeling of SP. As a result, the SP pattern is well explained by groundwater flow and infiltration processes. Thus, SP data is thought to be very useful for understanding of groundwater flow system on a mountain scale.

1. Introduction

Spatial distribution of self-potential or spontaneous potential (SP), which is time-invariant natural electrical potential, has been observed at the ground surface along a topographic slope. In general, a negative correlation between altitude and SP value along a slope is reported (e.g., [1–3]), which is called as terrain effect or topographic effect on SP. For example, Sasai et al. [3] have found relatively high SP values at the foothills of the Miyake-jima volcano, decreasing toward the middle altitude of the mountain. The SP profile is well correlated with the topographic profile, and the mean coefficient of topographic effect was −1.07 mV/m in the case of the Miyake-jima volcano.

Topographic effect on SP is thought to be caused by subsurface fluid flow along the slope through electrokinetic coupling (e.g., [4]). In rock-water systems, the flow of groundwater can drag excess electrical charges located in the close vicinity of the pore water/mineral interface in the electrical double layer with it, resulting in macroscopic electric current called as streaming current or drag current and the development of electric potential called as streaming potential in the direction of fluid flow (e.g., [5, 6]). Because the zeta potential, characteristic potential difference across the electrical double layer, is commonly negative for typical crustal rocks in water of pH higher than ~2 [6], the pore fluid flow carries positive charges with it; so a negative correlation between the hydraulic head and the streaming potential is expected. Yasukawa and Mogi [7] conducted numerical calculations of groundwater flow and associated streaming potential and confirmed a negative correlation between the altitude and SP value, which has observable amplitude on the slope surface. Although SP can be generated by other effects such as thermoelectric coupling [8], electrochemical effects (diffusion potential, redox potential, etc.; e.g., [8, 9]), SP along a mountain slopes is thought to be mainly caused by the electrokinetic effects.

On the basis of the topographic effect, a number of attempts to estimate hydrological features along mountain slopes have been made. For example, Fournier [10] assumed that main SP sources were aligned on water table and tried to estimate the water table elevation using SP data taken at the ground surface. Revil et al. [11] developed a sophisticated method for water table estimation; they inferred the water table elevations below a slope using an inverse technique with the observed SP data. Linde et al. [12] modeled both SP sources in the vadose zone and along the water table and tried to estimate the water table below a slope more precisely. The SP measurement is nonintrusive, quick, and inexpensive; therefore, these benefits tend to accelerate the applications of the SP method to groundwater studies, engineering, and environmental investigations.

There is still some debate, however, about the origin of topographic effect on SP along mountain slopes. Previous studies adopt very simple assumptions of hydrological process. Fournier [10] and Revil et al. [11] excluded the contribution of streaming current associated with downward flow of vadose-zone water, which was taken as the primary source for SP generation by Aubert and Atangana [13]. Although Linde et al. [12] considered both SP sources in the vadose and underlying water-saturated zones, they assumed a homogeneous vadose zone and uniform infiltration at the ground surface. However, numerical studies (e.g., [14]) suggest that complicated downward liquid flow in the vadose zone will make complicated SP pattern on the ground surface.

For small-scale slopes, less than 100 m in length, Ernstson and Scherer [15] suggest various causes of SP anomalies along the slopes: biological root activity, lateral water movements in alternating permeable and impermeable layers, discharge of perched-water aquifers, and infiltration and evapotranspiration together with groundwater recharge. However, the origins of SP along a larger-scale slope (e.g., a few km in length) have not been discussed enough. Detailed interpretation of topographic effect on mountain-scale SP pattern is useful to apply the SP method to estimate subsurface fluid motions, especially in volcanoes. As many former studies said [3, 16–19], SP anomalies of positive polarity are frequently found around volcano summits and near fumaroles. In order to discuss such SP anomalies, knowledge of SP distribution caused by topographic effect is necessary as a basis.

In order to discuss possible causes of SP along a mountain slope, comparison between the SP pattern and hydrological data along the slope is required, but few field studies were reported. In this study, we investigate self-potential distribution around the northern slope of Mt. Tsukuba, located in Japan, where extensive hydrological studies have been carried out. Mt. Tsukuba mainly consists of granitic basement, so that the SP distribution along Mt. Tsukuba will be a typical example of topographic effect on a plutonic mountain. We compare the SP data with hydrological results to discuss main causes of the SP anomalies.

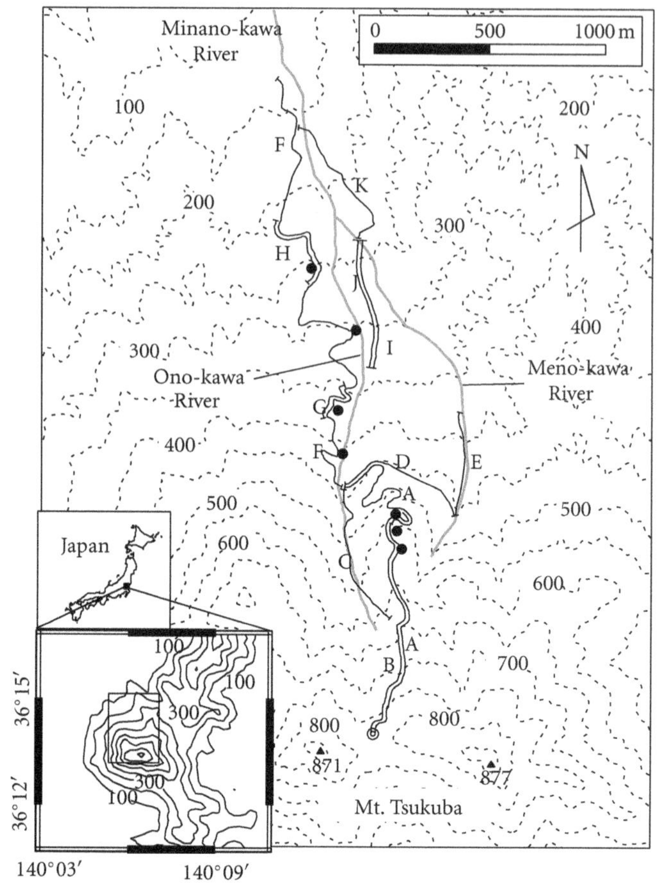

FIGURE 1: Location of Mt. Tsukuba and SP survey routes. Thick solid lines: the SP survey routes. A double circle: reference point for SP measurement. Solid circles: locations of DC resistivity survey. Thin solid lines: equi-topographic contour. Gray lines: rivers.

2. Self-Potential Measurement

Mt. Tsukuba is located near Tsukuba, central Japan (Figure 1). The maximum altitude is 877 m. The slope gradient is about 20 and 10 degrees at altitude around 500 and 300 meters, respectively. The basement of Mt. Tsukuba mainly consists of granitic rocks, and partly gabbroic rocks and debris [20]. There are many hydrological studies: for example, isotope analysis of river water [21], discharge analysis of river water [22], isotope analysis of spring water, and estimation of water table elevation [23].

SP survey was carried out along eleven routes (A to K) from August 20 to 22 in 2000 (Figure 1). The survey area covers the northern slope of Mt. Tsukuba with altitude range between 130 m and 780 m. The "leapfrog" method was used for SP observation with electrode spacing of 12–120 m (mainly 50 m). We used two types of paired electrodes and voltmeters as shown in Figure 2. Cu-$CuSO_4$ electrodes named AUE-type are suitable for the flat soil

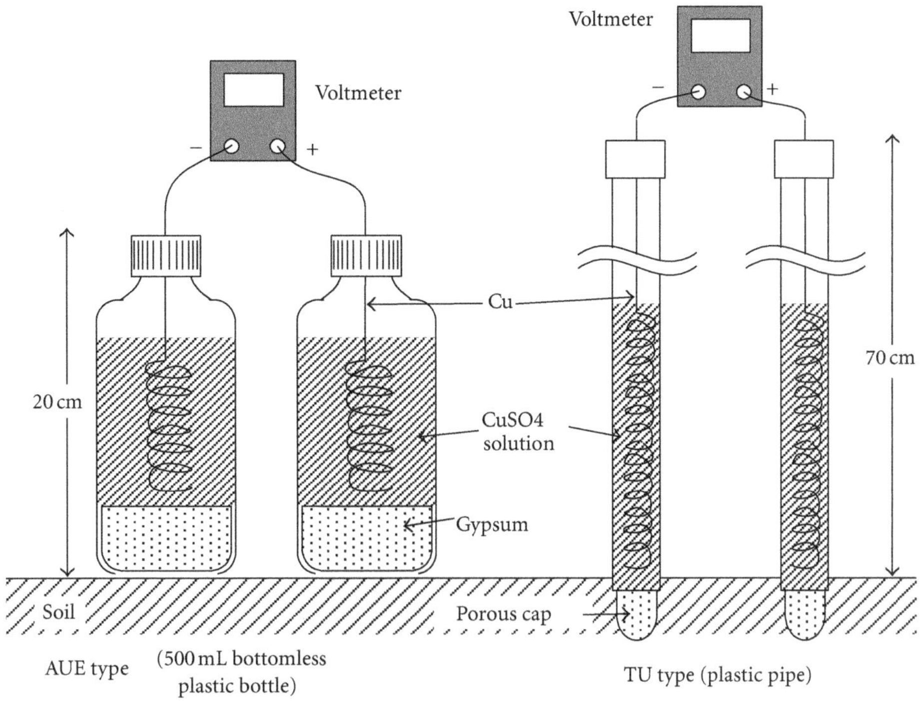

FIGURE 2: Schematic diagram of Cu-$CuSO_4$ electrodes used in this study.

surface, and those named TU-type are used for thick soil cases with stabbing. Electrode position at start or end of each survey route is common among several routes (Figure 1). All observed SP values are represented as relative ones to the reference station near the summit (shown by a double circle in Figure 1).

Stability of the electrode pair was checked by voltage measurement between the two electrodes set on the ground with 10 cm separation. Such offset check was done at start, end, and several points along each survey route. The offset voltages were within 5 mV in this study. We also confirmed accuracy of the observed SP values by making SP measurements along parallel additional routes (e.g., B, G, H, C, and J-K in Figure 1), whose electrode positions at start and end of profiles were common with main survey routes (A–F in Figure 1).

In addition to SP surveys, DC resistivity surveys were conducted at six sites in this area (Figure 1). A one-dimensional forward modeling algorithm [24] and a one-dimensional inversion algorithm [25] were applied to obtain near-surface resistivity values. The approximate sounding depth was 10–20 m.

3. Results

The observed SP anomalies are shown in Figure 3. Along the main SP profile (A–F), several sharp and local SP anomalies are recognized, in particularly at 770, 640, 520, and 360 m in altitude. Some of the amplitudes exceed 100 mV, and the typical altitudinal extent is about 50 m (corresponding to the horizontal spatial extent of about 150 m) or less. As shown in Figure 3(a), similar local SP anomalies are also observed along the neighboring survey routes, B, G, and H. Note that voltage difference between routes A and B was less than 10–20 mV, which is thought to be due to small difference (typically less than ~10 m) in the electrode position between these two survey routes except the start and end points.

The spectrum analysis gives us more quantitative features of the local and sharp SP anomalies. We applied the fast Fourier transform (FFT) to the obtained SP data along the main profile, A–F. Since the data spacing is not equal through the entire SP profile, we calculated SP values at equispaced points using the natural cubic splines. A linear increasing trend of SP toward the foothill (about −0.12 mV/m estimated by the least square regression) is subtracted before the FFT. Finally, the FFT was applied to SP values at 512 points equally spaced with altitude interval of about 1.286 m. The spectrum obtained by the FFT is shown in Figure 4. Three peaks of spectrum were clearly identified at altitudinal wavelength of 131.6 m, 65.8 m, and 43.9 m. To delineate characters of the spectrum peaks in Figure 4 more clearly, we first created a "regional" SP pattern with longer wavelength by applying the inverse FFT to a spectrum portion with longer wavelength (219–658 m) denoted by "A" in Figure 4 and adding the linear trend of SP described above to it. The resultant "regional" SP pattern is shown in Figure 5(a). Then, the "local" SP pattern with shorter wavelength is estimated by subtracting the regional SP pattern from the raw data. Figure 5(b) is the resultant local SP pattern with altitudinal wavelength shorter than 131.8 m. We found that the local SP anomalies have mainly positive peaks with altitudinal extent of 40–60 m. Note that large peaks are obvious at altitudes between 300 m and 700 m.

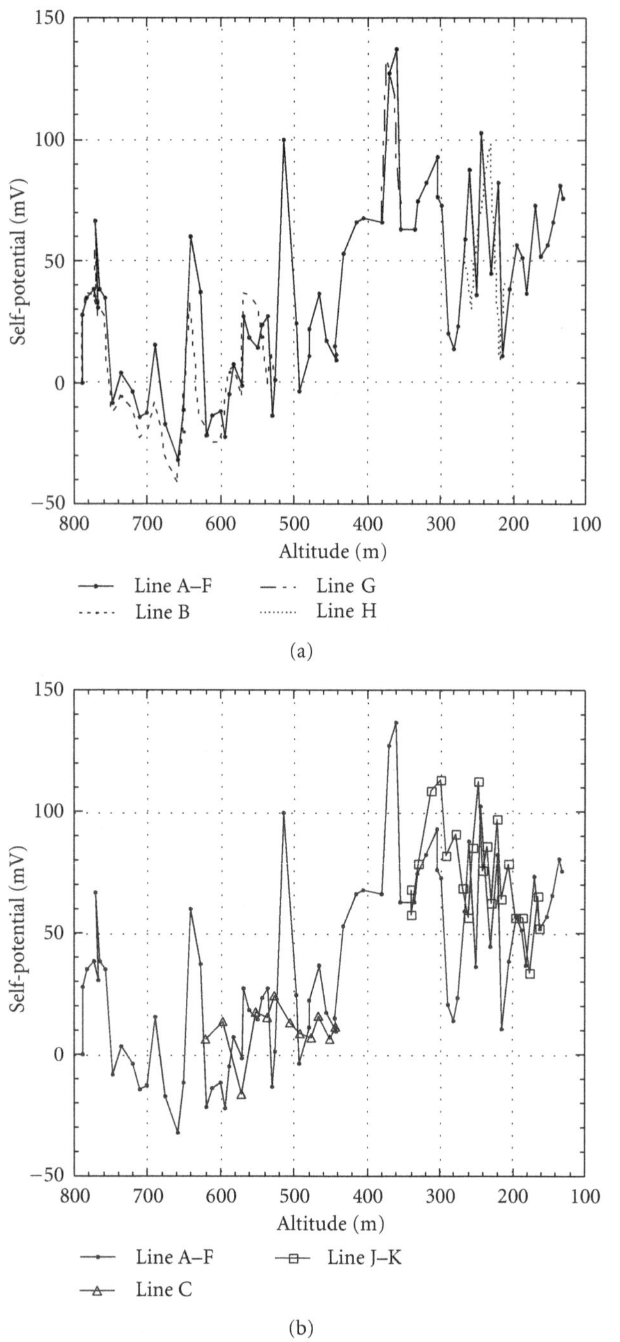

FIGURE 3: Observed SP profile along the northern slope of Mt. Tsukuba. (a) Those along the neighboring routes (A, F, B, G, and H) and (b) the parallel routes (C, J, and K) are plotted. The reference point is at the summit.

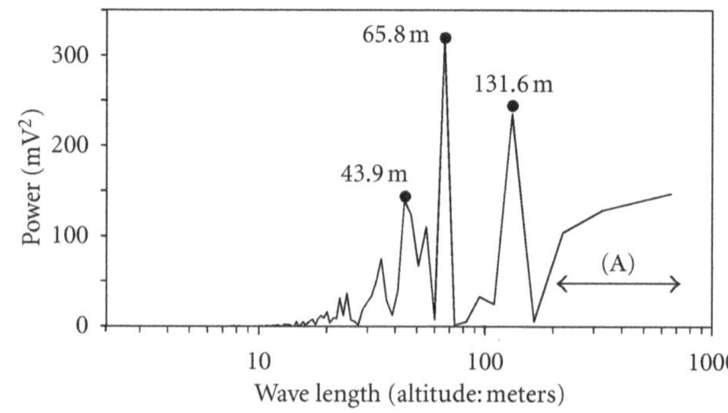

FIGURE 4: Power spectrum of spatial SP distribution against altitude. Three peaks are obvious. Spectrums with longer wavelength are denoted as "A."

Concerning the longer wavelength (more than ~200 m in altitude), SP patterns along parallel survey routes show similar features (Figure 3(b)). For example, SP profiles along routes A and C are similar to each other except for local SP anomalies described above. The "regional" SP pattern (Figure 5(a)) indicates gradual increase toward the foothill of Mt. Tsukuba. However, the increasing trend is not monotonic and includes two major perturbations. One is a decreasing trend toward lower altitude from 350 m to 250 m, and another is also a decreasing trend between 750 m and 650 m in altitude. The former decreasing trend is also recognized along parallel route J-K (Figure 3(b)), suggesting that this decreasing trend has substantial spatial extent in the same altitudes.

4. Discussion

4.1. Local SP Anomalies. "Local" SP anomalies of positive polarity (Figure 5(b)) have a limited spatial extent. The altitudinal extent of them is typically about 40–60 m, which corresponds to horizontal extent of 100–200 m (because of 10–20 degrees slope gradient). In order to estimate the approximate depth of electrical current source which generates such local SP anomalies, simple calculations were conducted. A point current source was assumed in a uniform resistivity (ρ) half-space. The depth of point source (d) was varied, and the surface potential field (V) was calculated using the following simple analytic solution:

$$V = \frac{\rho \cdot I}{2\pi\sqrt{r^2 + d^2}}, \quad (1)$$

where r is the horizontal distance between the source and the observation point, and I is the source current strength. The calculated potential profile with respective source depths is normalized by the SP value at $r = 0$ m (shown in Figure 6). For comparison between the calculated and observed SP, we plot two local SP anomalies (denoted by "A" and "B" in Figure 5(b)) in Figure 6. Zero level is fixed as an average SP values at the northern and southern edges of each anomaly, and SP values are normalized by the peak values. On the basis of Figure 6, the sources for local SP anomalies are thought to be located near the surface; we roughly estimate the depths to be shallower than ~10 meters, although the observed local SP values are substantially scattered.

A cause of local positive SP anomalies such as observed in the present study is generally inferred to be local springs, yielding a positive current source near the ground surface. The slope of Mt. Tsukuba is widely covered by debris or slope deposits, which consist of unconsolidated sediments and have high permeability. It is well known that perched

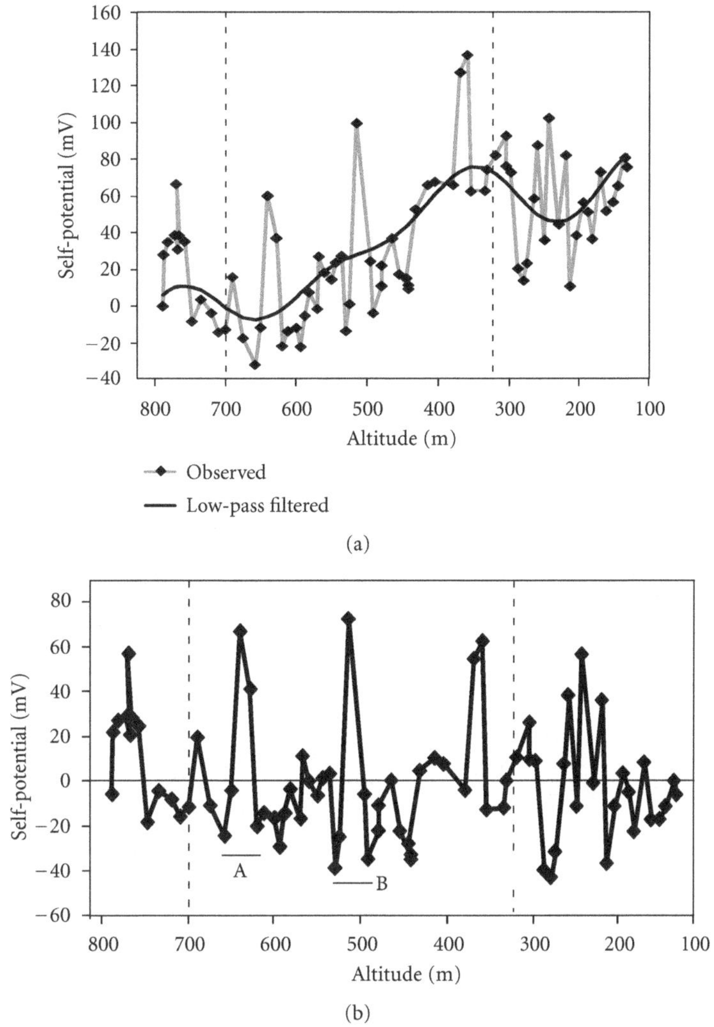

Figure 5: (a) The observed raw SP profile (gray line with diamonds) and low-pass filtered one called as "regional" pattern in this study (solid line). (b) Residual SP profile after subtracting regional SP from raw data, called as "local" anomalies.

water flow is present in a vadose zone developed in such slope deposits; therefore, we infer that the observed local SP anomalies reflect discharge of perched water. However, most of the anomalies are not located in the neighbor (within a few ten meters) of known springs. So instead of directly corresponding to surface springs, the current sources are thought to be located at underground transition zone of electrokinetic coupling coefficient where perched water flow loses its ability to carry electrical charges with it. In the area with altitude higher than about 450 m, the slope angle increases gently toward the summit (from 10 to 20 degrees), and larger gravels are conspicuous. May be, perched water flow becomes stream-like flow through underground channels (cavities), which does not transport sufficient electrical charges with it unlike the flow through porous media, at shallower depths than ~10 meters in the slope deposits. Such underground streams may start from the locations of local SP anomalies like "A" and "B" shown in Figure 5(b) and flow into nearby rivers. Anyway, such shallow phenomena seem to show seasonal variations. We will need to make repeated measurements to observe a correlation with precipitation, and so forth in the future.

4.2. Regional SP Pattern. The general trend of "regional" SP pattern indicates gradual increase toward the foothill of Mt. Tsukuba (Figure 5(a)). This is a typical example of the topographic effect on SP, interpreted by the mountain-scale groundwater flow from summit to foothill. The mean coefficient of topographic effect was about −0.12 mV/m, and the maximum was about −0.45 mV/m at altitude of 400 m. It is noteworthy that the measured streaming potential coefficient for intact rock samples of Inada granite is about 0.03×10^{-6} V/Pa for 0.001 mol/L KCl solution below 50°C [26, 27]. If we apply this data to the granitic basement of Mt. Tsukuba, topographic effect of −0.3 mV/m is possible for the case that water table is located close to the upper surface of the granitic basement (Figure 7(a)). Substantial part of the general trend is thought to be caused by the water flow in the granitic basement.

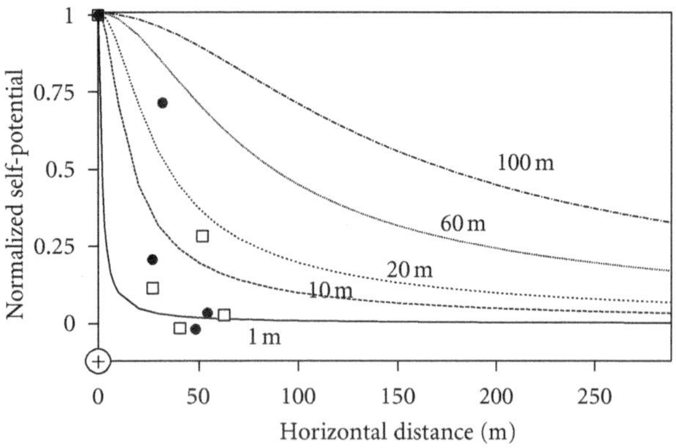

FIGURE 6: Calculated SP from various depths of a point source (1–100 m). The SP values are normalized by the respective value at 0 m in the horizontal distance where the source is located. The observed SP anomalies "A" and "B" in Figure 5(b) are shown by circles and rectangles, respectively.

There are two major perturbations from the general trend as mentioned in Section 3. Both of them seem to be present near geological contacts between different units along the mountain slope. The Holocene younger slope deposits (unconsolidated sediments) distribute below altitude of 320 m along the route F, while the Holocene older slope deposits distribute around altitude of 320–700 m (Figure 7(a)). Above 700 m in altitude, the gabbro basement rock is exposed with a cap-like distribution around the summit [20]. Two decreasing patterns toward the foothill ("RA" and "RB" in Figure 7(b)), which are inconsistent with the ordinal topographic effect, are located near these geological boundaries at the ground surface. In the following, we discuss causes of these two perturbations based upon the heterogeneity of subsurface structure and related groundwater flow.

4.2.1. Regional SP Perturbation at the Foot Hill. The decreasing pattern "RA" in Figure 7(b) is present near the boundary between older and younger slope deposits as mentioned above. One of the candidates for this perturbation is a difference in the coupling coefficient between the two deposits. Even though both slope deposits have the same material origin (gabbro and granite), the coupling coefficient may be different between them due to difference in the formation factor (porosity, tortuosity), and so forth (e.g., [6]). In addition to this, hydrological heterogeneities such as water table perturbation due to heterogeneous permeability distribution and heterogeneous infiltration of meteoric water can be candidates.

If the younger slope deposits (on the lower altitude side) are more permeable than the older slope deposits (on the higher altitude side), downward fluid flow crossing the boundary leaves positive charges there (with negative zeta potential) due to discontinuous pressure gradient normal to the boundary even if the coupling coefficient is homogeneous [28]. However, it depends on the hydrological boundary condition at the ground surface whether or not water flow though a heterogeneous permeability region

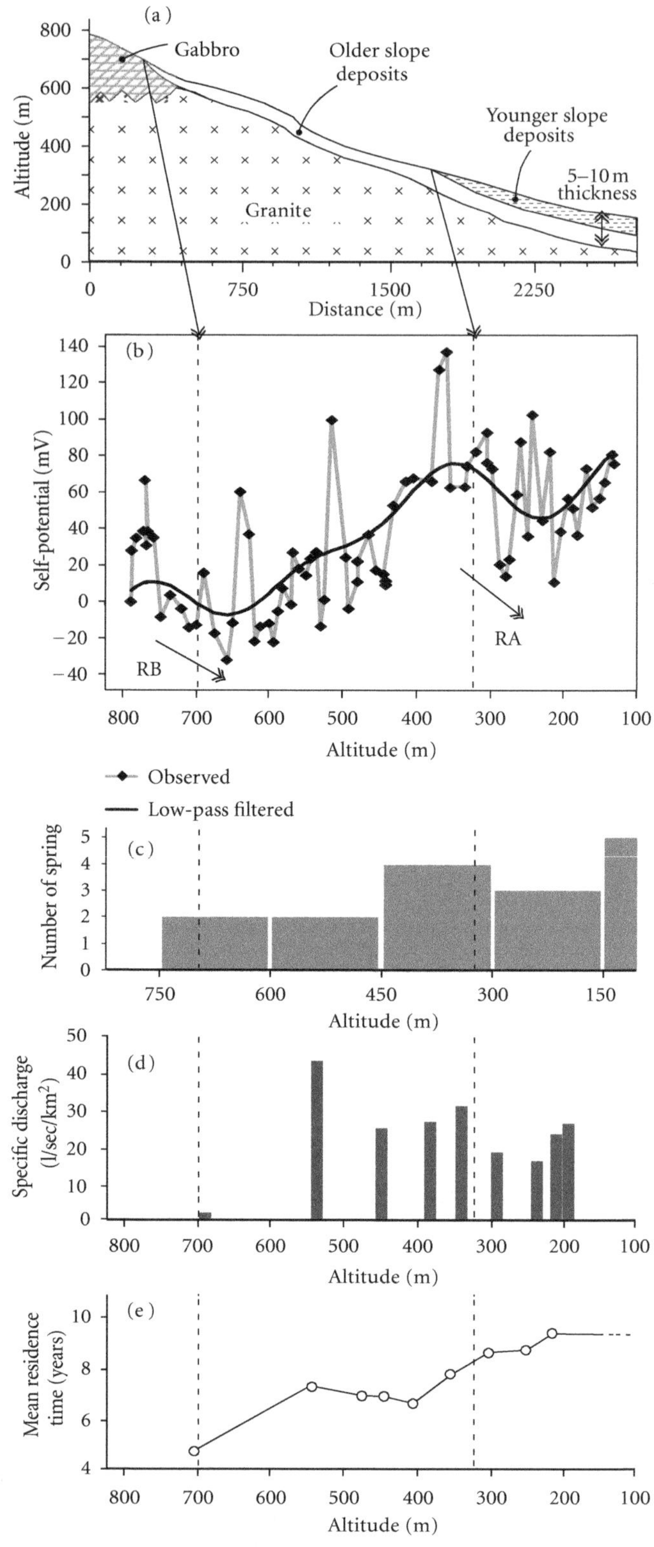

FIGURE 7: Comparison of the geological setting, SP distributions, and hydrological information along the northern slope of Mt. Tsukuba. (a) Geological structure [20]. Note that the surface layer is exaggerated vertically. (b) Raw and regional SP profiles are same as shown in Figure 5. (c) Histogram for number of springs around Mt. Tsukuba against altitude derived from [23]. (d) Specific discharge of the Ono-kawa river water [22]. (e) Mean residence time of stream water [21].

brings about surface SP anomalies. If it is pressure (hydraulic head) = constant, no SP anomaly will appear at the ground surface. Electrical charges appearing at the permeability boundary just work as sources to create potential distribution corresponding to the distorted pressure distribution at the boundary (see the tutorial by Jouniaux and Ishido [29], in this special issue "Electrokinetics in Earth Sciences" for more details). However, if the impermeable boundary condition (Neumann condition) is adopted, surface SP anomalies will appear corresponding to hydraulic head distribution along the surface boundary which is modified as a result of heterogeneous permeability distribution (T. Ishido, pers. comm., 2012). The effects of permeability heterogeneity on SP pattern are studied by Ozaki et al. [30]. In their numerical simulation using two-dimensional FDM (finite difference method), the Neumann boundary condition was adopted: no flow normal to all exterior boundaries of the model domain including the ground surface. In the present case (Figure 7(a)), water table (hydraulic head) will be lowered near the boundary between the younger and older slope deposits if the permeability is larger for the younger deposits. Since the lower water table corresponds to higher SP (e.g., [10, 11]), the observed decreasing trend "RA" toward the foothill cannot be explained by this mechanism.

Next, the possibility of heterogeneous infiltration is discussed hereinafter on the basis of a lot of hydrological information. Histogram of number of springs against altitude is made from a distribution of springs in the whole area of Mt. Tsukuba [23]. The histogram shown in Figure 7(c) indicates a clear correlation with the regional SP pattern. The higher SP value is, the more frequently springs are found. A distribution of specific discharge of river water also indicates a high correlation with the regional SP pattern around the foothill. Specific discharge is a flow rate of river water normalized by the catchment area and reflects the contribution of groundwater discharges into river water. Figure 7(d) is the observation result along the Ono-kawa River [22]. Lower specific discharges around 250 m altitude compared to those around 350 m altitude indicate that groundwater discharges into river water are relatively small in the catchment area around 250 m altitude. This means that the fraction of downward groundwater infiltration is relatively high around 250 m altitude compared to the area around 350 m altitude. This is consistent with the result of isotopic analysis [21]; the stream water shows relatively old and young ages at altitudes around 250 m and 350 m, respectively (Figure 7(e)). This means that the contribution of shallow groundwater originating from recent rainfalls to groundwater discharges is relatively small around 250 m altitude compared to that around 350 m altitude. These hydrological data imply that the vadose zone is relatively thick and thin at altitudes of 250 m and 350 m, respectively.

Our DC resistivity survey supports this implication; the average resistivity of the older slope deposits is 100–200 Ωm, while it is higher (200–500 Ωm) in the younger slope deposits, suggesting a highly unsaturated condition. Therefore, we may conclude that the SP decreasing pattern at the foothill ("RA" in Figure 7(b)) is mainly generated by vertical infiltration flow through the thicker vadose zone in the younger slope deposits and underlying layers. If the thickness of the vadose zone is comparable to that of the younger slope deposits (less than 10 meters, Figure 7(a)), the vertical potential gradient in the vadose zone needs to be around −3 mV/m to explain the regional negative SP at altitude of 250 m. This gradient (−3 mV/m) is an order of magnitude greater than the observed general topographic effect, which suggests that the effective streaming potential coefficient of the vadose zone in the younger slope deposits is significantly larger in magnitude than that of the surrounding area.

4.2.2. Regional SP Perturbation Near the Summit. The decreasing pattern "RB" in Figure 7(b) is present at the lower perimeter of the gabbro rock body exposed around the summit. Water table around the summit is estimated to be quite deep and almost flat based on a hydrological study with isotope analysis and numerical modeling of groundwater flow [23]. The water table is located below the gabbroic body in the granitic basement at higher elevation areas than about 500 m, and the thickness of vadose zone is more than 200 m around the boundary between the gabbroic rocks and the slope deposits at ~700 m altitude (Figure 7(a)). In order to discuss the main causes of the SP perturbation, we carried out a simple numerical modeling by assuming electrical current sources as shown in Figure 8.

The intensity of the current sources is given from the streaming current density J (in A/m^2) in the vadose zone (e.g., [11]):

$$J = -L\rho g \nabla h, \qquad (2)$$

where L is the electrokinetic coupling coefficient for the vadose zone (in A/Pa-m), ρ is the liquid water density (~10^3 kg/m^3), g is the gravitational acceleration (9.8 m/sec^2), and h is the hydraulic head (in meter). As for vertical infiltration of water in the vadose zone, ∇h can be assumed to be ~1. Although experimental data of L was reported for unsaturated sands (e.g., [31]), no data is available for liquid/gas two-phase flow in the granitic rocks. We simply assume $L = -2 \times 10^{-10}$ A/Pa-m for the vadose zone in the present calculation, which means $|J| = 2 \times 10^{-6}$ A/m^2 from (2). In all of the three models shown in Figure 8, the intensity of unit current source at the ground surface is given as $I = -0.002$ A for a surface area of 20 m (along the slope) × 50 m (perpendicular to the slope). Such current sources are assumed every 20 m between 0 m and 400 m along the slope, and current sources of the same intensity but with opposite sign ($I = +0.002$ A) are assumed every 20 m along the water table for charge conservation. As for the streaming current density in the underlying saturated region, we assume relatively large magnitude ($|J| = 4 \times 10^{-6}$ A/m^2, which may be too large considering a small ∇h value for the saturated region, T. Ishido, pers. comm., 2012). Thus, negative current sources (unit intensity of −0.004 A) are assumed every 20 m along the water table, and a positive current source of +0.08 A (=0.004 A × 20) is assumed for charge conservation on the slope surface (at 700 m foothill side from the origin of slope). A three-dimensional forward calculation code "3ddcxh" [32] is used to calculate the SP,

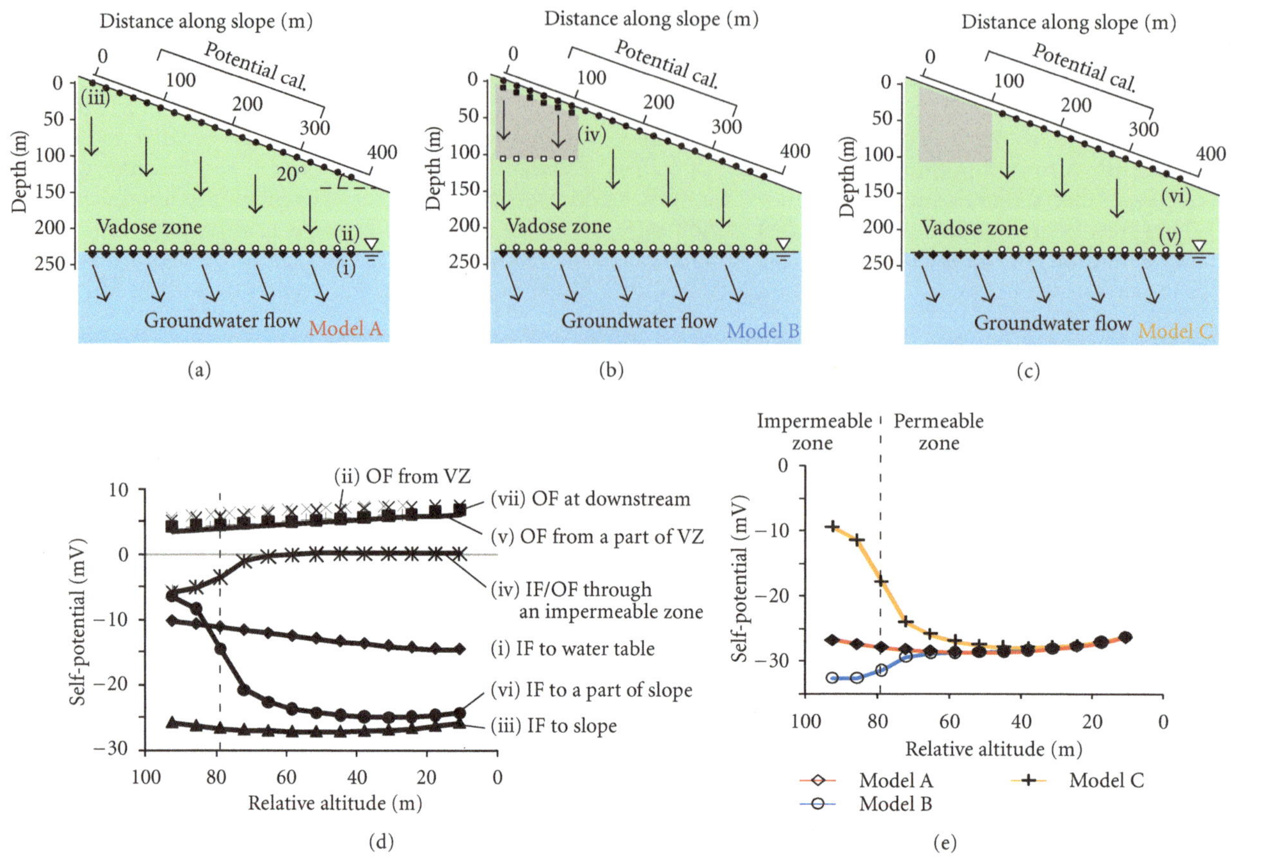

Figure 8: Numerical modeling of SP profiles along slopes. (a) Model A with homogeneous subsurface structure. (b) Model B with constant infiltration through both a less permeable zone (gray) and surroundings. (c) Model C with heterogeneous infiltration outside of an impermeable zone (gray). In all three models, diamonds, circles, and rectangular dots indicate electric current sources with intensity of 0.004, 0.002, and 0.001 A, respectively. Solid and open symbols mean negative and positive polarities, respectively. Arrows indicate assumed flow directions. SP profiles shown in (d) and (e) are calculated along the thick line portion on the slope for each case. (d) Calculated SP from respective current sources. The abbreviations are as follows: IF: inflow, OF: outflow, and VZ: vadose zone. (e) Resultant SP profiles from three models against altitude.

assuming a uniform subsurface resistivity of 200 Ωm (which is estimated from the DC resistivity survey).

As shown in Figure 8, we consider three models: (A) with homogeneous permeability and infiltration, (B) with a less permeable zone (gabbro) and homogeneous infiltration, and (C) with an impermeable zone (gabbro) and heterogeneous infiltration. The SP from model A (Figure 8(a)) is regarded as a reference without heterogeneity, which shows almost flat profile (Figure 8(e)). The reason why a negative correlation between SP and altitude is not seen here is due to the effect of intense negative current sources assigned to mimic downward saturated water flow along the water table (see the curve (i) in Figure 8(d)). In model B (Figure 8(b)), the streaming current density is assumed to be $|J| = 3 \times 10^{-6}$ A/m^2 for the gabbroic body, that is, one and half times that for the surrounding vadose zone (this assumption may be plausible considering that the liquid-phase saturation can be higher in the less permeable zone). This assumption produces a negative SP near the summit (see the curve (iv) in Figure 8(d)), which results in an opposite feature to the observation (Figure 8(e)). In model C (Figure 8(c)), no infiltration is assumed along the gabbroic rock surface, considering a situation that most of rainfall water is removed from the area due to rapid runoff along the almost impermeable rock surface and does not infiltrate downward. Due to this effect (see the curve (vi) in Figure 8(d)), the SP profile (shown in Figure 8(e)) shows a corresponding trend to the observed SP with decreasing pattern near the summit ("RB" in Figure 7(b)).

As a result of the above calculations, the heterogeneous infiltration at the surface is thought to be the most probable cause of the SP perturbation observed near the summit. A similar but more prominent feature observed near active volcano summits has been studied by numerical simulations (e.g., [14, 33]). Aizawa et al. [33] carried out a sophisticated modeling of both groundwater flow and SP and indicates that a heterogeneous infiltration due to a less permeable zone near the summit makes regional "sink" pattern of SP along the mountain slope. In their study, heterogeneous distributions of coupling coefficient and electrical conductivity are taken into account in addition to permeability heterogeneity. The presence of extensive alteration rocks around volcano summits is very influential on the SP distribution due to its high electrical conductivity rather than its impervious nature [14]. In such situations, we need to take into account various factors other than heterogeneous permeability and infiltration distributions.

In field observations, similar SP patterns are also found at small-scale mountain slopes. For example, in the Nagano prefecture, Japan, Abe [22] found the relatively positive SP anomaly near a ridge and negative "sink" SP pattern off the ridge. We suppose that the ridge areas have a thin permeable layer at the surface, and infiltration mainly occurs off the ridge having a thicker surface permeable layer. We believe that this phenomenon is rather common at many mountain slopes, so it is very important to take into account heterogeneous infiltration effects to discuss groundwater flow systems beneath mountain slopes.

5. Summary

On the basis of the proposed causes for local and regional SP patterns in this study, we summarize the inferred groundwater flow system in Mt. Tsukuba. As discussed in Section 4.2.1, the water table comes close to the surface at altitude of about 400 m, inferred by the high regional SP values, the younger mean residence time of stream water, and distribution of springs (Figures 7(b), 7(e), and 7(c)). The water table goes down at the foothill (lower than 350 m in altitude) covered by thick younger slope deposits with high permeability, inferred by the regional SP sink pattern "RA" and the specific discharge of stream (Figures 7(b) and 7(d)). The thickness of the vadose zone will be over 10 m estimated by the DC resistivity survey. Around the summit, as discussed in Section 4.2.2, the thicker vadose zone expands above the water table located in the granitic basement. The infiltration is not homogeneous with simple vertical flow but is concentrated outside of the gabbro rocks at the summit, inferred from the regional SP pattern "RB" near the summit (Figure 7(b)). In addition to the regional flow systems, local flow of perched water is developed in the vadose zone in the slope deposits as discussed in Section 4.1. The local positive SP anomalies (Figure 5(b)) are thought to be associated with outflow of perched water near the ground surface.

The implication for groundwater flow system in Mt. Tsukuba is not solely derived from hydrological information. The combination of hydrological and SP surveys gives us a chance of sophisticated interpretation of groundwater flow system. The major contribution of SP surveys comes from wider spatial coverage with high measurement density than the conventional hydrological surveys along the mountain slope. On the other hand, SP data alone is not interpreted uniquely among various possible patterns of groundwater flow. Therefore, SP surveys are brought into a complementary relation with hydrological information. In the future, coupled numerical examinations of both groundwater flow and SP should be performed for many cases along mountain slopes, which is very helpful in evaluation of groundwater flow systems for environmental, engineering, agricultural, and scientific issues.

Acknowledgments

The authors thank Dr. Y. Tanaka, Dr. Y. Sasai, and Dr. H. Murakami for their valuable comments about field experiments. They thank Yumiko Tsutsui, Takuya Omura, Takayuki Kimura, Chisa Kobayashi, and Manabu Shibata for their help on preexperiments. This paper has been greatly improved by constructive remarks from the Academic Editor Dr. Tsuneo Ishido and two anonymous reviewers. This work was partially supported by a Grant for Scientific Research from the Ministry of Education, Japan (no. 23501241).

References

[1] D. B. Jackson and J. Kauahikaua, "Regional self-potential anomalies at Kilauea Volcano (Hawaii)," *US Geological Survey Professional Paper*, vol. 1350, no. 2, pp. 947–959, 1987.

[2] F. S. Birch, "Testing Fournier's method for finding water table from self-potential," *Ground Water*, vol. 31, pp. 50–56, 1993.

[3] Y. Sasai, J. Zlotnicki, Y. NIshida et al., "Electromagnetic monitoring of miyake-jima volcano, izu-bonin arc, japan: a preliminary report," *Journal of Geomagnetism and Geoelectricity*, vol. 49, no. 11-12, pp. 1293–1316, 1997.

[4] T. Ishido, "Self-potential generation by subsurface water flow through electrokinetic coupling, in Detection of subsurface flow phenomena," *Lecture Notes in Earth Sciences*, vol. 27, pp. 121–131, 1989.

[5] A. Revil and P. Leroy, "Hydroelectric coupling in a clayey material," *Geophysical Research Letters*, vol. 28, no. 8, pp. 1643–1646, 2001.

[6] T. Ishido and J. Muzutani, "Experimental and theoretical basis of electrokinetic phenomena in rock-water systems and its applications to geophysics," *Journal of Geophysical Research*, vol. 86, no. 3, pp. 1763–1775, 1981.

[7] K. Yasukawa and T. Mogi, "Topographic effects on Sp anomaly caused by subsurface fluid flow—numerical approach," *Butsuri-Tansa*, vol. 51, pp. 17–26, 1998.

[8] R. F. Corwin and D. B. Hoover, "Self-potential method in geothermal exploration," *Geophysics*, vol. 44, no. 2, pp. 226–245, 1979.

[9] F. Timm and P. Möller, "The relation between electric and redox potential: evidence from laboratory and field measurements," *Journal of Geochemical Exploration*, vol. 72, no. 2, pp. 115–128, 2001.

[10] C. Fournier, "Spontaneous potentials and resistivity surveys applied to hydrogeology in a volcanic area: case history of the Chaine des Puys (Puy-de-Dome, France)," *Geophysical Prospecting*, vol. 37, no. 6, pp. 647–668, 1989.

[11] A. Revil, V. Naudet, and J. D. Meunier, "The hydroelectric problem of porous rocks: inversion of the position of the water table from self-potential data," *Geophysical Journal International*, vol. 159, no. 2, pp. 435–444, 2004.

[12] N. Linde, A. Revil, A. Bolève et al., "Estimation of the water table throughout a catchment using self-potential and piezometric data in a Bayesian framework," *Journal of Hydrology*, vol. 334, no. 1-2, pp. 88–98, 2007.

[13] M. Aubert and Q. Yéné Atangana, "Self-potential method in hydrogeological exploration of volcanic areas," *Ground Water*, vol. 34, no. 6, pp. 1010–1016, 1996.

[14] T. Ishido, "Electrokinetic mechanisms for the "W"-shaped self-potential profile on volcanoes," *Geophysical Research Letters*, vol. 31, no. 15, pp. L15616–5, 2004.

[15] K. Ernstson and H. U. Scherer, "Self-potential variations with time and their relation to hydrogeologic and meteorological parameters," *Geophysics*, vol. 51, no. 10, pp. 1967–1977, 1986.

[16] C. J. Zablocki, "Some electrical and magnetic studies of Kilauea Iki lava lake," Open-File Report—U. S. Geological Survey, Report OF 76-0304, Hawaii, USA, 1976.

[17] F. Massenet and P. Van Ngoc, "Mapping and surveillance of active fissure zones on a volcano by the self-potential method, Etna, Sicily," *Journal of Volcanology and Geothermal Research*, vol. 24, no. 3-4, pp. 315–338, 1985.

[18] J. Zlotnicki, M. Feuillard, and G. Hammouya, "Water circulations on La Soufriere volcano inferred by self- potential surveys (Guadeloupe, Lesser Antilles). Renewal of volcanic activity?" *Journal of Geomagnetism & Geoelectricity*, vol. 46, no. 9, pp. 797–813, 1994.

[19] T. Hashimoto and Y. Tanaka, "A large self-potential anomaly on Unzen volcano, Shimabara Peninsula, Kyushu Island, Japan," *Geophysical Research Letters*, vol. 22, no. 3, pp. 191–194, 1995.

[20] S. Igi, *Geologic Atlas of Japan Kanto Region*, Asakura Shoten, Tokyo, Japan, 1990.

[21] K. Sanjo, "Isotope hydrology of Mt. Tsukuba," *Science Reports—University of Tsukuba, Institute of Geoscience, Section A*, vol. 12, pp. 1–36, 1991.

[22] Y. Abe, *Estimation of groundwater flow using self-potential measurement in hillslopes of College of Natural Sciences*, Graduation thesis, First Cluster of Colleges, University Tsukuba, 2001.

[23] T. Matsumoto, *Age dating of spring and groundwater and its modeling in Mt. Tsukuba, central Japan*, M.S. thesis, University Tsukuba, 2011.

[24] Y. Murakami, "Computer program for processing and interpreting Schlumberger and Wenner resistivity data," Open File Report number 42, Geological Survey of Japan, 1987.

[25] K. Ushijima, H. Mizunaga, and A. Kato, "Interpretation of geoelectric sounding data by personal computer," *Butsuri-Tanasa*, vol. 40, pp. 423–435, 1987.

[26] T. Tosha, N. Matsushima, and T. Ishido, "Zeta potential measured for an intact granite sample at temperatures to 200°C," *Geophysical Research Letters*, vol. 30, no. 6, pp. 28–1, 2003.

[27] T. Ishido and N. Matsushima, "Streaming potential measured for an intact granite sample at temperatures to 200°C," in *Proceedings of the 32nd Workshop on Geothermal Reservoir Engineering, 2007*, SGP-TR-183, Stanford University.

[28] W. R. Sill, "Self-potential modeling from primary flows," *Geophysics*, vol. 48, no. 1, pp. 76–86, 1983.

[29] L. Jouniaux and T. Ishido, "Electrokinetics in earth sciences: a tutorial," *International Journal of Geophysics*, vol. 2012, Article ID 286107, 2012.

[30] Y. Ozaki, H. Mikada, T. Goto, J. Takekawa, M. Tsujimura, and F. Hachani, "Effect of heterogeneity of hydraulic conductivity on streaming potential," in *Proceedings of the 14th International Symposium on Recent Advances in Exploration Geophysics (RAEG'10)*, pp. 115–117, 2010.

[31] X. Guichet, L. Jouniaux, and J. P. Pozzi, "Streaming potential of a sand column in partial saturation conditions," *Journal of Geophysical Research B*, vol. 108, no. 2141, p. 2, 2003.

[32] K. Spitzer, "A 3-D finite-difference algorithm for DC resistivity modelling using conjugate gradient methods," *Geophysical Journal International*, vol. 123, no. 3, pp. 903–914, 1995.

[33] K. Aizawa, Y. Ogawa, and T. Ishido, "Groundwater flow and hydrothermal systems within volcanic edifices: delineation by electric self-potential and magnetotellurics," *Journal of Geophysical Research B*, vol. 114, no. 1, Article ID B01208, 2009.

12

Seasonal, Diurnal, and Solar-Cycle Variations of Electron Density at Two West Africa Equatorial Ionization Anomaly Stations

Frédéric Ouattara,[1] Doua Allain Gnabahou,[2] and Christine Amory Mazaudier[3]

[1] *Ecole Normale Supérieure de l'Université de Koudougou, P.O. Box 376, Koudougou, Burkina Faso*
[2] *Lycée Provincial de Koudougou, Direction Régional du Centre Ouest, P.O. Box 39, Koudougou, Burkina Faso*
[3] *LPP-UPMC, CNRS, 4 Avenue de Neptune, 94107 Saint Maur des Fossés, Paris, France*

Correspondence should be addressed to Frédéric Ouattara, fojals@yahoo.fr

Academic Editor: Yuichi Otsuka

We analyse the variability of foF2 at two West Africa equatorial ionization anomaly stations (Ouagadougou and Dakar) during three solar cycles (from cycle 20 to cycle 22), that is, from 1966 to 1998 for Ouagadougou and from 1971 to 1997 for Dakar. We examine the effect of the changing levels of solar extreme ultraviolet radiation with sunspot number. The study shows high correlation between foF2 and sunspot number (Rz). The correlation coefficient decreases from cycle 20 to cycle 21 at both stations. From cycle 21 to cycle 22 it decreases at Ouagadougou station and increases at Dakar station. The best correlation coefficient, 0.990, is obtained for Dakar station during solar cycle 22. The seasonal variation displays equinoctial peaks that are asymmetric between March and September. The percentage deviations of monthly average data from one solar cycle to another display variability with respect to solar cycle phase and show solar ultraviolet radiation variability with solar cycle phase. The diurnal variation shows a noon bite out with a predominant late-afternoon peak except during the maximum phase of the solar cycle. The diurnal Ouagadougou station foF2 data do not show a significant difference between the increasing and decreasing cycle phases, while Dakar station data do show it, particularly for cycle 21. The percentage deviations of diurnal variations from solar-minimum conditions show more ionosphere during solar cycle 21 at both stations for all three of the other phases of the solar cycle. There is no significant variability of ionosphere during increasing and decreasing solar cycle phases at Ouagadougou station, but at Dakar station there is a significant variability of ionosphere during these two solar-cycle phases.

1. Introduction

Many ionosphere studies concern ionosphere parameter variability [1, 2] and do not include the African sector [3]. Moreover, some papers deal with the comparison between ionospheric data and the International Reference Ionosphere (IRI) [4–9]. On the other hand, many studies have investigated the solar-cycle variation and/or geomagnetic activity variation of the critical frequency of the F2 layer ([10–16]. It is important to know that few studies integrate African sector data, as noted by Bilitza et al. [3]), and take into account long series of data. In fact we have in the African sector the works which treat the variability of equatorial F2 density [17–23] in the equatorial ionization anomaly (EIA) trough, and in the Asian sector we have the works of Le Huy et al. [24] and Pham Thi Thu et al. [25], which concern the variability of the EIA trough for South East Asia and the southern EIA crest in the Asian sector.

The present study relies on the use of long series of data (three solar-cycles of foF2) which are obtained from the African sector and particularly from the Sub-Saharan African sector. It is well-known that in Africa, and especially in Sub-Saharan Africa, there is a lack of data. In the past only a few ionosonde stations operated (see Figure 1). In Figure 1, green points indicate the stations that operated in 1960. We can see in Africa only four stations, with one station in West Africa (Dakar) and only two for the equatorial region (Dakar (lat: 14.8°N; long: 342.6°E) and Djibouti (lat: 11,5°N; long:

FIGURE 1: Ionosondes in operation in 1960.

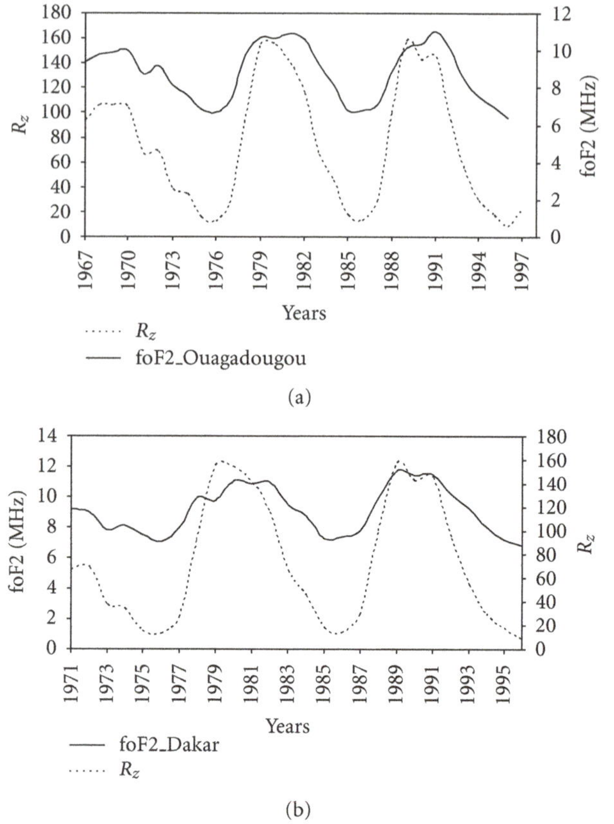

FIGURE 2: Yearly variation of foF2 and R_z. (a) for Ouagadougou station and (b) for Dakar station.

42,8°E)). After 1960, we can add Ouagadougou station (lat: 12.4°N; long: 358.5°E), Tamanrasset station (lat: 22.80°N; long: 354.47°E), Ibadan station (lat: 7.43°N; long: 356.10°E), and recently Korhogo station (lat: 9.3°N; long: 354.62°E).

The objective of this paper is to determine (1) foF2 variability of two West African EIA stations with solar-cycle, season, and time of day and (2) to point out foF2 longitudinal variations. Comparison between data and models will be done in another study. It will be important before testing models to know well the variability of station data.

The structure of the paper is as follows. After the treatment of data and methodology in Sections 2 and 3, we present and discuss our results and end the paper by conclusion as Section 4 of the paper.

2. Data Sets

For this study, F2 layer critical frequency of (foF2) data obtained from two African EIA ionosonde stations: (1) Ouagadougou (lat: 12.4°N; long: 358.5°E; dip: +1.45) and (2) Dakar (lat: 14.8°N; long: 342.6°E; dip: +5.53). These data covered three solar-cycles (cycles 20, 21 and 22) and are provided by the Ecole Nationale de Télécommunication de Bretagne (ENST-Bretagne).

Sunspot number (R_z) data obtained from the SPIDR web-site are also used in order to determine solar-cycle phases.

3. Methodology

Our database contains hourly foF2 values which are going from June 1966 to February 1998 for Ouagadougou and from January 1971 to February 1997 for Dakar. For this work are considered years with 75% monthly available data (i.e., 9/12 ratio of months per year). With this criterion, the available data go from 1971 to 1996 for Dakar and from 1965 to 1997 for Ouagadougou. For Ouagadougou, year 1986 must be excluded, for its available data are 5/12; but for the analysis of solar-cycle 22 data, weights have been used (1 for years with available number of months more than 75%, 0.75 for years with available number of months between 75% and 50%, and 0.5 when available number of months is less than 50%) in order to integrate year 1986.

It is important to note that, for the retained years, all hourly data are available during a day, and for most of the retained years (more than 98%) the number of months exceeds 75%. Therefore, daily values are an arithmetic mean over all hours, monthly values are an arithmetic mean over all days, and annual values are an arithmetic mean over all months.

As foF2 is greatly influenced by solar ultraviolet radiation, foF2 variability with solar-cycle-phase must show solar ultraviolet radiation variation with respect to solar-cycle phase. Solar-cycle phases are determined by considering the following conditions (see [15, 26, 27]: (1) minimum phase: $R_z < 20$, where R_z is the yearly average Zürich sunspot number, (2) ascending phase: $20 \leq R_z \leq 100$ and R_z greater than the previous year's value; (3) maximum phase: $R_z > 100$ (for small solar-cycles (solar-cycles with sunspot number maximum (R_z max) less than 100) the maximum phase is obtained by considering $R_z > 0.8 * R_z\,\text{max}$), and (4) descending phase: $100 \geq R_z \geq 20$ and R_z less than the previous year's value. Table 3 gives the years of the different solar-cycle phases and their R_z mean value.

foF2 variations are analysed by using (1) annual averaged data for solar-cycle variations, (2) monthly averaged data for seasonal variations, and (3) hourly averaged data for diurnal

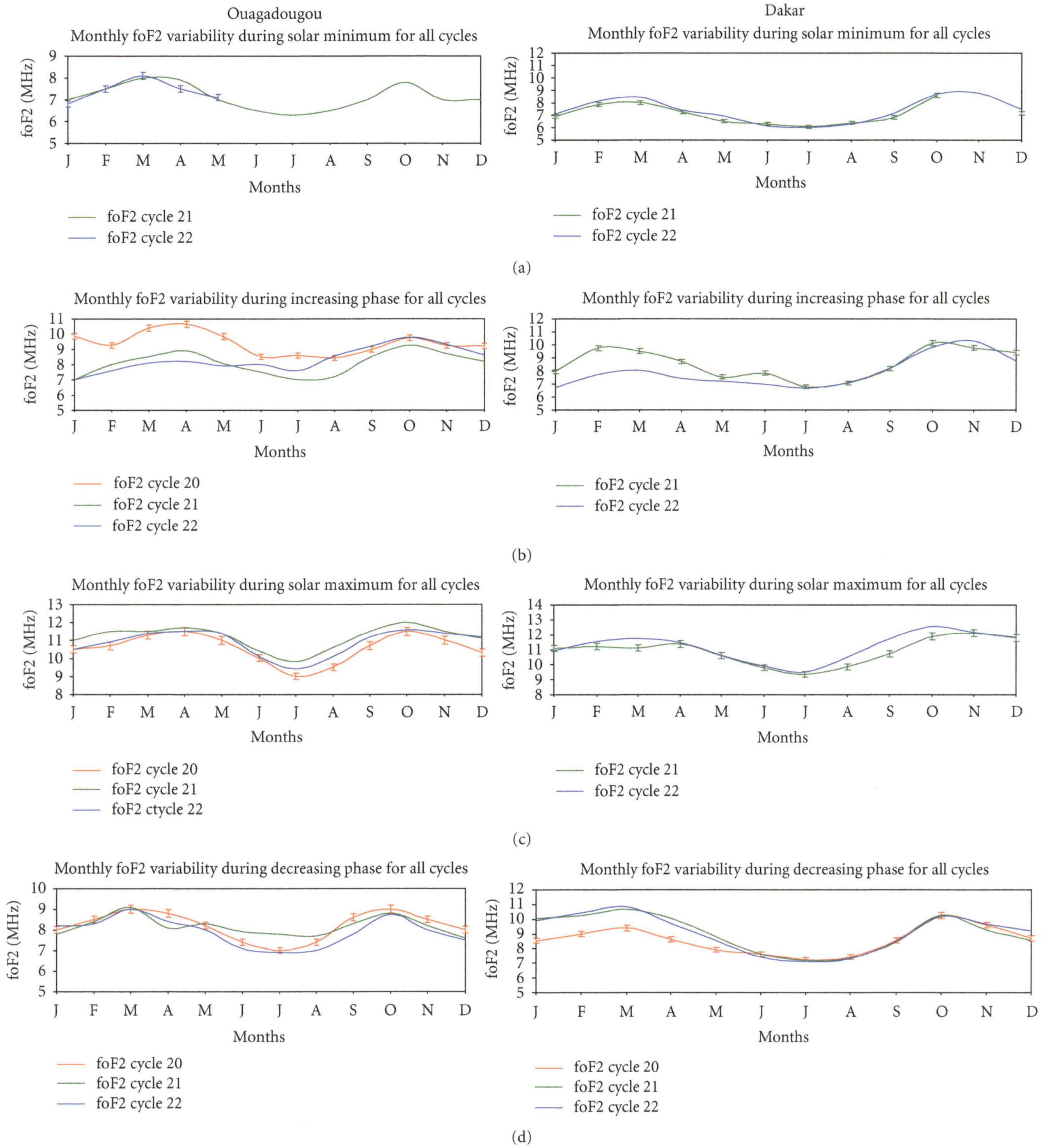

FIGURE 3: Monthly variations of foF2 at the stations of Ouagadougou and Dakar for cycles 20, 21, and 22 (a) during solar minimum, (b) for increasing solar-activity, (c) during solar maximum; and (d) for decreasing solar-activity.

Table 1: Correlation coefficients between foF2 and sunspot number (R_z) from cycle 20 to cycle 22.

Cycle	20	21	22	Location
foF2 correlation coefficient	0.977	0.973	0.948	Ouagadougou
	0.950	0.883	0.990	Dakar

Table 2: Predominance of March or September equinoctial peak for different solar-cycle phases.

Solar cycle phases	Solar cycles: Ouagadougou station 20	21	22	Dakar station 20	21	22	Nature of peak predominance
Minimum	X	X	X				March/April
				X	X	X	September/October
Increasing	X						March/April
		X	X	X	X	X	September/October
Maximum							March/April
	X	X	X	X	X	X	September/October
Decreasing	X	X	X		X	X	March/April
				X			September/October

variations. These analyses are made by taking into account solar-cycle phases.

As the solar-cycle 20 maximum ($R_z = 105.9$) is smaller than the maxima of cycles 21 and 22 (their maxima are comparable: 155.4 and 157.6, resp.; see Figure 2) we put error bars ($\sigma = \sqrt{\Delta}$, where Δ is the variance defined by $(1/N)\sum_{i=1}^{N}(x_i - \overline{x})^2$ with $\overline{x}$ mean value) of solar-cycle 20 data in Figure 3 in order to have a reference for the significance of solar-cycle differences. In case of lack of cycle 20 data, error bars of the solar-cycle 21 data are shown. Error bars of the solar minimum data are also shown in Figure 4 in order to have a reference for the significance of differences from the other solar-cycle phases.

For analyzing foF2 variability, we will use qualitative analysis based on examination of data plots (error bars will help us for this analysis) and quantitative analysis based on percentage deviation, expressed as $\sigma_{\text{rel}} = ((x_i^0 - x_i^m)/x_i^m) \times 100$, where x_i^0 and x_i^m are either (1) the monthly averaged foF2 data during the solar-cycle 20 (in case of lack of solar-cycle 20 data it expresses solar-cycle 21 data) and foF2 data during the other solar-cycles (in case of lack of solar-cycle 20 data it will be solar-cycle 22 data), respectively, for seasonal studies shown in Figure 3 or (2) solar minimum phase foF2 data and foF2 data of the other solar-cycle phase, respectively, for diurnal studies shown in Figure 4. For studying seasonal variation of foF2, we quantify the difference between each solar-cycle of foF2 data and its variability. As $\sigma_{\text{rel}} < 0$ shows higher foF2 and $\sigma_{\text{rel}} > 0$ lower foF2 than the reference less, the diurnal percentage deviation permits to study the variability of ionosphere.

4. Results and Discussions

4.1. Qualitative Analysis. Figure 2 shows the plot of R_z and foF2 for Ouagadougou station (panel a) and Dakar station (panel b) for available data of the three solar-cycles. It can be seen good correlation between R_z and foF2 for these two stations.

Table 1 shows the correlation coefficient between foF2 and R_z throughout the three solar-cycles (20, 21, and 22) for the two stations (Dakar and Ouagadougou). It can be seen for both stations the decrease of correlation coefficient from cycle 20 to cycle 21. The correlation coefficient decreases from cycle 21 to cycle 22 for Ouagadougou station and increases for Dakar station. Throughout the three solar-cycles, the best correlation is seen at Dakar station (0.990) even if the correlation is better at Ouagadougou station than Dakar station for cycles 20 and 21.

Figure 3 presents monthly mean variations of foF2 for the three solar-cycles during the four solar-cycle phases. The left panels concern Ouagadougou data and the right panels Dakar data. Panel (a) corresponds to solar-cycle minimum phase, panel (b) solar-cycle increasing phase, panel (c) solar-cycle maximum phase and panel (d) solar-cycle decreasing phase.

The red lines represent monthly mean variation of cycle 20 data, green lines those of cycle 21 and blue lines cycle 22 data.

The left panel (a) shows a lack of data during the minimum phase of solar-cycle 20 (absence of red line) for Ouagadougou, which operated since 1966 (with data available since 1967). The lack of data is also observed in the right panels (a), (b), and (c) (absence of red lines in these panels), because available data at Dakar station begins in 1971.

Figure 3 highlights the well-known seasonal variation of foF2 with two asymmetric peaks (error bars help us to see this asymmetry) at the equinoxes except at Ouagadougou during the decreasing phase and little bit during the maximum phase. From one cycle to another or from one phase to another, the predominance of the equinoctial peaks varies.

TABLE 3: Years of the different solar-cycle phases and their R_z mean value.

Solar-cycles	Solar-cycle phases				
	Minimum	Increasing	Maximum	Decreasing	Years and R_z mean
20	1964-1965	1966-1967	1968–1970	1971–1974	Years
	12.6	70.4	105.3	52	R_z mean
21	1975-1976	1977-1978	1979–1982	1983-1984	Years
	14.1	60	141.6	56.3	R_z mean
22	1985-1986	1987	1988-1989	1992–1994	Years
	15.7	29.4	136.5	59.6	R_z mean

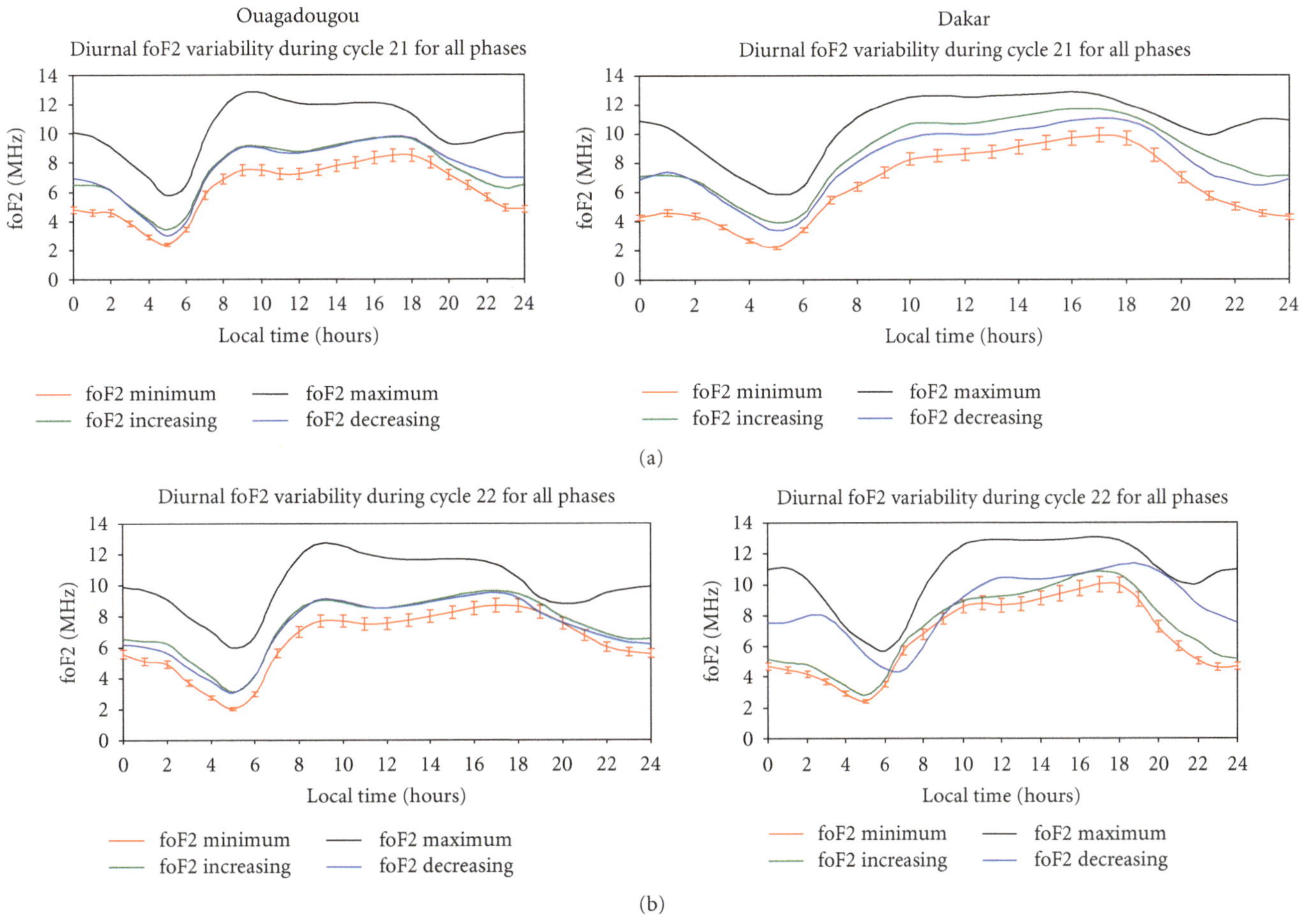

FIGURE 4: Diurnal variation of foF2 at Ouagadougou station (left panels) and at Dakar station (right panels). Panel (a) concerns solar-cycle 31 and panel (b) solar-cycle 22.

During June solstice, at Dakar for all solar-cycles, the density of ionization is the same. Whatever the station, the maximum of ionization appears always in October. The other maximums appear sometimes in March and sometimes in April. During solar-cycle maximum phase, foF2 profiles are regular and the density of ionization grows from cycle 20 to cycle 21.

Table 2 shows the peak predominance. It can be concluded that (1) during the minimum phase each station exhibits the same peak predominance (March/April for Ouagadougou station and September/October for Dakar station); (2) during the maximum phase, only September/October predominance is observed at both stations; (3) during the increasing phase we observe September/October predominance at both stations except for cycle 20 for Ouagadougou; (4) during the decreasing phase March/April predominance is observed for both stations except for solar-cycle 20 at Dakar. These observations (different monthly locations of peak predominance throughout the solar-cycle phases) suggest the necessity to analyse the variability of the ionosphere by taking into account each solar phase and not to consider only minimum and high solar-activity. The

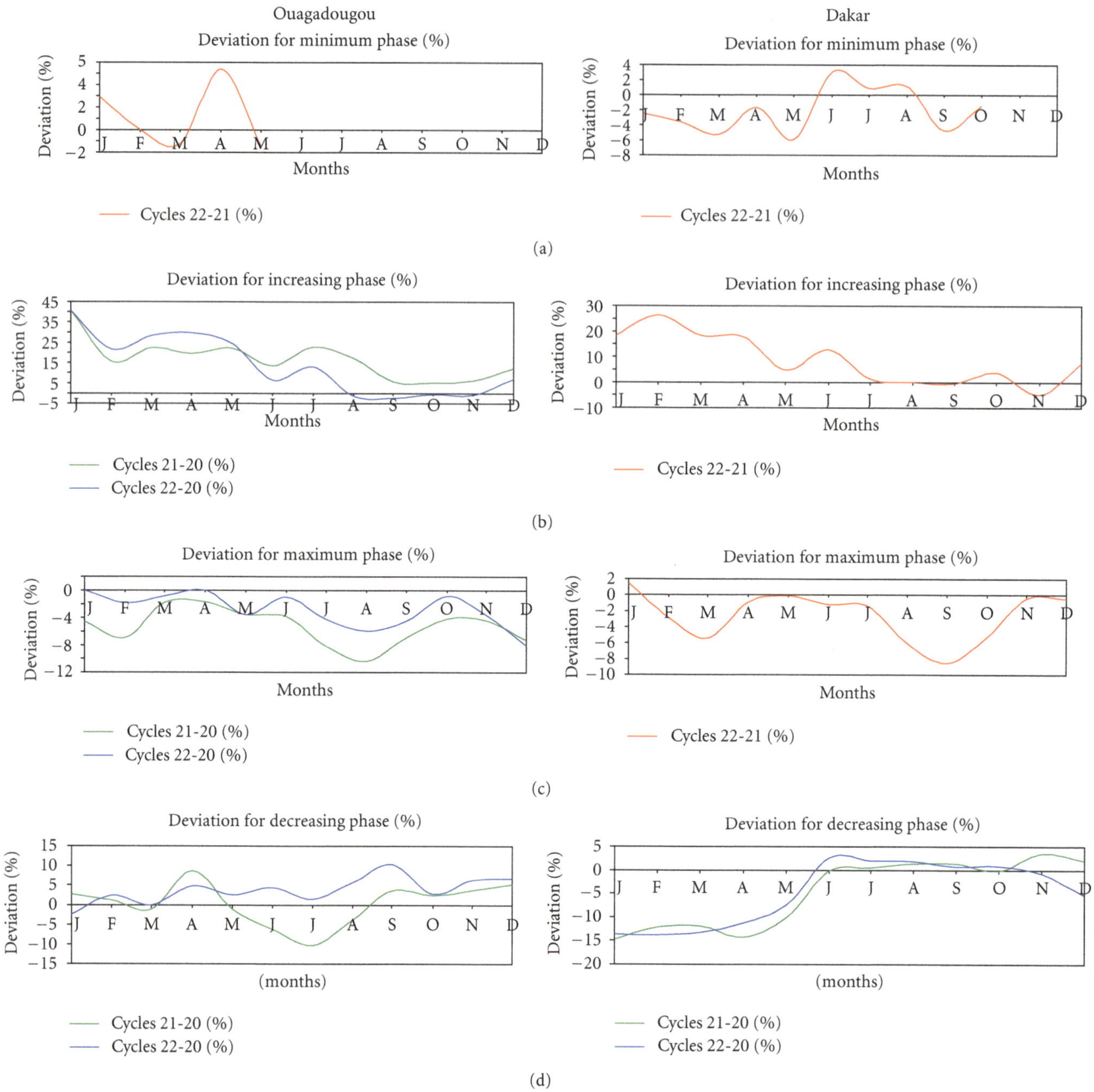

FIGURE 5: Percentage deviations of monthly average data from one solar-cycle to another with respect to month and solar-cycle phase. The left panels are for Ouagadougou and the right panels are for Dakar.

difference of predominant peak locations between the two stations may be due to longitudinal variation of the F2 layer critical frequency.

Figure 4 gives local time variations of foF2 for cycles 21 and 22. We do not consider here cycle 20 because only minimum and decreasing phase data are available. On the left we have Ouagadougou data and on the right Dakar data. Panels (a) and (b) show the local time variation of foF2 during cycles 21 and 22, respectively. Red lines correspond to the minimum phase, green lines the increasing phase, blue lines the decreasing phase, and black lines the maximum phase. foF2 during solar maximum phase of both solar-cycles 21 and 22 presents a secondary maximum during night time, which expresses the effect of the prereversal electric field in foF2 profiles. In fact, eastward daytime electric field at equatorial ionosphere (E and F regions) exhibits a significant increase just it reverses to its night time westward direction [28, 29]. The theoretical models of the low-latitude fields suggest that this enhancement is either caused mainly or entirely by F region winds (e.g., [30–33]) or produced solely by E region tidal winds (e.g., [34, 35]). The prereversal enhancement of the zonal electric field in the equatorial

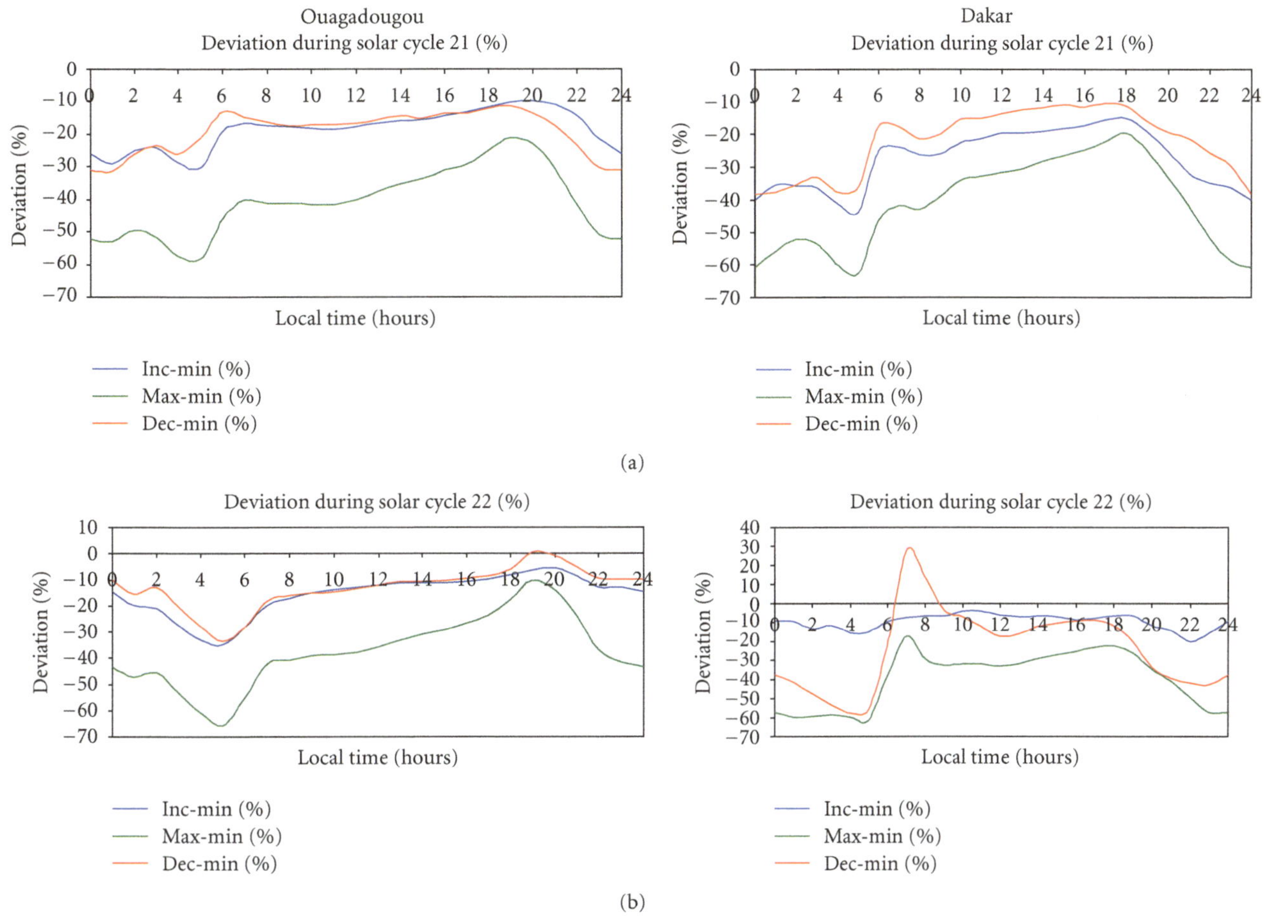

FIGURE 6: Percentage deviations (with sign reversed) of diurnal variations with respect to solar minimum conditions for three solar-cycle phases: increasing (blue), maximum (green), and decreasing (red).

ionosphere is well-known [33] and depends on season, level of magnetic activity, and phase of the solar-cycle [33, 36]. On a given day it might be absent, but it is a persistent feature of averaged data [33]. Fejer et al. [36] note that the occurrence of a sharp increase of the upward velocity in the dusk sector just before it reverses to its downward direction is the main characteristic of the equatorial F region vertical drift. They also pointed out that the evening upward velocity enhancement is responsible for the rapid rise of the equatorial F layer after sunset.

The analysis of Figure 4 points out for Ouagadougou station (left panels) the similarity of the increasing and decreasing phases for both cycles. The foF2 increases from the minimum phase to the maximum phase. The local time variation of foF2 during the all solar-activity phases present a noon bite out, with late-afternoon/evening maxima for all lines except for the maximum phase, where there is a secondary minimum in the early evening. All solar-activity phases foF2 start to increase before sunrise (before 0600 LT) and there is no difference of foF2 between increasing and decreasing phases for both solar-cycles.

Table 3 shows that for solar-cycle 21 solar-activity during increasing phase is higher than during decreasing phase and that it is the reverse for solar-cycle 22. Normally, foF2 of both stations and for both solar-cycles must vary with respect to solar-activity. Therefore the same local time variation of Ouagadougou foF2 data for both solar-cycles during solar increasing and decreasing phases must be explained by the bias of data classification with respect to the solar-cycle activity.

At Dakar (right panels) the lines for the increasing and decreasing phases are similar through a whole day for solar-cycle 21 but only at daytime for cycle 22. foF2 at this station also increases from solar minimum to solar maximum. It can be mentioned from this figure that an enhancement of foF2 occurs around midnight during solar maximum and that the foF2 enhancement can be seen before sunrise during solar decreasing phase. All solar-activity phases foF2 increase before sunrise (before 0600 LT) except during solar-activity maximum and decreasing phases of solar-cycle 22 where foF2 increases after sunrise (between 0600 LT and 0800 LT). The right panel (b) of Figure 4 shows that solar-activity increasing phase foF2 is higher than that of decreasing phase during both solar-cycles except between 0600 LT and 0900 LT during solar-cycle 22 where it is the reverse. By taking into account the results of Table 3, the local time variation of foF2 between 0600 LT and 0900 LT during solar-activity increasing and decreasing phases with no respect to solar-activity must be due to the difference between their foF2 data daytime increasing times.

The Ouagadougou data (left panels) show no difference of foF2 between the solar-activity decreasing and increasing phases. At Dakar (right panels); on the other hand, difference of foF2 between the increasing and decreasing phases is seen. The difference is larger during cycle 21 than that of cycle 22. For solar-cycle 22, foF2 is higher during the decreasing phase than during the increasing phase. This result has been pointed out by Özgüç et al. [12]. It can be also noted that at Dakar station, during solar-cycle 22 at daytime, foF2 does not change from minimum phase to increasing phase (see the error bars shown on the red curve).

The results of Figure 4 prevent us from treating together the variability of the foF2 for the increasing and decreasing phases as done by Bilitza et al. [3] for moderate solar-activity, for it depends on the station. This point of view has been considered in the work of Özgüç et al. [12] and Ataç et al. [13].

The difference between the Dakar and Ouagadougou electron densities shows the necessity to study separately the data from these kinds of stations, as foF2 shows longitudinal effects.

4.2. Quantitative Analysis. Figures 5 and 6 concern the evolution of the percentage deviation. The right panels concern Ouagadougou and the left panels Dakar.

In Figure 5 the green curves express the percentage deviation between the values of foF2 for solar-cycle 21 and these for solar-cycle 20. Blue curves give the percentage deviation between foF2 values of solar-cycle 22 and foF2 values of solar-cycle 20. The red curves express the same thing, but for solar-cycles 21 and 22.

Figure 5 shows for a given solar-cycle phase the same variability of percentage deviation graphs. Percentage deviation graphs decrease (1) from January to March/April during minimum phase, (2) from January to July/August during increasing phase, and (3) from January to August/September during maximum phase. The similar variation of the percentage deviation at both stations for the minimum, increasing, and maximum phases may be expressed the non longitudinal dependence of monthly solar ultraviolet radiation variability.

The phase-to-phase variability of the percentage variations shows the necessity to take into account the different solar-cycle phases in the study of ionosphere. The different variability of the percentage deviation values between the March and September equinox periods is related to the equinoctial asymmetry previously noted in the qualitative analysis. The qualitative analysis shows the phase-to-phase variability of foF2.

Figure 6 shows negative values of percentage deviation. We can assert that there is higher electron density during increasing, maximum, and decreasing solar-cycle phases than solar minimum phase. The quantity of foF2 is the highest during solar maximum phase at both stations and for both solar-cycles.

In Figure 6(a), the quantity of foF2 decreases from maximum phase to increasing phase for all stations and then decreases from increasing phase to decreasing phase at Dakar station; at Ouagadougou station this quantity is fairly constant during daytime.

In Figure 6(b), we observe the same phase-to-phase variation of foF2 at Ouagadougou station while at Dakar station the quantity of foF2 during the decreasing phase is higher than during the increasing phase.

In Figure 6, except during decreasing phase of solar-cycle 22 at Dakar station (blue curve in right panel (b)), all curves show double peaks: morning peak 0600 LT–0800 LT and afternoon or evening peak 1600 LT–2000 LT.

5. Conclusion

This study shows the correlation between foF2 and R_z for Ouagadougou and Dakar data. The correlation coefficient varies from one solar-cycle to another. The best correlation is observed at Dakar during solar-cycle 22. Seasonal variations of foF2 present asymmetric equinoctial peaks which vary among the different solar-cycle phases. The foF2 shows the phase-to-phase variability of the solar-activity due to solar ultraviolet radiation variability. At Ouagadougou station, foF2 during the solar-activity increasing phase is almost identical to that during the decreasing phases while at Dakar station foF2 is higher during the decreasing phase than during the increasing phase. It can be concluded that it is necessary to treat separately the variability of the ionosphere according to each type of solar-cycle phase.

Acknowledgments

Authors thank the Ecole Nationale de Télécomunication de Bretagne (ENST-Bretagne) and, SPIDR webmaster for providing data. Authors also thank Dr. Rolland Fleury and Dr. Patrick Lassudrie Duchesne from the ENST-Bretagne for their collaboration and Dr. Arthur Richmond from HAO at NCAR for his proofreading and advices. Thanks to International Journal of Geophysics editor and reviewers for their kindly remarks, suggestions, and propositions which allow them to improve the paper.

References

[1] H. Risbeth and M. Mendillo, "Patterns of F2 layer variability," *Journal of Atmospheric and Solar-Terrestrial Physics*, vol. 63, pp. 1661–1680, 2001.

[2] D. N. Fotiadis, G. M. Baziakos, and S. S. Kouris, "On the global behaviour of the day-to-day MUF variation," *Advances in Space Research*, vol. 33, no. 6, pp. 893–901, 2004.

[3] D. Bilitza, O. K. Obrou, J. O. Adeniyi, and O. Oladipo, "Variability of foF2 in the equatorial ionosphere," *Advances in Space Research*, vol. 34, no. 9, pp. 1901–1906, 2004.

[4] J. O. Adeniyi and I. A. Adimula, "Comparing the F2-layer model of IRI with observations at Ibadan," *Advances in Space Research*, vol. 15, no. 2, pp. 141–144, 1995.

[5] J. O. Adeniyi and S. M. Radicella, "Diurnal variation of ionospheric profile parameters B0 and B1 for an equatorial

station at low solar activity," *Journal of Atmospheric and Solar-Terrestrial Physics*, vol. 60, no. 3, pp. 381–385, 1998.

[6] J. O. Adeniyi and S. M. Radicella, "Variation of bottomside profile parameters B0 and B1 at high solar activity for an equatorial station," *Journal of Atmospheric and Solar-Terrestrial Physics*, vol. 60, no. 11, pp. 1123–1127, 1998.

[7] J. R. de Souza, G. J. Batley, M. A. Abdu, and I. S. Batista, "Comparison of low latitude F region peak densities, heights and equatorial ExB drift from IRI with obervational data and the Sheffield University plasmasphere ionosphere model," *Advances in Space Research*, vol. 18, pp. 41–44, 1996.

[8] I. S. Batista and M. A. Abdu, "Ionospheric variability at Brazilian low and equatorial latitudes: Comparison between observations and IRI model," *Advances in Space Research*, vol. 34, no. 9, pp. 1894–1900, 2004.

[9] K. O. Obrou, *Ionosphère équatoriale: contribution à l'amélioration du modèle International Reference Ionosphere (IRI)*, Thèse de Doctorat d'état es Sciences Physiques, UFR SSMT. Université de Cocody, 2008.

[10] P. A. Bradley, "Indices of ionospheric response to solar-cycle epoch," *Advances in Space Research*, vol. 13, no. 3, pp. 25–28, 1993.

[11] S. S. Kouris, P. A. Bradley, and I. K. Nissopoulos, "The relationships of foF2 and M(3000)F2 versus R_{12}," in *Proceedings of the COST 238/PRIME/ Workshop*, pp. 155–167, Eindhoven, The Netherlands, 1994.

[12] A. Özgüç, T. Ataç, and R. Pektaş, "Examination of the solar cycle variation of foF2 for cycles 22 and 23," *Journal of Atmospheric and Solar-Terrestrial Physics*, vol. 70, no. 2–4, pp. 268–276, 2008.

[13] T. Ataç, A. Özgüç, and R. Pektaş, "The variability of foF2 in different phases of solar cycle 23," *Journal of Atmospheric and Solar-Terrestrial Physics*, vol. 71, no. 5, pp. 583–588, 2009.

[14] J. Lei, J. P. Thayer, J. M. Forbes et al., "Ionosphere response to solar wind high-speed streams," *Geophysical Research Letters*, vol. 35, no. 19, Article ID L19105, 2008.

[15] F. Ouattara, C. Amory-Mazaudier, R. Fleury, P. Lassudrie Duchesne, P. Vila, and M. Petitdidier, "West African equatorial ionospheric parameters climatology based on ouagadougou ionosonde station data from june 1966 to february 1998," *Annales Geophysicae*, vol. 27, no. 6, pp. 2503–2514, 2009.

[16] D. H. Zhang, X. H. Mo, L. Cai et al., "Impact factor for the ionospheric total electron content response to solar flare irradiation," *Journal of Geophysical Research A*, vol. 116, no. 4, Article ID A04311, 2011.

[17] E. Sambou, P. M. Vila, and A. T. Kobea, "Non-trough foF2 enhancements at near-equatorial dip latitudes," *Annales Geophysicae*, vol. 16, no. 6, pp. 711–720, 1998.

[18] O. K. Obrou, D. Bilitza, J. O. Adeniyi, and S. M. Radicella, "Equatorial F2-layer peak height and correlation with vertical ion drift and M(3000)F2," *Advances in Space Research*, vol. 31, no. 3, pp. 513–520, 2003.

[19] D. Bilitza, O. K. Obrou, J. O. Adeniyi, and O. Oladipo, "Variability of foF2 in the equatorial ionosphere," *Advances in Space Research*, vol. 34, no. 9, pp. 1901–1906, 2004.

[20] O. K. Obrou, J. O. Adeniyi, A. T. Kobea, and B. Moukassa, "Electron density profile parameters B0 and B1 response during a magnetic disturbance at equatorial latitude," *Journal of Atmospheric and Solar-Terrestrial Physics*, vol. 67, no. 5, pp. 515–519, 2005.

[21] O. A. Oladipo, J. O. Adeniyi, S. M. Radicella, and O. K. Obrou, "Variability of equatorial ionospheric electron density at fixed heights below the F2 peak," *Journal of Atmospheric and Solar-Terrestrial Physics*, vol. 70, no. 7, pp. 1056–1065, 2008.

[22] F. Ouattara and J. L. Zerbo, "Ouagadougou station F2 layer parameters yearly and seasonal variations during severe geomagnetic storms generated by CMEs and fluctuating wind streams," *International Journal of Physical Sciences*, vol. 6, no. 20, pp. 4854–4860, 2011.

[23] F. Ouattara and C. Amory Mazaudier, "Statistical study of the diurnal variation of the Equatorial F layer at Ouagadougou from 1966 to 1998," submitted to *special issue of the Journal of Space Weather and Space Climate.*

[24] M. Le Huy, C. T. Nguyen, T. L. Tran et al., "The effect of the geomagnetic storm on the ionospheric total electron content in the Southeast Asian equatorial ionization anomaly region observed by the GPS data," *Journal of Sciences of the Earth*, vol. 29, no. 2, pp. 104–112, 2007.

[25] H. Pham Thi Thu, C. Amory-Mazaudier, and M. Le Huy, "Time variations of the ionosphere at the northern tropical crest of ionization at Phu Thuy, Vietnam," *Annales Geophysicae*, vol. 29, no. 1, pp. 197–207, 2011.

[26] F. Ouattara, "Solar magnetic fields components: phases, profiles and their relationships," *Journal of Science*, vol. 9, no. 2, pp. 9–16, 2009.

[27] J. L. Zerbo, F. Ouattara, C. Zoundi, and A. Gyébré, "Solar cycle 23 and geomagnetic activity since 1868," *La Revue CAMES: La Série A*, vol. 12, no. 2, pp. 255–262, 2011.

[28] B. G. Fejer, D. T. Farley, R. F. Woodman, and C. Calderon, "Dependence of equatorial F region vertical drifts on season and solar cycle," *Journal of Geophysical Research*, vol. 84, no. 10, pp. 5792–5796, 1979.

[29] B. G. Fejer, "The equatorial ionospheric electric fields. A review," *Journal of Atmospheric and Terrestrial Physics*, vol. 43, no. 5-6, pp. 377–386, 1981.

[30] H. Rishbeth, "The F-layer dynamo," *Planetary and Space Science*, vol. 19, no. 2, pp. 263–267, 1971.

[31] R. A. Heelis, P. C. Kendall, R. J. Moffeit, D. W. Windle, and H. Rishbeth, "Electrical coupling of the E- and F-regions and its effect on F-region drifts and winds," *Planetary and Space Science*, vol. 22, no. 5, pp. 743–756, 1974.

[32] N. Matuura, "Electric fields deduced from the thermosphere model," *Journal of Geophysical Research*, vol. 79, p. 4679, 1974.

[33] D. T. Farley, E. Bonell, B. G. Fejer, and M. F. Larsen, "The prereversal enhancement of the zonal electric field in the equatorial ionosphere," *Journal of Geophysical Research*, vol. 91, no. 13, pp. 723–728, 1986.

[34] E. K. Walton and S. A. Bowhill, "Seasonal variations in the low latitude dynamo current system near sunspot maximum," *Journal of Atmospheric and Terrestrial Physics*, vol. 41, no. 9, pp. 937–949, 1979.

[35] R. J. Stening, "A two-layer ionospheric dynamo calculation," *Journal of Geophysical Research*, vol. 86, p. 3543, 1981.

[36] B. G. Fejer, E. R. de Paula, S. A. Gonzalez, and R. F. Woodman, "Average vertical and zonal F region plasma drifts over Jicamarca," *Journal of Geophysical Research*, vol. 96, no. 8, pp. 901–906, 1991.

13

On a Correlation between the Ionospheric Electric Field and the Time Derivative of the Magnetic Field

R. R. Ilma,[1] M. C. Kelley,[1] and C. A. Gonzales[2]

[1] *School of Electrical and Computer Engineering, Cornell University, Ithaca, NY 14853, USA*
[2] *IBM Thomas J. Watson Research Center, Yorktown Heights, NY 10598, USA*

Correspondence should be addressed to R. R. Ilma, rri5@cornell.edu

Academic Editor: Yuichi Otsuka

A correlation of the ionospheric electric field and the time derivative of the magnetic field was noticed over thirty years ago and has yet to be explained. Here we report on another set of examples during the superstorm of November 2004. The electric field in the equatorial ionosphere, measured with the Jicamarca incoherent scatter radar, exhibited a 3 mV/m electric field pulse that was not seen in the interplanetary medium. It was, however, accompanied by a correlation with the time derivative of the magnetic field measured at two points in Peru. Our inclination was to assume that the field was inductive. However, the time scale of the pulse was too short for the magnetic field to penetrate the crust of the Earth. This means that the area threaded by $\partial B/\partial t$ was too small to create the observed electric field by induction. We suggest that the effect was caused by a modulation of the ring current location relative to the Earth due to the electric field. This electric field is required, as the magnetic field lines are considered frozen into the plasma in the magnetosphere. The closer location of the ring current to the Earth in turn increased the magnetic field at the surface.

1. Introduction

In his Ph.D. thesis, Gonzales [1] published Figure 1 using interplanetary and auroral indices, electric field measurements using the incoherent scatter radar (ISR) technique at Chatanika (Alaska) and Jicamarca (Peru), and $\partial B/\partial t$ measured on the ground at San Juan, Puerto Rico. The correlation between the latter three parameters was excellent. The ratio $E/(\partial B/\partial t)$ was 50 million meters. Here we report on an extensive set of observations we believe to be of the same type obtained during the superstorm of November 2004 (see [2], and companion papers).

2. Data Presentation

Various unusual phenomena that occurred in the November 2004 magnetic storm are documented in a series of papers, the first of which summarizes many of the observations from the interplanetary medium to the equatorial ionosphere [2]. The second part of the two-phase storm was monitored by the ISR chain in the American sector with unprecedented coverage. At Jicamarca, Peru, the eastward electric fields can be deduced from ISR drifts obtained from pulse-to-pulse experiments when the transmitting antenna was pointed perpendicular to the geomagnetic field **B** [3]. The radial velocity is a measurement of F-region vertical $\mathbf{E} \times \mathbf{B}$ drift with typical time and range resolutions of 5 minutes and 15 km, respectively. The uncertainty of these ion drifts is less than 1 m/s. In addition to the observed electric field, the horizontal component (H) of the geomagnetic field was measured at the stations indicated in Table 1 and included in the analysis. The magnetic latitudes of these stations were calculated using the International Geomagnetic Reference Field (IGRF) magnetic field model. these stations are located in the American Sector. The Jicamarca magnetometer provides measurements every second with an uncertainty of 0.1 nT and the Piura magnetometer every 10 seconds with 1 nT of resolution [4]. From all magnetic stations, the 1-minute average of measurements was utilized. Finally, to consider a magnetospheric perspective, the parallel component of the magnetic field measured by the geostationary satellite GOES 12 is also used in this study.

TABLE 1: Magnetometer stations.

Station name	Location	Geographic longitude, deg	Geographic latitude, deg	Magnetic latitude, deg
Jicamarca	Peru	283.13	−11.95	−1.73
Piura	Peru	279.36	−5.18	4.96
College	Alaska	212.14	64.87	65.38

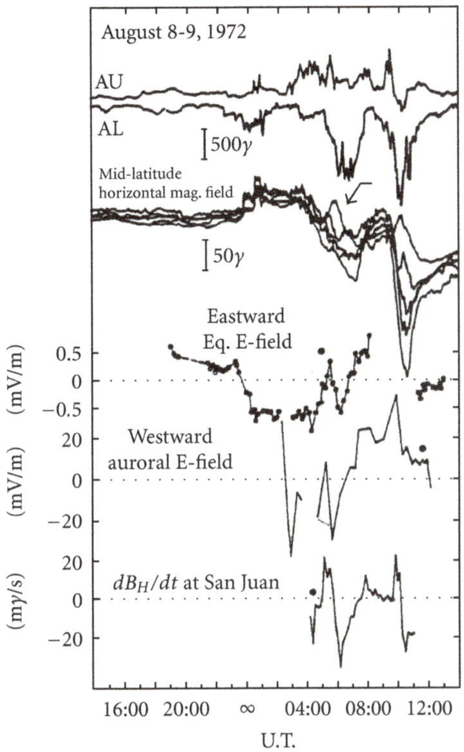

FIGURE 1: From top to bottom: The AU, AL indices; five superposed midlatitude magnetograms (Kakioka, Tashkent, Tangerang, San Juan, Honolulu) with the arrow pointing to the San Juan magnetogram; the zonal electric field at the equator (Jicamarca, Peru); the auroral zonal electric field at College, Alaska; the time derivative of the horizontal component of the magnetic field at San Juan, Puerto Rico. In the last three panels, the dots indicate local magnetic midnight. (After [5]. Reproduced with permission of the American Geophysical Union).

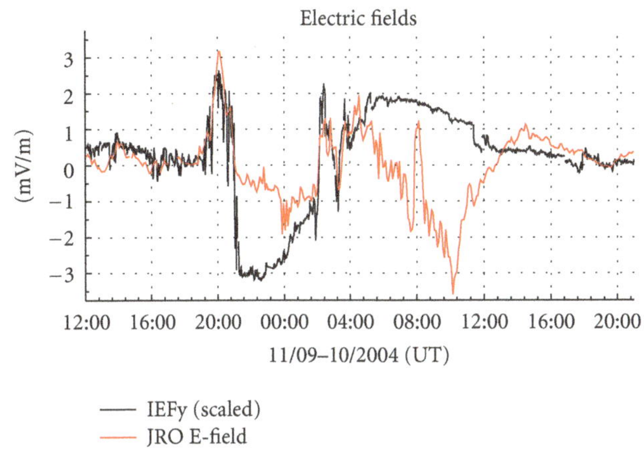

FIGURE 2: The black-line plot is the y-component of the electric field in the Earth's frame of rest delayed by the transit time of the solar wind to the front of the magnetosphere divided by ten for a 33 h period on November 2004. The red-line plot is the zonal component of the electric field in the equatorial ionosphere. (After [8]. Reproduced with permission of the European Geoscience Union).

3. Data Analysis

In the November 2004 event, the interplanetary electric field was correlated at a level of 85% with the zonal electric field measured at the Jicamarca Radio Observatory (JRO) [2]. The basic result is presented in Figure 2 where the delayed interplanetary electric field [6] is plotted, along with the zonal component of the electric field measured in the F region over JRO. The seeming lack of correlation after 0500 UT on 10 November 2004 is due to the local time dependence of the penetration electric field [6]. As was shown by [7], this deviation is expected theoretically and, in fact, shows that the two fields continue to be highly anticorrelated if local time is taken into consideration. But of interest here is the huge spike in the JRO electric field at 0800 UT. The change in the electric field is 3.0 mV/m, which, by equatorial standards, is a very large field. Curiously, there is no corresponding interplanetary electric field change. We conclude that this was due to a magnetospheric source.

In Figure 3, we compare the JRO electric field with the poleward component of the magnetic field measured on the ground at two locations in Peru: at Jicamarca and at Piura, about 1000 km north. The magnetic fields at Jicamarca and Piura are 90 degrees out of phase with the JRO electric field during the time of the large electric field change, as is clearly seen in Figure 5. In Figure 4, we plot the electric field along with the time derivative of the Jicamarca magnetic field. During the time period of 0720-0820 UT, which includes the largest electric field pulse, the two are in phase. The period of the pulse is about an hour. At other times, the fluctuations occur at higher frequencies and the phase shift is variable. During the period of 0700-0830 UT, the correlation between the JRO electric field and the time derivative of the local magnetic field is 71%. The peak time derivative of the magnetic field is 6 nT/min.

Figure 5 shows that the electric field and the magnetic field are 90 degrees out-of-phase. The large electric field pulse begins at 0740 UT. At 0800, GOES registered a dipolarization (see Section 4 for a definition) event in which the parallel component of B increased by over 40 nT. The time delay observed in the satellite data likely is due to the location of GOES relatively near the Earth. The electric field essentially

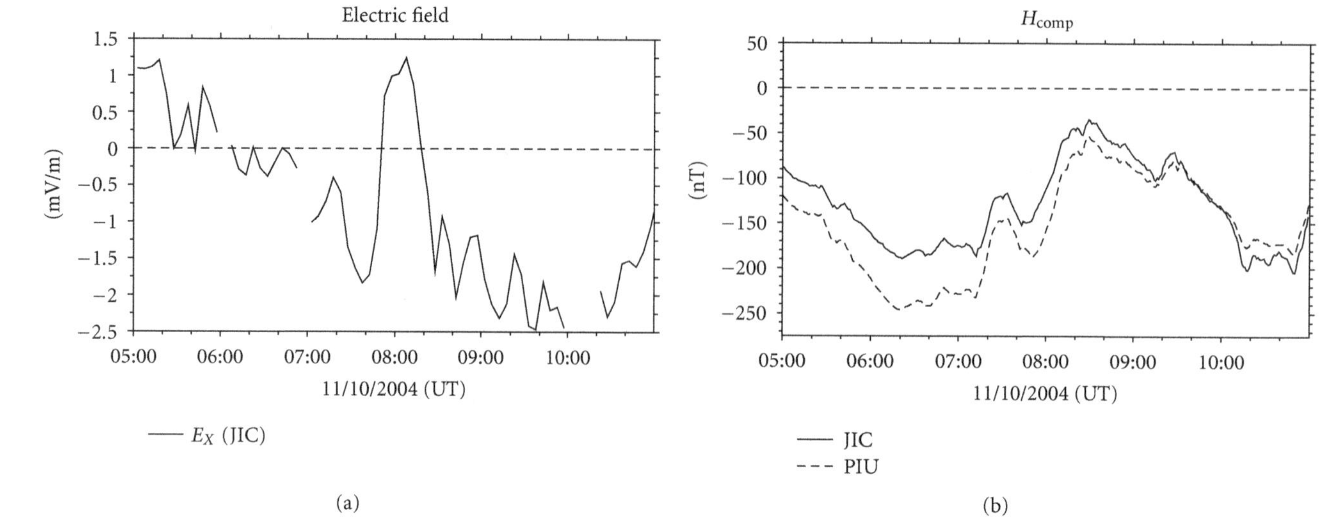

Figure 3: The eastward electric field at Jicamarca for a 6 hr period on 10 November 2004. (b) shows the horizontal magnetic field on (Jicamarca) and off (Piura) the equator.

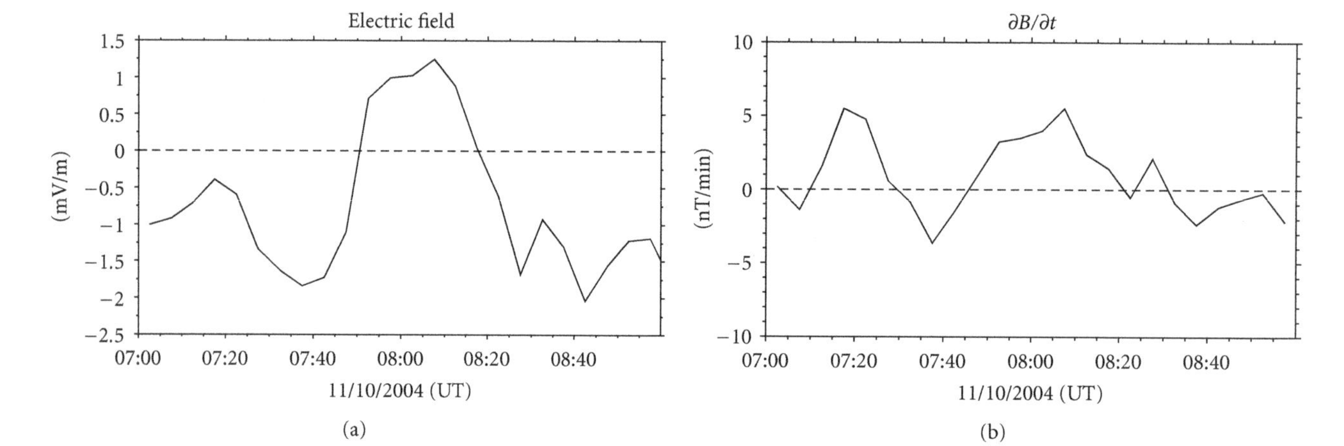

Figure 4: Comparison of the time-derivative of the horizontal component of the magnetic field simultaneously with the eastward electric field. Both quantities were measured at Jicamarca, Peru.

penetrates instantaneously [6]. In Figure 6, we investigate the 75° W GOES magnetic field component perpendicular to the equatorial plane.

4. Discussion

Maxwell's equations involve partial time and space derivatives, as opposed to the total time derivatives, which include advective terms. Comparison of the Jicamarca and equatorial magnetic fields indicates that the magnetic field of the most importance is northward at the nighttime equator. Using the integral form of Maxwell's equation, $\oint E \cdot dl = \iint (\partial B/\partial t) ds^2$, and assuming that the magnetic field entirely penetrates the Earth, $\partial B/\partial t \sim 0.07$ nT/s, and so we can estimate the induced electric field from

$$2\pi R_E E = \pi R_E^2 \frac{\partial B}{\partial t}, \quad (1)$$

and hence,

$$E = \frac{\partial B}{\partial t}\frac{R_E}{2} = 0.2\,\text{mV/m}, \quad (2)$$

which underestimates E by a factor of ten for the November 2004 event. However, this result requires the external magnetic field to penetrate the entire Earth. Penetration phenomena are referred to as magnetic field diffusion and are described by

$$\frac{\partial B}{\partial t} = \frac{\eta}{\mu_0}\nabla^2 B, \quad (3)$$

where η is the resistivity of the medium. The time constant for penetrating the magnetic field by a distance of L is thus the order of

$$\tau = \frac{\mu_0 L^2}{\eta}. \quad (4)$$

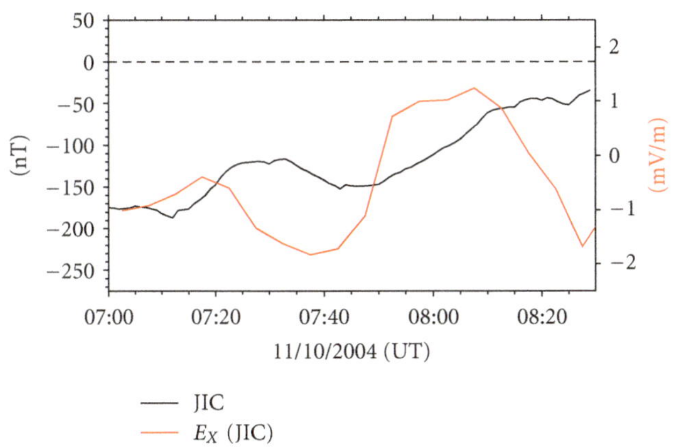

FIGURE 5: The electric and magnetic field time-series, 90 degrees out of phase over Jicamarca, Peru.

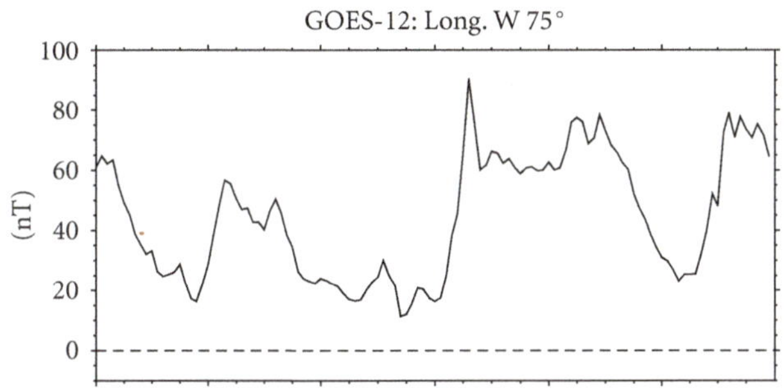

UT	07:20	07:40	08:00	08:20	08:40
X (GSE)	1.8	−0.7	−3	1.4	−1.1
Y (GSE)	−4.1	−3.3	−1	−4.1	−3.1
Z (GSE)	2.4	4	4	2.7	4.1

FIGURE 6: The GOES-12 (Long. W 75) parallel component of the magnetic field versus universal time (UT) and GSE coordinates.

Using typical crustal values of $\eta = 100\,\Omega/\text{m}$, we find that it takes 100 hours to penetrate $1R_E$ into the planet. Thus, the line integral of the electric field around the Earth at F-region heights must be equal to the surface integral of $\partial B/\partial t$ in the area between, say, $H = 300$ km and the surface. Thus, we have

$$2\pi R_E \sim (2\pi R)H\frac{\partial B}{\partial t}, \tag{5}$$

where we have approximated the area enclosed by the line integral. Substituting values and solving for E yields only 0.04 mV/m, which is almost two orders of magnitude too small. It was tempting, of course, to treat the electric field as an inductive effect, but the numbers just do not add up. As shown above, the solid earth is a good enough conductor that, on the time scale of the pulse, it cannot even penetrate the crust.

So what is happening? Without a penetrating electric field from the IMF, we could posit an electric field generated in the magnetosphere, say, during a magnetic substorm. In Figure 7, we compare the magnetic field measured in Alaska with the electric field in Peru for the event. Reference [9] showed that the key link between perturbations in the

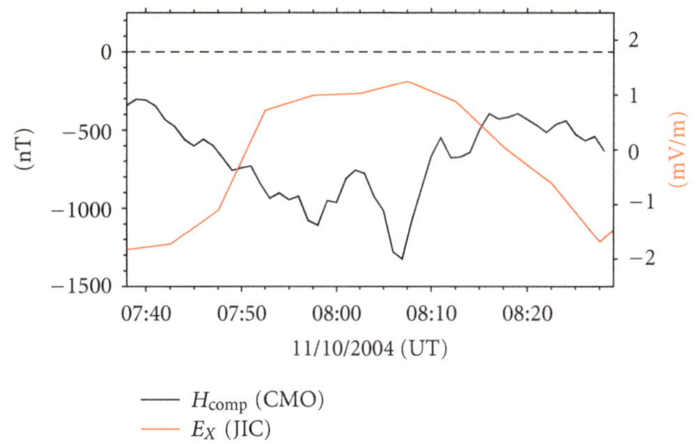

FIGURE 7: Time-derivative of the horizontal component of the magnetic field at the polar zone (College, Alaska; black solid curve) simultaneously with the eastward component of the electric field in the equatorial ionosphere (Jicamarca, Peru; red curve).

equatorial F-region electric fields and the interplanetary magnetic field is the electric field of magnetospheric origin. If there is a substorm-related, magnetospherically induced electric field oriented from dusk to dawn across the nightside, the ring current will move closer to the Earth, causing an increase in the magnetic field at the surface that will be proportional to the distance to the ring current. Since distance is the integral of the velocity and the electric field, and if the magnetospheric field penetrates to the equator, as it must, E and $\partial B/\partial t$ will be correlated. During a substorm, magnetic energy stored in the distorted magnetosphere is released into the atmosphere as accelerated electrons and Joule heating. At such a time, the magnetic field relaxes towards dipole, which is the lowest energy state. This is called dipolarization, which is associated with the release of stored magnetic energy into the ionosphere-atmosphere system. Such an event was reported by [10] using the THEMIS instrumentation. Since the magnetic field lines can be considered to be frozen into the plasma, an eastward electric field must be associated with this dipolarization. They found that the equatorial magnetospheric electric field was in the range of 2–6 mV/m. Observations of penetrating electric fields (PPE) at the equator showed that the PPE is at least 0.5 times the equatorial field [11]. The magnetic field at the center of a current loop is inversely proportional to the radius a, that is, $B = KI/a$. Thus,

$$\frac{\partial B}{\partial t} = -KI\frac{1}{a^2}\frac{\partial a}{\partial t} = B\frac{1}{a}\frac{\partial a}{\partial t}. \tag{6}$$

If $B = 100\,\text{nT}$, $a = R_e$, and $\partial a/\partial t = E/B_0 = 2\,\text{km/s}$, then $\partial B/\partial t = 2\,\text{nT/min}$. These parameters are conservative since a strong ring current could locate at $a = 3R_e$ and produce 400 nT at Earth. This mechanism seems to explain the observations. It is well known that inductive electric fields create problems for electric power systems (see [12], and references therein). We suggest that the time-varying electric fields potential reported here could be a source of power system disruptions at middle and low latitudes.

Acknowledgments

The authors gratefully acknowledge the Jicamarca staff for providing radar and magnetic data. Magnetic data at College, Alaska, was obtained from the International Real-time Magnetic Observatory Network (INTERMAGNET) database. Magnetic data from GOES 12 were obtained from the Space Physics Interactive Data Resource (SPIDR) database. M. Kelley and R. Ilma were supported by the National Science Foundation (NSF) under Grant ATM-0551107. The Jicamarca Radio Observatory is a facility of the Geophysical Institute of Peru and is operated with support from the NSF Cooperative Agreement ATM-0432565 through Cornell University.

References

[1] C. A. Gonzales, *Electric fields in the low latitude ionosphere and their relationship to magnetospheric and interplanetary phenomena*, Ph.D. thesis, Cornell University, Ithaca, NY, USA, 1979.

[2] M. C. Kelley, R. R. Ilma, M. Nicolls et al., "Spectacular low- and mid-latitude electrical fields and neutral winds during a superstorm," *Journal of Atmospheric and Solar-Terrestrial Physics*, vol. 72, no. 4, pp. 285–291, 2010.

[3] E. Kudeki, R.F. Woodman, and Z. Feng, "Incoherent scatter radar plasma density measurements at Jicamarca using a transverse-mode differentialphase method," *Geophysical Research Letters*, vol. 30, no. 1255, p. 4, 2003.

[4] D. Anderson, A. Anghel, J. Chau, and O. Veliz, "Daytime vertical E × B drift velocities inferred from ground-based magnetometer observations at low latitudes," *Space Weather*, vol. 21, no. 11001, p. 9, 2004.

[5] C. A. Gonzales, M. C. Kelley, B. G. Fejer, J. F. Vickrey, and R. F. Woodman, "Equatorial electric fields during magnetically disturbed conditions. II—implications of simultaneous auroral and equatorial measurements," *Journal of Geophysical Research*, vol. 84, pp. 5803–5812, 1979.

[6] M. C. Kelley and E. Dao, "On the local time dependence of the penetration of solar wind-induced electric fields to the magnetic equator," *Annales Geophysicae*, vol. 27, no. 8, pp. 3027–3030, 2009.

[7] R. W. Nopper and R. L. Carovillano, "Polar-equatorial coupling during magnetically active periods," *Geophysical Research Letters*, vol. 5, pp. 699–702, 1978.

[8] M. C. Kelley, R. R. Ilma, and G. Crowley, "On the origin of prereversal enhancement of the zonal equatorial electric field," *Annales Geophysicae*, vol. 27, pp. 2053–2056, 2009.

[9] M. C. Kelley, B. G. Fejer, and C. A. Gonzales, "An explanation for anomalous equatorial ionospheric electric fields associated with a northward turning of the interplanetary magnetic field," *Geophysical Research Letters*, vol. 6, pp. 301–304, 1979.

[10] A. Runov, V. Angelopoulos, M. I. Sitnov et al., "THEMIS observations of an earthward-propagating dipolarization front," *Geophysical Research Letters*, vol. 36, no. 14, Article ID L14106, 2009.

[11] C. A. Gonzales, M. C. Kelley, R. A. Behnke, J. F. Vickrey, R. Wand, and J. Holt, "On the longitudinal variations of the ionospheric electric field during magnetospheric disturbances," *Journal of Geophysical Research*, vol. 88, pp. 9135–9144, 1983.

[12] J. G. Kappenman, "An overview of the impulsive geomagnetic field disturbances and power grid impacts as130 sociated with the violent Sun-Earth connection events of 29-31 October 2003 and a comparative evaluation with other contemporary storms," *Space Weather*, vol. 3, article S08C01, 2005.

14

Frequency-Dependent Streaming Potentials: A Review

L. Jouniaux[1] and C. Bordes[2]

[1] *Institut de Physique du Globe de Strasbourg, UdS-CNRS UMR 7516, Université de Strasbourg, 5 rue René Descartes, 67084 Strasbourg, France*
[2] *Laboratoire des Fuides Complexes et leurs Réservoirs, UPPA-CNRS UMR5150, Université de Pau et des Pays de l'Adour, 64013 Pau, France*

Correspondence should be addressed to L. Jouniaux, laurence.jouniaux@unistra.fr

Academic Editor: Tsuneo Ishido

The interpretation of seismoelectric observations involves the dynamic electrokinetic coupling, which is related to the streaming potential coefficient. We describe the different models of the frequency-dependent streaming potential, mainly Packard's and Pride's model. We compare the transition frequency separating low-frequency viscous flow and high-frequency inertial flow, for dynamic permeability and dynamic streaming potential. We show that the transition frequency, on a various collection of samples for which both formation factor and permeability are measured, is predicted to depend on the permeability as inversely proportional to the permeability. We review the experimental setups built to be able to perform dynamic measurements. And we present some measurements and calculations of the dynamic streaming potential.

1. Introduction

Electrokinetics arise from the interaction between the rock matrix and the pore water. Therefore electrokinetic phenomena are often observed in aquifers, volcanoes, and hydrocarbon or hydrothermal reservoirs. Observations show that seismoelectromagnetic signals associated to earthquakes can be induced by electromagnetic induction [1, 2] or by electrokinetic effect [3, 4]. The electrokinetic phenomena are due to pore pressure gradients leading to fluid flow in the porous media or fractures and inducing electrical fields. These electrokinetic effects are associated to the electrical double layer which was originally described by Stern. The electrokinetic signals can be induced by global displacements of the reservoir fluids (streaming potential) or by the propagation of seismic waves (seismoelectromagnetic effect). As soon as these pressure gradients have a transient signature, the dynamic part of the electrokinetic coupling has to be taken into account by introducing the dependence on fluid transport properties.

It is generally admitted that two kinds of seismoelectromagnetic effects can be observed. The dominant contribution, commonly called "coseismic", is generated close to the receivers during the passage of seismic waves. The second kind, so-called "interfacial conversion" [5], is very similar to dipole radiation and is generated at physicochemical interfaces due to strong electrokinetic coupling discontinuities. This interface conversion is often perceived to have the potential to detect fine fluids transitions with higher resolution than seismic investigations, but in practice, signals are often masked by electromagnetic disturbances, especially when generated at great depth.

Nevertheless recent field studies have focused on the seismoelectric conversions linked to electrokinetics in order to investigate oil and gas reservoirs [6] or hydraulic reservoirs [5, 7–13]. It has been shown using these investigations that not only the depth of the reservoir can be deduced, but also the geometry of the reservoir can be imaged using the amplitudes of the electroseismic signals [14]. Moreover fractured zones can be detected and permeability can be measured using seismoelectrics in borehole [15–18]. This method is especially appealing to hydrogeophysics for the detection of subsurface interfaces induced by contrasts in permeability, in porosity, or in electrical properties (salinity and water content) [19–21].

The analytical interpretation of the seismoelectromagnetic phenomenon has been described by Pride [22], by connecting the theory of Biot [23] for the seismic wave propagation in a two-phase medium with Maxwell's equations, using dynamic electrokinetic couplings. The seismoelectromagnetic conversions have been modeled in homogeneous or layered saturated media [12, 21, 24–26] with applications to reservoir geophysics [27].

Theoretical developments showed that the electrical field induced by the P-waves propagation is related to the acceleration [12]. The electrokinetic coupling is created at the interface between grains and water, when there is a relative motion of electrolyte ions with respect to the mineral surface. Thus, seismic wave propagation in fluid-filled porous media generates conversions from seismic to electromagnetic energy which can be observed at the macroscopic scale, due to this electrokinetic coupling at the pore scale. The seismoelectric coupling is directly dependent on the fluid conductivity, the fluid density, and the electric double layer (the electrical interface between the grains and the water) (see [28], in this special issue "Electrokinetics in Earth Sciences" for more details). For more details on the surface complexation reactions see Davis et al. [29] or Guichet et al. [30]. It can be accurately quantified in the broad band by a dynamic coupling [22] which can be linked in the low-frequency limit to the steady-state streaming potential coefficient largely studied in porous media [30–44].

Laboratory experiments have also been investigated for a better understanding of the seismoelectric conversions [45–56]. These papers describe the laboratory studies performed to investigate this dynamic coupling. An oscillating pore pressure must be applied to a rock sample, and because of the relative motion between the rock and the fluid, an induced streaming potential can be measured. Depending on the oscillating frequency of the fluid, the fluid makes a transition from viscous dominated flow to inertial dominated flow. As the frequency increases, the motion of the fluid within the rock is delayed and larger pressure is needed. In order to know the dynamic coupling, both real and imaginary parts of the streaming potential must be measured.

2. From Dynamic Streaming Potential to Seismoelectromagnetic Coupling

The steady-state streaming potential coefficient is defined as the ratio of the streaming potential to the driving pore pressure:

$$\mathbf{C}_{s0} = \frac{\Delta V}{\Delta P} = \frac{\varepsilon\zeta}{\eta\sigma_f}, \tag{1}$$

which is called the Helmholtz-Smoluchowski equation, where σ_f, ε, and η are the fluid conductivity, the dielectric constant of the fluid, and the fluid dynamic viscosity respectively (see [28]). In this formula the surface electrical conductivity is neglected compared to the fluid electrical conductivity. The potential ζ is the electrical potential within the double layer on the slipping plane. Although the zeta potential can hardly be modeled for a rock and although it cannot be directly measured within a rock, the steady-state streaming potential coefficient can be measured in laboratory, by applying a fluid pressure difference (ΔP) and by measuring the induced streaming electric potential (ΔV) [30, 38, 39, 44, 57]. The electrical potential ζ itself depends on fluid composition and pH and the water conductivity [29–31, 38, 40, 42, 44, 58].

2.1. Packard's Model. Packard [59] proposed a model for the frequency-dependent streaming potential coefficient for capillary tubes, assuming that the Debye length is negligible compared to the capillary radius, based on the Navier-Stokes equation:

$$\mathbf{C}_{s0}(\omega) = \frac{\Delta V(\omega)}{\Delta P(\omega)} = \left(\frac{\varepsilon\zeta}{\eta\sigma_f}\right)\left(\frac{2}{a\sqrt{i\omega\rho_f/\eta}}\frac{J_1\left(a\sqrt{i\omega\rho_f/\eta}\right)}{J_0\left(a\sqrt{i\omega\rho_f/\eta}\right)}e^{-i\omega t}\right), \tag{2}$$

where ω is the angular frequency, a is the capillary radius, J_1 and J_0 are the Bessel functions of the first order and the zeroth order, respectively, and ρ_f is the fluid density.

The transition angular frequency for a capillary is

$$\omega_c = \frac{\eta}{\rho_f a^2}. \tag{3}$$

More recently Reppert et al. [60] used the low- and high-frequency approximations of the Bessel functions to propose the following formula, which corresponds to their equation 26 corrected with the right exponents −2 and −1/2:

$$\mathbf{C}_{s0}(\omega) = \left(\frac{\varepsilon\zeta}{\eta\sigma_f}\right)\left[1 + \left(\frac{-2}{a}\sqrt{\frac{\eta}{\omega\rho_f}}\left(\frac{1}{\sqrt{2}} - \frac{1}{\sqrt{2}}i\right)\right)^{-2}\right]^{-1/2} \tag{4}$$

with the transition angular frequency

$$\omega_c = \frac{8\eta}{\rho_f a^2} \tag{5}$$

and showed that this model was not very different from the model proposed by Packard [59].

The complete development relating Biot's theory and Maxwell's equations has been published by Pride in 1994 [22].

2.2. Pride's Model. Pride [22] derived the equations governing the coupling between seismic and electromagnetic wave propagation in a fluid-saturated porous medium from first principles for porous media. The following transport equations express the coupling between the mechanical and electromagnetic wavefields ([22] (174), (176), and (177)):

$$\begin{aligned}\mathbf{J} &= \sigma(\omega)\mathbf{E} + L(\omega)\left(-\nabla p + i\omega^2\rho_f\mathbf{u}_s\right),\\ -i\omega\mathbf{w} &= L(\omega)\mathbf{E} + \frac{k(\omega)}{\eta}\left(-\nabla p + i\omega^2\rho_f\mathbf{u}_s\right).\end{aligned} \tag{6}$$

In the first equation, the macroscopic electrical current density $\mathbf{J}$ is the sum of the average conduction and streaming current densities. The filtration velocity $\mathbf{w}$ of the second equation is separated into electrically and mechanically induced contributions. The electrical fields and mechanical forces that create the current density $\mathbf{J}$ and filtration velocity $\mathbf{w}$ are, respectively, $\mathbf{E}$ and $(-\nabla p + i\omega^2 \rho_f \mathbf{u_s})$, where p is the pore-fluid pressure, $\mathbf{u_s}$ is the solid displacement, and $\mathbf{E}$ is the electric field. The complex and frequency-dependent electrokinetic coupling $L(\omega)$, which describes the coupling between the seismic and electromagnetic fields [22, 60], is the most important parameter in these equations. The other two coefficients, $\sigma(\omega)$ and $k(\omega)$, are the electric conductivity and dynamic permeability of the porous material, respectively.

The seismoelectric coupling that describes the coupling between the seismic and electromagnetic fields is complex and frequency-dependent Pride [22]:

$$L(\omega) = L_0 \left[1 - i \frac{\omega}{\omega_c} \frac{m}{4} \left(1 - 2\frac{d}{\Lambda} \right)^2 \left(1 - i^{3/2} d \sqrt{\frac{\omega \rho_f}{\eta}} \right)^2 \right]^{-1/2}, \tag{7}$$

where L_0 is the low-frequency electrokinetic coupling, d is related to the Debye-length, Λ is a porous-material geometry term [65], and m is a dimensionless number (detailed in Pride [22]).

The transition angular frequency ω_c separating low-frequency viscous flow and high-frequency inertial flow is defined as

$$\omega_c = \frac{\phi \eta}{\alpha_\infty k_0 \rho_f}, \tag{8}$$

where ϕ is the porosity, k_0 is the intrinsic permeability, and α_∞ is the tortuosity.

2.3. Further Considerations. The low-frequency electrokinetic coupling L_0 is related to the steady-state streaming potential coefficient $\mathbf{C_{s0}}$ by

$$L_0 = -\mathbf{C_{s0}} \sigma_r, \tag{9}$$

where σ_r is the rock conductivity. The electrokinetic coupling $L(\omega)$ can be estimated by considering that steady-state models of $\mathbf{C_{s0}}$ can be applied to the calculation of L_0. When writting $\sigma_r = \sigma_f / F$ with surface conductivity neglected, the steady-state electrokinetic coupling can be written as

$$L_0 = -\frac{\varepsilon \zeta}{\eta F}. \tag{10}$$

We can see that the steady-state electrokinetic coupling is inversely proportional to the formation factor.

The transition angular frequency separating viscous and inertial flows in porous medium can be rewritten by inserting $\alpha_\infty = \phi F$ with F, as follows:

$$\omega_c = \frac{1}{F} \frac{\eta}{k_0 \rho_f}, \tag{11}$$

where F is the formation factor that can be deduced from resistivity measurements using Archie's law.

Since the permeability and the formation factor are not independent but can be related by $k_0 = CR^2/F$ [66] with C being a geometrical constant usually in the range 0.3–0.5 and R being the hydraulic radius, the transition angular frequency can be written as

$$\omega_c = \frac{\eta}{\rho_f C R^2}. \tag{12}$$

Equation (12) shows that the transition angular frequency in porous medium is inversely proportional to the square of the hydraulic radius.

Recently Walker and Glover [74] proposed a simplified equation of Pride's development assuming that the Debye length is negligible compared to the characteristic pore size, and assuming the following parameter:

$$m = 8 \left(\frac{\Lambda}{r_{\text{eff}}} \right)^2 \tag{13}$$

leading to

$$L(\omega) = L_0 \left[1 - 2i \frac{\omega}{\omega_c} \left(\frac{\Lambda}{r_{\text{eff}}} \right)^2 \right]^{-1/2} \tag{14}$$

with r_{eff} being the effective pore radius, and a transition angular frequency being

$$\omega_c = \frac{8\eta}{\rho_f r_{\text{eff}}^2}. \tag{15}$$

Garambois and Dietrich [12] studied the low-frequency assumption valid at seismic frequencies, meaning at frequencies lower than Biot's frequency separating viscous and inertial flows and gave the coseismic transfer function for low-frequency longitudinal plane waves. In this case, and assuming Biot's moduli $C \ll H$, they showed that the seismoelectric field $\mathbf{E}$ is proportional to the grain acceleration:

$$\mathbf{E} \simeq -\frac{L_0}{\sigma_r} \rho_f \ddot{\mathbf{u}} = \frac{\epsilon \zeta}{\eta \sigma_f} \rho_f \ddot{\mathbf{u}}. \tag{16}$$

Equations (16), (9), and (1) show that transient seismoelectric magnitudes will be affected by the bulk density of the fluid, and the streaming potential coefficient which is inversely proportional to the water conductivity and proportional to the zeta potential (which depends on the water pH).

2.4. The Electrokinetic Transition Frequency Compared to Hydraulic's One. The theory of dynamic permeability in porous media has been studied by many authors [61, 65, 75–77].

The frequency behavior of the permeability is given by Pride (1994) [22]

$$\frac{k(\omega)}{k_0} = \left[\left(1 - i \frac{\omega}{\omega_c} \frac{4}{m} \right)^{1/2} - i \frac{\omega}{\omega_c} \right]^{-1}. \tag{17}$$

Table 1: Measured or predicted transition frequency for dynamic streaming potential and permeability, for samples of porosity ϕ, formation factor F, permeability k_0, and half of the mean particle size r, from (SED) Smeulders et al. [61], (CKS) Charlaix et al. [62], (SG) Sears and Groves [63], (P) Packard [59], (TGR) Tardif et al. [64], and (RMLJ) Reppert et al. [60]. *Indicates predicted transition frequency from (3) and **indicates the transition frequency computed by the authors.

Sample	Particle size μm	ϕ [%]	F	k_0 [m^2]	f_c [Hz]	Source
Capillary	254 (radius)			10^{-8}	10–2.5* Hz	CKS
Capillary	508 (radius)				1.3–0.62* Hz	SG
Capillary G4	720 (radius)				0.31*–0.28** Hz	P
Capillary G2	826 (radius)				0.23*–0.21** Hz	P
Capillary 1	800–1100 (radius)				7.1 Hz	RMLJ
Glass beads	1.25–1.75	32	7.8	4.2×10^{-9}	4.8 Hz	SED
Glass beads	850 (r)	50	2.8	10^{-8}	6.2 Hz	CKS
Glass beads	580–700	31	8.7	9×10^{-10}	20 Hz	SED
Glass beads	450 (r)	50	3.2	2×10^{-9}	25 Hz	CKS
Glass beads	250 (r)	50	3	5×10^{-10}	108 Hz	CKS
Glass beads	200–270	31	9	1.4×10^{-10}	126 Hz	SED
Crushed glass	440 (r)	50	3	10^{-9}	44 Hz	CKS
Crushed glass	265 (r)	50	3.2	2×10^{-10}	45–103 Hz	CKS
Porous filter A	72.5–87				269 Hz	RMLJ
Porous filter B	35–50				710 Hz	RMLJ
Sand grains	1000–2000	31	9	26×10^{-10}	6.7 Hz	SED
Sand grains	150–300	29	10.7	10^{-10}	149 Hz	SED
Ottawa sand	200–250 (r)	31	4.7	1.2×10^{-10}	230–273 Hz	TGR

The transition angular frequency for a porous medium is the same as (8). Charlaix et al. [62] measured the behavior of permeability with frequency on capillary tube, glass beads, and crushed glass. The dynamic permeability is constant up to the transition frequency above which it decreases, and the more permeable the sample is, the lower the transition frequency is. Other measurements have been performed on glass beads and sand grains [61]. The transition frequency ($f_c = \omega_c/2\pi$) varies from 4.8 Hz to 149 Hz for samples having permeability in the range 10^{-8} to 10^{-10} m^2 (see Table 1), which are extremely high permeabilities.

The transition frequency indicates the beginning of the transition for both the permeability and the electrokinetic coupling. However the transition behavior and the cuttoff frequency are different between permeability and electrokinetic coupling ((7) and (17)), both depending on the pore-space geometry term m but in different manner.

We calculated the predicted transition frequency $f_c = \omega_c/2\pi$ with ω_c from (11) with $\eta = 10^{-3}$ Pa.s and $\rho_f = 10^3$ kg/m^3. The other parameters F and k_0 are measured from different authors cited in Bernabé [78] (see Table 2). We also calculated the parameters for four Fontainebleau sandstone samples. It has been shown for these samples that $F = \phi^{-2.01}$ (from Ruffet et al. [79]) and that $k_0 = a\phi^n$ with different values for n according to the porosity. The following laws were chosen: $k_0 = 1.66 \times 10^{-4}\phi^8$ for $\phi < 6\%$ and $k_0 = 2.5 \times 10^{-10}\phi^3$ for ϕ ranging between 8 and 25% [80]. We can see that the transition frequencies are of the order of kHz and MHz and no more from 0.2 to 150 Hz as measured or calculated on glass beads, sand grains, crushed glass, or capillaries. We plotted the results of the transition frequency as a function of the permeability on these various samples in Figure 1. Although the formation factor is not constant with the permeability, it is clear that the transition frequency is inversely proportional to the permeability as

$$\log_{10}(f_c) = -0.78\log_{10}(k) - 5.5, \quad (18)$$

and varies from about 100 MHz for 10^{-17} m^2 to about 10 Hz for 10^{-8} m^2, so by seven orders of magnitude for nine orders of magnitude in permeability.

3. Experimental Apparatus and Procedure

Several experimental setups were proposed to provide the sinusoidal pressure variations.

The first experimental apparatus proposed a sinusoidal motion delivered by a sylphon bellows which was driven by a geophone-type push-pull driver (Figure 2 from Packard [59]). The low-frequency oscillator (0.01 Hz to 1 kHz) was used for operation of the push-pull geophone driver. Similar setups were proposed by Thurston [81] (Figure 3) and Cooke [82], so that frequency of this kind of source was 1–400 Hz [82], 20–200 Hz [59], and 10–700 Hz [81]. The induced pressure was up to 2 kPa. More recently Schoemaker et al. [83] used a so-called Dynamic Darcy Cell (DCC) with a mechanical shaker connected to a rubber membrane leading to a frequency range for the oscillating pressure 5 to 200 Hz. The sinusoidal fluid flow was also applied by a displacement piston pump directly connected to the electrodes chambers (Figure 4 from [63, 88]). The piston was mounted on a Scotch Yoke drive attached to a controllable speed AC motor [84]. The frequency range of this source was then 0.4 Hz to

TABLE 2: Predicted transition frequency (from (11)), for dynamic streaming potential, for samples of porosity ϕ, formation factor F, and permeability k_0, from (1) calculated in the present study, and measured by (2) Taherian et al. [67], (3) Morgan et al. [68], (4) Fatt [69], (5) Wyble [70], (6) Dobrynin [71], (7) Chierici et al. [72], and (8) Yale [73].

Sample	ϕ [%]	F	k_0 [m^2]	f_c [Hz]
Fontainebleau sandstone[1]	20	25	2×10^{-12}	3.2 kHz
Fontainebleau sandstone[1]	15	45	8×10^{-13}	4.4 kHz
Fontainebleau sandstone[1]	10	102	2.5×10^{-13}	6.2 kHz
Sandstone-S22[2]	31.2	6	2.7×10^{-12}	9.7 kHz
Sandstone-S47[2]	20	14.4	8.5×10^{-13}	13 kHz
Boise[8]	26	12	9×10^{-13}	14.7 kHz
Berea sandstone500[8]	20	20	4.9×10^{-13}	16.2 kHz
Sandstone-S42[2]	19.7	14.7	6.7×10^{-13}	16.2 kHz
Sandstone-S45[2]	21	11.7	7.2×10^{-13}	18.8 kHz
Fahler 162[8]	3	294	2.7×10^{-14}	20 kHz
Sandstone-S43[2]	21.2	13	5.1×10^{-13}	23.5 kHz
Pliocene 41[7]	21	144.9	4.2×10^{-14}	26.1 kHz
Pliocene 35[7]	20	156.2	3.7×10^{-14}	27.5 kHz
Berea sandstoneC2H[3]	19.8	15.1	3.8×10^{-13}	27.7 kHz
Sandstone-S50[2]	18.3	17.2	3.1×10^{-13}	30 kHz
Triassic38[7]	21	12.6	4×10^{-12}	31.4 kHz
Triassic34[7]	20	13.9	3.5×10^{-13}	32.7 kHz
Berea sandstoneB2[3]	20.3	15.2	2.64×10^{-13}	39.7 kHz
Sandstone-S5[2]	26.4	8.7	4.1×10^{-13}	45 kHz
Sandstone-S35[2]	18.75	17.4	2×10^{-13}	46.5 kHz
Massillon DH[8]	16	23.8	1.3×10^{-13}	51.4 kHz
Cambrian 16[7]	14	312.5	9.5×10^{-15}	53.6 kHz
Fontainebleau sandstone[1]	5	412	6.5×10^{-15}	59.4 kHz
Berea sandstoneD1[3]	18.5	18.4	1.3×10^{-13}	66.5 kHz
Tensleep1[4]	15	18.9	1.2×10^{-13}	70.3 kHz
Tertiary 807[8]	22	14.9	1.5×10^{-13}	71.1 kHz
Cambrian 6[7]	8.1	90.9	2.3×10^{-14}	76.1 kHz
Torpedo[6]	20	41.7	4.5×10^{-14}	84.9 kHz
Miocene 7[7]	8.3	384.6	4.4×10^{-15}	94 kHz
Cambrian 14[7]	11	52.6	3.2×10^{-14}	94.5 kHz
Sandstone Triassic27[7]	18	20	7.2×10^{-14}	110.5 kHz
Sandstone-S9[2]	20.9	12	1×10^{-13}	126.2 kHz
Triassic26[7]	18	17.2	6.8×10^{-14}	135.7 kHz
Sandstone-S6[2]	22.8	10.6	8.3×10^{-14}	180.7 kHz
Berea 100H[8]	17	17.2	4.9×10^{-14}	188.4 kHz
Sandstone S15[2]	21.8	13.9	4.5×10^{-14}	256.7 kHz
Kirkwood[5]	15	40	1.2×10^{-14}	331.6 kHz
Indiana DV[8]	27	12	3×10^{-14}	440.3 kHz
Island Rust A1[3]	14.6	52.5	5.2×10^{-15}	579 kHz
Bradford[5]	11	90	2.5×10^{-15}	700.3 kHz
Austin chalk[3]	23.6	22.7	9.7×10^{-15}	763 kHz
Massillon DV[8]	19	27.8	6.9×10^{-15}	830.4 kHz
Sandstone-S34[2]	21.35	13.7	1.1×10^{-14}	1.06 MHz
Sandstone S44[2]	15.7	24.5	4.2×10^{-15}	1.5 MHz
Indiana L. SA1[3]	18	29.2	1.9×10^{-15}	2.9 MHz
Tennessee A1[3]	5.5	180.3	2.3×10^{-16}	3.8 MHz
AZPink (Coconino)[3]	10.3	62.4	6.3×10^{-16}	4.04 MHz
Leuders L.SA1[3]	15.2	41.5	7.1×10^{-16}	5.3 MHz

TABLE 2: Continued.

Sample	ϕ [%]	F	k_0 [m^2]	f_c [Hz]
Sandstone-S40[2]	10.9	130	1.9×10^{-16}	6.4 MHz
Sandstone-S23[2]	18.8	40.7	4.8×10^{-16}	8.1 MHz
Fahler 189[8]	1.9	714.3	2×10^{-17}	11.1 MHz
Penn blue A1[3]	3.9	219	6.2×10^{-17}	11.7 MHz
AZChoclate2[3]	9.5	159.3	5.8×10^{-17}	17.2 MHz
Fahler 161[8]	2.3	416.7	1×10^{-17}	38.2 MHz
Fahler 142[8]	7.6	164	2×10^{-17}	48.5 MHz
Sandstone S21[2]	12.1	65	3×10^{-17}	81.7 MHz
Fahler 154[8]	4.6	263.1	7×10^{-18}	86.4 MHz
Fahler 192[8]	4.4	128.2	9×10^{-18}	137.9 MHz

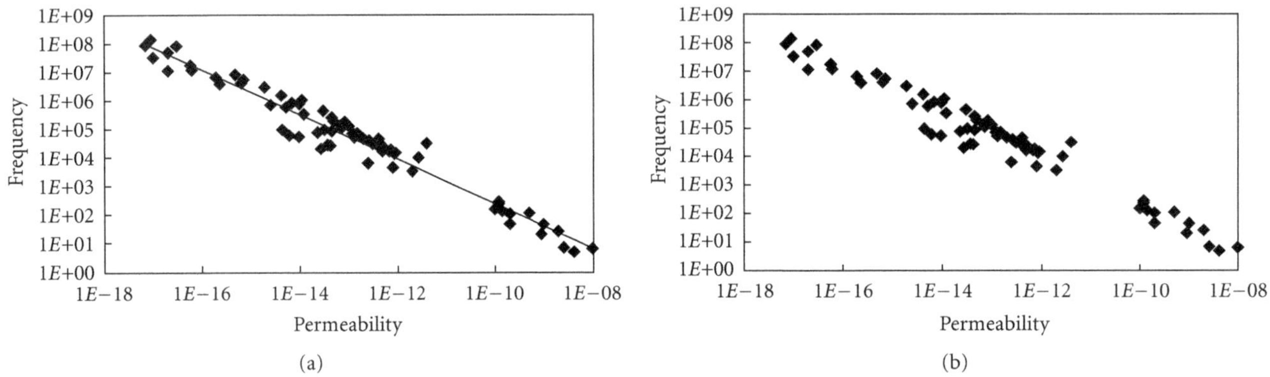

FIGURE 1: The transition frequency $f_c = \omega_c/2\pi$ (in Hz) predicted in the present study with ω_c from (11) with $\eta = 10^{-3}$ Pa.s and $\rho_f = 10^3$ kg/m^3 as a function of the permeability (in m^2). The transition frequency varies as $\log_{10}(f_c) = -0.78\log_{10}(k) - 5.5$. The parameters of the samples, F and k_0, are measured from different authors on various samples cited in Tables 1 and 2.

21 Hz and the pressure up to 15 kPa. Pengra et al. [85] used a piston rod attached to a loudspeaker driven by an audio power amplifier (Figure 5). They performed measurements up to 100 Hz, with an applied pressure of 5 kPa RMS. More recently it was proposed by Reppert et al. [60] to use an electromechanical transducer (Figure 6), and these authors covered a frequency range 1–500 Hz. The vibrating exciter proposed by Schoemaker et al. [86] was used from 5 Hz to 200 Hz. Recently Tardif et al. [64] used an electromagnetic shaker operating in the range 1 Hz to 1 kHz and provided measurements up to 200 Hz. Higher frequencies have been investigated [49, 50, 52, 54, 55] for the detection of the interfacial conversions.

The electromagnetic noise radiating from such equipment must be suppressed by shielding the setup and wires (shielded twisted cable pairs) [64, 86]. Moreover it is essential to have a rigid framework. A mechanical resonance can occur in the cell/transducer system (at 70 Hz in [85]), and the noise associated with mechanical vibration can be suppressed puting an additional mass to the frame [64].

Once the oscillatory pressure is applied, the pressure must be measured. Most of the setups include piezoelectric transducers to measure the pressure difference over the capillary or the porous sample. Reppert et al. [60] proposed to use hydrophones that have a flat response from 1 to 20 kHz. Tardif et al. [64] proposed to use dynamic transducers with a low-frequency limit 0.08 Hz and a maximum frequency of 170 kHz.

The electrodes are usually Ag/AgCl or platinum electrodes. The electrodes used by Schoemaker et al. [86] were sintered plates of Monel (composed of nickel and copper). The electrical signal must be measured using preamplifiers or a high-input impedance acquisition system. Since the impedance of the sample depends on the frequency, one must correct the measurements from this varying-impedance to be able to have a correct streaming potential coefficient [60]. Moreover the electrodes at top and bottom of the sample can behave as a capacitor, requiring a correction using impedance measurements too [86].

The sample is usually saturated and it is emphasized that the sample should be left until equilibrium with water. This equilibrium can be obtained by leaving the sample in contact with water for some time, and by flowing the water within the sample several times by checking the pH and the water conductivity until an equilibrium is reached [39]. The procedure including water flow is better because the properties of the water can be measured. When the properties of the water are measured only before saturating the sample, the resulting water once in contact with the sample is not known. Usually the water is more conductive

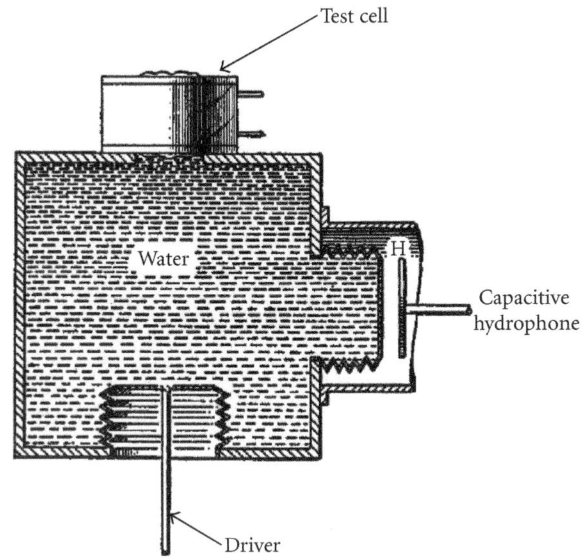

Figure 2: The sylphon bellows is driven by a geophone-type push-pull driver to apply a sinusoidal motion to the sample (modified from [59]).

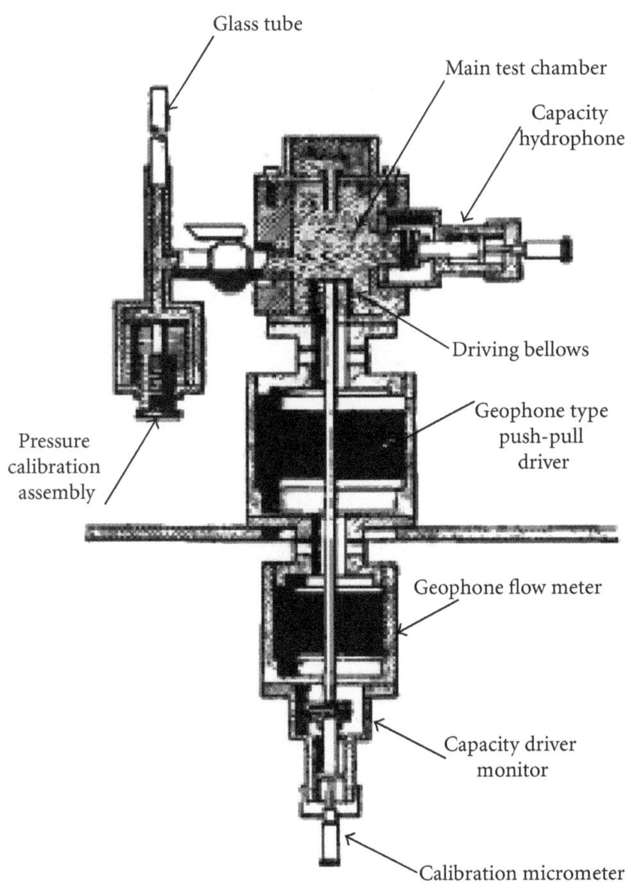

Figure 3: Experimental setup used by Thurston [87] (modified from [87]).

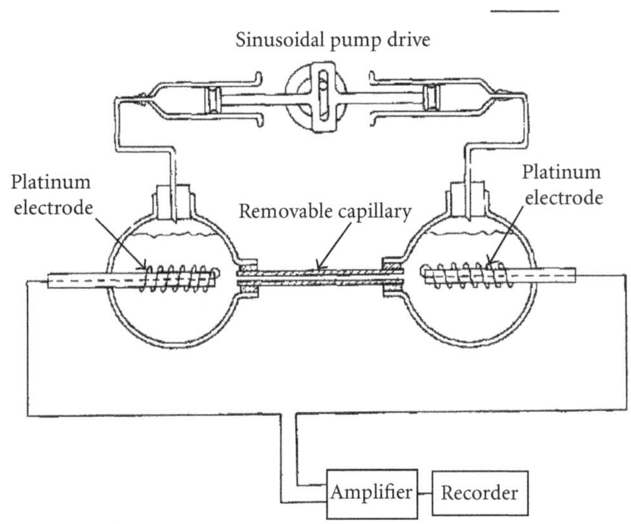

Figure 4: Experimental setup used by Groves and Sears (modified from [88]).

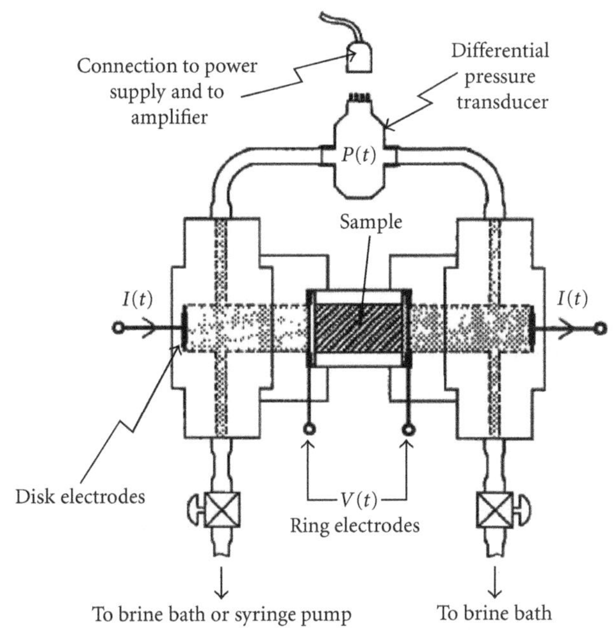

Figure 5: Experimental setup used by Pengra et al. [85] for streaming potential and electro-osmosis measurements (modified from [85]).

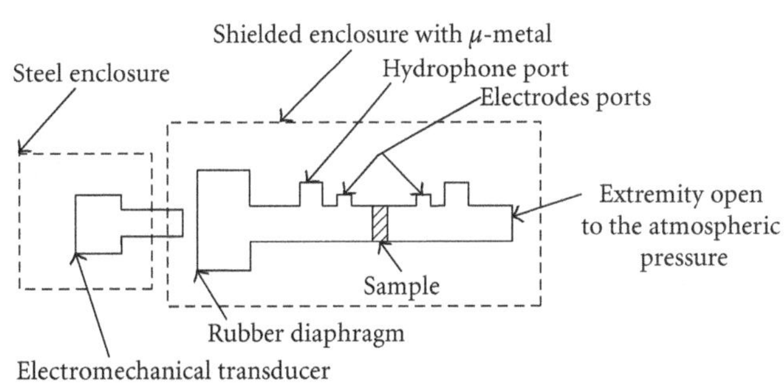

Figure 6: Experimental setup used by Reppert et al. [60] (modified from [60]).

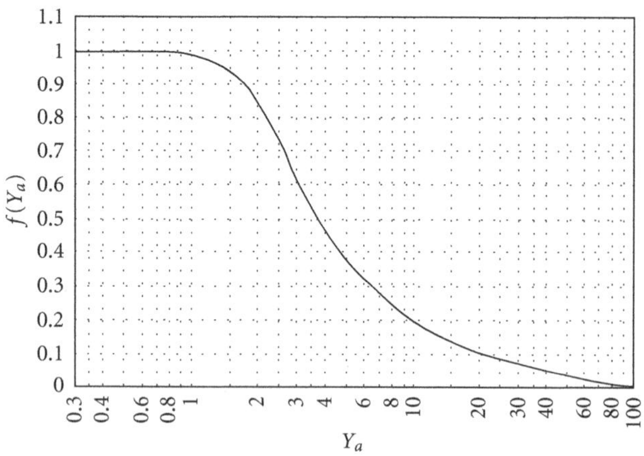

FIGURE 7: The absolute magnitude of the normalized streaming potential coefficient calculated by Packard [59] using (19) where $Y_a = a\sqrt{\omega\rho_f/\eta}$, equivalent to (2) (modified from [59]).

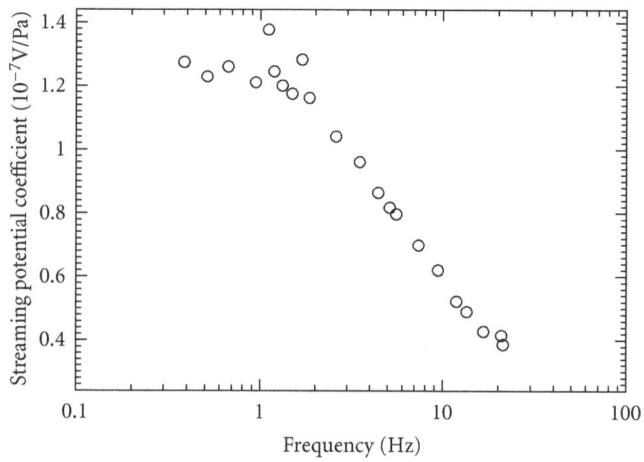

FIGURE 8: The streaming potential coefficient measured as a function of frequency by Sears and Groves [63] on a capillary coated with clay, incubated with BSA in 0.02 M Tris-HCl.

when in contact with the sample, and the pH can change. Recalling that the streaming potential is proportional to the zeta potential (which depends on pH) and inversely proportional to the water conductivity (1), it is essential to know properly the pH and the water conductivity.

4. Measurements and Calculations of the Dynamic Electrokinetic Coefficient

The absolute magnitude of the streaming potential coefficient normalized by the steady-state value was calculated by Packard [59] as

$$f(Y_a) = \left(\frac{-2}{Y_a} \frac{i\sqrt{i}J_1(\sqrt{i}Y_a)}{J_0(\sqrt{i}Y_a)} e^{-i\omega t} \right), \quad (19)$$

which is equal to (2), but expressed as a function of the parameter $Y_a = a\sqrt{(\omega\rho_f)/\eta}$, the transition frequency being obtained for $Y_a = 1$ (Figure 7). The streaming potential coefficient is constant up to the transition angular frequency and then decreases with increasing frequency.

Sears and Groves [63] measured the streaming potential coefficient on a capillary of radius 508 μm which was coated with clay-Adams Siliclad and then incubated with 1% bovine serum albumin and filled with 0.02 M Tris-HCl at pH 7.32. They reported the streaming potential and the pressure difference as a function of frequency in the range 0–20 Hz. We calculated the resulting streaming potential coefficient (see Figure 8) which decreases from about 1.3×10^{-7} to 4×10^{-8} V/Pa. These authors computed the zeta potential and concluded that the zeta potential is independent of the frequency with an average value of 28.8 mV. Moreover they concluded that the zeta potential is also independent of the capillary radius and capillary length.

The value of the streaming potential coefficient on Ottawa sand measured at 5 Hz by Tardif et al. [64] was -5.2×10^{-7} V/Pa using a 0.001 mol/L NaCl solution to saturate the sample. Values between 1 and 2×10^{-8} V/Pa were measured on samples saturated by 0.1 M/L NaCl brine [85]. A compilation of numerous streaming potential coefficients measured on sands and sandstones at various salinities in DC domain [44] showed that $\mathbf{C_{s0}} = -1.2 \times 10^{-8}\sigma_f^{-1}$, where $\mathbf{C_{s0}}$ is in V/Pa and σ_f in S/m. A zeta potential of -17 mV can be inferred from these collected data, assuming the other parameters (see (1)) independent of water conductivity. These assumptions are not exact, but the value of zeta is needed for numerous modellings which usually assume the other parameters independent of the fluid conductivity. Therefore an average value of -17 mV for such modellings can be rather exact, at least for medium with no clay nor calcite.

Reppert et al. [60] calculated the real part and the imaginary part of the theoretical Packard's streaming potential coefficient (2) for different capillary radii (see Figure 9). It can be seen that the larger the radius is, the lower the transition frequency is, as shown previously by the different theories. Recent developments by the group of Glover have been performed to build a new setup and to make further measurements on porous samples: two papers detail these studies in this special issue on Electrokinetics in Earth Sciences.

5. Conclusion

Since the theory of Pride in 1994 [22], the dynamic behavior of the streaming potential is known for porous media. However few experimental results are available, because of the difficulty to perform correct measurements at high frequency. Up to now, measurements of the frequency-dependence of the streaming potential have been performed up to 200 Hz on high-permeable samples. The main difficulty arises from electrical noise induced by mechanical vibration. Moreover it has been emphasized that the measurements must be corrected by impedance measurements as a function of frequency too because the impedance of the sample depends on frequency. Further theoretical

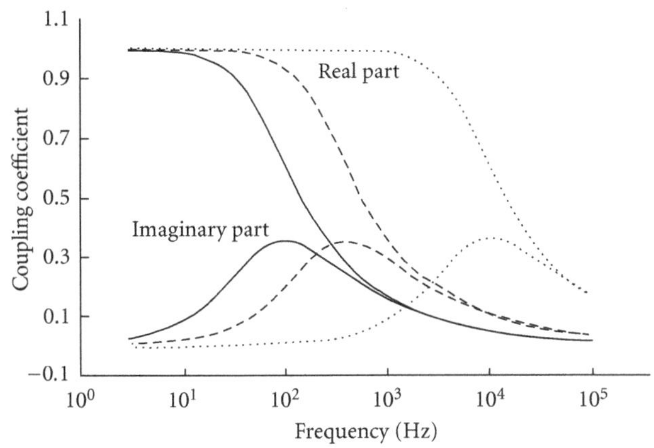

FIGURE 9: The real and imaginary part of Packard's model (2) calculated by Reppert et al. [60] for three capillary radii: 100 μm (continuous line), 50 μm (dashed line), and 10 μm (point line) (modified from [60]).

developments performed by Garambois and Dietrich [12] studied the low-frequency assumption valid at frequencies lower than the transition frequency. We show that this transition frequency, on a various collection of samples for which both formation factor and permeability are measured, is predicted to depend on the permeability as inversely proportional to the permeability.

Acknowledgments

This work was supported by the French National Scientific Center (CNRS), by the National Agency for Research (ANR) through TRANSEK, and by REALISE the "Alsace Region Research Network in Environmental Sciences in Engineering" and the Alsace Region. The authors thank two anonymous reviewers and the associate editor T. Ishido for very constructive remarks that improved this paper.

References

[1] Y. Honkura, Y. Ogawa, M. Matsushima, S. Nagaoka, N. Ujihara, and T. Yamawaki, "A model for observed circular polarized electric fields coincident with the passage of large seismic waves," *Journal of Geophysical Research B*, vol. 114, no. 10, Article ID B10103, 2009.

[2] M. Matsushima, Y. Honkura, N. Oshiman et al., "Seismoelectromagnetic effect associated with the İzmit earthquake and its aftershocks," *Bulletin of the Seismological Society of America*, vol. 92, no. 1, pp. 350–360, 2002.

[3] N. Takeuchi, N. Chubachi, S. Hotta, and K. Narita, "Analysis of earth potential difference signals by using seismic wave signals," *Electrical Engineering in Japan*, vol. 125, no. 4, pp. 52–58, 1998.

[4] M. A. Fenoglio, M. J. S. Johnston, and J. D. Byerlee, "Magnetic and electric fields associated with changes in high pore pressure in fault zones: application to the Loma Prieta ULF emissions," *Journal of Geophysical Research*, vol. 100, no. 7, pp. 12951–12958, 1995.

[5] J. C. Dupuis, K. E. Butler, A. W. Kepic, and B. D. Harris, "Anatomy of a seismoelectric conversion: Measurements and conceptual modeling in boreholes penetrating a sandy aquifer," *Journal of Geophysical Research B*, vol. 114, no. 10, Article ID B10306, 2009.

[6] A. Thompson, S. Hornbostel, J. Burns et al., "Field tests of electroseismic hydrocarbon detection," SEG Technical Program Expanded Abstracts, 2005.

[7] J. C. Dupuis and K. E. Butler, "Vertical seismoelectric profiling in a borehole penetrating glaciofluvial sediments," *Geophysical Research Letters*, vol. 33, no. 16, Article ID L16301, 2006.

[8] J. C. Dupuis, K. E. Butler, and A. W. Kepic, "Seismoelectric imaging of the vadose zone of a sand aquifer," *Geophysics*, vol. 72, no. 6, pp. A81–A85, 2007.

[9] M. H. P. Strahser, W. Rabbel, and F. Schildknecht, "Polarisation and slowness of seismoelectric signals: a case study," *Near Surface Geophysics*, vol. 5, no. 2, pp. 97–114, 2007.

[10] S. S. Haines, A. Guitton, and B. Biondi, "Seismoelectric data processing for surface surveys of shallow targets," *Geophysics*, vol. 72, no. 2, pp. G1–G8, 2007.

[11] M. Strahser, L. Jouniaux, P. Sailhac, P.-D. Matthey, and M. Zillmer, "Dependence of seismoelectric amplitudes on water content," *Geophysical Journal International*, vol. 187, no. 3, pp. 1378–1392, 2011.

[12] S. Garambois and M. Dietrich, "Seismoelectric wave conversions in porous media: field measurements and transfer function analysis," *Geophysics*, vol. 66, no. 5, pp. 1417–1430, 2001.

[13] S. S. Haines, S. R. Pride, S. L. Klemperer, and B. Biondi, "Seismoelectric imaging of shallow targets," *Geophysics*, vol. 72, no. 2, pp. G9–G20, 2007.

[14] A. H. Thompson, J. R. Sumner, and S. C. Hornbostel, "Electromagnetic-to-seismic conversion: a new direct hydrocarbon indicator," *The leading Edge*, vol. 26, no. 4, pp. 428–435, 2007.

[15] J. Singer, J. Saunders, L. Holloway et al., "Electrokinetic logging has the potential to measure the permeability," in *the 46th Annual Logging Symposium of the Society of Petrophysicists and Well Log Analysts*, 2005.

[16] C. C. Pain, J. H. Saunders, M. H. Worthington et al., "A mixed finite-element method for solving the poroelastic Biot equations with electrokinetic coupling," *Geophysical Journal International*, vol. 160, no. 2, pp. 592–608, 2005.

[17] O. V. Mikhailov, J. Queen, and M. N. Toksöz, "Using borehole electroseismic measurements to detect and characterize fractured (permeable) zones," *Geophysics*, vol. 65, no. 4, pp. 1098–1112, 2000.

[18] L. Jouniaux, "Electrokinetic techniques for the determination of hydraulic conductivity," in *Hydraulic Conductivity—Issues, Determination and Applications*, L. Elango, Ed., InTech, 2011.

[19] M. D. Schakel, D. M. J. Smeulders, E. C. Slob, and H. K. J. Heller, "Seismoelectric interface response: experimental results and forward model," *Geophysics*, vol. 76, no. 4, pp. N29–N36, 2011.

[20] M. Schakel and D. Smeulders, "Seismoelectric reflection and transmission at a fluid/porous-medium interface," *Journal of the Acoustical Society of America*, vol. 127, no. 1, pp. 13–21, 2010.

[21] S. Garambois and M. Dietrich, "Full waveform numerical simulations of seismoelectromagnetic wave conversions in fluid-saturated stratified porous media," *Journal of Geophysical Research B*, vol. 107, no. 7, article 2148, 2002.

[22] S. Pride, "Governing equations for the coupled electromagnetics and acoustics of porous media," *Physical Review B*, vol. 50, no. 21, pp. 15678–15696, 1994.

[23] M. A. Biot, "Theory of propagation of elastic waves in a fluid-saturated porous solid: I. Low frequency range," *Journal of the Acoustical Society of America*, vol. 28, no. 2, pp. 168–178, 1956.
[24] M. W. Haartsen and S. R. Pride, "Electroseismic waves from point sources in layered media," *Journal of Geophysical Research B*, vol. 102, no. 11, pp. 24745–24769, 1997.
[25] M. W. Haartsen, W. Dong, and M. N. Toksöz, "Dynamic streaming currents from seismic point sources in homogeneous poroelastic media," *Geophysical Journal International*, vol. 132, no. 2, pp. 256–274, 1998.
[26] Y. Gao and H. Hu, "Seismoelectromagnetic waves radiated by a double couple source in a saturated porous medium," *Geophysical Journal International*, vol. 181, no. 2, pp. 873–896, 2010.
[27] J. H. Saunders, M. D. Jackson, and C. C. Pain, "A new numerical model of electrokinetic potential response during hydrocarbon recovery," *Geophysical Research Letters*, vol. 33, no. 15, Article ID L15316, 2006.
[28] L. Jouniaux and T. Ishido, "Electrokinetics in Earth Sciences: a tutorial," *International Journal of Geophysics*, vol. 2012, Article ID 286107, 16 pages, 2012.
[29] J. A. Davis, R. O. James, and J. O. Leckie, "Surface ionization and complexation at the oxide/water interface. I. Computation of electrical double layer properties in simple electrolytes," *Journal of Colloid And Interface Science*, vol. 63, no. 3, pp. 480–499, 1978.
[30] X. Guichet, L. Jouniaux, and N. Catel, "Modification of streaming potential by precipitation of calcite in a sand-water system: laboratory measurements in the pH range from 4 to 12," *Geophysical Journal International*, vol. 166, no. 1, pp. 445–460, 2006.
[31] T. Ishido and J. Muzutani, "Experimental and theoretical basis of electrokinetic phenomena in rock-water systems and its applications to geophysics," *Journal of Geophysical Research*, vol. 86, no. 3, pp. 1763–1775, 1981.
[32] J. P. Pozzi and L. Jouniaux, "Electrical effects of fluid circulation in sediments and seismic prediction," *Comptes Rendus—Academie des Sciences, Serie II*, vol. 318, no. 1, pp. 73–77, 1994.
[33] L. Jouniaux and J. P. Pozzi, "Permeability dependence of streaming potential in rocks for various fluid conductivities," *Geophysical Research Letters*, vol. 22, no. 4, pp. 485–488, 1995.
[34] L. Jouniaux and J. P. Pozzi, "Streaming potential and permeability of saturated sandstones under triaxial stress: consequences for electrotelluric anomalies prior to earthquakes," *Journal of Geophysical Research*, vol. 100, no. 6, pp. 10197–10209, 1995.
[35] L. Jouniaux and J. P. Pozzi, "Laboratory measurements anomalous 0.1–0.5 Hz streaming potential under geochemical changes: implications for electrotelluric precursors to earthquakes," *Journal of Geophysical Research B*, vol. 102, no. 7, pp. 15335–15343, 1997.
[36] L. Jouniaux, S. Lallemant, and J. P. Pozzi, "Changes in the permeability, streaming potential and resistivity of a claystone from the Nankai prism under stress," *Geophysical Research Letters*, vol. 21, no. 2, pp. 149–152, 1994.
[37] L. Jouniaux, J. P. Pozzi, J. Berthier, and P. Massé, "Detection of fluid flow variations at the Nankai Trough by electric and magnetic measurements in boreholes or at the seafloor," *Journal of Geophysical Research B*, vol. 104, no. 12, pp. 29293–29309, 1999.
[38] L. Jouniaux, M. L. Bernard, M. Zamora, and J. P. Pozzi, "Streaming potential in volcanic rocks from Mount Pelée," *Journal of Geophysical Research B*, vol. 105, no. 4, pp. 8391–8401, 2000.
[39] X. Guichet, L. Jouniaux, and J. P. Pozzi, "Streaming potential of a sand column in partial saturation conditions," *Journal of Geophysical Research B*, vol. 108, no. 3, article 2141, 2003.
[40] M. Z. Jaafar, J. Vinogradov, and M. D. Jackson, "Measurement of streaming potential coupling coefficient in sandstones saturated with high salinity NaCl brine," *Geophysical Research Letters*, vol. 36, no. 21, Article ID L21306, 2009.
[41] L. Jouniaux, A. Maineult, V. Naudet, M. Pessel, and P. Sailhac, "Review of self-potential methods in hydrogeophysics," *Comptes Rendus—Geoscience*, vol. 341, no. 10-11, pp. 928–936, 2009.
[42] J. Vinogradov, M. Z. Jaafar, and M. D. Jackson, "Measurement of streaming potential coupling coefficient in sandstones saturated with natural and artificial brines at high salinity," *Journal of Geophysical Research B*, vol. 115, no. 12, Article ID B12204, 2010.
[43] M. D. Jackson, "Multiphase electrokinetic coupling: insights into the impact of fluid and charge distribution at the pore scale from a bundle of capillary tubes model," *Journal of Geophysical Research*, vol. 115, Article ID B07206, 2010.
[44] V. Allègre, L. Jouniaux, F. Lehmann, and P. Sailhac, "Streaming potential dependence on water-content in Fontainebleau sand," *Geophysical Journal International*, vol. 182, no. 3, pp. 1248–1266, 2010.
[45] N. Migunov and A. Kokorev, "Dynamic properties of the seismoelectric effect of water-saturated rocks," *Izvestiya, Earth Physics*, vol. 13, no. 6, pp. 443–445, 1977.
[46] R. Chandler, "Transient streaming potential measurements on fluid-saturated porous structures: an experimental verification of Biot's slow wave in the quasi-static limit," *Journal of the Acoustical Society of America*, vol. 70, no. 1, pp. 116–121, 1981.
[47] S. A. Mironov, E. I. Parkhomenko, and G. Y. Chernyak, "Seismoelectric effect in rocks containing gas or fluid hydrocarbon (english translation)," *Izvestiya, Physics of the Solid Earth*, vol. 29, no. 11, pp. 1000–1006, 1994.
[48] Y. G. Jiang, F. K. Shan, H. M. Jin, and L. W. Zhou, "A method for measuring electrokinetic coefficients of porous media and its potential application in hydrocarbon exploration," *Geophysical Research Letters*, vol. 25, no. 10, pp. 1581–1584, 1998.
[49] Z. Zhu, M. W. Haartsen, and M. N. Toksöz, "Experimental studies of electrokinetic conversions in fluid-saturated borehole models," *Geophysics*, vol. 64, no. 5, pp. 1349–1356, 1999.
[50] Z. Zhu, M. W. Haartsen, and M. N. Toksöz, "Experimental studies of seismoelectric conversions in fluid-saturated porous media," *Journal of Geophysical Research B*, vol. 105, no. 12, pp. 28055–28064, 2000.
[51] Z. Zhu and M. N. Toksoöz, "Crosshole seismoelectric measurements in borehole models with fractures," *Geophysics*, vol. 68, no. 5, pp. 1519–1524, 2003.
[52] B. Chen and Y. Mu, "Experimental studies of seismoelectric effects in fluid-saturated porous media," *Journal of Geophysics and Engineering*, vol. 2, no. 3, pp. 222–230, 2005.
[53] C. Bordes, L. Jouniaux, M. Dietrich, J. P. Pozzi, and S. Garambois, "First laboratory measurements of seismo-magnetic conversions in fluid-filled Fontainebleau sand," *Geophysical Research Letters*, vol. 33, no. 1, Article ID L01302, 2006.
[54] G. I. Block and J. G. Harris, "Conductivity dependence of seismoelectric wave phenomena in fluid-saturated sediments," *Journal of Geophysical Research B*, vol. 111, no. 1, Article ID B01304, 2006.
[55] Z. Zhu, M. N. Toksöz, and D. R. Burns, "Electroseismic and seismoelectric measurements of rock samples in a water tank," *Geophysics*, vol. 73, no. 5, pp. E153–E164, 2008.

[56] C. Bordes, L. Jouniaux, S. Garambois, M. Dietrich, J. P. Pozzi, and S. Gaffet, "Evidence of the theoretically predicted seismo-magnetic conversion," *Geophysical Journal International*, vol. 174, no. 2, pp. 489–504, 2008.

[57] V. Allègre, L. Jouniaux, F. Lehmann, and P. Sailhac, "Reply to comment by A. Revil and N. Linde on 'Streaming potential dependence on water-content in Fontainebleau sand'," *Geophysical Journal International*, vol. 186, no. 1, pp. 115–117, 2011.

[58] B. Lorne, F. Perrier, and J. P. Avouac, "Streaming potential measurements: 1. Properties of the electrical double layer from crushed rock samples," *Journal of Geophysical Research B*, vol. 104, no. 8, pp. 17857–17877, 1999.

[59] R. G. Packard, "Streaming potentials across glass capillaries for sinusoidal pressure," *The Journal of Chemical Physics*, vol. 21, no. 2, pp. 303–307, 1953.

[60] P. M. Reppert, F. D. Morgan, D. P. Lesmes, and L. Jouniaux, "Frequency-dependent streaming potentials," *Journal of Colloid and Interface Science*, vol. 234, no. 1, pp. 194–203, 2001.

[61] D. M. J. Smeulders, R. L. G. M. Eggels, and M. E. H. Van Dongen, "Dynamic permeability: reformulation of theory and new experimental and numerical data," *Journal of Fluid Mechanics*, vol. 245, pp. 211–227, 1992.

[62] E. Charlaix, A. P. Kushnick, and J. P. Stokes, "Experimental study of dynamic permeability in porous media," *Physical Review Letters*, vol. 61, no. 14, pp. 1595–1598, 1988.

[63] A. R. Sears and J. N. Groves, "The use of oscillating laminar flow streaming potential measurements to determine the zeta potential of a capillary surface," *Journal of Colloid And Interface Science*, vol. 65, no. 3, pp. 479–482, 1978.

[64] E. Tardif, P. W. J. Glover, and J. Ruel, "Frequency-dependent streaming potential of Ottawa sand," *Journal of Geophysical Research B*, vol. 116, no. 4, Article ID B04206, 2011.

[65] D. L. Johnson, J. Koplik, and R. Dashen, "Theory of dynamic permeability in fluid saturated porous media," *Journal of Fluid Mechanics*, vol. 176, pp. 379–402, 1987.

[66] M. S. Paterson, "The equivalent channel model for permeability and resistivity in fluid-saturated rock-A re-appraisal," *Mechanics of Materials*, vol. 2, no. 4, pp. 345–352, 1983.

[67] M. R. Taherian, W. E. Kenyon, and K. A. Safinya, "Measurement of dielectric response of water-saturated rocks," *Geophysics*, vol. 55, no. 12, pp. 1530–1541, 1990.

[68] D. Morgan, D. Lesmes, F. Samstag et al., *Laboratory reports for geophysics 612: Rock physics*, Texas A&M University, 1990.

[69] I. Fatt, "Effect of overburden and reservoir pressure on electric logging formation factor," *Bulletin of American Association of Petroleum Geologists*, vol. 41, pp. 2456–2466, 1957.

[70] D. Wyble, "Effect of applied pressure on the conductivity, porosity and permeability of sandstones," *Transactions of the AIME*, vol. 213, pp. 430–432, 1958.

[71] V. Dobrynin, "Effect of overburden pressure on some properties of sandstones," *Society of Petroleum Engineers Journal*, no. 2, pp. 360–366, 1962.

[72] G. Chierici, G. Ciucci, F. Eva, and G. Long, "Effect of the overburden pressure on some petrophysical parameters of reservoir rocks," in *Proceeding of the 7th World Petroleum Congress*, vol. 2, pp. 309–338, 1967.

[73] D. Yale, *Network modelling of flow, storage and deformation in porous rocks*, Ph.D. thesis, Stanford University, 1984.

[74] E. Walker and P. W. J. Glover, "Permeability models of porous media: Characteristic length scales, scaling constants and time-dependent electrokinetic coupling," *Geophysics*, vol. 75, no. 6, pp. E235–E246, 2010.

[75] J. L. Auriault, L. Borne, and R. Chambon, "Dynamics of porous saturated media, checking of the generalized law of Darcy," *Journal of the Acoustical Society of America*, vol. 77, no. 5, pp. 1641–1650, 1985.

[76] P. Sheng and M. Y. Zhou, "Dynamic permeability in porous media," *Physical Review Letters*, vol. 61, no. 14, pp. 1591–1594, 1988.

[77] P. Sheng, M. Zhou, E. Charlaix, A. Kushnick, and J. Stokes, "Scaling function for dynamic permeability in porous media, reply," *Physical Review Letters*, vol. 63, p. 581, 1988.

[78] Y. Bernabé, "Pore geometry and pressure dependence of the transport properties in sandstones," *Geophysics*, vol. 56, no. 4, pp. 436–446, 1991.

[79] C. Ruffet, Y. Guéguen, and M. Darot, "Complex conductivity measurements and fractal nature of porosity," *Geophysics*, vol. 56, no. 6, pp. 758–768, 1991.

[80] T. Bourbié, O. Coussy, and B. Zinszner, *Acoustic of Porous Media*, Institut Francais du petrole Publications, Editions Technip, 1987.

[81] G. Thurston, "Periodic fluid flow through circular tubes," *Journal of the Acoustical Society of America*, vol. 24, no. 6, pp. 653–656, 1952.

[82] C. E. Cooke, "Study of electrokinetic effects using sinusoidal pressure and voltage," *The Journal of Chemical Physics*, vol. 23, no. 12, pp. 2299–2303, 1955.

[83] F. C. Schoemaker, D. M. J. Smeulders, and E. C. Slob, "Simultaneous determination of dynamic permeability and streaming potential," *SEG Technical Program Expanded Abstracts*, vol. 26, no. 1, pp. 1555–1559, 2007.

[84] C. M. Cerda and K. Non-Chhom, "The use of sinusoidal streaming flow measurements to determine the electrokinetic properties of porous media," *Colloids and Surfaces*, vol. 35, no. 1, pp. 7–15, 1989.

[85] D. B. Pengra, S. X. Li, and P. Z. Wong, "Determination of rock properties by low-frequency AC electrokinetics," *Journal of Geophysical Research B*, vol. 104, no. 12, pp. 29485–29508, 1999.

[86] F. C. Schoemaker, D. M. J. Smeulders, and E. C. Slob, "Electrokinetic effect: theory and measurement," *SEG Technical Program Expanded Abstracts*, vol. 27, no. 1, pp. 1645–1649, 2008.

[87] G. Thurston, "Apparatus for absolute measurement of analogous impedance of acoustic elements," *Journal of the Acoustical Society of America*, vol. 24, no. 6, pp. 649–656, 1952.

[88] J. N. Groves and A. R. Sears, "Alternating streaming current measurements," *Journal of Colloid And Interface Science*, vol. 53, no. 1, pp. 83–89, 1975.

15

A "Non-Invasive" Technique for Qualifying the Reinforced Concrete Structure

Antonella Guida, Antonello Pagliuca, and Alessandro Tranquillino Minerva

Department of European and Mediterranean Cultures, School of Architecture, University of Basilicata, Via Lazazzera, 75100 Matera, Italy

Correspondence should be addressed to Antonella Guida, arch.antonellaguida@gmail.com

Academic Editor: Raffaele Solimene

In recent years, a lot of studies on built heritage emphasize the need to use appropriate techniques to evaluate the current condition of the structure before designing an intervention. The research focuses on the restoration of reinforced concrete buildings that begin to show signs of decay and deterioration. To verify the state of a building, it's possible to use the "destructive" methods (that require a local removal of material) and "nondestructive" tests. The combined results from different "nondestructive" tests are very interesting instruments to assess the concrete strength. This methodological approach can help to reduce the possible errors when using the sclerometer and ultrasonic tests separately; in this way, the combined method called "SonReb" (SONic + REBound) was developed. This paper would highlight the importance with respect to cultural heritage buildings and on the studied structure and contribute to developed engineering strategies to maintenance and restoration. The above-defined methodology has been tested on a postwar building which is located in Gravina in Puglia (Italy): the "Centrone" theatre; it was built using a mixed structure, that is, reinforced concrete and bearing masonry built of local stone. The building was used until the 1990s and now is abandoned. The analysis and qualification of the masonry structures of built heritage show how this approach is useful for classifying the pathological events on each building and to implement the innovative solutions to improve the durability of a restoration intervention.

1. Introduction

In the last decades, the architectural heritage of the modern movement seems to be more at risk than during any other period. This built inheritance embodies the dynamic spirit of the industrial age. At the end of the 1980s, many modern masterpieces had already been demolished or changed beyond recognition. This was mainly due to the fact that many were not considered to be elements of heritage that their original functions have substantially changed and that their technological innovations have not always endured long-term stresses.

A thorough research of built heritage enables understanding of the evolution of design philosophies and underlying cultural meanings and messages, artistic and functional qualities, and engineering achievements. A detailed knowledge of building materials, construction techniques, environmental services, external impacts, and internal impacts of use or disuse assist in identifying the problems affecting buildings and defining a methodological approach for interventions.

The research objective is to provide new qualitative information on the strength of reinforced concrete structures of a building by using innovative, non-invasive testing techniques. The case of study is the *"Centrone"* theatre in Gravina in Puglia (Italy).

The confluence of the local architectural styles (vernacular and academic) and the emerging aesthetic of reinforced concrete is explored to outline the context that influenced the building design.

The research methodology includes (a) the context in which the buildings were designed, (b) their history, (c) building technologies used, (d) non-invasive testing of the reinforced concrete structures, (e) the analysis of the test results, and (f) the conclusions.

Testing methods could be "destructive", as they require a local removal of material, or "nondestructive", that is, they do not affect the structure. A sclerometer test, an ultrasonic test and their combined use, called SonReb *(SONic + REBound)*, are "nondestructive" tests on reinforced concrete. The combined tests are a very useful method for assessing the concrete strength and to reduce the possibility of errors that can happen if the tests are not combined, as it has been noticed that the humidity content of a structural element can influence the sclerometer index and the ultrasound speed [1]. The combined method requires shorter time to obtain the results.

2. A *"Critical"* Approach to Building Restoration

The conservation intervention on a historic building, regardless of its architectural and/or artistic value, is generally more appropriate if information on its construction, evolution to date, materials, construction techniques, and structure is available. The conservation of built heritage highlights different issues such as the vast number of buildings needing attention and the urgency of cases that have to be resolved with limited economic resources and time.

To undertake a suitable intervention, three questions have to be answered: whether, where, and how to restore. The fourth question could be added, in which the economic aspect dominates: when to restore. To be able to respond adequately to these questions, it is necessary to proceed by developing specifications step by step, starting from a detailed knowledge of the entire building, the level and causes of degradation, and by finishing with the "operational" description of the proposed interventions.

Within the preliminary data acquisition phase, the direct survey of the building characteristics and condition and the mapping of any noticed changes are accompanied by the research on the project documentation and the events that have affected the structure during its construction and throughout the building life.

These approaches, often coupled with normative models which translate living conditions into objective parameters and standards, hinder the interventions such as a "simple" maintenance or a restoration or produce the result that is not logically related to the structural, typological, functional, and technological characteristics of the artefacts.

An "appropriate" restoration should plan the reuse of built heritage and aim to achieve building performance comparable to new buildings. The reuse interventions should be integrated with the conservation and not imposed.

The above methodology highlights how the approach could be helpful for the classification of pathological events within a building and for the application of innovative solutions to increase the durability of restoration interventions.

3. The Investigative Techniques for Concrete

The investigative techniques for concrete are also classified in two defined macrocategories ("destructive" and "nondestructive"). The former are based on the extraction of concrete samples to undertake compressive tests and represent the most reliable instrument for assessing the mechanical properties of concrete. The second investigative typology, the nondestructive tests, can be further subdivided in: really "nondestructive" investigations and "partially destructive" investigations. The latter include (a) the penetration test with a Windsor gun (ASTM C 83) which enables the identification of compressive resistance of concrete by measuring the depth of penetration of the special metal pins projected with a Windsor gun into concrete, (b) the extraction test (pull-out) (UNI 10157 : 1992-ASTM C 900-06) which enables the assessment of compressive resistance of concrete by measuring the force used by a hydraulic jack for extracting a special plug inserted into concrete.

The really nondestructive investigations include, among others, (a) endoscopy that enables a direct observation of form and appearance of an investigated object, (b) thermography that assists in recognising potential structural anomalies by using the capacity of materials to transfer heat, (c) magnetometry which enables localising metal bars in reinforced concrete, (d) the Ground Penetrating Radar (GPR), especially with regard the investigation of the internal reinforcement bars, (e) measurement of the electric potential of concrete which enables defining the level of corrosion of metal reinforcements in concrete, (f) ultrasound investigation that allows qualitative assessment of the concrete resistance by using the capacity of the concrete components to transfer ultrasound waves, (g) sclerometric tests that assess the concrete resistance by reading the bounce results, and finally, (h) so-called "SonReb" (SONic + REBound) that enables assessing the concrete resistance by combining the speed of ultrasound waves and the index of surface bounce through a synergic use of the two previous investigations.

3.1. The "SonReb" Method for Determining the Resistance of Concrete. The SonReb method, as mentioned above, allows a qualitative determination of the concrete resistance [2] through the cross-examination of the values of the speed of ultrasound waves and the values of sclerometric bounce. This investigation method is standardised by RILEM Recommendations [3] 43 CND-EN 13791 : 2007, the EC regulation 1-2010 UNI EN 12504-2 : 2001, ASTM C597, UNI EN 12504-4 : 2005, the Test Report CUR 69, the standards UNI 7997, UNI 9524 and UNI 83308.

This method is used for assessing the resistance of concrete, enabling the elimination of errors, at least partly, that appear when the two investigation methods are separately applied. This method, in fact, allows reducing the errors made when the sclerometric and ultrasound tests are undertaken separately [4].

It has been noticed, for example, that the humidity content leads to an underestimation of the sclerometric index and in an inversely proportional way leads to an overestimation of the ultrasound speed; similarly, the sclerometric index rises in a directly proportional way to the increase of the age of concrete [1], while the ultrasound speed decreases in an inversely proportional way to it [5].

In fact, the risk that can come up in the separate use of sclerometric and/or ultrasound tests is related, for example,

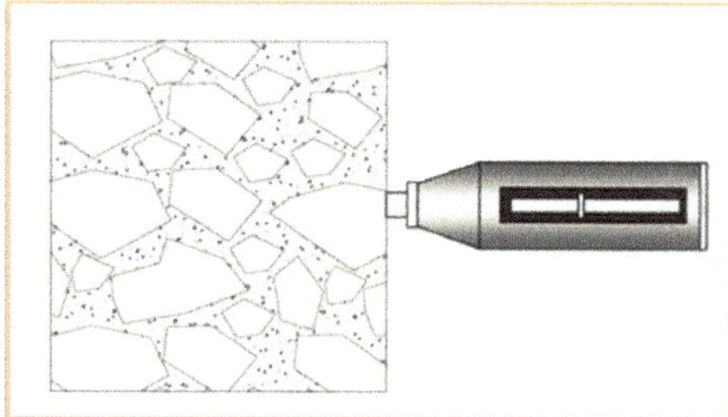

FIGURE 1: Sclerometer beats against a large piece of aggregate.

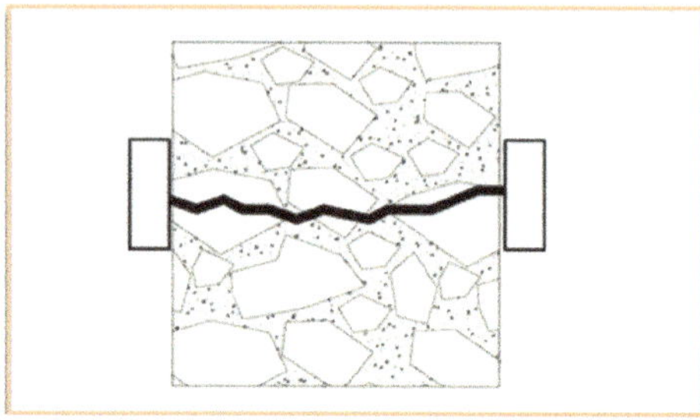

FIGURE 2: Ultrasound equipment: the impulse transmission.

to the chance nature of the position of aggregates in relation to the external surface of an investigated component; the sclerometer can easily beat against a large piece of aggregate, probably obtaining a higher value of the bounce index S in relation to the one returned when the blow is executed on a homogenous concrete surface (Figure 1).

Again as an example, the ultrasound test (Figure 2) can equally be affected by the chance nature of the disposition of aggregates in a cement mix when the gaps between the aggregates are arranged in the way that induces the rise of the void index; in this case, the speed of wave spread V decreases in relation to the value that would be obtained when the wave spreads through an area of "*homogenous*" concrete.

The application of the "SonReb" method asks for the appraisal [6] of local values of the ultrasound speed V and of the bounce index S from which it is possible to obtain the resistance of concrete R_c through expressions such as:

$$R_{c,\text{sonReb}} = a \cdot S^b \cdot V^c. \quad \text{(a)}$$

In the scientific literature this formula has assumed different forms, each one expressing the experiments undertaken directly on site or in a laboratory on standardised samples.

For example:

(i) Gasparik [7] (1992)

$$R_{c,2} = 8,06 \cdot 10^{-8} \cdot S^{1,246} \cdot V^{1,85}, \quad \text{(b)}$$

(ii) RILEM (1993) [2], NDT 4

$$R_{c,1} = 9,27 \cdot 10^{-11} \cdot S^{1,4} \cdot V^{2,6}, \quad \text{(c)}$$

(iii) Di Leo and Pascale [6] (1994)

$$R_{c,3} = 1,2 \cdot 10^{-9} \cdot S^{1,058} \cdot V^{2,446}, \quad \text{(d)}$$

in which:

(i) R_c is the resistance of a cube under compression [N/mm^2];

(ii) S is the sclerometric index;

(iii) V is the ultrasound speed [m/s].

The formula (c) depicts the correlation curve applied in the investigated case studies. In (c) R_c is expressed in MPa and the ultrasound speed in m/s; this relationship is related to a standard concrete whose properties are described in RILEM 43-CND. When a different type of concrete is employed the following relationship is applicable [8]:

$$R_{\text{sonReb}} = R'_{\text{sonReb}} \cdot \left(C_c \cdot C_d \cdot C_a \cdot C_f \cdot C_P \cdot C_m\right), \quad (1)$$

where R'_{sonReb} is the value obtained from (1), while C_c (cement type), C_d (cement content), C_a (aggregate types), C_f (proportions of fines), C_p (maximum aggregate size), and C_m (errors) are coefficients of influence that permit the extension of (b) to the cases of a nonstandard concrete (as defined above).

Hence, if the concrete has the same characteristics as the one whose experimental curves are available, the graph directly provides the estimated resistance of concrete. Otherwise, as usually happens, when the concrete has a different composition from the one presented by the curve, the corrective coefficients, that take into account the type of cement and the related dosage, the nature and dimensions of aggregates, and the potential additives need to be applied to obtain an approximately true estimate of the concrete resistance [1].

It is evident that the above equations could not have universal validity except for the fact that the values of S and V depend on the characteristics of concrete, even when the specific indications on the limitations of applicability are missing.

However, a qualitative appraisal of the resistance of concrete can be made even by using the graphs (Figure 3) which show a series of isoresistance curves in the plane V-S (obtained from the above analytic expressions) and which refer to the tests undertaken on standardised samples in a laboratory.

Although it appears absolutely necessary to analyse a wider range of cases, the above methodological investigation approach, based on the comparative analysis of the two described test campaigns, suggests several important considerations regarding the modality of investigation and the interpretation of results, demonstrating the need for establishing general investigation criteria: more than defining in a strict manner the number of tests that should be undertaken, it would be necessary to preset the level of significance to be achieved.

4. The Case of Study: "Centrone" Theatre in Gravina in Puglia (Italy)

The above-defined methodology has been tested on a postwar building which is located in Gravina in Puglia

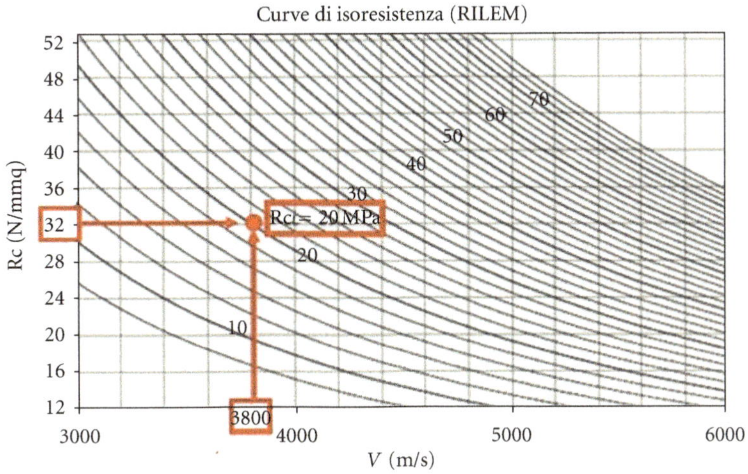

FIGURE 3: Example of isoresistance curves—RILEM NDT 4 recommendations.

FIGURE 4: Historical image of vault construction.

FIGURE 5: Historical image of iron trusses.

(Italy): the "Centrone" theatre; it was built using a mixed structure, that is, reinforced concrete and bearing masonry built of local stone.

The building is not very old, but it is very interesting for the local civic history; in fact, it represents different values. Values that are set in different reels from the life of the building and revolved the surrounding environment in which it is inserted.

Historical value related to the historical development of the building and to the importance that it has in terms of expression of the local civic history; social value that is related to the social development of the city; cultural value that is related to the cultural growth of society; value of identity that has the building in terms of physics and metaphysics belong to a place then; but not least; the economic value related to the ability to reuse the building readapting it to new needs.

The building was built between 1946 and 1948; the foundations were built on pillars and vaults (Figure 4) and the pitched roof using iron trusses (Figure 5).

The building is a rectangular block (Figure 6), developed on three levels and covered using plaster and stone; the main façade is symmetrical, tripartite and follows the ground inclination; in the opposite side, the building shows two projecting polygonal elements realized using the local stone.

In the central part of the building; approximately 10 m wide; is located the entrance, highlighted by the massive presence of two Doric columns that support a loggia. The building is organized in different spaces: the central part, where at ground floor there are the foyer entrance and ticket office, while on the first floor there is the Room "Italy" (Figure 7) that contains 110 seats; at second floor there are three flats for the managers of the theatre. The mail hall contains the stalls with 800 seats (a big space 32.5 m long and 21 m wide); it is partly covered by a balcony (at the first level) with 300 seats and a gallery on the second level with 300 seats.

FIGURE 6: The "Centrone" theatre in Gravina in Puglia (Italy).

The pathologies that interest the building are not particularly serious, since it does not affect in any way the

FIGURE 7: The room "Italy" on the first floor of the theatre.

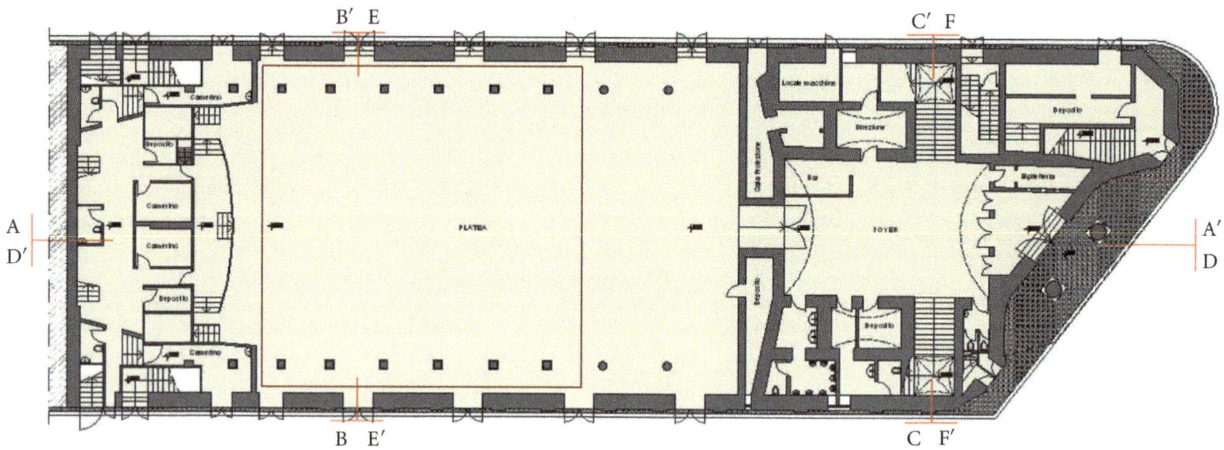

FIGURE 8: The specification of the location of the investigation points.

static system of structure and are easily solvable with non-invasive interventions. Inside are not detectable serious and manifest diseases, except for some water infiltrations from the roof that are determined as chromatic alterations and plaster detachment.

The condition of the pillars is generally acceptable, since they do not show obvious pathologies; however, you should note that some have a bad state of preservation of the plaster due to moisture that caused severe swelling, internal cavities and an evident surface lack of homogeneity.

4.1. On-Site Tests: Formulation. Essential elements for the organisation of an investigation campaign are the selection of the components that will be examined (which have to be representative of the whole structure), the investigation methodology that will be used, and the number and location of the investigation points; these requirements, in fact, are fundamental to guarantee a certain level of "reliability" of the obtained results and a "trustworthiness" with regard to the qualitative indications related to the characteristics of the material. In the case of the "Centrone" theatre, the identification of the components to be investigated was made by selecting the structural components which make the internal core of the load-bearing structure for two reasons: (1) because they enable to investigate a concrete in the state of "natural aging" (i.e., without considering the aggressive atmospheric agents which could have modified the condition of the material) and (2) for easiness of selecting the testing points. The structure of the whole building consists of a dual typology of structural components: those inserted in the external fabric (therefore difficult to investigate) and those located within the building.

The selection of the location of the investigation points (Figure 8) was random (with the aim to guarantee the representativeness of the investigation for the entire structure); in addition, "*homogenous*" areas [1] (which have the same characteristics) were noted and considered by making the obtained results "qualitative" and representative for all the

TABLE 1: The value of rebound index.

Structural element		Sclerometer value										Average value
Element code	Size [cm]	Rebound value										
P1 Dx T	43 × 43	27	28	26	29	25	25	31	26	28	28	28
P2 Dx T	43 × 43	34	38	36	42	40	36	37	38	41	38	38
P3 Dx T	43 × 43	32	33	30	31	32	32	34	31	31	31	32
P4 Dx T	43 × 43	35	33	34	30	36	34	34	34	36	38	38
P5 Dx T	43 × 43	35	33	34	34	33	32	35	37	34	34	34
P6 Dx T	43 × 43	35	37	38	38	36	34	35	38	36	36	36
P1 Sx T	43 × 43	41	37	45	47	37	38	42	39	40	41	41
P2 Sx T	43 × 43	43	42	42	40	44	42	46	44	44	40	43
P3 Sx T	43 × 43	34	34	37	35	34	33	34	32	36	35	34
P4 Sx T	43 × 43	32	28	32	32	34	34	34	34	30	32	32
P5 Sx T	43 × 43	30	28	28	32	30	28	32	32	30	30	30
P6 Sx T	43 × 43	34	35	34	36	35	37	38	35	38	35	36
P1 Dx I p	43 × 43	23	21	21	22	22	25	25	26	24	22	23
P2 Dx I p	43 × 43	33	32	33	32	30	30	30	33	31	30	31
P3 Dx I p	43 × 43	37	40	38	41	43	38	38	38	41	38	39
P4 Dx I p	43 × 43	38	38	38	38	37	39	37	38	39	39	38
P5 Dx I p	43 × 43	37	39	35	38	37	38	38	40	38	40	38
P6 Dx I p	43 × 43	36	38	37	37	38	38	37	38	36	38	37
P1 Sx I p	43 × 43	26	22	24	22	24	20	23	22	22	22	23
P2 Sx I p	43 × 43	40	41	37	40	40	39	38	38	37	38	39
P3 Sx I p	43 × 43	40	37	35	37	36	35	35	35	35	35	36
P4 Sx I p	43 × 43	35	35	35	37	35	37	36	38	36	38	36
P5 Sx I p	43 × 43	40	40	37	35	36	35	38	36	36	36	37
P6 Sx I p	43 × 43	36	38	36	38	38	38	40	40	38	38	38

components that show the same characteristics; a further condition is that the selected surfaces do not show any obvious condition of degradation. The methodology used to undertake the tests was exclusively selected in terms of the possibility to undertake the tests. Regarding the sclerometric tests, they were undertaken by placing the instrument always orthogonally in relation to the investigated surface; the preference was given to the direct method of investigation.The considered reference (in terms of the number of tests with the aim to guarantee the reliability of the results) for the investigation campaign was taken from the framework of the Italian investigation procedures.

4.2. On-Site Tests: Analysis and Results. Sclerometer tests were carried out following the directions contained in UNI EN 12504-2 (2001). The instrument used is the concrete GEI model. On each pillar identified, it was carried out 10 measurements (Table 1) of the value of rebound (for each test, ten values were taken, from which the average value was considered as illustrative).

Three different areas were selected for each investigated component; at 0.70 m, at 1.40 m, and at 2.10 m from the floor. This selection, in fact, was motivated by the need to investigate the columns at the points of major stress (base and middle), having assumed (1) a uniform distribution of the loads and imposed loads (due to the homogenous structure) and (2) a homogenous condition of the loads (there are no visible cracks to induce any different considerations).

Ultrasonic tests were carried out according to the UNI EN 12504-4 (2005). Ultrasonic equipment used is the DSP model of Ultrasonic UTD 1004. On each pillar identified, it was carried out 10 measurements (Table 2) of the value of rebound (for each test ten values were taken, from which the average value was considered as illustrative).

These data were compared with the method "SonReb" (Tables 3, 4, 5, and 6) to improve—as said before—the qualitative interpretation of results; the analysis showed satisfactory results of the state of concrete condition, as well as the importance to compare the ultrasonic and sclerometer tests.

The obtained results show a significant homogeneity of values measured at each investigated level; more precisely, the sclerometric tests show an average value (of the ten measurements made at each point) of the bounce index uniformly distributed on the surfaces; whereas, the ultrasound tests returned average values of the ultrasound between a minimum value of $3050\,m/s^2$ and a maximum value of $3660\,m/s^2$. Comparing the results by using the method SonReb, the obtained values of the concrete resistance are from a minimum value of 18 MPa to a maximum value of

TABLE 2: The value of ultrasonic speed.

P1 Sx T average ultrasonic speed: 2862,9 m/s						
Height: 0,70 m Size 0,46 × 0,46	*X* direction	2039	2036	2035	2038	2041
		2032	2034	2045	2048	2029
	Y direction	3656	3669	3686	3676	3664
		3668	3672	3675	3663	3665
Height: 1,40 m Size 0,43 × 0,43	*X* direction	2043	2041	2042	2045	2046
		2049	2068	2061	2043	2048
	Y direction	3665	3681	3687	3690	3684
		3689	3676	3679	3695	3700
Height: 2,10 m Size 0,40 × 0,40	*X* direction	2052	2045	2049	2046	2048
		2058	2060	2039	2059	2056
	Y direction	3685	3684	3682	3679	3669
		3689	3694	3692	3686	3696
P1 Dx T average ultrasonic speed: 2873,2 m/s						
Height: 0,70 m Size 0,46 × 0,46	*X* direction	3367	3355	3362	3360	3355
		3394	3397	3389	3362	3392
	Y direction	2520	2521	2531	2513	2512
		2515	2510	2533	2511	2525
Height: 1,40 m Size 0,43 × 0,43	*X* direction	3250	3257	3260	3252	3247
		3302	3315	3333	3341	3348
	Y direction	2490	2499	2510	2512	2505
		2496	2493	2495	2491	2510
Height: 2,10 m Size 0,40 × 0,40	*X* direction	3322	3320	3333	3327	3325
		3139	3152	3134	3152	3157
	Y direction	2526	2516	2545	2544	2558
		2556	2559	2562	2568	2564
P6 Sx T average ultrasonic speed: 2134,1 m/s						
Height: 0,70 m Size 0,46 × 0,46	*X* direction	2160	2084	2053	2016	2066
		2161	2016	2103	2105	2150
	Y direction	2160	2154	2111	2102	2013
		2163	2136	2165	2130	2122
Height: 1,40 m Size 0,43 × 0,43	*X* direction	2160	2136	2165	2154	2155
		2155	2153	2151	2160	2165
	Y direction	2152	2154	2150	2153	2103
		2100	2155	2015	2140	2150
Height: 2,10 m Size 0,40 × 0,40	*X* direction	2150	2099	2088	2066	2200
		2035	2150	2156	2171	2015
	Y direction	2030	2105	2145	2144	2014
		2156	2148	2154	2148	2154
P6 Dx T average ultrasonic speed: 2560,9 m/s						
Height: 0,70 m Size 0,46 × 0,46	*X* direction	2562	2563	2545	2568	2567
		2575	2574	2586	2544	2578
	Y direction	2543	2546	2566	2564	2568
		2545	2544	2543	2549	2548
Height: 1,40 m Size 0,43 × 0,43	*X* direction	2566	2531	2533	2540	2545
		2562	2545	2565	2555	2551
	Y direction	2556	2565	2559	2557	2549
		2552	2553	2545	2551	2553

Table 2: Continued.

P6 Dx T average ultrasonic speed: 2560,9 m/s						
Height: 2,10 m	X direction	2565	2596	2586	2567	2568
Size 0,40 × 0,40		2564	2572	2575	2574	2573
	Y direction	2564	2561	2563	2559	2568
		2567	2586	2584	2583	2569
P1 Sx I p average ultrasonic speed: 2582,0 m/s						
Height: 0,70 m	X direction	1899	2130	2156	2102	2105
Size 0,46 × 0,46		2111	2182	2221	2310	2229
	Y direction	3200	3180	3196	3102	3104
		3016	3038	3086	3048	3062
Height: 1,40 m	X direction	1700	1750	1890	1806	1830
Size 0,43 × 0,43		1834	1905	1942	1972	1973
	Y direction	3109	3140	3149	3131	3158
		3160	3263	3264	3285	3189
Height: 2,10 m	X direction	2005	2103	1999	2150	2015
Size 0,40 × 0,40		2060	2160	2163	2184	2150
	Y direction	3005	3158	3139	3190	3100
		3089	3048	3069	3099	3105
P1 Dx I p average ultrasonic speed: 3006,5 m/s						
Height: 0,70 m	X direction	3000	3002	3005	3019	2993
Size 0,46 × 0,46		3010	3015	3019	2921	2996
	Y direction	3012	3015	3018	3016	3018
		3001	3021	3029	3028	3020
Height: 1,40 m	X direction	3000	3052	3013	3002	3015
Size 0,43 × 0,43		2999	3050	3015	3016	3019
	Y direction	2996	3047	3036	3033	3031
		3002	3052	3035	3063	3043
Height: 2,10 m	X direction	2899	2915	3005	3001	3045
Size 0,40 × 0,40		3015	3062	3036	3035	3046
	Y direction	3051	3015	3063	3026	3033
		3018	3017	3021	3012	3026

Table 3: Left side pillar (Sx): Ground floor.

Pillar	Average rebound value	Average ultrasonic speed
P1 Sx T	41	2862,9
P2 Sx T	43	3297,1
P3 Sx T	34	3345,3
P4 Sx T	32	3181,2
P5 Sx T	30	3192,1
P6 Sx T	36	2134,1

Table 4: Right side pillar (Dx): Ground floor.

Pillar	Average rebound value	Average ultrasonic speed
P1 Dx T	28	2873,2
P2 Dx T	38	3103,6
P3 Dx T	32	3199,3
P4 Dx T	38	3313,2
P5 Dx T	34	3082,1
P6 Dx T	36	2560,9

22 MPa, that take in a significant consideration the contribution of the covered concrete conditions.

5. Conclusions

The tests carried out are the basis of a diagnostic project that is possible to implement and monitor to guarantee a deeper knowledge, with the goal of attaining a level of thorough understanding aimed at the "preservation and improvement" of a building.

The recovery project, that takes particular care in the methodological application of the diagnostic phase, cannot leave out of consideration the necessity of a careful and timely monitoring of building conditions.

TABLE 5: Left side pillar (Sx): First floor.

Pillar	Average rebound value	Average ultrasonic speed
P1 Sx I p	23	2582,0
P2 Sx I p	39	3476,1
P3 Sx I p	36	3358,2
P4 Sx I p	36	3319,6
P5 Sx I p	37	3450,3
P6 Sx I p	38	3346,1

TABLE 6: Right side pillar (Dx): First floor.

Pillar	Average rebound value	Average ultrasonic speed
P1 Dx I p	23	3006,5
P2 Dx I p	31	3210,2
P3 Dx I p	39	3120,9
P4 Dx I p	38	3204,6
P5 Dx I p	38	3098,1
P6 Dx I p	37	3301,5

The carried out tests are the first and simplest analysis for a qualitative assessment; it is necessary to classify the structure regarding the following consolidation procedures. The recovery and conservation project, as well as an "indispensable" transformation of an old building, must be in that evaluations, of feasibility and suitability, both economic and practice which is the basis for a "suitable choice" of recovery intervention, that permit to annul the "cancellation" of the Built Heritage.

References

[1] A. Masi, *La Stima della Resistenza del Calcestruzzo in situ Mediante Prove Distruttive e non Distruttive*, Il Giornale delle Prove non Distruttive Monitoraggio Diagnostica no. 1, 2005.

[2] RILEM Draft Recommendation, 43-CND. Combined nondestructive testing of concrete. Draft recommendation for in situ concrete strength determination by combined nondestructive methods. Materials and Structures no. 26, 1993.

[3] RILEM (The International Union of Testing and Research Laboratories for Materials and Structures) is an organisation which enables exchanges through an international network of testing engineers, researchers, academics, educators and practitioners.

[4] F. Braga, M. Dolce, A. Masi, and D. Nigro, *Valutazione delle Caratteristiche Meccaniche dei Calcestruzzi di Bassa Resistenza Mediante Prove non Distruttive*, L'Industria Italiana del Cemento no. 3, 1992.

[5] P. Bocca and S. Cianfrone, "Le prove non distruttive sulle costruzioni: una metodologia combinata," L'Industria Italiana del Cemento no. 6, 1986.

[6] A. Di Leo and G. Pascale, *Prove Non Distruttive Sulle Costruzioni in Cemento Armato*, Convegno Sistema Qualità e Prove non Distruttive per l'affidabilità e la Sicurezza delle Strutture Civili, Bologna, Italy, 1994.

[7] J. Gasparik, *Prove Non Distruttive Nell'edilizia*, Università di Brescia, 1992.

[8] R. Giochetti and L. Lacquaniti, *Controlli non Distruttivi su Impalcati da Ponte in Calcestruzzo Armato*, Nota Tecnica 04, Università degli Studi di Ancona, Facoltà di Ingegneria, Istituto di Scienza e Tecnica delle Costruzioni, 1980.

16

Lg Coda Variations in North-Central Iran

Mojtaba Naghavi,[1] Zaher Hossein Shomali,[1] and Mehdi Zare[2]

[1] *Institute of Geophysics, University of Tehran, Tehran, 14155-6466, Iran*
[2] *International Institute of Earthquake Engineering and Seismology (IIEES), Tehran, 19395-3913, Iran*

Correspondence should be addressed to Mojtaba Naghavi, mmnaghavi@ut.ac.ir

Academic Editor: P. Talwani

Ground motion records in north-central Iran have been used in order to obtain the Lg coda Q, using stack spectral ratio method. The lateral variations in Q_0 and its frequency dependence are estimated in the 0.3–7.0 Hz frequency range which led to $Q = 267(\pm 32) \times f^{0.71(\pm 0.14)}$. The observed variations in quality factor show low values of Q_0 in western part of the study region where bounded by relatively high values in southern and northern parts. Since the seismicity of the study area is quite shallow the obtained results can be attributed to the upper 30 km of the crust. The Damavand volcano and its surrounding region also exhibit variations in the value of Q_0 which result in low and intermediate values of Q_0 in western and eastern parts, respectively. Current seismicity in Damavand is mostly confined to its southwestern part, whereas our results proved to possess low values of Q_0. In general, most of Q factor variations can be attributed to the lateral heterogeneity as well as the severity of the crustal velocity gradient, and as expected the north-central Iran is well inferred as a tectonically active region.

1. Introduction

The energy of a seismic wave decays while passing through a "real" medium such as the earth which is not completely elastic. The decay in energy due to nonelastic phenomena is called intrinsic attenuation and is characterized with the Q parameter, the large values of which represent small values of attenuation and as Q approaches zero the pertaining attenuation will become quite strong. Therefore, Q could be considered as a measure of elasticity of the media.

In the present study, the Q factor is obtained for the Central Alborz in Iran using the Lg coda method. The Lg phase was originally observed using earthquakes occurred in California and has similar characteristics as SH phase with an average group velocity in the order of 3.5 km/s [1]. The regional Lg phase, comprised of multiple shear wave reverberations trapped in the crust, is often used to determine the quality factor (Q) variations for event-station distances less than 1500 km. The Lg phase is typically the most prominent short-period seismic phase observed over continental paths at distances greater than 200 km [2]. Lg phase can be treated as a sum of higher modes of surface waves, or multiple supercritically reflected *S* waves trapped in the crust [2]. The average group velocity of Lg phase is in order of 3.3∼3.6 km/s, while in stable shields show faster velocities than young active regions [3]. In this study, the Lg coda Q is estimated using ground motions recorded at 35 short-period seismic stations in north-central Iran, using stack spectral ratio (SSR) procedure developed by Xie and Nuttli [4]. This method is widely applied to Lg code studies in the crustal structures in different parts of the world [3, 5–7].

Frequency dependency of Lg phase, due to inelasticity and wave scattering through the crust, significantly depends on tectonic environment and local seismicity. Stable continents are characterized by low-frequency dependence and high-quality factor. In the case of active regions, that is, regions with complex structures in crust and upper mantle, the energy attenuation results in a reduced quality factor and an increased frequency dependence [8]. The method used in

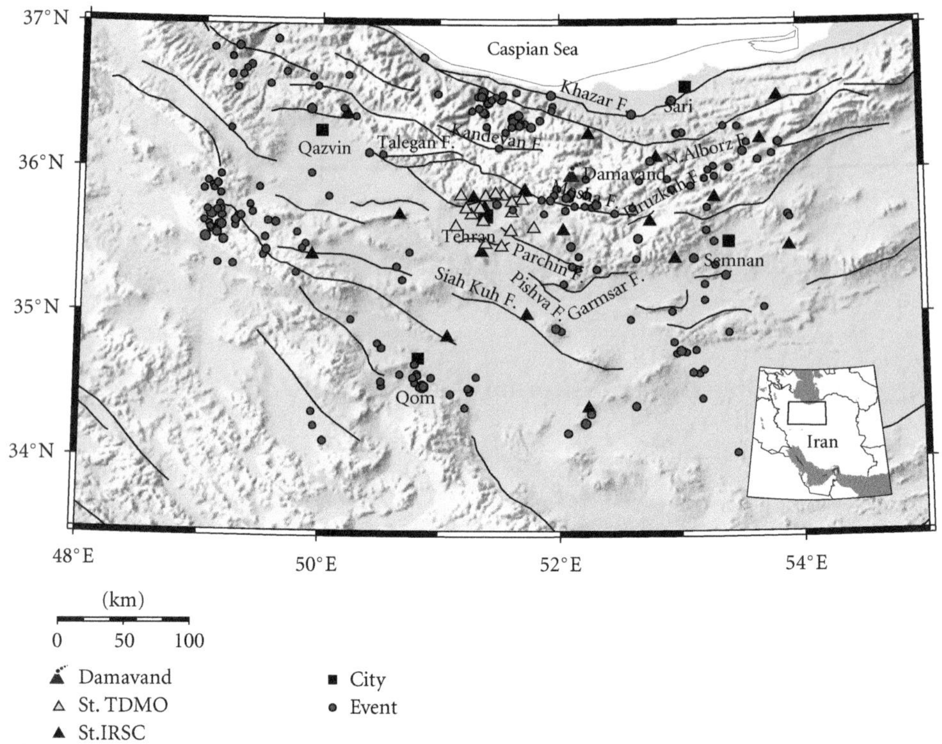

Figure 1: Topographic map of the study area accompanied with the distribution of recording stations and the earthquakes used for the study. Solid lines illustrate the major faults, rectangles represent significant cities, the circles show earthquake epicenters, and triangles indicate seismic stations.

this research to estimate the attenuation dose not distinguish between intrinsic and scattering attenuation but produces an apparent attenuation for the study area.

2. Tectonic Settings of the Studied Area

The study area is located in the northern seismotectonic province of Iran (Alborz). The term "Alborz" is usually used for the mountain range in northern Iran (south of the Caspian Sea) and some of surrounding regions. The Damavand peak is the most prominent geological as well as topographic feature in the region.

The Alborz mountain range, in northern Iran, consists of several sedimentary and volcanic layers and numerous active faults. This mountain belt has an eastwest trend and its width and length are approximately 100 km and 600 km, respectively. The entire system is limited to the Talesh mountains in west and Kopeh Dagh mountains in the east [9, 10]. This area has experienced several destructive earthquakes [11] and accommodates the general motion between the central Iran and southern Caspian Sea [12]. Damavand volcano is the prevalent geo-seismological feature in this mountain range.

Central Alborz is an active mountain-belt and is a part of Alpine-Himalayan mountain belt. The tectonic activity in the Alborz mountain range is predominately affected by two mechanisms, northsouth shortening between Eurasia plate and central Iran and northwest ward motion of the south Caspian basin [13].

The study region is located in central Alborz, bounded by 37°N to the north, 34°N to the south, 49°W to the west, and 54°W to the east. The area of study, distribution of earthquakes as well as the recording stations used in this research are shown in Figure 1.

3. Data Set and Data Processing

Two main seismological networks are used in this study. The first network, Iranian Seismological Center (IRSC) is the national short period seismic network operated by the Institute of Geophysics, University of Tehran (IGUT; http://geophysics.ut.ac.ir/En/) consisting of 54 seismic stations across the country. These stations are equipped with three component seismometers with an eigen-frequency of 1–1.25 Hz (negative feedback SS-1 Ranger Seismometers). The second seismic network is the Tehran city Seismic Network (TCSN), operated by Tehran Disaster Management and Mitigation Organization (TDMMO; http://www.tdmmo.ir/), equipped with three component short-period seismometers established in June 2004 in order to monitor stress field variations and seismicity of Tehran city and its vicinity (Figure 1).

The data used in this study consists of 1020 velocity time histories recorded from 205 earthquakes occurred between 2000 and 2009 with magnitudes between M_L 3.5 and 6.5. The

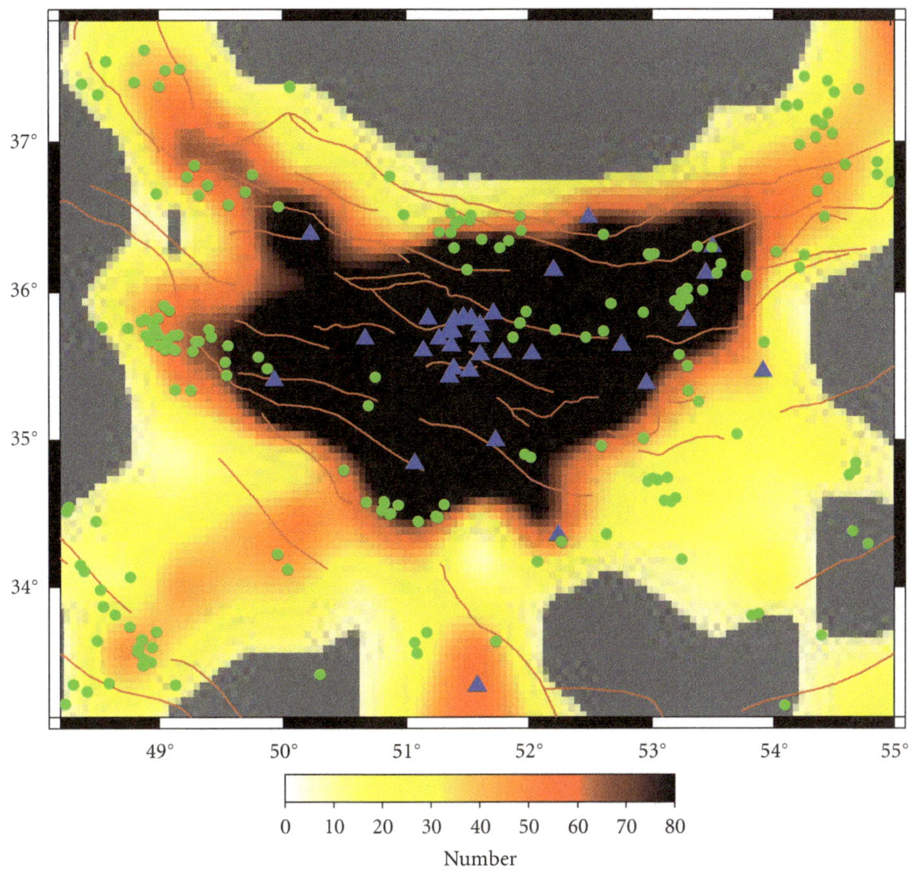

FIGURE 2: The path-coverage density variation in the study area. Circles and triangles represent earthquake epicenters and seismic stations, respectively.

data come from 19 stations of IRSC and 16 stations of TCSN (triangles in Figure 1). Data acquisition is performed at a sampling rate of 50 and 75 samples per second for IRSC and TCSN networks, respectively. Figure 1 illustrates the stations as well as the recorded seismicity used in this study.

The initial data set consisted of 6470 waveforms. Since the Lg phase is predominantly observed on the vertical components of stations with epicentral distances greater than 140 km, as the first step of data processing, the records of such stations were chosen and their Z-components were selected. As the next step, only the waveforms with high signal-to-noise ratios were selected for the following calculations. Iaspe91 [14] model was also used in order to pick P and S phases on the selected waveforms. As the final step, these waveforms were individually addressed by accurate (manual) phase picking and thus the final data set was prepared.

The path coverage of the data used in this study is shown in Figure 2. The path coverage exhibits large number of crossing paths in the central part of the study area. The resolution is decreased towards the marginal parts of the model.

4. Analysis Method

The group velocity of Lg phase may vary from one region to another depending on different geological settings of the uppermost crust. If there are layered sediments along the propagation path, the velocity will decrease significantly.

In this method, as the very first step, the beginning of P and S phases are marked accurately on each waveform. Then by windowing the waveform, the Lg window spectral ratios are found for each subsequent window (Figure 3) and the quality factor is determined based on frequency estimation by stack spectral ratio and linear regression.

The investigation method is based on the stack spectral ratio procedure developed by Xie and Nuttli [4]. This method can be used to obtain frequency-dependent $Q(f)$ estimates from Lg coda. The use of the stacking procedure is a major development in obtaining stable estimates of $Q(f)$, while there is no need to remove the response of seismograph.

Having estimated the arrivals of P and S waves, the Lg coda time series are obtained using the average Lg-phase group velocity. The time series are divided into several

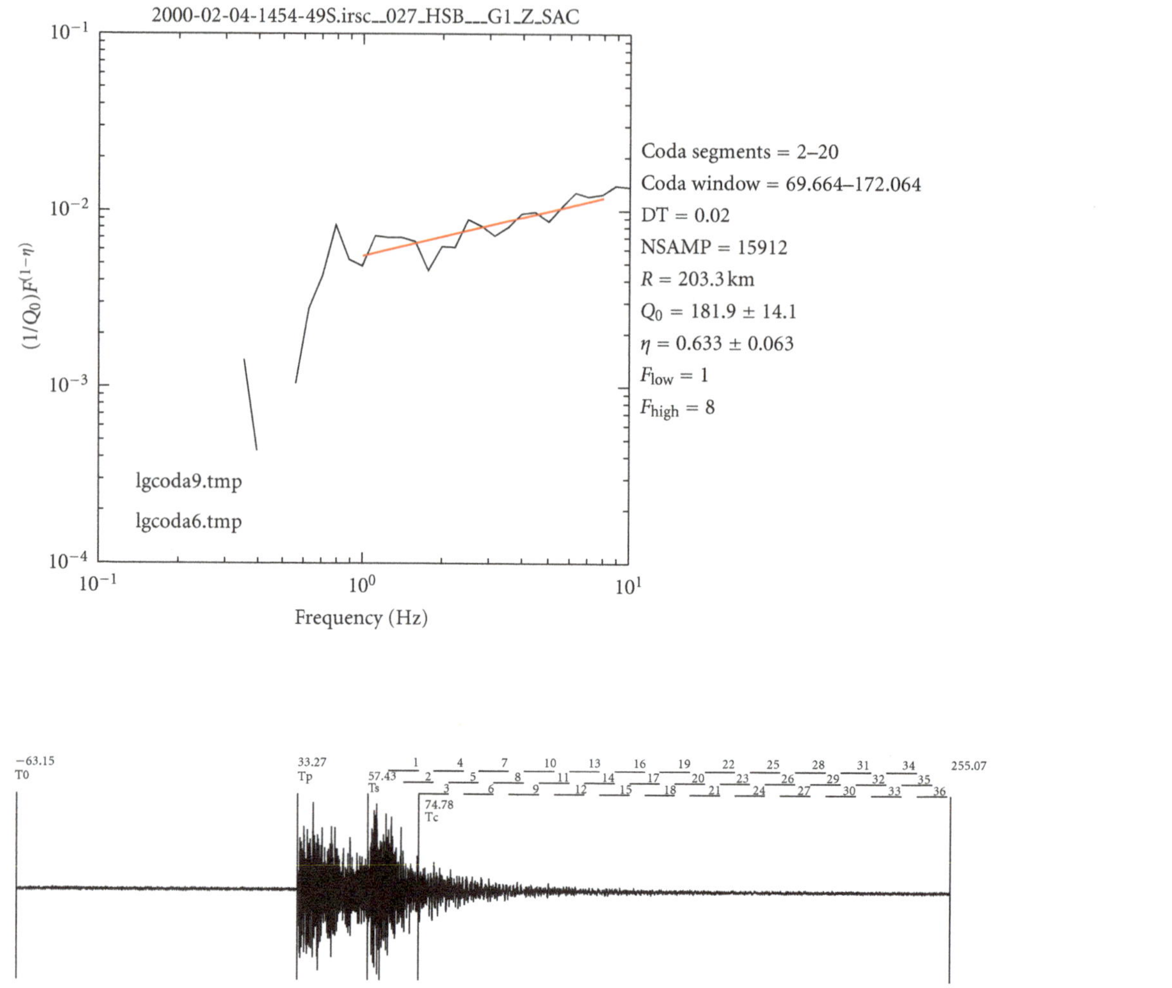

FIGURE 3: Logarithmic $(1/Q_0)F^{1-\eta}$ versus frequency diagram for earthquakes with epicentral distance of 203 km. Windows are shown on the waveform. Average Q_0 and η are estimated using a linear regression (solid line). The y-intercept of this line gives the quality factor Q_0. Estimated errors are also indicated in the corresponding frequency ranges.

overlapping windows, and the spectral ratio is obtained for each two successive windows. Following Xie and Nuttli [4], the Lg coda spectrum is defined in a fixed width window with a centered time as given below:

$$S(f, t_i) = A_0 G(r, t_i) e^{-\pi f t_i / Q(f)}, \tag{1}$$

where $S(f, t_i)$, f, r, $G(r, t_i)$, and A_0 represent the spectra of the windows, frequency, epicentral distance, geometrical spreading function, and total effects of site, source, and instrument, respectively. We focused on events with epicentral distances greater than 140 km for two reasons. Firstly, in many cases with short epicentral distances, it was difficult to determine the presence of Lg because of interference with that of the higher amplitude S arrivals. Secondly, by using events with distances greater than 140 km, we can assume a constant, frequency-independent geometrical spreading of order of 0.5 [15–18].

The spectral ratio for two successive windows centered at t_1 and t_2 can be written as the following:

$$\text{RATIO}_{12} = \frac{(S(f, t_2)/G(r, t_2))}{(S(f, t_1)/G(r, t_1))} = e^{-\pi f (t_2 - t_1)/Q(f)}. \tag{2}$$

This ratio is not stable due to variations of amplitude spectrum in different successive windows. To improve the estimation of Q the following conditions should be provided.

(1) Applying an appropriate function to window the seismogram. This, prior to the calculation of Fourier transform would result in a smooth amplitude spectrum.

(2) Selecting a wide enough time difference $t_2 - t_1$.

(3) Calculating the mean-value for a large number of spectral ratios which could be expressed as:

$$\text{LNRATIO} = \frac{\ln \text{RATIO}_{ij}}{-\pi \left(t_i - t_j\right)}, \tag{3}$$

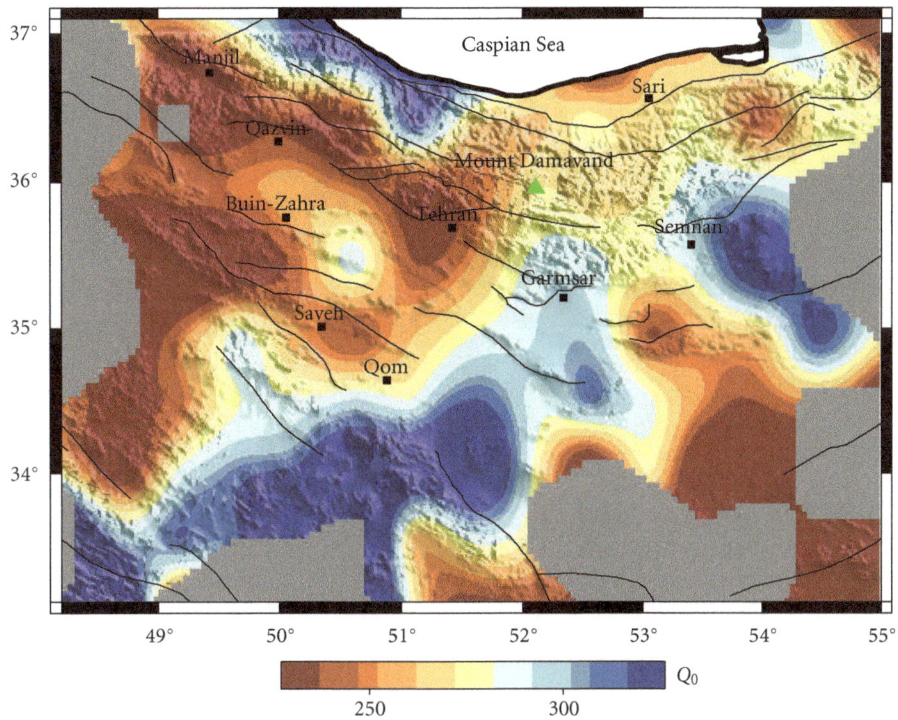

FIGURE 4: Q_0 variations in the study area. Triangles and squares show Damavand volcano and main cities, respectively. Major faults are marked by black lines.

$Q(f)$ can be then expressed by (4)

$$Q(f) = Q_0 f^{\eta}. \tag{4}$$

By working on (4) and adding ε as the random error we obtain

$$\ln\left(\frac{f}{Q(f)}\right) = (1-\eta) \times \ln(f) - \ln Q_0 + \varepsilon. \tag{5}$$

Plotting $(1/Q_0)F^{1-\eta}$ versus f on a logarithmic scale by using a linear regression, the mean value of Q_0 is estimated (Figure 3). A Comparison between the obtained results from theoretical and empirical shows that the stack spectral ration estimates a stable Q in which the error value pertaining to Q_0 is one order less than Q_0 itself.

5. Estimating the Quality Factor for the Study Area

In this study to calculate the lateral variations in quality factor over the region of study we use back projection method to create a tomographic image of the distribution of Q_0 in the studied area.

In this technique, the study area is divided into a grid of rectangular cells, from which the quality factor is calculated for each record and each grid/cell. The median quality factor value between the lowest and the highest values over all the paths crossing each grid/cell is used to start the inversion process. For each record, a residual term is calculated and a new estimate of the quality factor for each grid/cell is found by back-projecting [6, 19] the residual term into the inverse of the quality factor. The quality factor is updated through the iterative application of the inversion until the residual term falls below acceptable minimum threshold.

An estimation of quality factor is determined for each ray-path which corresponds to each pair of station-event. We divided the study area to different grids and then we tested various grid-size in the process, and the results presented in this research (a grid by 0.05°× 0.05° units) is considered robust in relation to the resolution of the final image and the quality of the data. The final results, Q_0 variation in the study area, are shown in Figure 4. Note that the area with less number of ray paths is masked.

Frequency dependency of Q-factor resulted from this study is given below:

$$Q = 267(\pm 32) \times f^{0.71(\pm 0.14)}, \quad (0.3 \le f \le 7.0\,\text{Hz}). \tag{6}$$

The weighted least-squares of $Q(f)$ fit to the Q_0 estimates from this study (north-central Iran) are compared in Figure 5 with other regions including: Alaska [20], south-central Alaska [21], the northeastern United States (NEUS) the Basin and Range province (BRP) [22], the Tibetan Plateau [23], and South-Central Alaska [24].

Figure 5 shows apparent $Q(f)$ values, from several previous researches estimating over a variety of tectonic regions. Tectonically active regions such as the north-central Iran, the BRP, and south-central Alaska generally have low values of Q_0 and high-frequency dependence, while high Q_0 and low η values are observed in tectonically stable continental regions

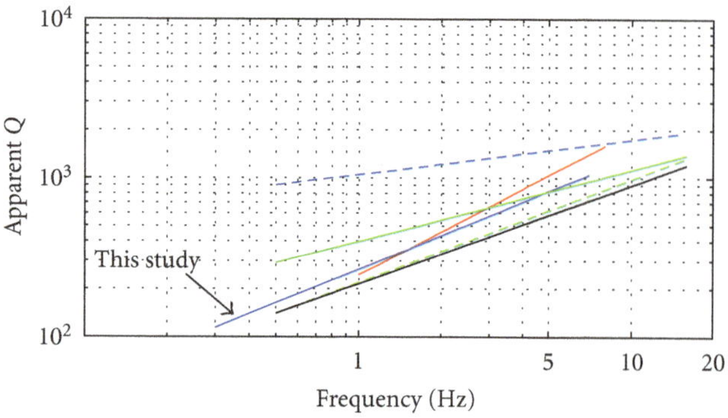

$Q = 245 \times f^{0.9}$ → AEIC-continental Alaska
$Q = 267 \times f^{0.71}$ → central-north Iran (Alborz)
$Q = 399 \times f^{0.45}$ → TP—Tibetan Plateau
$Q = 220 \times f^{0.66}$ → AKLg-south-central Alaska
$Q = 1052 \times f^{0.22}$ → NEUS—northeastern united states
$Q = 235 \times f^{0.56}$ → BRP—basin and range province

Figure 5: Comparison between the proposed formulas for other tectonic regions in the world. It can be seen that the proposed formulas has an acceptable conformity with other functions of quality factor.

like the NEUS. The quality factors determined from this study, as expected the north-central Iran, are more indicative of a tectonically active region.

6. Discussion and Conclusions

The quality factor is related to seismicity and the characteristics in seismotectonic and thermal variations. Therefore based on the results obtained in this research most of the studied region is considered as an active region from the seismicity point of view. This implies a great inhomogeneous region in Alborz active seismic zone and at its vicinity. Alborz mountain range is part of Alpide belt. It forms a barrier between the south Caspian basin and the central plateau of Iran.

Damavand volcano and its surrounding regions are characterized by relatively low Q_0 in the eastern area and a high gradient of Q_0 can be observed in the region. A change (strong gradient) in Q_0 value is observed in the results ranging from anomalously low Q_0 values in the western side although the current seismicity is mostly confined to the eastern side. Most of Damavand's recent earthquakes have occurred in the central and southwestern parts of the Damavand volcano which appear with relatively low Q_0 in the final image. Figure 4 shows that western part of the studied region has been surrounded by low Q_0 anomaly through two zones with relatively high Q_0 in the south and in the north. It seems that this model is related to high rate of probability in seismicity and therefore further investigation is required.

According to the results obtained in this study, Damavand volcano and its surrounding region are characterized with a change (strong gradient) in Q_0 value, ranging from anomalously low Q_0 values in the western side although the current seismicity is mostly confined to the eastern side.

Considering the distribution of the past earthquakes and comparing it with the shape of quality factor's change (Figure 4), it was observed that the Q_0 value in the eastern part is larger than that of the western part.

According to the results, in Damavand region, Q_0 has a relatively higher gradient than that of the surrounding region. It sharply declines moving from east to west. The Q_0 map (Figure 4) correlates well with the large-scale tectonic units of the studied area and also several clear trends corresponding to different characteristics of seismic activity and attenuation field. Most of Q factor variations can be attributed to the lateral heterogeneity as well as the severity of the crustal velocity gradient. It should be noted that that the Q_0 results, such as the one shown in Figure 4, are taken as mean values of each propagation path. Since the seismicity in the area is quite shallow (earthquake depths mostly are less than 30 km), the Q_0 results can be attributed to the average of upper part of the 30 km of the crust of the study area.

It is worth noting that our results impose reliable constraints and interpretations where density of crossing rays allows. On the other hand, we must be cautious interpreting results where the crossing rays are relatively low, in particular at the edge of the network.

Acknowledgments

The authors sincerely thank Iranian Seismological center (IRSC) and TCSN for providing the earthquake data needed for the research.

References

[1] F. Press and M. Ewing, "Two slow surface wave across north America," *Bulletin of the Seismological Society of America*, vol. 42, pp. 219–228, 1952.

[2] J. Xie, Z. Wu, R. Liu, D. Schaff, Y. Liu, and J. Liang, "Tomographic regionalization of crustal Lg Q in eastern Eurasin," *Geophysical Research Letters*, vol. 33, Article ID L03315, 2006.

[3] B. J. Mitchell, Y. Pan, J. Xie, and L. Cong, "Lg coda Q variation across Eurasia and its relation to crustal evolution," *Journal of Geophysical Research B*, vol. 102, no. 10, pp. 22767–22779, 1997.

[4] J. Xie and O. W. Nuttli, "Interpretation of high-frequency coda at large distances: stochastic modelling and method of inversion," *Geophysical Journal*, vol. 95, no. 3, pp. 579–595, 1988.

[5] S. Baqer and B. J. Mitchell, "Regional variation of Lg coda Q in the continental United States and its relation to crustal structure and evolution," *Pure and Applied Geophysics*, vol. 153, no. 2–4, pp. 613–638, 1998.

[6] L. Cong and B. J. Mitchell, "Lg coda Q and its relation to the geology and tectonics of the Middle East," *Pure and Applied Geophysics*, vol. 153, no. 2–4, pp. 563–585, 1998.

[7] J. L. De Souza and B. J. Mitchell, "Lg coda Q variations across South America and their relation to crustal evolution," *Pure and Applied Geophysics*, vol. 153, no. 2–4, pp. 587–612, 1998.

[8] M. Campillo, "Propagation and attenuation characteristics of the crustal phase Lg," *Pure and Applied Geophysics*, vol. 132, no. 1-2, pp. 1–19, 1990.

[9] M. Alavi, "Tectonostratigraphic synthesis and structural style of the Alborz mountain system in Northern Iran," *Journal of Geodynamics*, vol. 21, no. 1, pp. 1–33, 1996.

[10] A. M.C. Şengör, D. Altiner, A. Cin, T. Ustaömer, and K. J. Hsü, "Origin and assembly of the Tethyside orogenic collage at the expense of Gondwana Land," *Geological Society Special Publication*, vol. 37, no. 1, pp. 119–181, 1988.

[11] M. Berberian and R. S. Yeats, "Contribution of archaeological data to studies of earthquake history in the Iranian Plateau," *Journal of Structural Geology*, vol. 23, no. 2-3, pp. 563–584, 2001.

[12] J. Jackson, K. Priestley, M. Allen, and M. Berberian, "Active tectonics of the South Caspian basin," *Geophysical Journal International*, vol. 148, no. 2, pp. 214–245, 2002.

[13] J. F. Ritz, H. Nazari, A. Ghassemi et al., "Active transtension inside central Alborz: a new insight into northern Iran-southern Caspian geodynamics," *Geology*, vol. 34, no. 6, pp. 477–480, 2006.

[14] B. L. N. Kennett and E. R. Engdahl, "Traveltimes for global earthquake location and phase identification," *Geophysical Journal International*, vol. 105, no. 2, pp. 429–465, 1991.

[15] R. L. Street, R. B. Herrmann, and O. W. Nuttli, "Spectral characteristics of the Lg wave generated by central United States earthquakes," *Geophysical Journal of Royal Astronomical Society*, vol. 41, pp. 51–63, 1975.

[16] R. B. Herrmann and A. Kijko, "Modeling some empirical vertical component Lg relations," *Bulletin of the Seismological Society of America*, vol. 73, pp. 157–171, 1983.

[17] R. W. Burger, P. G. Somerville, L. S. Barker, R. B. Herrrnann, and D. V. Helmberger, "The effect of crustal structure on strong ground motion attenuation relations in eastern North America," *Bulletin of the Seismological Society of America*, vol. 77, pp. 120–139, 1987.

[18] G. M. Atkinson and R. F. Mereu, "The shape of ground motion attenuation curves in southeastern Canada," *Bulletin of the Seismological Society of America*, vol. 82, no. 5, pp. 2014–2031, 1992.

[19] J. Xie and B. J. Mitchell, "A back-projection ethodforimaginglarge-scale lateral variations of Lg coda Q with application to ontinental Africa," *Geophysical Journal International*, vol. 100, pp. 161–181, 1990.

[20] G. J. Steensma and N. N. Biswas, "Frequency dependent characteristics of coda wave quality factor in central and southcentral Alaska," *Pure and Applied Geophysics*, vol. 128, no. 1-2, pp. 295–307, 1988.

[21] R. Hansen, D. McNamara, E. Van Ark, and D. Christensen, "Lg ProPagation Continental Alaska," *Eos Transactions of American Geophysical Union*, vol. 79, article 79, 1998.

[22] H. M. Benz, A. Frankel, and D. M. Boore, "Regional Lg attenuation for the continental United States," *Bulletin of the Seismological Society of America*, vol. 87, no. 3, pp. 606–619, 1997.

[23] D. E. McNamara, T. J. Owens, and W. R. Walter, "Propagation characteristics of Lg across the Tibetan Plateau," *Bulletin of the Seismological Society of America*, vol. 86, no. 2, pp. 457–469, 1996.

[24] D. E. McNamara, "Frequency dependent Lg attenuation in south-central Alaska," *Geophysical Research Letters*, vol. 27, no. 23, pp. 3949–3952, 2000.

17

A Short Walk along the Gravimeters Path

Iginio Marson

Department of Civil Engineering and Architecture, University of Trieste, Via Valerio 6, 34127 Trieste, Italy

Correspondence should be addressed to Iginio Marson, marson@units.it

Academic Editor: Umberto Riccardi

The history of gravity measurements begun in 1604 with Galileo Galilei experiments on the acceleration due to the gravity force of the earth, *g*, along inclined planes. In his memory, the most used unit to measure *g* is the gal (10^{-2} m/s^2). The paper takes the interested reader through a walk along some of the most important achievements in gravity measurements and gives some perspectives for future developments in terrestrial gravity.

1. Introduction

The gravity force of the earth is a physical quantity of interest for geodesy, geophysics, and metrology. It is the main force which is responsible of the structure and shape of the earth. The combined effect of gravitational attraction and centrifugal force acts to distribute less dense material in the outermost layers of the earth and denser one in the innermost ones.

The study of the shape of the earth and of its deformation is the basic topic of geodesy. To describe the shape of the earth, geodesists have chosen the equipotential surface which coincides with the surface of the mean sea level in a given epoch, the geoid. The surface of the mean sea level in a given epoch is determined by the gravity field of the earth. From the gravity field it is possible to determine the morphology of the geoid over the entire planet. Space geodesy and in particular satellite altimetry provides a direct measurement of the sea level surface over the oceans and hence the determination of the geoid. The determination of the geoid over the land is, however, possible only by means of the gravity field. The geoid is the reference surface for elevation measurements and it is therefore important for both cartography and navigation purposes. Oceanic currents are also partly controlled by the morphology of the geoid and hence a knowledge of the geoid is of importance for oceanography, hydrology, and for environmental studies related with possible changes of the mean sea level.

Geophysics is more concerned with studying anomalies in the gravity field which are due to anomalous distribution of masses within the earth. Geodynamical processes like plate tectonics, mountain building, convective motions in the earth mantle, and volcanism have all strongly changed the mass distribution within the earth from that which would have been produced simply by the combined effects of gravitational attraction and centrifugal force. The thickness and composition of the earth lithosphere are highly variable over the whole planet. This diversification generates important anomalies in the gravity field. Geophysics is particularly interested in the study of the gravity anomalies because they permit the determination of the geological structures (faults, crust-mantle boundary, density anomalies in the mantle, etc.) which are responsible of the anomalies. Gravity is therefore a basic tool for studies of structural geology. Some geological structures within the earth crust (such as faults, synclines, antisynclines, salt domes) are frequently associated with potential reservoirs for oil and gas. As a result, the study of the Earth's gravity field also plays an important role in the prospection for oil as well as for geothermic.

In Metrology, a knowledge of the gravity field is used as a transfer standard in the determination of metrological standards involving force and pressure measurements. Therefore gravity is required in connection with applications of the electrical standards to a nonartifact standard of mass (i.e., a new Kilogram standard) as well as the calibration of pressure gauges and weighting devices. Metrological use of gravity as a constant, since it varies with both position and time, requires that it is measured wherever it is needed, and, if the accuracy requirements are higher than 10^{-8}, it will be necessary to measure it also at the time it is needed. The definition of reference gravity networks to provide reference sites for metrological, geodetic, and geophysical purposes at different scales (from national to continental to global) is a topic for geodetic metrology. The main activities of this branch of metrology are the determination of gravity datum and the verification of the gravity scale. The accuracy requirements differ considerably and are strongly related to the purpose of the research. Metrology, related to standards of force and pressure, needs to know g with an accuracy of the order of 10^{-6}. The same order of magnitude is also required by geophysical studies aimed at the definition of deep and broad geological structures (for instance, the crust-mantle boundary). For shallower structures (faults within the earth crust) the accuracy requirements are about one order of magnitude higher, 10^{-7}. This order of magnitude is also required for geoid determination. Modern gravimetry allows for the determination of gravity anomalies with accuracy of the order of 10^{-8} which are of importance for civil engineering and archeology. This high accuracy allows also for the determination of time-dependant gravity anomalies related with motion of fluids within the earth crust, such as geothermal fluids, and changes in the water table or density variations due to geodynamical processes. Force metrology, as it will applied to the replacement of the kilogram, will require an accuracy between 10^{-8} and 10^{-9}. Stationary gravimeters of high accuracy, 10^{-10}, are used to measure temporal variations of the gravity field at a given site with the principal target of this research being the determination of the Earth's response to tidal forces. Different applications require not only different accuracy but also different special and temporal sampling. Typically, the spacial density of measuring sites ranges from one site every 10 km for deep geological structures, to one site every 0.5 km for prospection for oil and gas, to one site every 20 m for civil engineering.

Anomalies in the earth gravity field can be also determined by studying the perturbations in the orbits of artificial satellites. The long-wavelength part of the earth's gravity field from space as determined from various tracking measurements of earth orbiting satellites was of considerably varying quality and of incomplete geographical coverage. Consequently the accuracy and resolution of the earth gravity field models were poor, with most of the satellite contributions limited to wavelengths of 700 km or longer. Only broad geophysical features of the earth's structure could be detected. An impressive progress in satellite gravimetry is due to the GRACE twin satellites mission, Gravity Recovering and Climate Experiment, launched by NASA in March 2002, which has made detailed measurements of earth's gravity field (http://www.csr.utexas.edu/grace/). More recently, the ESA mission GOCE, Gravity Field and Steady-State Ocean Circulation Explorer that was launched on March 17, 2009, aims at the determination of gravity-field anomalies with an accuracy of the order of 10^{-6} and of the geoid with an accuracy of about 2 cm and a spatial resolution better than 100 km (http://www.esa.int/SPECIALS/GOCE/).

Space geodesy techniques, principally GPS has made it possible to determine the position of a moving object with high (submeters) accuracy. This capability has made a main contribution to the solution of the key problem in the measurement of gravity from a moving platform, that is, the separation of gravitational and inertial accelerations. Ship-borne gravimetry has been carried out for several decades. In the case of sea-surface gravity measurements, the separation between gravitational and inertial accelerations is facilitated by the different wavelengths of the two signals and by the periodic character of the inertial accelerations which therefore can be removed through the use of filters. In the case of airborne gravimetry, however, the speed of the aircraft, being considerably greater than the speed of a ship, reduces the gap between the wavelengths of the two signals so that the use of filters is no longer effective. Here differential GPS has helped us to tackle this problem by providing direct measurements of the position of the aircraft as a function of time, from which, by double differentiation, the inertial acceleration can be computed. Recent experiments have demonstrated that airborne gravimetry can be used to measure the gravity field with an accuracy of about 10^{-6} making it possible to also use this technique for applications such as oil prospecting. Table 1 summarizes some application fields together with the related accuracy target.

2. A Historical Walk along the Gravimeters Path

In principle any physical process which involves gravity can be used to measure g. The early theory of gravity based on Aristotle (384–322 BC) principle that the velocity of a free falling body is proportional to its weight was changed by Galileo in 1604 who discovered, by means of experiment with inclined planes, that in the free fall the acceleration is constant and independent from the mass.

The measurement of g, an absolute value of an acceleration, implies the measurement of length and time: period and equivalent length of a pendulum or time required by a free falling object to travel a given distance, hence the need of metrological standard for length and time for high precision absolute measurement of g. For the measurement of difference of gravity, however, it is possible to measure only time, for example, the change of the period of a pendulum of known length or the extension of a mechanical or electromagnetic spring supporting a fixed mass. Table 2, partially compiled after Torge [1], reports the progress made in measuring the gravity acceleration.

2.1. Yesterday. The pendulum method for measuring g, introduced by Huygens in 1656 was first used because of its mechanical simplicity and its nearly amplitude-independent (isochronous) period of oscillation. For an ideal, small amplitude of swing pendulum, the period of oscillation is only a function of the length of the pendulum and g. In an ideal (mathematical) pendulum, its length is simply the distance between the point of suspension and the point at which the entire mass is concentrated. In a real (physical) pendulum, which consists of an extended mass swinging

TABLE 1: Field of application, accuracy target, type of instruments and their availability.

Application	Accuracy target	Instruments/availability
Isostasy	10^{-5}	Relative gravimeters (pendulum)/1817
Prospection for oil and gas	10^{-6}—10^{-7}	Relative gravimeters/1950
Structural geology	10^{-6}	Relative gravimeters/1950
Microgravity prospection	10^{-8}	Relative gravimeters/1980
Metrology (forces)	10^{-6}—10^{-7}	Absolute gravimeters/1980
Metrology (kilogram)	10^{-8}—10^{-9}	Absolute gravimeters/1990
Geodetic metrology	10^{-8}	Absolute gravimeters/1980
Geodesy	10^{-7}—10^{-8}	Rel + Abs gravimeters/1980
Earth tide and earth dynamic	10^{-10}—10^{-11}	Superconducting gravimeters/1970

TABLE 2: Progress in measuring g(partially compiled after Torge 1989, page 14, figure 1.1 [1]

Type of instrument	Time span	Uncertainty range	Rate
Reversible pendulum	1817 to1945	4×10^{-5} to 2×10^{-6}	One order of magnitude/130 years
Spring type gravimeters	1935 to 2010	4×10^{-7} to 3×10^{-9}	Two orders of magnitude/75 years
Superconducting gravimeters	1965 to 2010	5×10^{-10} to 5×10^{-12}	Two orders of magnitude/45 years
Absolute gravimeters	1970 to 2010	8×10^{-7} to 2×10^{-9}	Two orders of magnitude/40 years

about a certain point of suspension, its length depends on the mass distribution of the pendulum relative to the point of suspension. To achieve an accuracy of the order of 10^{-6}, the length of the pendulum must be measured with an accuracy of the order of 10^{-6} and the period of oscillation with an accuracy two times better than this. Because of the isochronous character of the oscillation, the accuracy in the measurement of the period can be increased by measuring the time for a number of oscillations. However, in spite of its conceptual simplicity, the main problem with pendulum measurements remains with the determination of the "length". Nevertheless, the measurement of g begun with, and exclusively depended on, pendulum measurements until the end of nineteenth century. Figure 1 shows an example of a pendulum.

Pendulum measurement played a role in the discussion between Cassini and Newton concerning the flattening of the earth. In 1672 the French astronomer Jean Richer had measured the period of a pendulum at Paris and at Cayenne, which is almost on the equator. The gravitational force was found to be less in Cayenne than in Paris, a discrepancy that G. D. Cassini, the first of a dynasty that was to dominate French observational astronomy throughout the eighteenth century, took to be evidence that there was less mass of the earth under the equator than under more northerly latitudes; that is, that the earth is not quite a sphere but is a prolate ellipsoid. (Rowlinson 2007 [2]).

Until the end of the eighteen century, all the pendulum measurements were absolute, that is, both period of oscillation and length of the pendulum were measured at a given site and g computed from these two quantities. The measurements were rather lengthy and complicated and were made mainly in laboratory conditions. Of fundamental importance for the history of gravimetry remains the absolute measurement of g made by Kuhnen and Furtwangler [3] between 1898 and 1904 in Potsdam which was used as the base for the Potsdam Gravity System introduced in 1908 and extended worldwide. In spite of the duration of the measurements, the Potsdam datum had a significant systematic error which was discovered only several years later. In 1817, Kater invented the reversible pendulum [4]. Consider a pendulum which can be hung from either of two points. If the period of swing is the same about each of the two points of support, then the distance separating the two points of suspension is equal to the length of a simple pendulum having the same period. Thus the metrology is reduced to measuring the period of oscillation and the distance between the two supports. Pendulum measurements are affected by several sources of errors, the main ones being: the vibration of the pendulum support, the change in the pendulum length and the influences of nongravitational forces (friction, magnetic and electrostatics). A comprehensive overview of the development of instruments, measuring techniques and application is given in Nabighian et al. [5].

During the first half of the twentieth century several improvements were made on pendulums: apparatus with two or four pendulums, generally made by quartz to minimize temperature dependent effects, were manufactured to reduce sources of errors. Pendulums both as absolute and relative measurements have been used in various configurations until about 1940, both for laboratory and field observations in land, on ships and underwater. Kater has also the merit to have recognized that for the geophysical needs of gravimetry, detection of anomalies in the gravity field of the Earth, it is sufficient to measure the changes in gravity with the position, hence diminishing constraints in the physical parameters to be measured and increasing the accuracy. For instance, provided that the "length" of the pendulum remains constant during the measurements, the gravity differences between two observations points are related to the difference in the periods of oscillations. Relative gravimetry was born after this achievement.

FIGURE 1: The Mendenhall pendulum.

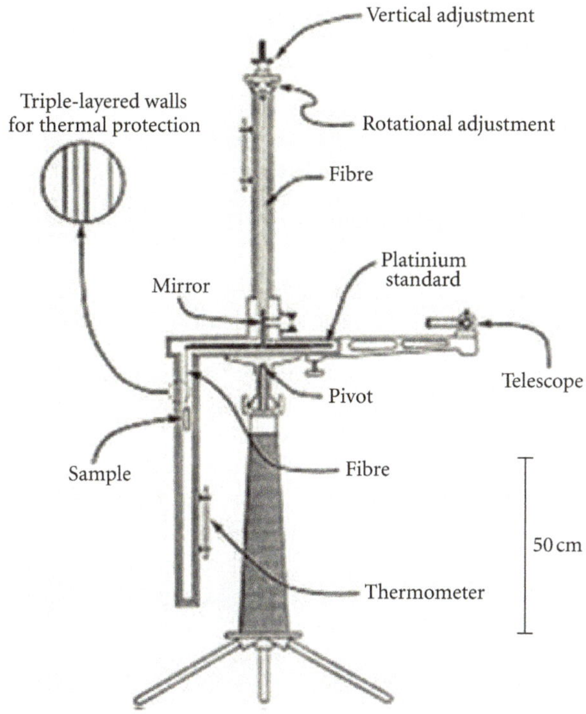

FIGURE 2: Eotvos Torsion balance.

A significant breakthrough in gravity measurements resulted from the introduction of spring-type relative gravimeters. These instruments are basically weighing devices in which the force on a mass due to gravity is balanced by the restoring force of a spring. Historically the spring-type gravimeters can be divided in two types: linear, in which the equilibrium conditions are between two forces (elastic and gravitational) and astatic, in which a condition close to equilibrium is reached by equating the momentum of the gravitational and elastic forces. The great majority of the modern gravimeters belongs to the second category. The advantage of astatic gravimeters over linear ones is that their sensitivity can be greatly increased through the use of astatization where a small change in force results in a large change in position. Gravimeter springs are normally made of metal or fused silica. Metal leaf, coiled, and helical springs have been used in different type of gravimeters. A key problem with metal springs is the strong dependency of their elasticity on temperature. Usually this type of gravimeters requires a well-defined operating temperature. A natural counterpart to metal is fused silica. Springs made of fused silica are rather strong and slightly less sensitive to temperature. On the other hand this type of gravimeters shows a large linear drift due to evaporation of silica in its container under vacuum. Sam Worden developed in in the late 1940's the Worden Gravity Meter based on this principle. Another problem which affects spring-type gravimeters is related to pressure compensation, that is obtained either by a pressure compensation box mounted on the beam opposite to the mass with respect to the rotation axes or by putting the instrument in a sealed container. As a third category of "spring" some gravimeters use gas at a particular temperature and pressure as the spring element. Gas-type gravimeters were employed in the '30ies mainly as ship-borne gravimeters because the high degree of damping which resulted from the use of a gas spring which made this kind of instruments less sensitive to vibrations and hence better suited for measurements on the sea surface.

A peculiar kind of gravity meter, used between the two World Wars, is the torsion balance developed by Baron Roland von Eotvos in 1896. The torsion balance consists of a torsion platinum-iridium fiber suspended vertically with a horizontal aluminum bar suspended from its lower end. The proof mass is positioned on one end of the bar, while on the other end there is an identical proof mass suspended by another fiber several centimeters below the horizontal plane of the bar. If the gravity field of the earth at the measurement point is distorted by an anomalous mass at depth, the balance bar rotates since the horizontal component of gravity at one test mass is different from that at the other test mass. The magnitude of the torque can be determined by measuring the angle through which the balance has been rotated by the torque. Figure 2 shows a schematic of a Torsion balance.

A second break-through in relative gravimetry resulted from the introduction of the zero length spring. For this type of spring, the restoring force is proportional to the entire length of the spring. Its importance results from the fact that when employed in an astatic scheme, with the point of application of the spring and the pivot of the arm holding the test mass located on the same vertical line, the equilibrium condition is independent from the deflection angle of the arm, hence making the sensitivity (in principle) infinite. Since 1934 when LaCoste [6] made their first gravity meter based on the zero length spring (which was actually born as long period seismometer), this type of gravimeter has dominated the scene of relative gravimetry for something like

50 years. Figure 3 shows a schematic diagram of a LaCoste-Romberg gravity meter.

The principal problem with spring-type gravimeters is the elastic variability of the spring. All these kinds of gravimeters suffer because of anelastic drift which depends on time as well as temperature changes and the effects of mechanical shocks and vibrations. The discovery of superconductivity in certain materials at low temperature resulted in a new kind of "spring-type" gravimeters which is, at least in principle, drift free. In the superconducting gravimeter, the test mass, a metallic sphere made from or coated with a layer of superconducting metal, is levitated by the magnetic field generated by current flowing in a superconducting coil. The instrument, a stationary one, is installed at a fixed site and measures gravity variations with time. The measured changes in current necessary to keep the test mass in the reference position are proportional to the variations in the gravity field as results of tides, variations of barometric pressure and local atmospheric, mass movements, or density variations within the earth. Superconducting gravimeters, invented by Prothero and Goodkind [7] in the mid '60ties, provide a very high sensitivity of the order of 10-10 and, owing to the superconducting properties, can be nearly drift free.

Relative gravimeters all require a calibration in order to determine what changes of spring length or of electrical current correspond to a given gravity change. Relative gravimeters have been widely employed to make measurements on land, on the sea surface, at the sea bottom and, recently, from the air. All geophysical problem as well as several geodetic problems can be solved using relative techniques. The basic problem of geodetic metrology (definition of gravity datum and of gravity scale), however, cannot be solved using relative gravimetry. It was just after the second world war, when techniques for high-accuracy measurements of time and space intervals became available, that measurements of g through studying the motion of a free falling object became possible. A pendulum's advantage with respect to studying a free falling object is that, in the case of pendulum, the accuracy of the time measurement can be simply increased by increasing the number of oscillations. This is not possible in the case of a free fall experiment; on the contrary, in this case, it is necessary to measure with high accuracy small time intervals. Fortunately our ability to do this has increased dramatically over the past 3-4 decades. As a result, the second half of the 20th century has seen the development of free fall methods for the absolute determination of g. During this period, a number of absolute determinations used the method of geometrical optics to define the position of the falling object, while during the past 30 years the method of laser interferometry has been used. Using laser interferometry to measure distances has been a key point in the development of today's transportable absolute gravity meters: it permits the measurement of distances of the order of 1 m with the accuracy of 1 nm during the subsecond timescale associated with a single drop. A comprehensive overview of this historical period of absolute gravimeters is given in Marson and Faller 1986 [8].

After the first instruments which used geometrical optics (Martsyniak1956 [9], Preston-Thomas et al. 1960 [10] and Thulin, 1960 [11]), Faller, 1962 had developed the first free fall interferometric instrument (Faller 1965 [12]). Meanwhile Sakuma at Bureau International des Poides et Mesures, Sevres, France [13] was working on a rise and fall interferometric instrument. When lasers with frequency stabilization became available, Hammond and Faller (1967) [14] developed the first transportable absolute gravity meter of the free fall type. A few years later, Hammond (1970) [15] modified and made improvements of this transportable meter. Both, the initial and modified instruments used multiple stations coupled with laser interferometry. Based on the principle of Sakuma's instrument the Italian Metrology Institute "G. Colonnetti" was developed in 1974 [16] in cooperation with BIPM, a portable version of the stationary apparatus (Figure 4). In this instrument, laser interferometry was applied to measure position in space. As in Sakuma's instrument, the moving object followed a rise-and-fall trajectory.

For several years, the advantages and disadvantages of these different approaches, direct free fall and rise and fall, have been debated. The free fall type of ballistic gravimeter is in principle simpler from a mechanical point of view with respect to the rise-and-fall one. This is mainly because the verticality of the trajectory in the second type strongly depends on being able to successfully throw the object upwards starting with a large release velocity, whereas in the free fall the release velocity is essentially zero. The symmetry of the trajectory on the other hand eliminates certain frequency-dependent errors in the fringe signal. Moreover, the symmetry also compensate the air drag effect due to residual pressure, diminishing the level of vacuum required for the experiment. Due to drag g seems larger at the rise and smaller at the fall. A rise and fall experiment requires a particular care in setting and monitoring the verticality of the trajectory, while the free fall one is less sensitive to this. Hence a free fall experiment can be made automatic and controlled by computer in a easier way with respect to a rise and fall one. In a measuring session, free fall instruments can perform thousands of drops while rise and fall ones almost hundred. The first are computer controlled, the second are under the control of human beings with all the plus and minus related to it.

In order to establish the accuracy of absolute gravimeters, and eventually detect sources of systematic errors, eight intercomparison campaigns have been realized at the BIPM laboratories in Sevres France. Seven instruments took part in the first campaign in 1981. All instruments were developed in specialized institutes, mainly metrological institutes or laboratories related to them, and were operated by the scientists and technicians who developed them and hence knew "every single screw". Twenty two instruments participated in the 8th campaign, the last at BIPM. The rapid increase of absolute gravimeters is due to the excellent work of Faller who developed, with his PhD students Niebauer [17] and Rinker [18], 6 absolute gravimeters, known as JILAg, within a scientific framework with six geodetic/geophysics institute in the USA and Europe and made available to a wider body of scientists these kind of instruments [19] (Figure 5).

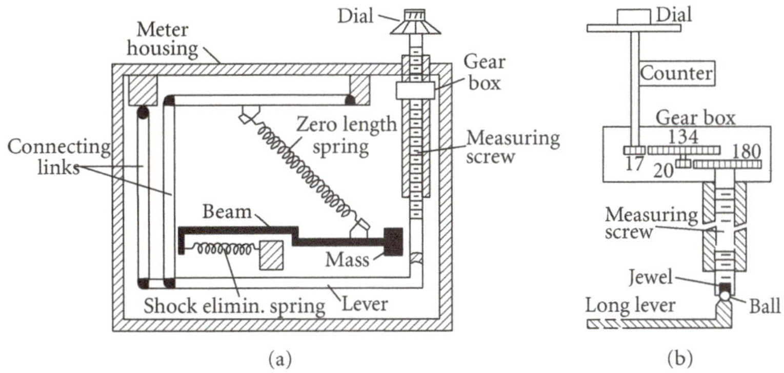

FIGURE 3: Schematic of a La Coste-Romberg gravity meter.

FIGURE 4: The IMGC absolute gravity meter.

FIGURE 5: The JILAg absolute gravity meter.

FIGURE 6: AXIS FG5 absolute gravimeter.

After Faller, T. Niebauer developed, on a commercial base (AXIS, Microg Solutions, MicrogLaCoste), a successful series of instruments, FG5, with a worldwide distribution. All these instruments are based on the free fall principle (Figure 6).

2.2. Today. The terrestrial relative gravity world is dominated for several decades, since 1939, by LaCoste&Romberg gravity meters (Mod D and Mod G) and has witnessed in the last years the growth of Scintrex Autograv (CG3, CG5) (http://www.scintrexltd.com/gravity.html), Microg LaCoste (gPhone), http://www.microglacoste.com and ZLS (Burris), http://www.zlscorp.com. The long lasting superiority of LaCoste instrument was also due to the introduction of a capacitive transducer to monitor electronically the position of the beam in a much more impersonal way than the optical readout. It opened the way to the introduction of "feedback" electronics in order to keep always the beam of instruments at the same position (zero method). This technique was introduced first on the tidal LaCoste (ET meter) used for tidal registrations. The first attempt based on the use of a step motor to drive the micrometer was not very successful. Later an electrostatic restoring force has been used successfully. All instruments have nowadays electronic levels, electronic read-out, and computer-controlled data acquisition. The reading sensitivity is of 1×10^{-9}, the field repeatability around 5×10^{-9}. The development of computer-controlled data acquisition and to certain extent also error compensation has certainly increased the instrumental reliability. As counter fact, the interaction between operator and gravity meter, which was inherent with the original instruments, has been lost.

Since the early days at the beginning of '80ties, absolute gravimetry has been subjected to continuous upgrade and developments, mainly due to the continuously increased R&D at Micro-g Solutions first and, since 2005, at Micro-g LaCoste. The original FG5 instrument, already an excellent instrument, has been followed by the FG5-X with a new dropping chamber and an extended free fall length of 36 cm within 260 ms, by FG-L, a simplified version of FG5 and A10, a smaller version of FG5 intended for field measurements. Declared accuracy of the order of 10^{-9} for FG5 and FG5-X and 10^{-8} for FG-L and A10. Reading the report of the Intercomparison Campaign of Absolute Gravimeters, held in Sevres, France at the BIPM laboratories in 2005, Jiang et al. 2011 [20] we clearly see that the absolute gravity world is dominated by free-fall-type instruments (FG5, A10, JILAg) with just few exceptions. One of these is the new instrument developed by the Italian Institute for Metrological Research-INRIM and indicated as IMGC-02, a memory of the original IMGC instrument. The instrument features a completely redesigned launching chamber, now automatic, a Mach-Zendher interferometer and computer-controlled data acquisition [21, 22] (Figure 7).

The repeatability, of the order of 10^{-9} and the total uncertainty, 8×10^{-9}, are similar to those of the free fall instruments. But what is very important to stress once again is the symmetry of the trajectory and its property of cancelling most, if not all, errors dependent from the frequency of the fringe signal.

Recently, J. E. Faller together with A. L. Vitoushkin 2005 [23], have developed a new absolute gravity meter which uses a cam to create the repeated necessary motions of hold, release, catch, and lift. By using a second cam that drives an auxiliary mass, it is possible to have the center of mass of the free fall system unchanged throughout the measurement cycle avoiding therefore any soil recoil effect. The instruments are very small, weighs less than 45 kg and have an accuracy comparable to that of the FG5.

The intercomparison campaigns are of high importance for the metrology of absolute gravimetry: they represent the best tool to identify and possibly correct for systematic errors and state the effective level of accuracy. Since 1981 the Bureau International des Poides et Mesures, Sevres, France has hosted eight campaigns (1981, 1985, 1989, 1994, 1997, 2001, 2005, 2009) while two campaigns (2003 and 2011) were held in Walferdange, Luxemburg at the European Center for Geodynamics and Seismology (ECGS). The increase in the numbers of instruments which participate in the campaigns is impressive [20] as well as the results of the last intercomparisons which indicate an agreement between the various instruments of about 2×10^{-9} (Figure 8).

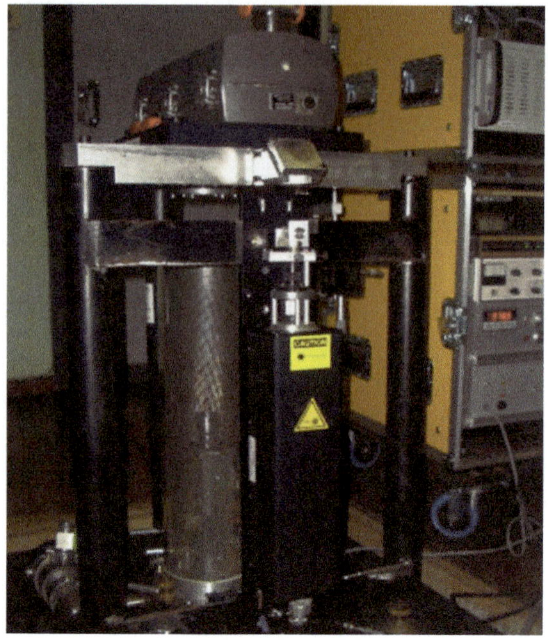

Figure 7: IMGC-02 absolute gravity meter.

Figure 8: Impressive image of absolute gravimeters in "a raw" at Walferdange.

As far as Superconducting gravity meters is concerned, GWR Instruments, Inc. is the exclusive manufacturer of this kind of instruments. The levitation of a spherical test mass in an ultrastable magnetic field generated by persistent currents in two niobium coils that are superconducting below a temperature of 9.3 K replaces the mechanical spring. As a result, the ultrastable magnetic field, weak gradient, and operation at cryogenic temperatures eliminate the sources of noise and drift commonly found in mechanical spring gravity meters so that sensitivity of the instrument reaches the level of 10^{-12}. GWR produces two types of SG: the GWR Observatory

Superconducting Gravimeter and *i*Grav SG Meter-Simplified Superconducting Gravimeter for Portable Operation.

After 30 years of continuous development of gravity meters we now have at our disposal a powerful pool of high precision and high accuracy instruments to study our dynamic Earth. It is almost impossible to overview the high amount of scientific papers which report field of applications and results of gravity experiments. List of relevant papers on the use of superconducting gravimeters can be found at http://www.gwrinstruments.com/published-papers.html and on the use of absolute gravity meters at http://www.microglacoste.com/grav_bib.php.

2.3. The Future of Terrestrial Gravity

2.3.1. New Instruments. Research and development aimed at reducing the size and cost of portable absolute and superconducting gravity meters is certainly the most obvious target. The most challenging task, however, concerns the development of absolute gravimeters based on a new vision of the interferometric principle: atomic interferometry, Peters et al. 1999 [24], Peters et al. 2001 [25], Schmidt et al. 2011 [26]. To measure g a cold cloud of Rb atoms is prepared in an ultra high-vacuum chamber to be used as a test mass. During their free fall, these atoms undergo three stimulated Raman transitions, Weiss et al. [27], Moler et al. [28] that separate, redirect, and recombine the atomic wave function, resulting in an atomic interferometer, respectively. The total phase shift between the two paths of this atomic interferometer depends on g and scales with the square of the time interval between two consecutive Raman pulses. This gravity phase shift is canceled by chirping the frequency difference of the two Raman beams to compensate the time-dependent Doppler shift. The value of g is therefore derived from a frequency chirp. In the paper by Louchet-Chauvet et al. [29], the authors report an intercomparison experiment between three instruments based on three different methods: free fall, symmetrical rise and fall, cold atomic interferometry. Apart from the results of the intercomparison, which showed some discrepancies and indicated that there is still work to be done, the important fact is that it shows that absolute g instruments based on atomic interferometry are in our near future and can achieve accuracies comparable to those of absolute gravimeters based on optical interferometry. The fundamental step forward in this new kind of instruments is that there is no need of dropping or launching mechanisms and there are no solid bodies in free fall. On the other hand, very specialized scientists to set up and perform the experiment are required.

A challenging topic for new instruments regards the development of vertical gradiometers based on the superconducting principle, van Kann et al. [30], ballistic methods, Brown et al. [31], or atom interferometry, McGuirk et al. [32]

2.3.2. Metrology. The kilogram is the last remaining base unit of the International System of Units which is still defined by a material artifact: it is equal to the mass of the international prototype of the kilogram which since 1889 has been kept at the BIPM. The main disadvantage of this definition is that it relies on the long-term stability of the mass of a material artifact. To ensure the long-term stability of the mass unit, it is advisable to link the unit of mass to a fundamental constant. Therefore, the General Conference for Weights and Measures recommended at its 21st meeting in 1999 that "national laboratories continue their efforts to refine experiments that link the unit of mass to fundamental or atomic constants with a view to a future redefinition of the kilogram". The redefinition of the kilogram would also benefit three other base units: the ampere, the mole, and the candela, the definitions of which depend on the kilogram. The watt balance experiment links the Planck constant h to the mass of the international prototype of the Kilogram (IPK), that is, it measures the ratio h/m_{IPK}. The physical principle is balancing an electrical power with a mechanical power, the weight of a reference mass. Hence the need of high-precision measurement of g which can be achieved only by means of the modern absolute gravimeters based on different principles: free fall rise-and-fall and atom interferometry Louchet-Chauvet et al. [29]

2.3.3. Moving Platform. As already noted, the achievements of satellite positioning systems (GPS, GNNS, GALILEO) made it possible to measure the gravity field by means of airborne gravimeters. Baumann et al. have shown [33] the potential applications of ballistic absolute gravimeters also in these kind of experiments. A step forward in this direction could be also the development of absolute vertical gradiometers due to their intrinsic property of reducing, if not eliminating, inertial accelerations.

3. Conclusions

The second half of the last century has witnessed a rapid increase of precision and accuracy of the gravity meters, from 10^{-6} at the beginning of the century to 10^{-9}, both in absolute and relative instruments, and up to 10^{-12} for superconducting gravimeters in specific measurement conditions.

New type of gravimeters has been developed among which the most relevant ones are based on cold-atom interferometry which opens up a new way to perform the free fall experiment. The combination of different instruments allows for the "best as possible" measurement of g when high accuracy is needed. The colocation of superconducting relative gravimeters of high sensitivity with absolute gravity meters has opened up new perspectives in the study of geodynamical processes in the earth (evolution of the lithosphere, isostatic responses to mass changes, density changes due to thermal cooling and heating, lithospheric slab pull).

Disclosure

In the paper the author has mentioned some commercial identities manufacturers of gravity meters. The author declares that he/she, as author of the paper, does not have any direct financial relation with the commercial identity mentioned in the paper that might lead to a conflict of interest.

References

[1] W. Torge, *Gravimetry*, Walter De Gruyter, Berlin, Germany, 1989.

[2] J. S. Rowlinson, "Flattening the earth," *Notes and Records of The Royal Society*, vol. 61, no. 1, pp. 67–68, 2007.

[3] F. Kuhnen and P. Furtwangler, *Bestimmung der absoluter Grösse der Schwerkraft zu Potsdam mit Reversionspendeln*, vol. 27 of *Veröffentlichung des Königl. preussischen geodätischen Institutes*, 1906.

[4] H. Kater, "An account of experiments for determining the length of the pendulum vibrating seconds in the latitude of London," *Philosophical Transactions of the Royal Society of London*, vol. 108, pp. 33–102, 1818.

[5] M. N. Nabighian, M. E. Ander, V. J. S. Grauch et al., "Historical development of the gravity method in exploration," *Geophysics*, vol. 70, no. 6, pp. 63–89, 2005.

[6] L. J. B. LaCoste, "A new type long period vertical seismograph," *Physics*, vol. 5, no. 7, pp. 178–180, 1934.

[7] W. A. Prothero and J. M. Goodkind, "A superconducting gravimeter," *Review of Scientific Instruments*, vol. 39, no. 9, pp. 1257–1262, 1968.

[8] I. Marson and J. E. Faller, "g-the acceleration of gravity: its measurement and its importance," *Journal of Physics E*, vol. 19, no. 1, pp. 22–32, 1986.

[9] A. Y. Martsyniak, "Opredelenje absoliutnoi velichiny uskoreniia sily trax zhesti po paderniiu zhezla v vakuume," *Izmer. Tekh. Poverochn Delo*, vol. 5, p. 11, 1956.

[10] H. Preston-Thomas, L. G. Turnball, E. Green, T. M. Dauphice, and E. H. Kalra, "An absolute measurement of the acceleration due to gravity at ottawa," *Canadian Journal of Physics*, vol. 38, pp. 824–852, 1960.

[11] A. Thulin, "Acceleration due to gravity at the national bureau of standards," *NBS Monograph*, vol. 107, pp. 1–20, 1960.

[12] J. E. Faller, "Results of an absolute determination of the acceleration of gravity," *Journal of Geophysical Research*, vol. 70, pp. 4035–4038, 1965.

[13] A. Sakuma, "État actuel de la nouvelle détermination absolue de la pesanteur au Bureau International des Poids et Mesures," *Bulletin Géodésique*, vol. 69, no. 1, pp. 249–260, 1963.

[14] J. A. Hammond and J. E. Faller, "12.5-Laser-interferometer system for the determination of the acceleration of gravity," *IEEE Journal of Quantum Electronics*, vol. 3, no. 11, pp. 597–602, 1967.

[15] J. A. Hammond, "A laser-interferometer system for the absolute determination of the acceleration of gravity," *JILA Report*, no. 103, 1970.

[16] G. Cerutti, L. Cannizzo, A. Sakuma, and J. Hostache, "Transportable apparatus for absolute gravity measurement," *VDI Berichte*, no. 21, pp. 49–51, 1974.

[17] T. M. Niebauer, *New absolute gravity instruments for physics and geophysics [Ph.D. thesis]*, University of Colorado, Boulder, Colo, USA, 1987.

[18] R. L. Rinker, *Super spring—a new type of low-frequency vibration isolator [Ph.D. thesis]*, University of Colorado, Boulder, Colo, USA, 1983.

[19] W. Torge, R. H. Röder, M. Schnüll, H. G. Wenzel, and J. E. Faller, "First results with the transportable absolute gravity meter JILAG-3," *Journal of Geodesy*, vol. 61, no. 2, pp. 161–176, 1987.

[20] Z. Jiang, O. Francis, L. Vitushkin et al., "Final report on the Seventh International Comparison of Absolute Gravimeters (ICAG 2005)," *Metrologia*, vol. 48, no. 5, pp. 246–260, 2011.

[21] G. D'Agostino, *Development and metrological characterization of a new transportable absolute gravimeter [Ph.D. thesis]*, 2005.

[22] G. D'Agostino, S. Desogus, A. Germak et al., "The new IMGC-02 transportable absolute gravimeter: measurement apparatus and applications in geophysics and volcanology," *Annals of Geophysics*, vol. 51, no. 1, pp. 39–49, 2008.

[23] J. E. Faller, "The measurement of little g: a fertile ground for precision measurement science," *Journal of Research of the National Institute of Standards and Technology*, vol. 110, no. 6, pp. 559–581, 2005.

[24] A. Peters, K. Y. Chung, and S. Chu, "Measurement of gravitational acceleration by dropping atoms," *Nature*, vol. 400, no. 6747, pp. 849–852, 1999.

[25] A. Peters, K. Y. Chung, and S. Chu, "High-precision gravity measurements using atom interferometry," *Metrologia*, vol. 38, no. 1, pp. 25–61, 2001.

[26] M. Schmidt, A. Senger, M. Hauth, C. Freier, V. Schkolnik, and A. Peters, "A mobile high-precision absolute gravimeter based on atom interferometry," *Gyroscopy and Navigation*, vol. 2, no. 3, pp. 170–177, 2011.

[27] D. S. Weiss, B. C. Young, and S. Chu, "A precision measurement of h/M_{Cs} based of photon recoil using laser cooled atoms and atom interferometry," *Applied Physics B Lasers and Optics*, vol. 59, no. 3, pp. 217–256, 1994.

[28] K. Moler, D. S. Weiss, M. Kasevich, and S. Chu, "Theoretical analysis of velocity-selective Raman transitions," *Physical Review A*, vol. 45, no. 1, pp. 342–348, 1992.

[29] A. Louchet-Chauvet, S. Merlet, Q. Bodart et al., "Comparison of 3 absolute gravimeters based on different methods for the e-MASS project," *IEEE Transactions on Instrumentation and Measurement*, vol. 60, no. 7, pp. 2527–2532, 2011.

[30] F. J. van Kann, M. J. Buckingham, C. Edwards, and R. Matthews, "Performance of a superconducting gravity gradiometer," *Physica B*, vol. 194, no. 1, pp. 61–62, 1994.

[31] J. M. Brown, T. M. Niebauer, F. J. Klopping, and A. T. Herring, "A new fiber optic gradiometer for 4-D absolute differential gravity," *Geophysical Research Letters*, vol. 27, no. 1, pp. 33–36, 2000.

[32] J. M. McGuirk, G. T. Foster, J. B. Fixler, M. J. Snadden, and M. A. Kasevich, "Sensitive absolute-gravity gradiometry using atom interferometry," *Physical Review A*, vol. 65, no. 3, article 033608, 14 pages, 2002.

[33] H. Baumann, E. E. Klingelé, and I. Marson, "Absolute airborne gravimetry: a feasibility study," *Geophysical Prospecting*, vol. 60, no. 2, pp. 361–372, 2012.

18

Rayleigh Waves in a Rotating Orthotropic Micropolar Elastic Solid Half-Space

Baljeet Singh,[1] Ritu Sindhu,[2] and Jagdish Singh[2]

[1] *Department of Mathematics, Post Graduate Government College, Sector 11, Chandigarh 160 011, India*
[2] *Department of Mathematics, Maharshi Dayanand University, Rohtak, 124 001 Haryana, India*

Correspondence should be addressed to Baljeet Singh; bsinghgc11@gmail.com

Academic Editor: Rudolf A. Treumann

A problem on Rayleigh wave in a rotating half-space of an orthotropic micropolar material is considered. The governing equations are solved for surface wave solutions in the half space of the material. These solutions satisfy the boundary conditions at free surface of the half-space to obtain the frequency equation of the Rayleigh wave. For numerical purpose, the frequency equation is approximated. The nondimensional speed of Rayleigh wave is computed and shown graphically versus nondimensional frequency and rotation-frequency ratio for both orthotropic micropolar elastic and isotropic micropolar elastic cases. The numerical results show the effects of rotation, orthotropy, and nondimensional frequency on the nondimensional speed of the Rayleigh wave.

1. Introduction

Material response to external stimuli depends heavily on the motions of its inner structures. Classical elasticity does not include this effect, where only translation degrees of freedom of material point of body are considered. Eringen [1] developed the linear micropolar theory of elasticity, which included the intrinsic rotations of the microstructure. It provides a model which can support body and surface couples and display high frequency optical branch of the wave spectrum. For engineering applications, it can model composites with rigid chopped fibres, elastic solid with rigid granular inclusions, and other industrial materials such as liquid crystals.

The assumptions of isotropy in a solid medium may not capture some of significant features of the continuum responses of soils, geological materials, and composites. Iesan [2–4] studied some static problems in orthotropic micropolar elasticity. Kumar and Choudhary [5, 6] studied the mechanical sources and dynamic behaviour of orthotropic micropolar elastic medium. Kumar and Chaudhary [7] studied the plane strain problem in a homogeneous orthotropic micropolar elastic solid. Kumar and Ailawalia [8] studied the response of a micropolar cubic crystal due to various sources. Kumar and Gupta [9] studied the propagation of waves in transversely isotropic micropolar generalized thermoelastic half-space. Singh [10] investigated the two-dimensional plane wave propagation in an orthotropic micropolar elastic solid.

Surface waves in elastic solids were first studied by Rayleigh [11] for an isotropic elastic solid. The extension of surface wave analysis and other wave propagation problems to anisotropic elastic materials has been the subject of many studies; see, for example, [12–21]. The aim of the present paper is to study the propagation of Rayleigh wave in a rotating orthotropic micropolar elastic solid half space. The frequency equation of the Rayleigh wave is obtained. The speed of Rayleigh wave is computed with the help of approximated frequency equation. The effects of orthotropy, non-dimensional frequency, and rotation are shown graphically on the non-dimensional speed of the Rayleigh wave.

2. Formulation of the Problem and Solution

We consider a homogeneous and orthotropic medium of an infinite extent with Cartesian coordinate system (x, y, z). We restrict our study to the plane strain parallel to xy-plane, with the displacement vector $\mathbf{u} = (u_1, u_2, 0)$ and microrotation vector $\phi = (0, 0, \phi_3)$. Following Eringen [22] and Schoenberg and Censor [23], the field equations in xy-plane for homogeneous and rotating orthotropic micropolar solid in absence of body forces and couples are written as

$$A_{11}u_{1,11} + (A_{12} + A_{78})u_{2,12} + A_{88}u_{1,22} - K_1\phi_{3,2} = \rho\left[\frac{\partial^2 u_1}{\partial t^2} - \Omega^2 u_1 - 2\Omega\frac{\partial u_2}{\partial t}\right], \quad (1)$$

$$(A_{12} + A_{78})u_{1,12} + A_{77}u_{2,11} + A_{22}u_{2,22} - K_2\phi_{3,1} = \rho\left[\frac{\partial^2 u_2}{\partial t^2} - \Omega^2 u_2 + 2\Omega\frac{\partial u_1}{\partial t}\right], \quad (2)$$

$$B_{66}\phi_{3,11} + B_{44}\phi_{3,22} - \chi\phi_3 + K_1u_{1,2} + K_2u_{2,1} = \rho j\ddot{\phi}_3, \quad (3)$$

where

$$K_1 = A_{78} - A_{88}, \qquad K_2 = A_{77} - A_{78}, \qquad \chi = K_2 - K_1. \quad (4)$$

We consider the following surface wave solutions

$$\{u_1, u_2, \phi_3\} = \{\overline{u}_1(y), \overline{u}_2(y), \overline{\phi}_3(y)\}e^{ik(x-ct)}, \quad (5)$$

where k is the wave number, c is phase velocity of the wave, and $\omega = kc$ is the angular frequency. Making use of (5) in (1) to (3), we obtain three homogeneous equations in $\overline{u}_1(y)$, $\overline{u}_2(y)$, and $\overline{\phi}_3(y)$, which have nontrivial solutions if

$$\alpha D^6 - \beta D^4 + \gamma D^2 - \delta = 0, \quad (6)$$

where $D = d/dy$ and α, β, γ, and δ are given in Appendix.

Let m_1, m_2, and m_3 be the roots of auxiliary equation (6). Then, the general solutions of (6) are written as

$$u_1 = (A_1e^{-m_1y} + A_2e^{-m_2y} + A_3e^{-m_3y} + A_4e^{m_1y} + A_5e^{m_2y} + A_6e^{m_3y})e^{ik(x-ct)},$$

$$u_2 = (\zeta_1A_1e^{-m_1y} + \zeta_2A_2e^{-m_2y} + \zeta_3A_3e^{-m_3y} + \zeta_1A_4e^{m_1y} + \zeta_2A_5e^{m_2y} + \zeta_3A_6e^{m_3y})e^{ik(x-ct)},$$

$$\phi_3 = (\eta_1A_1e^{-m_1y} + \eta_2A_2e^{-m_2y} + \eta_3A_3e^{-m_3y} + \eta_1A_4e^{m_1y} + \eta_2A_5e^{m_2y} + \eta_3A_6e^{m_3y})e^{ik(x-ct)}, \quad (7)$$

where

$$m_1^2 + m_2^2 + m_3^2 = \frac{\beta}{\alpha}, \quad (8)$$

$$m_1^2m_2^2 + m_2^2m_3^2 + m_3^2m_1^2 = \frac{\gamma}{\alpha}, \quad (9)$$

$$m_1^2m_2^2m_3^2 = \frac{\delta}{\alpha}, \quad (10)$$

and the expressions for ζ_1, ζ_2, and ζ_3 and η_1, η_2, and η_3 are given in the Appendix.

With the use of the radiation conditions $u_1 \to 0$, $u_2 \to 0$, $\varphi_3 \to 0$ as $y \to \infty$, we obtain the particular solutions for medium $(y > 0)$ as

$$u_1 = (A_1e^{-m_1y} + A_2e^{-m_2y} + A_3e^{-m_3y})e^{ik(x-ct)}, \quad (11)$$

$$u_2 = (\zeta_1A_1e^{-m_1y} + \zeta_2A_2e^{-m_2y} + \zeta_3A_3e^{-m_3y})e^{ik(x-ct)}, \quad (12)$$

$$\phi_3 = (\eta_1A_1e^{-m_1y} + \eta_2A_2e^{-m_2y} + \eta_3A_3e^{-m_3y})e^{ik(x-ct)}. \quad (13)$$

3. Boundary Conditions

The mechanical boundary conditions at $y = 0$ are the vanishing of normal force stress tangential force stress; and tangential couple stress that is,

$$t_{22} = 0, \qquad t_{21} = 0, \qquad m_{23} = 0, \quad (14)$$

where

$$t_{22} = A_{12}\frac{\partial u_1}{\partial x} + A_{22}\frac{\partial u_2}{\partial y},$$

$$t_{21} = A_{78}\frac{\partial u_2}{\partial x} + A_{88}\frac{\partial u_1}{\partial y} + (A_{88} - A_{78})\phi_3, \quad (15)$$

$$m_{23} = B_{44}\frac{\partial \phi_3}{\partial y}.$$

The solutions given by (11) to (13) satisfy the boundary conditions (14) at $y = 0$, and we obtain the following frequency equation:

$$\frac{A_{12}}{A_{22}}\frac{A_{78}}{A_{88}}\sum\frac{m_1}{k}\frac{\eta_1}{k}(\zeta_2 - \zeta_3) + i\frac{A_{12}}{A_{22}}\sum\frac{m_1}{k}\frac{m_2}{k}\left(\frac{\eta_1}{k} - \frac{\eta_2}{k}\right) + i\frac{A_{12}}{A_{22}}\frac{K_1}{A_{88}}\sum\frac{m_1}{k}\frac{\eta_1}{k}\left(\frac{\eta_2}{k} - \frac{\eta_3}{k}\right) + i\frac{A_{78}}{A_{88}}\sum\frac{m_1}{k}\frac{m_2}{k}\zeta_3\left(\frac{\eta_2}{k}\zeta_1 - \frac{\eta_1}{k}\zeta_2\right) + \frac{K_1}{A_{88}}\sum\frac{m_1}{k}\frac{m_2}{k}\frac{\eta_3}{k}\left(\frac{\eta_1}{k}\zeta_2 - \frac{\eta_2}{k}\zeta_1\right) + \sum\frac{m_1}{k}\frac{m_2}{k}\frac{m_3}{k}\left(\frac{\eta_1}{k}\zeta_2 - \frac{\eta_2}{k}\zeta_1\right) = 0. \quad (16)$$

4. Particular Case

The frequency equation (16) reduces to the frequency equation for an isotropic rotating micropolar elastic case, if we take

$$A_{11} = A_{22} = \lambda + 2\mu + \kappa,$$

$$A_{77} = A_{88} = \mu + \kappa, \qquad A_{12} = \lambda, \qquad A_{78} = \mu, \quad (17)$$

$$B_{44} = B_{66} = \gamma, \qquad -K_1 = K_2 = \frac{\chi}{2} = \kappa.$$

5. Numerical Results and Discussion

From relations (8) to (10), we obtain the following approximated roots:

$$\frac{m_1^2}{k^2} \cong \frac{\left(A_{11} - \rho c^2 \Omega^*\right)}{A_{88}}, \quad (18)$$

$$\frac{m_2^2}{k^2} \cong \frac{\left(A_{77} - \rho c^2 \Omega^*\right)}{A_{22}}, \quad (19)$$

$$\frac{m_3^2}{k^2} \cong \frac{\left(B_{66} + \rho j c^2 + \left(\chi / k^2\right)\right)}{B_{44}}. \quad (20)$$

With the help of (18) to (20), the frequency equation (16) reduces to the approximated frequency equation for an orthotropic rotating micropolar elastic case. The approximated frequency equation is used to compute the non-dimensional speed of the Rayleigh wave in orthotropic micropolar solid half-space for the following arbitrary physical constants:

$$\begin{aligned} A_{11} &= 11.65 \times 10^{10}\,\mathrm{Nm}^{-2}, \\ A_{22} &= 11.71 \times 10^{10}\,\mathrm{Nm}^{-2}, \\ A_{12} &= 7.69 \times 10^{10}\,\mathrm{Nm}^{-2}, \\ A_{77} &= 1.99 \times 10^{10}\,\mathrm{Nm}^{-2}, \\ A_{78} &= 1.98 \times 10^{10}\,\mathrm{Nm}^{-2}, \\ A_{88} &= 2.01 \times 10^{10}\,\mathrm{Nm}^{-2}, \end{aligned} \quad (21)$$

$$B_{44} = 0.036 \times 10^{10}\,\mathrm{N}, \qquad B_{66} = 0.037 \times 10^{10}\,\mathrm{N},$$

$$\rho = 2.19 \times 10^3\,\mathrm{Kg\,m}^{-3}, \qquad j = 0.000196\,\mathrm{m}^2.$$

The non-dimensional speed of the Rayleigh wave is also computed for isotropic micropolar elastic case with following relevant parameters [24]:

$$\begin{aligned} \lambda &= 7.59 \times 10^{10}\,\mathrm{Nm}^{-2}, & \mu &= 1.89 \times 10^{10}\,\mathrm{Nm}^{-2}, \\ \kappa &= 0.0149 \times 10^{10}\,\mathrm{Nm}^{-2}, & \gamma &= 0.0268 \times 10^{9}\,\mathrm{N}, \\ \rho &= 2.19 \times 10^3\,\mathrm{Kg\,m}^{-3}, & j &= 0.000196\,\mathrm{m}^2. \end{aligned} \quad (22)$$

The non-dimensional speed c^* $(= \rho c^2/A_{22})$ of Rayleigh wave is computed for orthotropic micropolar elastic case and isotropic micropolar elastic case for different values of non-dimensional frequency ω^* $(= \omega^2/[\chi/\rho j])$ and rotation-frequency ratio Ω/ω.

The non-dimensional speed c^* is plotted against the rotation-frequency ratio Ω/ω, when non-dimensional frequency $\omega^* = 5, 10$, and 20. The speed c^* decreases with the increase in value of rotation-frequency ratio Ω/ω. For each value of rotation-frequency ratio Ω/ω, the speed c^* increases with the increase in value of non-dimensional frequency ω^*. The effect of non-dimensional frequency ω^* on non-dimensional speed c^* decreases with the increase in values of Ω/ω. The variations showing the effect of orthotropy on non-dimensional speed of Rayleigh wave are shown in Figure 1, where solid lines and dotted lines correspond to the orthotropic micropolar elastic case and isotropic micropolar elastic case, respectively.

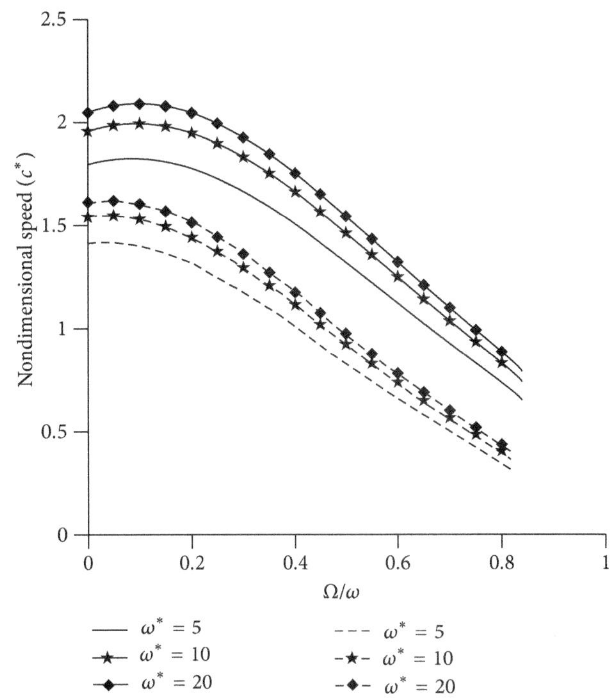

Figure 1: Variations of the nondimensional speed of Rayleigh wave against rotation-frequency ratio (solid lines:orthotropic micropolar elastic case, dotted lines:isotropic micropolar elastic case).

The speed c^* is also plotted against the non-dimensional frequency ω^*, when rotation-frequency ratio $\Omega/\omega = 0.2, 0.4$, and 0.6. The speed c^* of Rayleigh wave increases sharply with the increase in value of non-dimensional frequency ω^* for both orthotropic micropolar elastic case and isotropic micropolar elastic case. Beyond $\omega^* = 4$, it increases slowly in the both cases. Here, the speed c^* decreases with the increase in values of Ω/ω in each case. The effect of rotation on non-dimensional speed c^* increases with the increase in value of non-dimensional frequency ω^*. The variations showing the effect of orthotropy on non-dimensional speed of Rayleigh wave are shown in Figure 2, where solid lines and dotted lines correspond to the orthotropic micropolar elastic case and isotropic micropolar elastic case, respectively.

6. Conclusion

The propagation of Rayleigh wave is studied in an orthotropic micropolar elastic solid half-space, where we obtained the required approximated frequency equation of Rayleigh wave. The non-dimensional speed c^* $(= \rho c^2/A_{22})$ is computed for certain ranges of non-dimensional frequency ω^* and rotation-frequency ratio Ω/ω. The comparison of solid and dotted line curves in the figures reveals the effect of orthotropy,

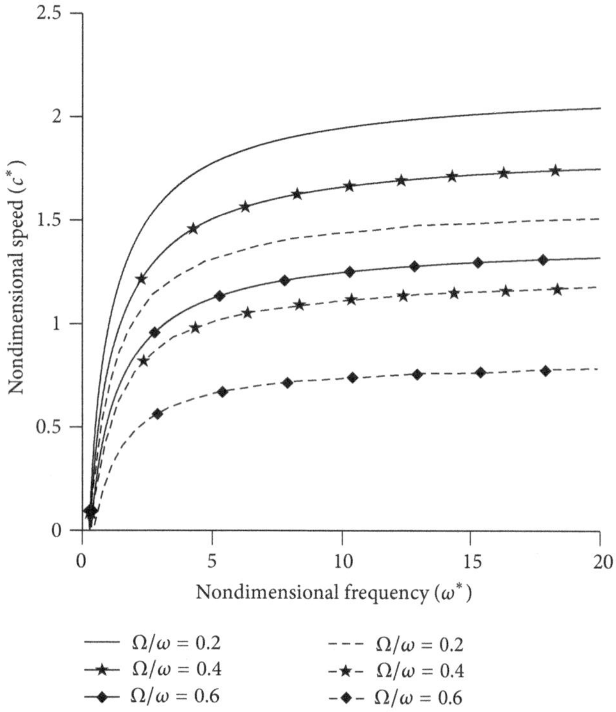

FIGURE 2: Variations of the non-dimensional speed of Rayleigh wave against non-dimensional frequency ω^* (solid lines:orthotropic micropolar elastic case, dotted lines:isotropic micropolar elastic case).

rotation, and non-dimensional frequency ω^* on the non-dimensional speed c^* of the Rayleigh wave in an orthotropic micropolar elastic solid half-space.

Appendix

The values of α, β, γ, and δ are given as

$$\begin{aligned}
\alpha &= A_{22}A_{88}B_{44}, \\
\beta &= k^2\left(A_{22}A_{88}L_4 + A_{88}B_{44}L_3 + A_{22}B_{44}L_1 - B_{44}L_2^2\right) - A_{22}K_1^2, \\
\gamma &= k^4\left(- L_2^2L_4 + L_3L_4A_{88} + L_1L_4A_{22} + L_1L_3B_{44} - 4\left(\rho c^2\right)^2\left(\frac{\Omega}{\omega}\right)^2 B_{44}\right) + k^2\left(-K_2^2A_{88} + 2K_1K_2L_2 - K_1^2L_3\right), \\
\delta &= k^6\left[L_1L_3L_4 - 4\left(\rho c^2\right)^2\left(\frac{\Omega}{\omega}\right)^2 L_4\right] - k^4L_1K_2^2, \\
L_1 &= A_{11} - \rho c^2\Omega^*, \qquad L_2 = A_{12} + A_{78}, \\
L_3 &= A_{77} - \rho c^2\Omega^*, \\
L_4 &= B_{66} + \rho jc^2 + \frac{\chi}{k^2}, \qquad \Omega^* = 1 + \frac{\Omega^2}{\omega^2}.
\end{aligned} \tag{A.1}$$

The values of ζ_1, ζ_2, and ζ_3 and η_1, η_2, and η_3 are obtained as

$$\begin{aligned}
\zeta_i = i\Bigg[&\left(\frac{m_i}{k}\left(\frac{m_i}{k}\frac{A_{12}+A_{78}}{A_{88}} - \frac{2\rho c^2}{A_{88}}\frac{\Omega}{\omega}\right) - \frac{K_2}{K_1}\left(\frac{m_i^2}{k^2} - \frac{A_{11}-\rho c^2\Omega^*}{A_{88}}\right)\right) \\
&\times\left(\frac{K_2}{K_1}\left(\frac{m_i}{k}\frac{A_{12}+A_{78}}{A_{88}} + \frac{2\rho c^2}{A_{88}}\frac{\Omega}{\omega}\right) + \frac{m_i}{k}\frac{A_{22}}{A_{88}}\left(\frac{m_i^2}{k^2} - \frac{A_{77}-\rho c^2\Omega^*}{A_{22}}\right)\right)^{-1}\Bigg], \\
\frac{\eta_i}{k} = &\left(\left[\left(\frac{m_i^2}{k^2}\frac{\left(A_{12}+A_{78}\right)^2}{A_{22}A_{88}} - 4\frac{\left(\rho c^2\right)^2}{A_{22}A_{88}}\left(\frac{\Omega}{\omega}\right)^2\right) - \left(\frac{m_i^2}{k^2} - \frac{A_{11}-\rho c^2\Omega^*}{A_{88}}\right)\left(\frac{m_i^2}{k^2} - \frac{A_{77}-\rho c^2\Omega^*}{A_{22}}\right)\right]\right) \\
&\times\left(\left[\frac{K_2}{A_{88}}\left(\frac{m_i}{k}\frac{A_{12}+A_{78}}{A_{22}} + 2\frac{\rho c^2}{A_{22}}\frac{\Omega}{\omega}\right) + \frac{m_i}{k}\frac{K_1}{A_{88}}\left(\frac{m_i^2}{k^2} - \frac{A_{77}-\rho c^2\Omega^*}{A_{22}}\right)\right]\right)^{-1}.
\end{aligned} \tag{A.2}$$

References

[1] A. C. Eringen, "Linear theory of micropolar elasticity," *Journal of Mathematics and Mechanics*, vol. 15, pp. 909–924, 1966.

[2] D. Iesan, "The plane micropolar strain of orthotropic elastic solids," *Archives of Mechanics*, vol. 25, no. 3, pp. 547–561, 1973.

[3] D. Iesan, "Torsion of anisotropic elastic cylinders," *Zeitschrift für Angewandte Mathematik und Mechanik*, vol. 54, no. 12, pp. 773–779, 1974.

[4] D. Iesan, "Bending of orthotropic micropolar elastic beams by terminal couples," *Analele Ştiinţifice ale Universităţii Al. I. Cuza din Iaşi*, vol. 25, pp. 411–418, 1974.

[5] R. Kumar and S. Choudhary, "Mechanical sources in orthotropic micropolar continua," *Proceedings of the Indian Academy of Sciences*, vol. 111, no. 2, pp. 133–141, 2002.

[6] R. Kumar and S. Choudhary, "Dynamical behaviour of orthotropic micropolar elastic medium," *Journal of Vibration and Control*, vol. 8, no. 8, pp. 1053–1069, 2002.

[7] R. Kumar and S. Choudhary, "Response of orthotropic micropolar elastic medium due to time harmonic source," *Sadhana*, vol. 29, pp. 112–116, 2004.

[8] R. Kumar and P. Ailawalia, "Deformation in micropolar cubic crystal due to various sources," *International Journal of Solids and Structures*, vol. 42, no. 23, pp. 5931–5944, 2005.

[9] R. Kumar and R. R. Gupta, "Propagation of waves in transversely isotropic micropolar generalized thermoelastic half

space," *International Communications in Heat and Mass Transfer*, vol. 37, no. 10, pp. 1452–1458, 2010.

[10] B. Singh, "Wave propagation in an orthotropic micropolar elastic solid," *International Journal of Solids and Structures*, vol. 44, no. 11-12, pp. 3638–3645, 2007.

[11] L. Rayleigh, "On waves propagated along the plane surface of an elastic solid," *Proceedings of the London Mathematical Society*, vol. 17, pp. 4–11, 1885.

[12] D. L. Anderson, "Elastic wave propagation in layered anisotropic media," *Journal of Geophysical Research*, vol. 66, pp. 2953–2963, 1961.

[13] D. M. Barnett and J. Lothe, "Free surface (Rayleigh) waves in anisotropic elastic half-spaces: the surface impedance method," *Proceedings of the Royal Society A*, vol. 402, pp. 135–152, 1985.

[14] P. Chadwick and G. D. Smith, "Foundations of the theory of surface waves in anisotropic elastic materials," *Advances in Applied Mechanics*, vol. 17, pp. 303–376, 1977.

[15] M. Destrade, "The explicit secular equation for surface acoustic waves in monoclinic elastic crystals," *Journal of the Acoustical Society of America*, vol. 109, no. 4, pp. 1398–1402, 2001.

[16] M. A. Dowaikh and R. W. Ogden, "On surface waves and deformations in a compressible elastic half-space," *Stability & Applied Analysis of Continuous Media*, vol. 1, pp. 27–45, 1991.

[17] M. J. P. Musgrave, "The propagation of elastic waves in crystals and other anisotropic media," *Reports on Progress in Physics*, vol. 22, pp. 74–96, 1959.

[18] R. W. Ogden and P. C. Vinh, "On Rayleigh waves in incompressible orthotropic elastic solids," *Journal of the Acoustical Society of America*, vol. 115, no. 2, pp. 530–533, 2004.

[19] D. Royer and E. Dieulesaint, "Rayleigh wave velocity and displacement in orthorhombic, tetragonal, hexagonal and cubic crystals," *Journal of the Acoustical Society of America*, vol. 76, no. 5, pp. 1438–1444, 1984.

[20] T. C. T. Ting, "An explicit secular equation for surface waves in an elastic material of general anisotropy," *Quarterly Journal of Mechanics and Applied Mathematics*, vol. 55, pp. 297–311, 2002.

[21] T. C. T. Ting, "Explicit secular equations for surface waves in monoclinic materials with the symmetry plane at $x_1 = 0$, $x_2 = 0$ or $x_3 = 0$," *Proceedings of the Royal Society A*, vol. 458, no. 2021, pp. 1017–1031, 2002.

[22] A. C. Eringen, "Theory of micropolar elasticity," in *Fracture*, vol. 2, pp. 621–729, Academic Press, New York, NY, USA, 1968.

[23] M. Schoenberg and D. Censor, "Elastic waves in rotating media," *Quarterly of Applied Mathematics*, vol. 31, no. 1, pp. 115–125, 1973.

[24] R. D. Gauthier, "Experimental investigation on micropolar media," in *Mechanics of Micropolar Media*, O. Brulin and R. K. T. Hsieh, Eds., World Scientific, Singapore, 1982.

3D Imaging of Dead Sea Area Using Weighted Multipath Summation: A Case Study

Shemer Keydar,[1] Benjamin Medvedev,[1,2] Abdallah Al-Zoubi,[3] Michael Ezersky,[1] and Emad Akkawi[3]

[1] *Project Department, The Geophysical Institute of Israel, Lod 71100, Israel*
[2] *Schlumberger, Via dell'Unione Europea, 4 San Donato Milanese, Milan, Italy*
[3] *Engineering Faculty, Al-Balqa' Applied University, Salt 19117, Jordan*

Correspondence should be addressed to Shemer Keydar; shemer@gii.co.il

Academic Editor: Umberta Tinivella

The formation of sinkholes along the Dead Sea is caused by the rapid decline of the Dead Sea level, as a possible result of human extensive activity. According to one of the geological models, the sinkholes in several sites are clustered along a narrow coastal strip developing along lineaments representing faults in NNW direction. In order to understand the relationship between a developing sinkhole and its tectonic environment, a high-resolution (HR) three-dimensional (3D) seismic reflection survey was carried out at the western shoreline of the Dead Sea. A recently developed 3D imaging approach was applied to this 3D dataset. Imaging of subsurface is performed by a spatial summation of seismic waves along time surfaces using recently proposed multipath summation with proper weights. The multipath summation is performed by stacking the target waves along all possible time surfaces having a common apex at the given point. This approach does not require any explicit information on parameters since the involved multipath summation is performed for all possible parameters values within a wide specified range. The results from processed 3D time volume show subhorizontal coherent reflectors at approximate depth of 50–80 m which incline on closer location to the exposed sinkhole and suggest a possible linkage between revealed fault and the sinkholes.

1. Introduction

During the last thirty years, hundreds of sinkholes have appeared along the Dead Sea (DS) shoreline in both Israel and Jordan (Figure 1) [1–3]. The process began in the southern part of the DS coast and slowly spread northward along the western coast. The eastern coast which is usually steeper has been less affected at the flat-lying region close to the Lisan Peninsula. The sinkholes have already caused considerable damage to infrastructure, and there is obvious potential for further collapses beneath main highways and other infrastructure.

In order to understand the relationship between a developing sinkhole and its tectonic environment, a numerous number of high resolution seismic reflection surveys were carried out using common midpoint (CMP) technique [3, 5, 6]. The high-resolution seismic reflection method is intended for the study faults and conduits in the shallow subsurface (down to 100–200 m deep), which could bring the fresh water to the salt layer presumably from below. The conventional processing technique, the so-called common midpoint (CMP) method, essentially consists of a "stack" (summation) of properly corrected traces thus increasing the signal-to-noise ratio. However, in the shallow subsurface, the conventional CMP method causes a loss of information because of the problem of "stretching" of data caused by NMO time correction formula. That is why a new 2D and 3D free stretch imaging approach had been applied in the framework of the MERC project M27-050 in order to examine tectonic hypothesis connecting the sinkholes with tectonic faults [7–9]. This approach involves zero-offset common shot point (CSP) stacking and diffraction imaging method [9]. On the basis of those methods in combination with a new weighted multipath summation technique [10, 11], a package

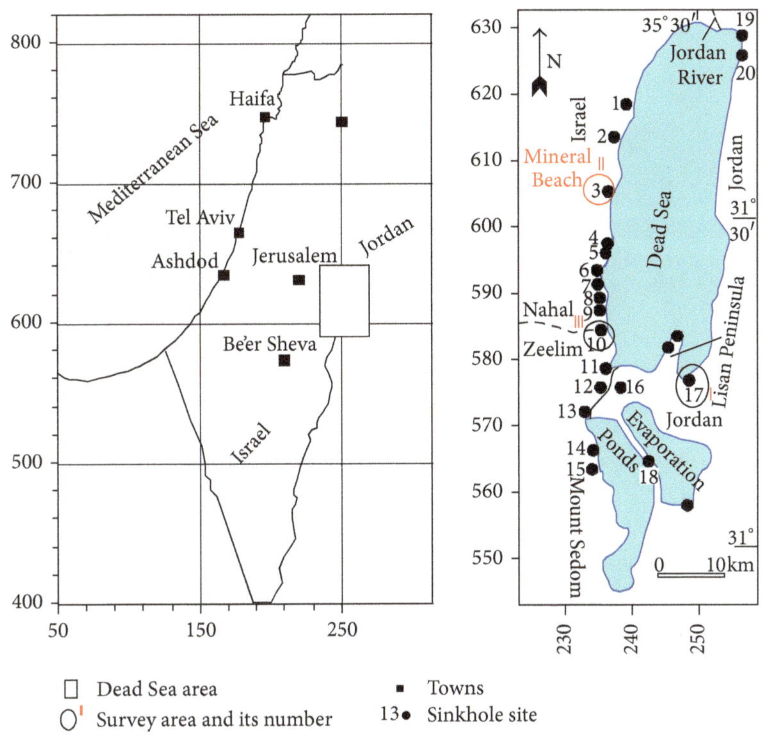

FIGURE 1: Sinkhole sites along the Dead Sea shore: (1) Palms, (2) Samar Spring, (3) Mineral Beach, (4) Ein Gedi and Nahal Arugot, (5) Yesha, (6) Zeruya, (7) Nahal Hever northern, (8) Nahal Hever southern, (9) Asa'el, (10) Nahal Zeelim, (11) Mezada, (12) Rahaf, (13) Mor, (14) Ein Boqeq, (15) Newe Zohar, (16) Lisan Peninsula, (17) Ghor Al-Haditha, (18) Dam-2. I–III: sites under investigation. (coordinates are in km, new Israel Mercator grid).

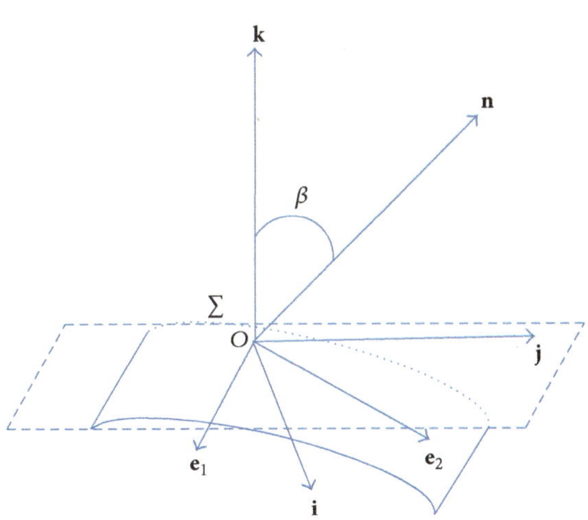

FIGURE 2: Scheme of wavefront arriving at a Common Shot Point.

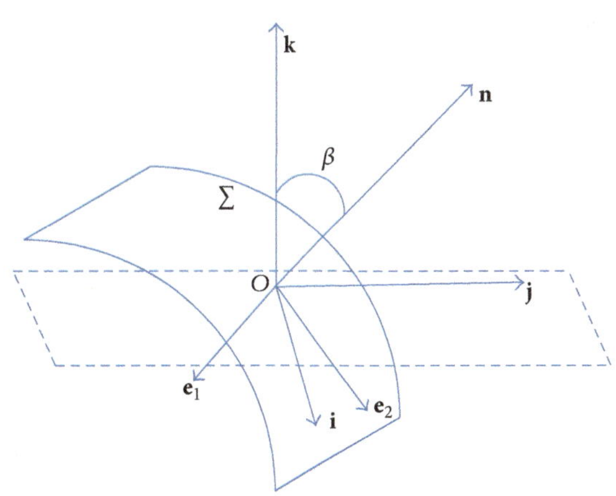

FIGURE 3: The same as Figure 2 in case of one direction is normal to vector **k**.

of programs was written. The CSP stack and diffraction method are complementary to each other and reveal useful information about the subsurface. The diffraction method serves as a tool for detection of faults and voids, while the CSP stacking contains information about the structure of the subsurface. These methods were applied along the Dead Sea's shorelines. We have carried out several new seismic surveys along line crossing the sinkhole lineaments and reinterpreted a number of previous reflection sections. The 2D imaging technique has been presented in [7]. This paper examines the new 3D imaging approach and its application to 3D data that was acquired at the western shoreline of the Dead Sea. The results of the imaging allow better understanding between

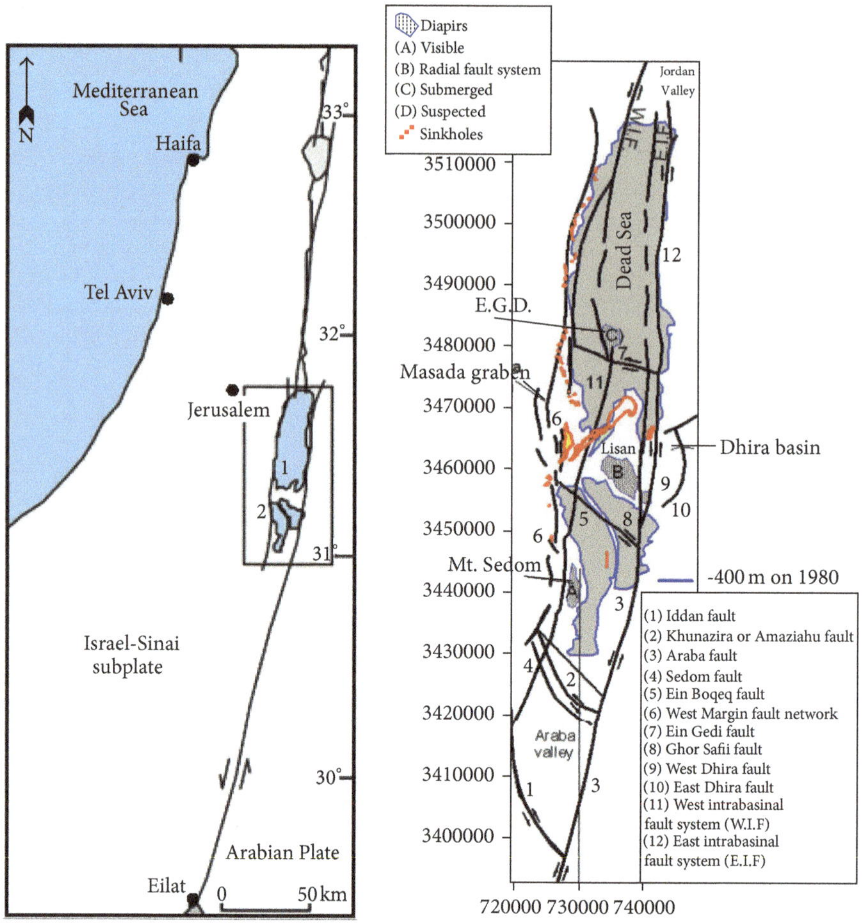

(a) (b)

FIGURE 4: Dead Sea tectonic setting: (a) Dead Sea transform; (b) faults through the DS area [4].

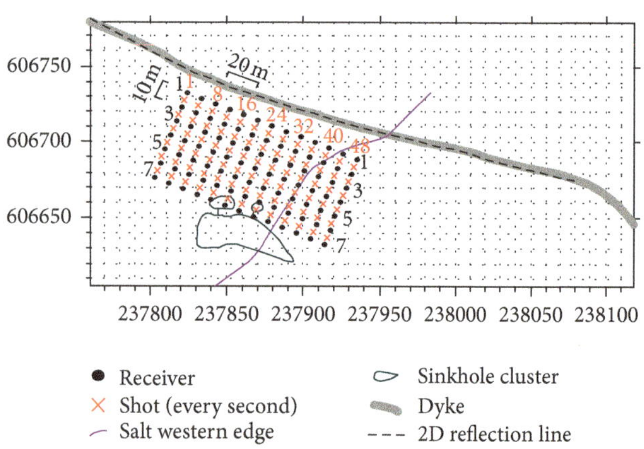

FIGURE 5: The geometry of 3D seismic survey.

a developing sinkhole and its tectonic environment. We give here a short description of 3D zero-offset stacking and multipath weighted summation. For a more comprehensively discussion of the methods, one can read in the relevant papers.

2. The 3D Zero Offset Stacking Method

The basis of the 3D zero-offset stacking method is a new normal moveout (NMO) time correction formula for three-dimensional media as function on wavefront parameters [11, 12]. One of these parameters on which the proposed time correction formula is based is the emergence angle β, defined as the acute angle between the wavefront normal and the normal to the acquisition plane at the CSP. Let us consider a central ray and its associated wavefront arriving at a common shot point (CSP) (Figure 2).

An ellipsoidal wavefront emerges at the CSP location O. It has two independent vectors $\mathbf{e}_1$ and $\mathbf{e}_2$ tangent to the wavefront and the normal $\mathbf{n}$. The emergence angle β is the angle between $\mathbf{n}$ and the normal to the acquisition plane, $\mathbf{k}$. Additional fundamental parameters on which the time correction formula is based are the curvatures of the wavefront. It is well known from the differential geometry of surfaces that at each point of the wavefront two principal curvatures can be defined, being the eigenvalues of the 2×2 matrix of second derivatives. These curvatures are the minimal and maximal curvatures and are related to the mean and Gaussian curvatures. The principal curvatures are associated

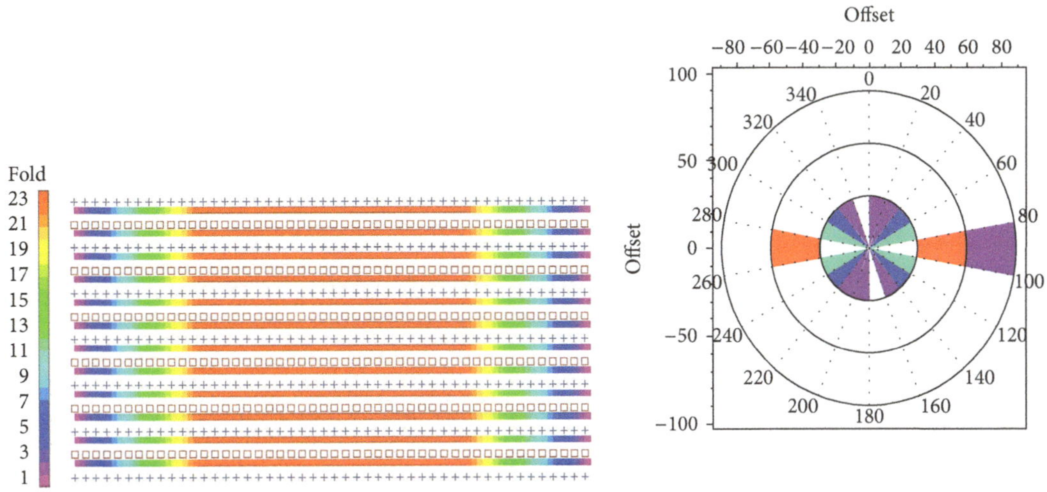

FIGURE 6: Fold scheme. (a) The shots locations are marked by "+" and receivers locations are marked by squares, (b) fold as function of azimuth.

to two perpendicular principal directions, which together with the wavefront normal form a rectangular system of axes. The curvature of the wavefront is completely described by the two principal curvatures and the angle the principal directions make with the general coordinate system. This is denoted in Figure 2 as **i**, **j**, and **k**. The principal curvatures can be expressed in terms of principal radii of curvature, which play an important role in the traveltime correction expressions derived in this paper. Figure 3 illustrates the case where one of the principal directions is normal to **k**. Based on those principles and eikonal equation only, the following 3D travel-time correction formula for an arbitrary system of coordinates was obtained:

$$t = \frac{1}{V}\left(a_{31}x + a_{32}y\right) + \frac{1}{2VR_1}\left(a_{11}x + a_{12}y\right)^2 + \frac{1}{2VR_2}\left(a_{21}x + a_{22}y\right)^2 + \cdots, \quad (1)$$

$$a_{31} = \sin\beta\,\sin\phi, \qquad a_{32} = -\sin\beta\cos\phi,$$

$$a_{11} = \cos\phi, \qquad a_{12} = \sin\phi,$$

$$a_{21} = -\cos\beta\sin\phi, \qquad a_{22} = \cos\beta\cos\phi,$$

where β is the emergence angle of the wavefront at point O, ϕ azimuth angle, defined as the angle between the axis $O\widetilde{X}$ and the principal direction of the wavefront at point O, $\tilde{x}$ and $\tilde{y}$ are the offsets along the axis $O\widetilde{X}$ and $O\widetilde{Y}$. For a given source receiver, gather the moveout equations that express the moveout correction with respect to a zero-offset trace by five parameters measured at the central point. The parameters are principal radii of the wavefront R_1, R_2, the azimuth angle ϕ, the emergence angle β, and the reference velocity V.

In the following, we give a short description of the weighted multipath summation.

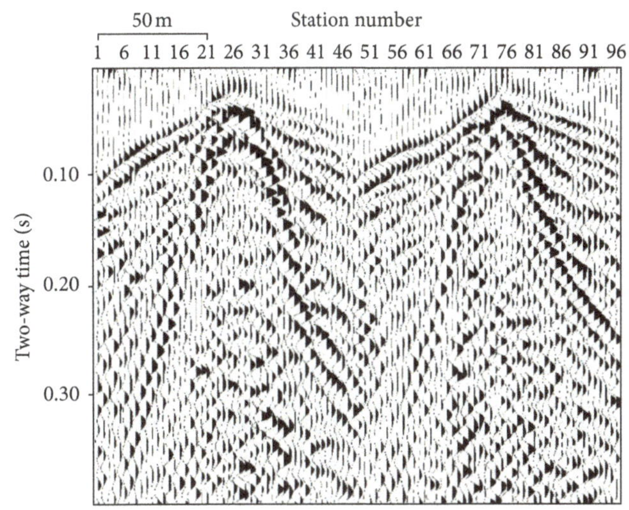

FIGURE 7: Typical common shot gathering including 96 channels.

3. Multipath Summation

The summation of 3D stacking is performed along time surfaces defined by (1). The summation can be implemented using either of the two approaches. In the first conventional approach, the target waves are stacked along the time surfaces defined by wavefront parameters and velocity. These optimal parameters usually are estimated using optimization problem which consists of finding parameters which maximize some correlation functional. We use an alternative, more formal recently proposed multipath summation with a proper weighting [10, 13–15]. The weighted multipath summation allows us to replace the complex optimization problem of estimating the optimal parameters, by summation along all possible surfaces that are created from small variations

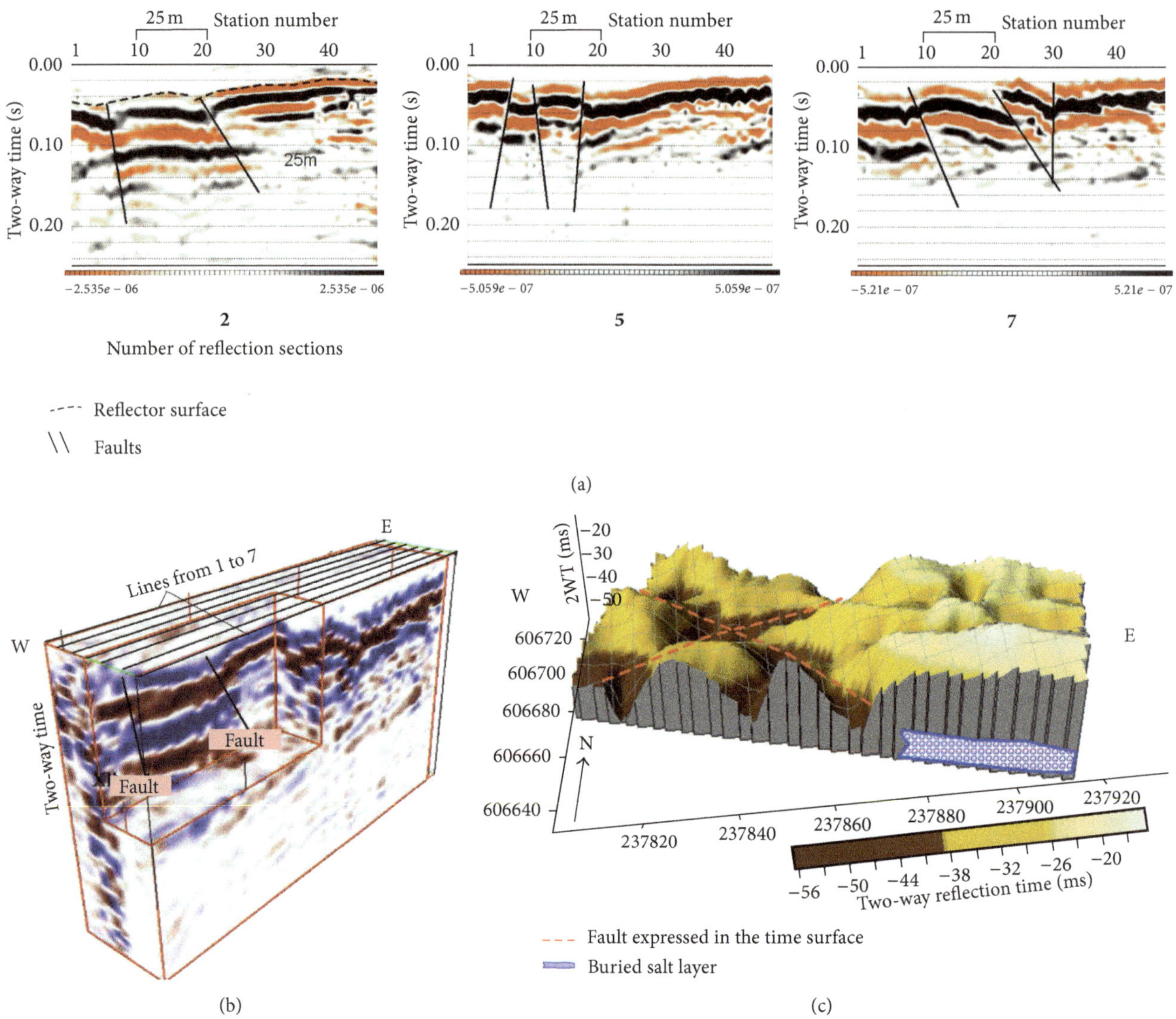

Figure 8: Typical two-way time sections (a) constructed from 3D time cube obtained using 3D imaging approach (b); two-way time map of reflector in the 3D presentation (c).

of those parameters. The weighted multipath summation (WMPS) can be described by the following expressions:

$$I_{\text{WMPS}} = \frac{\sum_P I_P W(I_P)}{\sum_P W(I_P)},$$
$$I_P = \sum_P \sum_S \sum_R U_{SR}(T_{SR}(P)), \qquad (2)$$

where $W(I_P) = \exp(\lambda I_P)$ is an "optimal" weighting function and λ is an undimensional large number. I_P is the image obtained for fixed parameters P, namely, for radii of curvatures azimuth and emergence angles. $U_{SR}(T_{SR})$ is a seismic trace for a given source-receiver pair. T_{SR} is travel time from shot S to receiver R. The summation (2) with the proper weight $W(I_P)$ is performed for all possible parameter values within a specified range. The image defined by I_P means that for every point to be imaged, seismic amplitudes are stacked together along all possible time surfaces defined by (1). The constructive and destructive interference of the amplitudes contributed by each time surface produces an image close to that obtained by stacking with the "correct" parameter [16].

4. Case Study

4.1. Dead Sea Area—Brief Tectonic Setting. Dead Sea (DS) is a pull apart basin or strike-slip one undergone to NW-SE compression and NE-SW tensile stress (Figure 4(a)). DS consists of two basins: the northern one (deeper) (denoted by 1 in Figure 4(a)) and the southern one (shallower) (denoted by 2 in Figure 4(a)). And it is extended in approximately N-S direction.

Numerous faults have been detected through DS area (Figure 4(b)) and have mainly extended in the same direction

[4, 17, 18]. The Dead Sea is located nowadays at elevation of −426 m below the sea level (b.s.l).

4.2. Study Site. Mineral beach study site (Figure 1) is located between the Dead Sea shoreline at the east and route number 90 (the main road along the western DS shore) at the west. The area is characterized by N-S normal faults [6]. Around Mineral Beach, sinkholes develop in both mudflat (southern boreholes Mn-2, Mn-5) and alluvial fan (northern borehole Mn-4) areas. Boreholes show that salt layer at the northern part of the area is located at 10 m deeper (−434 m elevation) than that at the southern one (−424 m). At the northern part of the area, salt is overlain by sandy-gravel sediments, whereas the southern part is composed of DS mud (clay) overlaying the salt.

The field acquisition covers 120 m by 60 m and consists of seven receiver lines in a 10 m interval between them. Shooting was carried out using six shot lines (288 shots in total) and each shot includes 96 channels in 2.5 m interval between them. The distance between shot lines is 10 m (Figure 5).

We used truck mounted accelerated weight (Digipulse) as an energy source and single 10 Hz geophone per station. In order to image the new developing fault in details, the survey was designed with a full azimuth cover for offsets less than 30 m (Figure 6).

The data was recorded using 0.5 msec sample rate and 0.5 sec record length. Typical common shot gathers are seen on Figure 7.

The data was processed using the new 3D stacking algorithm. The range of variation of parameters in (2) was as follows: radii of curvature from 3 to 600 meter, emergence angle from −5 to +5 degree, and velocity from 300 m/s to 1000 m/s. The value of λ was 900. The results from processed 3D imaging are presented in Figure 8(a) as a time sections along reflection lines and in Figure 8(b) as a 3D cube time section.

A time map of the first reflector is presented in Figure 8(c). On all the figures, are clearly seen subhorizontal coherent reflectors at approximate depth of 50–80 m, which incline on closer location to the exposed sinkhole. In addition, faults are seen on all sections (Figure 8(a) sections denoted by 2, 5, and 7). These faults are clearly seen on reflector presented in Figure 8(c). The results are consistent with the results that were obtained from a previous 2D study [3] at the same site. This work provides the first 3D HR imaging on the edge of a sinkhole and nearby fault. The results of the seismic interpretation of 3D image suggest a possible linkage between revealed fault sinkholes and field observation.

5. Conclusions

In order to understand the relationship between a developing sinkhole and its tectonic environment, a high-resolution (HR) three-dimensional (3D) seismic reflection survey was carried out at the western shoreline of the Dead Sea. The 3D image of the subsurface was obtained by the use of a recently developed 3D imaging approach. The core of this approach is a new 3D NMO time correction surface formula. Imaging of subsurface is performed by a spatial summation of seismic waves along these time surfaces using recently proposed multipath summation with proper weights. The multipath summation is performed by stacking the target waves along all possible time surfaces having a common apex at the given point. This approach does not require any explicit information on parameters since the involved multipath summation is performed for all possible parameters values within a wide specified range. The results from processed 3D time volume show subhorizontal coherent reflectors at approximate depth of 50–80 m which incline on closer location to the exposed sinkhole and suggest a possible linkage between revealed fault and the sinkholes.

Acknowledgments

This study has been performed within the framework of MERC project M27-050 fund sponsored by the USAID. Thanks are due to the support of the Israel Ministry of Infrastructure. The authors are also grateful to the Geophysical Institute of Israel for permission to publish this paper.

References

[1] Y. Arkin and A. Gilat, "Dead Sea sinkholes: an ever-developing hazard," *Environmental Geology*, vol. 39, no. 7, pp. 711–722, 2000.

[2] A. Frumkin and E. Raz, "Collapse and subsidence associated with salt karstification along the Dead Sea," *Carbonates and Evaporites*, vol. 16, no. 2, pp. 117–130, 2001.

[3] Y. Yechieli, M. Abelson, A. Bein, O. Crouvi, and V. Shtivelman, "Sinkhole "swarms" along the Dead Sea cost: reflection of disturbance of lake and adjacent groundwater systems," *Bulletin of the Geological Society of America*, vol. 118, no. 9-10, pp. 1075–1087, 2006.

[4] Z. Ben-Avraham, "Geophysical framework of the Dead Sea: structure and Tectonics," in *The Dead Sea: the Lake and Its Setting*, T. M. Niemi, Z. Ben-Avraham, and J. Gat, Eds., pp. 22–35, Oxford University Press, Oxford, UK, 1997.

[5] M. Abelson, G. Baer, V. Shtivelman et al., "Collapse-sinkholes and radar interferometry reveal neotectonics concealed within the Dead Sea basin," *Geophysical Research Letters*, vol. 30, no. 10, pp. 52–1, 2003.

[6] M. Abelson, Y. Yechieli, O. Crouvi et al., "Evolution of the Dead Sea sinkholes," *Special Paper of the Geological Society of America*, no. 401, pp. 241–253, 2006.

[7] S. Keydar, B. Medvedev, M. Ezerky, and L. Sobolevsky, "Imaging shallow subsurface of Dead Sea area by Common Shot Point stacking and diffraction method using weighted multipath summation," *Journal of Civil Engineering and Science*, vol. 1, no. 2, pp. 75–79, 2012.

[8] S. Keydar, L. Bodet, C. Camerlynck et al., "A new approach for shallow subsurface imaging and its application to the Dead Sea sinkhole problem," in *Proceedings of the 73rd EAGE Conference and Exhibition*, pp. 1–4, Vienna, Austria, April 2011.

[9] S. Keydar, B. Medvedev, A. Al-Zoubi, and M. Ezersky, "Another look of imaging of shallow subsurface: real examples from the Dead Sea sinkhole development areas," in *EGU General Assembly*, vol. 14 of *Geophysical Research Abstracts, vol. 14*, p. 1432, Vienna, Austria, April 2012.

[10] S. Keydar, "Homeomorphic imaging using path integrals," in *Proceedings of the 66th EAGE Conference & Exhibition*, pp. 7–10, Paris, France, June 2004.

[11] S. Keydar and M. Mikenberg, "Prestack time migration using the Kirchhoff sum along a new approximation of the reflection travel time curve," in *Proceedings of the 72nd European Association of Geoscientists and Engineers Conference and Exhibition (EUROPEC '10)*, pp. 4916–4920, Barcelona, Spain, June 2010.

[12] S. Keydar and M. Mikenberg, "A new time correction formula in three-dimensional media as a function of wavefront attributes," *Journal of Seismic Exploration*, vol. 17, no. 4, pp. 349–369, 2008.

[13] S. Keydar and V. Shtivelman, "Imaging zero-offset sections using multipath summation," *First Break*, vol. 23, pp. 21–24, 2005.

[14] E. Landa, S. Fomel, and T. J. Moser, "Path-integral seismic imaging," *Geophysical Prospecting*, vol. 54, no. 5, pp. 491–503, 2006.

[15] J. Schleicher and J. C. Costa, "Migration velocity analysis by double path-integral migration," *Geophysics*, vol. 74, no. 6, pp. WCA225–WCA231, 2009.

[16] V. Shtivelman, S. Keydar, and M. Mikenberg, "Imaging near-surface inhomogeneities using weighted multipath summation," *Near Surface Geophysics*, vol. 7, no. 3, pp. 171–177, 2009.

[17] A. Al-Zoubi and U. S. Ten Brink, "Salt diapirs in the Dead Sea basin and their relationship to Quaternary extensional tectonics," *Marine and Petroleum Geology*, vol. 18, no. 7, pp. 779–797, 2001.

[18] A. Frumkin, M. Ezersky, A. Al-Zoubi, E. Akkawi, and A.-R. Abueladas, "The Dead Sea hazard: geophysical assessment of salt dissolution and collapse," *Geomorphology*, vol. 134, pp. 102–117, 2011.

20

Seasonal and Local Time Variations of *E*-Region Field-Aligned Irregularities Observed with 30.8-MHz Radar at Kototabang, Indonesia

Y. Otsuka

Solar-Terrestrial Environment Laboratory, Nagoya University, Nagoya, Japan

Correspondence should be addressed to Y. Otsuka, otsuka@stelab.nagoya-u.ac.jp

Academic Editor: R. Dabas

A VHF backscatter radar with operating frequency 30.8 MHz has been operated at Kototabang (0.20°S, 100.32°E; dip latitude 10.36°S), Indonesia, since February 2006. We analyzed *E*-region field-aligned irregularities (FAIs) observed by this radar through a year of 2007 and found that the *E*-region FAI observed at Kototabang can be classified into two groups. One is "descending FAI". Altitude of the FAI echo region descends with time from 102 km to 88 km altitude during 0700–1000 and 1900–0000 LT in June solstice season. The other is "low-altitude FAI", which is observed in an altitude range from 88 to 94 km mainly during nighttime. The observed Doppler velocity show distinct local time and altitude dependence. The seasonally averaged zonal velocity above (below) approximately 94 km altitude is westward (eastward) during daytime and eastward (westward) during nighttime. Meridional/vertical velocity perpendicular to the geomagnetic fields is upward during daytime and downward during nighttime. The direction of the FAI velocity above approximately 94 km altitude is consistent with that of the background $\mathbf{E} \times \mathbf{B}$ plasma drifts reported previously.

1. Introduction

VHF, UHF, and L-band radars are used for observations of the Bragg scatter echo from field-aligned irregularities (FAIs) with a spatial scale of one half the radar wavelength. Regarding the *E*-region FAIs at mid-latitudes, using a powerful radar, middle and upper atmosphere (MU) radar at Shigaraki, Japan, Yamamoto et al. [1] have discovered two types of the *E*-region FAI echoes: quasiperiodic (QP) and continuous echoes. The QP echoes appearing intermittently at altitudes of sporadic *E* (*Es*) layer (above 100 km) with periods of 5–20 min. Morphological features of the QP echoes have been reported by Yamamoto et al. [2]. Occurrence rate of the QP echoes shows distinct seasonal variation with a maximum in summer and rises abruptly after the local sunset and decreases after midnight. Numerous experimental and theoretical studies have been made to disclose the mechanisms generating the QP echoes (e.g.,[3–7]). On the other hand, the continuous FAI echoes appear continuously in time at low-altitude of the *E*-region (90–100 km), mainly around sunrise and occasionally at postmidnight [8].

However, studies of the *E*-region FAIs at low latitudes are limited compared to those at the mid-latitude regions. Using the Piura VHF radar in Peru (5.2°S, 80.6°W; dip latitude 7°N) and the Gadanki MST radar in India (13.5°N, 79.2°E; dip latitude 6.3°N), Chau et al. [9] and Patra et al. [10] have reported statistical characteristics of the *E*-region FAIs, respectively. Using the Gadanki radar, Patra et al. [11] and Pan and Rao [12] have observed QP echoes at altitudes lower than those of the usual QP echoes.

The Equatorial Atmosphere Radar (EAR) was installed at Kototabang (0.20°S, 100.32°E; dip latitude 10.36°S) in West Sumatra, Indonesia, on 2001 March [13]. *E*-region FAI observations have been performed using the EAR and Fukao et al. [13] have shown that the *E* region echoes above 100 km are quite similar to those at other low-latitude and mid-latitude locations. Patra et al. [14] have performed simultaneous wind observations made using a meteor radar installed at the EAR site and compared the meteor with the FAI Doppler velocities. Their results show that the low-altitude FAI velocities are controlled mainly by neutral winds.

As described above, the EAR is a powerful tool to observe *E*-region FAIs at Kototabnag, Indonesia. However, the EAR is usually operated in a mode of troposphere and stratosphere measurements, and the ionospheric FAI measurements are made only approximately 4 weeks a year, on average although the routine measurements have started since July 2010. Consequently, a climatological study of the *E* region FAIs has not been made yet. To make continuous observations of FAIs over Indonesia, a VHF radar with an operating frequency of 30.8 MHz has been installed at the EAR site. The present paper reports seasonal and local time variations of FAIs observed with the 30.8-MHz radar at Kototabang, Indonesia, which is located at magnetically low-latitude.

2. Observations

A VHF backscatter radar with an operating frequency of 30.8 MHz was installed at Kototabang, Indonesia, on February 2006. Specification of the radar system is shown in Table 1. Peak and average transmitting powers are 20 kW and 1.5 kW, respectively. The antenna is composed of a linear array of 18 three-element Yagi antennas and aligned in zonal direction with a total length of 117 m. The VHF radar has a capability to steer the radar beam to 9 directions between ±54° in azimuth around geographic south (126°–234°). Zenith angle of all the radar beams is set at 20°. Half-power full-beam widths in azimuth and zenith directions are 12° and 40°, respectively. The perpendicularity between the radar beam and geomagnetic field line can be achieved within the half-power full-beam width.

The VHF radar was operated routinely in a mode which consisted of *E* and *F* region FAI measurements. For the *E*-region FAI measurements through a year of 2007, five beams with azimuth of 125.8°, 153.0°, 180.0°, 207.0°, and 234.2° were steered to reveal spatial and temporal variations of the FAI echoes. This arrangement covered approximately 90 km in zonal direction at 100 km altitude. This coverage is almost same as the EAR multibeam measurements conducted by Patra et al. [14]. The range and time resolutions of the VHF radar measurement were 1.2 km and about 56 seconds, respectively. Details of the observational mode are listed in Table 2.

In this study, the altitude of the target from which the radar echo comes is defined as the altitude at which center of the radar beam is perpendicular to the geomagnetic field. It is important to note that the radar beam of this radar is rather broad in both azimuth and elevation so that the altitude shown in this study may be inaccurate. Furthermore, since the operating frequency of 30.8 MHz is relatively low compared to other coherent radar for the FAI measurements, the altitude of the radar echo may be underestimated due to the refraction of the radio wave.

TABLE 1: Specifications of the VHF radar at Kototabang.

Location	0.20° S, 100.32° E
Dip latitude	10.36° S
Operating frequency	30.8 MHz
Antenna	Linear array of 18 three-element Yagi antennas
Gain	22 dBi
Beam width	12° in azimuth (half-power full-width) 40° in zenith (half-power full-width)
Beam steering	Active phase control Azimuthal beam directions between +54° and −54° at 13° step
Transmitter	Solid-state pulse transmitter
Peak power	20 kW
Average power	1.5 kW
Pulse width	1–200 μs
Subpulse width	1, 2, 4, 8, 16 μs
Receiver	Single coherent receiver
A/D converter	16 bit

TABLE 2: Observational mode for *E*-region FAI.

Parameter	Value
Azimuth of beams	234.2°, 207.0°, 180.0°, 153.0°, 125.8° from due north
Interpulse period	1.6 ms
Transmitted pulse	64 μs (8 μs × 8-bit complementary)
Number of coherent integration	2
Number of FFT points	256
Number of spectral average	3
Range resolution	1.2 km
Sampling intervals	1.2 km
Range	90.0–141.6 km
Time resolution	56 seconds.

3. Results

Figure 1 shows range-time-SNR plots of the *E*-region FAI echoes observed in the five radar beams on July 29, 2007. Color scale represents signal-to-noise ratio (SNR) of the FAI echoes. The vertical scale at the right of each figure shows the altitude at which the radar beam is perpendicular to the geomagnetic field. The echo region of the *E*-region FAI descended approximately from 100 km to 90 km altitude between 0700 and 1100 LT and then disappeared. Around the sunset (1800 LT), the FAI appeared again at an altitude of approximately 100 km and descend with time to approximately 90 km until 2300 LT, indicating semidiurnal variation of the descending FAIs. Between 0000 and 0600 LT, the other descending FAI was observed. Zenith angle of the five radar beams ranges between 21° and 33°, assuming that the FAI echo comes from the point where the radar beam is perpendicular to the geomagnetic field. The displacement of the FAI echo toward the radar is, in general, a combination of the horizontal displacement toward the radar and downward displacement. However, for the current event as shown in Figure 1, altitude of the FAI echo region is almost identical among the five radar beams. This feature was seen in most *E*-region FAI events and indicates that the FAI regions move in the vertical direction rather than the horizontal direction. In the current study, we interpret the FAI displacement as the vertical displacement of the FAI region.

Figure 2 shows the Doppler velocity in the five radar beams on July 29, 2007. In this study, the Doppler velocity was obtained from the first moment of the FAI echo spectrum. We adopt the data with SNR larger than 0 dB in order to exclude noisy data. From the figure, we find that the Doppler velocity varies with local time and altitude. The Doppler velocity in an upper part of the *E*-region FAI echoes (approximately above 94 km altitude) between 0700 and 1200 LT is negative (away from the radar) on westward beam and positive (toward the radar) on eastward beam, indicating westward velocity of the FAI. On the other hand, between 0000 and 0600 LT, the Doppler velocity in an upper part of the *E*-region FAI echoes is toward the radar on westward beam and away from the radar on eastward beam, indicating eastward velocity.

Therefore, the zonal velocity was westward during daytime and eastward during nighttime. The Doppler velocity in the lower *E* region (approx. below 94 km altitude) between 1800 and 2300 LT is toward the radar on all of the five beams, indicating northward/downward velocity perpendicular to the magnetic fields.

By analyzing the *E*-region FAI data obtained with the VHF radar at Kototabang through a year of 2007, we investigate seasonal and local time variations of the FAI echo occurrence rate. The one-year interval was divided into three seasons, namely, June solstice season (May–August), December solstice season (November–February), and equinox (March, April, September, and October). Figure 3 shows local time and altitude variations of FAI occurrence rate in each hourly and altitude bin. We defined the occurrence rate as the ratio of the time period duration of the FAI occurrence to the hourly observation intervals for each altitude bin. The altitude bin corresponds to the sampling interval of 1.2 km in range. From Figure 3, we find two types of the *E*-region FAI. One is "descending FAI", which can be seen most clearly in June solstice season. Altitude of the FAI echo region descends with time from 102 km to 88 km altitude between 0700 and 1000 LT. In winter, the descending FAI is less pronounced. Another descending FAIs can be seen between 1900 and 0000 LT in the June solstice season, indicating semidiurnal variation of the FAI appearance. The other type of the FAI is "low-altitude FAI." The FAI echo was observed at around 88–94 km mainly during nighttime, and the altitude of the echo region is mostly constant with time through a night.

Figure 4 shows averaged Doppler velocities of the *E*-region FAIs observed in the southward beam with the VHF radar at Kototabang in 2007 for different seasons. All of

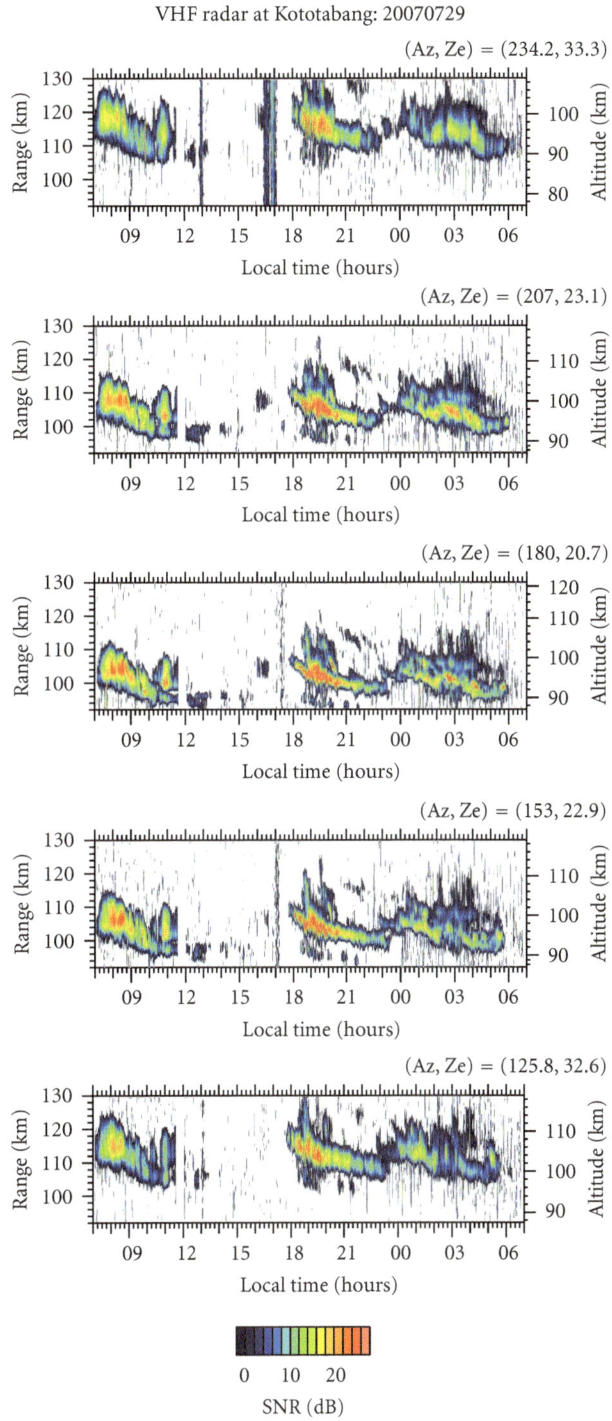

FIGURE 1: Range-time-SNR plot of the *E*-region Field-Aligned Irregularity (FAI) observed on beams with azimuth 125.8°, 153.0°, 180.0°, 207.0° and 234.2° (top to bottom panels) with the 30.8 MHz radar at Kototabang, Indonesia on July 29, 2007. Left vertical axis shows the range from the radar and the right axis shows the altitude at which the radar beam is perpendicular to the geomagnetic field.

the FAI observed at Kototabang in 2007 were classified into each bin of an hour and a sampling interval in range (1.2 km), and the data within each bin are averaged to obtain averaged Doppler velocity. The results show that direction of the Doppler velocity is different between daytime and nighttime; the velocity is southward/upward during daytime and northward/downward during nighttime.

Zonal velocity of the FAI is obtained from combining the Doppler velocities on eastward and westward beams. Figure 5 shows local time variation of the averaged zonal velocity (positive eastward). The velocity depends on not only local time but also altitude. In the lower (higher) altitudes of the FAI echo regions, the velocity is eastward (westward) during daytime and westward (eastward) during nighttime.

4. Discussion and Conclusions

Based on the seasonal and local time variations of the FAI echo occurrence rate, the *E*-region FAI observed at Kototabang can be classified into two groups. One is "descending FAI." Altitude of the FAI echo region descends with time from 102 km to 88 km altitude during 0700–1000 and 1900–0000 LT. The descending FAIs are frequently observed in the June solstice season compared with other seasons. The other is "low-altitude FAI." The FAI echo appears around 88–94 km and the altitude of the echo region is mostly constant with time through a night. The altitude of the low-altitude FAI is consistent with that of the continuous FAIs observed at Shigaraki, Japan (e.g., [8]). However, the continuous FAIs in Japan occur after midnight and continue until the morning, whereas the low-altitude FAIs at Kototabang appear after sunset and continue through a night with maximum occurrence rate between sunset and midnight. This difference could be attributed to latitudinal dependence of the *E*-region FAIs.

From the spectral characteristics, the FAIs observed by the VHF radar at Kototabang are considered to be type-2 echo. The Doppler velocity (V_d) observed by the radar in the meridional beam is given by (e.g., [15])

$$V_d = \frac{1}{1+\psi}\frac{E_{E-W}}{B} + \frac{\psi}{1+\psi}U_{N-S}\sin\theta, \quad (1)$$

where E_{E-W} is zonal component of the electric field, B is the magnetic field, U_{N-S} is meridional component of the neutral wind, θ is zenith angle of the radar beam. $\psi = \nu_e\nu_i/\omega_e\omega_i$, where ν_e (ν_i) is the electron (ion) collision frequency and ω_e (ω_i) is the electron (ion) cyclotron frequency. Here, the contribution of vertical neutral winds is neglected. ψ decreases with altitude because collision frequency, which proportional to the neutral density decreases with altitude. At an altitude around 95 km, ψ is equal to unity, then $1/(1+\psi) = \psi/(1+\psi)$ [16]. Therefore, at higher altitudes above 95 km ($\psi \ll 1$), the Doppler velocities are dominated by the $\mathbf{E}\times\mathbf{B}$ drift, whereas at lower altitudes ($\psi \gg 1$), it is dominated by the meridional neutral wind.

As shown in Figures 4 and 5, the velocities of the observed FAIs above approximately 94 km altitude is westward and southward (upward) during daytime, and eastward and northward (downward) during nighttime. The direction of the FAI velocity is consistent with the $\mathbf{E}\times\mathbf{B}$ plasma drift measured by the Jicamarca incoherent scatter radar [17].

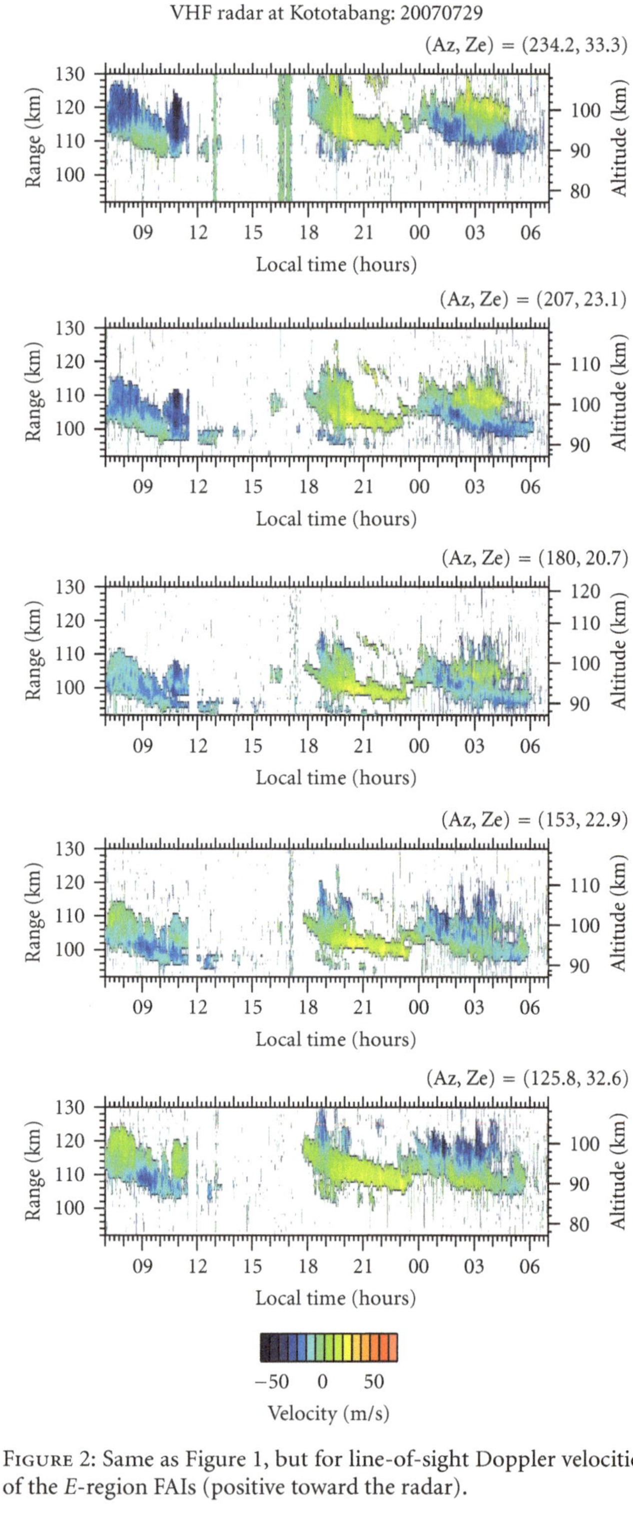

FIGURE 2: Same as Figure 1, but for line-of-sight Doppler velocities of the *E*-region FAIs (positive toward the radar).

This feature is different from the mid-latitude FAIs observed by the MU radar in Japan [1, 2]. The mid-latitude FAIs show westward velocities. This is probably because the velocity of the mid-latitude FAIs corresponds to the $\mathbf{E}\times\mathbf{B}$ drift caused by the polarization electric fields through the *E* and *F* region electrodynamical coupling processes associated with medium-scale traveling ionospheric disturbances (MSTIDs)

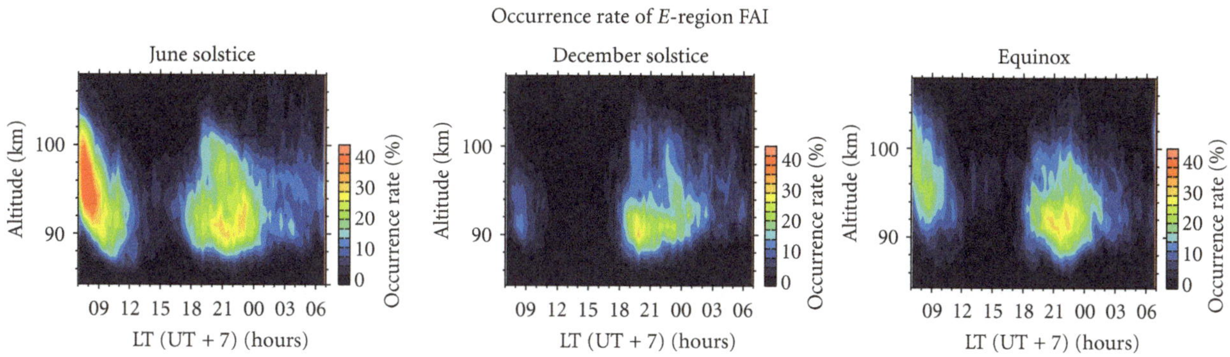

FIGURE 3: Local time and altitude variations of FAI occurrence rate at Kototabang, Indonesia, for (a) June solstice, (b) December solstice, and (c) equinox seasons in 2007.

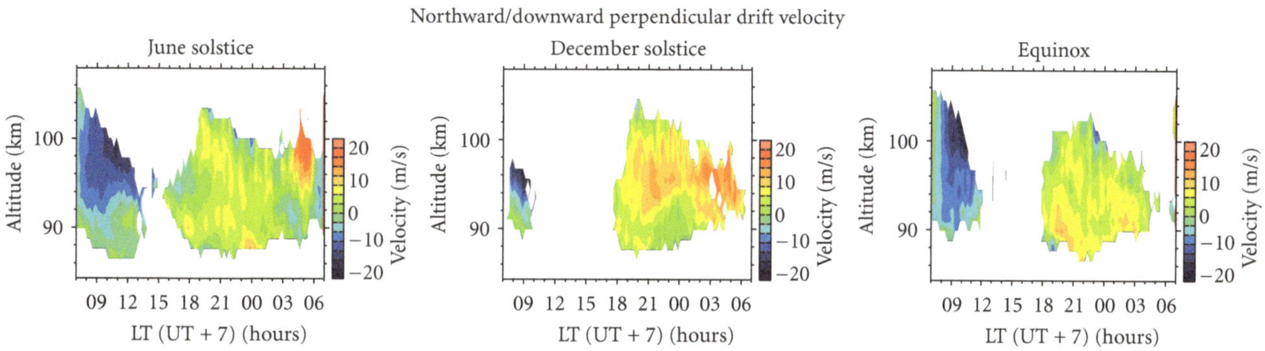

FIGURE 4: Local time and altitude variations of averaged Doppler velocity of the E-region FAIs in the southward/downward perpendicular direction to the magnetic fields (a) June solstice, (b) December solstice, and (c) equinox seasons in 2007.

in the F region [18, 19]. At mid-latitudes, both E-region FAIs and MSTIDs have band-like structures elongated from NW to SE in the northern hemisphere (from NE to SW in the southern hemisphere). At Kototabang, occurrence rate of such mid-latitude MSTIDs is low [20]. Therefore, the morphological features of the low-latitude FAIs are different from those at mid-latitudes although the underlying basic plasma instability processes are similar.

In the lower altitude of the FAI echo regions (below 94 km altitude), the Doppler velocity is westward during nighttime. This is probably due to westward neutral wind in consistence with the work of Patra et al. [14]. Altitude gradient of the plasma density can be considered to be upward at this altitude range (88–94 km altitude). This region is stable for the gradient drift instability. However, if neutral wind is eastward, it can make this region unstable via the gradient drift instability. Other processes may operate as the generation mechanism of the FAIs in the low-altitude E-region. One of possible mechanisms is the Kelvin-Helmholtz instability [21]. This instability could generate turbulence of the neutral atmosphere. Since the ion-neutral collision frequency is high in the lower E-region, the plasma moves at the almost same velocity as the neutral particles, so that the ionospheric irregularities are produced. When Richardson number is less than 0.25, this process operates and generates structures similar to the Kelvin-Helmholtz billows. In terms of this process, the mid-latitude quasi-periodic (QP) echo has been explained. On the other hand, at low latitude, using the Gadanki radar, Sripathi et al. [22] have observed the daytime QP echo with periods in the range 1–3 minutes and suggested that the Kelvin-Helmholtz instability seems to be the most likely mechanism to explain the daytime QP echoes.

The Doppler velocity of the low-altitude FAIs shows semidiurnal signature; the westward velocity becomes large at sunset and postmidnight, and weak around 2200 LT. This semidiurnal signature is most clearly seen in June solstice and equinox. Chau et al. [9], who have investigated the E-region FAI observed with the VHF radar at Piura, report that the Doppler velocity of the FAI shows semidiurnal pattern, particularly between April and July. Furthermore, the current study shows that the descending FAI echoes appear at 0700–1100 and 1900–2400 LT, indicating semidiurnal variation. The descending rate of the echo region is approximately 3-4 km/h, which is consistent with downward phase velocity of the semidiurnal tide and sporadic E (Es) layer [23]. Es layer could be formed by the convergence of ions due to the vertical shear in zonal wind which is eastward below the Es layer and westward above the Es layer [24]. The descending FAIs observed at Kototabang could be accompanied by the Es layer.

At mid-latitudes, the E-region FAIs do not occur during daytime except the continuous FAI echo which appears at low-altitudes of the E-region (90–100 km) mostly around sunrise [1]. On the other hand, the present observations

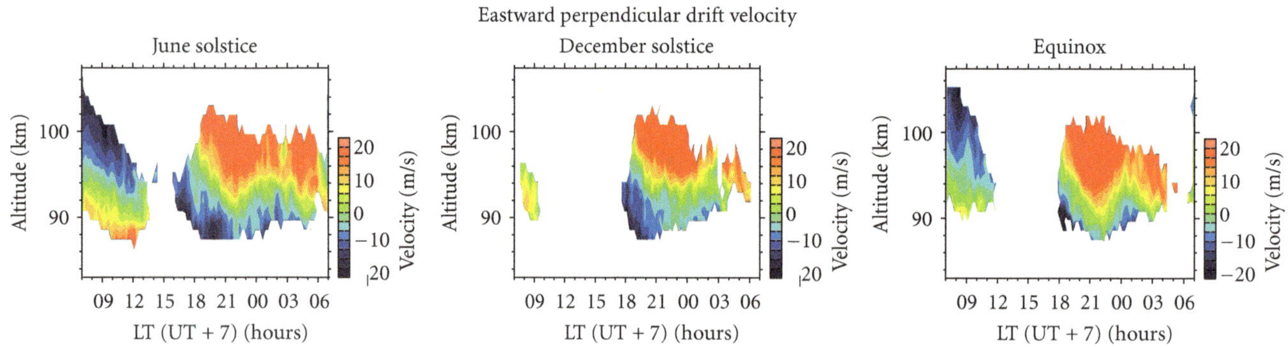

FIGURE 5: Same as Figure 4 but for the Doppler velocity toward the east.

show that the *E*-region FAIs over Kototabang, located at geomagnetically low-latitudes, occur even after sunrise until noon. The daytime *E*-region FAIs are frequently observed by the Gandanki MST radar in India, but they do not occur at Piura in Peru [9, 25]. These results suggest that the occurrence of the daytime *E*-region FAIs at low-latitudes have a longitudinal dependence. Occurrence rate of the nighttime *E*-region FAIs at mid-latitudes shows a distinct seasonal variation with a maximum in summer [2, 26]. This seasonal variation is consistent with that of *Es* layer. This could be because high plasma density and its gradient in the *Es* layer are needed for generating the FAIs. The nighttime *E*-region FAI over Kototabang shows highest occurrence rate around June solstice, but its seasonal variation is not so distinct compared to that at mid-latitudes. This feature is similar to that of the nighttime *E*-region FAIs at Gadanki in India and Piura in Peru. Chau et al. [9] have shown that the nighttime FAIs over Piura occur frequently in summer. Phanikumar et al. [25] have shown that the FAI occurrence rate at Gadanki is high in both summer and winter and relatively less in equinoxes and pointed out that the FAI occurrence is consistent with inhomogeneity of the plasma density in the *Es* layer. The FAI occurrence rate at Kototabang also needs to be compared with the occurrence rate of the *Es* layer to reveal the mechanisms generating the *E*-region FAIs.

Acknowledgments

This work was supported by Grants-in-Aid for Scientific Research (20684021) and Strategic Funds for the Promotion of Science and Technology of the Ministry of Education, Culture, Sports, Science and Technology of Japan. The 30.8-MHz radar is operated by STEL, Nagoya University, in collaboration with LAPAN and RISH, Kyoto University.

References

[1] M. Yamamoto, S. Fukao, R. F. Woodman, T. Ogawa, T. Tsuda, and S. Kato, "Midlatitude E region field aligned irregularities observed with the MU radar," *Journal of Geophysical Research*, vol. 96, no. A9, pp. 15943–15949, 1991.

[2] M. Yamamoto, S. Fukao, T. Ogawa, T. Tsuda, and S. Kato, "A morphological study on mid-latitude E-region field-aligned irregularities observed with the MU radar," *Journal of Atmospheric and Terrestrial Physics*, vol. 54, no. 6, pp. 769–777, 1992.

[3] S. Fukao, M. Yamamoto, R. T. Tsunoda, H. Hayakawa, and T. Mukai, "The SEEK (Sporadic-E Experiment over Kyushu) Campaign," *Geophysical Research Letters*, vol. 25, no. 10, pp. 1761–1764, 1998.

[4] R. F. Woodman, M. Yamamoto, and S. Fukao, "Gravity wave modulation of gradient drift instabilities in midlatitude sporadic E irregularities," *Geophysical Research Letters*, vol. 18, no. 7, pp. 1197–1200, 1991.

[5] R. T. Tsunoda, S. Fukao, and M. Yamamoto, "On the origin of quasi-periodic radar backscatter from midlatitude sporadic E," *Radio Science*, vol. 29, no. 1, pp. 349–365, 1994.

[6] R. B. Cosgrove and R. T. Tsunoda, "A direction-dependent instability of sporadic-E layers in the nighttime midlatitude ionosphere," *Geophysical Research Letters*, vol. 29, no. 18, 2002.

[7] T. Yokoyama, M. Yamamoto, and S. Fukao, "Computer simulation of polarization electric fields as a source of midlatitude field-aligned irregularities," *Journal of Geophysical Research A*, vol. 108, no. 2, article no. 1054, 2003.

[8] T. Ogawa, M. Yamamoto, and S. Fukao, "Middle and upper atmosphere radar observations of turbulence and movement of midlatitude sporadic E irregularities," *Journal of Geophysical Research*, vol. 100, no. A7, pp. 12173–12188, 1995.

[9] J. L. Chau, R. F. Woodman, and L. A. Flores, "Statistical characteristics of low-latitude ionospheric field-aligned irregularities obtained with the Piura VHF radar," *Annales Geophysicae*, vol. 20, no. 8, pp. 1203–1212, 2002.

[10] A. K. Patra, S. Sripathi, V. Sivakumar, and P. B. Rao, "Statistical characteristics of VHF radar observations of low latitude E-region field-aligned irregularities over Gadanki," *Journal of Atmospheric and Solar-Terrestrial Physics*, vol. 66, no. 17, pp. 1615–1626, 2004.

[11] A. K. Patra, S. Sripathi, V. Sivakumar, and P. B. Rao, "Evidence of kilometre-scale waves in the lower E region from high resolution VHF radar observations over Gadanki," *Geophysical Research Letters*, vol. 29, no. 10, article 1499, 2002.

[12] C. J. Pan and P. B. Rao, "Low altitude quasi-periodic radar echoes observed by the Gadanki VHF radar," *Geophysical Research Letters*, vol. 29, no. 11, article 1530, 2002.

[13] S. Fukao, H. Hashiguchi, M. Yamamoto et al., "Equatorial atmosphere radar (EAR): system description and first results," *Radio Science*, vol. 38, no. 3, article 1053, 2003.

[14] A. K. Patra, T. Yokoyama, M. Yamamoto, T. Nakamura, T. Tsuda, and S. Fukao, "Lower E region field-aligned irregularities studied using the Equatorial Atmosphere radar and

meteor radar in Indonesia," *Journal of Geophysical Research A*, vol. 112, no. 1, Article ID A01301, 2007.

[15] T. Tanaka and S. V. Venkateswaran, "Characteristics of field-aligned E-region irregularities over Iioka (36°N), Japan-II," *Journal of Atmospheric and Terrestrial Physics*, vol. 44, no. 5, pp. 395–406, 1982.

[16] A. K. Patra, "On the interpretation of the VHF radar echoes from the equatorial lower E region," *Geophysical Research Letters*, vol. 29, no. 14, article 1695, 2002.

[17] B. G. Fejer, E. R. dePaula, S. A. Gonzalez, and F. Woodman, "Average vertical and zonal F region plasma drifts over Jicamarca," *Journal of Geophysical Research*, vol. 96, no. A8, pp. 13901–13906, 1991.

[18] Y. Otsuka, F. Onoma, K. Shiokawa, T. Ogawa, M. Yamamoto, and S. Fukao, "Simultaneous observations of nighttime medium-scale traveling ionospheric disturbances and e region field-aligned irregularities at midlatitude," *Journal of Geophysical Research A*, vol. 112, no. 6, Article ID A06317, 2007.

[19] Y. Otsuka, K. Shiokawa, T. Ogawa, T. Yokoyama, and M. Yamamoto, "Spatial relationship of nighttime medium-scale traveling ionospheric disturbances and F region field-aligned irregularities observed with two spaced all-sky airglow imagers and the middle and upper atmosphere radar," *Journal of Geophysical Research A*, vol. 114, no. 5, Article ID A05302, 2009.

[20] K. Shiokawa, Y. Otsuka, and T. Ogawa, "Quasi-periodic southward-moving waves in 630-nm airglow images in the equatorial thermosphere," *Journal of Geophysical Research*, vol. 111, article A06301, 2006.

[21] M. F. Larsen, "A shear instability seeding mechanism for quasiperiodic radar echoes," *Journal of Geophysical Research A*, vol. 105, no. 11, Article ID 1999JA000290, pp. 24931–24940, 2000.

[22] S. Sripathi, A. K. Patra, V. Sivakumar, and P. B. Rao, "Shear instability as a source of the daytime quasi-periodic radar echoes observed by the Gadanki VHF radar," *Geophysical Research Letters*, vol. 30, no. 22, article 2149, 2003.

[23] N. Christakis, C. Haldoupis, Q. Zhou, and C. Meek, "Seasonal variability and descent of mid-latitude sporadic E layers at Arecibo," *Annales Geophysicae*, vol. 27, no. 3, pp. 923–931, 2009.

[24] J. D. Whitehead, "The formation of the sporadic-E layer in the temperate zones," *Journal of Atmospheric and Terrestrial Physics*, vol. 20, no. 1, pp. 49–58, 1961.

[25] D. V. Phanikumar, A. K. Patra, C. V. Devasia, and G. Yellaiah, "Seasonal variation of low-latitude E-region plasma irregularities studied using Gadanki radar and ionosonde," *Annales Geophysicae*, vol. 26, no. 7, pp. 1865–1876, 2008.

[26] C. Haldoupis and K. Schlegel, "Characteristics of midlatitude coherent backscatter from the ionospheric E region obtained with Sporadic E Scatter experiment," *Journal of Geophysical Research*, vol. 101, no. A6, pp. 13387–13397, 1996.

21

Multitemporal High-Resolution Satellite Images for the Study and Monitoring of an Ancient Mesopotamian City and its Surrounding Landscape: The Case of Ur

Giacomo Di Giacomo and Giuseppe Scardozzi

Istituto per i Beni Archeologici e Monumentali (IBAM), Consiglio Nazionale delle Ricerche (CNR), Strada per Monteroni c/o Campus Universitario, 73100 Lecce, Italy

Correspondence should be addressed to Giuseppe Scardozzi, g.scardozzi@ibam.cnr.it

Academic Editor: Nicola Masini

The paper concerns the use of multitemporal high-resolution satellite images for the study of the ancient city of Ur, in southern Mesopotamia, inaccessible to scholars from 2003. The acquired dataset is composed by two Gambit KH-7 (1966) and one Corona KH-4B (1968) declassified spy space photos and by few images taken by the recent satellites for civilian use QuickBird-2 (2002, 2004, 2007), Ikonos-2 (2008), and WorldView-1 (2008). The processing of all these images and the integration with ASTER and SRTM DEMs allowed the acquisition of new data about the topographical layout of the city and its monuments and ancient roads; the georeferencing of all archaeological remains and traces visible on the images allowed the upgrade of the archaeological map of Ur. The research also provided important data concerning the reconstruction of the surrounding landscape, where a lot of traces of old channels and riverbeds of the Euphrates were identified in areas much modified and altered during the last decades by urbanization and agricultural works. Moreover, the multitemporal images allowed the monitoring of the conservation of the archaeological area, particularly before and after second Gulf War.

1. Introduction

The research activity concerning Ur was gained in the context of the "Virtual Museum of Iraq" Project, promoted by the Italian Ministry of Foreign Affairs and carried out by the National Research Council (CNR) of Italy. The project was designed to the construction of a Virtual Museum of the ancient civilizations that flourished in the territory of modern Iraq; the result was a rich website (http://www.virtualmuseumiraq.cnr.it/), free to the general public, based on the archaeological collection of one of the most important museums in the world. In fact, in 2003, during the 2nd Gulf War, the National Museum of Baghdad has been looted, stripping it of a priceless collection of cultural artefacts that are important historical treasures not only for Iraqi people but also for all humanity. The "Virtual Museum of Iraq" Project explored different new integrated digital technologies for virtual heritage, particularly focusing on the use of image-based 3D modelling and satellite remote sensed data [1–7]. Specifically, the present work concerns the case of the ancient city of Ur, in southern Mesopotamia (Figure 1), that is exemplificative of the contribution of high-resolution satellite images for the reconstruction of the ancient topography and the monitoring of the site.

2. Methodology: The Research Activity concerning Ur in the "Virtual Museum of Iraq" Project

In the "Virtual Museumof Iraq" Project, the activities of the Institute for Archaeological and Monumental Heritage (IBAM-CNR) were focused on the realization of 3D reconstruction of settlements, monuments, and objects of the most important archaeological Iraqi sites and on the contextualization of finds stored into the Baghdad Museum,

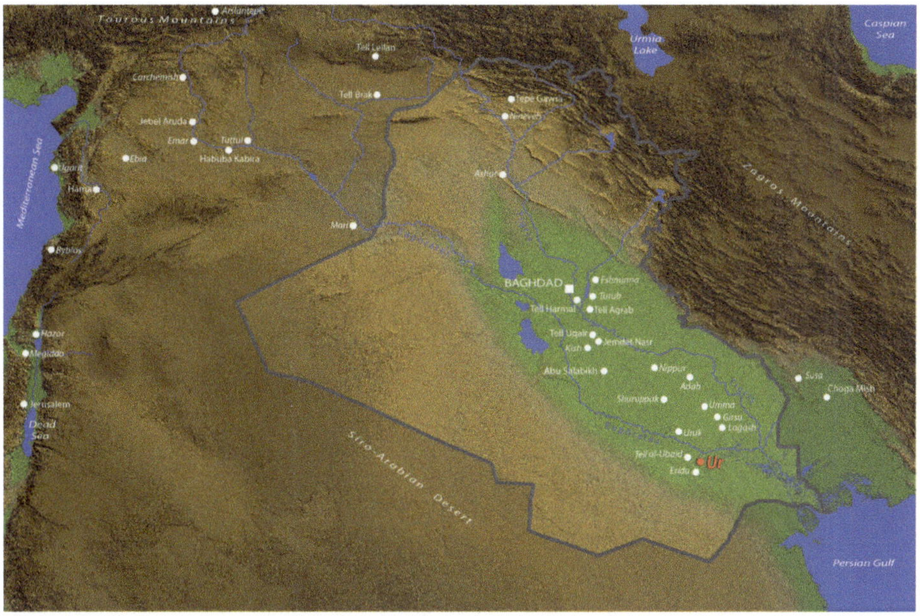

Figure 1: Ur in the southern Mesopotamia (Digital Elevation Model processed from the Shuttle Radar Topography Mission).

linking them to their sites of provenance. This process of contextualization enables the "visitors" of the Virtual Museum to view the original territorial contexts of the finds and it was implemented on three temporal levels: the ancient times, the moment of the discovery, and modern times. The virtual reconstructions, including those based on excavation photos, yield excellent results for the first two levels. Concerning the third temporal level, allowing a virtual visit to the place of provenance in its current situation, the photographic images, both aerial and from the ground, allow the user to view the archaeological areas with a very high detail. Considerable problems arise, however, in situations where the contexts are not easily accessible and adequate photographic documentation is not available. This is the case of much ancient cities and settlements of Iraq; generally, only few terrestrial images and some oblique aerial photographs of small sectors of the archaeological areas are available, together with some documents (graphic and photographic) of the excavations. An important contribution to the resolution of this problem, as well as to the recovery of documentation that serves to complement any photographic material that may be available, has been provided by high-resolution satellite images.

For each chronological section of the Virtual Museum, two or three archaeological sites more representative were chosen; they were documented by cataloguing the most significant monuments in order to provide the "visitors" of the Virtual Museum with a complete image of the characteristics of the contexts where the artefacts were found. In particular, the ancient city of Ur was chosen both for the Sumerian and the Neo-Babylonian "Halls" of the Museum.

In the section "Archaeological Sites" of the Iraq Virtual Museum, the contribution of high-resolution multitemporal satellite images was very important. First of all, the images were used to document the modern situation of ancient sites and for monitoring their conditions before and after the first and second Gulf War. These images make it possible to observe and document the archaeological areas as they are today, both as a substitute for a real visit (currently impossible) and in preparation for a future visit. The detailed examination of these images also constitutes an important opportunity to acquire new data about the topographical layout of the sites and their principal monuments [8, 9]. Moreover, satellite orthoimages were used for the production of new archaeological map of the ancient sites investigated in the research project [10].

In the case of Ur, we have a rich data set, starting from three panchromatic photos taken in the 1960s by the United States reconnaissance satellites Gambit KH-7 (1966; one vertical and one oblique, with a ground geometric resolution of 0.6 m) and Corona KH-4B (1968; ground geometric resolution of about 1.83 m). The documentation of the USA spy satellites covers the entire site and its surroundings; it is a fundamental tool for archaeological research, very useful when old aerial photos are scarce. In fact, thanks to their high geometric resolution, they allow the examination in detail of ancient urban areas and territories that later are changed or also partially destroyed; so, it is possible to detect and examine ancient structures, paleoenvironmental elements, and archaeological traces of buried features not yet visible [11].

The satellite documentation of the 1960s was augmented by the use of five high-resolution images taken between 2002 and 2008 by the commercial satellites QuickBird-2 (2002, 2004, 2007; *Standard Ortho-ready* products), Ikonos-2 (2008; *Carterra Geo Ortho kit* product), and WorldView-1 (2008; *Standard Ortho-ready* product). These images constitute an up-to-date documentation, very useful for the study and presentation of the archaeological site, through which it is possible also to compare with the aerial photos taken by

Figure 2: The new general plan of Ur (drawn up by F. Ghio): 1: City Wall; 2: North Harbour; 3: Palace of Ennigaldi-Nanna; 4: Harbour Temple; 5: Houses on City Wall; 6: Kassite Fort; 7: enclosure of the Sacred Area in the neo-Babylonian period; 8: enclosure of the Sacred Area during the 3rd Dynasty; 9: Nanna Court; 10: Etemenniguru; 11: Ziggurat; 12: Nanna Temple; 13: Boat Shrine; 14: Ningal Temple; 15: Giparku; 16: Edublamakh; 17: Ganunmakh; 18: Ekhursag; 19: Mausolea of the 3rd Dynasty; 20: Royal Graves; 21: Nimintabba Temple; 22: Houses (EM District); 23: West Harbour; 24: Houses (AH District); 25: neo-Babylonian Houses; 26: Enki Temple.

Royal Air Force pilots during the main excavations of Ur in 1930s. The high geometric, spectral, and radiometric resolutions of the recent satellite images allowed the detailed analysis of the archaeological site; particularly, their very high ground geometric resolution (between 1 and 0.5 m in panchromatic mode; between 4 and 2.40 m in multispectral mode) makes them similar, in terms of definition, to vertical aerial photographs on a middle scale that in the case of Ur are scarce. The images made it possible to obtain a detailed view from above of the current state of the ancient remains and their layouts, given the possibility of noticing even quite small details; they enabled us to identify elements that are not easily perceptible on the ground and to recognize the traces of structures that are still buried or have become buried again since their original excavation. Moreover, it is also possible to identify some paleoenvironmental features [12–18].

Very useful in the study of the site was also the integration of the images with ASTER and SRTM DEMs, from which were also extracted the contour lines of the new archaeological map of Ur elaborated during the research (Figure 2). The 3D models of the terrain show the tell of Ur, composed from successive stratifications, that rises approximately 20 m above the surrounding plain. The DEMs also allowed the 3D visualization of the site and its surroundings, and they made it possible to associate the archaeological traces and anomalies to the morphology of the ground thanks to the draping of the satellite images on the models (see Figure 7).

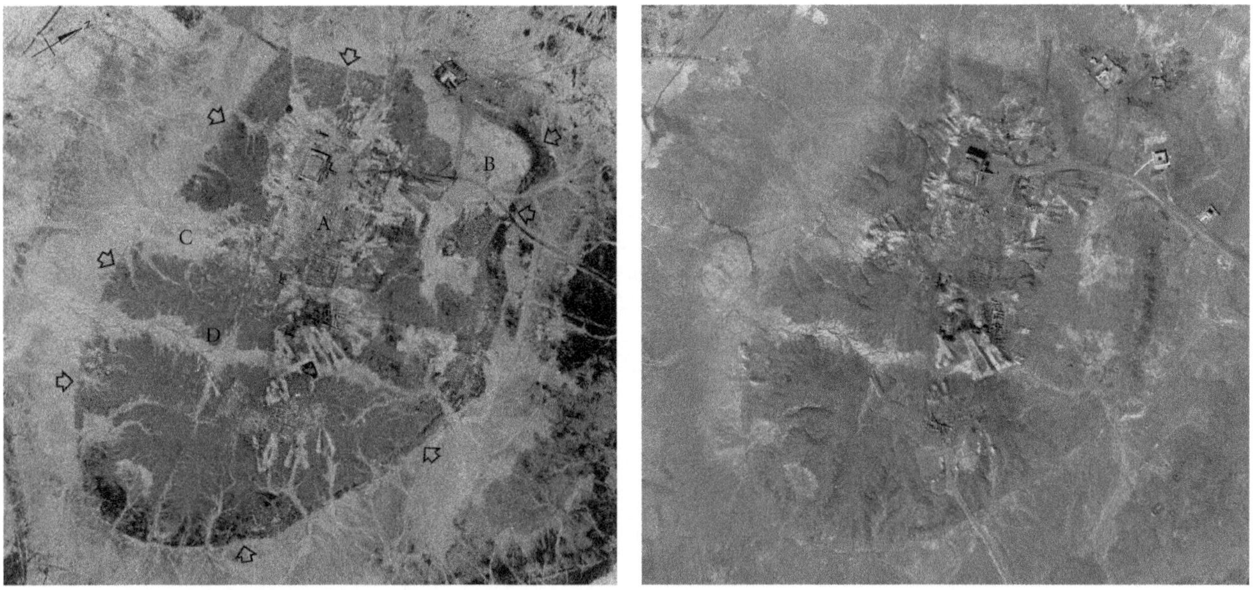

(a) (b)

FIGURE 3: The hill of Ur in a Gambit KH-7 photograph taken in April 1966 and in a WorldView-1 image acquired in March 2008: A: the sacred area with the Ziggurat and the other main monuments; B: the North Harbour; C: the West Harbour, a depression that probably corresponds to an ancient channel. The arrows indicate traces and remains of the city walls.

3. Data Set

The data set used in this research is very detailed and covers a time span ranging from 1966 to 2008. Technical developments in the fields of electronics, optics, and space engineering of the last 50 years have made a substantial contribution to the study of the archaeological site of Ur, allowing multitemporal analysis, using historical images taken by spy satellites during the 1960s, and then multispectral investigation, thanks to the images of modern satellites for civilian use.

The oldest images coming from the historical archives of the USGS (United States Geological Service-http://www.usgs.gov/) provide datasets of declassified images from spy satellites constellation Corona and Gambit. The Corona program was the first United States strategic reconnaissance satellites project operated by the Central Intelligence Agency in cooperation with US Air Force. The Corona satellites were originally used for photographic surveillance of the Soviet Union (USSR) and of China, but during the operational phase satellites acquired images of many other areas of the world. The Corona program began in June 1959 and ended in May 1972. Satellite series were designated KH-1, KH-2, KH-3, KH-4, KH-4A, and KH-4B; KH is an acronym for "Key Hole" and the incrementing number indicates changes in the instrumentation (e.g., the camera type). 144 Corona satellites were launched and 102 of them returned usable photographs.

The altitude of these satellites was 165 to 460 kilometers above the surface of the Earth, so the cameras could acquire images with a ground resolution down to 7.5 meters; satellites used special 70 millimeter film with 60-centimeters focal length cameras. The two KH-4A and KH-4B systems improved this resolution to 2.75 m and 1.8 m, respectively, because they operated at lower orbital altitudes. The KH-4 satellites used two panoramic cameras with 30-degree separation angle to produce stereoscopic imagery; in particular, the KH-4B system used two convergent rotating panoramic (black and white) cameras to record dual images of specific areas on the Earth with ground resolution of about 1.83 m and focal length of 60 centimetres and a Dual Improved Stellar Index Camera for target location identification and assistance in mapping. During 16 successful missions (from September 1967 to May 1972) it has produced 188,000 frames: an image taken on May 4, 1968, was used for this work, but this is not the oldest image available.

In fact, the dataset also includes two images taken in 1966 from a satellite of Gambit KH-7 series. The KH-7 was the first satellite for high-quality reconnaissance images. The system was designed to work with Corona program to take higher resolution images of a smaller area and the operational phase started in July 1963 and ended in June 1967. Gambit KH-7 successfully returned film on 34 of 38 missions and the usable images came from 30 missions. Satellites returned a total of 19,000 frames, with an initial ground resolution of 1.2 m; by 1966, best resolution improved to 0.6 m. These satellites, equipped with a Kodak advanced lens drive scanning Optical Bar camera system with its forward rotating primary mirror and folding optics capable of creating images on large format on a 9 inch wide, provided key cartographic information for large-scale (1 : 50,000) maps for US Department of Defense; priority targets for systems included Soviet and Chinese nuclear installations, but photos of other areas were taken. Most of the imagery from the KH-7 satellites was declassified

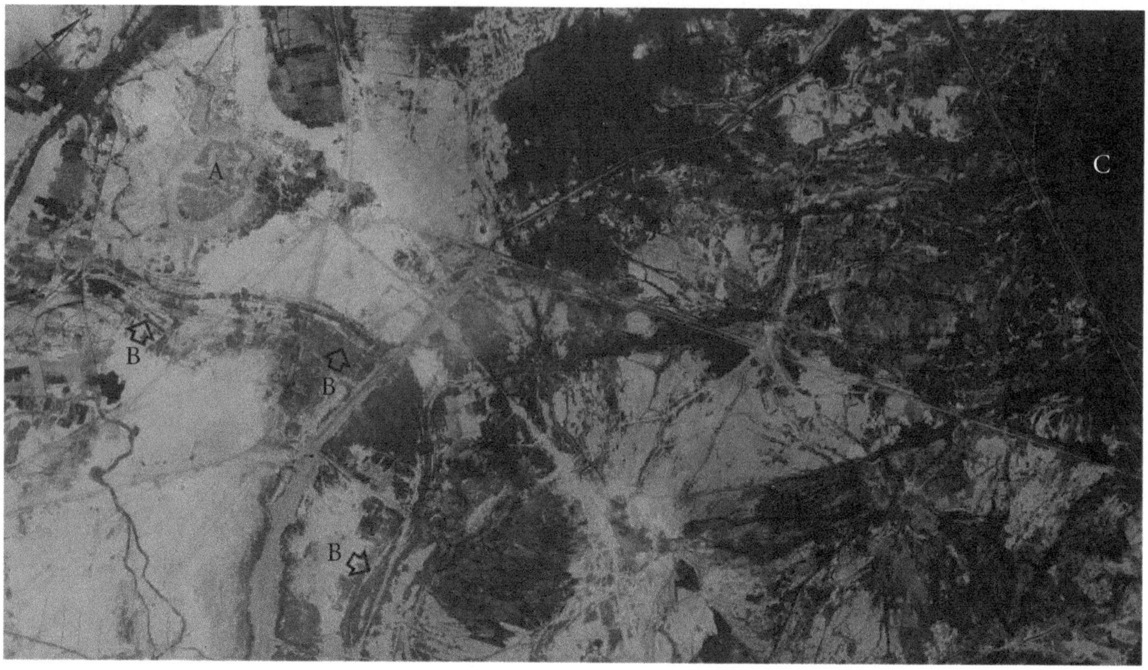

Figure 4: The territory of Ur (A) in a Gambit KH-7 photograph taken in April 1966: the traces of a paleoriver bed of the Euphrates (B) and the present river bed (C) are visible.

in 2002, and one oblique taken in February 18, 1966 and one vertical taken in April 23, 1966 (Figures 3 and 4; see also Figure 9) are part of the data set used for the research concerning Ur. In particular, the vertical image, as all the images acquired at nadir on the later KH-7 missions, can be compared favourably with the best commercial systems like Ikonos-2, QuickBird-2, and WorldView-1.

Ikonos-2 is a commercial earth observation satellite, owned by GeoEye, and was the first to collect publicly available high-resolution imagery at 1 m in panchromatic mode and 4 m in multispectral mode (http://www.geoeye.com/). Ikonos-2 was launched in 2001 and imagery began being sold on January 1, 2000. Ikonos-2 is the first next-generation commercial satellite, capable of acquiring images in stereo-mode, by rotating the camera along track; this satellite uses telemetry to transfer data to the ground and to receive navigation data from remote control center. The system, based on a fully digital image acquisition, has a primary mirror aperture of 0.7 m and a folded optical focal length of 10 m using 5 mirrors. The detectors at the focal plane include a panchromatic and a multispectral sensor, with 13500 pixels and 3375 pixels, respectively (cross-track). The sensor collects data with a 11-bit (0–2047) sensitivity that are delivered in an unsigned 16-bit (0–65535) data format. Sometimes data are rescaled down to 8 bit (0–255) to decrease file size, but this causes a loss of data. Ikonos-2 revisit time is 3 to 5 days for off-nadir and 144 days for true-nadir. An image taken by Ikonos-2 on August 9, 2008 was used for this work.

QuickBird-2 is another high-resolution commercial earth observation satellite, owned by DigitalGlobe and launched in 2001 (http://www.digitalglobe.com/). Until 2007 it has been the platform that has taken Earth images with the best geometric resolution; in fact the system collects 61-centimeter class panchromatic and 2.5-meter multispectral data over a large field of view with rapid target selection. QuickBird-2 was equipped with a sensor implemented by Ball Aerospace and Technologies Corporation, named Ball Global Imaging System 2000 (BGIS 2000), designed to be an agile, stable, and highly accurate Earth remote sensing platform. The imaging system returns 11-bit digital high resolution imagery from space, with 23 meters of horizontal accuracy. The operational orbit of the satellite is set to an altitude of 450 km with 98 degree sun synchronous inclination. Revisit frequency is 1 to 3.5 days depending on latitude. In the research concerning Ur three Quickbird-2 images were used; they were taken on February 27, 2002, on April 04, 2004, and on March 20, 2007.

WorldView-1, another satellite of DigitalGlobe company, was successfully launched on September 18, 2007. It was equipped with only a panchromatic imaging system, which can acquire half-meter resolution imagery with a dynamic range of 11 bit per pixel. Operating at an altitude of 496 kilometers, WorldView-1 has an average revisit time of 1.7 days and is capable of collecting up to 750,000 square kilometers per day of half-meter imagery. The satellite is also equipped with state-of-the-art geolocation capabilities and exhibits stunning agility with rapid targeting and efficient in-track stereo collection. Only one image, captured by this satellite on March 10, 2008, was used for the case study concerning Ur (Figure 3).

These eight images taken over 40 years from different satellites, with different characteristics and in different seasons, make up the data set that was used for this case study.

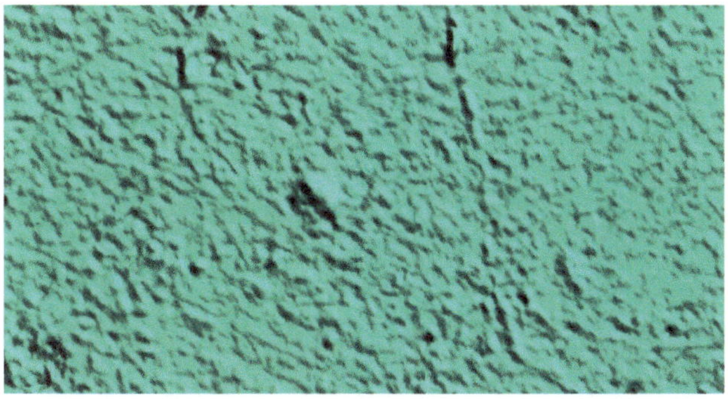

(a)

(b)

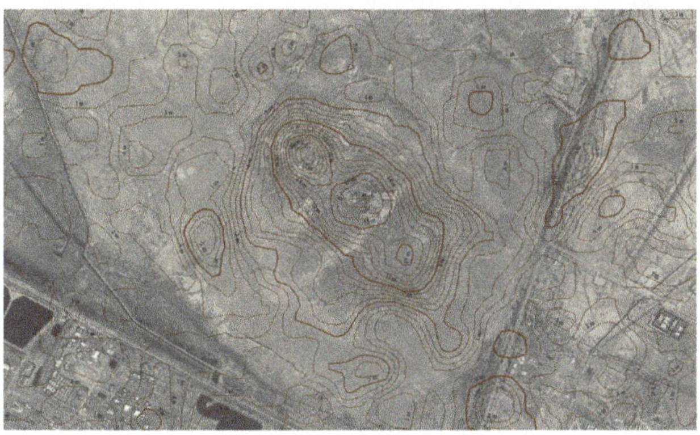

(c)

FIGURE 5: The hill of Ur in ASTER (a) and SRTM (b) DEMs; contour lines (c) extracted from the SRTM DEM, draped with the QuickBird-2 panchromatic image of 2004.

Depending on the type of the image, it was possible to perform various analyses and processing, from georeferencing, reading, and interpretation of the older images to the application of algorithms for postprocessing of the digital images taken by modern satellites.

4. Data Processing

In this case study, and also during all activities of the "Virtual Museum of Iraq" Project, the analysis of satellite images represented an opportunity to acquire new data on the ancient topography of the archaeological sites that were examined. This documentation was very important specifically for contexts that have been studied many years ago or that are characterized by a history of the research centered mainly on individual complexes, monuments, or wealth of finds. In addition, these studies have often not interested the general layout of the settlement. However, is important to specify that in the analysis of remote sensing data verification on the ground of presences, traces, and anomalies is fundamental; it is in order to clarify their real pertinence to archaeological elements, their interpretation and, if possible, also their dating, avoiding misunderstandings and mistakes. But in the study cases of the project, the ground control with a survey of the sites was impossible, so we could often formulate only hypotheses and the only comparison was with published studies.

The preliminary data processing has been the developing of three-dimensional models of the area of Ur (Figure 5), starting from the radar data taken by the Shuttle Radar Topography Mission and from an optical stereo pair taken by the radiometer ASTER (Advanced Spaceborne Thermal Emission and Reflection) of the Terra satellite; this stereo pair was already processed by the data center of Land Processes (LP) of Distributed Active Archive Centers (DAAC), which is a part of NASA's Earth Observing System Data and Information System (EOSDIS).

The Shuttle Radar Topography Mission (SRTM) obtained elevation data on a near-global scale to generate the most complete high-resolution digital topographic database of Earth. SRTM consisted of a specially modified radar system that flew onboard the Space Shuttle during an 11-day mission in February of 2000. SRTM is an international project spearheaded by the National Geospatial-Intelligence Agency (NGA) and NASA. The processing method of SRTM data is divided in three stages: the first processing stage involves importing and merging the 1-degree tiles into continuous elevational surfaces in ArcGRID format. The second process fills small holes iteratively and the cleaning of the surface to reduce pits and peaks. The third stage then interpolates through the holes using a range of methods. The method used is based on the size of the hole and the landform that surrounds it. In the processing Arc/Info AML model was used. The final product is a shape file that contains three-dimensional data of Earth surface that can be used as a Digital Elevation Model to extract contours.

Instead the ASTER Digital Elevation Model product is generated using bands 3 N (nadir-viewing) and 3B (backward-viewing) of an ASTER Level-1A image acquired in the Visible Near Infrared (VNIR). The VNIR subsystem of ASTER includes two independent telescopes that facilitate the generation of stereoscopic data. There is a time lag of approximately one minute between the acquisition of the nadir and backward images. Starting in early summer of 2006, LP DAAC has implemented a new production software for efficiently creating quality DEMs starting from the ASTER stereo pairs [19, 20]. Based on an automated stereocorrelation method, the new software generates a relative DEM without any ground control points (GCPs). It utilizes the ephemeris and altitude data derived from both the ASTER instrument and the Terra spacecraft platform. The ASTER DEM is a single-band product with 30-meters

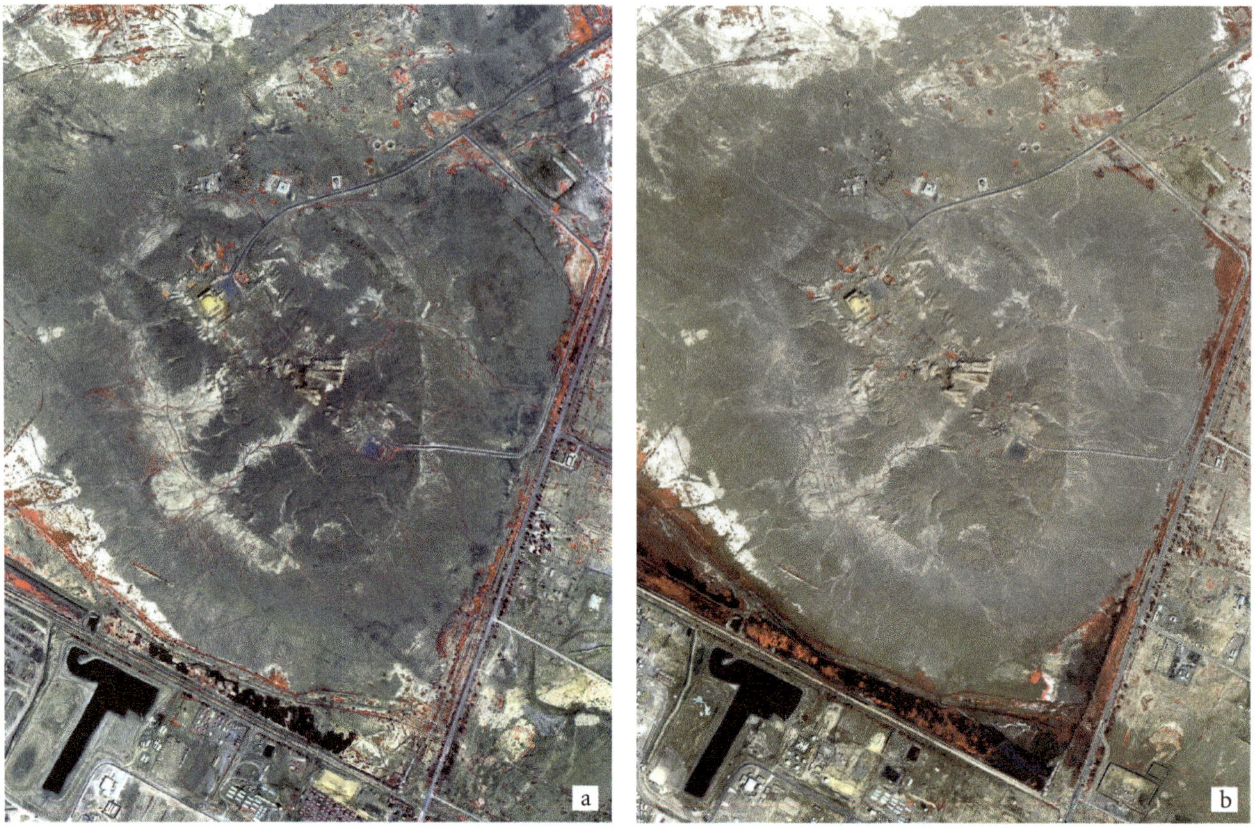

(a) (b)

Figure 6: Ur in a pan-sharpened QuickBird-2 image of April 2004 (colour composite RGB bands 4-3-1) and in a pan-sharpened QuickBird-2 image of March 2007 (colour composite RGB bands 4-3-2).

horizontal postings that is geodetically referenced to the UTM coordinate system and referenced to the Earth's geoid using the EGM96 geopotential model.

Both the ASTER DEM and the SRTM DEM were alternatively used to orthorectify satellite images through the Rational Polynomial Coefficients released by the satellite company: Rational Polynomial satellite sensor models are simpler empirical mathematical models relating image space (line and column position) to latitude, longitude, and surface elevation. The name Rational Polynomial derives from the fact that the model is expressed as the ratio of two cubic polynomial expressions. Actually, a single image involves two such rational polynomials, one for computing line position and one for the column position. The coefficients of these two rational polynomials are computed by the satellite company from the satellite's orbital position and orientation and the rigorous physical sensor model.

The aberrations on the images caused by satellite absolute position at taking time has been reduced (as much as possible in the absence of Ground Control Points) using three-dimensional models SRTM and ASTER in association with ortho-rectifying algorithms implemented in software like ENVI or PCI Geomatica. So it was possible to upgrade the archaeological plan of the site and to create an up-to-date vector documentation.

Once georeferenced and geometrically corrected, image processing was carried out on some satellite images, panchromatic and multispectral, acquired from Ikonos-2 and QuickBird-2 platforms. In particular, work was oriented in two directions: on the one side, we proceeded to elaborate the multispectral bands and to produce high-resolution colour images from the fusion with panchromatic images through various algorithms of pan-sharpening; on the other side, the georeferencing and orthorectification of panchromatic images for cartographic purposes were carried out. The multispectral processing of the images and the subsequent analysis of data from the near infrared and from the bands of the visible spectrum, was taken up in successive steps: the first phase of the work is consisted in the fusion of the different bands forming the data set of a raw satellite image. In fact the sensors of commercial satellites QuickBird-2 and Ikonos-2 acquire at the same time the same image in two different geometric resolutions: the panchromatic ones can reach a resolution at nadir of about 0.60 and 0.80 m, while the multispectral ones can reach a resolution between 2.40 and 3.20 m. The acquisition in multispectral mode for this type of satellites is in turn subdivided in four bands of the spectrum: red, green, blue (bands of the spectrum visible to the naked eye), and near infra-red (NIR-spectrum invisible to the naked eye). The images with different geometric resolution do not allow their direct overlay, unless an appropriate breakdown of the pixels is carried out, and a resampling through several algorithms: the most performance for images analysis aimed to the archaeological

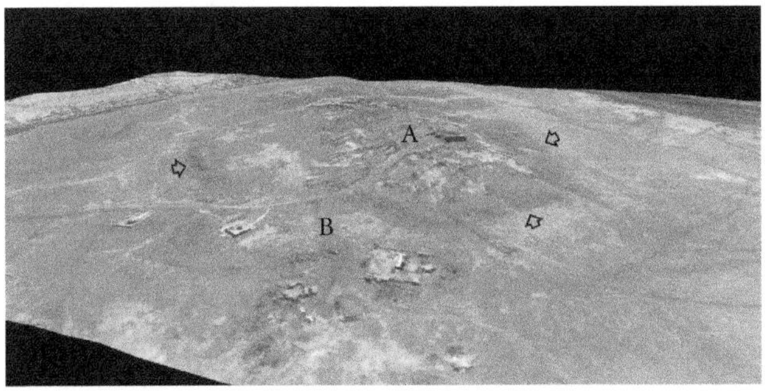

Figure 7: The QuickBird-2 panchromatic image of April 2004 draped on the DEM based on SRTM data: a view from the north, with the Ziggurat on the top of the hill (A), and the North Harbour area (B); the arrows indicate traces and remains of the city walls.

research were those of Zhang, implemented in the PCI Geomatica software, and the Gram-Schmiedt algorithm, implemented in ENVI [21]. Using these algorithms it is possible to obtain high-resolution images (between 0.70 and 1 m) in the four bands acquired by QuickBird-2 and Ikonos-2 sensors (Figure 6; see also Figures 10–11 and 13).

The orderly overlaying of the single bands thus allows to visualize the image in true colours or in false colours, highlighting each time the different types of anomalies: for example, the use of red and near infra-red bands resulted being the most suitable to spot the anomalies and traces that are an outcome of vegetation growth. In fact, in the visible spectrum the red wavelength is very sensitive to the energy absorbed by chlorophyll during photosynthesis, while in the near infra-red the energy released by the internal structure of the leaves is highlighted. These data were also analyzed by dedicated software like PCI Geomatica 10.0 and ENVI 4.7 that implements algorithms for image enhancement like for example, the NDVI, useful for analyzing the growth rate, of vegetation in a defined area [22]; these are interpreted in order to acquire new data on the potential presence of buried structures that have affected vegetation growth.

The availability of modern satellite images with high geometric resolution represents a tool with enormous potential for the study of ancient urban and territorial contexts. These images provide plan views and, together with the photos of the 1960s spy satellites, make a complete multitemporal documentation of the archaeological areas. The manual vectorization of all archaeological remains and traces visible in the multi-temporal remote sensing data so allowed the creation of a new archaeological map (see Figure 2), ready for input in a GIS, with new data on the ancient topography of the site, monuments, and roads; in the map, the contour lines were extracted from the DEM based on SRTM and ASTER data. In addition, the three-dimensional models were used, together with QuickBird-2 panchromatic image of April 2004, to create a realistic 3D image of the ancient area, as it now appears (Figure 7).

5. Ancient Topography of the Site and Monitoring of the Archaeological Area

The processing and visual analysis of all acquired photos and satellite images available were very important for the study of Ur due to the inaccessibility of the archaeological area for scholars and the unavailability of vertical aerial photos and cartographies. This remote-sensed documentation allowed the acquisition of new data about the topographical layout of the city and its monuments, the ancient road network, and the paleoenvironmental context; these data are critical for the reconstruction of the historic landscape in which the settlement was located. In fact, a lot of traces and anomalies linked to archaeological and paleoenvironmental features were identified [1, 2]. In particular, the satellite images taken in winter and early spring, as most of those used in this study, are very useful for the visual photo-interpretation; in fact in these periods the more humid climate allows some types of traces (crop marks and damp marks) to be better visible, while in southern Iraq summer and early autumn are too dry and the archaeological traces are scarce clear. Furthermore, the draping of satellite images on DEMs (together with the only oblique image and in which the scarce air-photo available) highlighted the anomalies due to microrelief and allowed to connect the traces with the morphology of the site.

The integration of the documentation from old excavations and the remote-sensed data allowed the update of the archaeological map of Ur (see Figure 2), where all ancient remains and traces visible in the satellite images (orthorectified using the SRTM DEM and the Rational Polynomial Coefficients) were georeferenced. From the new digital archaeological map of Ur it is also possible to extract diachronic maps that document the historical development of the city during its main phases (Figure 8): (i) the Sumerian period, from the 1st to the 3rd Dynasty; (ii) the Isin and Larsa Dynasties; (iii) the Kassite period; (iv) the Neo-Babylonian age.

First of all, satellite images provide plan views of the entire archaeological area of Ur, very useful for a site where excavations were focused only on single monumental complexes or buildings and in which the studies had not interested the general layout of the settlement; they make available a complete image of the site, where different monuments and archaeological areas are together integrated and correctly localized. So, it is possible to understand the mutual relationships between various sectors of the urban area and different buildings.

In particular, it is possible to see the whole perimeter of the city with its circuit of walls and the position, plan, and reciprocal relationships of the buildings that were excavated between 1919 and 1934 (see Figure 3). In the highest part of the tell, the sacred area with the Ziggurat and the other main monuments are visible, while at the northern end and in the south-eastern part of the city it is possible to distinguish the depressions corresponding to the two harbours, today silted up. It is important also to note the presence of another depression stretching southwards from the northern harbour and passing to the east of the sacred area; it then curves towards the south-west, terminating

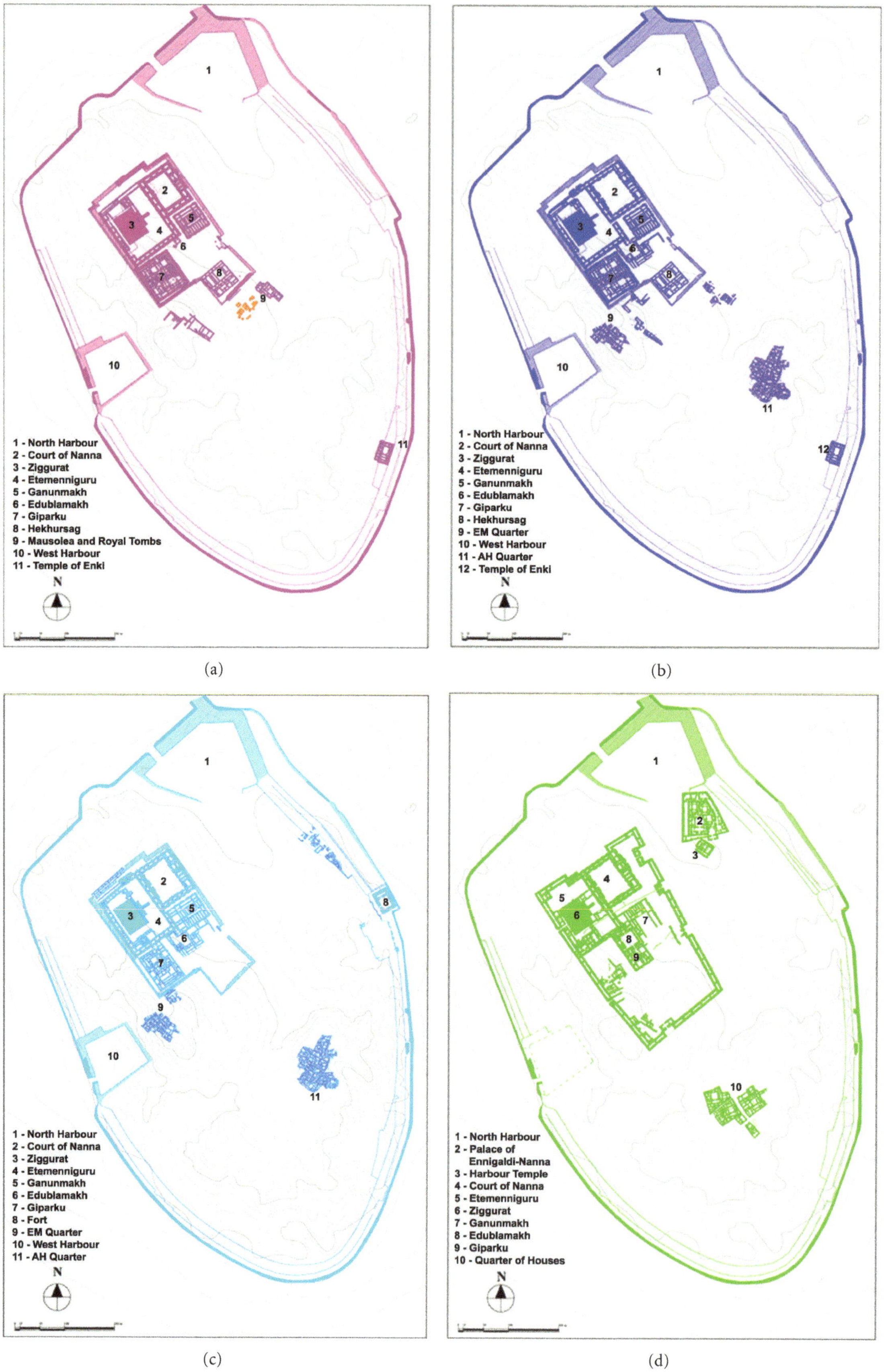

FIGURE 8: Maps of the main phases of Ur: (a) the Sumerian period, from the 1st to the 3rd Dynasty; (b) the Isin and Larsa Dynasties; (c) the Kassite period; (d) the Neo-Babylonian age.

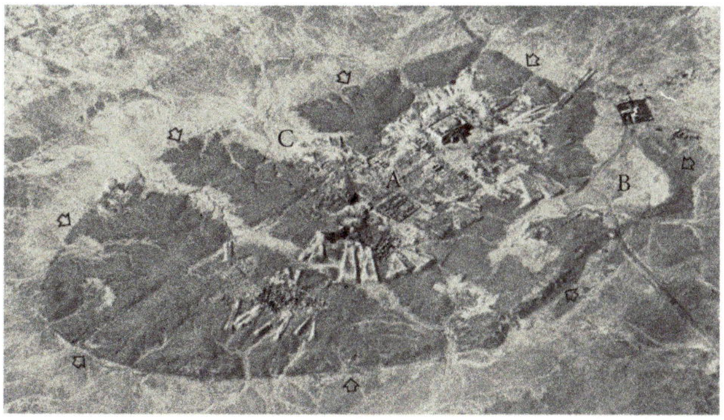

FIGURE 9: An oblique view of Tell al Muqayyar (Ur) in a Gambit KH-7 photograph acquired in February 1966: A: the sacred area on the top of the hill; B: the North Harbour; C: the West Harbour. The arrows indicate traces and remains of the city walls.

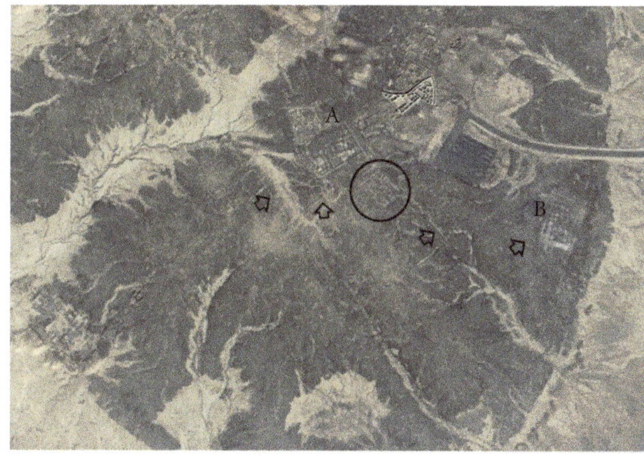

FIGURE 11: Traces of structures and streets between the Neo-Babylonian Houses (A) and the Enki Temple (B) in a pan-sharpened QuickBird-2 image of February 2002 (colour composite RGB bands 3-2-1): the arrows indicate buried streets and walls along their sides, while the circle highlights buried structures (maybe other Neo-Babylonian houses).

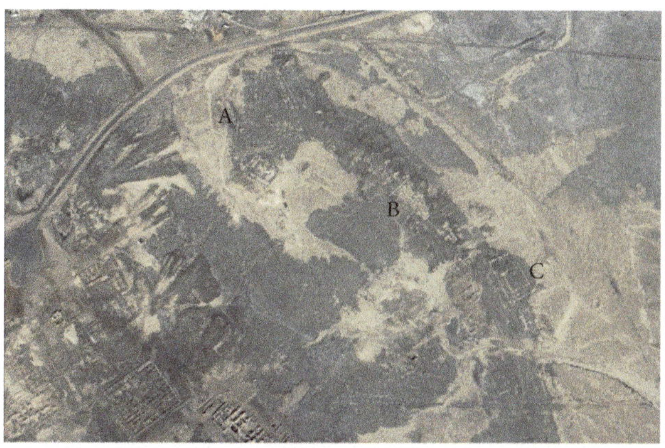

FIGURE 10: Traces and remains of the Palace of Ennigaldi-Nanna (A), of few houses built against the city walls (B), and of the Kassite Fort (C) in a pan-sharpened QuickBird-2 image of February 2002 (colour composite RGB bands 3-2-1).

to the south of the south-western harbour: this depression probably corresponds to the channel that crossed the city from the northern harbour and which was built or restored by Hammurabi. The remains and traces of the defensive city walls were also well visible in a Gambit KH-7 oblique photo of February 1966 (Figure 9) that shows very well the morphology of the site.

In the Northern part of the urban area, the satellite images show in detail the depression corresponding to the main harbour of Ur, which, besides being buried, is also partially covered by some modern constructions; particularly in the Gambit KH-7 vertical photo of April 1966, it is even possible to make out a few traces of the inner edge of the northern and south-western docks. Further south may be seen the remains and traces of the Palace of Ennigaldi-Nanna, today partly reburied after its excavation (Figure 10). The scant remains of the so-called Harbour Temple are visible immediately to the south. In the north-eastern sector of Ur, the satellite images show quite clearly the traces of the fortifications, only partially excavated. Also visible are the scant remains and traces of a few houses built against the walls and of the quadrangular Kassite Fort, which stood in this stretch of the fortifications.

In the south-eastern sector of the city, satellite images clearly show the path of the walls; the scarce traces of the small Temple of Enki and the remains of the so-called AH Quarter are visible nearby (Figure 11). The last residential complex dates back to the period of the Isin and Larsa dynasties; the view from above shows the plan of the restored area, with the houses built up against each other and separated by narrow winding streets. Traces of the structures, partly reburied after the excavations, of another residential district, dateable to the Neo-Babylonian period, are visible to the south-west; the houses were inserted in a regular urban layout, marked by broad straight roads. The satellite image makes it possible to identify even traces of buried or semiexposed structures, identifiable with other houses and parallel streets: so, it is possible to integrate the excavation areas, showing how the regular urban planning of the Neo-Babylonian period concerned a broad section of the south-eastern part of Ur.

In the central sector of the city, it is possible to read the entire plan of the monumental sacred area and verify the preservation of the buildings excavated in 1920s and 1930s. The Ziggurat dominates this area, also thanks to its impressive visual impact, which is made even more evident by the modern restoration; on the other hand, traces of the Etemenniguru are scarce. The remains of the Edublamakh, the Ganunmakh, the Giparku, and the Hekhursag are clearly visible, while only some structures pertaining to the so-called Court of Nanna are preserved. It is possible to observe the structures and the traces of two other excavated monuments, the Temple of Nimin-Tabba and the so-called EM District, a

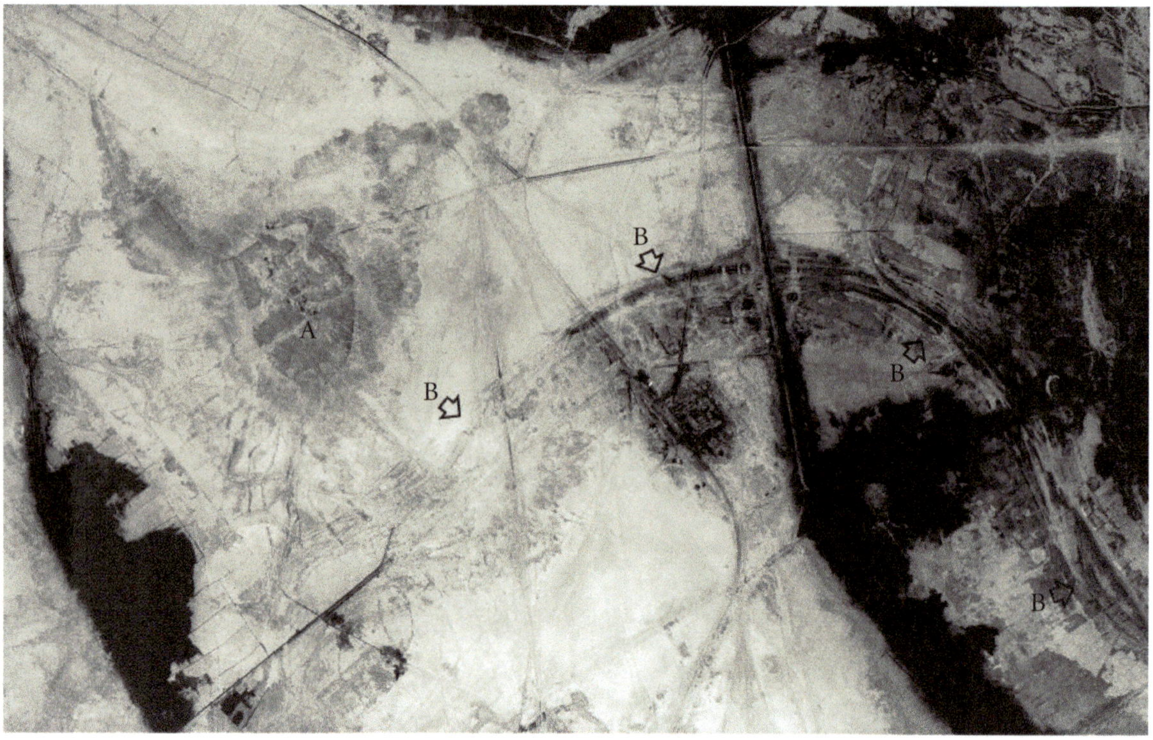

FIGURE 12: Ur (A) and the surrounding territory in a Corona KH-4B photo of May 1968: a lot of traces (B) of the paleoriver bed of the Euphrates are visible.

residential sector poorly conserved and dateable to the Isin-Larsa period. The broad pit resulting from the excavation of the so-called Royal Cemetery Clearly is visible to the south of the Hekhursag; the necropolis is characterized by the presence of numerous burials of high dignitaries and kings of Ur.

The remote-sensed documentation also provided important data concerning the reconstruction of the landscape surrounding Ur, where a lot of traces of old channels and paleoriver beds of the Euphrates were identified in areas much modified and altered during the last decades by urbanization and agricultural works. For this reason, the high-resolution USA reconnaissance photographs taken in 1966 by Gambit KH-7, which document the region of Nasiriyah, are very useful (see also Figure 4). In particular, they show the traces of a paleo-river bed of the Euphrates to the south-west, to the south, and to the south-east of the hill where the city was built; in fact, in ancient times the bed of the river ran near Ur, that, how we have seen, had two harbours and was an important trading centre. So, the Gambit image documents the paleohydrology of the area, in a territory that today is partially modified and altered by modern structures and above all by the military base and the airport built about 500 m south-west of the ancient city, exactly over the paleoriver bed, which was visible also in the Corona KH-4B image of 1968 (Figures 12 and 13). The oblique Gambit KH-7 photo, taken from the east in 1966, shows the modern route of the Euphrates, very far from the ancient city (Figure 14).

FIGURE 13: Ur (A) and the surrounding territory in a pan-sharpened Ikonos-2 image of 2008 (colour composite RGB bands 4-3-2): only some traces (B) of the paleo river bed of the Euphrates are visible.

FIGURE 14: Gambit KH-7 oblique photo of 1966: A: Ur; B: paleo river bed of the Euphrates; C: present river bed of the Euphrates; D: Nasiriyah.

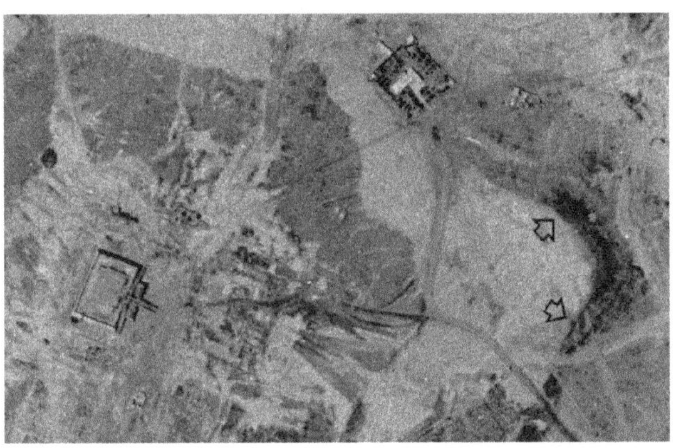

(a)

(b)

FIGURE 15: The area of the North Harbour in a Gambit KH-7 photo of 1966 (a) and in the WorldView-1 image of 2008 (b): the arrows show the traces and remains of the northern and north-eastern docks.

Lastly, the multitemporal satellite images allowed the monitoring of the conservation of archaeological area during the last four decades and particularly before and after second Gulf War. In fact, in time of war, satellite images provided a crucial and invaluable tool for monitoring archaeological sites as well as their preservation or fdamage. Moreover, for a documentation of the situation before the first Gulf War and to study the area in a period chronological nearer the times of the archaeological excavations, the photos taken from the United States reconnaissance satellites are very useful. For example, the comparison between the vertical Gambit KH-7 image of 1966 and the WorldView-1 image of 2008 shows that during the 1960s some buildings and structures excavated between 1919 and 1934 were better preserved than today (see Figure 3): the Palace of Ennigaldi-Nanna and the nearby residence of the high priestess of Nanna/Sin, the quadrangular Kassite fortress, the so-called AH Quarter and the North Harbour (Figure 15).

6. Conclusions

Through this experience, we could appreciate, one more time, the usefulness of historical and recent satellite images for remote analysis of archaeological sites. The potential of such data in research studies becomes even more evident when one considers that the revisit time of the last built satellites has further lowered: this feature allows not only to acquire information programming also the time of taking the image (useful for example to analyze the shadows and calculate the height above ground of archaeological evidence) but also to implement systems for remote monitoring of archaeological sites. The completeness of the data set available for the study of Ur also made it possible not only to create an archaeological map in which all the evidence brought to light by the excavations and the visible traces are displayed in their correct relation to each other but also to contextualize the surrounding area, also reconstructed by using old satellite images. The analysis of historical and recent images, together with the availability of altimetric data acquired by SRTM mission and ASTER images, has helped to achieve a much more accurate and precise classification of the entire archaeological area, as well as to simplify the presentation of data, which are now easily understandable to nonspecialists, who may visit the site of the Virtual Museum of Iraq (http://www.virtualmuseumiraq.cnr.it/), while for a viewing of the maps created ad hoc for three other ancient centers of Iraq (Hatra, Khorsabad, and Seleucia), they can see the "Ancient cities of Mesopotamia" webGIS [7].

Acknowledgments

The "Virtual Museum of Iraq" Project was coordinated by Dr. Massimo Cultraro of the CNR-IBAM. The work concerning the satellite remote-sensed data was carried out by Drs. Laura Castrianni, Imma Ditaranto, Fabrizio Ghio, Giuseppe Pellino, Ilaria Miccoli, and Veronica Randino in the Ancient Topography, Archaeology and Remote Sensing Laboratory of the CNR-IBAM. The work concerning the 3D reconstruction was directed by Dr. Francesco Gabellone and carried out by Ivan Ferrari, Francesco Giuri, and Massimo Limoncelli in the Information Technologies Laboratory of the CNR-IBAM. The present work was made in strict cooperation between the authors; Sections 1, 2, and 5 are written by Giuseppe Scardozzi, while Giacomo Di Giacomo is the author of Sections 3, 4, and 6.

References

[1] F. Gabellone and G. Scardozzi, "Integrated technologies for the reconstructive study of Mesopotamian cultural heritage: the cases of Ur, Uruk and Nimrud," in *Proceedings of the International Congress on Cultural Heritage and New Technologies, Workshop 11"Archäologie und Computer"*, pp. 1–46, Vienna, Austria, October 2006.

[2] F. Gabellone and G. Scardozzi, "From the object to the territory: image-based technologies and remote sensing for the reconstruction of ancient contexts," *Archeologia e Calcolatori*, supplement 1, pp. 123–142, 2007.

[3] M. Cultraro, F. Gabellone, and G. Scardozzi, "Integrated methodologies and technologies for the reconstructive study of Dur-Sharrukin (Iraq)," in *Proceedings of the 21st International Symposium on AntiCIPAting the Future of the Cultural Past (CIPA '07)*, vol. 1, pp. 253–258, Athens, Greece, 2007.

[4] M. Cultraro, F. Gabellone, and G. Scardozzi, "From remote sensing to 3D modelling and virtual reconstructions of the Iraqi archaeological sites: the case of Hatra," in *Proceedings of the 1st International Workshop on Remote Sensing for Archaeology and Cultural Heritage Management*, R. Lasaponara and N. Masini, Eds., pp. 239–242, Rome, Italy, 2008.

[5] M. Cultraro, F. Gabellone, and G. Scardozzi, "The virtual musealization of archaeological sites: between documentation and communication," in *Proceedings of the Workshop on 3D Virtual Reconstruction and Visualization of Complex Architectures*, pp. 1–6, Trento, Italy, February 2009.

[6] M. Cultraro, F. Gabellone, and G. Scardozzi, "The virtual musealization of archaeological sites: between documentation and communication," in *Proceedings of 14th International Congress on Cultural Heritage and New Technologies*, pp. 294–308, Vienna, Austria, November 2009.

[7] G. Di Giacomo and G. Scardozzi, "Un webGIS per la conoscenza delle antiche città della Mesopotamia," in *Proceedings of the 2nd Congreso Internacional de Arqueología e Informática Gráfica, Patrimonio e Innovación*, pp. 135–139, Seville, Spain, 2010.

[8] G. Scardozzi, "Multitemporal satellite high resolution images for the knowledge and the monitoring of the Iraqi archaeological sites: the case of Seleucia on the Tigris," *European Journal of Applied Remote Sensing*, vol. 45, no. 4, pp. 143–160, 2009.

[9] G. Scardozzi, "Multitemporal satellite images for knowledge of the assyrian capital cities and for monitoring landscape transformations in the upper course of tigris river," *International Journal of Geophysics*, vol. 2011, Article ID 917306, 17 pages, 2011.

[10] G. Scardozzi, "The contribution of high resolution satellite images to the production of base-maps and cartographies for archaeological research in Turkey and Iraq," in *Remote Sensing for Environmental Monitoring, GIS Applications, and Geology IX*, vol. 7478 of *Proceedings of SPIE*, pp. 1–12, Berlin, Germany, September 2009.

[11] G. Scardozzi, "Old high resolution satellite images for landscape archaeology: case studies from Turkey and Iraq," in *Remote Sensing for Environmental Monitoring, GIS Applications, and Geology*, vol. 7110/03 of *Proceedings of SPIE*, pp. 1–14, Cardiff, UK, September 2008.

[12] G. Bitelli and V. A. Girelli, "Metrical use of declassified satellite imagery for an area of archaeological interest in Turkey," *Journal of Cultural Heritage*, vol. 10, supplement 1, pp. e35–e40, 2009.

[13] A. Beck, G. Philip, M. Abdulkarim, and D. Donoghue, "Evaluation of Corona and Ikonos high resolution satellite imagery for archaeological prospection in western Syria," *Antiquity*, vol. 81, no. 311, pp. 161–175, 2007.

[14] J. Casana and J. Cothren, "Stereo analysis, DEM extraction and orthorectification of CORONA satellite imagery: archaeological applications from the Near East," *Antiquity*, vol. 82, no. 317, pp. 732–749, 2008.

[15] N. Galiatsatos, D. N. M. Donoghue, and G. Philip, "High resolution elevation data derived from stereoscopic CORONA imagery with minimal ground control: an approach using Ikonos and SRTM data," *Photogrammetric Engineering and Remote Sensing*, vol. 74, no. 9, pp. 1093–1106, 2008.

[16] M. Jahjah, C. Ulivieri, A. Invernizzi, and R. Parapetti, "Archaeological remote sensing application pre-post war situation of Babylon archaeological site-Iraq," *Acta Astronautica*, vol. 61, no. 1–6, pp. 121–130, 2007.

[17] R. Goossens, A. De Wulf, J. Bourgeois, W. Gheyle, and T. Willems, "Satellite imagery and archaeology: the example of CORONA in the Altai Mountains," *Journal of Archaeological Science*, vol. 33, no. 6, pp. 745–755, 2006.

[18] J. A. Ur, "Sennacherib's northern Assyrian canals: new insights from satellite imagery and aerial photography," *Iraq*, vol. 67, pp. 317–345, 2005.

[19] T. Toutin, "Three-dimensional topographic mapping with ASTER stereo data in rugged topography," *IEEE Transactions on Geoscience and Remote Sensing*, vol. 40, no. 10, pp. 2241–2247, 2002.

[20] T. Toutin, "ASTER DEMs for geomatic and geoscientific applications: a review," *International Journal of Remote Sensing*, vol. 29, no. 7, pp. 1855–1875, 2008.

[21] R. Lasaponara and N. Masini, "Pan-sharpening techniques to enhance archaeological marks: an overview," in *Satellite Remote Sensing. A New Tool for Archaeology*, R. Lasaponara and N. Masini, Eds., pp. 87–109, Springer, 2012.

[22] R. Lasaponara, N. Masini, and ", "Image enhancement, feature extraction and geospatyal analysis in an archaeological perspective," in *Satellite Remote Sensing. A New Tool For Archaeology*, R. Lasaponara and N. Masini, Eds., pp. 17–63, Springer, 2012.

22

An Experimental and Modeling Study on the Response to Varying Pore Pressure and Reservoir Fluids in the Morrow A Sandstone

Aaron V. Wandler, Thomas L. Davis, and Paritosh K. Singh

Department of Geophysics, Colorado School of Mines, 1500 Illinois Street, Golden, CO 80401, USA

Correspondence should be addressed to Aaron V. Wandler, awandler@mymail.mines.edu

Academic Editor: Marek Grad

In mature oil fields undergoing enhanced oil recovery methods, such as CO_2 injection, monitoring the reservoir changes becomes important. To understand how reservoir changes influence compressional wave (P) and shear wave (S) velocities, we conducted laboratory core experiments on five core samples taken from the Morrow A sandstone at Postle Field, Oklahoma. The laboratory experiments measured P- and S-wave velocities as a function of confining pressure, pore pressure, and fluid type (which included CO_2 in the gas and supercritical phase). P-wave velocity shows a response that is sensitive to both pore pressure and fluid saturation. However, S-wave velocity is primarily sensitive to changes in pore pressure. We use the fluid and pore pressure response measured from the core samples to modify velocity well logs through a log facies model correlation. The modified well logs simulate the brine- and CO_2-saturated cases at minimum and maximum reservoir pressure and are inputs for full waveform seismic modeling. Modeling shows how P- and S-waves have a different time-lapse amplitude response with offset. The results from the laboratory experiments and modeling show the advantages of combining P- and S-wave attributes in recognizing the mechanism responsible for time-lapse changes due to CO_2 injection.

1. Introduction

Currently, there are 128 enhanced oil recovery (EOR) projects worldwide using CO_2 injection as a tertiary recovery method, with 113 taking place in the United States [1]. Often the criterion for successful seismic monitoring of CO_2 projects is the ability to collect high-resolution seismic data, since seismic methods have shown to be very effective in detecting time-lapse changes [2–6]. The effectiveness of seismic monitoring depends upon the changes in acoustic and elastic impedance caused by injection and production. These changes are then observed in the amplitude and travel time of seismic waves.

Time-lapse seismic methods utilizing compressional waves (P-wave) and shear waves (S-wave) can ideally improve the identification of flooded or pressurized compartments within the reservoir during a CO_2 EOR operation. The use of P-wave data is not a definitive method for pressure detection, but the addition of S-wave data may provide a distinction between a gas-bearing zone and a zone of high pore pressure [7], this is a motivation for using S-waves as a time-lapse tool. An example provided by Xue et al. [8] explains if an observed time-lapse P-wave anomaly is caused by a decrease in velocity without an associated S-wave anomaly, the zone has most likely been flushed with CO_2 without any changes in pore pressure. However, if the S-wave data show a similar anomaly, due to a reduction in velocity, the zone may have undergone a pore pressure buildup in addition to being flushed with CO_2. Identifying the impact of changing stresses within the reservoir is as important as recognizing the effects of saturation.

Saturation effects are typically estimated from fluid substitution modeling (e.g., Gassmann [9] and Kuster and Toksoz [10]). Since fluid substitution methods do not directly account for changing rock properties with respect to varying pore pressure, additional information must be used to account for the effects of pore pressure. This additional information may be obtained from, for example, reservoir flow simulation or experimental core studies. The purpose of this paper is to show how the results from rock physics experiments that include saturation and pore pressure variation can be used to estimate amplitude changes in a reservoir undergoing CO_2 injection.

The contribution of this study applies directly to a major ongoing research project at Colorado School of Mines. The Reservoir Characterization Project (RCP) is a consortium at Colorado School of Mines whose research goal is to monitor CO_2 EOR projects to effectively manage and enhance the recovery process. The RCP has used P-, S-, and converted-wave (PS) time-lapse seismic acquisition to monitor CO_2 floods for EOR. The RCP conducted the first ever time-lapse, multicomponent seismic survey in conjunction with a CO_2 huff-n-puff project at Vacuum Field, New Mexico. Multicomponent time-lapse seismic data was used to monitor the EOR project at Weyburn Field, Saskatchewan, Canada. The current research project is a time-lapse study of an incised valley fill system undergoing a water alternating gas (i.e., CO_2) injection scheme for EOR at Postle Field, Texas County, Oklahoma, located in the Anadarko Basin. The results from the experiment and analysis reflect the use of reservoir core samples and actual reservoir fluids. The aim of this work is to provide insight related directly to this specific type of environment.

1.1. Previous Work. Studies on the effects of saturation, pressure, and pore fluid have been documented as early as 1966 [12]. King [12] reported on the change in velocities of five sandstone samples saturated with air, an NaCl solution, and kerosene as a function of hydrostatic confining pressure and internal pore pressure. Furthermore, Domenico [13] showed that S-wave velocity is more sensitive to increasing differential pressure than P-wave velocity in three sandstone cores of low, intermediate, and high porosity.

A series of papers have been published on the results of the P-wave velocity response to the migration of CO_2 in water-saturated sandstones using an experimental setup that simulates cross-well seismic profiling. Xue et al. [8] mapped the velocity changes caused by CO_2 injection in water-saturated Tako and Shirahama sandstones at constant pore pressure. In a similar experiment, Xue and Ohsumi [14] mapped the velocity changes caused by the injection of gas, liquid, and supercritical phase CO_2 into a water-saturated Tako sandstone. They investigated the changes in velocity with increasing pore pressure and concluded that CO_2 injection causes larger velocity changes than the change associated with increasing pore pressure in their core samples. In addition to the P-wave velocity, strain was measured as a function of hydrostatic pressure in a dry sample. Their observations showed how strain normal and parallel to the bedding plane are not equal and more deformation occurred normal to the bedding plane. To improve upon their cross-well seismic profiling experiment and the accuracy of the P-wave velocity changes relative to the CO_2 injection, Xue and Lei [15] applied a difference tomography method to the measured slowness. This technique correlated areas of higher porosity with greater velocity reduction. Velocity reduction when CO_2 was in liquid or supercritical phase was twice as much as when in the gas phase. From the same dataset as Xue and Lei [15], Shi et al. [16] focused on P-wave velocity reductions due to supercritical CO_2 displacement of pore water, where the velocity reductions deviated significantly from the Gassmann-predicted velocities for both patchy and uniform saturations. Lei and Xue [17] expanded on the results of Xue and Lei [15] and Shi et al. [16] and showed that P-wave velocity decreased, and the attenuation coefficient increased, due to gaseous, liquid, and supercritical CO_2 partially replacing pore water. Kim et al. [18] also investigated the use of P-wave velocity and resistivity as a way to monitor saturation of a sandstone during a CO_2 injection process. This was accomplished by injecting supercritical CO_2 into a water-saturated Berea sandstone sample at simulated reservoir conditions.

The effects of saturation, pore pressure, and effective stress, studied on sandstone core samples from the Northwest Shelf of Australia, show that the V_P/V_S ratio is an indicator for the saturation state between oil-saturated and the dry state [19]. Siggins [20] also studied the response of CO_2-saturated sandstones at in-situ reservoir conditions for two synthetic sandstones and one reservoir sandstone/mudstone from the Otway Basin, C seam, Waarre Formation. His findings were both P-wave and S-wave velocities decreased when the samples were saturated with liquid CO_2, compared to dry rock. Also, the synthetic sandstones showed better agreement with Gassman's fluid substitution theory than did the reservoir sandstone. Similarly, Siggins et al. [21] compared synthetic sandstones and Waarre C formation reservoir sandstones when dry and saturated with gaseous and liquid-phase CO_2. Larger changes were observed in both P-wave and S-wave velocity when CO_2 was in the liquid phase.

The effects of saturation and pore pressure have also been studied with respect to the effective pressure law and the effective pressure coefficient [22, 23] and how fractures change the constituency of the rock and the effective pressure coefficient [24]. The work by Xu et al. [24] on effective pressure shows the importance of why assuming the effective pressure coefficient to be unity could lead to significant errors when predicting time-lapse changes.

The RCP has used experimental rock physics to understand how the changes in confining pressure, pore pressure, and fluid saturation affect P-wave and S-wave velocities. This knowledge has then been applied to the interpretation of multicomponent time-lapse surface seismic monitoring. Capello [25] used rock physics as a feasibility study for seismic monitoring of a CO_2 huff-n-puff flood in the Permian San Andres Formation, a dolomitized carbonate deposit at Vacuum Field. Also using samples from the San Andres Formation, Duranti [26] investigated the coexistence of equant pores and fractures and found that a dual porosity rock physics model was more successful in predicting the changes in differential pressure and saturation observed in the experimental data.

Brown [27] integrated experimental rock physics measurements from the carbonate Marly beds of the Mississippian Charles Formation to improve the interpretation of time-lapse seismic data for the monitoring of a CO_2 flood at Weyburn Field. Yamamoto [28] included the rock physics results obtained by Brown [27] to improve the flow modeling

TABLE 1: Depth, density, porosity, and permeability of core samples from three different lithological zones.

Sample	Depth (m)	Density (gm/cm^3)	Porosity (%)	Permeability (mD) Preexperiment	Permeability (mD) Postexperiment
Cemented	1875	2.394	11.54	10.6	6.67
High permeability no. 1	1878	2.105	19.96	270	192
High permeability no. 2	1879	2.067	22.64	123	105
Lower permeability no. 1	1888	2.179	18.73	11.0	11.0
Lower permeability no. 2	1888	2.104	20.84	68.8	26.8

of CO_2 by minimizing the difference between calculated and observed acoustic impedance while simultaneously matching the production history.

Lastly, Rojas [29] measured dry rock velocities for the purpose of analyzing the different responses from Gassmann fluid substitution. This study on tight gas sands linked V_P/V_S sensitivity to pore pressure changes, lithology, and fluid content to aid in identifying zones with a higher reservoir potential of the Late Cretaceous Williams Fork Formation at Rulison Field, CO.

Perhaps the earliest and most cited laboratory experiment related to monitoring a CO_2 EOR operation is that by Wang and Nur [30]. They measured ultrasonic P-wave and S-wave velocities on seven sandstones and one unconsolidated sand saturated with n-hexadecane and then flushed with CO_2. Their measurements were done at constant confining pressure over a range of pore pressures and temperatures above and below the critical point of CO_2. Their findings show a significant decrease in P-wave velocity which was attributed to CO_2 saturation. The S-wave velocity exhibited a greater effect to pore pressure than saturation and was shown to be less sensitive to saturation.

Outside of research done on the Tako, Shirahama, Waarre C Formation, Williams Fork Formation sandstones and the carbonates of San Andres and Charles Formation, most of the experimental studies are not done on reservoir samples obtained from well core. This is evident in the statement made by Siggins et al. [21], "literature on laboratory measurements of the effects of CO_2 on the seismic properties of reservoir rocks is relatively sparse."

2. Postle Field Overview

The geologic environment of the producing reservoir at Postle Field is the Upper Morrow sandstone Formation, which is a Pennsylvanian-age siliciclastic incised valley fill system. Typical Morrow sediments consist of shales punctuated by valley fill sand deposits [31]. The Morrow Formation is encased between two high-stand limestones [32]. The upper boundary is a conformable surface at the base of the Atoka Thirteen Fingers Limestone. The lower boundary of the Morrow Formation is an unconformable surface above the Mississippian Chester Limestone. The deposition of Postle Field during Pennsylvanian time was a relative sea-level low-stand shoreline where the Morrow Formation is much sandier [33]. The primary producing unit in the Morrow Formation is the Upper Morrow A, A1, and A2 sands. In Postle Field, the producing formation that is undergoing the CO_2 flood is the Morrow A sandstone.

Postle Field is a large mature oil field undergoing a water alternating gas (WAG) flood for enhanced oil recovery of the Morrow A sandstone. The volume of water to CO_2 injected into the reservoir is referred to as the WAG ratio and is expressed in reservoir volumes at reservoir conditions. The average WAG ratio implemented at Postle Field is 0.35 to 1. The reservoir pressure is maintained above the minimum miscibility pressure of 14.48 MPa by injecting water and CO_2 at 29.65 MPa and by keeping the bottom hole pressure of a producing well higher than 6.89 MPa [34].

3. Sample Selection

We retrieved five core plug samples from three different lithological zones within the Morrow A sandstone interval. The three zones were a low-porosity and low-permeability cemented zone, a higher-porosity and high-permeability zone, and a higher-porosity and lower-permeability zone. Each sample was cut parallel to the bedding plane. The cylindrical dimensions of the samples were roughly 3.81 cm in length and 2.54 cm in diameter. The depths from where the core samples were retrieved, as well as density, porosity, and permeability are listed in Table 1.

3.1. Mineralogy. We used QEMSCAN (quantitative evaluation of minerals by scanning electron microscopy) to quantitatively evaluate mineral content on each of the three lithological zones. The cemented zone (Figure 1) consists of subrounded quartz grains cemented by ankerite and kaolinite clay. This cemented zone shows a lower amount of grain-to-grain contact which suggests the possibility of early cementation. The lower amount of grain-to-grain contact is likely related to the presence of ankerite which may indicate that the pore water was rich in iron. In addition, there are occurrences of angular overgrowths over subrounded quartz grains. Even though the QEMSCAN measured 14% porosity, most of the porosity for this cemented zone is interstitial between the fine-grained kaolinite clay.

The higher-porosity and high-permeability zone (Figure 2) consists mostly of subrounded quartz grains with lesser amounts of kaolinite clay cement, where most of the porosity (25%) occurs between quartz grains. Unlike the other zones, only a trace amount of Ankerite occurs as cement.

The higher-porosity and lower-permeability zone (Figure 3) consists mostly of subrounded quartz grains with lesser

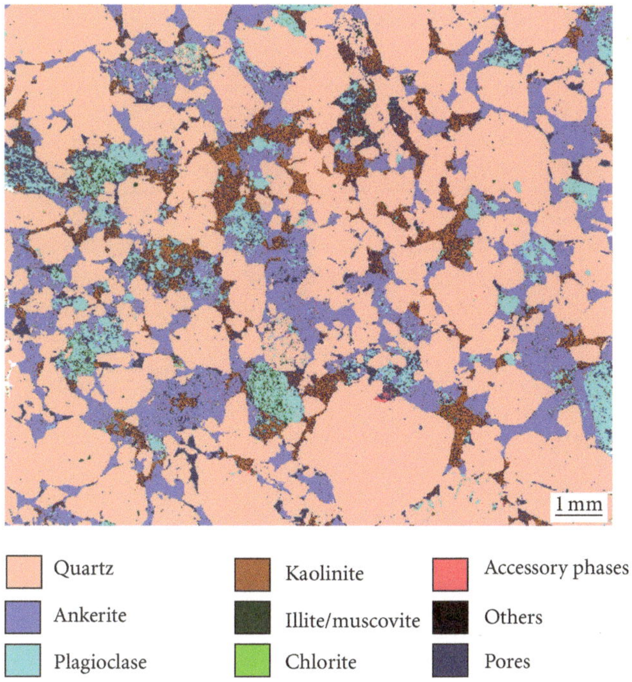

FIGURE 1: QEMSCAN image from the cemented zone.

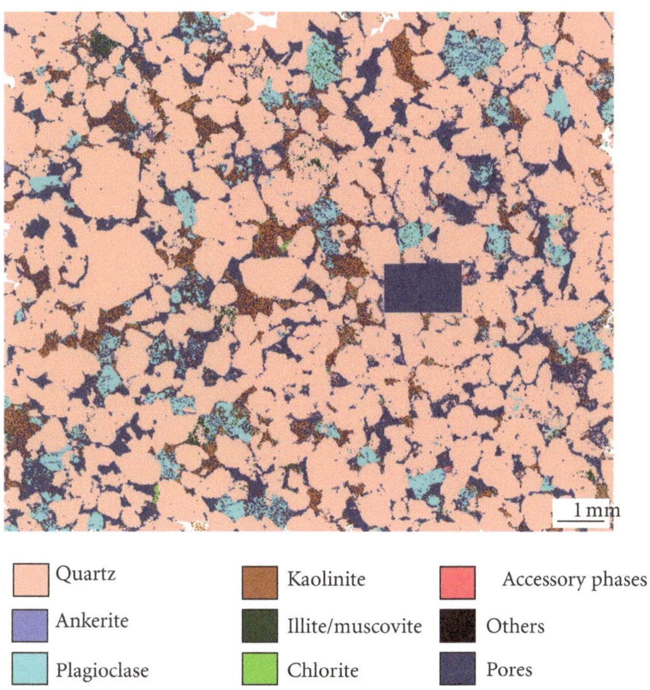

FIGURE 2: QEMSCAN image from the high-permeability zone. (The solid blue rectangle is a file conversion artifact.)

amounts of ankerite and kaolinite clay cement. This zone is more similar to the cemented zone with respect to the presence of ankerite and interstitial porosity. Much like the cemented zone, most of the porosity (20%) is interstitial between the fine-grained kaolinite and chlorite.

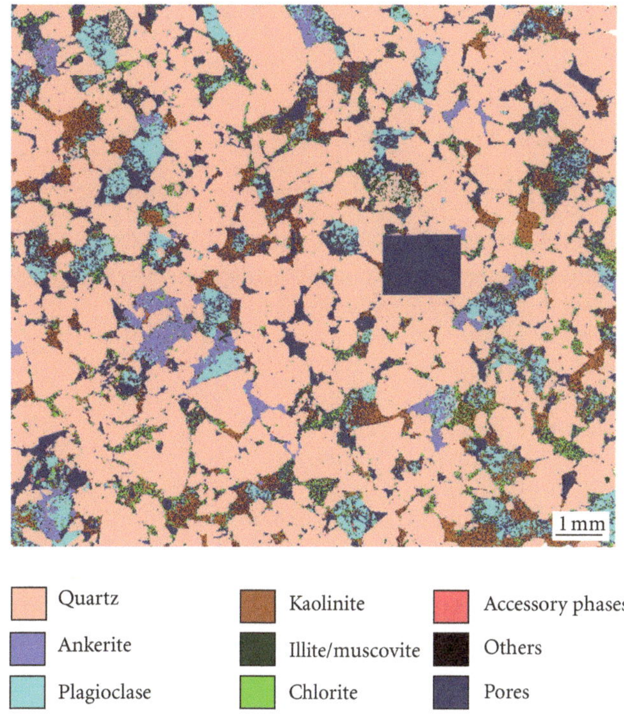

FIGURE 3: QEMSCAN image from the lower-permeability zone. (The solid blue rectangle is a file conversion artifact.)

Table 2 shows the mineral volume percentage related to each of the zones as described above.

4. Methodology

4.1. Experimental Description. For the laboratory core experiments, we used the direct pulse transmission technique to record compressional and shear-wave velocities at ultrasonic frequencies (≈1 MHz). The experimental setup included a pressure vessel to encase and apply an isostatic confining pressure to the core sample; heat tapes wrapped around the pressure vessel to control the temperature; independent pumps for regulating confining pressure and pore pressure; a pulse generator; a digital oscilloscope for recording wave forms. A transducer was placed at each flat end of the core plug sample and was capable of recording a compressional wave and two shear waves in orthogonal directions. The two shear waves were oriented along the fast- and slow-velocity directions relative to the core sample at bench-top conditions. Each transducer casing had a fluid line for injecting the different fluids through the sample. In addition to recording the different wave modes, average axial strain was measured along the major axis of the sample. Uncertainties from the experimental setup correspond to a velocity error of 0.3% to 0.4%. However, the actual error is higher due to hand-picking first arrivals.

4.2. Testing Sequence. We started the testing sequence with dry rock measurements where we incrementally increased

Table 2: QEMSCAN mineral volume (%) of the three zones.

Zone	Quartz	Ankerite	Plagioclase	Kaolinite	Illite	Chlorite	Accessory Phases	Others
Cemented	66	19	6	7	trace	trace	trace	1
High permeability	84	trace	6	8	1	trace	trace	trace
Lower permeability	75	4	8	8	1	3	trace	trace

the confining pressure from 3.45 MPa to 41.37 MPa at the estimated reservoir temperature of 63.9°C.

Following the dry rock measurements, we vacuum saturated the core sample with brine then recorded measurements at different pore pressures. The fluids flushed through the samples following brine were live oil, live oil with a 0.334 mol fraction of CO_2, and pure CO_2. The pore pressure range for the brine-saturated samples was 29.65 MPa to 3.45 MPa, to simulate the stress path from an injection well to a production well. For the other fluids, live oil, live oil with a 0.334 mol fraction of CO_2, and pure CO_2, the pore pressure range was 29.65 MPa to 6.89 MPa. For these fluids, we did not record data below 6.89 MPa since these pressures would be below the bubble point. For the fluid-saturated and fluid-flushed cases, a constant confining pressure of 44.82 MPa is maintained to stimulate the stress of the overburden, and all the measurements were recorded at 63.9°C.

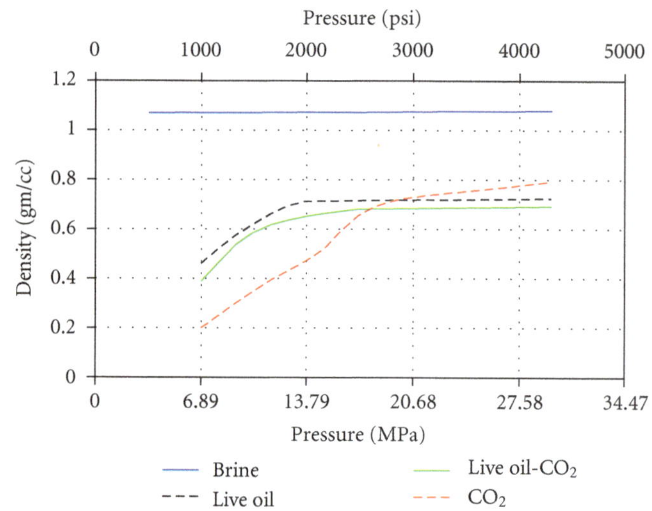

Figure 4: Fluid density as a function of pressure, modeled using the FLAG calculator [11].

4.3. Fluid Properties. The fluids used to saturate and flush through the core samples were reproduced to simulate the actual reservoir fluids at Postle Field. These fluids consist of brine with an NaCl concentration of 142,000 ppm, live oil with a gas oil ratio (GOR) of 103 L/L, live oil with a 0.334 mole fraction of CO_2, and pure CO_2. The live oil has an API gravity of 38.6 and a gas gravity of 0.786. This information was obtained from a reservoir fluid study provided by the current field operator. Using the Fluid Acoustics for Geophysics (FLAG) calculator [11], the live oil bubble point is estimated to be 13.33 MPa at 63.9°C. The mixture consisting of live oil and a mole fraction of 0.334 CO_2 has an estimated bubble point of 17.14 MPa at 63.9°C. The GOR and volume of gas for the live oil with a 0.334 mole fraction of CO_2 was calculated assuming a gas gravity of 1.53 and a molecular weight of 44.01 gm/mole for CO_2. This approach was used by Wang et al. [35] based on experimental measurements of Postle Field oil and mixtures of approximately 10 mole% and 50 mole% CO_2. Figures 4 and 5 show the properties of the fluids over the pore pressure range used in the experiment calculated with the FLAG calculator [11]. The phase of CO_2 for these pressures is supercritical fluid with the exception of the lowest pressure, 6.89 MPa, where CO_2 is in the gas phase.

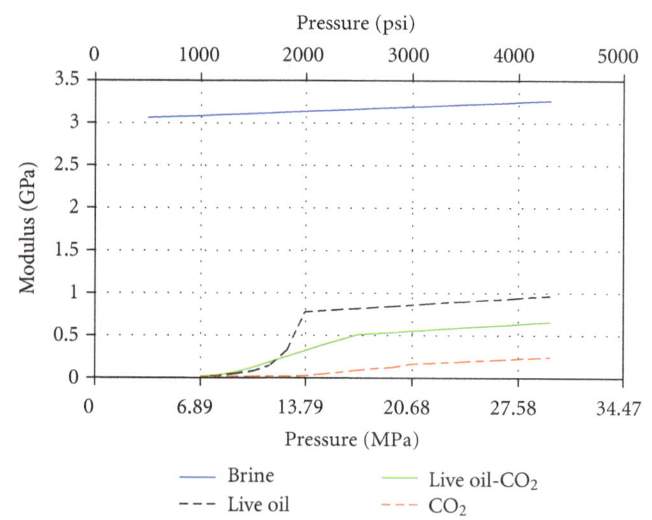

Figure 5: Bulk modulus of fluids as a function of pressure, modeled using the FLAG calculator [11].

5. Results and Discussion

First, we measured compressional wave (P-wave) and shear wave (S-wave) velocities in air-saturated samples (dry rock) while loading and unloading an isostatic confining pressure. Following the dry rock testing, we measured P- and S-wave velocities, as a function of pore pressure, on brine-saturated core samples which were then flooded with live oil, live oil with a 0.334 mol fraction of CO_2, and pure CO_2, with a constant confining pressure of 44.82 MPa.

5.1. Core-Measured Velocities. The measured velocities, P- and S-wave, are shown in Figures 6–10 for the dry rock and fluid-saturated core samples. The S-wave velocity shown in these figures is the average of the two orthogonally measured

Table 3: Percentage change of P-wave velocity with increasing differential pressure from 15.17 MPa to 37.92 MPa.

Sample	Dry Rock Loading (%)	Dry Rock Unloading (%)	Brine Sat. (%)	Oil Flushed (%)	Oil-CO_2 Flushed (%)	CO_2 Flushed (%)
Cemented	6.5	−3.6	0.94	2.4	—	3.2
High permeability no. 1	5.4	−6.1	2.5	—	−0.24	1.4
High permeability no. 2	8.0	−5.1	2.4	1.8	4.6	5.9
Lower permeability no. 1	4.0	−3.3	2.8	1.9	3.0	4.6
Lower permeability no. 2	5.4	−4.1	2.9	2.0	—	3.4

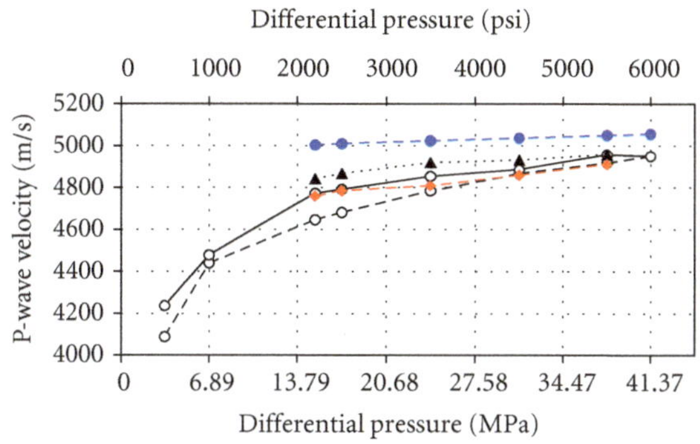

(a) P-wave velocities for the cemented sample versus differential pressure

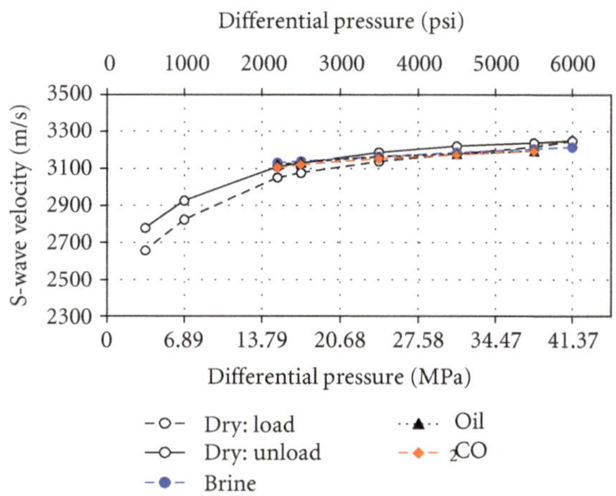

(b) S-wave velocities for the cemented sample versus differential pressure

Figure 6: P- and S-wave velocities for the cemented sample versus differential pressure for dry rock, brine saturated, and when flushed with live oil and CO_2.

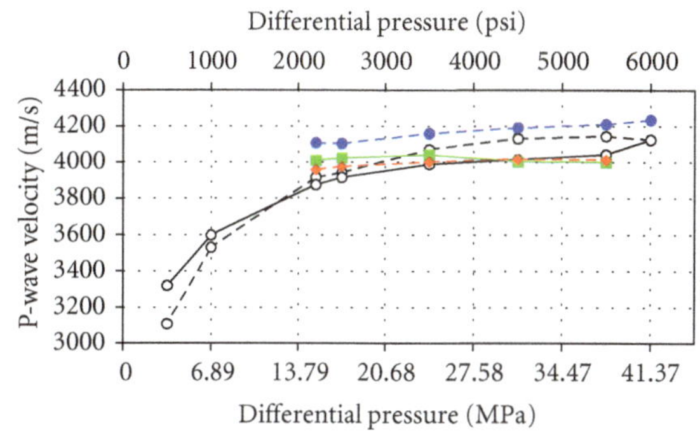

(a) P-wave velocities for the high-permeability number 1 sample versus differential pressure

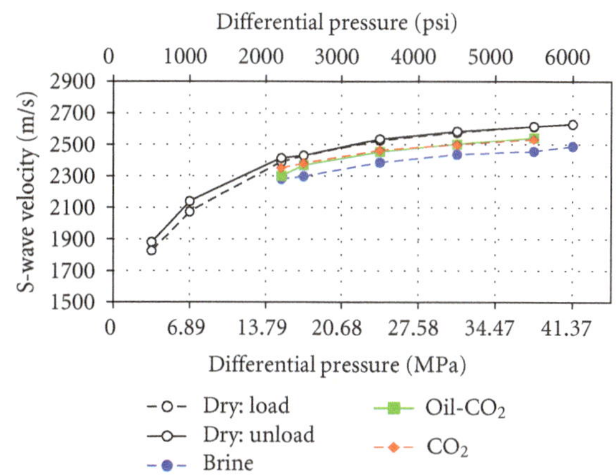

(b) S-wave velocities for the high-permeability number 1 sample versus differential pressure

Figure 7: P- and S-wave velocities for the high-permeability number 1 sample versus differential pressure for dry rock, brine saturated, and when flushed with the oil-CO_2 mixture, and CO_2.

S-waves. Differential pressure, P_d, is defined as confining pressure, P_c, minus pore pressure, P_p: (1);

$$P_d = P_c - P_p. \quad (1)$$

In general, the high-permeability samples show a greater sensitivity to both confining pressure and pore pressure. Both P-wave and S-wave velocities decrease with decreasing differential pressure (i.e., increasing pore pressure). The P-wave velocities show a larger decrease in velocities due to pressure when flooded with CO_2. Whereas pore fluid has little bearing on S-wave velocities. Tables 3 and 4 show the percentage change in velocity for dry rock and fluid-saturated core samples over the differential pressure range of 15.17 MPa to 37.92 MPa.

In terms of different lithological characteristics of the core samples with respect to velocity, the higher the permeability of the sample, the lower the P-wave and S-wave velocities. This relationship cannot be extended to porosity since the porosities of the lower-permeability samples were very to the high-permeability samples.

Table 4: Percentage change of S-wave velocity with increasing differential pressure from 15.17 MPa to 37.92 MPa.

Sample	Dry Rock Loading (%)	Dry Rock Unloading (%)	Brine Sat. (%)	Oil Flushed (%)	Oil-CO_2 Flushed (%)	CO_2 Flushed (%)
Cemented	6.5	−4.3	2.3	2.3	—	2.9
High permeability no. 1	10.2	−8.3	7.7	—	10.5	7.7
High permeability no. 2	8.3	−6.4	8.7	9.7	10.3	9.1
Lower permeability no. 1	6.4	−5.2	5.0	5.9	5.3	5.7
Lower permeability no. 2	6.9	−5.5	3.4	3.9	—	4.4

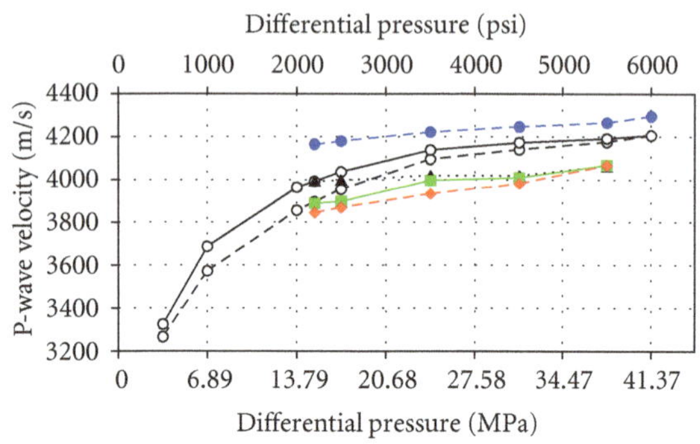

(a) P-wave velocities for the high-permeability number 2 sample versus differential pressure

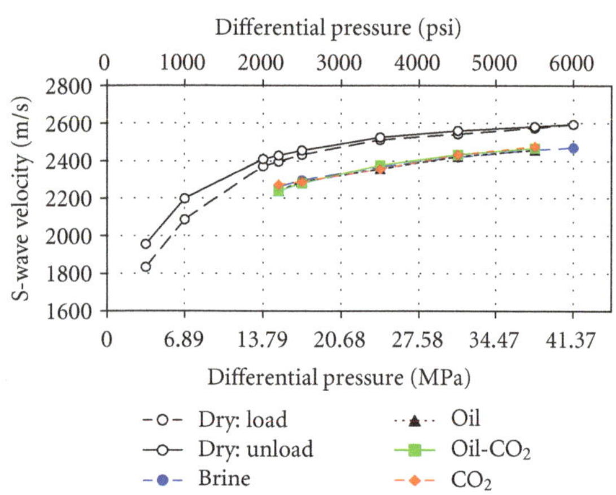

(b) S-wave velocities for the high-permeability number 2 sample versus differential pressure

Figure 8: P- and S-wave velocities for the high-permeability number 2 sample versus differential pressure for dry rock, brine saturated, and when flushed with live oil, the oil-CO_2 mixture, and CO_2.

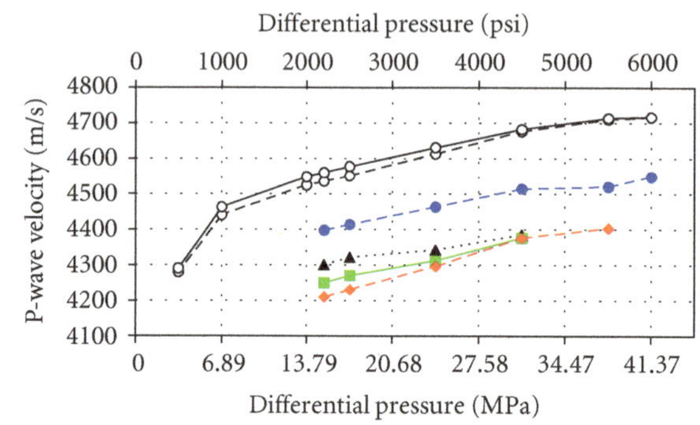

(a) P-wave velocities for the low-permeability number 1 sample versus differential pressure

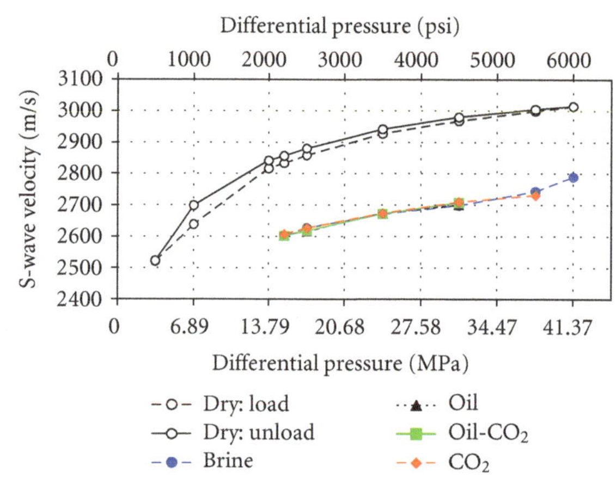

(b) S-wave velocities for the low-permeability number 1 sample versus differential pressure

Figure 9: P- and S-wave velocities for the low-permeability number 1 sample versus differential pressure for dry rock, brine saturated, and when flushed with live oil, the oil-CO_2 mixture, and CO_2. (Note dry rock velocities were recorded at 20°C, fluid-saturated velocities were recorded at 63.9°C.)

The P-wave velocities show a response to both pressure and fluid type. The P-wave velocities systematically decrease, where brine saturated is the fastest and CO_2 flushed is usually the slowest. The S-wave velocities' response is mainly dominated by pressure and generally the fluid effect is negligible.

Some specific observations with respect to each of the core samples are as follows.

5.1.1. Cemented Sample. The S-wave velocity for the brine-saturated sample has a slightly higher velocity than when CO_2 flushed (Figure 6(b)). More commonly the opposite behavior is observed.

5.1.2. High Permeability Number 1. The dry rock P-wave velocities (Figure 7(a)) show an unusual pattern where the velocity due to loading is higher than the velocity while unloading. This loading and unloading cycle may have caused permanent damage to the core leading to this response. Additionally, the P-wave velocities of the mixture consisting of live oil with a mole fraction of 0.334 CO_2 and

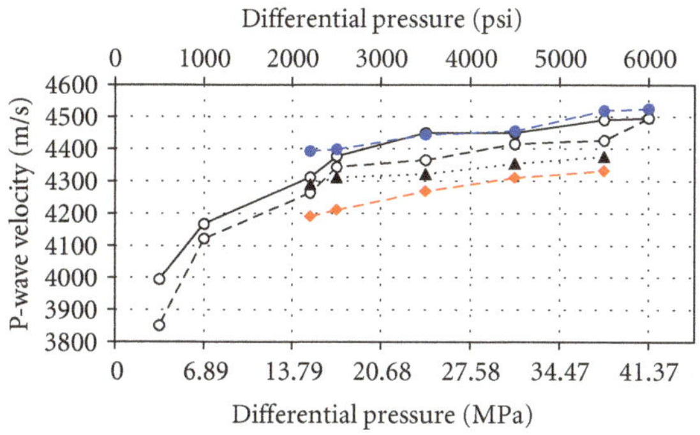

(a) P-wave velocities for the low-permeability number 2 sample versus differential pressure

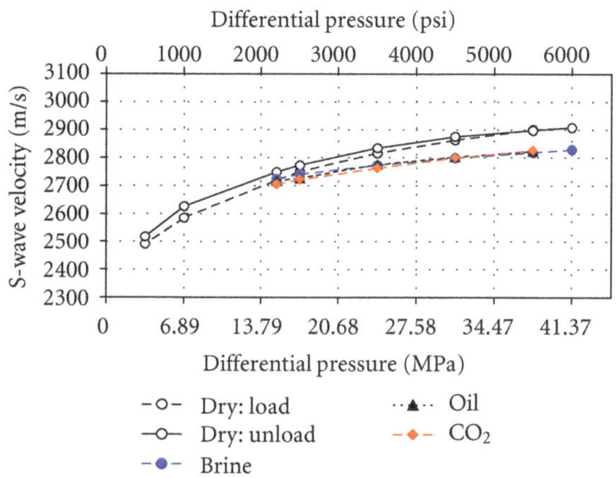

(b) S-wave velocities for the low-permeability number 2 sample versus differential pressure

FIGURE 10: P- and S-wave velocities for the low-permeability number 2 sample versus differential pressure for dry rock, brine saturated, and when flushed with live oil and CO_2.

pure CO_2 become nearly equal at a differential pressures of 31.03 MPa and 37.92 MPa (equivalent to pore pressures of 13.79 MPa and 6.89 MPa). The estimated bubble point for the oil-CO_2 mixture is 17.14 MPa, suggesting the dominant phase of the mixture at these pressures could be gas.

In regards to the S-wave velocities (Figure 7(b)), the dry rock velocities do not show the same pattern as the P-wave velocities during the loading and unloading cycle. The dry rock S-wave velocities exhibit a response consistent with the other samples, where the velocity slightly increases during the unloading cycle. This sample showed the most S-wave velocity response to fluid; where the brine-saturated velocity is slower than when flushed with the oil-CO_2 mixture and CO_2.

5.1.3. High Permeability Number 2. Much like the high-permeability number 1 sample, the high-permeability number 2 sample (Figure 8(a)) shows a very similar P-wave velocity response to live oil with a mole fraction of 0.334 CO_2 and pure CO_2. These velocities nearly converge at differential pressures of 31.03 MPa and 37.92 MPa. Also, the velocity of the sample when flushed with live oil begins to converge with the oil-CO_2 mixture and CO_2 at differential pressures

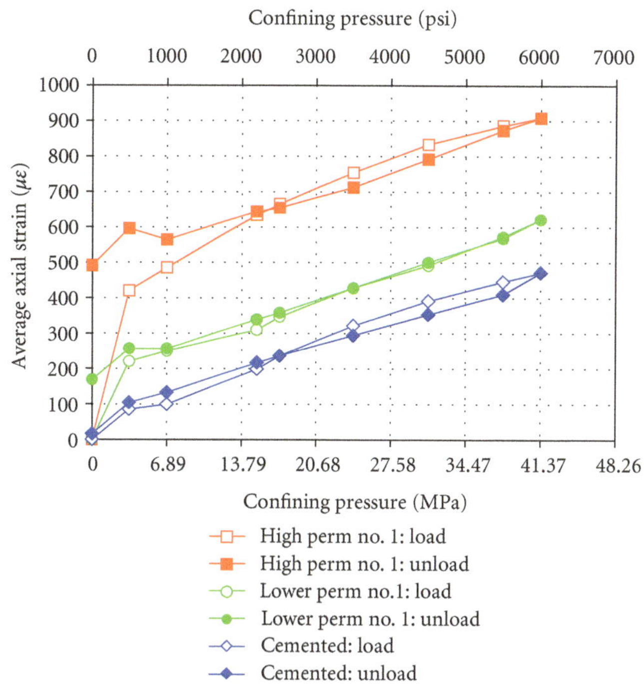

FIGURE 11: Average axial strain as a function of isostatic confining pressure for three dry rock core samples.

of 31.03 MPa and above. Again, this is most likely due to the effect of gas coming out of solution (live oil has a bubble point estimated to be 13.33 MPa at 63.9°C).

5.1.4. Low Permeability Number 1. The most obvious difference between the low-permeability number 1 sample and the other core samples is the much higher dry rock velocity for both P- and S-waves when compared to the fluid-saturated and fluid-flushed velocities (Figure 9). These higher dry rock velocities are the result of being recorded at 20°C, while the fluid measurements were recorded at the estimated reservoir temperature of 63.9°C.

5.1.5. Low Permeability Number 2. The low-permeability number 2 sample S-wave velocities (Figure 10(b)) show a similar trend as the cemented sample where the brine-saturated velocity is at some pressures marginally higher than when the sample was flushed with live oil and CO_2. Again, typically the opposite effect is observed.

5.2. Average Axial Strain. Average axial strains measured during the dry rock loading and unloading cycle for three samples, each from a different lithological zone, are shown in Figure 11. The isostatic confining pressure has a greater effect on the high-permeability sample and least affects the cemented sample. We used these results to correct for sample length, porosity, and density at each differential pressure.

5.3. Log Facies Model. One method to link the core sample information to well log response is through log facies modeling. We generated a log facies model using 12 wells

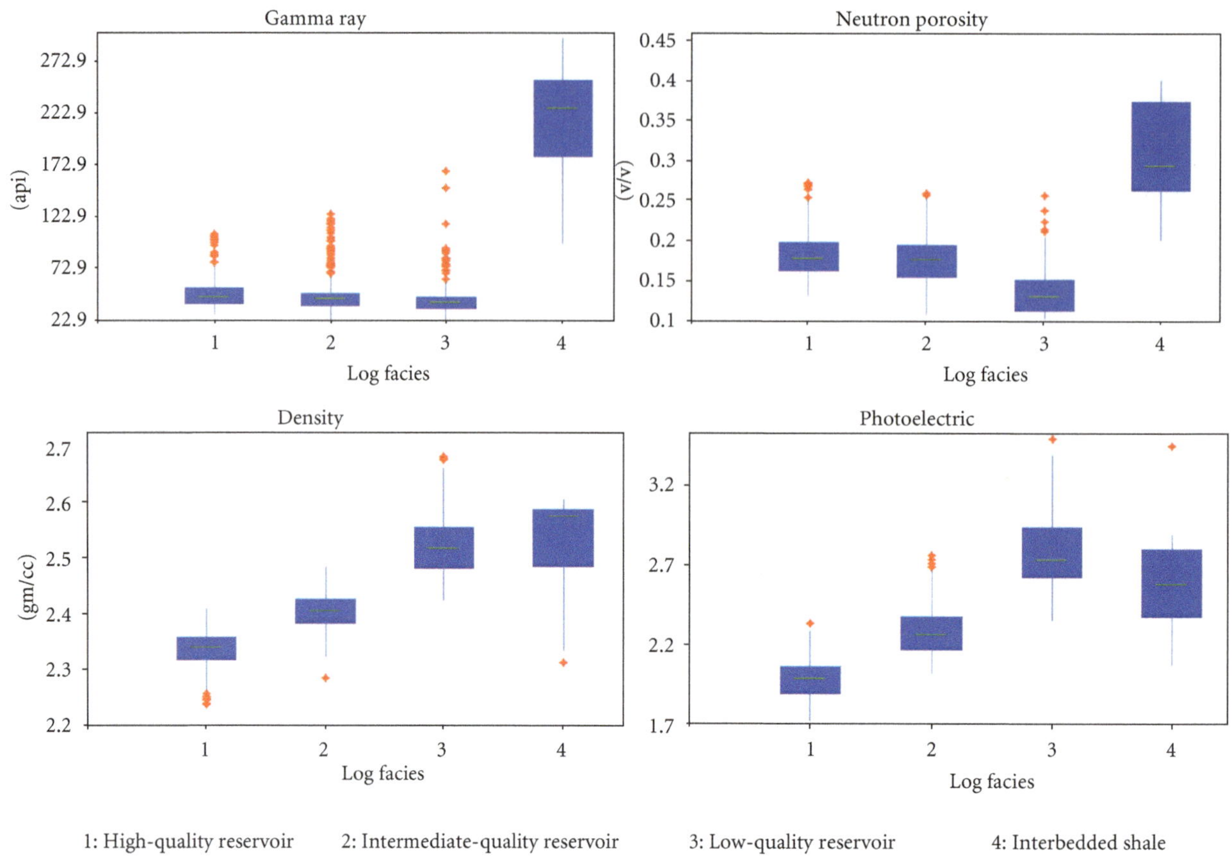

FIGURE 12: Box and whisker plot of the four clusters for the input logs used to generate the log facies model.

logged with the same technology within Postle Field. These well logs were corrected for environmental effects related to pressure, temperature, and borehole diameter. The well logs used as inputs for the log facies model were gamma ray, neutron porosity, bulk density, and photoelectric logs. The reservoir zones from each well log were spliced together in series to capture the variability of the reservoir. To identify the different log responses within the reservoir zone, principal component analysis was applied to maximize variability followed by cluster analysis to group data by similar log response. Figure 12 shows the box and whisker plot for each of the four clusters and the log response associated with each cluster. For each box, the central green line is the median of the data, the edges of the box are the 25th and 75th percentiles, and vertical lines (whiskers) extend to the extreme data points not considered outliers, the outliers are plotted as red diamonds. Based on the trends of the four box and whisker plots for each log response, log facies number 1 is interpreted as high-quality reservoir, log facies number 2 as intermediate-quality reservoir, log facies number 3 as low-quality reservoir, and log facies number 4 represents an interbedded shale.

Once the log facies are established, a supervised classification method is then applied to wells with additional data such as sonic or core data. This is shown in Figure 13 where the log facies model has been applied to a well with core-measured porosity and permeability. From Figure 13, an inference can be made that the dark blue log facies (number 1) relates to high-permeability zones, the light blue log facies (number 2) correlates to intermediate-permeability zones, and the green log facies (number 3) corresponds to zones of low permeability.

Unfortunately, the well which the core samples used for the laboratory velocity measurements did not have the necessary input logs for supervised classification. To relate the core samples to the log facies model, we compared the permeability of the core samples to the permeability associated with each log facies taken from the well in Figure 13. Figure 14 shows this comparison where the core samples used in the laboratory velocity measurements are shown as stars and each box and whisker plot represents the variability in permeability for each of the reservoir log facies. This assumes each core sample used in the laboratory velocity measurements corresponds to a log facies. Hence, the cemented sample is associated with the low-quality reservoir log facies, the high-permeability number 1 and number 2 samples are related to the high-quality log facies and the lower-permeability number 1 and number 2 samples correspond to the intermediate-quality log facies.

5.3.1. Integrating the Log Facies Model and Core-Measured Velocity. We achieved only limited success using Gassmann's relation [9] and the effective medium theory of Mori and Tanaka [36, 37] to predict velocity changes related to fluid

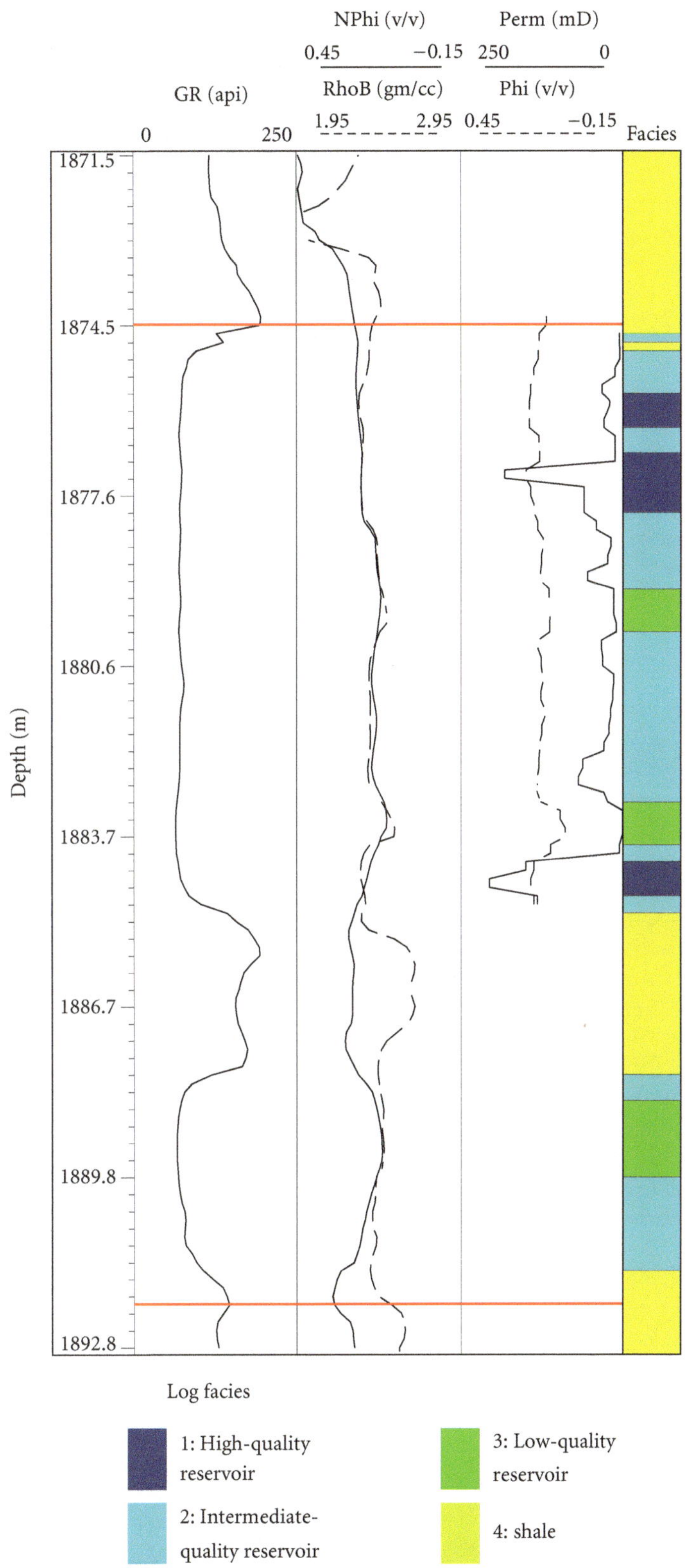

Figure 13: Log facies model applied to a well with core-measured porosity and permeability. The first track shows the gamma ray log. Track 2 plots the neutron porosity and bulk density. Track 3 represents core-measured permeability and porosity. The reservoir zone (Morrow A sandstone) is shown between the horizontal red lines.

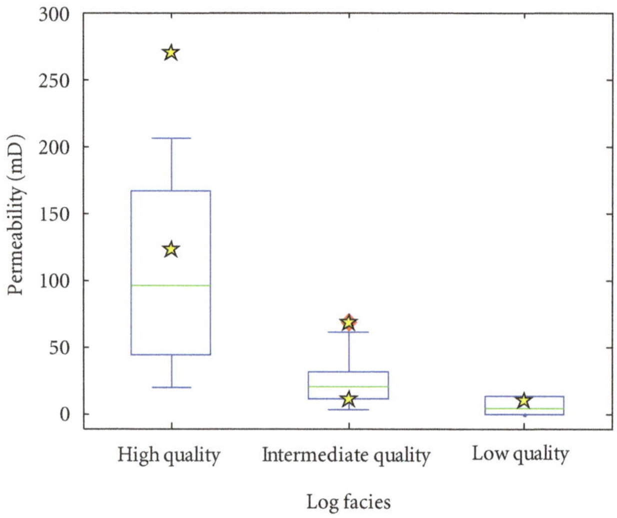

FIGURE 14: Permeability of the core samples used in the laboratory velocity measurements (stars) superimposed on the box and whisker plot of the permeability related to each log facies (from the well in Figure 13).

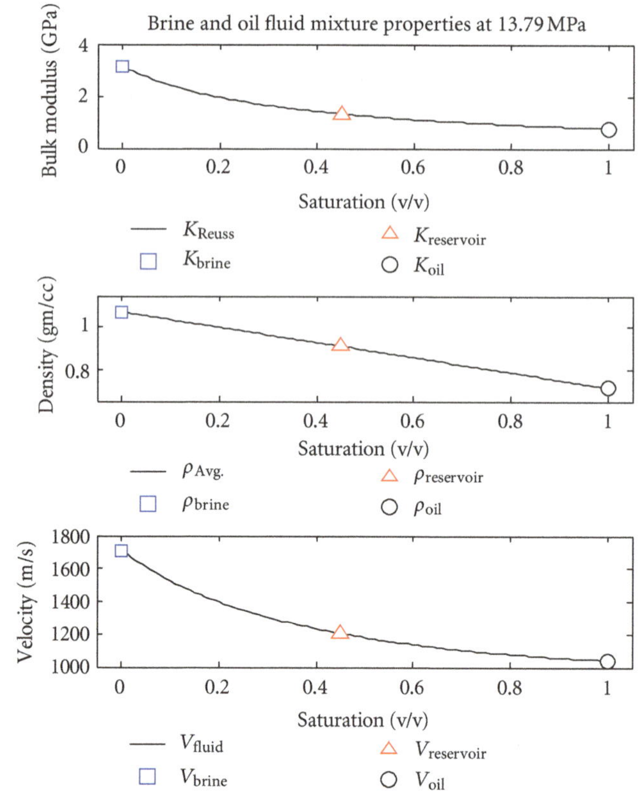

FIGURE 15: Calculated fluid properties at reservoir conditions for live oil, brine, and a mixture with a brine saturation of 45%.

saturation and differential pressure in the core samples. Generally, Gassmann's relation is not expected to model fluid effects at ultrasonic frequencies. This result only means Gassmann's relation is not appropriate for this laboratory data but may be suitable at the field scale [38]. Therefore, we apply an alternative approach which assumes a direct relationship between the log facies model and the core-measured velocities. This direct relationship is based on estimating the reservoir condition in the laboratory velocity measurements, then applying the changes associated with fluid type and pressure observed in the laboratory measurements directly to the only cross-dipole velocity well logs recorded in the Postle Field study area.

The reservoir condition from petrophysical analysis estimates a water saturation of 45% [34]. Figure 15 shows the properties of live oil, brine, and the reservoir condition with a water saturation of 45%. To estimate the reservoir condition the difference in velocity between oil and brine is taken and is 76% less than the velocity of brine.

This estimated velocity at reservoir conditions now allows for changes to be applied to the log response based on laboratory velocity measurements made on the core samples. For example, to estimate the log response at complete brine saturation, the log would be multiplied by the percent change between brine-saturated velocity and the estimated reservoir velocity from the experimental core measurement.

Since very little fluid effect is observed in the S-wave velocity core measurements, it is assumed that the fluid effect is negligible and the only changes in S-wave velocity are due to pressure. For estimating changes in the S-wave log response, the S-wave velocity is multiplied by the percent change from the reservoir condition to the desired pore pressure.

The core samples we selected to estimate the changes in P- and S-wave velocity log response were the cemented sample which is associated to the low-quality reservoir log facies, the high-permeability sample number 2 which is associated to the high-quality reservoir log facies, and the lower-permeability sample number 2 which is associated to the intermediate-quality reservoir log facies. Changes in bulk density were calculated using the density transformation equation (2) where ρ_2 is bulk density saturated with the new fluid, ρ_1 is the initial bulk density, ρ_{f1} is the density of the initial fluid, and ρ_{f2} is the density of the new fluid:

$$\rho_2 = \rho_1 + \phi\left(\rho_{f1} - \rho_{f2}\right). \qquad (2)$$

Figure 16 shows the changes in log response for the conditions of complete brine saturation at pore pressures of 6.89 MPa and 29.65 MPa. Figure 17 shows the changes in log response for the conditions of complete CO_2 saturation for pore pressures of 6.89 MPa and 29.65 MPa.

5.4. Time-Lapse Amplitude Difference. We generated synthetic seismic gathers using the full waveform modeling method described in Singh and Davis [39]. To understand the P- and S-wave response at the Morrow A sandstone level, we created 1D models for the brine- and CO_2-saturated cases using the modified well logs (Figures 16 and 17). The logs were blocked at 1.5 m and simplified to have three main reflectors: the Atoka limestone above the reservoir, the Morrow A sandstone (reservoir), and a limestone layer below the reservoir. Above the Atoka limestone, the log response was smoothed to remove the effect of multiples.

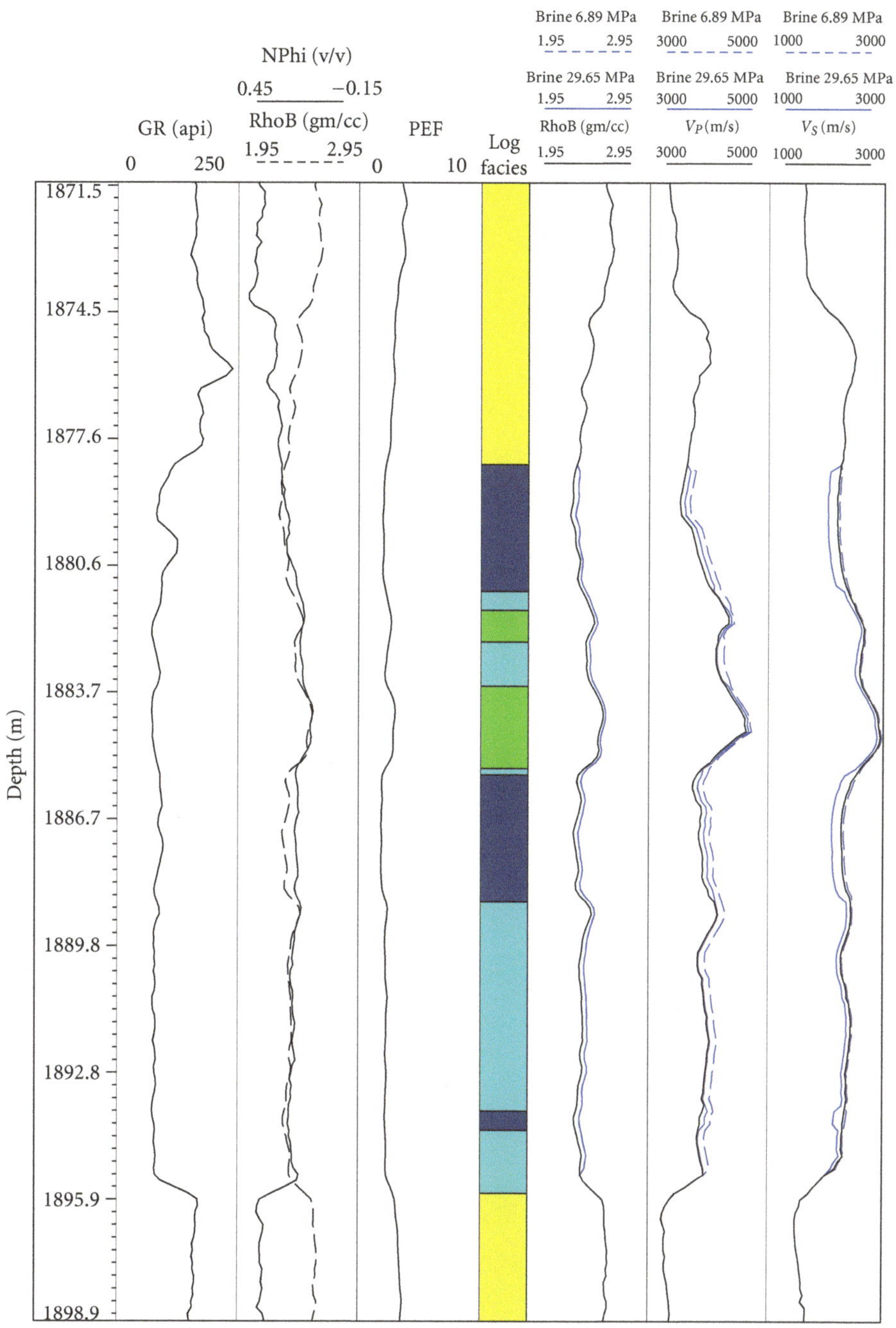

FIGURE 16: Estimated changes in the Morrow A sandstone for P-wave velocity, S-wave velocity, and bulk density log response for conditions of complete brine saturation at pore pressures of 6.89 MPa (blue dashed lines) and 29.65 MPa (solid blue lines). The recorded in-situ P-wave velocity, S-wave velocity, and bulk density logs are shown in black.

From the 1D models, we generated 2D multicomponent gathers simulating vertical and horizontal sources, and receivers with vertical and horizontal components. We extracted statistical wavelets from the processed P- and S-wave seismic volumes recorded at Postle Field for use in the full waveform modeling. The P-wave wavelet was zero phase with a peak frequency of 18 Hz and the S-wave wavelet was also zero phase with a peak frequency of 13 Hz. The gathers (Figure 18) are modeled with a maximum offset of 1829 m, which is consistent with the usable offset from the seismic data recorded at Postle Field.

Seismic acquisition parameters and examples of P- and S-wave seismic data recorded at Postle Field can be found in Singh and Davis [39]. Using nearly identical wavelets and the same cross-dipole sonic logs used in this study, Singh and Davis [39] also address the tuning and interference effect at the Morrow A sandstone level.

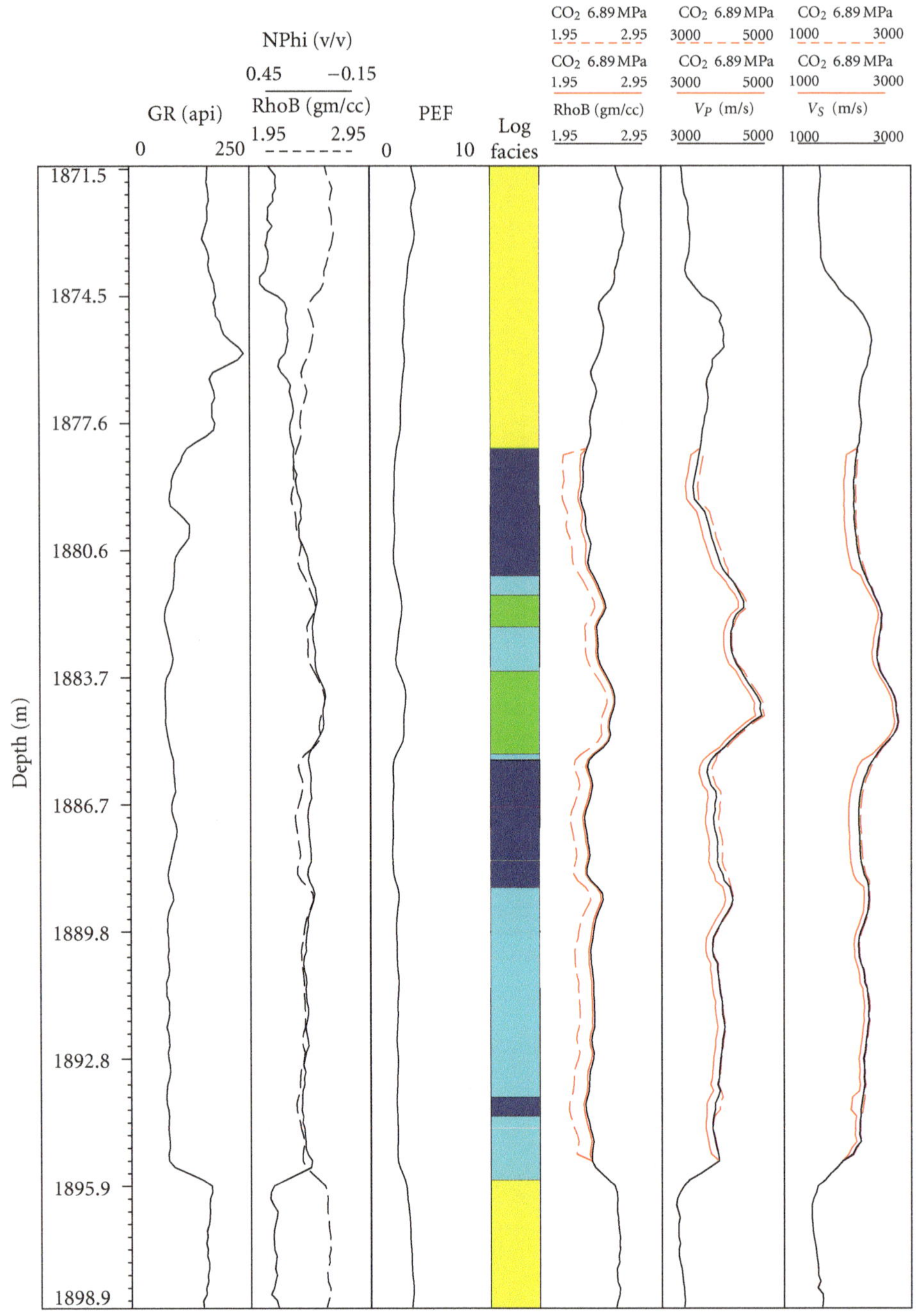

Figure 17: Estimated changes in the Morrow A sandstone for P-wave velocity, S-wave v.elocity, and bulk density log response for conditions of complete CO_2 saturation at pore pressures of 6.89 MPa (red dashed lines) and 29.65 MPa (solid red lines). The recorded in-situ P-wave velocity, S-wave velocity, and bulk density logs are shown in black.

To evaluate the changes in seismic response, we analyzed the amplitude difference of the P- and S-wave gathers for the brine- and CO_2-saturated cases. The amplitude difference for the brine-saturated case represents a pore pressure increase from 6.89 MPa to 29.65 MPa in the Morrow A sandstone. Similarly, the amplitude difference for the CO_2-saturated case represents a pore pressure increase from 6.89 MPa to 29.65 MPa in the Morrow A sandstone. The time shift for the brine and CO_2 cases was minor, calculated to be less than 0.5 milliseconds. The P-wave amplitude differences are shown in Figure 19 and the S-wave amplitude differences are shown in Figure 20.

For both the brine and CO_2 cases, the P-wave modeling shows a negative amplitude difference for an increase in pressure (Figure 19), this occurs because an increase in pore pressure results in a lower acoustic impedance. For both cases, the amplitude difference is greatest at zero offset and becomes less with increasing offset. The brine-saturated case shows a large amplitude difference starting from the reflection at the Morrow A sandstone level. The CO_2 case

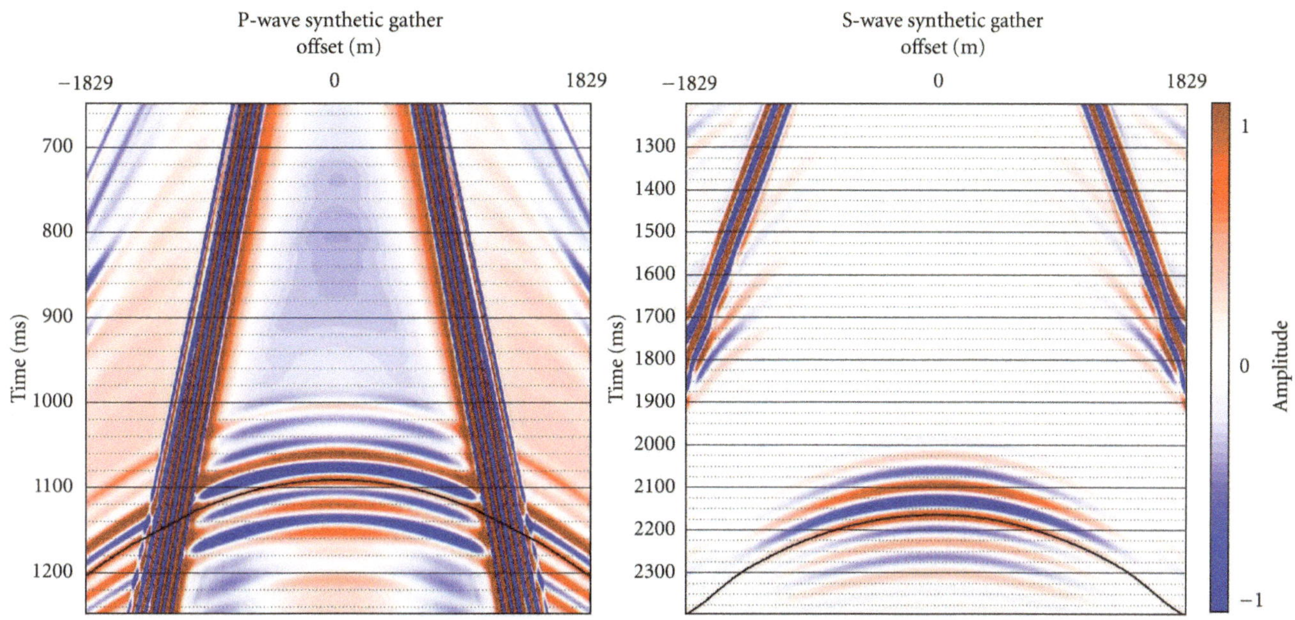

FIGURE 18: Full waveform synthetic gathers for P-wave (left) and S-wave (right) reflectivity. The Morrow A sandstone top is indicated by the black curve.

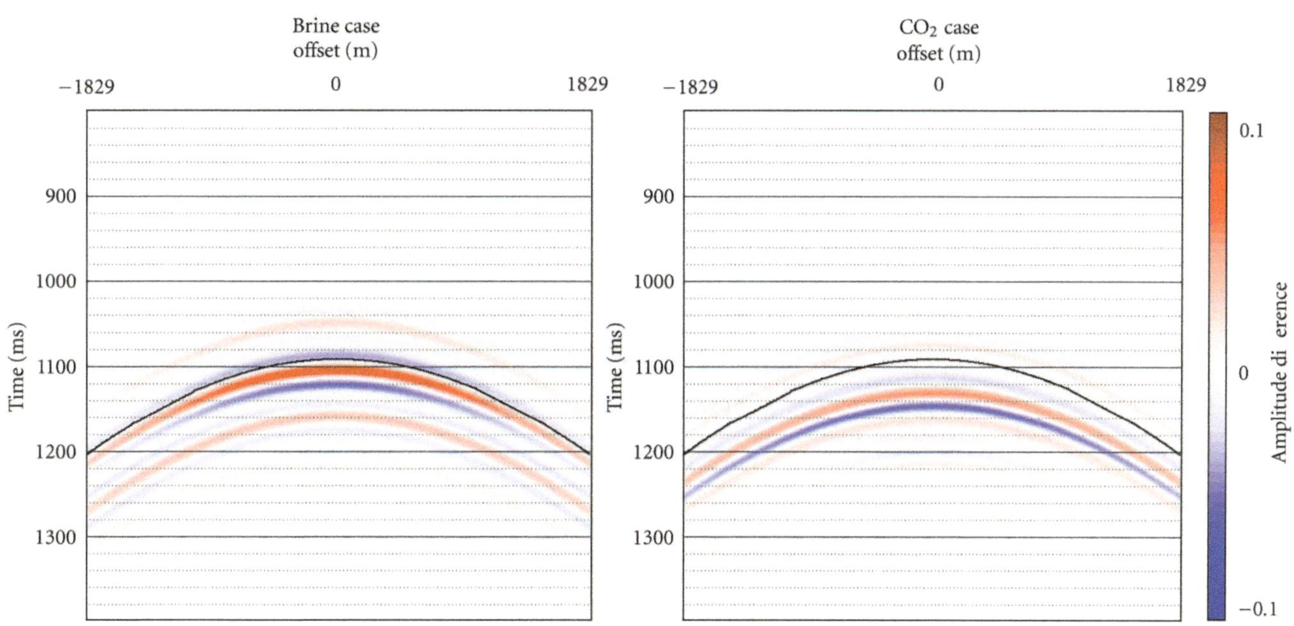

FIGURE 19: Comparison of the P-wave amplitude difference for the brine- and CO_2-saturated cases with a pore pressure increase from 6.89 MPa to 29.65 MPa. The Morrow A sandstone top is indicated by the black curve.

shows a very weak amplitude decrease at the Morrow A sandstone reflection and a larger amplitude difference from the reflectors below the Morrow A sandstone. A change in waveform around the Morrow A sandstone for the CO_2 case affects the signature of the amplitude difference. These observations suggest near-angle P-wave stacks would be better suited for time-lapse analysis with larger amplitude differences occurring below the reservoir.

Increasing pore pressure lowers the elastic impedance and hence the S-wave amplitude difference is negative with increasing pore pressure for both the brine and CO_2 cases (Figure 20). The most notable observation is the amplitude difference with increasing offset between the two cases. The amplitude difference is highest at zero offset for the brine-saturated case and decreases with increasing offset. However, the CO_2-saturated case shows the amplitude difference to

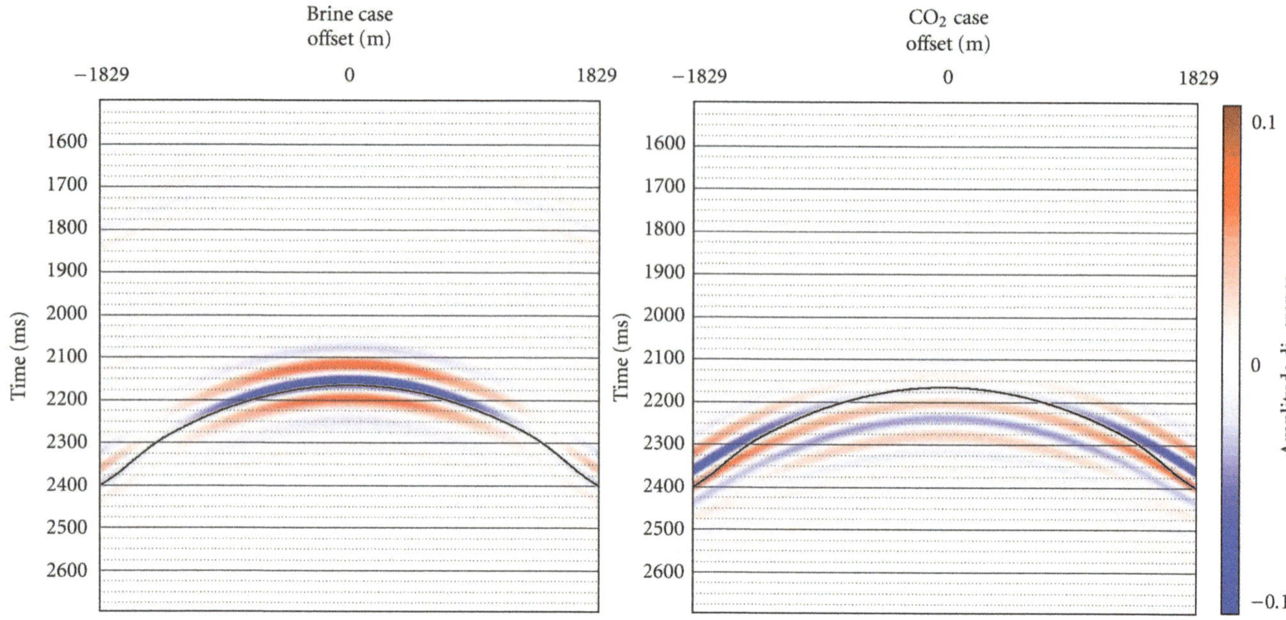

FIGURE 20: Comparison of the S-wave amplitude difference for the brine- and CO_2-saturated cases with a pore pressure increase from 6.89 MPa to 29.65 MPa. The Morrow A sandstone top is indicated by the black curve.

increase with increasing offset. Analysis on S-wave angle stacks, rather than full offset stacks, may be more effective in detecting changes in saturation and, therefore, useful for CO_2 monitoring.

The amplitude difference of the full waveform seismic modeling shows the effects fluid saturation and pressure have on P- and S-wave reflections. P-wave reflections show the near offsets to have the most amplitude difference. While S-wave reflections show a maximum amplitude difference at near offsets for the brine case and at far offsets for the CO_2 case. Also, the S-wave amplitude difference is greater than the P-wave amplitude difference for these two cases.

To determine if the time-lapse changes are mainly due to time shift or only an amplitude difference, we compared the root mean square (RMS) amplitude difference for the two cases. Since we did not take into account any time shift when we differenced the modeled synthetic seismic gathers, representing the minimum and maximum pressures for the brine and CO_2 cases (Figures 19 and 20), we calculated the RMS amplitude difference at the Morrow A sandstone top. We examined a 20 ms window centered at the Morrow A sandstone top for the P-wave gathers and a 40 ms window for the S-wave gathers. These RMS amplitude differences are shown in Figure 21 for both the brine and CO_2 cases. We observe that the RMS amplitude difference at the Morrow A top is comparable to the amplitude difference of the synthetic gathers shown in Figures 19 and 20. The RMS amplitude difference with offset also agrees with the amplitude difference of the synthetic gathers. A comparison of the RMS amplitude difference and the differenced synthetic seismic gathers demonstrate that the effect of time shift is almost negligible at the Morrow A top. The time-lapse changes at the Morrow A top (observed in Figures 19 and 20) are mainly due to amplitude differences. Figure 21 also confirms that S-wave gathers show a larger time-lapse change as compared to the P-wave gathers.

6. Conclusions

Laboratory core measurements show the effects of differential pressure and fluid saturation through the P- and S-wave velocity response of five Morrow A sandstone samples. The samples were saturated with brine and flushed with live oil, an oil-CO_2 mixture, and pure CO_2. The P-wave velocity shows a sensitivity to both fluid saturation and pressure. Whilst the S-wave velocity shows a dependence to changes in differential and a nearly negligible sensitivity to fluid saturation. Core samples taken from the high-permeability zone show the most velocity change to confining pressure and pore pressure. The core sample from the cemented zone shows the least amount of velocity change to confining pressure and pore pressure. The overall velocity decrease was greater for S-waves as shown in Tables 3 and 4. Average axial strain also showed the high-permeability sample to be the most stress-sensitive and the cemented sample the least sensitive to confining pressure.

We attempted modeling the core-measured velocities using Gassmann's relation [9] and the effective medium theory of Mori and Tanaka [36, 37]. Neither of these models successfully predicted the measured P-wave or S-wave velocities. The P- and S-wave velocity well logs were modified to estimate changes due to pressure and fluid saturation based on the laboratory core-measured velocities. This approximation was implemented through a log facies model where each log facies was associated with a core sample.

The P-wave amplitude difference for the brine- and CO_2-saturated cases, modeling a pore pressure increase from 6.89 MPa to 29.65 MPa, displayed a maximum amplitude

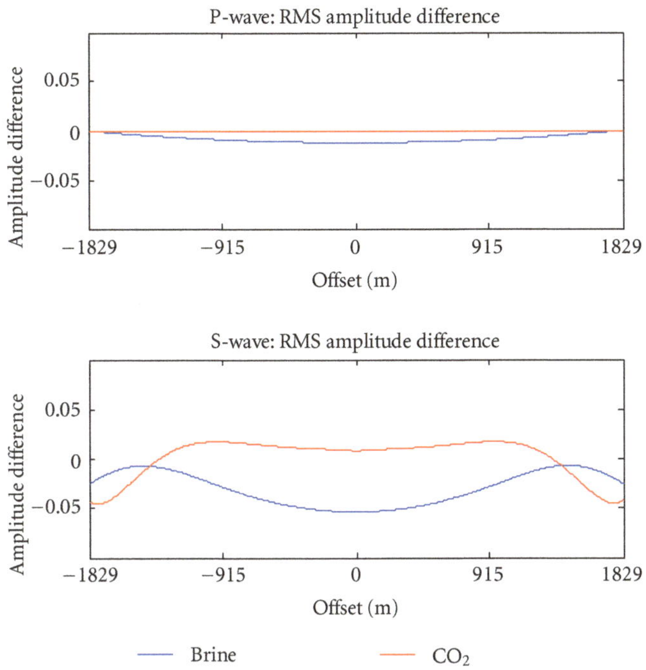

FIGURE 21: RMS amplitude difference at the Morrow A sandstone top for the synthetic seismic modeling of the P- and S-wave gathers.

difference at the near offsets. Whereas the S-wave shows a maximum amplitude difference at the near offsets for the brine case and a maximum amplitude difference at far offsets for the CO_2 case.

The full waveform modeling results show how fluid and pressure changes in the Morrow A sandstone reservoir affect the characteristics of P- and S-wave amplitude differences. In addition, the full waveform modeling shows an advantage to using angle-limited stacks as compared to full offset stacks for time-lapse analysis. These results demonstrate the benefit and establish motivation for using S-wave seismic data in time-lapse studies involving changes in pressure and fluid saturation.

Acknowledgments

The authors would like to thank the sponsors of The Reservoir Characterization Project at Colorado School of Mines, specifically Whiting Petroleum: the current operator of Postle Field; the Center for Rock Abuse at Colorado School of Mines for the use of their Rock Squeezing Laboratory, especially Professor Michael Batzle and Weiping Wang; Tom R. Bratton for his help and guidance generating the log facies model; BP America for the use of the seismic modeling code. A. V. Wondler would like to thank BP for the BP Colorado School of Mines Ph.D. Fellowship.

References

[1] L. Kootungal, "2010 world wide EOR survey," *Oil and Gas Journal*, vol. 108, no. 14, pp. 41–53, 2010.

[2] A. Tura and D. Lumley, "Estimating pressure and saturation changes from time-lapse AVO data," in *the 69th Annual International Meeting, SEG, Expanded Abstracts*, pp. 1655–1658, 1999.

[3] M. Landrø, "Discrimination between pressure and fluid saturation changes from time-lapse seismic data," *Geophysics*, vol. 66, no. 3, pp. 836–844, 2001.

[4] T. L. Davis, M. J. Terrell, R. D. Benson, R. Cardona, R. R. Kendall, and R. Winarsky, "Multicomponent seismic characterization and monitoring of the CO_2 flood at Weyburn Field, Saskatchewan," *Leading Edge*, vol. 22, no. 7, pp. 696–697, 2003.

[5] R. J. Pawar, N. R. Warpinski, J. C. Lorenz et al., "Overview of a CO_2 sequestration field test in the West Perl Queen reservoir, New Mexico," *Environmental Geosciences*, vol. 13, no. 3, pp. 163–180, 2006.

[6] A. Robinson and T. L. Davis, "Improving efficiency of CO_2 sequestration using short-term seismic monitoring: Applications from the Postle field EOR programme," *First Break*, vol. 29, no. 1, pp. 75–78, 2011.

[7] A. Nur and Z. Wang, Eds., *Seismic and Acoustic Velocities in Reservoir Rocks, Experimental Studies*, vol. 1, chapter 1, Society of Exploration Geophysicists, 1998.

[8] Z. Xue, T. Ohsumi, and H. Koide, "An experimental study on seismic monitoring of a CO_2 flooding in two sandstones," *Energy*, vol. 30, no. 11-12, pp. 2352–2359, 2005.

[9] F. Gassmann, "Uber die elastizitat poroser medien," *Vierteljahrsschrift der Naturforschenden Gesellschaft in Zrich*, vol. 96, pp. 1–23, 1951.

[10] G. T. Kuster and M. N. Toksoz, "Velocity and attenuation of seismic waves in two-phase media: Part I. Theoretical formulations," *Geophysics*, vol. 39, no. 5, pp. 587–606, 1974.

[11] "Fluid Acoustics for Geophysics calculator," Rock Physics Lab, University of Houston and Center for Rock Abuse, Colorado School of Mines, 2009.

[12] M. King, "Wave velocities in rocks as a function of changes in overburden pressure and pore fluid saturants," *Geophysics*, vol. 31, pp. 50–73, 1966.

[13] S. N. Domenico, "Rock lithology and porosity determination from shear and compressional wave velocity," *Geophysics*, vol. 49, no. 8, pp. 1188–1195, 1984.

[14] Z. Xue and T. Ohsumi, "Seismic wave monitoring of CO_2 migration in water-saturated porous sandstone," *Exploration Geophysics*, vol. 35, pp. 25–32, 2004.

[15] Z. Xue and X. Lei, "Laboratory study of CO_2 migration in water-saturated anisotropic sandstone, based on P-wave velocity imaging," *Exploration Geophysics*, vol. 37, pp. 10–18, 2006.

[16] J. Q. Shi, Z. Xue, and S. Durucan, "Seismic monitoring and modelling of supercritical CO_2 injection into a water-saturated sandstone: Interpretation of P-wave velocity data," *International Journal of Greenhouse Gas Control*, vol. 1, no. 4, pp. 473–480, 2007.

[17] X. Lei and Z. Xue, "Ultrasonic velocity and attenuation during CO_2 injection into water-saturated porous sandstone: measurements using difference seismic tomography," *Physics of the Earth and Planetary Interiors*, vol. 176, no. 3-4, pp. 224–234, 2009.

[18] J. Kim, Z. Xue, and T. Matsuoka, "Experimental study on CO_2 monitoring and saturation with combined P-wave velocity and resistivity," in *International Oil and Gas Conference and Exhibition (IOGCEC '10)*, pp. 312–319, June 2010.

[19] A. F. Siggins and D. N. Dewhurst, "Saturation, pore pressure and effective stress from sandstone acoustic properties," *Geophysical Research Letters*, vol. 30, no. 2, pp. 61-1–61-4, 2003.

[20] A. F. Siggins, "Velocity-effective stress response of CO_2-saturated sandstones," *Exploration Geophysics*, vol. 37, pp. 60–66, 2006.

[21] A. F. Siggins, M. Lwin, and P. Wisman, "Laboratory calibration of the seismo-acoustic response of CO_2 saturated sandstones," *International Journal of Greenhouse Gas Control*, vol. 4, no. 6, pp. 920–927, 2010.

[22] N. I. Christensen and H. F. Wang, "The influence of pore pressure and confining pressure on dynamic elastic properties of Berea sandstone," *Geophysics*, vol. 50, no. 2, pp. 207–213, 1985.

[23] R. Hofmann, X. Xu, M. Batzle, M. Prasad, A. K. Furre, and A. Pillitteri, "Effective pressure or what is the effect of pressure?" *Leading Edge*, vol. 24, no. 12, pp. 1256–1260, 2005.

[24] X. Xu, R. Hofmann, M. Batzle, and T. Tshering, "Influence of pore pressure on velocity in low-porosity sandstone: implications for time-lapse feasibility and pore-pressure study," *Geophysical Prospecting*, vol. 54, no. 5, pp. 565–573, 2006.

[25] M. A. Capello, *Geology and rock physics of the San Andres Formation in Vacuum Field, New Mexico*, M.S. thesis, Collororado School of Mines, 1995.

[26] L. Duranti, *Time-lapse multicomponent seismic analysis of reservoir dynamics*, Ph.D. dissertation, Colorado School of Mines, 2001.

[27] L. T. Brown, *Integration of rock physics and reservoir simulation for the interpretation of time-lapse seismic data at Weyburn Field, Saskatchewan*, M.S. thesis, Colorado School of Mines, 2002.

[28] H. Yamamoto, *Using time-lapse seismic measurments to improve ow modeling of CO_2 injection in the Weyburn Field: a naturally fractured, layered reservoir*, Ph.D. dissertation, Colorado School of Mines, 2004.

[29] E. Rojas, *Elastic rock properties of tight gas sandtones for reservoir characterization at Rulison Field, Colorado*, M.S. thesis, Colorado School of Mines, 2005.

[30] Z. Wang and A. M. Nur, "Effects of CO_2 flooding on wave velocities in rocks with hydrocarbons," *SPE Reservoir Engineering*, vol. 4, no. 4, pp. 429–436, 1989.

[31] S. Sonnenberg, "Tectonic and sedimentation model for Morrow sandstone deposition, Sorrento Field, Denver Basin, Colorado," *Mountain Geologist*, vol. 22, pp. 180–191, 1985.

[32] J. Benton, "Subsurface stratigraphic analysis, Morrow (Pennsylvanian), North Central Texas County, Oklahoma," *The Shale Shaker Digest VII*, vol. 21–23, pp. 1–28, 1973.

[33] T. D. Jobe, *Optimizing geo-cellular reservoir modeling in a braided river incised valley fill: Postle Field, Texas County, Oklahoma*, M.S. thesis, Colorado School of Mines, 2010.

[34] A. E. Heris, *Integrated ow simulation and time-lapse seismic reservoir characterization in an enhanced oil recovery project, Postle Field, Texas County, Oklahoma*, Ph.D. dissertation, Colorado School of Mines, 2011.

[35] W. Wang, M. Batzle, and A. Wandler, "Effects of CO_2 on brine, gas, and oil properties," in *2009 Annual Meeting of Fluid/DHI*, Rock Physics Lab, University of Houston and Center for Rock Abuse, Colorado School of Mines, 2009.

[36] T. Mori and K. Tanaka, "Average stress in matrix and average elastic energy of materials with misfitting inclusions," *Acta Metallurgica*, vol. 21, no. 5, pp. 571–574, 1973.

[37] Y. Benveniste, "A new approach to the application of Mori-Tanaka's theory in composite materials," *Mechanics of Materials*, vol. 6, no. 2, pp. 147–157, 1987.

[38] G. Mavko, "Expert answers," *CSEG Recorder*, vol. 30, no. 5, pp. 8–12, 2005.

[39] P. Singh and T. Davis, "Advantages of shear wave seismic in Morrow sandstone detection," *International Journal of Geophysics*, vol. 2011, Article ID 958483, 16 pages, 2011.

Frequency-Dependent Streaming Potential of Porous Media—Part 2: Experimental Measurement of Unconsolidated Materials

P. W. J. Glover,[1] E. Walker,[1] J. Ruel,[2] and E. Tardif[2]

[1] *Department of Geology and Engineering Geology, Laval University, Québec City, QC, Canada G1K 7P4*
[2] *Department of Mechanical Engineering, Laval University, Québec City, QC, Canada G1K 7P4*

Correspondence should be addressed to P. W. J. Glover, paglover@ggl.ulaval.ca

Academic Editor: Tsuneo Ishido

Frequency-dependent streaming potential coefficient measurements have been made upon Ottawa sand and glass bead packs using a new apparatus that is based on an electromagnetic drive. The apparatus operates in the range 1 Hz to 1 kHz with samples of 25.4 mm diameter up to 150 mm long. The results have been analysed using theoretical models that are either (i) based upon vibrational mechanics, (ii) treat the geological material as a bundle of capillary tubes, or (iii) treat the material as a porous medium. The best fit was provided by the Pride model and its simplification, which is satisfying as this model was conceived for porous media rather than capillary tube bundles. Values for the transition frequency were derived from each of the models for each sample and were found to be in good agreement with those expected from the independently measured effective pore radius of each material. The fit to the Pride model for all four samples was also found to be consistent with the independently measured steady-state permeability, while the value of the streaming potential coefficient in the low-frequency limit was found to be in good agreement with other steady-state streaming potential coefficient data.

1. Introduction

There have only been 10 measurements of the frequency-dependent streaming potential coefficient of porous geological and engineering materials. A review of the existing measurements was carried out by Glover et al. [1]. These previous measurements can be divided into two groups: (i) transient measurements with a percussive source and (ii) harmonic measurements with a vibrating source.

While the first of these approaches mimics many of the possible applications more closely [2–4], it cannot provide the streaming potential coupling coefficient as a function of frequency without using the frequency domain filtering and Fourier techniques. Such techniques can only be used in a linear system. Although the equations that describe the streaming potential coefficient are linear below the transition frequency and there is no evidence that they become non-linear above that frequency, it has not yet been shown that such an approach can be made to work for streaming potential coupling coefficient measurements on rocks. The processing of such data has, however, been discussed at length in Reppert and Morgan [5]; they mention that inertial effects may be seen if the transient signal has strong enough high-frequency components.

The second approach is capable of providing the streaming potential coupling coefficient at each frequency directly. Its disadvantage is that a high-quality harmonic driving pressure is required to create the time-varying flow. Various authors have shown that measurements on a range of materials are possible in the range 1 Hz to 600 Hz [6–10], but before the recent paper of Tardif et al. [11] only one measurement had been made on a geological material [10].

This paper reports research that uses the electromagnetic drive concept proposed by Glover et al. [1] to create an apparatus for measuring the frequency-dependent streaming potential coupling coefficient of unconsolidated materials such as sands, gravels, and soils. Unconsolidated materials were chosen because it is easier to arrange a sample holder with no need for a sleeve and a sleeve pressure. The experimental data have been analyzed using a number of theoretical models. The first part of this paper is dedicated to describing these models, followed by experimental measurements on samples of sand and glass beads. The theoretical models have been compared with the measured data in order to obtain the transition frequency, which has then been used to calculate the effective pore radius of the sands and glass bead packs using the theory in Glover and Walker [12].

2. Theoretical Models

The steady-state streaming potential coefficient (the streaming potential per driving fluid pressure difference) has long been described by the Helmholtz-Smoluchowski (HS) equation, and is given in the form most convenient for application to rocks (e.g., [13]);

$$C_{so} = \frac{\Delta V}{\Delta P} = \frac{\varepsilon \zeta}{\eta_f \sigma}, \quad \text{where} \quad \sigma = \sigma_f + \frac{2\Sigma_s}{\Lambda}. \tag{1}$$

In this equation ΔP (Pa) is the fluid pressure difference, ε (F/m) is the dielectric constant of the fluid, η_f (Pa.s) is the dynamic viscosity of the fluid, ζ (V) is the zeta potential, ΔV (V) is the streaming potential, σ_f (S/m) is the electrical conductivity of the bulk fluid, Σ_s (S) is the specific electrical conductance of the surface (i.e., that due to the double layer), σ (S/m) is the electrical conductivity of the mobile fluid, and Λ (m) is a characteristic length associated with the microstructure of the pore network [14–18]. The steady-state streaming potential is independent of the sample geometry.

The importance of considering the surface conductance when applying the HS equation to geological materials has been discussed by a number of authors including [1, 14–17]. Recently several modified versions of the classical HS equation have been published that take into account the variability of the streaming potential coupling coefficient as a function of grain size [17, 19–21], pore size [17], and pore throat size [17].

There are several theoretical models for the frequency-dependent streaming potential coupling coefficient. The models fall into three categories: (i) models based only on vibrational mechanics [22], (ii) models based on flow in capillary bundles [6], and (iii) those that have been developed for porous media [23].

2.1. Vibrational Mechanics Models. If we apply the amplitude of the critically damped second-order vibrational behaviour [22] to the frequency-dependent streaming potential coupling coefficient, we get

$$\frac{C_s(\omega)}{C_{so}} = \left(1 + \left(\frac{\omega}{\omega_t}\right)^2\right)^{-1/2}, \tag{2}$$

where $C_s(\omega)$ (in V/MPa) is the streaming potential coupling coefficient at a frequency ω (in Hz), C_{so} (in V/MPa) is the steady-state streaming potential coupling coefficient, and ω_t (in Hz) is the transition frequency. This model only exists in the real domain and is a special case (for $\xi = 1$) of the general second-order vibrational behaviour with variable damping that is given by

$$\frac{C_s(\omega)}{C_{so}} = \left(\left(1 - \left(\frac{\omega}{\omega_t}\right)^2\right)^2 + \left(2\xi\left(\frac{\omega}{\omega_t}\right)\right)^2\right)^{-1/2}, \tag{3}$$

for the frequency-dependent streaming potential coupling coefficient. In this equation ξ is the damping factor of the system. This equation provides the modulus or magnitude of the vibration as is usually measured experimentally either by a peak-to-peak or an RMS measurement.

Both equations can be fitted to experimental data where $C_s(\omega)/C_{so}$ is plotted as a function of frequency. For (2) the transition frequency is the only fitting variable, while (3) has two fitting variables: the transition frequency and the damping factor.

These vibrational mechanics models are purely formal and contain no underlying physics. They are interesting in that they can show that a system is behaving in a certain manner, but no inference can be made, for example, about what controls the damping coefficient. This lack of specificity often allows such models to apparently fit the data better than other models which include more of the underlying physics.

2.2. Capillary Tube Models. The capillary tube model was introduced by Packard [6] together with a small number of experimental measurements. It is given by

$$\frac{C_s(\omega)}{C_{so}} = \left(\frac{-2}{kr}\frac{J_1(kr)}{J_0(kr)}\right), \tag{4}$$

where

$$k^2 = \frac{-i\omega\rho_f}{\eta_f}, \tag{5}$$

where $i = \sqrt{-1}$, ρ_f (kg/m^3) is the density of the bulk fluid, ω (rad/s) is the angular frequency, r (m) is the radius of the capillary in which the flow takes place, and J_0 and J_1 are Bessel functions of the zeroth and first order, respectively. Note that k has dimensions of inverse length; hence, kr is dimensionless, and the real part of the Bessel function expression $\Re[(2J_1(kr)/krJ_0(kr))]$ varies between unity at low frequencies and zero at high frequencies, and its imaginary part $\Im[(2J_1(kr)/krJ_0(kr))]$ is zero at both low and high frequencies, but attains a peak value at a frequency known as the transition frequency.

Reppert et al. [9] have provided a simplification of (4):

$$\frac{C_s(\omega)}{C_{so}} = \left(1 - \frac{2}{r}\sqrt{\frac{\eta_f}{\omega\rho_f}}\left\{\frac{1}{\sqrt{2}} - \frac{1}{\sqrt{2}}i\right\}\right), \tag{6}$$

(their Equations 26 and 38). However, a recent study showed that the simplification is incorrect [11]. The correct simplification, which is consistent with the simplified model solution shown in their Figure 4 [9] (T. Ishido, pers. comm., 2011), is

$$\frac{C_s(\omega)}{C_{so}} = \left(1 + \left[\frac{-2}{r}\sqrt{\frac{\eta_f}{\omega\rho_f}}\left\{\frac{1}{\sqrt{2}} - \frac{1}{\sqrt{2}}i\right\}\right]^{-2}\right)^{-1/2}. \tag{7}$$

When (4) or (7) is fitted to experimental data where the viscosity and density of the process fluid is known, the only fitting variable is the effective capillary radius r.

2.3. Porous Media Models. An extremely important study by Pride [23] has provided a model of streaming potential coupling coefficient for porous media. It takes the form

$$C_s(\omega) = \left(\frac{\varepsilon\zeta}{\eta_f\sigma}\left\{1 - 2\frac{\delta}{\Lambda}\right\}\right) \times \left(1 - i\frac{\omega}{\omega_t}\frac{m^*}{4}\left\{1 - 2\frac{\delta}{\Lambda}\right\}^2\left\{1 - i^{3/2}\delta\sqrt{\frac{\omega\rho_f}{\eta_f}}\right\}^2\right)^{-1/2}, \quad (8)$$

where

$$\omega_t \equiv \frac{\phi}{\tau_e\kappa_{\mathrm{DC}}}\frac{\eta_f}{\rho_f}, \quad (9)$$

and

$$m^* \equiv \frac{\phi\Lambda^2}{\tau_e\kappa_{\mathrm{DC}}}. \quad (10)$$

Equation (8) can be rewritten in the same form as the previous models as

$$\frac{C_s(\omega)}{C_{so}} = \left(1 - i\frac{\omega}{\omega_t}\frac{m^*}{4}\left\{1 - 2\frac{\delta}{\Lambda}\right\}^2\left\{1 - i^{3/2}\delta\sqrt{\frac{\omega\rho_f}{\eta_f}}\right\}^2\right)^{-1/2}, \quad (11)$$

noting that the steady-state term in this model includes an additional factor $C_{so} = \varepsilon\zeta/(\eta_f\sigma)\{1 - 2(\delta/\Lambda)\}$.

In (8)–(11), τ_e (unitless) is the electrical tortuosity of the pore network, ϕ (unitless) is the porosity of the sample, κ_{DC} (m^2) is the steady-state fluid permeability, and δ (m) is the Debye length. The parameter ω_t (rad/s) is the transition frequency, at which the quadrature component of the dispersive system is greatest. While this equation may seem complex, it should be noted that for most geological media the pore fluid is sufficiently saline for the Debye length to be much smaller than the characteristic length scale (i.e., $\delta \ll \Lambda$), which allows significant simplifications to be made.

Recently, such a simplification of (11) has been proposed [19]:

$$\frac{C_s(\omega)}{C_{so}} = \left(1 - i\frac{m^*}{4}\frac{\omega}{\omega_t}\right)^{-1/2}, \quad (12)$$

where

$$\omega_t = \frac{\phi}{\tau_e\kappa_{\mathrm{DC}}}\frac{\eta_f}{\rho_f} = \frac{8}{r_{\mathrm{eff}}^2}\frac{\eta_f}{\rho_f}, \quad (13)$$

where r_{eff} (m) is the effective pore radius of the rock. Equation (12) makes the assumption that $\delta \ll \Lambda$, which is valid for the majority of porous rocks that are saturated with saline fluids with a concentration of 10^{-3} mol/dm^3 or more. If we take $m^* = 8/3$ as suggested by [19], we get

$$\frac{C_s(\omega)}{C_{so}} = \left(1 - i\frac{2}{3}\frac{\omega}{\omega_t}\right)^{-1/2}, \quad (14)$$

which is dependent solely on the transition frequency.

If either the full Pride model (11) or its simplifications (12) and (14) are fitted to experimental data where the viscosity and density of the pore fluid and the microstructural parameters $(\delta, \Lambda, \phi, \tau_e, \kappa_{\mathrm{DC}})$ of the porous medium are known, the only fitting variable is the transition frequency.

2.4. Physical Interpretation of Theoretical Models. Until recently only the Packard model [6] and its simplification [9] had been tested against data from a few measurements on capillary tubes and filter material [9, 10]. Recently [11] published some data for Ottawa sand which suggested that the vibrational mechanics models [22] fitted their data best. They reported that the best fit of all of the models was a damped second-order vibration mechanics model (3). However, there were errors in their data processing. A correct analysis of their data shows that their data follow the Pride model best, which agrees very well with the results of this study.

Most of the theoretical models have a real and imaginary part. In this paper we have analysed these two contributions separately, comparing the measured data with the overall magnitude and each of the complex components of each model. We have taken an RMS measurement approach which provides the magnitude of the variation and maximises the precision with which it can be measured.

It is interesting to consider the physical meaning of the real and imaginary contributions to the streaming potential coupling coefficient. Currently we do not have sufficient information to answer this question with authority. However, there are some indications. The streaming potential coupling coefficient is defined as the ratio of the streaming potential to the pressure drop across the sample. However, it is the fluid velocity that separates the charge and causes the streaming potential. This implies that the frequency dependence of the streaming potential coupling coefficient depends on the frequency dependence of the dynamic fluid permeability. The dynamic fluid permeability at low frequencies is controlled by viscous flow that is represented by the real part of the dynamic permeability. However, when a critical frequency is reached, the inertial acceleration of the fluid begins to control the flow (e.g., [9, 24]). The inertial acceleration is represented by the imaginary part of the dynamic permeability. Hence, we might expect the real and imaginary parts of the streaming potential coupling coefficient to be influenced by the same transition from viscous-dominated to inertial-dominated fluid flow. In this scenario, the transition frequency is the same as the critical frequency at which viscous-dominated fluid flow becomes inertially dominated. It is becoming clear, however, that the frequency dependence of fluid flow and of the streaming potential coupling coefficient are subtly different. The normalised dynamic permeability of a bundle of capillary tubes approximately follows a Debye model for all frequencies, while the normalised streaming potential coupling coefficient follows the same Debye model up to the transition frequency then deviates from it considerably [25].

3. Experimental Tests

3.1. Sample Material and Basic Characteristics. Experimental tests were carried out on samples of Ottawa sand and on packs of glass beads using the apparatus described in the as- sociated paper [1]. Their main properties are shown in Table 1.

Ottawa Sand. Ottawa sand was obtained from Fisher Scientific and washed repeatedly in distilled water in order to remove any rock powder before being dried in a vacuum oven prior to use. The pore throat and grain size distributions of the sample material were measured using mercury injection porosimetry and are shown as Figure 1 in [11] together with the grain size distribution obtained by laser diffraction measurements using a Malvern Mastersizer 2000. The porosity of sand samples was also measured using a helium pycnometer. The complex electrical properties of a saturated sample of the sand were measured at 60 frequencies between 1 Hz and 1 MHz using a Solartron 1260A Impedance Analyzer. The absolute value of the complex electrical conductivity at the frequency for which the out-of-phase conductivity was minimum has been used together with the modal grain size from the laser diffraction measurements and the helium porosity in order to calculate the mean pore size of the sand samples using the method of [12].

The quasi-steady-state permeability was obtained by calculating the volume of fluid flowing through the sample per second at 10 Hz using the measured piston displacement and also measuring the pressure required to move this fluid. The permeability at 10 Hz was $\kappa_{10} = 1.19 \times 10^{-10}\,\text{m}^2$. We have taken this value to represent the steady-state permeability in the absence of steady-state permeability on the sample.

The measurements shown in this paper are the same as those reported in [11]. However, it should be noted that there were major errors in the analysis of the data in [11] which are corrected in this work. Consequently, this work represents the correct treatment of the Ottawa sand data, while the previous paper should be referred to for a detailed petrophysical description of the material as well as measurements of the petrophysical properties of the Ottawa sand that have been made by other authors.

Glass Beads. Three sizes of soda lime glass bead (nominally 0.5, 1, and 2 mm in diameter) were obtained from Endecotts Ltd.. Samples of the beads were washed repeatedly in distilled water before being dried in a vacuum oven prior to use. No mercury porosimetry was carried out on the beads because they are too expensive to be disposed of after only one use. The grain size distribution was obtained by laser diffraction measurements and using the detailed calibration information provided by Endecotts Ltd., which is based on a sieve analysis. The porosity of the glass bead samples was measured using a helium pycnometer. The complex electrical properties of a saturated sample of the sand were measured at 60 frequencies between 1 Hz and 1 MHz using a Solartron 1260A Impedance Analyzer, and the results were used to calculate the mean pore size of each bead pack using the Glover and Walker method [12] in the same way as for the Ottawa sand.

The steady-state permeability of each bead pack was measured using a gravitational pressure head. The results are shown in Table 1. These values are in very good agreement with the value predicted from the grain size and electrical measurements using the RGPZ method [26] (their Equation (10)).

Fluids. The fluid used in the experiments was 10^{-3} mol/L NaCl with a measured density of 997 kg/m^3, which agrees well with the equation of state of NaCl solutions (e.g., [27]) and a viscosity of 8.94×10^{-4} Pa.s calculated using the model of [28]. The electrical conductivity of the fluid was measured at $\sigma_f = [1.23 \pm 0.05]\times10^{-2}$ S/m at 25°C before use, which is in good agreement with the model of Sen and Goode [29] at $[1.23 \pm 0.05]\times10^{-2}$ S/m. For the Ottawa sand, the conductivity of the fluid emerging from the apparatus during the measurement was extremely close to the original conductivity of the fluid ($[1.21 \pm 0.05]\times10^{-2}$ S/m at 25°C). In the case of the glass beads, the conductivity of the fluid slowly increased to 1.48×10^{-2} S/m, 1.38×10^{-2} S/m, and 1.33×10^{-2} S/m for 0.5, 1, and 2 mm diameter beads, respectively, while being circulated through the sample for 24 hours before the electrical and electrokinetic measurements were made. This amounts to an increase of concentration from 1×10^{-3} mol/L to 1.23×10^{-3} mol/L, 1.15×10^{-3} mol/L, and 1.10×10^{-3} mol/L, respectively, which we associate with dissolution of the grains during the attainment of physicochemical equilibrium between the grains and the fluid. For the Ottawa sand experiment, the pH of the fluid during the measurement was pH 6. In the case of the glass beads, the initial pH of the fluid was pH 6.9, which reduced during the recirculation of the fluids. The pH of the fluid was measured on samples of fluid emerging from the apparatus during the electrokinetic experiment. The stable values were pH 6.4, 6.6, and 6.7 for the 0.5, 1, and 2 mm diameter beads, respectively. We note again that the changes in the fluid conductivity and pH are not as great as some authors have experienced (e.g., Leroy et al. [30]), and we associate this with the repeated washings that we subjected the material to initially. However, we note later that modelling of the steady-state streaming potential coupling coefficient is extremely sensitive to the fluid conductivity and the pH. Hence it is extremely important for these parameters to be measured on the fluid emerging from the experimental apparatus during the electrokinetic measurement.

3.2. Experimental Methodology. The cell was loaded with either a sample of sand or beads in layers of 1 cm with light tamping between the layers in the case of the sand, and with agitation after each layer in the case of the glass beads.

Table 1: Physical properties of the sample material.

Property	Unit	Ottawa sand	Glass beads			Comment
			0. 5 mm	1 mm	2 mm	
Modal grain radius (laser diffraction), r_g	μm	251 ± 56	—	—	—	Using a Malvern Mastersizer 2000.
Modal grain radius (image analysis), r_g	μm	245 ± 46	249 ± 2.32	492 ± 11	995 ± 7	Image analysis using Sigma Scan 4.
Modal grain radius (Hg injection), r_g	μm	211 ± 59	—	—	—	Calculated from pressure data using Mayer-Stowe theory.
D10 grain radius (sieving), r_g	μm	—	251.1 ± 0.200	493 ± 0.838	989.6 ± 0.989	
Effective pore radius, r_p	μm	67.6 ± 16.2	70.8 ± 0.70	139 ± 3.41	287 ± 4.03	Using the method of Glover and Walker [12].
Effective pore throat radius, r_{pt}	μm	40.7 ± 9.0	42.7 ± 0.42	84 ± 2.06	173 ± 2.43	Using Glover and Déry [17], for random packing.
Modal pore throat radius (Hg injection), r_{pt}	μm	38.5 ± 7.5	—	—	—	Using a Micromeritics AutoPore IV
Porosity (gravimetry)	—	0.314	0.383	0.380	0.382	Please see text.
Porosity (helium expansion)	—	0.325	0.391	0.383	0.385	Using a real gas expansion pycnometer.
Porosity (mercury injection)	—	0.304	—	—	—	Using a Micromeritics AutoPore IV.
Measured permeability, κ_{meas}	m^2	1.19×10^{-10}	1.62×10^{-10}	5.95×10^{-10}	27.2×10^{-10}	The measured permeability at 5 Hz for the Ottawa sand and under steady-state conditions for the glass beads.
Predicted permeability using the RGPZ method, κ_{RGPZ}	m^2	1.22×10^{-10}	1.52×10^{-10}	5.80×10^{-10}	24.8×10^{-10}	Permeability predicted from electrical data and the grain diameter using the method of Glover et al. [26].
Electrical conductivity, σ_r	S/m	2.63×10^{-3}	3.58×10^{-3}	3.33×10^{-3}	3.32×10^{-3}	For a 0.001 mol/L NaCl at 25°C.
Electrical conductivity of the fluid, σ_f	S/m	1.23×10^{-2}	1.48×10^{-2}	1.38×10^{-2}	1.33×10^{-2}	Measured on the recycled fluid at the sample outlet after equilibration.
Formation factor, F	—	4.676	4.13	4.18	4.16	Calculated from the conductivities of the saturated rock and the saturating fluid.
Connectedness, G	—	0.214	0.242	0.239	0.241	Calculated from the formation factor [33, 34].
Cementation exponent, m	—	1.372	1.48	1.48	1.48	Calculated from the formation factor.
Electrical tortuosity, τ_e	—	1.519	1.583	1.589	1.588	Calculated from the porosity and m.
Theta factor, Θ	—	3.705	3.53	3.57	3.55	From the method of [12].
Predicted transition frequency, ω_t	Hz	256.48	213.36	57.32	12.61	At 24°C using the method of [19].

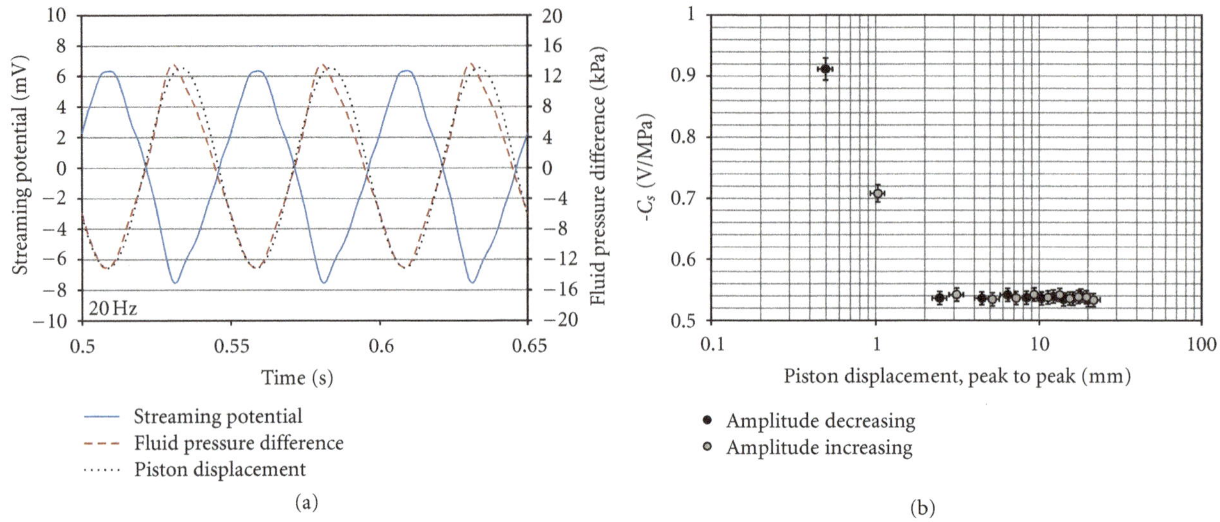

FIGURE 1: (a) The measured waveforms (streaming potential, fluid pressure and piston position (LVDT)) at 20 Hz. (b) The calculated streaming potential coupling coefficient as a function of piston amplitude at 10 Hz for Ottawa sand saturated with 10^{-3} mol/L NaCl solution at 24°C ($\phi = 0.325$, $\kappa_{10} = 1.19 \times 10^{-10}$ m^2). The errors in the frequency were calculated from the analysis of a train of approximately 500 cycles, while the errors in the streaming potential coupling coefficient were calculated from the errors in the RMS streaming potential and the measured RMS pressure difference (500 cycles).

The system was fully saturated with the process fluid, using back-pressure where necessary to remove all air bubbles. Once saturated, the steady-state permeability of the glass beads was measured using gravity-driven flow. The process fluid was then recycled through the sample for 24 hours to ensure full physicochemical equilibrium. During this time the permeability of the Ottawa sand was measured at a frequency of 10 Hz using the pressure transducers and calculating the flow by measuring the piston displacement with the LVDT.

Figure 1 shows the typical measured waveforms for 20 Hz, noting that there is a tendency that higher frequencies provide better-quality waveforms. High-quality measurements were possible between 5 and 200 Hz for the Ottawa sand and between 10 Hz and 500 Hz for the glass bead packs. Both the streaming potential and the dynamic pressure can be measured with acceptable levels of noise. Figure 1(a) shows that the fluid pressure and piston displacement are in phase with each other and in antiphase with the streaming potential, as theory requires. The compressive part of each cycle is sinusoidal as expected, whereas the backstroke is slightly distorted due to the inflow for new fluid through the check valve. This slight asymmetry might be corrected using digital filtering of the measured data or by imposing a background DC fluid flow and a back-pressure instead of using check valves. This latter approach would also remove any tendency for the fluid to cavitate.

In the case of Ottawa sand, tests were made up to 600 Hz, when the sample tube failed. It was observed that the seal between the piston and the tube let in air at frequencies higher than 200 Hz. Although the data for frequencies greater than 200 Hz seem to behave well, we have not reported them because the presence of air bubbles may make the measurements unreliable. We corrected the air leakage for the glass bead pack measurements simply by lubricating the piston seal.

This paper contains results for three diameters of glass bead (0.5, 1, and 2 mm). We also attempted to make measurements on glass bead packs with a 0.25 mm and 3.35 mm nominal diameter. Unfortunately we could not generate sufficient pressure to produce a streaming potential of sufficient size to measure the 3.35 mm beads with accuracy, and the experiment with the 0.25 mm beads did not provide data of sufficient quality to report.

The frequency-dependent streaming potential coefficients were calculated using the methods described in Reppert et al. [9] and Reppert and Morgan [5], the most important step being the renormalisation of the data taking into account of the frequency-dependent impedance of the sample and measuring circuit, which is shown for each sample at each of the measurement frequencies in Figure 2. This procedure ensures that the streaming potential coefficient is calculated with the correct sample conductivity, that is, that which relates to the frequency of the data. Without such a step, the data seem to fit better the vibrational mechanics models, as was erroneously reported by us in [11]. Inclusion of the correction results in the data fitting the Pride model [23] better than the other models. This is a satisfying result as the Pride model was specifically conceived for porous media.

3.3. Displacement Tests. One of the characteristics of an electromagnetic shaker is that the piston amplitude decreases with frequency for any set driving current [1]. However, the differential fluid pressure generated by the piston increases with frequency [1]. If the driving current is kept constant throughout a suite of tests at different frequencies, the combination of these two effects is to generate smaller

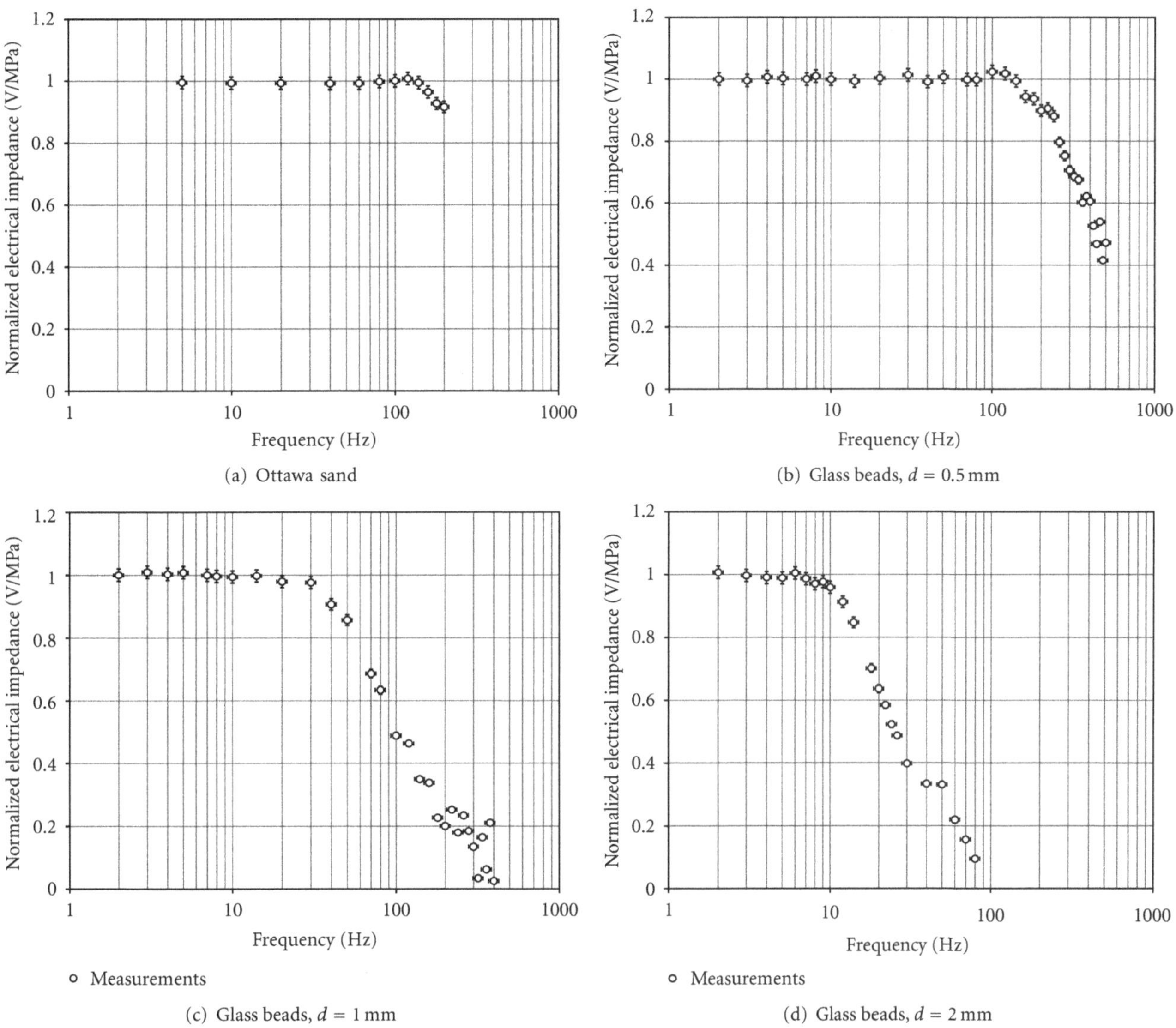

FIGURE 2: Normalised electrical impedance data for (a) Ottawa sand, and glass beads with (b) d = 0.5 mm, (c) d = 1 mm, and (d) d = 2 mm.

differential fluid pressures at high frequencies, lowering the signal-to-noise ratio. One solution to this problem is to vary the drive current in order to generate a differential fluid pressure that can be measured with good signal-to-noise ratio and to keep that differential pressure constant for as great a range of frequencies as possible. Such a procedure implies that the piston amplitude will be different for each frequency. Current understanding of frequency-dependent streaming potentials does not indicate that there is a piston amplitude below which the coupling is not fully developed or any other reason why the streaming potential coupling coefficient should vary with piston amplitude. However, it is important experimentally to know if we will introduce any systematic errors by allowing the piston amplitude to vary and also to know if there is a piston amplitude below which measurements become unreliable due to the signal-to-noise ratio.

We have carried out tests to examine the measured streaming potential as a function of the piston amplitude using a sample of the Ottawa sand and an arbitrary frequency of 10 Hz. Dynamic fluid pressure, dynamic streaming potential, and instantaneous piston position measurements were made while decreasing the shaker driving current in increments (and hence the piston amplitude) until the measured values were below the noise threshold. Measurements were then made while incrementally increasing the driving current until the maximum displacement was reached.

The results are shown in Figure 1(b). The initial peak-to-peak displacement was just over 20 mm, and this decreased incrementally until it was 0.5 mm, then increased again to a maximum about 21.7 mm. In the range about 2 mm to 21.7 mm, the measured streaming potential and the calculated streaming potential coupling coefficient remained stable with $C_s = 0.5374 \pm 0.0029$ V/MPa, which represents a

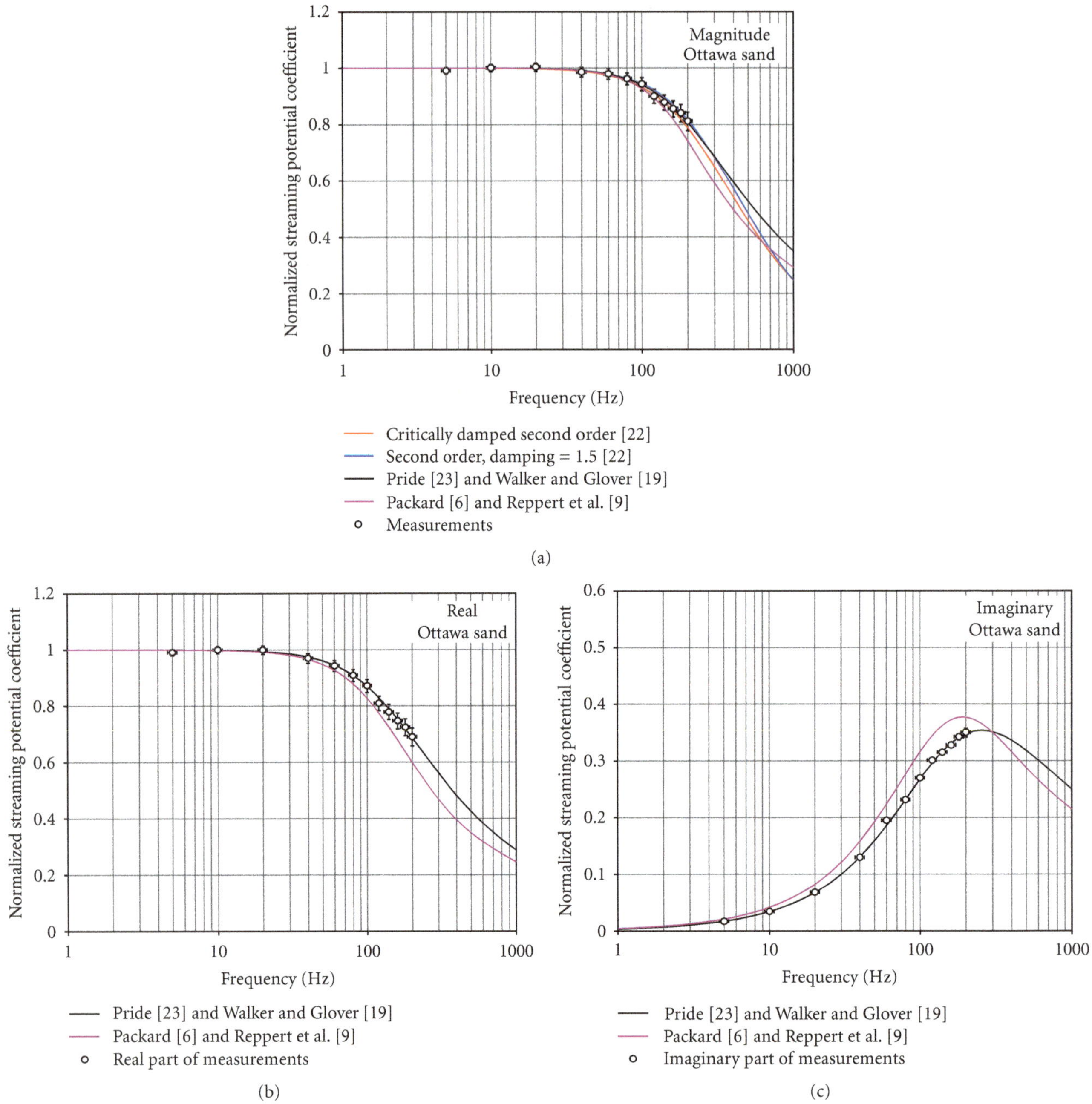

FIGURE 3: The calculated streaming potential coupling coefficient (normalised to the value at 5 Hz, which was 0.518 V/MPa) as a function of frequency for Ottawa sand saturated with 10^{-3} mol/L NaCl solution at 24°C ($\phi = 0.325$, $\kappa_{10} = 1.19 \times 10^{-10}\,\text{m}^2$). (a) Magnitude with six models shown fitting the data [6, 9, 19, 22, 23], (b) real component for models [6, 9, 19, 23], and (c) imaginary component for models [6, 9, 19, 23].

variability of about 0.5%. The values during reduction and augmentation of piston displacement were $C_s = 0.5369 \pm 0.0026$ V/MPa and 0.5379 ± 0.0032 V/MPa, respectively, so there is no sensitivity to whether the piston amplitude is increasing or decreasing. We can remark, therefore, that in the range 2 mm to 21.7 mm the streaming potential and the calculated streaming potential coupling coefficient measured by this instrument are independent of piston amplitude and direction of piston amplitude change. Hence the piston amplitude may be varied to optimize the measurement conditions.

The measured streaming potential was affected at piston amplitude less than about 2 mm. Under these conditions the measured pressure difference is very small for our high permeability sample, and it is difficult to distinguish the measurements from the background noise. We believe that the observed increase in the streaming potential coupling coefficient for displacements less than 2 mm is due to the

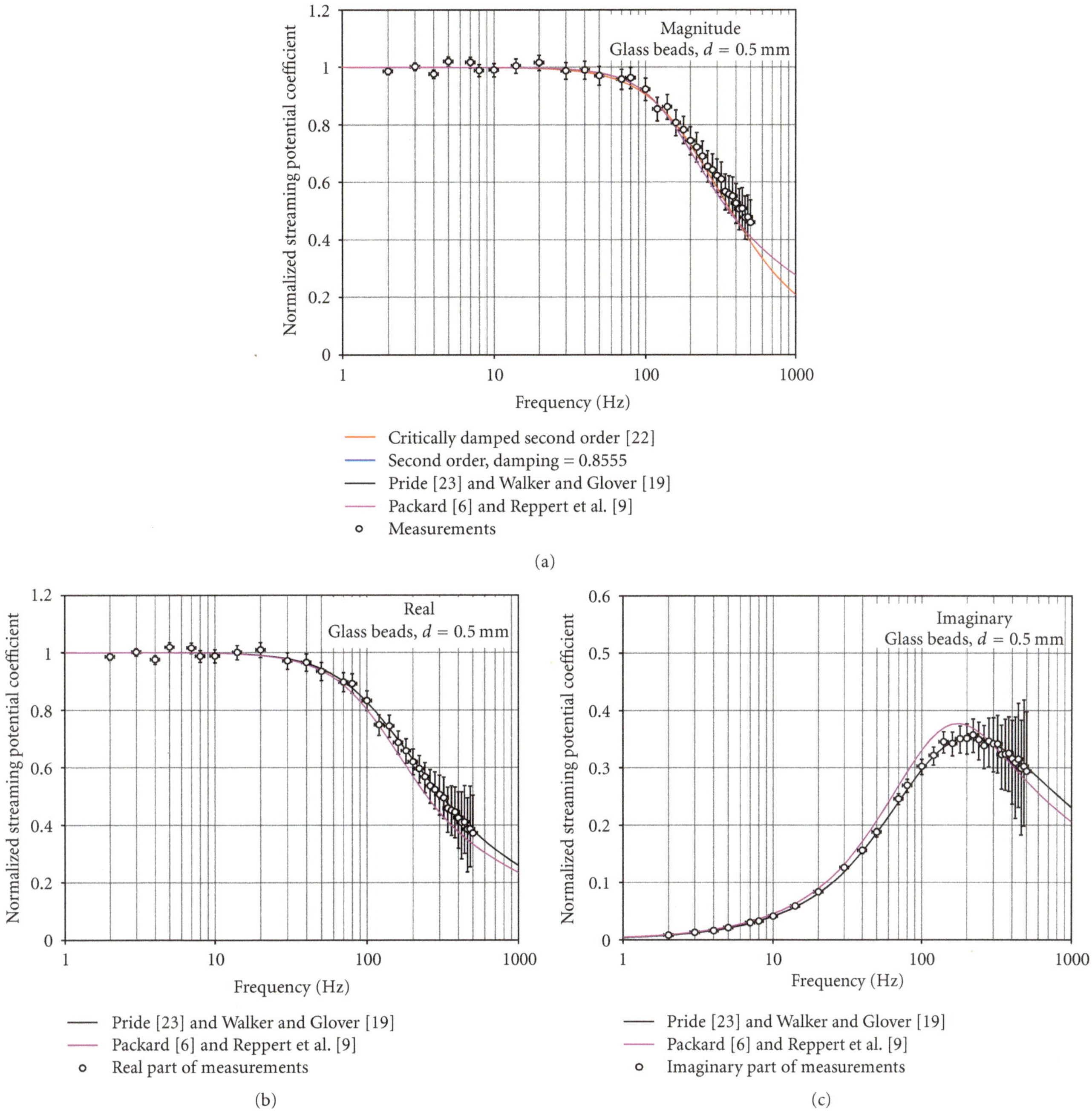

FIGURE 4: The calculated streaming potential coupling coefficient (normalised to the value at 2 Hz, which was 1.37 V/MPa) as a function of frequency for a pack of nominally 0.5 mm diameter glass beads saturated with 10^{-3} mol/L NaCl solution at 24°C ($\phi = 0.383$, $\kappa_{DC} = 1.62 \times 10^{-10}\,\text{m}^2$). (a) Magnitude with six models shown fitting the data [6, 9, 19, 22, 23], (b) real component for models [6, 9, 19, 23], and (c) imaginary component for models [6, 9, 19, 23].

difficulty in measuring these small pressures. There was no evidence for turbulent fluid flow at large piston amplitudes.

3.4. Initial Frequency-Dependent Streaming Potential Coupling Coefficient Results. Figures 3, 4, 5 and 6 show the normalised measured streaming potential coupling coefficient for Ottawa sand and the three grades of glass bead as a function of frequency. The normalised streaming potential coupling coefficient was calculated by dividing the measured streaming potential coupling coefficient at a given frequency by that measured at the lowest frequency available (5 Hz for Ottawa sand and 2 Hz for the glass beads). If one can assume that the streaming potential coupling coefficient at this low frequency approximates to that during steady-state flow, it is possible to say that the data shown in Figures 3–6 represent the right-hand side of (2), (3), (4), (11) and (14), that is, the frequency-dependent terms.

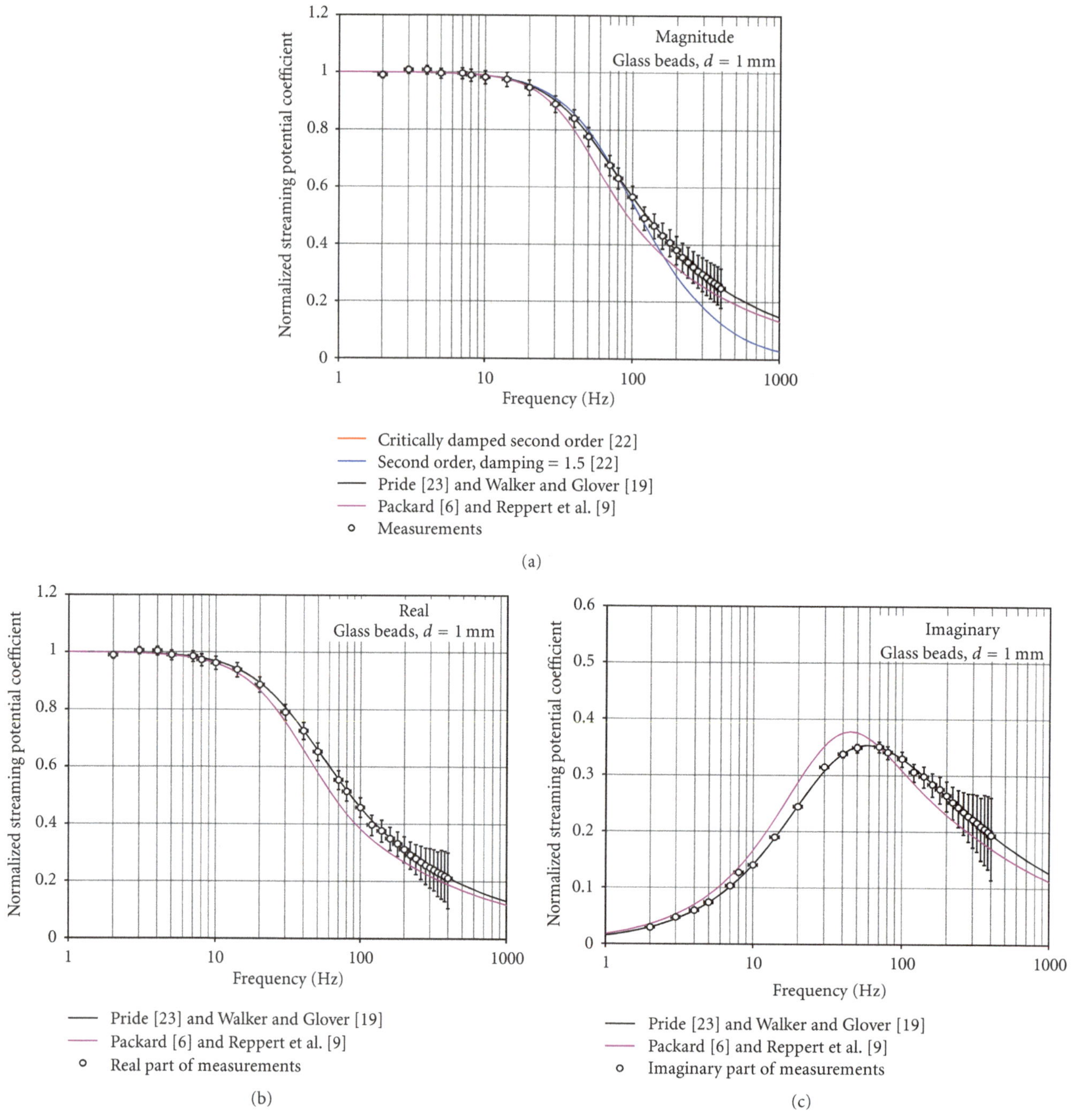

FIGURE 5: The calculated streaming potential coupling coefficient (normalised to the value at 2 Hz, which was 1.61 V/MPa) as a function of frequency for a pack of nominally 1 mm diameter glass beads saturated with 10^{-3} mol/L NaCl solution at 24°C ($\phi = 0.38$, $\kappa_{DC} = 5.95 \times 10^{-10}$ m^2). (a) Magnitude with six models shown fitting the data [6, 9, 19, 22, 23], (b) real component for models [6, 9, 19, 23], and (c) imaginary component for models [6, 9, 19, 23].

It should be noted in these figures that the error bars become larger at the higher frequencies. This is due to the difficulty in measuring small streaming potentials at frequencies greater than the transition frequency.

4. Analysis and Modelling of the Experimental Results

4.1. Steady-State Streaming Potential Coupling Coefficient. The majority of the data analysis will concentrate on the frequency-dependent part of the streaming potential coupling coefficient. However, we should say a few words about the steady state streaming potential coupling coefficient. Although this was not measured in our apparatus, we can perhaps use the streaming potential coupling coefficient at the lowest frequency as a reasonable indication of that under true steady-state conditions considering that Figures 3–6 show that these values tend towards the steadystate value. The lowest frequency for the Ottawa sand was 5 Hz while those for the glass beads was 2 Hz. We will call the

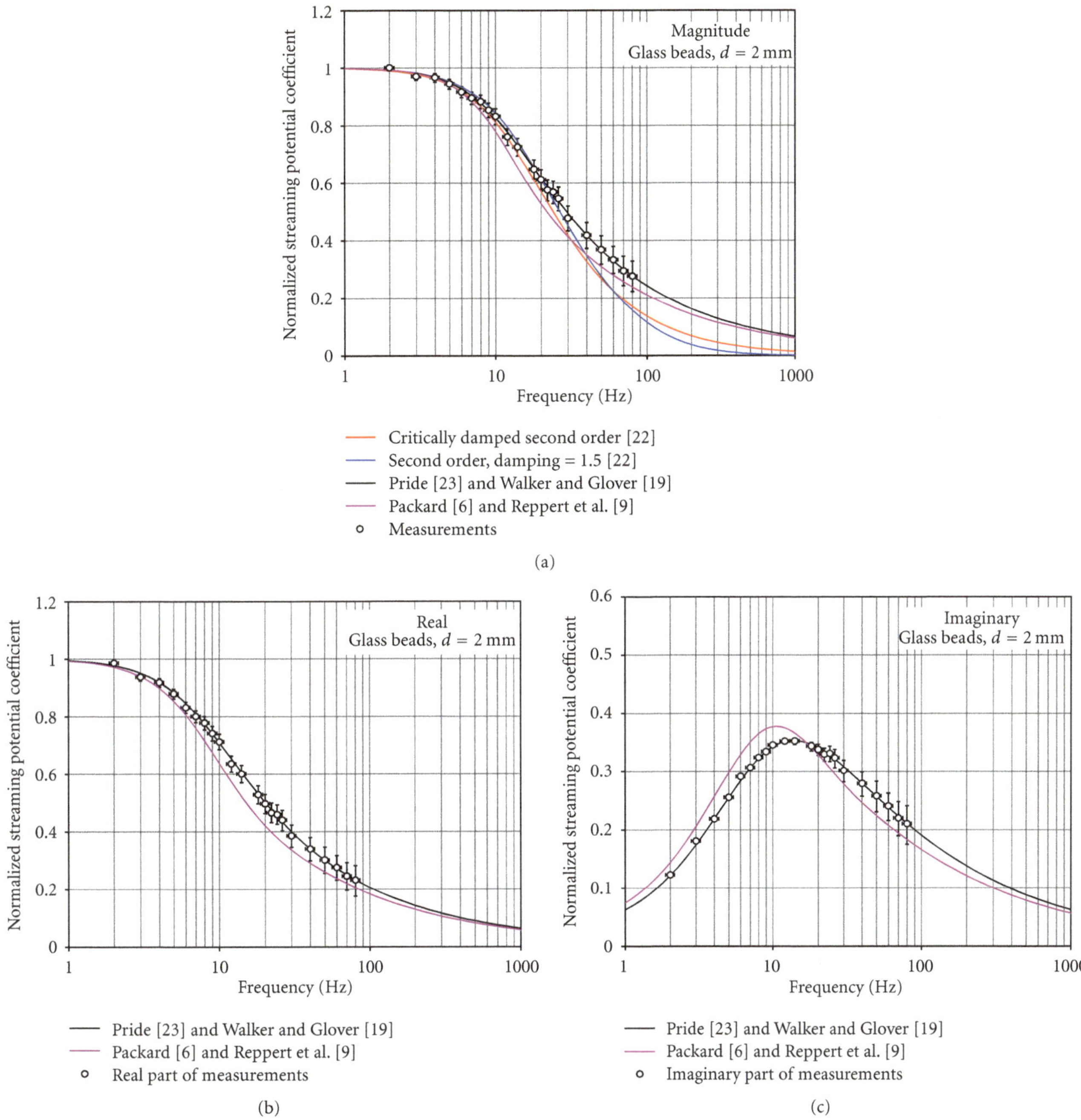

FIGURE 6: The calculated streaming potential coupling coefficient (normalised to the value at 2 Hz, which was 1.80 V/MPa) as a function of frequency for a pack of nominally 2 mm diameter glass beads saturated with 10^{-3} mol/L NaCl solution at 24°C (ϕ = 0.382, κ_{DC} = 27.2×10^{-10} m^2). (a) Magnitude with six models shown fitting the data [6, 9, 19, 22, 23], (b) real component for models [6, 9, 19, 23], and (c) imaginary component for models [6, 9, 19, 23].

streaming potential coupling coefficients at these frequencies the quasi-steady-state values. The quasi-steady-state values of the streaming potential coupling coefficient for each of the samples are shown in Table 2. This table also shows some electrokinetic modelling that we have carried out using the same approach as Glover and Déry [17] and Glover et al. [16]. In this modelling we kept the following parameters constant: Γ_o = 5 sites/nm^2, pK_{me} = 7.5, pK_{-} = 8, while the formation factor, porosity, cementation exponent, grain diameter fluid concentration, and pH were set to the values related to each sample (Table 1). It can be seen from Table 2 that the modelled values of the steady-state streaming potential coupling coefficient slightly overestimate the measured values. We need to put the measured and modelled values in the context of the measurements made by others and the experimental errors. Figure 7 shows the measured values and the model curves in the context of a database of other steady-state streaming potential coupling coefficient measurements

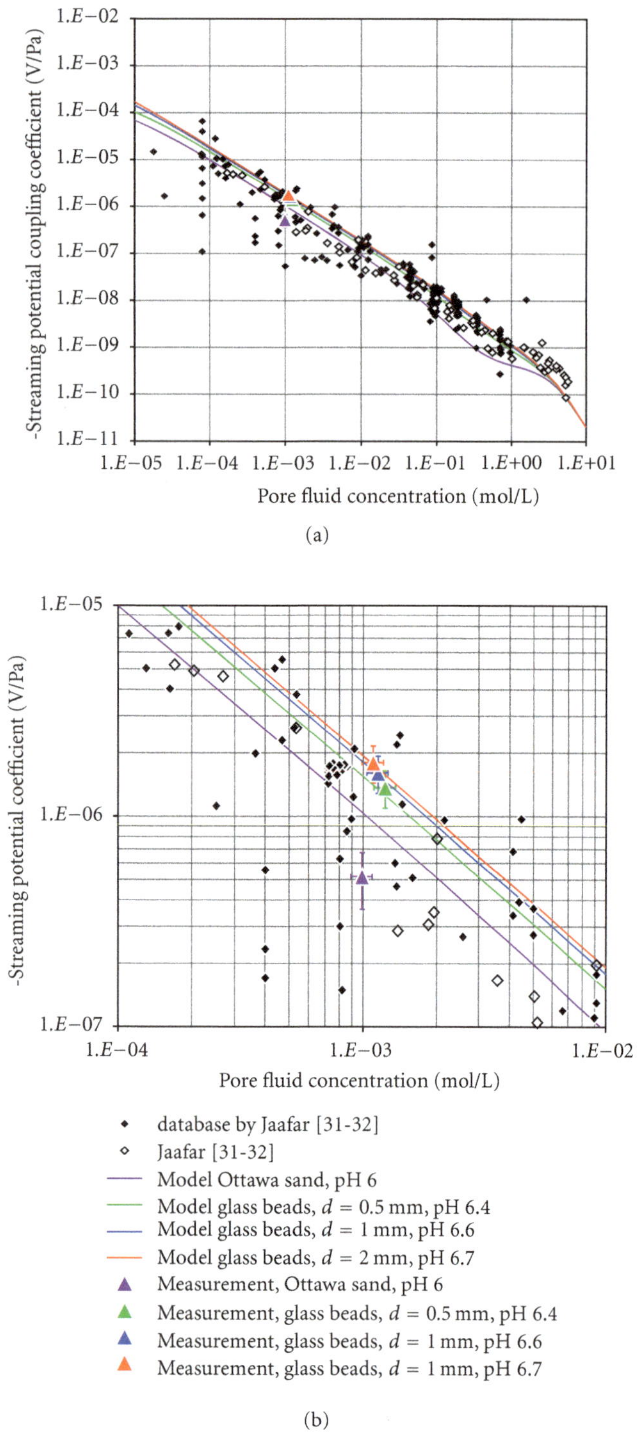

FIGURE 7: (a) The steady-state streaming potential coupling coefficient measured in this work shown with a compilation of silica-based earth materials measured by (open symbols) or compiled by (solid symbols) Jaafar [31, 32]. The lines represent the theoretically modelled streaming potential coupling coefficient using the method of Glover et al. [16] and Glover and Déry [17] and with the parameters $\Gamma_o = 5\ \text{sites/nm}^2$, $pK_{me} = 7.5$, $pK_{-} = 8$, and $T = 24.0°C$, with the porosity, cementation exponent, grain size, and pH for the individual samples (given in Table 1). (b) The same diagram as (a) on an expanded scale.

(from [31, 32]). It should be noted that the experimental values are plotted here at the fluid concentrations calculated from the fluid electrical conductivity that was measured on the fluid leaving the apparatus during the experiment rather than that of the original stock fluid, which was 0.001 mol/L. The error bars are approximate and represent a conservative assessment of the measurement errors (30% for the streaming potential coupling coefficient and 10% for the pore fluid concentration). It can be seen that the measured values are in fact in extremely good agreement with the existing data and the model. The model is extremely sensitive to the values of fluid concentration and pH used, and hence it is extremely important to have an accurate measurement of the pH value and the electrical conductivity of the fluid that emerges from the experimental apparatus during the experiment in order that the measurements can be compared with the model at an acceptable level of precision.

4.2. Frequency-Dependent Streaming Potential Coupling Coefficient. Figures 3–6 show the experimental data together with theoretical curves that (i) are based on standard equations in vibration mechanics, for example, [22], (ii) have been developed for capillary tubes [6, 9], and (iii) have been derived for porous media [19, 23]. In each figure there are three parts. The first shows the behaviour of the magnitude (absolute value, modulus) of the complex variable, while the other two parts show the real and imaginary components of the complex variable. The vibrational mechanics models [22], the Pride model [23], and its simplification [19] may all be fitted to the data to obtain the transition frequency ω_t, from which the characteristic pore radius of the sand can be calculated using (13), while the Packard model [6] and its simplification [9] may be used to obtain the effective capillary radius of the sand directly. A single curve is included for the Packard capillary tube model [6] and the Reppert et al. simplification [9] because they are indistinguishable at the scale of the figures and which indicates that the corrected Reppert et al. simplification (i.e., (8)) performs extremely well. The Pride model [23] and its simplification by Walker and Glover [19] are also represented by a single curve for the same reason.

Figures 3–6 show that all of the models describe the data fairly well. However, the following discussion shows that some of these fits do not use parameters that are consistent with other physical properties of the samples. The best fit for all the samples is provided by the Pride model [23] and its Walker and Glover simplification [19]. These models are specifically designed for porous media but do not implicitly take into account the surface conduction that occurs naturally in geological porous media. In fitting this model, we used the independently measured quasi-steady-state permeability k_{DC}, the electrical tortuosity τ_e that was calculated from the electrical impedance measurements, and the porosity by helium pycnometry ϕ, all of which are given in Table 1. The fitting variables were the transition frequency and the characteristic length scale of the pore space Λ. The values of the fitting variables are given in Table 2. The transition frequency can be used to derive an effective

Table 2: Summary of results.

Property	Unit	Ottawa sand	Glass beads 0. 5 mm	Glass beads 1 mm	Glass beads 2 mm	Comment
Steady-state electrokinetic modelling						
Measured steady-state streaming potential coupling coefficient, C_{so}	V/MPa	0.518 ± 0.155	1.37 ± 0.411	1.61 ± 0.483	1.80 ± 0.54	Value at lowest frequency measured.
Modelled steady-state streaming potential coupling coefficient, C_{so}	V/MPa	1.05	1.26	1.57	1.76	Using [16, 17].
Modelled zeta potential, ζ	mV	−15.9	−29.6	−29.6	−29.6	Using [16, 17].
pH for electrokinetic modelling	—	6	6.7	6.7	6.7	
Transition frequencies						
Transition frequency, critically damped 2nd order vibrational model, ω_t	Hz	230	234	54	13	Using [22].
Transition frequency, 2nd order model with variable damping, ω_t	Hz	748.8	636.9	176.4	41.7	Using [22].
Damping factor, ξ	—	1.5	1.5	1.5	1.5	Using [22].
Transition frequency from the Pride model, ω_t	Hz	256.58	213	58.79	13.85	Using [23].
Transition frequency from the Glover and Walker simplification, ω_t	Hz	256.58	213.36	57.32	12.61	Using [19].
Predicted effective pore radius						
Calculated effective pore radius from independent measurement, r_p	μm	67.6 ± 16.2	70.8 ± 0.70	139 ± 3.41	287 ± 4.03	From Table 1
From the Packard model, equivalent capillary tube radius, r_{eff}	μm	67.5	72	145	302	Using [6].
From the critically damped 2nd order vibrational model	μm	70.46	69.85	145.41	296.35	Using (12)
From the Pride model	μm	66.71	73.21	139.36	287.11	Using (12)
From the Glover and Walker simplification	μm	66.71	73.15	141.13	300.90	Using (12)
Characteristic length scale, Λ	μm	62.40	67.76	131.45	280.24	Using (10)
Predicted permeabilities						
Measured permeability, κ_{meas}	m^2	1.19×10^{-10}	1.62×10^{-10}	5.95×10^{-10}	27.2×10^{-10}	See text.
Predicted permeability using the RGPZ method, κ_{RGPZ}	m^2	1.22×10^{-10}	1.52×10^{-10}	5.80×10^{-10}	24.8×10^{-10}	Using [26].
From the critically damped 2nd order vibrational model	m^2	1.33×10^{-10}	1.48×10^{-10}	6.32×10^{-10}	26.4×10^{-10}	Using (12)
From the Pride model	m^2	1.19×10^{-10}	1.62×10^{-10}	5.80×10^{-10}	24.8×10^{-10}	Using (12)
From the Glover and Walker simplification	m^2	1.19×10^{-10}	1.62×10^{-10}	5.95×10^{-10}	27.2×10^{-10}	Using (12)

pore size for the sample r_{eff} using (13), which is also given in Table 2. When this is done, it is clear that both Λ and r_{eff} agree well with the independently obtained pore size of the sample r_p (Tables 1 and 2). We conclude that the Pride model and its simplification perform extremely well when compared to experimental data.

While not as effective as the Pride model and its simplification, the Packard model [6] and its Reppert et al. simplification [9] also provide a fairly good fit to the data. However, the discrepancies between these models and the data clearly show that a model which is based on a bundle of capillary tubes is not as effective as the Pride model in describing a porous medium when the fluid flow and electrical flow have a tortuosity which is significantly different from unity such as in our samples where the electrical tortuosity is approximately 1.5 (see Table 1). Many rocks have tortuosities much higher than this. Hence, one would expect the Packard model and Reppert et al. models to perform worse for these rocks, overestimating the effective capillary radius and hence the predicted permeability of the sample. The advantage with these models is that when used with a single pore fluid at a constant temperature and pressure, they use a single variable: the effective capillary radius, which is shown in Table 2 for comparison with the independently measured value r_p.

The critically damped second-order vibrational mechanics model also provides a reasonable fit to the data, giving transition frequencies and effective pore radii that are consistent with the independently obtained measurements (Table 2). However, the variably damped second-order vibrational mechanics model has difficulty fitting the data. Here there are two fitting parameters, the transition frequency and the damping coefficient. A large number of different combinations of these parameters provide curves that seem to fit the experimental data approximately. Figures 3–6 show one particular combination where the damping coefficient $\xi = 1.5$ and the transition frequencies are given in Table 2. The transition frequencies are clearly much too large and predict effective pore radii that are badly underestimated. Increasing the damping coefficient further allows the frequency roll-off to approach the experimental data, but only at the expense of even higher predicted transition frequencies. We conclude, therefore, that this model is of no practical use when describing frequency-dependent streaming potential coefficients of porous media.

4.3. Sensitivity of the Pride Model to Steady-State Permeability. The full Pride model calculates the transition frequency from the sample porosity, electrical tortuosity, and permeability as well as the density and viscosity of the pore fluid (9). It also calculates a parameter which we have called m^* (10) from the porosity, characteristic length scale of the pores Λ, the electrical tortuosity, and the permeability of the sample. It became clear in our modelling that the fit to the experimental data depends strongly on the value of m^*, which in turn depends upon the steady-state permeability of the rock k_{DC}, its electrical tortuosity τ_e, and its porosity ϕ as well as the characteristic length scale of its pores Λ. While τ_e, ϕ, and Λ vary from sample to sample, they can

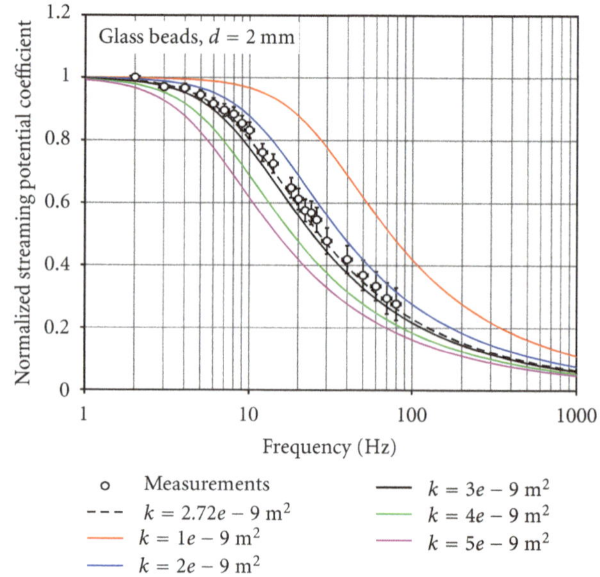

FIGURE 8: The calculated streaming potential coupling coefficient (normalised to the value at 2 Hz, which was 1.80 V/MPa) as a function of frequency using the Pride model [23] for a pack of nominally 2 mm diameter glass beads saturated with 10^{-3} mol/L NaCl solution at 24°C ($\phi = 0.382$, $\kappa_{DC} = 27.2 \times 10^{-10}$ m^2). Each solid curve shows the results of the model for a different value of permeability. The dashed line is for the permeability of the sample that was measured independently.

be measured independently. This is also true of the steady-state permeability of the rock. However, the steady-state permeability of the rock can vary over many orders of magnitude which makes the Pride model extremely sensitive to this parameter. Figure 8 shows the Pride model for a range of different permeabilities from 1×10^{-9} m^2 to 5×10^{-9} m^2 and the other parameters matching those for our 2 mm glass bead dataset. It is worthwhile noting that this is a very small permeability range considering (i) the precision that is usual when measuring permeability in the laboratory and (ii) the natural variability of permeability from sample to sample even in isotropic, homogeneous clean reservoir rocks. While this sensitivity might be considered to be a problem when forward modelling, it is a huge advantage when backward modelling as it should allow the predicted permeability to have a very high precision. Hence, although these measurements are currently difficult to make, they have the potential of providing an extremely precise method of obtaining the permeability of porous media.

4.4. Transition Frequencies and Pore Sizes. The transition frequencies and capillary radii calculated from each of the 5 models are given in Table 2. All of the transition frequencies, except those from the second order vibrational mechanics model with variable damping, are in broad agreement and vary according to the pore size of the sample. Figure 9 shows all the existing data as compiled by Tardif et al. [11] together with the Ottawa sand data of [11] and the new glass bead data presented in this paper. The dashed lines represent the

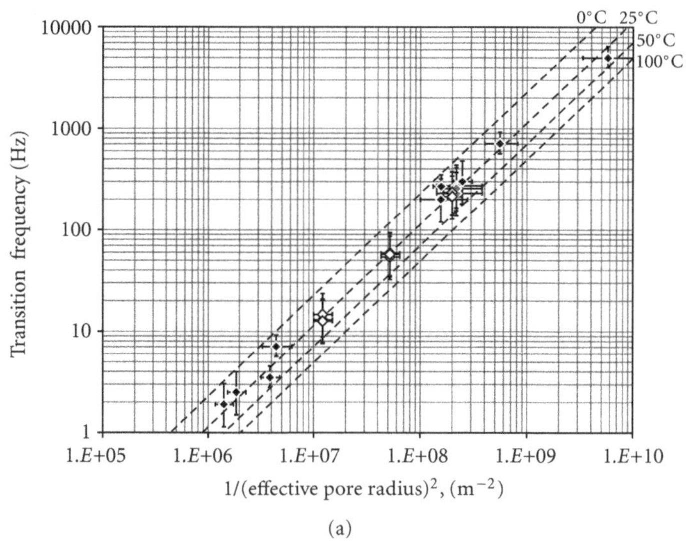

FIGURE 9: The electrokinetic transition frequency as a function of the inverse square characteristic pore size (a) in full and (b) at expanded scale. Black symbols, previous data for capillary tubes, filters frits, and rocks, Figure 6 of [19]; grey symbols, Ottawa sand measured at 24.0°C ± 0.8°C; open symbols, glass bead packs. The dashed lines represent the theoretical result [19] at four different temperatures.

result of (13) [19] using the fluid viscosities and densities for the pore fluid at various different temperatures. It is clear that the new data are compatible with the independently obtained effective pore radius measurements.

The transition frequency can be used to predict the effective pore radius of each sample using (13). Table 2 shows that all the predictions are in good agreement with the value of effective pore radius derived from the measured grain diameter using the method of Glover and Walker [12] and with the equivalent capillary radius derived from the Packard model [6].

The transition frequency can be used to predict the steady-state permeability of the sample using (13). Table 2 shows that all the predictions are in good agreement with the measured values of permeability and the permeability predicted using the RGPZ method [26]. It should be noted that the predictions for the porous media models [19, 23] are exact because the permeability is an input to these models. Hence, these models can only be used to predict permeability if they are fitted to the experimental data with the permeability as a variable parameter.

5. Conclusions

We have used the electromagnetic drive approach to create an experimental apparatus to measure the dynamic streaming potential coupling coefficient of disaggregated porous media between 1 Hz and 1 kHz. The apparatus has been used to measure samples of Ottawa sand and glass bead packs. Measurements were made on Ottawa sand between 5 Hz and 200 Hz, and on glass bead packs between 2 Hz and 500 Hz. In most cases the full variation either side of the transition frequency was captured. Measurements were possible up to 1 kHz, but in practice the streaming potential values became so small at high frequencies that they were unreliable; only those with reasonably small errors have been included in this paper.

Analysis of the steady-state part of the measured data, shows that the measured steady-state streaming potential coupling coefficient is compatible with the latest theoretical models of electrokinetics.

The dynamic experimental data, in the form of normalised streaming potential coupling coefficient, have been fitted with five theoretical models that were derived (i) from vibrational mechanics theory, (ii) for bundles of capillary tubes, and (iii) for porous media. The Pride model and its simplification, which were developed for porous media, fitted the data best and provided transition frequencies, characteristic length scales, and effective pore radii that were consistent with independently measured values for the samples. The Packard model and its simplification, which were developed for capillary tubes, also performed well. The second order vibrational mechanics model with variable damping only fitted the data when unreasonable transition frequencies were used, but the critically damped second order vibrational model performed reasonably well.

We found that the Pride model and its simplification models are extremely sensitive to the steady-state permeability which may cause difficulties in forward modelling given that this parameter is rarely known precisely and that there is often a large range of permeabilities even in isotropic, homogeneous, clean reservoir rock. However, the sensitivity is an advantage in reverse modelling as it should allow precise permeability determinations to be made by fitting this model to experimental dynamic streaming potential coupling coefficient data.

Acknowledgments

This work has been made possible thanks to funding by the Natural Sciences and Engineering Research Council of Canada (NSERC) Discovery Grant Programme. The authors would also like to thank Guillaume Lalande and the members of the mechanical engineering workshop for their help.

References

[1] P. W. J. Glover, J. Ruel, and E. Tardif, "Frequency-dependent streaming potential of porous media: 1. Experimental approaches and apparatus design," *International Journal of Geophysics*. In press.

[2] C. Bordes, L. Jouniaux, S. Garambois, M. Dietrich, J. P. Pozzi, and S. Gaffet, "Evidence of the theoretically predicted seismo-magnetic conversion," *Geophysical Journal International*, vol. 174, no. 2, pp. 489–504, 2008.

[3] C. Bordes, L. Jouniaux, M. Dietrich, J. P. Pozzi, and S. Garambois, "First laboratory measurements of seismo-magnetic conversions in fluid-filled Fontainebleau sand," *Geophysical Research Letters*, vol. 33, no. 1, Article ID L01302, 2006.

[4] R. Chandler, "Transient streaming potential measurements on fluid-saturated porous structures: an experimental verification of biot's slow wave in the quasi-static limit," *Journal of the Acoustical Society of America*, vol. 70, no. 1, pp. 116–121, 1981.

[5] P. M. Reppert and F. D. Morgan, "Streaming potential collection and data processing techniques," *Journal of Colloid and Interface Science*, vol. 233, no. 2, pp. 348–355, 2001.

[6] R. G. Packard, "Streaming potentials across glass capillaries for sinusoidal pressure," *The Journal of Chemical Physics*, vol. 21, no. 2, pp. 303–307, 1953.

[7] C. E. Cooke, "Study of electrokinetic effects using sinusoidal pressure and voltage," *The Journal of Chemical Physics*, vol. 23, no. 12, pp. 2299–2303, 1955.

[8] A. R. Sears and J. N. Groves, "The use of oscillating laminar flow streaming potential measurements to determine the zeta potential of a capillary surface," *Journal of Colloid and Interface Science*, vol. 65, no. 3, pp. 479–482, 1978.

[9] P. M. Reppert, F. D. Morgan, D. P. Lesmes, and L. Jouniaux, "Frequency-dependent streaming potentials," *Journal of Colloid and Interface Science*, vol. 234, no. 1, pp. 194–203, 2001.

[10] P. M. Reppert, *Electrokinetics in the earth*, Ph.D. thesis, Massachusetts Institute of Technology, 2000.

[11] E. Tardif, P. W.J. Glover, and J. Ruel, "Frequency-dependent streaming potential of Ottawa sand," *Journal of Geophysical Research B*, vol. 116, no. 4, Article ID B04206, 2011.

[12] P. W. J. Glover and E. Walker, "A grain-size to effective pore-size transformation derived from electrokinetic theory," *Geophysics*, vol. 74, no. 1, pp. E17–E29, 2009.

[13] F. D. Morgan, E. R. Williams, and T. R. Madden, "Streaming potential properties of Westerly granite with applications," *Journal of Geophysical Research B*, vol. 94, no. 9, pp. 12449–12461, 1989.

[14] P. M. Reppert and F. D. Morgan, "Temperature-dependent streaming potentials: 1. Theory," *Journal of Geophysical Research B*, vol. 108, no. 11, pp. 3–12, 2003.

[15] P. M. Reppert and F. D. Morgan, "Temperature-dependent streaming potentials: 2. Laboratory," *Journal of Geophysical Research B*, vol. 108, no. 11, pp. 4–13, 2003.

[16] P. W. J. Glover, E. Walker, and M. D. Jackson, "Streaming-potential coefficient of reservoir rock: A theoretical model," *Geophysics*, vol. 77, no. 2, pp. D17–D43, 2012.

[17] P. W. J. Glover and N. Déry, "Streaming potential coupling coefficient of quartz glass bead packs: dependence on grain diameter, pore size, and pore throat radius," *Geophysics*, vol. 75, no. 6, pp. F225–F241, 2010.

[18] D. L. Johnson, J. Koplik, and R. Dashen, "Theory of dynamic permeability and tortuosity in fluid-saturated porous media," *Journal of Fluid Mechanics*, vol. 176, pp. 379–402, 1987.

[19] E. Walker and P. W. J. Glover, "Characteristic pore size, permeability and the electrokinetic coupling coefficient transition frequency in porous media," *Geophysics*, vol. 75, no. 6, pp. E235–E246, 2010.

[20] A. Revil and L. M. Cathles III, "Permeability of shaly sands," *Water Resources Research*, vol. 35, no. 3, pp. 651–662, 1999.

[21] A. Revil, P. A. Pezard, and P. W. J. Glover, "Streaming potential in porous media 1. Theory of the zeta potential," *Journal of Geophysical Research B*, vol. 104, no. 9, pp. 20021–20031, 1999.

[22] W. T. Thomson and M. D. Dahleh, *Theory of Vibration with Applications*, Prentice Hall, 1998.

[23] S. Pride, "Governing equations for the coupled electromagnetics and acoustics of porous media," *Physical Review B*, vol. 50, no. 21, pp. 15678–15696, 1994.

[24] E. Charlaix, A. P. Kushnick, and J. P. Stokes, "Experimental study of dynamic permeability in porous media," *Physical Review Letters*, vol. 61, no. 14, pp. 1595–1598, 1988.

[25] P. W. J. Glover and E. Walker, "Modelling the frequency dependence of hydraulic flow and streaming potential coupling coefficients in capillary bundles and porous rocks," in *Proceedings of the Annual Congress of European Geosciences Union*, paper A-02892, Vienna, Austria7, April 2008.

[26] P. W. J. Glover, I. I. Zadjali, and K. A. Frew, "Permeability prediction from MICP and NMR data using an electrokinetic approach," *Geophysics*, vol. 71, no. 4, pp. F49–F60, 2006.

[27] S. N. Lvov and R. H. Wood, "Equation of state of aqueous NaCl solutions over a wide range of temperatures, pressures and concentrations," *Fluid Phase Equilibria*, vol. 60, no. 3, pp. 273–287, 1990.

[28] S.L. Phillips, H. Ozbek, and R.J. Otto, "Basic energy properties of electrolytic solutions database," in *Proceedings of the 6th International CODATA Conference*, Santa Flavia, Italy, May 1978, http://www.osti.gov/bridge/purl.cover.jsp; jsessionid =3954E775156A8BC0FA35DB5CE5B402D4? purl=/ 6269880-iPJPhB/ DB5CE5B402D4?purl =/6269880-iPJPhB/.

[29] P. N. Sen and P. A. Goode, "Influence of temperature on electrical conductivity on shaly sands," *Geophysics*, vol. 57, no. 1, pp. 89–96, 1992.

[30] P. Leroy, A. Revil, A. Kemna, P. Cosenza, and A. Ghorbani, "Complex conductivity of water-saturated packs of glass beads," *Journal of Colloid and Interface Science*, vol. 321, no. 1, pp. 103–117, 2008.

[31] M. Z. Jaafar, J. Vinogradov, and M. D. Jackson, "Measurement of streaming potential coupling coefficient in sandstones saturated with high salinity NaCl brine," *Geophysical Research Letters*, vol. 36, no. 21, Article ID L21306, 2009.

[32] J. Vinogradov, M. Z. Jaafar, and M. D. Jackson, "Measurement of streaming potential coupling coefficient in sandstones saturated with natural and artificial brines at high salinity," *Journal of Geophysical Research B*, vol. 115, no. 12, Article ID B12204, 2010.

[33] P. W. J. Glover, "What is the cementation exponent? A new interpretation," *The Leading Edge*, vol. 28, no. 1, pp. 82–85, 2009.

[34] P. W. J. Glover, "A generalized Archie's law for n phases," *Geophysics*, vol. 75, no. 6, pp. E247–E265, 2010.

New Bouguer Gravity Maps of Venezuela: Representation and Analysis of Free-Air and Bouguer Anomalies with Emphasis on Spectral Analyses and Elastic Thickness

Javier Sanchez-Rojas[1, 2]

[1] *Institut für Geowissenschaften Abteilung Geophysik, Christian-Albrechts-Universität zu Kiel, Otto-Hahn-Platz 1, 24118 Kiel, Germany*
[2] *Departamento de Geofísica, Fundación Venezolana de Investigaciones Sismológicas FUNVISIS, 1070 Caracas, Venezuela*

Correspondence should be addressed to Javier Sanchez-Rojas, jsanchez@geophysik.uni-kiel.de

Academic Editor: Joerg Schleicher

A new gravity data compilation for Venezuela was processed and homogenized. Gravity was measured in reference to the International Gravity Standardization Net 1971, and the complete Bouguer anomaly was calculated by using the Geodetic Reference System 1980 and 2.67 Mg/m^3. A regional gravity map was computed by removing wavelengths higher than 200 km from the Bouguer anomaly. After the anomaly separation, regional and residual Bouguer gravity fields were then critically discussed in term of the regional tectonic features. Results were compared with the previous geological and tectonic information obtained from former studies. Gravity and topography data in the spectral domain were used to examine the elastic thickness and depths of the structures of the causative measured anomaly. According to the power spectrum analysis results of the gravity data, the averaged Moho depths for the massif, plains, and mountainous areas in Venezuela are 42, 35, and 40 km, respectively. The averaged admittance function computed from the topography and Free-Air anomaly profiles across Mérida Andes showed a good fit for a regional compensation model with an effective elastic thickness of 15 km.

1. Introduction (Earlier Gravity Mapping in Venezuela)

Gravity surveys have been carried out in Venezuela since 1945 with the intensification of oil exploration. Thus, gravity surveying was first confined to the oil-producing sedimentary basins (i.e., Maracaibo and Eastern Venezuela basins). The first precise gravimetric survey covering the whole country was carried out as late as 1970 under the framework of the Latin American Gravity Standardization Network [1], which established a national network. The Venezuelan National Geographic Institute published the first gravity map of the entire country after 1988 [2]. Since then, several Venezuelan universities and official institutes have cooperated with international institutes to improve the coverage of the national network. In addition, the Venezuelan National Oil Company (PDVSA) has released a large amount of data for educational purposes. In this context, Izarra et al. [3] presented the last gravity data compilation from Simon Bolívar University in 2005. However, due to the extent of the country and the presence of inaccessible areas such as the Amazon forest and Mérida Andes, general improvement in the data coverage has been slow.

The purpose of this work is to present a new Bouguer anomaly map of Venezuela using the data available thus far. This study has four main aims: (1) to study the correlation between the Bouguer anomalies and the known regional geology by means of regional and residual gravity anomaly separation; (2) to estimate the continental crust thickness (Moho) using a spectral technique and to correlate it with Moho estimations derived from independent geophysical techniques (i.e., P-wave velocity models, which stem from wide-angle refraction seismic and the results from receiver function methods); (3) to use the coherence-admittance techniques to determine the spatial variations in elastic thickness (*Te*) in the mountainous areas; (4) to provide a new updated gravity database for future investigations.

2. Geological Setting

Different theories have been proposed about the origin of the Caribbean Plate [4, 5]. Nowadays, it is commonly accepted that extensive volcanism in the mid to late Cretaceous (~90–110 Ma) resulted in the formation of the Caribbean Large Igneous Province and a thickening of large areas of the Caribbean crust. Around this time, the Caribbean plate (CA) initiated an eastward migration relative to South America (SA) [6–10]. The resulting variety in geologic features between the CA and SA margins includes accretionary wedges, fold and thrust belts, and extensional and foreland basins [11]. Current relative plate motions make this margin predominantly strike-slip; it is limited on the east by subduction of oceanic (Atlantic) SA beneath the Lesser Antilles island arc [12] and in the west by partial subduction of the Caribbean plate under South America (Figure 1). On the continental deformation front observed along the Venezuela coastline, the important (large oil-producing) Barinas-Apure and Eastern basins have been formed. Barinas-Apure is mainly a foreland basin generated by the flexural response to the Mérida-Andes Mountain load [13]. On the other hand, the Eastern basin can be considered to be the result of (a) flexural loading of the Cordillera de la Costa range, (b) large and continuous deposition of continental material from the Guyana shield, and (c) subduction dynamics in the east [14]. The Barinas-Apure and Eastern Venezuelan sedimentary basins are separated by a geomorphologic-structural high called El Baúl High, formed by an igneous-metamorphic complex. This massif was described by Bellizzia [15], who differentiated several granitic, volcanic, and metasedimentary units and subunits. For instance, the Maracaibo basin is a foreland basin separated from Barinas-Apure by the emplacement of the Mérida Andes. This basin is bounded on the north by the Oca-Ancon fault system, Sierra de Perijá to the west, and the Mérida Andes mountains to the south and east. Mérida Andes (MA) is a NE2-SW trending mountain chain about 420 km long and with a maximum height of 5 km. This uplifted block was formed as a consequence of the convergence of the Panama arc and western South America [16–19]. According to Schubert [20] and Kohn et al., [21] this convergence has the main period of shortening during the Oligocene-Miocene, and evidence suggest that there is still significant present-day deformation.

Due to the large amount of interactions and its complexity between the CA-SA plates, the location of the plate boundary is controversial. Thus, this boundary has been interpreted as a 300 km wide plate boundary zone [22] that forms an orogenic float. This orogenic float is represented by the Mérida Andes; it is linked to the north with the Northern Cordilleras of Venezuela (i.e., Cordillera de la Costa range) and goes as far as the Trinidad-Tobago islands to the east. The boundary zone is limited by the Southern Caribbean thrust system on the north and the Mérida Southern Foothills, Guárico Frontal thrust belt, and Serranía thrust belt on the south [19]. The Cordillera de la Costa range is formed by different geological provinces containing (from north to south) late Jurassic to early Cretaceous basic and ultrabasic rocks, Precambrian and Paleozoic basement rocks, Jurassic to Cretaceous lower crust-upper mantle fragments, volcano-sedimentary sequences and basaltic to rhyolitic rocks, and late Cretaceous-Paleocene molassic sediments and flysh sequences [23–28].

The stable South American crust is represented by the Guyana Shield. This massif outcrops as sialic Precambrian continental crust composed mainly of metasedimentary and metaigneous rocks at amphibolite to granulite facies that have been intruded by granites [29]. The reported ages of these crystalline rocks range from 3600 to 800 Ma [30].

3. Gravity Database

A new gravity data compilation is presented here that includes data from early compilations by Simon Bolivar University [3] and data from the National Geophysical Data Center (NGDC). In order to improve the quality and coverage of onshore data, new high-resolution gravity surveys from the Venezuelan Foundation for Seismological Research (FUNVISIS) were included. Offshore data coverage was improved by new measurements from several marine surveys provided by the Marine Geoscience Data System (MGDS) [31].

(1) The Simon Bolivar University database consists of datasets from different Venezuelan institutions, oil companies, universities, and international surveys. Most of the data are evenly spaced; however, the compilation includes data with high accuracy (1×10^{-5} m/s^2 or less for gravity and 0.5 m or less for height) and an average station spacing of less than 100 m. This dataset has coordinates measured through the use of precise leveling methods, which were mainly collected by oil companies. In contrast, for data collected by universities, in some cases, the stations coordinates and heights were derived by reading from topographic maps. These data have elevation errors of several meters and errors of higher than 1×10^{-5} m/s^2 for the observed gravity. In addition, this dataset does not cover oil-producing areas and rugged mountainous terrains with limited accessibility. Thus, it cannot be completely discharged.

(2) The NGDC dataset was made available by the National Geophysical Data Center (NGDC), which belongs to the US Department of Commerce. This dataset includes offshore gravity data collected during research cruises from 1953 to the present from several oceanographic institutions and government agencies worldwide. As a consequence, the resolution and accuracy of this data are very variable.

(3) The MGDS dataset was made available by the Marine Geoscience Data System, which serves different communities of National Science Foundation-funded researchers and provides direct access to data

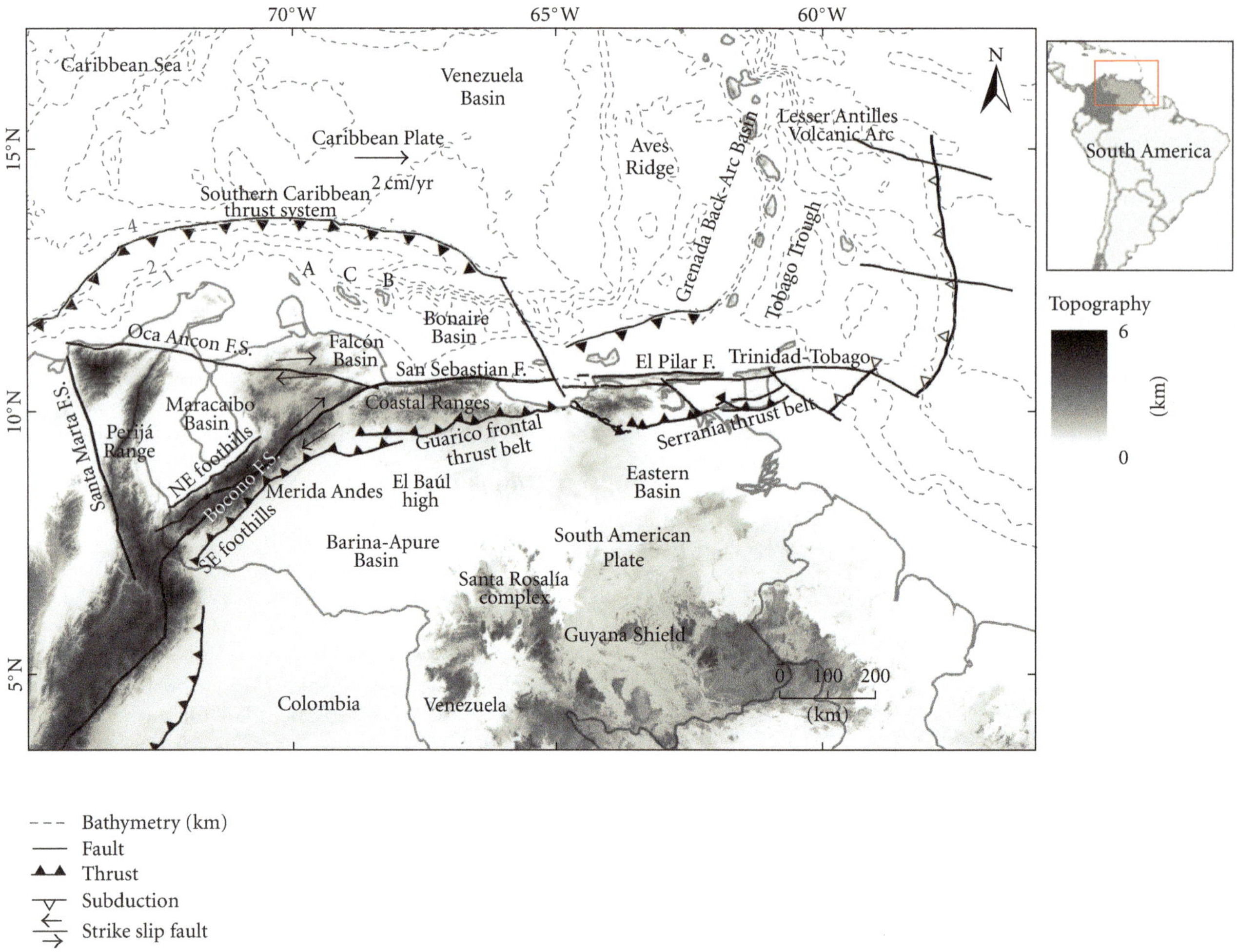

Figure 1: Study area and general tectonic and kinematic features of the northern margin of South America setting (modified from F. E. Audemard and F. A. Audemard [19]) on a digital elevation map. Arrows indicate relative regional motions. Texts indicate the name of geological or tectonic features. A, B, and C mean Aruba, Bonaire, and Curazao (ABC islands).

(http://www.marine-geo.org/). The collected data include research cruises from 1977 to 2004.

(4) The FUNVISIS dataset includes more than 4000 observations collected by a Scintrex CG-5 gravimeter combined with GPS system. The resolution and accuracy of this gravity data are $0.5 \times 10^{-5}\,\text{m/s}^2$ for gravity and ± 1 m for station heights with average station spacing of 500 m.

Presently, the gravity compilation contains about 80,000 onshore observations and more than 40,000 offshore stations (Figure 2). The average station interval is less than 1 km, which results in an average station density of 1 station/km^2 or higher.

3.1. Data Processing. A comparison of the different datasets showed that the gravity datasets and surveys in the compilation refer to different datum levels and exhibit variable quality and accuracy. Therefore, data homogenization focusing on the gravity datum and calibration and the coordinate determination and anomaly equations was required for anomaly reduction. After processing, the data were manually edited to remove erroneous measurements. Stations with outlier gravity values were removed after interpolating a high resolution gravity map. Additionally, stations with erroneous coordinates and/or heights have been removed after comparing the station heights with those obtained by a high resolution digital terrain model (90 m spacing). Stations with heights differences of 50 m or higher were also removed. In total, more than 2000 stations were eliminated from the original database [3].

The Bouguer anomaly was calculated using the following assumptions.

(i) The horizontal coordinates and elevations of the gravity stations based on the Geodetic Reference System 1980.

(ii) Absolute gravity datum is referred to International Gravity Standardization Net 1971 (IGSN71).

(iii) Theoretical gravity calculated at the normal ellipsoid: Somigliana's formula [32].

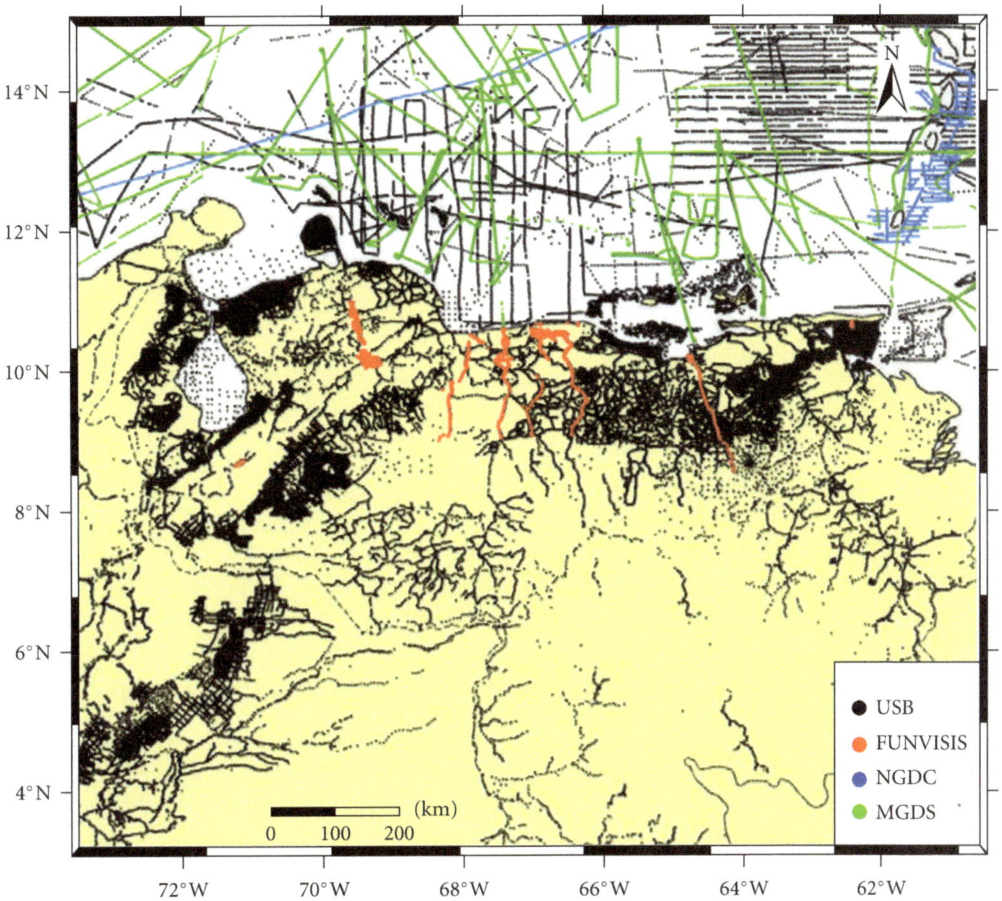

Figure 2: Station distribution map of gravity stations used for the new gravity anomaly map of Venezuela.

(iv) Topography correction calculated for a spherical cap of up to 167 km radius (Hayford zone O2) [33] assuming a constant density of 2670 kg/m^3. This value is close to the mean density of the surface rocks in the investigated area and the standard value used in the Bouguer anomaly correction. The digital terrain model used was based on the Shuttle Radar Topography Mission (SRTM) with a grid spacing of 90 m (onshore), and Gtopo30 (offshore).

(v) Bouguer correction equation for spherical Earth [34].

(vi) Height correction estimated by a Taylor series expansion of normal gravity up to 2nd order [35].

Taking into account all the different sources of errors in the databases, the accuracy of the computed Bouguer anomaly values was estimated to be about $\pm 1-5 \times 10^{-5}$ m/s^2 (1×10^{-5} m/s^2 = 1 mGal). Additional details about the Venezuelan gravity network and the dataset can be found in Drewes et al. [36] and Izarra et al. [3].

4. Gravity Maps

The calculated anomaly map consists of Bouguer anomalies (BA) onshore (correction density of 2.67 Mg/m^3) and Free-Air anomalies (FAA) offshore (Figure 3). Anomaly values range from –225 to 225 $\times$ 10^{-5} m/s^2. Offshore, in the Venezuela Basin (VB), magnitudes of the FAA barely reach low positive values. In contrast, the South Caribbean accretionary prism shows a broad gravity low with a WNW-ESE trend. A gathering of local anomaly highs separated by relative low gravity values appears as local highs along the Leeward Antilles (Aruba, Bonaire, and Curazao). The highest FAA values (more than 200 $\times$ 10^{-5} m/s^2) are caused by the subduction in northeastern Venezuela (i.e., Lesser Antilles volcanic arc), which extends further eastward. In the area of the Grenada basin (GrB) and Tobago trough (TT), gravity values decrease down to -80×10^{-5} and -45×10^{-5} m/s^2, respectively. Onshore, the most relevant BA is the prominent anomaly low observed in the Eastern Venezuelan basin (EVB) mostly caused by a large amount of infill sediments. The BA has NE-SW trending positive values in the area between the Guárico basin (GB) and Guyana shield (i.e., Precambrian rocks).

4.1. Regional and Residual Gravity Maps. In order to analyze the anomalies, a set of wavelength filters were applied to progressively separate local effects from regional effects within the gravity field. Here, the concept of the Butterworth band-pass filter was applied in the frequency domains to separate regional and residual fields. The Butterworth filter is a spectral domain filter with a roll-off and requires an

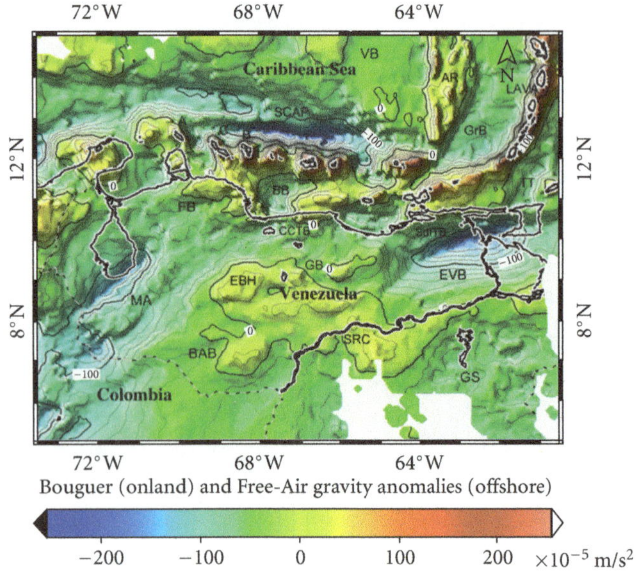

FIGURE 3: Gravity anomaly map of Venezuela. Complete Bouguer anomalies are given on land, and Free-Air anomalies are offshore. The Bouguer reduction density was 2.67 Mg/m^3. Contour interval: 20×10^{-5} m/s^2. Geographical coordinates. Geological features labeled are BB: Bonaire basin, BAB: Barinas Apure basin, CCTB: Coastal Cordillera thrust belt, EBH: El Baúl high, EVB: Eastern Venezuela basin, FB: Falcon basin, GB: Guárico basin, GS: Guyana shield, GrB: Granada basin, LAVA: Lesser Antilles volcanic arc, MA: Mérida Andes, SCAP: South Caribbean accretionary prism, SdITB: Serranía del Interior thrust belt, SRC: Santa Rosalía complex, and TT: Tobago through.

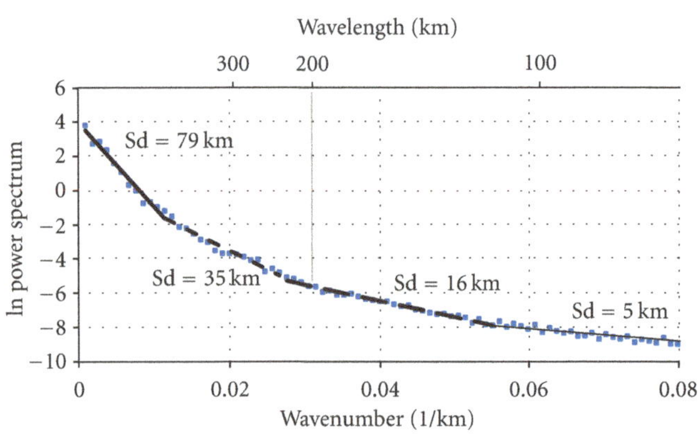

FIGURE 4: Graph representing the natural logarithm of the radially averaged power density spectrum versus the radial wavenumber and wavelength for the Bouguer anomaly map (blue points). The straight lines present the power spectra of mass points located at a maximum estimated depth. Four sources were recognized with estimated depths (Sd) of 79, 35, 16, and 5 km.

order (n) to implement the transition between the passing and rejected portions of the data spectrum; the higher the order, the steeper the transition. Based on the analysis of the radial power spectrum (Figure 4), the wavelength cutoff adopted here retained short-wavelength components for the observed data of 200 km, and the order was set at 3.

Figure 5(a) shows a map of the regional gravity field obtained by applying this filter to the Fourier transform of the demeaned and detrended observed gravity data after conversion to the spatial domain taking the inverse Fourier transform. The residual gravity field (Figure 5(b)) was the result of subtracting the regional gravity field from the observed Bouguer anomaly. The gravity map of the regional component of the gravity field (Figure 5(a)) showed anomalies between +125 and -125×10^{-5} m/s^2. These anomalies are characterized by a WNW-ESE relative gravity low. The residual gravity map (Figure 5(b)) ranges from +125 to −120 $\times 10^{-5}$ m/s^2. The distribution of anomalies is similar to the observed Bouguer anomaly map, and their main features are the strong gradients associated with the Caribbean-South America boundary fault system. The maximum gravity lows are related to basins in the study area (e.g., Granada and Eastern basins), and the maximum gravity highs related to several local geological features such as volcanic and metamorphic outcrops in the ABC islands. These gravity highs and lows show predominant orientation following the Antilles arc region.

5. Power Spectral Analysis

The application of Fourier analysis to the interpretation of potential field data is common and is frequently used to obtain the regional/residual field components of the gravity field. Here, the complete Bouguer gravity anomaly was used to get estimated depths for the structures that cause the measured anomaly. The methodology assumed that for large samples, the logarithm of the power spectrum (E) of the gravity field of a monopole source versus the wave number (radial/distance) may show a linear relationship. The slope of the straight line is proportional to the depth to the top of the corresponding body causing the gravity anomaly. Thus, if k denotes the wavenumber and $S(k)$ denotes the power spectrum of the gravity field, the depth (d) to the source can be estimated from the relation $S(k) = f(k)$ by employing the formula

$$\ln S(k) = -2 * k * d. \quad (1)$$

The power spectrum analysis was carried out through the 2D fast Fourier transformation of the gravity field. Due to the two-dimensional character of the dataset, radial averaging of the power spectrum was performed to obtain a one-dimensional representation [39–43]. Confidence limits for the depth estimations were calculated from the standard errors of the slopes of the best fitting lines for the linear segments.

The gravity data of the study area, covering a surface area of 900×900 km^2, was interpolated to produce a grid with a node spacing of 4 km. Results of the power spectrum analysis for this dataset showed four tendencies for the correlation between the energy (E) and wavenumber (radial/distance) (Figure 6(a)). The most regional part of the spectrum resulted in greater depths of about 79 km. The local part of the spectrum resulted in depths of ~16 km. The main aim of the spectral analysis was focused on the intermediate depths

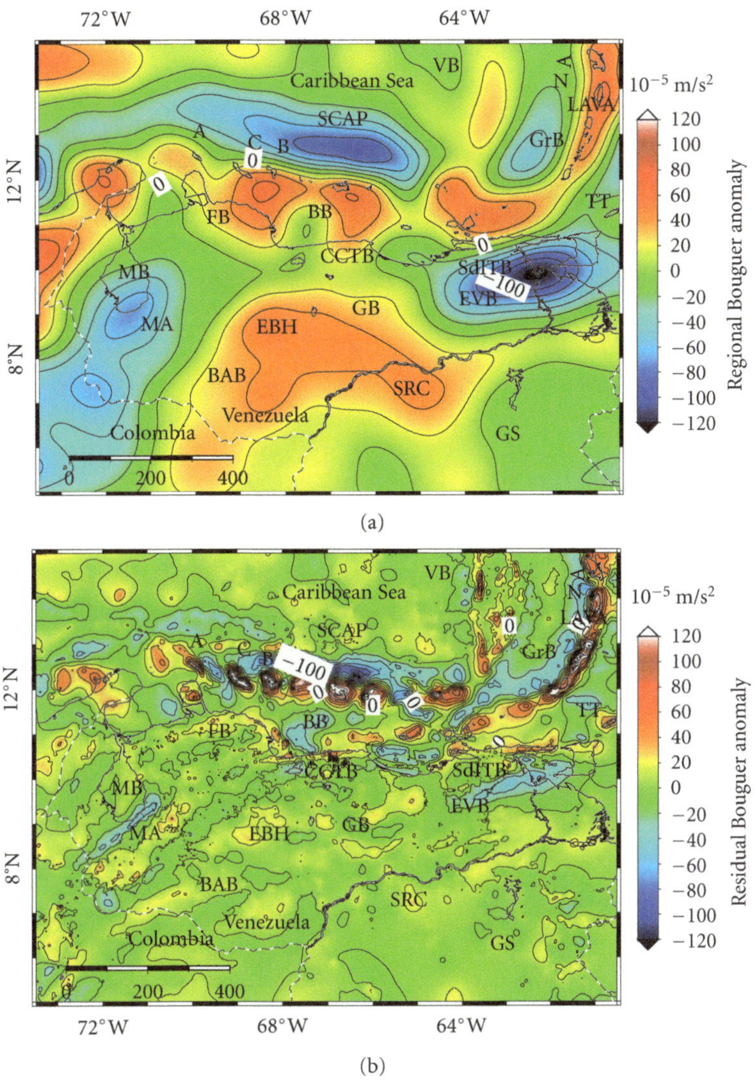

Figure 5: (a) Regional gravity map obtained by applying a Butterworth filter defined by results of the spectral analysis (Figure 4) of the gravity anomaly map. (b) Residual gravity map obtained by subtracting the regional gravity map from the observed gravity map in Figure 3. Both maps: contour interval = 5×10^{-5} m/s^2. Graphical indications as described in Figure 3.

of about 35 km because this depth could correspond to the crust-mantle boundary (Moho). The effect of shallower sources (~5 km) may account for local wavelengths of the gravity field in the area of study. The characteristics and origin of these sources were not investigated because they were outside the main aim of this work.

In order to perform the same analysis over all anomalies in the study area, a data block (window), measuring 300 × 300 km each, was used for the calculation of Moho depths across the N-S and E-W directions. The window size of 300 km^2 corresponds to six-times the expected source depth (Moho) to assure a depth-estimation error of <10% according to Regan and Hinze [44]. Figures 6(a)–6(d) show results for windows over the anomalies observed in the Barinas-Apure basin, Mérida Andes, Maracaibo basin, Guárico basin, Eastern basin, and Leeward Antilles. The final results are shown in Figures 7 and 8.

6. Effective Elastic Thickness Determination

The correlation between gravity observations and sea floor topography has been well documented. This relationship can be analyzed by using a linear transfer function called admittance. The basic techniques for determining the admittance function between the gravity and bathymetry data have been discussed in detail by McKenzie and Bowin [45] and Watts [46].

The method assumes that Free-Air gravity anomalies are caused by topography, and its compensation and attempts to determine a function when convolved with topography produce the gravity anomaly. The advantage of this method is that the admittance function may be derived from the observed data independent of a particular isostatic model and can be interpreted in terms of simple models of isostasy.

The isostatic response method simply involves deriving a filter that, convolved with the bathymetry in space domain,

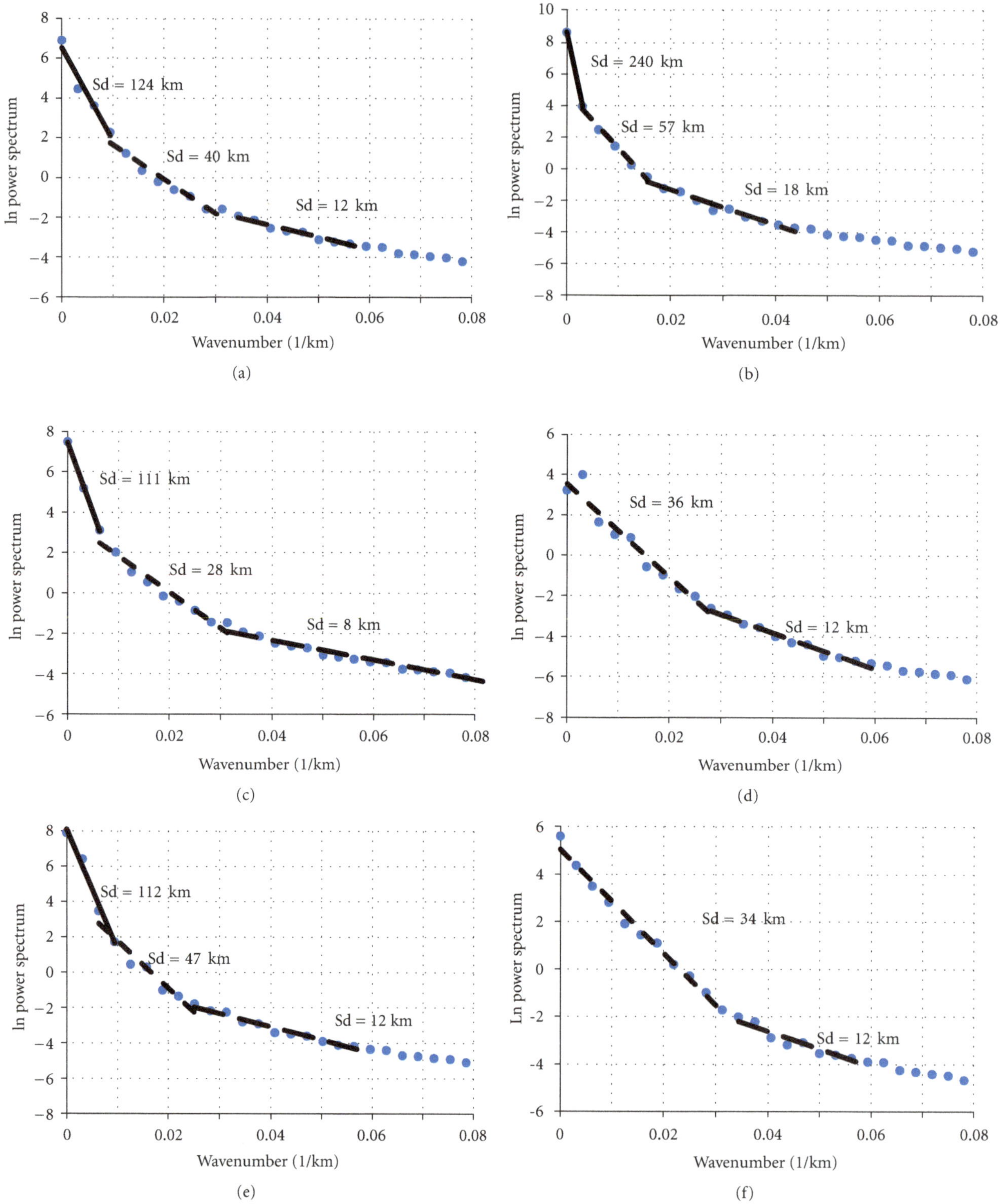

FIGURE 6: Results for the power spectrum analysis of the Bouguer gravity field using 2D fast Fourier transform. Black lines represent the trends of the linear regression separating the spectrum of mass points (blue points) located at a maximum estimated depth: (a) Barinas-Apure basin, (b) Mérida Andes, (c) Maracaibo basin, (d) Guárico basin, (e) Eastern basin, and (f) Leeward Antilles. Estimated depth (Sd).

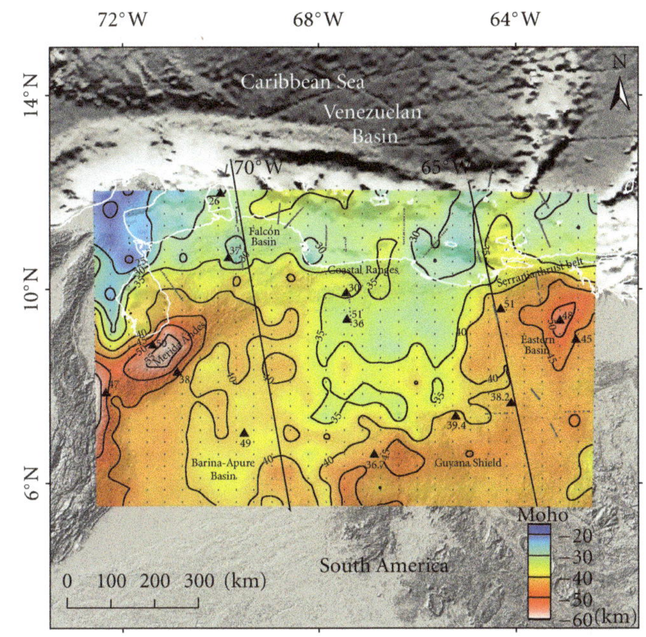

Figure 7: Map of source-depth estimation (Moho) obtained by spectral analysis of gravity data. Triangles are the receiver function Moho depth estimations in kilometers [37]. Gray lines are Moho depths estimated from seismic studies [38]. Location of profiles are shown as black lines labeled by 70 W and 65 W.

produces a series that resembles the observed gravity, again in the space domain. This process can be represented in the space domain by using the convolution operator*:

$$g(x) = f(x) * h(x), \tag{2}$$

where $h(x), f(x)$, and $g(x)$ are the series representing the topography, filter, and Free-Air gravity, respectively. The above convolution in the space domain is equivalent to multiplication in the spatial frequency domain,

$$G(k) = Z(k)H(k), \tag{3}$$

where $G(k), Z(k)$, and $H(k)$ are the discrete Fourier transforms of $g(x), f(x)$, and $h(x)$, respectively. $Z(k)$ is known as the admittance or transfer function and the wavenumber $k = 2\pi/\lambda = n2\pi/L$, where $n = 0, 1, 2, \dots, L/2\Delta\zeta, L$ is the length of the Fourier series, $\Delta\zeta$ is the distance between two consecutive sampling points, and λ is the wavelength. Equation (3) can be rewritten as:

$$Z(k) = \frac{G(k)}{H(k)}. \tag{4}$$

However, the function $Z(k)$ obtained in this way is influenced by noise in the gravity field, particularly at short wavelengths. The noise present in any data may be due to measurement errors, the data reduction procedure, or variations in the structure of the linear system under consideration. In the presence of noise, a better estimate of $Z(k)$ [45] can be obtained from

$$Z(k) = \frac{C(k)}{Eb(k)}, \tag{5}$$

where

$$C(k) = \frac{1}{N}\sum_{r=1}^{N} G_r(k)H_r^*(k),$$
$$Eb(k) = \frac{1}{N}\sum_{r=1}^{N} H_r(k)H_r^*(k). \tag{6}$$

N is the number of profiles, and $C(k)$ is the complex cross-spectrum of bathymetry and gravity. $Eb(k)$ is the power spectrum of bathymetry. *denotes the complex conjugate.

The quality and reliability of the admittance amplitudes were controlled by means of four additional parameters: the coherence, phase of admittance, and the coherent and incoherent energies:

$$\phi(k) = \tan^{-1}\left[\frac{\mathrm{Im}[Z(k)]}{\mathrm{Re}[Z(k)]}\right],$$
$$\gamma^2(k) = \frac{N(C(k)/EbEg) - 1}{N - 1}, \tag{7}$$
$$\text{Coherent energy} = \gamma^2(k)Eg(k),$$
$$\text{Incoherent energy} = (1 - \gamma^2(k))Eg(k),$$

where $\varphi(k)$ is phase of admittance. $\mathrm{Im}[Z(k)]$ and $\mathrm{Re}[Z(k)]$ are the real and imaginary parts of $Z(k)$. $\gamma^2(k)$ is the coherence. $Eg(k)$ is the power spectrum of gravity.

The observed admittance curve was compared with a set of theoretical admittance curves for the Airy and Flexure models, which contain the mean crust (Tc) and elastic plate thickness (Te) parameters. The final values of Te and Tc were obtained through the selection of the lowest mean square error between the observed admittance curve and each of the theoretical curves.

The theoretical curves for the admittance of the Free-Air anomaly for Airy isostatic compensation ($Z(k)^{\mathrm{Airy}}$) and for the flexure or plate isostatic compensation model ($Z(k)^{\mathrm{Flex}}$) were calculated following the method given by McKenzie and Fairhead [47],

$$Z(k)^{\mathrm{Airy}} = 2\pi G\rho_c\left(1 - e^{(-kTc)}\right), \tag{8}$$

$$Z(k)^{\mathrm{Flex}} = 2\pi G\rho_c\left(1 - \frac{e^{(-kTc)}}{A}\right), \tag{9}$$

where

$$A = 1 + \frac{Dk^4}{g(\rho_m - \rho_c)},$$
$$D = \frac{ET_e^3}{12(1 - \upsilon^2)}. \tag{10}$$

G is the gravitational constant, $E =$ is Young's modulus, υ is Poisson's ratio, g is acceleration due to gravity, ρ_c and ρ_m are the average crustal density and density of material below

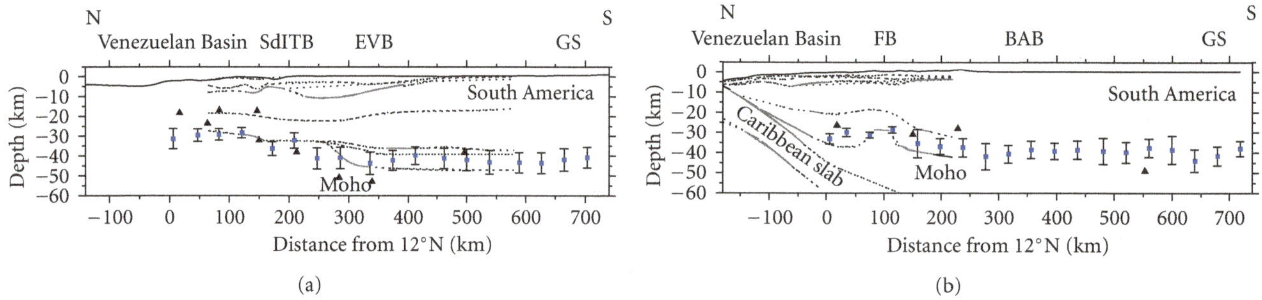

Figure 8: Moho depths cross-sections along N-S (blue boxes) with associated errors. (a) Profile 70 W. (b) Profile 65 W. Triangles are Receiver Function Moho depth estimations [37]. Gray lines are Moho depths estimated from seismic studies [38]. See Figure 7 for location. BAB: Barinas Apure basin, EVB: Eastern Venezuela basin, FB: Falcon basin, GS: Guyana shield, SdITB: Serranía del Interior thrust belt.

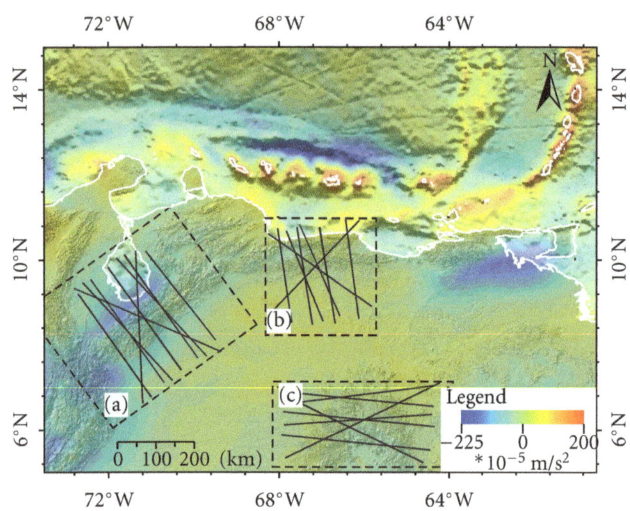

Figure 9: FAA and topography/bathymetry maps. Black-dotted boxes labeled as (a), (b), and (c) are selected regions for the estimation of elastic thickness: (a) Mérida Andes, (b) the Cordillera de la Costa range, and (c) the Guyana shield. Black straight lines are location of the selected profiles.

the assumed flexed elastic plate, Tc is the effective depth of compensation, and Te is the effective elastic thickness.

For the admittance analysis, in the present work, three areas were selected: Mérida Andes (Zone A), the Cordillera de la Costa range (Zone B), and the Guyana shield (Zone C). For each area, a maximum of nine profiles were extracted from the Free-Air anomaly map and topography map, each of ~500 km (Figure 9). Elevation data profiles were extracted from the Shuttle Radar Topography Mission (SRTM) with a grid spacing of 90 m. Each dataset along each profile was regularly spaced at intervals of 0.5 km. The coherence, phase of admittance, and energies (7) for each area were plotted with respect to wavenumber and are shown in Figure 10.

The observed admittance was computed from (5) for $2n$ samples, where n was taken as 9, which correspond to half of the longitude of the regularly spaced profiles. Theoretical admittance curves were computed (9) for Te values between 5, 10, 15, 20, 25, and 30 km. Te values were obtained after studying the square medium errors (RMS) of the correlation between observed and theoretical admittance curves. A best fit was considered whenever RMS dropped below a minimum value with the correlated theoretical Te value.

Finally, the observed and theoretical admittance curves were plotted with respect to wavenumber (Figure 11).

7. Results and Discussion

7.1. Regional and Residual Gravity Anomalies and Their Correlation with Major Tectonic Provinces. The 3 min resolution maps shown in Figure 5 can be used to identify gravity anomalies associated with geological bodies approximately 50 km or greater in size. These maps are therefore suitable for correlating gravity with tectonic provinces having dimensions of several hundred kilometers. In addition, in the southern part of Venezuela, large areas have poor coverage of gravity observations, especially in the Amazon region (i.e., Guyana shield).

Considering major topography and gravity changes, the study areas were separate in major topography/gravity provinces. The offshore provinces encompass the Venezuelan basin, South Caribbean accretionary prism, Aves Ridge, Grenada back arc basin, Lesser Antilles arc, Bonaire basin, and Tobago trough. The onshore provinces include the Guyana shield, Barinas-Apure basin, Maracaibo basin, Falcón basin, East Venezuela basin, Perijá-Mérida, and coastal ranges.

Large areas of negative residual anomalies onshore (Figure 5(b)) are associated with Mesozoic and Cenozoic sedimentary basins (Figure 1). These intracratonic Mesozoic basins such as the Barinas-Apure, Maracaibo, Falcón, and East Venezuela basins have distinct gravity signatures. For instance, the Barinas-Apure basin is characterized by long wavelength anomalies with low amplitude (-20 to $+20 \times 10^{-5}$ m/s^2). Another negative residual anomaly, near inland, but not as large in magnitude, is a FAA low through Bonaire basin. This basin has a quasirectangular shape with low negative residual anomalies which are emphasized by the surrounding gravity highs associated with the coastal ranges and the Leeward Antilles arc.

The negative Bouguer anomaly observed in the Eastern Venezuelan basin (Figure 3) is characterized by a SW-NE

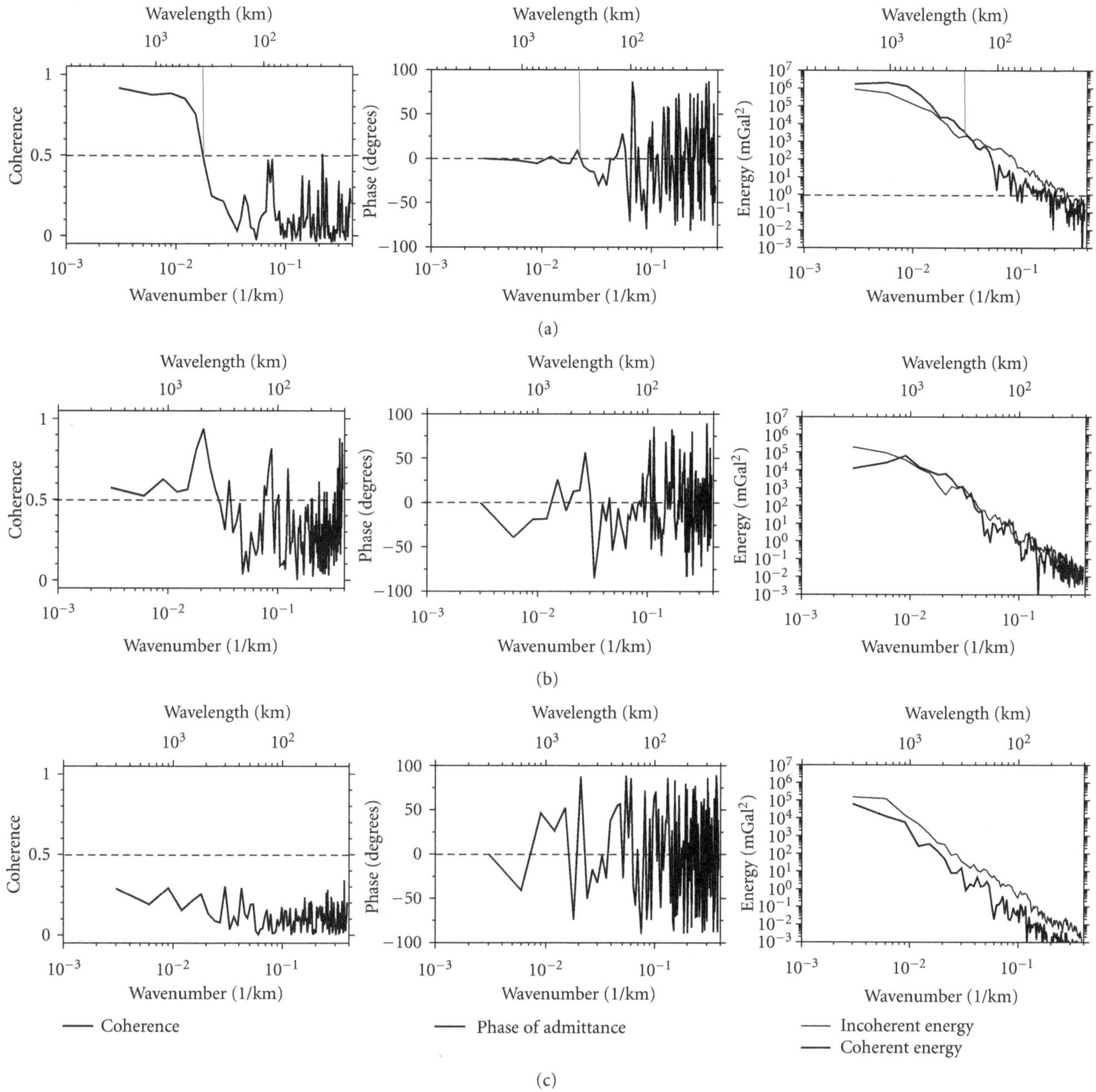

Figure 10: Observed admittance values computed from FAA and topography for areas (a), (b), and (c) in Figure 8. Red lines indicate cutoff values for coherence, phase, and coherent (thick line) and incoherent (thin line) energies. (a) For Mérida Andes, the coherence is high for $k < 0.04$, the estimated phase is close to zero for $k < 0.08$, and the coherent energy is higher than the incoherent energy for values of $k < 0.08$.

trending of a negative residual anomaly ($\sim -50 \times 10^{-5}\,\mathrm{m/s^2}$) (Figure 5(b)). This gravity low is associated with Cenozoic sediments, which accommodates the result of thrust-sheet loading (i.e., Serranía thrust belt) that forced the American continental lithosphere to flex downward between the Guyana shield and El Pilar fault [48]. Schmitz el al. [49] used seismic refraction data to estimate a maximum of 10–13 km of sedimentary infill for this basin. According to Jácome et al. [14], the amount of sediments is the result of multiple processes: (a) flexural loading of the Cordillera de la Costa range, (b) large and continuous deposition of continental material from the Guyana shield as the main source, and (c) the subduction dynamics of oceanic South America underneath the Caribbean plate. The spectral analysis of the Bouguer anomaly (Figure 6(d)) estimated an anomaly source located 12 km deep that may correspond to the top of the basin basement.

In the Mérida Andes, the residual gravity map (Figure 5(b)) showed a SW-NE trending of positive residual anomalies ($\sim 40 \times 10^{-5}\,\mathrm{m/s^2}$) flanked on both sides by negative anomalies (-20 to $-40 \times 10^{-5}\,\mathrm{m/s^2}$) that are associated with Barinas-Apure basin in the south and the

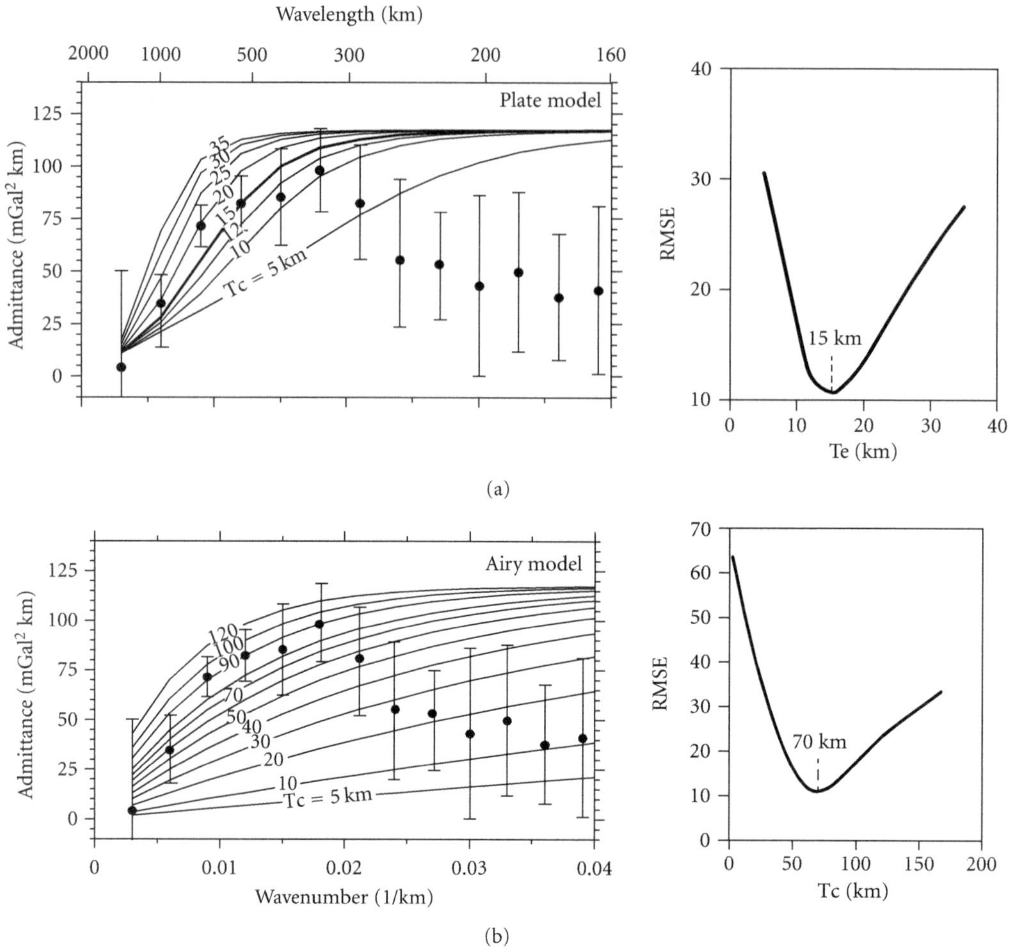

FIGURE 11: Observed admittance values computed from the FAA and topography profiles in Figure 9 for low wavenumbers, where $k = 0.0$–0.04. Observed (solid circles) and theoretical (solid curves) admittance for the (a) Flexure model, where Te is the effective elastic thickness, and (b) Airy isostatic model, where Tc is the mean thickness of the crust as a varying parameter of the model. Admittance errors were computed from the coherence [46]. The best fitting curve for Mérida Andes was the simple flexure model of the lithosphere with $Te = 17 \pm 3$ km.

Maracaibo basin. These gravity highs are associated with exposed Archean to Lower Proterozoic high-grade metamorphic basements within the mountain range. Relative maximums follow dikes (diabase) that intruded the Precambrian and Paleozoic rocks of Mérida Andes, which formed as a consequence of the convergence of the Panama arc and western South America, which initiated the formation of this mountain range [16–19]. Negative residual anomalies flanking the Mérida Andes showed similar shapes but very different amplitudes. These anomalies are mainly caused by strong flexes of the crust and the infill sediments of the Barinas-Apure basin in the south, and the Maracaibo basin in the north, in association with a chain-scale thrust and back thrust systems along the Mérida foothills. The northern flank has lower amplitude (< -40–40×10^{-5} m/s^2) with a relative maximum located in the northwest. In contrast, the southern flank barely reaches -40–40×10^{-5} m/s^2 in the most southeastern area of the anomaly. These anomalies were interpreted by F. E. Audemard and F. A. Audemard [19] to be the consequence of the rheological characteristics of two different continental crusts (the crust underneath the Maracaibo basin and the South American crust underneath the Barinas-Apure basin). In other words, the Maracaibo crust underwent a more recent tectonic and thermal event due to continental rifting during the Jurassic than the long-cooled Precambrian crust of the SA craton in the south. In summary, the Andes of Venezuela is a floating orogen involving incipient, gently NW-dipping continental subduction that generates a shallow foreland basin on the Barinas-Apure basin side, while it strongly flexes the Maracaibo crust on the forearc-equivalent side, where a deep flexural basin develops in association with a chain-scale backthrust along the Maracaibo foothills on the northwest [19].

Other regions where positive residual anomalies were observed were the Lesser Antilles arc and the Aves ridge. They were characterized by shorter wavelengths (<100 km) but high amplitudes (>100×10^{-5} m/s^2) in the residual anomalies map. These pronounced gravity highs and steep gradients were the combined effect of the steep slope of the bathymetric and high-density volcanic rock outcroppings in these localities, which displayed a wide range of compositions.

The regional gravity anomaly map provided a general view of the areal extent of the gravity response from the major geological units and reflects the gross crustal structure of the area. The calculated regional anomaly map displayed several well-defined gravity highs and lows of varying dimensions and relief. For instance, under the Guyana shield, the regional gravity field showed no evidence of significant crustal thickness variations (Figure 5(b)). This result is consistent with seismic studies that indicate a crust thickness of ~45 km [38]. In this region, regional and residual gravity maps suggest that the observed anomalies can be explained by upper crustal lateral density variations alone rather than by changes in the thicknesses of the lower crust and Moho. This is supported by the small range of ~ 20 to $\sim -20 \times 10^{-5}$ m/s^2 for the regional field in this area.

A conspicuous gravity high ($\sim 40 \times 10^{-5}$ m/s^2) in central Venezuela trending from NE to SW separates the Barinas-Apure basin and Eastern Venezuela basin. This anomaly extends from the El Baúl high to granitic rocks of the Santa Rosalía complex in the Precambrian Guyana shield, which crop out in the southwestern margin of the Orinoco River. In addition, Viscarret et al. [50] determined U-Pb zircon ages and interpreted the El Baúl massif to be a Paleozoic basement belt, which shows more affinity with the Mérida Andes than with rocks of the Guyana shield. Therefore, this long wavelength anomaly is very likely created by the overlapping of anomalies caused by the El Baúl high and the Santa Rosalía complex.

After the separation of the regional component of the BA, the Eastern Venezuelan basin still has the most relevant anomaly. This pronounced negative Bouguer gravity anomaly has been studied by several authors [3, 51–54]. The negative gravity anomaly is roughly parallel to the arc platform extension, but it does not extend west of Gulf of Paria. According to gravity and recent seismic refraction studies, this negative gravity anomaly is indicative of the very large load on the South American lithosphere here as well as the large amount of sediments in the basin.

Regional gravity anomalies of the South Caribbean accretionary prism and Leeward Antilles arc (i.e., ABC Islands) form a positive-negative gravity pair characteristic of subduction zones. The high-density rock in the Leeward Antilles islands are characterized by residuals with very high amplitudes of $100–120 \times 10^{-5}$ m/s^2, but regional field anomalies also have very high amplitudes. Whereas, the gravity anomalies caused by strong density contrast in the masses that form these volcanic bodies, that are or are being accreted to the continent, produce local anomalies that are very emphasized by the steep bathymetry. This interpretation is similar to the one by Bonini et al. [51], based on gravimetric data.

7.2. Effective Elastic Thickness and Moho Depth. The admittance and coherence results presented here for the Mérida Andes (Figure 11) best fit a flexural model with $Te = 15$ km. The Airy model fits the observed admittance when the mean crustal thickness is 70 km. This value is incompatible with the estimates of the typical continental crust thickness (35–40 km). Consequently, the Airy model cannot be accepted as a possible isostatic compensation mechanism. A simple plate flexure model with $Te = 15$ km is more reasonable.

Along the Cordillera de la Costa range, the admittance calculation agrees more closely with the theoretical values calculated for a plate thickness of 10 km. Te is around ~10 km north of the Eastern Venezuelan basin (i.e., along the Cordillera de la Costa range), indicating that low rigidity amplified the subsidence of the basin.

Coherence, phase of admittance, and coherent and incoherent energies calculated for the Cordillera de la Costa range and Guyana shield showed poor quality and reliability for Te estimations. In other words, the phase of the admittance was not close to zero, the coherence was less than 0.5, and incoherent energy was higher than coherent energy for all wavenumbers in the selected profiles (Figure 9). The admittance estimates for the set of profiles crossing Mérida Andes showed a noticeable data scattering at wavelengths longer than 160 km. This scattering could be caused by the wide variability of the thickness of sediments and also by very short wavelength anomalies that could be considered as noise.

Tassara et al. [55] used a wavelet formulation of the classical spectral isostatic analysis to invert satellite-derived gravity and topography/bathymetry for Te over South America. According to their calculations, Mérida Andes, which is the region located between the coastline (i.e., Cordillera de la Costa range) and Colombian Eastern Cordillera, exhibits variable Te from 35 to 40 km with an uncertainty of ±10 km. Beneath the Cordillera de la Costa range, Te decreases to ~35 ± 10 km. This high Te value in the Cordillera de la Costa range may reflect the combined effect of the strength of the upper continental and the partially subducted Caribbean oceanic lithospheres. Beneath the Eastern Venezuela basin, the elastic thickness ranges between 40 and 50 km with an uncertainty of ±10 km. The Guyana shield shows Te as low as 10 km, although with an uncertainty of up to 25 km.

The Te values estimated in this study did not match the elastic plate thicknesses estimated by Tassara et al. [55]. These discrepancies between the admittance results are the most probable cause for the presence of a considerable amount of loading from the top, which corresponds to stacking of thrust sheets and uplifts of basement rocks in coastal ranges, and from the bottom, that is, the partially subducted Caribbean slab. Another factor to consider is the size of the window used to compute admittance. According to Pérez-Gussinyé et al. [56] windows that are too small introduce spurious spatial variations, and windows that are too large tend to average the spatially varying Te values and smooth the true structure. These discrepancies could be caused by the large and tectonically heterogeneous area required by the admittance technique, which would tend to bias their result towards a weaker plate [57].

The results obtained in the spectral analysis confirm the previously established value of 35 km as the mean reference Moho depth [38]. In addition, the crust thickness is not homogeneous. For instance, the Moho topography shows a NE-SW depression beneath the Mérida Andes with a maximum depth of ~55 ± 5 km; this value gradually decreases toward the south and north of the main strike of this feature. The maximum crustal thickness in Mérida Andes (Figure 7) was partially constrained by the receiver function analysis [37], which shows Moho depths of 50 and 47 km near this maximum. The Maracaibo basin seems to have a crustal thickness of 35–40 km. The estimated Moho geometry under the area of the Falcón basin agreed well with the refraction seismic data modeled along the 70°W profile, which indicates a crustal thickness of ~27 km [58]. The crustal thickness beneath the Eastern Venezuela basin ranges from ~50 to ~40 km. The Guyana shield is regionally underlain by a crust of ~45 km. Northwards beneath the Guárico and Barinas-Apure basins, the crustal thickness reaches ~40 km. Along the coastline, the Caribbean Mountain System crustal thickness oscillates around 30 km and decreases slightly toward the Leeward Antilles (Figures 7 and 8).

Figure 8 show that there are significant differences among Moho values estimated with spectral analysis and Moho estimated with receiver function. On the other hand, Moho depth values in Figure 7 and closely match the values estimated by Schmitz et al. [38] and also produces the best fits to the observed Bouguer gravity data [54]. According to Schmitz et al. [38] the mismatch between seismic Moho and receiver function Moho is caused by the resolution of these technics.

8. Conclusions

A new Bouguer gravity map of Venezuela was developed based on an up-to-date dataset available in the country. All data were reprocessed and homogenized according to gravity processing standards, and the effects of the topography were corrected with digital topographic maps. The final dataset was comprised of more than ~80,000 observation points that are now available for future detailed interpretations and future crustal investigations, such as 2D and 3D gravity modeling.

The results obtained in the Bouguer gravity map presented in this work can be greatly improved by adding more gravity data in areas where scarce and poor quality data are the only data available, such as in mountain ranges (e.g., Andes) and the Amazon forest (e.g., Guyana shield).

The Free-Air gravity and topography admittance analysis of the data windows over the Mérida Andes provided elastic thicknesses in the range of 30–35 km.

Finally, Te values presented in this work using the admittance method were found to be lower than Te calculations carried out in previous studies on the scale of South America and surrounding plates. These differences in the Te calculation could be caused by tectonical heterogeneities and problems associated with the admittance technique (window size, data coverage and presence of noise), but these arguments are still in discussion. It is possible that satisfactory results for this area that has a very complicated structure could only be obtained by a more sophisticated approach.

Acknowledgments

The author is thankful to USB (Simón Bolívar University), FUNVISIS (Venezuelan Foundation for Seismological Research), and Christian-Albrechts University for the data and resources provided for the development of this work. He gratefully acknowledges the contributions from Christian-Albrechts University colleagues Proffessor Dr. Hans-Jürgen Götze and to Dr. S. Schmidt for the software packages used during this work. He also acknowledges useful comments from Dr. Michael Schmitz (FUNVISIS), Oscar H. Lücke (CAU), and Benjamin D. Gutknecht (CAU). His special thanks go to Joerg Schleicher (editor) and two anonymous reviewers for very helpful and constructive comments on the paper. The authors expresses thanks to FONACIT (Venezuelan National Fund for Science, Technology and Innovation) for financial support. Figures were prepared using the Generic Mapping Tool (GMT) software package (Wessel, P. and Smith, W. H. F., 1991). This is a contribution to project GEODINOS (FONACIT 2002000478).

References

[1] R. K. McConnell, P. J. Winter, and R. F. Geller, "Latin American Gravity Standardization Network 1977 (LAGSN77)," ed. Panama: Earth Physics Branch, Canada, and Servicio Geodesico Interamericano, 1979.

[2] V. R. Graterol, "Mapa de anomalía de Bouguer de la Republica de Venezuela," ed. Caracas: Universidad Simón Bolívar; Dirección de Cartografía Nacional de Venezuela; Energy, Mines and Resources Canada; Defense mapping agency, Inter American geodetic survey, 1988, p. Mapa de anomalía de Bouguer de la Republica de Venezuela (P=2.67 g/cc).

[3] C. Izarra, M. I. Jácome, M. Schmitz et al., "Analyzing gravity anomalies over the Caribbean and northern Venezuela tectonic plate boundary," in *Proceedings of the 6th International Symposium on Andean Geodynamics (ISAG '05)*, Barcelona, Spain, 2005.

[4] J. L. Pindell, "Evolution of the Gulf of Mexico and the Caribbean," in *Caribbean Geology, An Introduction*, S. K. Donovan, Ed., pp. 13–39, University of the West Indies, Kingston, Jamaica, 1994.

[5] M. Meschede and W. Frisch, "A plate-tectonic model for the Mesozoic and early Cenozoic history of the Caribbean plate," *Tectonophysics*, vol. 296, no. 3-4, pp. 269–291, 1998.

[6] J. L. Pindell and J. F. Dewey, "Permo-Triassic reconstruction of western Pangea and the evolution of the Gulf of Mexico/Caribbeanregion," *Tectonics*, vol. 1, no. 2, pp. 179–211, 1982.

[7] K. Burke, "Tectonic evolution of the Caribbean," *Annual Review of Earth and Planetary Sciences*, vol. 16, pp. 201–230, 1988.

[8] J. L. Pindell, S. C. Cande, W. C. Pitman et al., "A plate-kinematic framework for models of Caribbean evolution," *Tectonophysics*, vol. 155, no. 1–4, pp. 121–138, 1988.

[9] J. L. Pindell and S. Barrett, "Geologic evolution of the Caribbean: a plate-tectonic perspective," in *The Geology of North America*, G. Dengo and J. E. Case, Eds., vol. H of *The Caribbean Region*, pp. 405–432, Geological Society of America, 1990.

[10] J. L. Pindell, L. Kennan, W. V. Maresch et al., "Plate kinematics and crustal dynamics of circum-Caribbean arc-continent

interactions: tectonic controls on basin development in proto-Caribbean margins," in *Caribbean-South American Plate Interactions, Venezuela, GSA Special Paper 394*, H. G. A. Lallemant, V. B. Sisson, and H. G. A. Lallemant, Eds., pp. 7–52, Geological Society of America, Boulder, Colo, USA, 2005.

[11] M. Ostos, F. Yoris, and H. G. A. Lallemant, "Overview of the Southeast Caribbean-south American plate boundary zone," *Special Paper of the Geological Society of America*, no. 394, pp. 53–89, 2005.

[12] J. C. Weber, T. H. Dixon, C. DeMets et al., "GPS estimate of relative motion between the Caribbean and South American plates, and geologic implications for Trinidad and Venezuela," *Geology*, vol. 29, no. 1, pp. 75–78, 2001.

[13] L. Chacín, M. I. Jácome, and C. Izarra, "Flexural and gravity modelling of the Mérida Andes and Barinas-Apure Basin, Western Venezuela," *Tectonophysics*, vol. 405, no. 1–4, pp. 155–167, 2008.

[14] M. I. Jácome, N. Kusznir, F. Audemard, and S. Flint, "Formation of the Maturín foreland basin, eastern Venezuela: thrust sheet loading or subduction dynamic topography," *Tectonics*, vol. 22, no. 5, pp. 1–1, 2003.

[15] M. Bellizzia, "Geología del Macizo de El Baúl, Estado Cojedes," in *Proceedings of the 3rd Congreso Geológico Venezolano*, 1961.

[16] J.-F. Stephan, *Evolution geodinamique du domaine Caraibe, Andes et chaine Caraibe sur la transversale de Barquisimeto (Venezuela)*, Ph.D. thesis, Paris, France, 1982.

[17] F. E. Audemard, *Tectonics of Western Venezuela*, Ph.D. thesis, Rice University, Houston, Tex, USA, 1991.

[18] B. Colletta, F. Roure, B. de Toni, D. Loureiro, H. Passalacqua, and Y. Gou, "Tectonic inheritance, crustal architecture, and contrasting structural styles in the Venezuela Andes," *Tectonics*, vol. 16, no. 5, pp. 777–794, 1997.

[19] F. E. Audemard and F. A. Audemard, "Structure of the Mérida Andes, Venezuela: relations with the South America-Caribbean geodynamic interaction," *Tectonophysics*, vol. 345, no. 1–4, pp. 299–327, 2002.

[20] C. Schubert, "Basin formation along the Bocono-Moron-El Pilar fault system, Venezuela," *Journal of Geophysical Research*, vol. 89, no. 7, pp. 5711–5718, 1984.

[21] B. P. Kohn, R. Shagam, P. O. Banks et al., "Mesozoic-Pleistocene fission-track ages on rocks of the Venezuelan Andes and their tectonic implications," in *The Caribbean South American Plate Boundary and Regional Tectonics*, W. E. Bonini, R. B. Hargraves, and R. Shagam, Eds., vol. 162, pp. 365–384, Geological Society of America Memoirs, 1984.

[22] H. G. avé Lallemant, "Transpression, displacement partitioning, and exhumation in the eastern Caribbean/South American plate boundary zone," *Tectonics*, vol. 16, no. 2, pp. 272–289, 1997.

[23] C. Beck, *La chaine Caraïbe au méridien de Caracas: géologie, tectogenese, place dans l'évolution géodynamique Mesozoique-Cenozoique des Caraïbes Méridionales*, Ph.D. thesis, L'Universite des Sciences et Techniques de Lille, 1985.

[24] A. B. Bellizzia, "Sistema Montañoso del Caribe, una cordillera alóctona en la parte Norte de América del Sur," in *Proceedings of the the 6 Congreso Geológico Venezolano, Memorias, Sociedad Venezolana de Geólogos*, 1986.

[25] E. Navarro, M. Ostos, and F. Yoris, "Revisión y redefinición de unidades litoestratigráficas y síntesis de un modelo tectónico para la evolución de la parte Norte-Central de Venezuela durante el Jurásico Medio-Paleógeno," *Acta Cientifica Venezolana*, vol. 39, pp. 427–436, 1988.

[26] T. W. Donnelly, D. Beets, M. J. Carr et al., "History and tectonic setting of Caribbean magmatism," in *The Caribbean Region*, G. Dengo and J. E. Case, Eds., vol. H of *The Geology of North America*, pp. 339–374, The Geological Society of America, 1990.

[27] M. Ostos, "Tectonic evolution of the south-central margin of the Caribbean based on geochemical dates (Evolución tectónica del margen Sur-Central del Caribe basado en dates geoquímicos)," *GEOS*, vol. 30, pp. 1–294, 1990.

[28] G. Giunta, L. Beccaluva, M. Coltorti, F. Siena, and C. Vaccaro, "The southern margin of the Caribbean Plate in Venezuela: tectono-magmatic setting of the ophiolitic units and kinematic evolution," *Lithos*, vol. 63, no. 1-2, pp. 19–40, 2002.

[29] G. Feo-Codecido, F. D. J. Smith, N. Aboud et al., "Basement and Paleozoic rocks of the Venezuelan Llanos basins," in *The Caribbean-South American Plate Boundary and Regional Tectonics*, R. B. Hargrave, Ed., pp. 175–187, The Geological Society of America, Boulder, Colo, USA, 1984.

[30] C. G. de Juana, J. Iturralde, and X. Picard, *Geología de Venezuela y de sus Cuencas Petrolíferas*, Ediciones Foninves, Caracas, Venezuela, 1980.

[31] S. M. Carbotte, R. Arko, D. N. Chayes et al., "New integrated data management system for Ridge2000 and MARGINS research," *Eos, Transactions, American Geophysical Union*, vol. 85, no. 51, pp. 553–559, 2004.

[32] H. Moritz, "Geodetic reference system 1980," *Bulletin Géodésique*, vol. 54, no. 3, pp. 395–405, 1980.

[33] C. Kwang-Sun, L. Young-Cheol, and L. Mu-Taek, "Precise terrain torrection for gravity measurement considering the earth's curvature," *Journal of the Korean Earth Science Society*, vol. 28, pp. 825–837, 2007.

[34] J. H. Karl, "The Bouguer correction for the spherical earth," *Geophysics*, vol. 36, pp. 761–762, 1971.

[35] W. A. Heiskanen and H. Moritz, "Physical geodesy," *Bulletin Géodésique*, vol. 86, pp. 491–492, 1967.

[36] H. Drewes, W. Torge, R. H. Röder, C. Badell, D. Bravo, and O. Chourio, "Absolute and relative gravimetric surveys of national and geodynamic networks in Venezuela," *Journal of South American Earth Sciences*, vol. 4, no. 4, pp. 273–286, 1991.

[37] F. Niu, T. Bravo, G. Pavlis et al., "Receiver function study of the crustal structure of the southeastern Caribbean plate boundary and Venezuela," *Journal of Geophysical Research*, vol. 112, no. 11, Article ID B11308, pp. 1–15, 2007.

[38] M. Schmitz, J. Avila, M. Bezada et al., "Crustal thickness variations in Venezuela from deep seismic observations," *Tectonophysics*, vol. 459, no. 1–4, pp. 14–26, 2008.

[39] A. Spector and F. S. Grant, "Statistical methods for interpreting aeromagnetic data," *Geophysics*, vol. 35, pp. 293–302, 1970.

[40] D. C. Mishra and P. S. Naidu, "Two-dimensional power spectral analysis of aeromagnetic fields," *Geophysical Prospecting*, vol. 22, no. 2, pp. 345–353, 1974.

[41] B. K. Bhattacharyya and L. K. Leu, "Spectral analysis of gravity and magnetic anomalies due to rectangular prismatic bodies," *Geophysics*, vol. 42, no. 1, pp. 41–50, 1977.

[42] K. Dimitriadis, G. A. Tselentis, and K. Thanassoulas, "A BASIC program for 2-D spectral analysis of gravity data and source-depth estimation," *Computers and Geosciences*, vol. 13, no. 5, pp. 549–560, 1987.

[43] G. A. Tselentis, J. Drakopoulos, and K. Dimitriadis, "A spec-tral approach to moho depths estimation from gravity measurements in Epirus (NW Greece)," *Journal of Physics of the Earth*, vol. 36, no. 6, pp. 255–266, 1988.

[44] R. D. Regan and W. J. Hinze, "The effect of finite data length in the spectral analysis of ideal gravity anomalies," *Geophysics*, vol. 41, no. 1, pp. 44–55, 1976.

[45] D. McKenzie and C. Bowin, "The relationship between bathymetry and gravity in the Atlantic Ocean," *Journal of Geophysical Research*, vol. 81, pp. 1903–1915, 1976.

[46] A. B. Watts, "An analysis of isostasy in the World's Oceans 1. Hawaiian-Emperor Seamount Chain," *Journal of Geophysical Research*, vol. 83, pp. 5989–6004, 1978.

[47] D. McKenzie and D. Fairhead, "Estimates of the effective elastic thickness of the continental lithosphere from Bouguer and free air gravity anomalies," *Journal of Geophysical Research B*, vol. 102, no. 12, pp. 27523–27552, 1997.

[48] F. Roure, J. O. Carnevali, Y. Gou, and T. Subieta, "Geometry and kinematics of the North Monagas thrust belt (Venezuela)," *Marine and Petroleum Geology*, vol. 11, no. 3, pp. 347–362, 1994.

[49] M. Schmitz, A. Martins, C. Izarra, M. I. Jácome, J. Sánchez, and V. Rocabado, "The major features of the crustal structure in north-eastern Venezuela from deep wide-angle seismic observations and gravity modelling," *Tectonophysics*, vol. 399, no. 1–4, pp. 109–124, 2005.

[50] P. Viscarret, J. Wright, and F. Urbani, "New U-Pb zircon ages of El Baúl Massif, Cojedes State, Venezuela," *Revista Tecnica de la Facultad de Ingenieria Universidad del Zulia*, vol. 32, no. 3, pp. 210–221, 2009.

[51] W. E. Bonini, C. Acker, and G. Buzan, "Gravity studies across the western Caribbean Mountains," in *Proceedings of the 5th Congreso Latinoamericano de Geología*, Caracas, Venezuela, 1977.

[52] R. M. Russo and R. C. Speed, "Oblique collision and tectonic wedging of the South American continent and Caribbean terranes," *Geology*, vol. 20, no. 5, pp. 447–450, 1992.

[53] J. C. VanDecar, R. M. Russo, D. E. James et al., "Aseismic continuation of the Lesser Antilles slab beneath continental South America," *Journal of Geophysical Research*, vol. 108, 2003.

[54] J. Sanchez, H. J. Götze, and M. Schmitz, "A 3-D lithospheric model of the Caribbean-South American plate boundary," *International Journal of Earth Sciences*, vol. 100, no. 7, pp. 1697–1712, 2011.

[55] A. Tassara, C. Swain, R. Hackney, and J. Kirby, "Elastic thickness structure of South America estimated using wavelets and satellite-derived gravity data," *Earth and Planetary Science Letters*, vol. 253, no. 1-2, pp. 17–36, 2007.

[56] M. Pérez-Gussinyé, A. R. Lowry, A. B. Watts, and I. Velicogna, "On the recovery of effective elastic thickness using spectral methods: examples from synthetic data and from the Fennoscandian Shield," *Journal of Geophysical Research*, vol. 109, no. 10, Article ID B10409, 2004.

[57] D. W. Forsyth, "Subsurface loading and estimates of the flexural rigidity of continental lithosphere," *Journal of Geophysical Research*, vol. 90, pp. 12623–12632, 1985.

[58] M. J. Bezada, M. Schmitz, M. I. Jácome, J. Rodríguez, F. Audemard, and C. Izarra, "Crustal structure in the Falcón Basin area, northwestern Venezuela, from seismic and gravimetric evidence," *Journal of Geodynamics*, vol. 45, no. 4-5, pp. 191–200, 2008.

25

On the Potential of Least Squares Response Method for the Calibration of Superconducting Gravimeters

Mahmoud Abd El-Gelil,[1] Spiros Pagiatakis,[2] and Ahmed El-Rabbany[3]

[1] *Geomatics Division, Faculty of Engineering & the Built Environment, University of Cape Town, Cape Town 7700, South Africa*
[2] *Department of Earth and Space Science and Engineering, York University, Toronto, ON, Canada M3J 1P3*
[3] *Department of Civil Engineering, Ryerson University, Toronto, ON, Canada M5B 2K3*

Correspondence should be addressed to Mahmoud Abd El-Gelil, mahmoud.abdel-gelil@uct.ac.za

Academic Editor: Umberto Riccardi

One of the most important operating procedures after the installation of a superconducting gravimeter (SG) is its calibration. The calibration process can identify and evaluate possible time variability in the scale factor and in the hardware anti-aliasing filter response. The SG installed in Cantley, Canada is calibrated using two absolute gravimeters and the data are analysed in the time and frequency domains to estimate the SG scale factor. In the time domain, we use the weighted linear regression method whereas in the frequency domain we use the least squares response method. Rigorous statistical procedures are applied to define data disturbances, outliers, and realistic data noise levels. Using data from JILA-2 and FG5-236 separately, the scale factor is estimated in the time and frequency domains as $-78.374 \pm 0.012\,\mu$Gal/V and $-78.403 \pm 0.075\,\mu$Gal/V, respectively. The relative accuracy in the time domain is 0.015%. We cannot identify any significant periodicity in the scale factor. The hardware anti-aliasing filter response is tested by injecting known waves into the control electronics of the system. Results show that the anti-aliasing filter response is stable and conforms to the global geodynamics project standards.

1. Introduction and Background

Nowadays, most systems and devices are controlled by digital computers where digital output is the norm. The superconducting gravimeter (SG) is no exception; its output is digitized by an A/D converter. The SG is a relative gravimeter and must be calibrated accurately to estimate its scale factor before further use of its records and data interpretation. The voltage output from the SG has no physical meaning until it is converted to the proper units of μGal by using specific calibration techniques. Therefore, physical experiment or externally induced acceleration can be used to represent or simulate the input or the physics (gravity changes) behind this voltage variation.

Reference [1] reported that a highly accurate scale factor is essential to constrain mantle anelasticity. In [2], it is mentioned that an accurate scale factor is needed, for ocean tide loading, removal of synthetic tide, and amplitude determination of the Earth's normal modes in the sub-seismic band. To meet the above-mentioned application requirements, it is very important to estimate the scale factor with relative accuracy of less than 0.1% [3]. However, it is really a challenge to calibrate a device using another one, which is much less precise: the absolute gravimeter (AG) precision is at the μGal level while the SG is at the nGal level. In other words, the SG precision is about three orders of magnitude higher than that of the AG.

2. Methods of SG Calibration

Various methods have been developed and implemented to calibrate the SG, specifically the amplitude calibration. One such method is called "relative method." It uses another gravimeter, for example an absolute gravimeter or a relative gravimeter with a well-known scale factor. According to this method, both gravimeters have to run in parallel that is, at the same time and as close as possible to one another so they can sense the same gravity variations. Then, the least squares regression method is applied directly to both records to estimate the scale factor. The accuracy of this method is dependent on the data length and the precision of the gravimeters used in the experiment.

More often than not, the SG is calibrated using the relative method. Reference [3] ran the SG model TT70 and JILA-5 at the Strasbourg station for one day and estimated the calibration parameter with relative precision of less than 1.0% using linear regression. Later on, a compact SG (GWR C026) was installed at Strasbourg station. Reference [4] used over three years of AG and SG data to check the stability of the scale factor in the time domain. Reference [5] calibrated the SG at Membach, Belgium, using 47 days of data involving a large number of AG drops. A precision of about 0.1% was obtained through linear regression and tidal amplitude factors comparison. In addition to this high precision, the scale factor was checked in the frequency domain. Instead of running only one AG parallel to the SG, [6] used four FG5 AGs operating simultaneously for an average of a one-week period. The calibration factor was estimated from this experiment within 0.1%. Reference [7] also proved that this precision can be achieved with FG5 AG, recording for at least five days. The Japanese stations at Esashi, Canberra, and Matsushiro have been calibrated using the relative calibration method (e.g., [8–11]).

An alternative calibration method is known as the "absolute calibration" method. This method either uses a moving mass close to the SG or fixes the SG itself on a moving platform subjected to a known acceleration. The induced acceleration from the moving mass can be determined using Newton's law while the second method, which fixes the SG on a moving platform, uses the second time derivative of the vertical displacement of the platform. The accuracy of this method is limited to the accuracy of the gravitational constant, the homogeneity of the moving mass, and the geometry of the position of the mass with respect to the SG sphere [2]. Furthermore, the time span of the experiment is short; therefore, the regression of the induced acceleration and the output voltage from the SG has to be extrapolated to estimate the scale factor at zero frequency "DC" [12].

The SG absolute calibration method was first introduced by [13] by moving a sphere filled with mercury vertically under the SG; they achieved a precision of 0.2%. Reference [14] applied the same concept by moving a 273 kg annular mass around the SG and then used Newton's law to calculate the Newtonian attraction, which can easily be linked to the SG output; the scale factor was estimated with 0.3% relative accuracy. Reference [15] developed and applied the acceleration platform, which was used at the Frankfurt station. So far, the reported results from the platform provide a calibration precision of 0.01% [12]. This method is difficult to use because it requires special installations that are not easy nor are they available at Cantley station. Instead we used the relative method.

The SG at Cantley, Canada, was calibrated twice since the first installation by different researchers. Reference [16] estimated the scale factor with 0.4% precision by applying the theoretical tide method. Reference [17] estimated the scale factor to be $-78.3 \pm 0.1\ \mu$Gal/V using JILA-2 gravimeter, with a relative precision of 0.13%. Hence, the SG at any station must continuously be calibrated whenever data are available using different methods. In order to contribute to the previous calibration effort of the SG at Cantley station, we recalculated the SG scale factor based on long series from the superconducting and absolute gravimeters from 1997 to 2009. The calibration was carried out in the time and frequency domains. To the best of our knowledge, the Least Squares Response Method (LSRM) has never been used to calibrate the SG in the frequency domain.

The least squares response method (LSRM) developed in [18], which will be revisited again in the methodology section, is used to estimate the scale factor in the frequency domain. The LS spectra of both SG and AG are used to define common peaks. These common peaks are identified through the product spectrum of the two factor spectra. Then, the frequencies of the statistically significant peaks at 95% CL are used to estimate the amplitude and phase by suppressing its period in the time series (AG and SG). Finally, at each frequency the amplitude and phase difference of the scale factor is estimated, and then the least squares fitting method is used to estimate the scale factor in different frequency bands.

In addition to the frequency domain analysis mentioned above, we estimate the scale factor in the time domain by separating the gravity data into blocks based on the duration of the experiment. Then, the scale factor of each block is estimated using the weighted least squares regression and its value corresponds to the middle of the block. The weighted average of all individual scale factors and its standard error are finally calculated.

3. Methodology

3.1. Least Squares Spectral Analysis. We use the Least Squares Spectral Analysis (LSSA) to estimate the spectra of the absolute and superconducting gravimeter series and subsequently produce their product spectrum. The LSSA has its roots in [19, 20]. Its advantages over classical Fourier analysis have already been presented in the literature and need not be repeated here (see, e.g., [21–23]). For the sake of comprehensiveness, we only present the fundamental formulas and emphasize the statistical properties of this method.

Consider a time series $f(t_i)$ observed at discrete times t_i, $i = 1, 2, \ldots, n$, not necessarily evenly spaced, which is essentially equivalent to the presence of gaps in the time series. This time series also has an associated covariance matrix C_f that carries information on the uncertainty of the observed values as well as on their correlation (usually not available). The LSSA spectrum is described by the percentage variance $s(\omega_j)$ of the spectral content of a specific cyclic frequency ω_j, which is the ratio of the quadratic norm of the specific signal to the total quadratic norm of the series:

$$s(\omega_j) = \frac{f^T C_f^{-1} \hat{P}(\omega_j)}{f^T C_f^{-1} f}, \quad \text{where } f^T C_f^{-1} f \neq 0, \qquad (1)$$

where $\hat{P}(\omega_j)$ is the projection of $f(t_i)$ onto the model space spanned by the trigonometric base functions of cyclic frequency ω_j; $s(\omega_j) \in (0,1)$. For cyclic frequency ω_j : $\hat{P}(\omega_j) = \hat{c}_1 \cos(\omega_j t) + \hat{c}_2 \sin(\omega_j t)$, where $\hat{c}_1$ and $\hat{c}_2$ are estimated coefficients from the least squares fitting. For more details about the geometrical interpretation of the least squares transform and functional analysis see, for example, [24, 25]. Reference [22] showed that the probability density function (PDF) of the LS spectrum $s(\omega_j)$ is a *beta* distribution defined by two parameters α and β, where $\alpha = 1$ and β depends on the number of data points (length of the series) and the number of unknown parameters estimated by the LS estimation (degrees of freedom of the LS system).

3.2. Least Squares Product Spectrum. The LS product spectrum has been used as an effective and rigorous approach to identify common spectral peaks among the time series obtained from superconducting gravimeters for the purpose of detecting very weak signals buried in noise, such as the Slichter triplet [26]. In our case, two different time series (AG and SG) are used to obtain their product spectrum. It should be pointed out that both series do not necessarily need to have the same sampling rate, only their corresponding spectra must be of the same band and resolution (number of spectral values in the band) to make possible the mutual multiplication. This procedure can be applied to two or more time series to define the common features (multichannel system or multi-input single-output system).

The PDF of the product spectrum of two series can be derived using standard statistical approaches (e.g., [27]). This PDF is very important for the definition of the significance of the common peaks at different confidence levels. Moreover, the mean and the variance of the product spectrum can easily be defined. Let a random variable X denote the LS spectrum, which is distributed as $beta(x;\alpha,\beta)$, with $\alpha = 1$ and β dependent upon the degrees of freedom of the linear LS system as follows [22, Theorem 3]:

$$\beta = \frac{m-u-2}{2}, \tag{2}$$

where m is the number of data points in the time series and u is the number of unknown parameters that are estimated simultaneously with the LS spectrum. This *beta* distribution has finite first and second moments [22]:

$$\mu = \frac{2}{m-u}, \qquad \sigma^2 = \frac{4(m-u-2)}{(m-u+2)(m-u)^2}. \tag{3}$$

For the derivation of the PDF of the product LS spectrum (H) of two random variables X_1 and X_2, that is, the derivation of the PDF of the random variable $H = X_1 X_2$, we follow the same procedure as in [26] by deriving the PDF of the new random variable $\ln H$. However, the derivation of the PDF in [26] was applied for n random variables, where n was larger than 10, which resulted in a PDF approximately following the Gaussian distribution. In this paper, we deal with two random variables both having the same *beta* distribution but different parameters. These two random variables are the percentage variances $s(\omega_j)$ of both AG and SG series. First, we take its natural logarithm so the new random variable becomes

$$\ln H = \ln X_1 + \ln X_2. \tag{4}$$

Before deriving the PDF of $\ln H$, we must project the PDF of X to $\ln X$. If X is a random variable of continuous type, such that $X \sim beta(x;1,\beta) = \beta(1-x)^{\beta-1}$ [26] in the set $\mathcal{A} = \{x/x : 0 < x < 1\}$, we must find the PDF of the new random variable $Y = \ln X$. We introduce the transformation $y = v(x) = \ln x$, which gives $x = w(y) = e^y$. We now have the mapping of $\mathcal{A}$ into $\mathcal{B} = \{y/y : -\infty < y < 0\}$, and the Jacobian of transformation is $J = dx/dy = e^y$. The PDF of $Y = \ln X$ is then $g(y) = beta(e^y)|J|$, or $Y = \ln X \sim g(y) = \beta e^y (1-e^y)^{\beta-1}$, $\mathcal{B} = \{y/y : -\infty < y < 0\}$ with μ_y and σ_y^2, which can be obtained by applying the well-known definition of the mean and the variance of PDF, see, for example, [27]:

$$\mu_y = \int_{-\infty}^{0} y g(y)\,dy = -\frac{1}{\beta}(\gamma\beta + \beta\psi(\beta) + 1), \qquad \sigma_y^2 = \int_{-\infty}^{0} y^2 g(y)\,dy - \mu^2 = \frac{\pi^2\beta^2 - 6\beta^2\psi(1,\beta) + 6}{6\beta^2}, \tag{5}$$

where γ is Euler's constant, $\psi(\beta)$ is the logarithmic factorial function, and $\psi(n,\beta)$ is the nth derivative of $\psi(\beta)$ [28]. Now, the mean and the variance of the new random variable $\ln H$ can be defined as follows: the mean is the sum of the two individual means and the variance is the sum of the two variances (e.g., [27]):

$$\mu = -\sum_{i=1}^{2} \frac{1}{\beta_i}(\gamma\beta_i + \beta_i\psi(\beta_i) + 1), \qquad \sigma^2 = \sum_{i=1}^{2} \frac{\pi^2\beta_i^2 - 6\beta_i^2\psi(1,\beta_i) + 6}{6\beta_i^2}. \tag{6}$$

We can now obtain the PDF of the sum of the natural logarithms of the two spectra $Z = \ln X_1 + \ln X_2 = Y_1 + Y_2$, where Y_1 and Y_2 are in the set $\mathcal{B} = \{y/y : -\infty < y < 0\}$ and z is in the set $\mathcal{C} = \{z/z : -\infty < z < 0\}$, by

$$f(z) = \int_{z}^{0} \beta_1\beta_2 e^{z}(1-e^{z-y_2})^{\beta_1-1}(1-e^{y_2})^{\beta_2-1}dy_2, \tag{7}$$

where y_2 is the integration variable. Since $f(z)$ is the sum of two independent random variables Y_1 and Y_2, both following $\beta_i e^{y_i}(1-e^{y_i})^{\beta_i-1}$, $\mathcal{B} = \{y/y : -\infty < y < 0\}$ distribution, then $f(z)$ follows the same distribution but with different parameters [27]. So, the general formula of $f(z)$ is $f(z) = ce^{az}(1-e^{bz})^d$. From the PDF characteristics (e.g., $\int_{-\infty}^{0} f(z)dz = 1.0$ and $\mu_z = \int_{-\infty}^{0} z f(z)dz = \mu_{y_1} + \mu_{y_2}$), two out of the four unknown parameters (c and d) can be defined analytically giving the final PDF:

$$f(z) = \frac{be^{az}\left(1-e^{bz}\right)^{\beta_1+\beta_2-1}}{B(a/b, \beta_1+\beta_2)}, \tag{8}$$

where a and b are two parameters dependent on β_1 and β_2, both of which are smaller than unity and can be estimated from the numerical integration of (7). Notice that B is the *beta* function. The above PDF underlines the product LS spectrum and can be used to identify the significance of product peaks of both series at any confidence level (e.g., 90%, 95%, or 99%).

3.3. Least Squares Response Method. After producing the product LS spectrum of AG and SG records and using its PDF as a guide, we can identify statistically significant peaks at any confidence level. In this paper, we use the 95% confidence level to identify significant peaks. These peaks (actually their periods) are suppressed separately in the AG and SG series to estimate their amplitudes and phases using weighted least squares as follows:

$$\mathrm{AG}(t) = D_G + \mathrm{LT}_G + \sum_{i=1}^{m} A_{Gi}\cos(\omega_i t - \phi_{Gi}), \quad (9)$$

$$\mathrm{SG}(t) = D_S + \mathrm{LT}_S + \sum_{i=1}^{m} A_{Si}\cos(\omega_i t - \phi_{Si}), \quad (10)$$

where D_G and D_S are the offsets or averages in the AG and SG records, respectively, LT_G and LT_S are the parameters of AG and SG linear trends, if any, m is the number of common significant peaks, A_{Gi} and A_{Si} are the amplitudes of AG and SG constituents for the ith frequency, respectively, and ϕ_{Gi}, ϕ_{Si} are their associated phases. All the above parameters have an associated covariance matrix estimated from the LS estimation process. Only the statistically significant amplitudes and phases are used to estimate the scale factor. The magnitude and phase difference of the scale factor is then estimated for each ith frequency from

$$\mathrm{SF}(\omega_i) = \frac{A_{Gi}}{A_{Si}}, \quad (11)$$

$$\Delta\phi(\omega_i) = \phi_{Gi} - \phi_{Si}. \quad (12)$$

By applying the covariance law to (11) and (12), we can calculate the standard deviations σ_{SF} and $\sigma_{\Delta\phi}$ of the scale factor and phase difference, respectively:

$$\sigma_{\mathrm{SF}} = \left[\frac{1}{A_{Si}^2}\left(\sigma_{A_{Gi}}^2 + \sigma_{A_{Si}}^2\left\{\frac{A_{Gi}^2}{A_{Si}^2}\right\}\right)\right]^{1/2}, \quad (13)$$

$$\sigma_{\Delta\phi} = \left[\sigma_{\phi_{Gi}}^2 + \sigma_{\phi_{Si}}^2\right]^{1/2}. \quad (14)$$

We note here that the scale factor is calculated only at specific cyclic frequencies at which both AG and SG data have statistically significant amplitudes.

4. SG and AG Data at Cantley Station

The SG at Cantley station started recording gravity data on November 7, 1989. Shortly after the installation the SG was drifting excessively (0.5 μGal per day), and several years later a new gravimeter sensor was installed and started

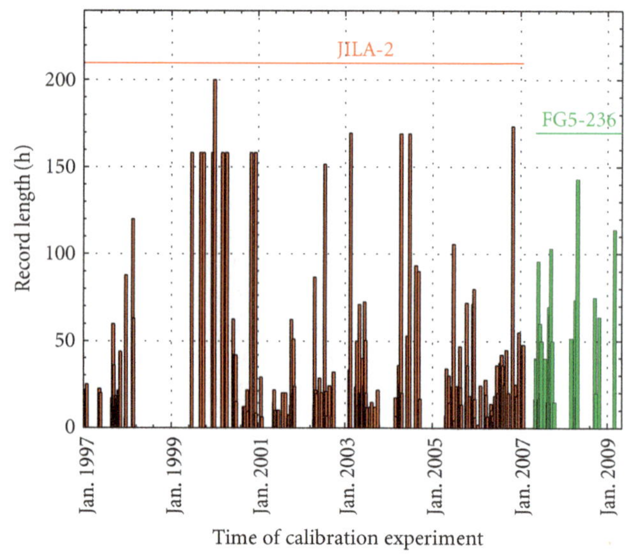

FIGURE 1: Absolute gravity records since January 1, 1997. The y-axis shows the duration of each calibration experiment in hours. The vertical red bars show the length of the experiments taken by JILA-2 while the vertical green bars show the length of the experiment taken by FG5-236.

recording in January 1997 and contributed to GGP during phase I and II. The SG has been operating on the pier of the Canadian Absolute Gravity Site (CAGS) close to JILA-2 and most recently FG5-236 Absolute Gravimeter. The JILA-2 gravimeter contributed calibration data until January 2007, and then it was replaced by FG5-236 in May 2007. In this work, we use AG data from both JILA-2 and FG5-236 for the calibration. Having the AG and SG running in parallel provides a good opportunity to examine the drift of the SG and check the signals coming out of both [17].

Over a period of 13 years (1997–2009), there have been 180 AG calibration experiments. The experiment durations range from a few to 200 hours with 7200 h of total length. The observation periods were most likely synchronised with the maximum Earth tide at Cantley station. Also, it is worth mentioning that the AG observations are unevenly spaced but the average sampling interval is very close to 60 s, which is not a problem when applying the LSSA method. Figure 1 shows the data availability and their duration per calibration experiment. Over 0.52 million sets are available from the AG. These records are distributed along the entire 13-year period and contaminated by all kinds of noises, for example, earthquakes, snow storms, and others.

5. Data Preprocessing and Filtering

Data preprocessing is one of the most important steps for obtaining a high precision scale factor. Three major factors or parameters have to be checked and tested before we estimate the scale factor namely, data offsets, data outliers, and the variance of the original data. The SG data are not free from small or large offsets, which need to be identified before the calibration process. Large offsets can be defined easily even

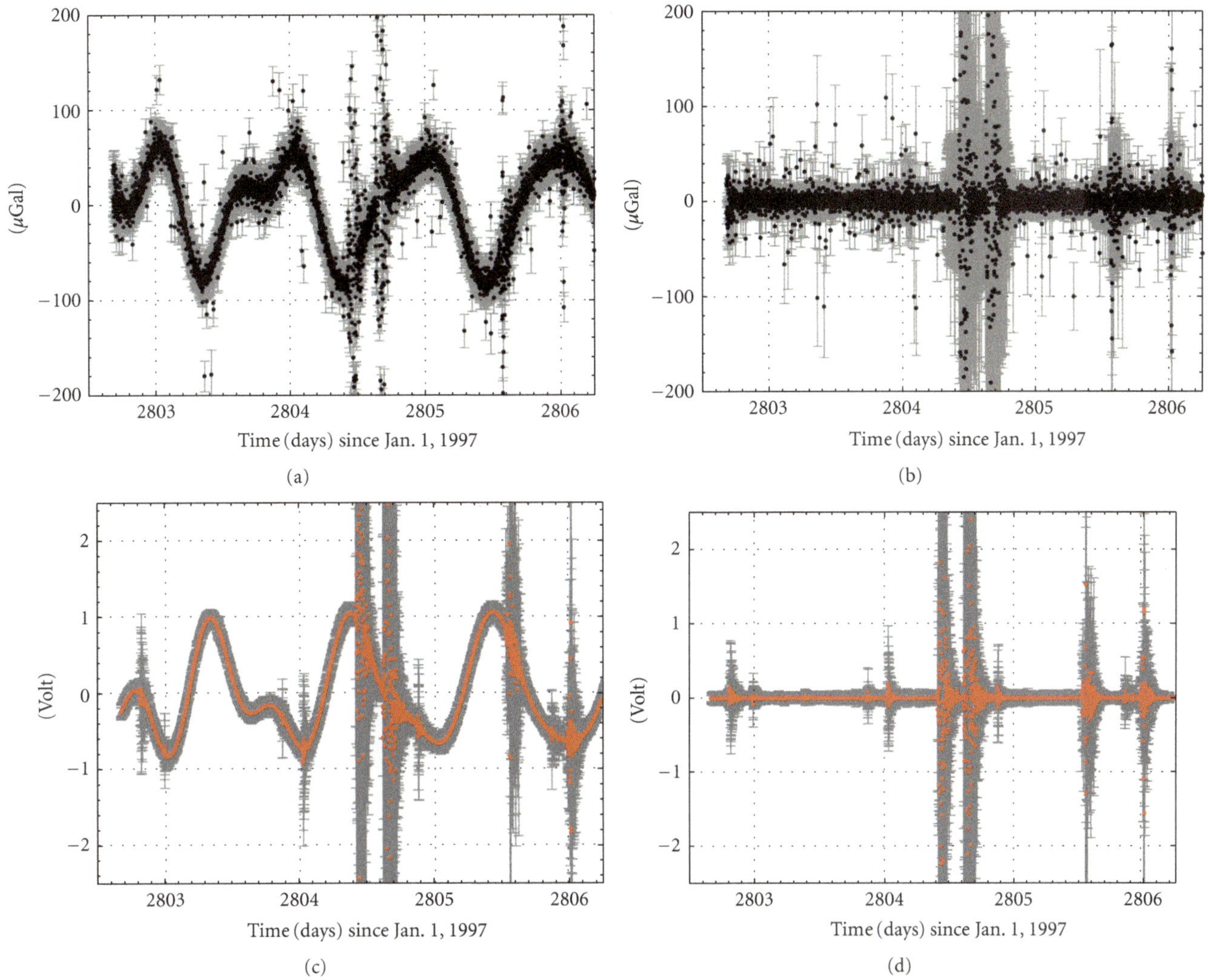

FIGURE 2: AG and SG before and after removing Earth tide, ocean loading, pressure, and polar motion. The x-axis is the time in hours since January 1, 1997. The top panels (a and b) are the AG while the bottom panels (c and d) are the SG.

by visual inspection but small offsets are not. To identify small offsets, all known signals must be removed first and then check the significance of the existing offsets as well as the residual outliers. Also, with this approach, the realistic noise level of the residuals can be estimated. The procedure below is followed when dealing with the SG data.

(1) For each AG block or experiment, the corresponding SG data segment is retrieved but extended 15 s beyond each end of the AG record to accommodate the filtering process. However, if there are no SG data corresponding to the AG records for any reason, the AG record is discarded.

(2) A nominal scale factor (from previous calibration) is used to transform the SG data from volts to μGal. Then, the Earth tide, ocean loading, pressure, and polar motion are removed from the raw SG data. Earth and ocean tides are removed based on the synthetic tide generated from "GWAVE" [29] while the pole tide is computed using the daily polar motion data from the International Earth Rotation and Reference Systems Service (IERS). A constant atmospheric admittance ($-0.3\,\mu$Gal/mbar) is used to model the pressure effect. Hence, the resulting residual series is practically free from most of the known signals.

(3) A Parzen window function with lag = 15 s (cutoff frequency $f_c \approx 0.0433$ Hz and bandwidth $b \approx 0.124$ Hz) is used to remove the high frequency component and provide a measure of the noise level. This is achieved by propagating the variance of the SG residual series segment within the window to the filtered value [23, 26]. Each filtered value is estimated exactly at the time AG measurements are available. In other words, the Parzen window is centered at the same time marks as the AG measurements.

(4) For each SG segment corresponding to the AG experiment, the LSSA v5.02 software [22, 23] is used for the following purposes: (a) offset detection and

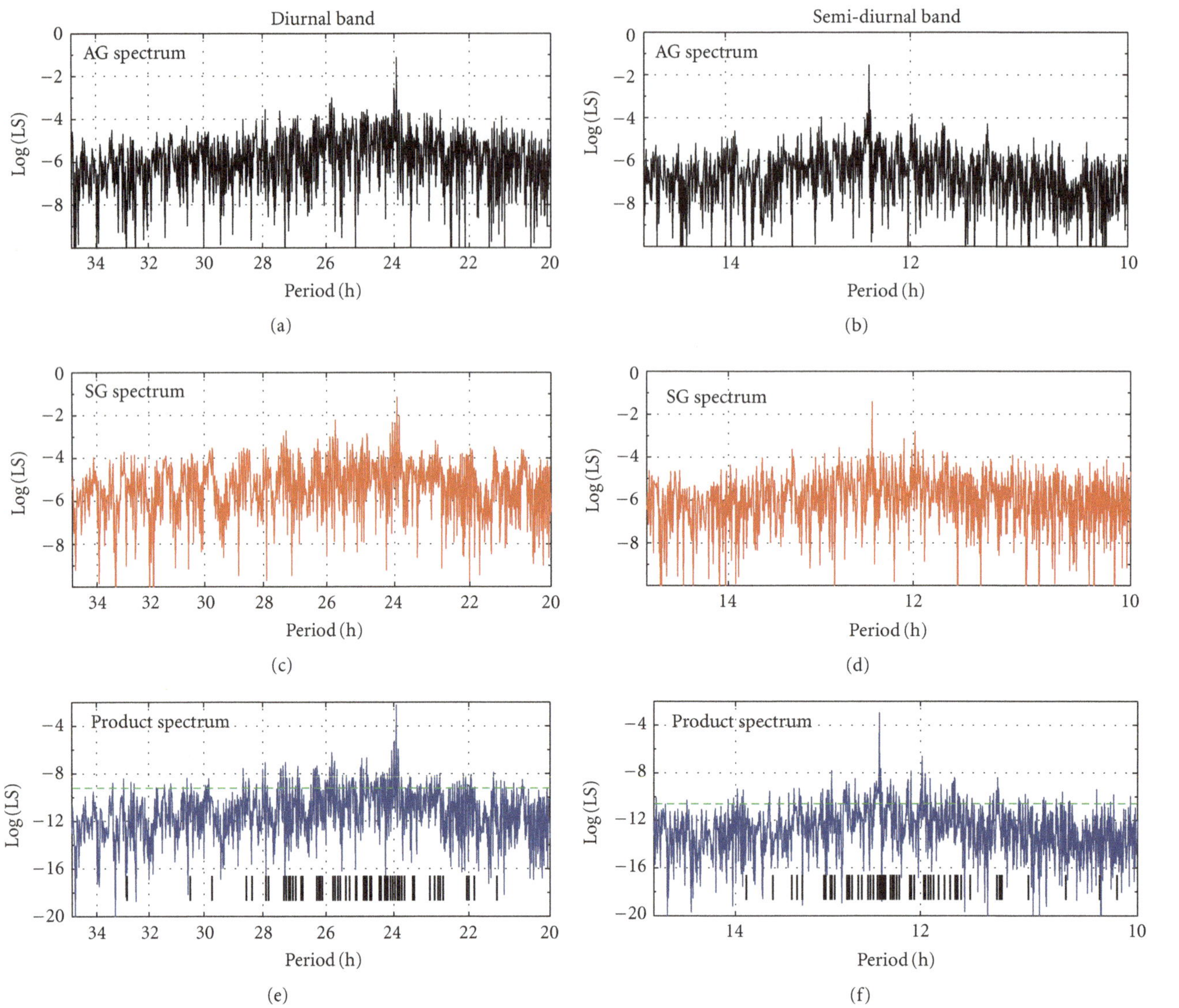

FIGURE 3: LS spectra of AG and SG and their product spectrum. The left panels (a, c, e) present the diurnal band while the right panels (b, d, f) present the semidiurnal band. The horizontal dashed green lines show the 95% CL while the small vertical lines in the product spectrum panels (e and f) show the common peaks.

identification of the time they occur, (b) residual outlier detection and flagging, (c) detection of any periodicities or deterministic signals followed by their suppression, and (d) test on the variance factor using Chi-square test [24].

(5) Based on the previous step, (a) the times of statistically significant offsets are identified for later analysis, (b) outliers more than 4.215 σ (95.0% CL for in-context test with 2000 data points [24]) are rejected. Their corresponding values in the AG record are excluded as well, and (c) the standard deviation estimated in step (3), which is associated with the filtered value, is rescaled (adjusted) using the a posteriori variance factor $\hat{\sigma}_o^2$ when the Chi-Square test fails in step (4d).

(6) Finally, all removed signals from steps (2) and (4c) above are restored in the residual series and translated back into volts using the same scale factor as in step (2) as well the standard deviation estimated in step (5c).

The AG data are provided to us as follows: time stamp, gravity set, and standard deviation. The standard deviation of each gravity value is a formal estimate. The standard deviation is between 1.0 and 3.0 μGal, which is rather too optimistic. Hence, the procedure below is followed to calibrate the standard deviation and outlier detection.

(1) Similar to step (2) in the SG data preprocessing, all known signals, such as Earth and ocean tide, atmospheric pressure, and polar motion are removed from the AG records.

(2) At each data point, we retrieve all AG data that extend 5.0 min at either side to estimate the standard deviation of the gravity value at the selected point. Then, the mean and the trend of this small data segment are subtracted and the estimated standard deviation of the residual values in the segment replaces the original value.

(3) Residual outliers are rejected along with their corresponding values in the SG record. The Chi-Square tests on the variance factor for all segments passed. This indicates that the standard deviation estimated in step 2 represents a realistic noise level of the AG records.

(4) All removed signals are restored in the residual AG series, as with the SG.

Figure 2 shows a segment from AG and SG before and after calibrating the standard deviation. The top-left panel (a) shows the AG with an uncalibrated error (provided with the original data), which is very optimistic, while the top-right panel (b) shows the AG after removing all signals with the calibrated standard deviation (gray error bars) based on the realistic noise level. The lower panels of Figure 2 represent the SG data in volts.

6. Scale Factor Determination

6.1. Scale Factor in the Frequency Domain. Due to the large number of AG data points (~0.52 million), a lowpass filter is applied to decimate the data (AG and SG) and reduce the high frequency noise. A Parzen window with lag = 450 s ($f_c \approx 1.40$ mHz and $b \approx 4.13$ mHz) is applied to produce data points with an average sampling interval of 15 min. The LS spectrum in the band 6.00000–0.00025 cpd (0.25–4015.00 d) of the filtered data is estimated using the LSSA v5.02 software. The identification of all significant peaks in the product LS spectrum at 95% CL is done rigorously using the probability density function (PDF) of the product LS spectrum defined by (8) (dashed lines in Figure 3). In fact there are many peaks that are significant, but they are close to each other. When these significant peaks are suppressed, we find that most of the estimated amplitudes and phases are highly correlated (from the variance-covariance matrix). Due to the high noise level in the AG data and the huge number of tidal constituents especially in the diurnal, semidiurnal and terdiurnal bands, it becomes difficult to resolve these peaks. Thus, we filter out these peaks and we use only the ones that satisfy the following two criteria: (1) highest peak and (2) smallest correlation with others.

Next, all significant peaks identified in the previous step are suppressed using (9) and (10) to estimate their amplitudes and phases simultaneously with other systematic noise (e.g., trend and offsets). We use (11)–(14) to estimate the magnitude and phase difference of the scale factor associated with the covariances. We note here that the scale factor is calculated only at specific frequencies ω_i at which both AG and SG time series have significant amplitudes. Then, the weighted LS regression is used to define the best fit to all admittance points to obtain a smooth response function [18].

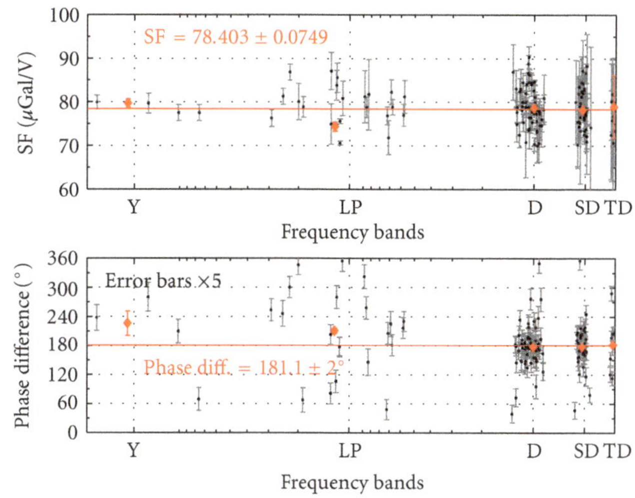

FIGURE 4: Amplitude of the scale factor (upper panel) and its phase lag (lower panel). The red diamonds show the scale factor estimates in certain bands (TD, SD, D, LP, and Y).

The estimated amplitude of the scale factor in the frequency domain is 78.403 ± 0.0750 μGal/V, while the phase difference is 181.1° ± 2.0°. The relative accuracy is 0.096%, which is higher than that in the time domain. The 180.0° phase difference (negative sign for all the SGs in GGP network) indicates how the feedback system in SG responds to gravity variation. Our estimation of the phase difference indicates that SG lags behind AG by 1.1° ± 2.0°, which is not significant. To check the scale variability we select five bands (ter-diurnal (TD), semidiurnal (SD), diurnal (D), long period (LP), which includes Mf, Mm, and Mtm, and other seasonal (Y), which includes Ssa, Sa, and others) to estimate the scale factor amplitude and phase difference at the midpoint of these bands. The red diamonds in Figure 4 and Table 1 represent these values.

6.2. Scale Factor in the Time Domain. All AG and their corresponding SG calibration experiments are used to estimate the scale factor in the time domain. For each individual experiment block the scale factor (SF) can be estimated by fitting a straight line through AG and SG residuals by minimizing the square of the weighted residuals AG_r (in the least squares sense) as follows:

$$AG_r = AG + SF \times SG, \quad (15)$$

where AG and SG are, respectively, the AG and the SG data after subtracting the mean value. The estimated scale factor and its associated standard error correspond to the middle of the block (see Figure 5(c)). Figure 5(a) shows one of this regression lines for one calibration experiment. Then, the least squares weighted average is applied to determine the scale factor based on 11 years of data from JILA-2 and two years from FG5-236. Using only JILA-2 data, the SG scale factor is estimated in the time domain as

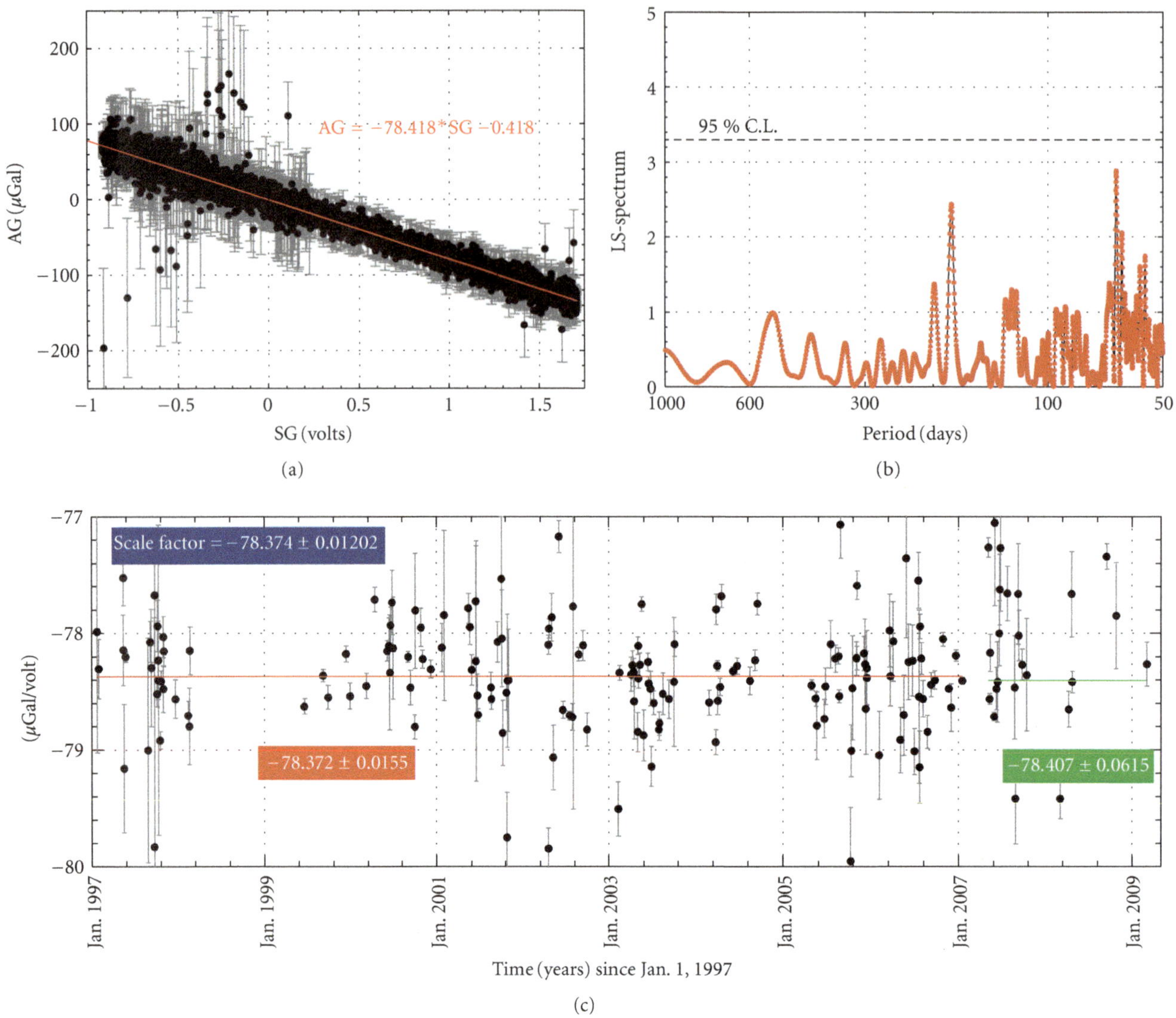

Figure 5: Scale factor in time domain. Panel (a) is the regression of one block and, (b) the scale factor in the frequency domain, while (c) is the average scale factor (solid line) and the estimated scale factor of each block.

Table 1: Amplitudes and phase differences in different bands (TD: tridiurnal, SD: semidiurnal, D: diurnal, LP: long period, and Y: yearly).

Band	Period (days)		Amplitude	Phase difference (°)
	From	To	(μGal/V)	AG−SG
TD	0.320	0.368	78.962 ± 7.150	182.22 ± 1.03
SD	0.457	0.593	78.149 ± 0.120	177.84 ± 0.60
D	0.802	1.396	78.647 ± 0.107	177.24 ± 0.55
LP	5.729	34.85	74.389 ± 1.018	221.60 ± 0.98
Y	91.33	437.9	79.662 ± 0.976	226.00 ± 45.00

−78.372 ± 0.0155 μGal/V. The FG5-236 short time series (May 2007–February 2009) is used to reestimate the scale factor in the time domain to be −78.407 ± 0.0615 μGal/V, which is statistically compatible with the one estimated from JILA-2 data. Combining both datasets the final scale factor becomes −78.374 ± 0.0120 μGal/V with a relative accuracy of 0.015%, which is comparable with the absolute method of calibration of SG when using a platform. This is the first time to achieve such high precision for the scale factor due to the fact that there are very long AG records for calibration. Further analysis of the scale factor in the time domain reveals that there is no statistically significant drift. However, the LSSA is applied to investigate the existence of any periodicity in the scale factor. The frequency domain representation of

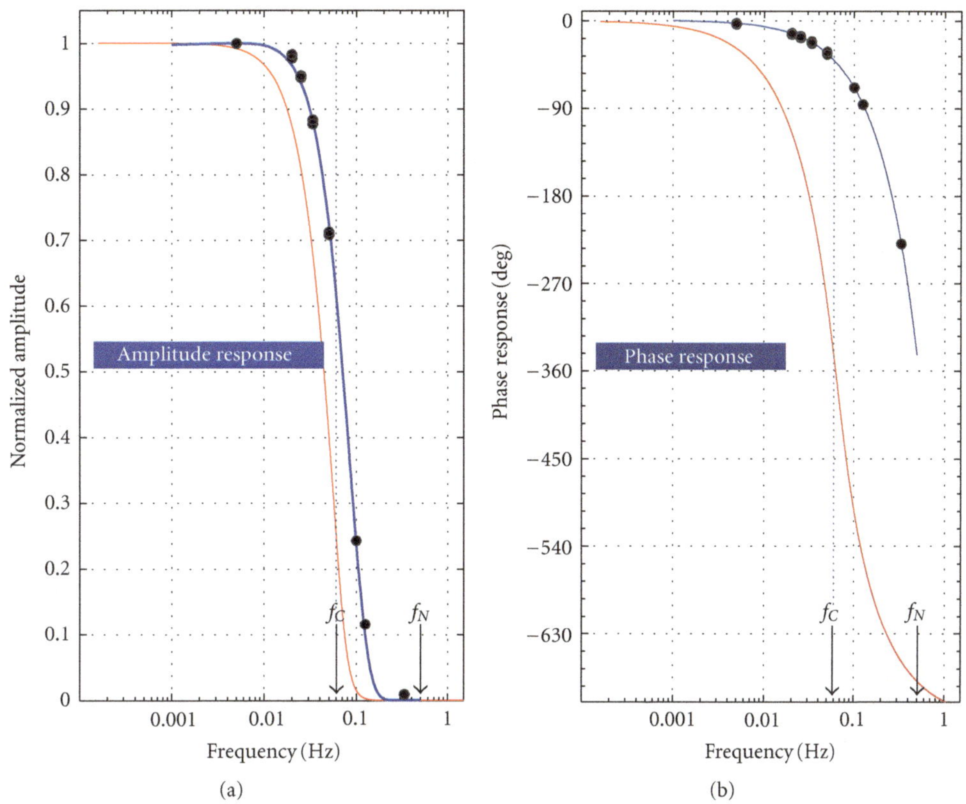

Figure 6: Hardware filter response with $f_c = 0.0615$ Hz and $f_N = 0.5$ Hz. Panel (a) shows the amplitude response, while panel (b) shows a smooth phase response without jumps by adding multiples of 360. The black dots show the real transfer function from the injected experiment with a blue line representing the best fitting of the amplitude and phase response, while the red line shows the theoretical Bessel filter response.

this series shows no evidence of any periodicities in the scale factor as there is no single peak greater than the 95% CL (Figure 5(b)).

7. Calibration of SG Electronics

The SG electronic package includes electronics that sense and control all subsystems (e.g., gravimeter, tilt, and temperature). An important feature of the gravimeter control is the gravity circuit card, which is mounted in the analog control chassis. On this card, there is an eight-pole Bessel hardware filter with a corner frequency of 0.0615 Hz. This filter, known as GGP1, is used as antialiasing hardware filter for digitizing the continuous gravity signal to 1 s sampling interval with a Nyquist frequency $f_N = 0.5$ Hz. The GGP1 filter conforms with recommendations put forth by the GGP. The red line in Figure 6 shows the theoretical filter response in the frequency domain represented by the gain and phase response.

This analog Bessel filter is characterized by almost constant group delay across the entire passband (0–50 mHz), thus preserving the wave shape of filtered signals. In addition, the filter does not attenuate the signals up to the frequency of 2.8 mHz (as seen from Figure 6). However, this has to be tested frequently by injecting a series of sine and/or square waves to the sensor electronics and study its actual response [30].

In January 22, 2008 the original filter (provided with the instrument) at Cantley station was replaced by the GGP1 filter. Six months later, the transfer function (gain and phase) of this filter was tested by injecting 14 sine and square waves with periods ranging from 3 s to 200 s and amplitudes ranging from 0.013 V to 0.6386 V, where the lowest amplitude is equivalent to 1 μGal. The black dots in Figures 6(a) and 6(b), show the gain and phase response, respectively.

It is obvious from Figure 6(a), that any noise at frequencies higher than 0.160 Hz (<6 s) is filtered out totally while the real cutoff frequency from this experiment is 0.070 Hz (14.3 s), which is close to the theoretical $f_c = 0.0615$ Hz. Also, the filter preserves the signals at frequencies lower than 0.005 Hz (200 s). Based on this test (as seen in Figure 6(b)), the estimated phase delay at $f_c = 0.0615$ Hz is 14.4 s, which is close to the specified filter parameters required by GGP.

8. Summary and Concluding Remarks

In this paper, the scale factor of the Cantley SG was determined in both time and frequency domains. The time domain estimation gave better accuracy ($-78.374 \pm$

0.0120 μGal/V) than the frequency domain ($-78.403 \pm 0.0750\,\mu$Gal/V). A couple of reasons may have contributed to this result: (1) the frequency domain estimation considers the noise level of both gravimeters SG and AG, while the time domain calculation deals only with AG noise; (2) using the whole dataset may hide real existing offsets and reduce the accuracy level in the frequency domain, while in the time domain the data are already segmented into blocks. However, these two estimates are statistically compatible (the compatibility of two means was statistically tested using t-test at 95% CL).

In the frequency domain, the scale factor in diurnal and semidiurnal bands is more accurate than in the other bands as these two bands have the strongest Earth tide signals. In addition, the scale factor determined in the time domain is closer to diurnal and semidiurnal bands estimation although not commensurate with the accuracy levels.

Finally, we could not identify any drift or periodicities in the scale factor. Also, the new installed GGP filter was tested to conform to the published specifications. Last but not least, we recommend the scale factor posted at the GGP website be changed to the new and accurate one estimated in this research from the time domain.

Acknowledgments

This research was financially supported by the Natural Sciences and Engineering Research Council (NSERC) of Canada, Ryerson University Post-Doctoral Fellowship, and the Start-Up Grant from University of Cape Town for the first author. The authors thank the manager of Cantley, Canada station for making the SG and AG data available for this research.

References

[1] T. Sato, M. Ooe, K. Nawa, K. Shibuya, Y. Tamura, and K. Kaminuma, "Long-period tides observed with a superconducting gravimeter at Syowa Station, Antarctica, and their implication to global ocean tide modeling," *Physics of the Earth and Planetary Interiors*, vol. 103, no. 1-2, pp. 39–53, 1997.

[2] J. Hinderer, D. Crossley, and R. Warburton, "Superconducting gravimetry," in *Treatise on Geophysics*, T. Herring and G. Schubert, Eds., vol. 3, Elsevier, Amsterdam, The Netherlands, 2007.

[3] J. Hinderer, N. Florsch, J. Makinen, H. Legros, and J. E. Faller, "On the calibration of a superconducting gravimeter using absolute gravity measurements," *Geophysical Journal International*, vol. 106, no. 2, pp. 491–497, 1991.

[4] M. Amalvict, J. Hinderer, J. Boy, and P. Gegout, "A three year comparison between a superconducting gravimeter (GWR C026) and an absolute gravimeter (FG5 206) in Strasbourg (France)," *Journal of the Geodetic Society of Japan*, vol. 47, pp. 410–416, 2001.

[5] O. Francis, "Calibration of the C021 superconducting gravimeter in membach (Belgium) using 47 days of absolute gravity measurements, gravity, geoid and marine geodesy," in *Proceedings of the International Association of Geodesy Symposia*, vol. 117, pp. 212–219, 1997.

[6] O. Francis and T. Van Dam, "Evaluation of the precision of using absolute gravimeters to calibrate superconducting gravimeters," *Metrologia*, vol. 39, no. 5, pp. 485–488, 2002.

[7] O. Francis, T. M. Niebauer, G. Sasagawa, F. Klopping, and J. Gschwind, "Calibration of a superconducting gravimeter by comparison with an absolute gravimeter FG5 in Boulder," *Geophysical Research Letters*, vol. 25, no. 7, pp. 1075–1078, 1998.

[8] T. Sato, Y. Tamura, S. Okubo, and S. Yoshida, "Calibration of scale factor of superconducting gravimeter at Esashi using an absolute gravimeter FG5," *Journal of the Geodetic Society of Japan*, vol. 42, pp. 225–232, 1996.

[9] M. Amalvict, H. McQueen, and R. Govind, "Absolute gravity measurements and calibration of SG-CT031 at Canberra, 1999-2000," *Journal of the Geodetic Society of Japan*, vol. 47, pp. 334–340, 2001.

[10] Y. Imanishi, T. Higashi, and Y. Fukuda, "Calibration of the superconducting gravimeter T011 by parallel observation with the absolute gravimeter FG5 #210—a Bayesian approach," *Geophysical Journal International*, vol. 151, no. 3, pp. 867–878, 2002.

[11] Y. Tamura, T. Sato, Y. Fukuda, and T. Higashi, "Scale factor calibration of a superconducting gravimeter at Esashi Station, Japan, using absolute gravity measurements," *Journal of Geodesy*, vol. 78, no. 7, pp. 481–488, 2004.

[12] R. Falk, M. Harnisch, G. Harnisch, I. Nowak, B. Richter, and P. Wolf, "Calibration of superconducting gravimeters SG103, C023, CD029, and CD030," *Journal of the Geodetic Society of Japan*, vol. 47, no. 1, pp. 22–27, 2001.

[13] R. Warburton, C. Beaumont, and J. Goodkind, "The effect of ocean tide loading on tides of the solid earth observed with the superconducting gravimeter," *Geophysical Journal of the Royal Astronomical Society*, vol. 43, pp. 707–720, 1975.

[14] V. Achilli, P. Baldi, G. Casula et al., "A calibration system for superconducting gravimeters," *Bulletin Géodésique*, vol. 69, no. 2, pp. 73–80, 1995.

[15] B. Richter, "Calibration of superconducting gravimeters," in *Proceedings of the Workshop on Non Tidal Gravity Changes Intercomparison between Absolute and Superconducting Gravimeters*, vol. 3, pp. 99–107, Conseil de l'Europe, Cahiers du Centre Europeen de Geodynamique et de Seismologie, Luxembourg, 1991.

[16] D. Bower, J. Liard, D. Crossley, and R. Bastien, "Preliminary calibration and drift assessment of the superconducting gravimeter GWR12 through comparison with absolute gravimeter JILA2," in *Proceedings of the Workshop on Non Tidal Gravity Changes Intercomparison between Absolute and Superconducting Gravimeters*, vol. 3, pp. 129–142, Conseil de l'Europe, Cahiers du Centre Europeen de Geodynamique et de Seismologie, Luxembourg, 1991.

[17] J. Merriam, S. Pagiatakis, and J. Liard, "Reference level stability of the Canadian superconducting gravimeter installation," *Journal of the Geodetic Society of Japan*, vol. 47, no. 1, pp. 417–423, 2001.

[18] M. Abd El-Gelil, S. Pagiatakis, and A. El-Rabbany, "Frequency-dependent atmospheric pressure admittance of superconducting gravimeter records using least squares response method," *Physics of the Earth and Planetary Interiors*, vol. 170, no. 1-2, pp. 24–33, 2008.

[19] P. Vaníček, "Approximate spectral analysis by least-squares fit—successive spectral analysis," *Astrophysics and Space Science*, vol. 4, no. 4, pp. 387–391, 1969.

[20] P. Vaníček, "Further development and properties of the spectral analysis by least-squares," *Astrophysics and Space Science*, vol. 12, no. 1, pp. 10–33, 1971.

[21] M. Craymer, *The least squares spectrum, its inverse transform and autocorrelation function: theory and some applications in geodesy [Ph.D. thesis]*, University of Toronto, Toronto, Canada, 1998.

[22] S. D. Pagiatakis, "Stochastic significance of peaks in the least-squares spectrum," *Journal of Geodesy*, vol. 73, no. 2, pp. 67–78, 1999.

[23] S. Pagiatakis, "Application of the least-squares spectral analysis to superconducting gravimeter data treatment and analysis," in *Proceedings of the workshop on High precision gravity measurements with application to geodynamics and Second GGP*, vol. 17, pp. 103–113, Cahiers Du Centre Europeen De Geodynamique et de Seismologie (ECGS), 2000.

[24] P. Vaníček, *Geodesy: The Concepts*, North Holland, Amsterdam, The Netherlands, 1986.

[25] H. Yin and S. Pagiatakis, "Least squares spectrum analysis and its application to superconducting gravimeter data analysis," *Geo-Spatial Information Science*, vol. 7, no. 4, pp. 279–283, 2004.

[26] S. D. Pagiatakis, H. Yin, and M. A. El-Gelil, "Least-squares self-coherency analysis of superconducting gravimeter records in search for the Slichter triplet," *Physics of the Earth and Planetary Interiors*, vol. 160, no. 2, pp. 108–123, 2007.

[27] R. V. Hogg and A. T. Craig, *Introduction to Mathematical Statistics*, Prentice-Hall, Upper Saddle River, NJ, USA, 6th edition, 2005.

[28] M. Abramowitz and I. A. Stegun, *Handbook of Mathematical Functions*, Dover Publications, 1965.

[29] J. B. Merriam, "An ephemeris for gravity tide predictions at the nanogal level," *Geophysical Journal International*, vol. 108, no. 2, pp. 415–422, 1992.

[30] M. Van Camp, H. G. Wenzel, P. Schott, P. Vauterin, and O. Francis, "Accurate transfer function determination for superconducting gravimeters," *Geophysical Research Letters*, vol. 27, no. 1, pp. 37–40, 2000.

26

Review of Recent Developments and the Future Prospective in West African Atmosphere/Land Interaction Studies

Yongkang Xue,[1,2] Aaron Boone,[3] and Christopher M. Taylor[4]

[1] *Department of Geography, University of California, Los Angeles, CA 90095, USA*
[2] *Department of Atmospheric and Oceanic Sciences, University of California, Los Angeles, CA 90095, USA*
[3] *Meteo-France, CNRM, Toulouse, France*
[4] *Centre for Ecology & Hydrology, Crowmarsh Gifford, Wallingford, Oxon OX10 8BB, UK*

Correspondence should be addressed to Yongkang Xue, yxue@geog.ucla.edu

Academic Editor: Andrea Sealy

This paper reviews West African land/atmosphere interaction studies during the past decade. Four issues are addressed in this paper: land data development, land/atmosphere interactions at seasonal-interannual scales, mesoscale studies, and the future prospective. The development of the AMMA Land Surface Model Intercomparison Project has produced a valuable analysis of the land surface state and fluxes which have been applied in a number of large-scale African regional studies. In seasonal-interannual West African climate studies, the latest evidence from satellite data analyses and modeling studies confirm that the West African region has a climate which is particularly sensitive to land surface processes and there is a strong coupling between land surface processes and regional climate at intraseasonal/seasonal scales. These studies indicate that proper land surface process representations and land status initialization would substantially improve predictions and enhance the predictability of West African climate. Mesoscale studies have revealed new understanding of how soil moisture heterogeneity influences the development of convective storms over the course of the diurnal cycle. Finally, several important issues regarding the future prospective are briefly addressed.

1. Introduction

West Africa is a diverse climatic region that includes a semiarid tropical zone in its northern section and a humid tropical climate zone in its southern section. Today, it is a bioclimatic zone of predominantly annual grasses with shrubs and trees, receiving a mean annual rainfall of between 150 and 2500 mm per year (Figure 1(a)). According to the United Nations' definition [1], it includes 16 African countries and covers an area in excess of approximately 6 million km^2. Its northern boundary is along the southern margin of the Sahara Desert, with the Atlantic Ocean to its west and to its south. There is no consensus on its eastern boundary, which is commonly specified using a line running from Mount Cameroon to Lake Chad. The elevation of the vast majority of this land is less than 300 meters above sea level, though isolated points of higher elevation exist in several countries along the southern shore of the region.

West Africa has experienced significant climate anomalies in the past century. Its northern section, which is referred to as the Western Sahel, suffered the most severe and longest droughts in the world during the twentieth century. Annual rainfall has persistently remained below the long-term average since the late 1960s, with devastating droughts during the late 1960s and the early 1980s (Figures 1(b) and 1(c)) [2–5]. Starting from the late 1980s, however, there has been some rainfall recovery relative to the very dry period (Figures 1(b) and 1(d)). A study by Lau et al. [6] finds that there has been an increase in the "continentality" of the West African monsoon (WAM), reflected by a shift of the WAM convection towards land from the later 1980s to the 1990s and the 2000s. The hydrologic variability in West Africa shows one of the strongest interdecadal signals worldwide during the 20th century [7]. There has been rapid population growth in the region over the past 50 years [8]. The severe droughts and resulting famines have devastating

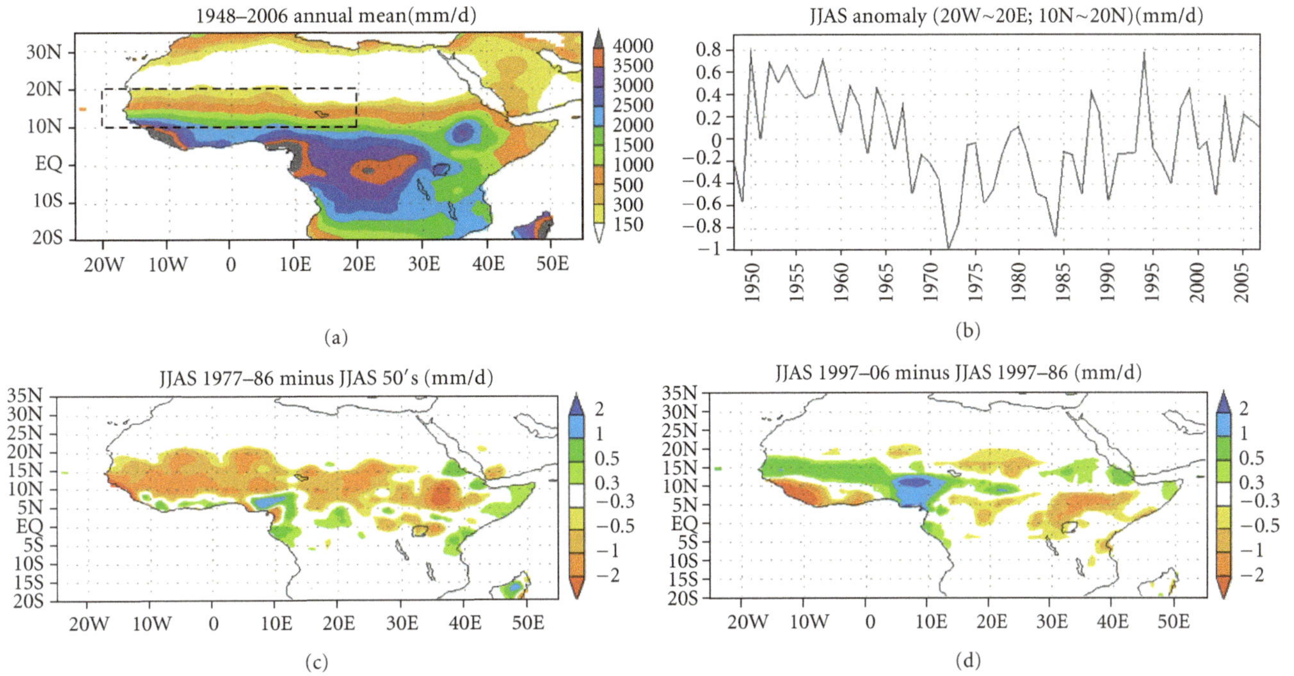

Figure 1: (a) Observed climate annual mean precipitation (mm day^{-1}); (b) time series of June-July-August-September (JJAS) mean precipitation (mm day^{-1}) averaged over 20°W to 20°E and 5°N to 20°N (the box outlined in (a)); (c) observed JJAS precipitation difference (mm day^{-1}) between the 1980s and the 1950s; (d) observed JJAS precipitation difference (mm day^{-1}) between 1996–2006 and 1976–1987. Data Sources: National Center for Environmental Prediction Climate Prediction Center Merged Analysis of Precipitation (CMAP, [23]).

environmental and socioeconomic consequences, which, along with the scientific challenge to understand causes of such extraordinary drought and climate variability, have presented a strong motivation for the scientific community to study West African climate variability and causes of Sahel drought, most of which occurred in West Africa. The term "Sahel drought" will be used interchangeably with "West African drought" in this paper. The Sahel drought is known to be strongly influenced by sea surface temperature (SST) anomalies, both globally and in regions adjacent to the African continent [9, 10]. Meanwhile, land surface processes, including the land use and land cover (LULC) change, have also been extensively investigated as one of the major factors influencing the West African climate system, which is the focus of this paper.

There is a steep latitudinal gradient in climate, soils, vegetation, fauna, land use, and human utilization, with strong correlations among them [11]. The continental-scale land mass and relatively flat orography suggests that land-surface/atmosphere interactions play a major role in the regional climate. From the mid-1970s through the 1990s, numerous studies have been conducted to explore the impact of the land surface on the West African climate. Modeling studies with different levels of complexity in their atmospheric models and land surface parameterizations have consistently suggested that the Sahel is one of the regions in the world where the land/atmosphere feedback plays an important role in regional climate variability, and LULC may be an important contributing factor to the persistent Sahel drought [12–20]. Comprehensive reviews on the relevant land/West African climate research mainly before the 21 century were summarized in Xue et al. [21].

There are several important characteristics in land/West African climate research reviewed in [21]. (1) Most studies test the sensitivities of regional climate to a few land parameters, such as surface albedo and soil moisture, explicitly or implicitly with LULC change embedded in the experimental designs. Only recently, encouraged by those early sensitivity studies, have more sophisticated vegetation models been introduced in West African climate studies (2) The Sahel drought was the main motivation for those studies. Most studies were conducted using either 3 D or 2 D coupled atmosphere/land surface models to estimate the possible land contribution to the drought. (3) There were only a few measurements over very few sites. Due to the limited availability of observational data over the West African region at that time, validation of the land models for regional climate and hydrological cycle study was largely missing. However, these research efforts have established a consensus in the public and part of the meteorological community that, in addition to SST, land surface processes are an important component affecting the West African regional climate, thereby stimulating more extensive research in this field.

In the past decade, important developments for obtaining a better understanding of the West African monsoon (WAM)/land interaction at intraseasonal/interannual/interdecadal scales have been made, in particular under the African Monsoon Multidisciplinary Analysis (AMMA) Project [22]. We will discuss a few major issues and achievements in the West African climate and land process interaction study in the past decade in this paper. In Section 2, new land data development for the WAM study and its application are presented. Section 3 focuses on the land/West African climate interactions studies at seasonal, interannual, and decadal scales. Section 4 discusses mesoscale interactions, and Section 5 presents a summary and future prospective in this field.

2. Land Data Development for the West African Monsoon Study and Its Application

Modeling the West African monsoon has proven difficult mainly due to both the paucity of observations at sufficient space-time resolutions, and because of the complex interactions of the relevant processes between the biosphere, atmosphere, and hydrosphere over this region. Distinct from the previous observational efforts which encompassed only one or a few sites [24], the AMMA program has organized comprehensive activities in data collection in order to increase the understanding and improve the modeling of the WAM/land interactions over a wide range of spatial and time scales [7]. A comprehensive review of AMMA field campaigns' key accomplishments and lessons learned from the extraordinary efforts made by a community of several hundred researchers has been discussed by Lebel et al. [25].

The application of observations from many local sites for large-scale atmosphere/land interaction study is always a challenge due to scale discrepancy. The AMMA Land Surface Model (LSM) Intercomparison Project (ALMIP; [26]) addresses this issue. In ALMIP, a set of state-of-the-art off-line land surface models (uncoupled from their host atmospheric models) have been forced with the best quality and highest space and time resolution data available to better understand the key processes and their corresponding scales. The forcing consists of a combination of short-term numerical weather forecast low-level atmospheric state variables, satellite-based rainfall (such as the Tropical Rainfall Measuring Mission (TRMM, [27]), and downwelling radiation flux products (Land-Satellite Application Facility; [28]), and data from the AMMA observational field campaigns. The ALMIP LSM outputs for the period 2002–2007 have been compared with satellite-derived surface brightness temperature [29], satellite-based soil water storage estimates [30], and observed turbulent fluxes [26]. In the evaluation, the site turbulence measurements were first upscaled to the ALMIP mesoscale grids for comparison [31].

Overall, there is a good LSM intermodel agreement. For example, the predicted latent fluxes agree well with a coefficient of variation ranging from approximately 5 to 15% over most of West Africa, with the best agreement over the semiarid Sahel where the vast majority of rainfall is subsequently reevaporated at the land surface. The ALMIP LSM-predicted surface fluxes and thermal and hydrological state variables at the land surface are akin to a multimodel surface reanalysis product. Studies [32] have shown the advantages and improved realism of using a multi-LSM model average of simulated surface properties. The ensemble signal to noise ratio of the surface turbulent fluxes is fairly large (approximately 0.7 to 0.9) over most of the region [33].

Because the ALMIP data cover most of West Africa, this data set can be considered as one of the best currently available proxies for large-scale estimates of the land surface state over West Africa for different applications. For example, ALMIP results have been used for a number of mesoscale to regional scale case studies within AMMA, such as the study of feedbacks between dust emissions and the atmosphere [34, 35], and the regional atmospheric nitrogen balance [36]. Results have also been used to examine the influence of initial soil moisture on medium range weather prediction at the European Centre for Medium-Range Weather Forecasts (ECMWF; [37]). In addition, ALMIP results have been recently used for evaluating the land surface component of numerical weather prediction (NWP) and mesoscale models [38], regional climate models (RCMs, [39]), and general circulation models (GCMs, [40]). Moreover, regional scale water budget estimates have been made [41] using ALMIP surface evaporation fluxes.

The application of the ALMIP products in the West African Monsoon Modeling and Evaluation (WAMME) project [42] clearly demonstrated its utility in the large-scale African land/atmosphere interaction studies [33]. It was found that simulated net radiation within WAMME GCMs and RCMs generally agrees rather well with ALMIP. However, the partitioning of this energy into turbulent fluxes is different from ALMIP and is quite variable. This seems to be due, in part, to further northward placement of the monsoon in areas and positive precipitation bias with larger potential energy for evaporation in most of the WAMME models. Furthermore, ALMIP produces the maximum latent heat flux during the monsoon retreat as stored water is evaporated before solar radiation reaches the boreal winter minima; only 6 of the WAMME models have this feature, and it is generally rather weak. These features are summarized in the Hovmoller plot in Figure 2, where WAMME model simulated latent heat fluxes are compared to those from ALMIP for 2004 averaged over the region from $-10°$ to $10°$E latitude, and the southern coast of West Africa to $20°$N. Finally, in contrast to ALMIP results and local scale observations which indicate values on average of the meridional ratio of sensible heat flux to the net radiation to be approximately 0.35, the ratio becomes quite low (generally less than 0.10) from the Sahel to the southern coast of West Africa during the core monsoon season for nearly all of the WAMME models. All these findings should help identify weaknesses in current West African climate modeling, understand the WAM mechanisms, and contribute to further model development.

By and large, ALMIP provides a robust regional scale estimate of land surface state variables and fluxes from an ensemble of state-of-the art land surface models in its first phase experiment. Satellite-based data (GRACE, AMSR-E, etc.) were introduced for evaluation of ALMIP data. ALMIP data address many key science issues within AMMA, such as how coupled models simulate the land states and fluxes and obtain better estimates of the regional scale water budget. The development of the data was a major achievement in the

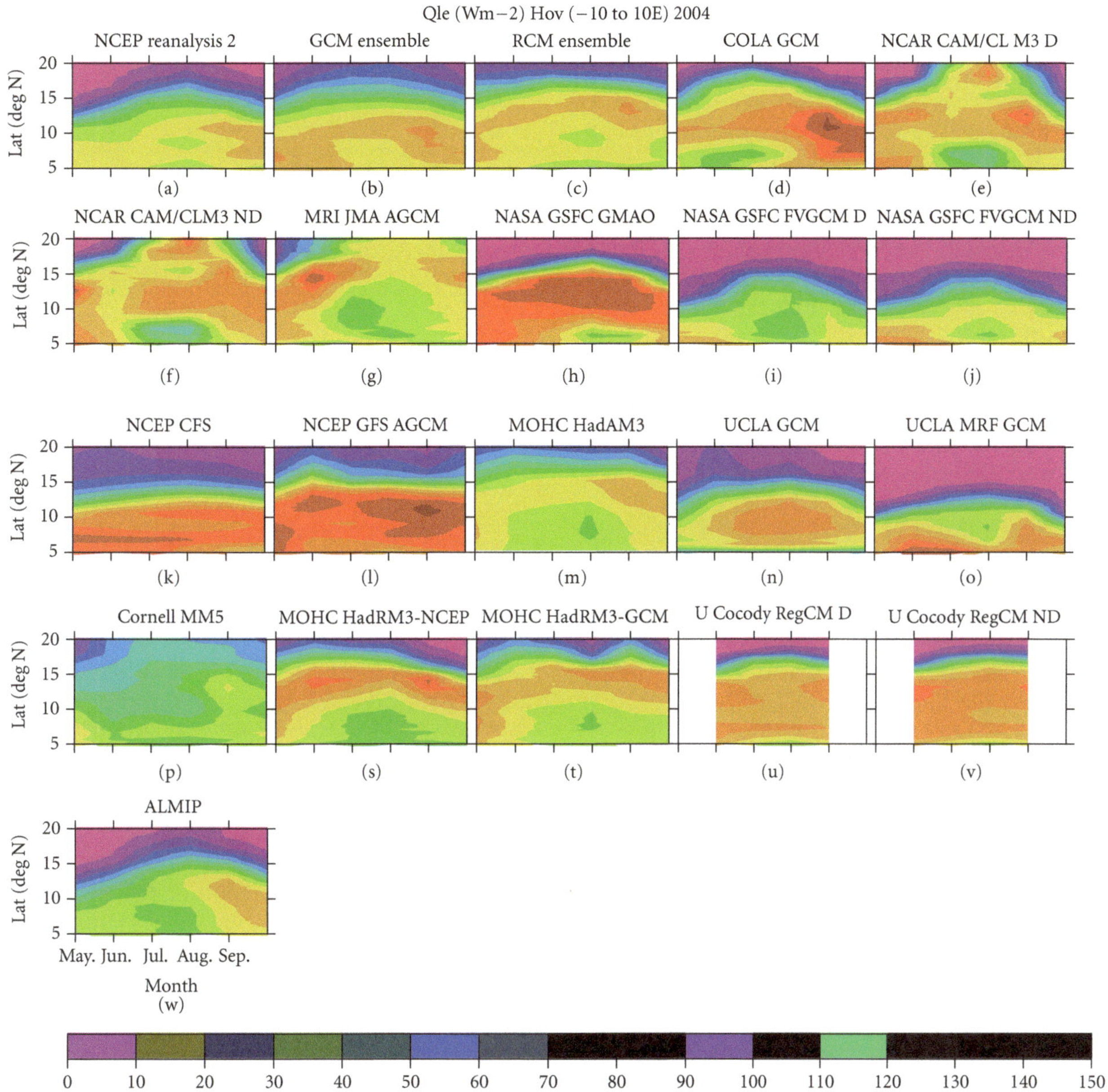

Figure 2: The WAMME latent heat flux area averaged from June through September (JJAS) for 2004 [33]. The corresponding ALMIP field is shown in panel (w).

past decade, which should pave the way for more scientific understanding in Sahel climate and its relationship with land surface processes and LULC change.

3. Seasonal and Interannual West African Climate Variability and Land Surface Processes

In the past decade, studies of West African climate/land interaction have focused primarily on predictability and prediction, that is, whether introducing adequate land surface processes improves the simulation of West African climate and enhances its predictability at different scales and possible mechanisms. The West African climate is dominated by the WAM system. Monsoon circulations are forced and maintained by sea-land thermal contrasts and by latent heat released into the atmosphere. Therefore, the WAM/land interaction was a major issue in these studies.

Insufficient data was a major obstacle to the evaluation of model-produced land/atmosphere interactions. Numerous remotely sensed products have become available during the past decade. Some derived satellite products, such as the Normalized Difference Vegetation Index (NDVI) and the fraction of photosynthetically active radiation (FPAR), have been considered to be observed measures of green vegetation condition. Analyses of these satellite products and climate data have confirmed a close relationship between vegetation conditions and precipitation at different scales [43–46], which early on was only suggested by modeling studies. For example, Martiny et al. [43] found that, in the southern Sahel, annual NDVI was highly correlated to annual precipitation of the concurrent year and the previous year. In another study with a statistical model of vegetation greenness [46], the results suggested a positive feedback between vegetation and Sahel rainfall at the monthly time scale and a vegetation memory operating at the annual time scale. This study suggests the previous year's vegetation growth affects greenness in the following year.

The statistical model also found that these vegetation-rainfall interactions increased the interannual variation in Sahelian precipitation, accounting for as much as 30% of the variability in annual precipitation in some sub-Sahel regions between 15° and 20°N. Furthermore, the seasonal differences had also been noted in some studies. For example, Liu et al. [45] found that, in the Sahel region the FPAR-lead correlations with precipitation are positive in spring and summer but become negative in fall and winter. Moreover, the trends and variability of these satellite-derived products and their major modes were also the subjects of some diagnostic studies [47–49]. Studies have revealed an increase in NDVI/LAI (leaf area index) and a decrease in surface albedo over the Sahel during the period 1982–1999 [50–52] that are interpreted as a vegetation recovery from the very dry periods of the 1980s.

Despite limitations in these types of studies due to data quality and data processing errors, short in duration, robustness in statistical approaches used, and uncertainty in using satellite products to represent real vegetation conditions, these results, nevertheless, shed light on the real vegetation/atmosphere interaction and provide a new and valuable reference for evaluating model-simulated vegetation/atmosphere interactions.

Meanwhile, modeling studies have shown more evidence that land/atmosphere interactions have profound effects on the WAM system at different scales. In a Global Land-Atmosphere Coupling Experiment (GLACE) study [53] investigating soil moisture/atmosphere coupling strength with multiple general circulation models (GCMs), each model conducted several ensembles of boreal summer (June through August) simulations: in one ensemble of 16 simulations, soil moisture varied with the simulations, while in another ensemble, the specified geographically varying time series of subsurface soil moisture was forced to be the same across the 16 simulations. If the model-simulated precipitation does not respond to soil moisture, the precipitation difference between these two ensembles should be very little. Soil moisture/atmosphere coupling strength of each model then was estimated based on each model's two ensembles with a statistical method. Figure 3 shows the global map of the difference averaged across twelve of the participating models, with higher values implying a higher impact of the soil moisture conditions on precipitation. This multiple model estimation of land atmosphere coupling strength reveals that the Sahel, along with a few other regions,

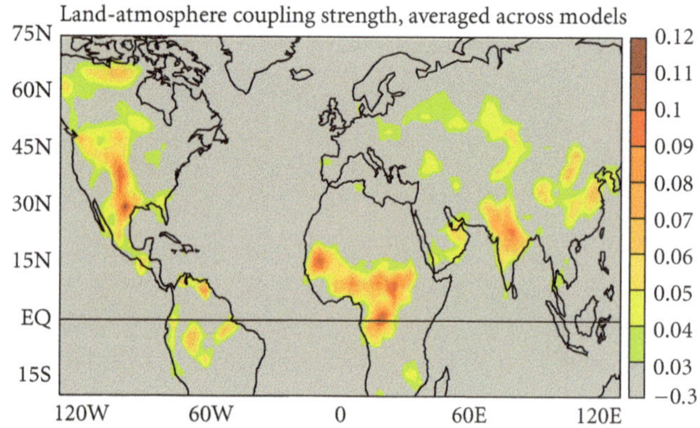

FIGURE 3: The soil moisture-atmosphere coupling strength diagnostic for boreal summer (the difference, dimensionless, describing the impact of soil moisture on precipitation), averaged across the 12 participating models. Courtesy by Dr. Randy Koster of NASA based on Koster et al. [53].

most of which lie along ecotones associated with semiarid areas, has the greatest soil moisture/climate coupling strength in the world.

The GLACE results are consistent with other soil moisture feedback studies. For instance, Douville et al. [54, 55] found that the proper simulation of the African monsoon required adequate soil moisture information, such as the soil moisture data from the Global Soil Wetness Project [56]. Using the Intergovernmental Panel on Climate Change IPCC models' simulation results, Notaro [57] also identifies North Africa as a region with strong soil moisture feedback. Furthermore, Philippon and Fontaine [58] proposed a soil moisture/monsoon regulation mechanism in which September to November Sahelian rainfall of the previous year could influence July to September Sahelian rainfall. However, other modeling studies [59, 60] showed that the root-zone soil moisture did not act as a memory of rainfall anomaly that would produce an effect into the following rainy season, and, therefore, it is not found to be related to the long-term persistence of the drought. A study using a regional climate model [61] found that the role of surface conditions in the precipitation dynamics is generally not overwhelming, even under conditions where soil moisture exerts a significant influence on surface evaporation. During the heart of the wet seasons in the Sahel, soil moisture is abundant and hardly constraining evaporation, precluding a strong land-atmosphere coupling. In addition, Koster et al. [53] also show that, although the average over 12 participating models' simulations indicate the Sahel area had higher coupling strength, substantial discrepancies among models existed. Further investigation on the soil moisture issues is warranted.

The impact of the vegetation biophysical process (VBP) on the WAM is another important subject under investigation. VBP refers to those land surface processes relevant to climate interactions associated with vegetation [62]. Xue et al. [62] analyzed two GCMs coupled to three different land models with varying degrees of physically based complexity in the representation of VBP. The importance of VBP effects

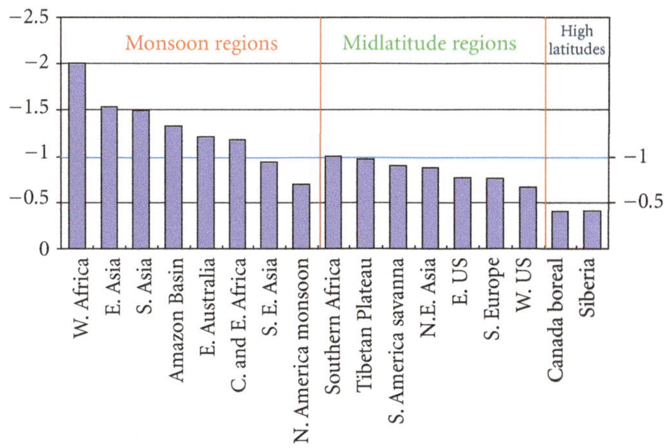

FIGURE 4: Reduction of absolute annual mean precipitation bias ($mm\,day^{-1}$) due to vegetation biophysical processes for different regions, indicating the impact of vegetation biological process on precipitation [62]. Differences of absolute bias of annual mean precipitation ($mm\,day^{-1}$) between LA-VBP and LA-VBP.

on climate were assessed based on the skill of simulations of observed global precipitation by two GCMs with different land models. Since a more realistic representation of VBPs in a GCM should improve precipitation simulations if VBP has effects on the real climate system, the statistically significant reduction of absolute bias and root-mean-square errors (RMSE) between simulated precipitation and the observation was adopted as criteria to identify VBP effects. Figure 4 shows the reduced absolute annual mean bias of 5-year precipitation simulations (or improved prediction skill) due to VBP process in the GCM. West Africa again shows the largest impact in the world. Further analyses revealed that, in West Africa, land surface processes have the most impact during the summer and the fall, with some impact during spring. The VBP's importance for the WAM has also been demonstrated by the application of satellite-derived vegetation products, such as LAI and vegetation coverage. By using remotely sensed LAI datasets, compared with using LAI based on a few ground surveys, the GCM produced substantial improvements in the near-surface climate in a few summer monsoon areas, including West Africa [63].

In addition to the seasonal mean, the land effect on the intraseasonal climate variability at large-scales was also a subject for some modeling studies [39, 64–66]. The studies showed that vegetation properties played a significant role in the intraseasonal variations of precipitation and circulation over West Africa [64], and intraseasonal soil moisture fluctuations were intense enough to feed back on the low-level vorticity structure [65]. In another study with a regional model, it has been shown that with a more realistic VBP representation, the seasonal timing and magnitude of mean monsoon precipitation more closely match observations, especially the timing of the monsoon advance and retreat across the Guinean Coast. In addition, the migration and position of the African Easterly Jet more closely match reanalysis winds [39].

The mechanisms for land surface/atmosphere interactions on the WAM, especially with respect to precipitation and the important WAM feature-African Easterly Jet (AEJ), have also been examined by a number of studies. When the atmospheric temperatures below the mid-troposphere are higher to the north (i.e., over the Sahara) and lower to the south, meridional temperature gradients are positive over tropical West Africa, and, according to basic atmospheric circulation thermodynamics principles, the thermal wind (and hence the jet) is easterly. The maximum latitudinal temperature gradient, which is related to the hot, dry surface conditions and a deep, well-mixed boundary layer in the Sahara heat low and cool, moist surface conditions associated with deep moist convection in the intertropical convergence zone, is associated with the strongest easterly thermal wind and therefore the AEJ in the mid-troposphere [67–69]. Wu et al. [70] suggest that it is not only the evaporation gradient that highly correlated with soil moisture, but a combination of vegetation properties and orography is also essential for the maintenance of the mean climatological AEJ. Therefore, it is the very particular combination of moisture gradients, vegetation distribution, and orography that produces the AEJ, not the soil moisture gradients alone.

Modeling studies further demonstrate the close relationship between the meridional thermal gradient, AEJ, and WAM precipitation and its onset. In the WAMME first experiment, 11 GCMs have been integrated from April through October with multiple members [42]. Common empirical orthogonal functions (CEOF) analysis based on observational data and multimodel simulation results was applied to characterize the major WAM features and has identified that the PC1 of precipitation and PC1 (Sahara mode) and PC2 (Sahel mode) of surface temperature characterize the WAM precipitation evolution and the northward movement of temperature gradient, respectively, and the WAM precipitation northward movement/retreat is closely associated with the weakening/enhancing of the Sahel mode (temperature PC2) and the enhancing/weakening of the Sahara mode (temperature PC1), which in turn enhances/weakening the meridional temperature gradient. The speed of the WAM evolution and the position of the WAM precipitation band are closely related to the development of this thermal gradient. The study identifies a close relationship between WAM onset and thermal low development in the Sahara: the timing of monsoon onset is about 10–15 days after the peak of the summer temperature anomaly in the Sahara.

Analyses further show that spatial distributions of surface sensible heat flux, surface temperature, and moisture convergence were closely associated with the simulated spatial distribution of precipitation; while surface latent heat flux is closely associated with the AEJ [42]. All these indicate the important role of the land surface energy partitioning in the WAM system, especially the thermal effect and surface induced large-scale circulation change as indicated by the moisture flux convergence change.

4. Mesoscale Circulation Variability and Land Surface Processes

Observations indicate high spatial heterogeneity of surface vegetation and soil conditions in the Sahel area [71]. Results from AMMA have provided new understanding of how this heterogeneity influences the development of convective storms over the course of the diurnal cycle. Because evaporative demand is high and the vegetation is sparse across the Sahel, particularly early in the wet season, the surface energy balance is strongly influenced by the availability of soil moisture close to the surface [72]. Rain events create strong mesoscale variability in soil moisture, and the resulting patterns of sensible and latent heating have been found to play an important role for the development of new storms during the afternoon in a number of studies.

Using land surface temperature data from satellite as a proxy for soil moisture (wet surfaces are up to 10 K cooler than dry surfaces), aircraft flights during AMMA were targeted at areas containing wet and dry soils. Taylor et al. [73] found that the temperature within the convective boundary layer during the afternoon was significantly correlated with the land surface temperature (i.e., anticorrelated with soil moisture) for surface features as small as 2.5 km and, furthermore, found a clear correlation between surface-induced temperature gradients and the low-level winds. Further evidence for mesoscale circulations induced by the land was presented over forest-crop transitions [74].

Numerical model experiments have shown how surface-driven convergence provides a favored location for the initiation of deep convection [75, 76]. The importance of this feedback mechanism has been highlighted in a recent study by Taylor et al. [77] analyzing nearly 4000 storms from satellite data of clouds and antecedent land surface temperature. Through detailed tracking of Mesoscale Convective Systems (MCS) with cloud-top temperature back to the location where the first deep cloud appeared, they were able to examine the relationship between the spatial structure of soil moisture and the probability of convective storm initiation. They found that, compared to a homogenous surface, storm initiation was twice as likely to occur 10 km upwind of a transition from dry to wet soils. This result is consistent with numerical modeling studies [75, 76], where contrasts in sensible heat flux drive a circulation which creates maximum low-level convergence where the surface-driven component opposes the large-scale flow (Figure 5). The study estimated that this effect was important for 1 in 8 MCS initiations in the region and implies a nontrivial impact of spatial scale on the sign of the soil moisture-precipitation feedback. Storms tend to initiate over dry soil (a locally negative feedback), but the occurrence of previous rain to create the heterogeneity is a prerequisite for this feedback. The MCS then travels, often many hundreds of kilometers, in turn creating new heterogeneity and favoring more initiations in subsequent days. This implies a positive feedback on the scale of the MCS, ~100 s km. Similar mesoscale processes are likely to be responsible for favored initiation of MCS around extensive wetland areas such as the Niger Inland Delta [78].

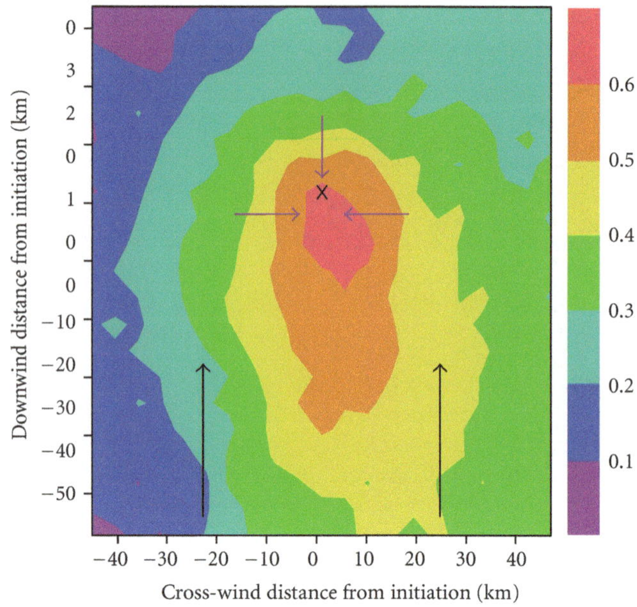

Figure 5: Favoured location (X) for initiation of convective storms [77]. Storms tend to develop over locally warm surfaces (shading denotes land surface temperature anomalies (K) based on a composite of nearly 4000 cases), but close to wetter (cooler) soils downwind. Arrows in mauve and black denote, respectively, the surface-induced, and large-scale components of the wind.

Once MCS has developed, various processes may govern the strength, and even the sign, of feedbacks between soil moisture and individual storms [22]. Gantner and Kalthoff [76] show that mature MCS with well-developed gust fronts can intensify over wet soils due to increased convective available potential energy, consistent with an observational case study from AMMA [72]. However, this positive feedback may reverse depending on the phase of the diurnal cycle, the convective inhibition and synoptic state, and the scale and orientation of the wet patch relative to the large-scale flow [75].

5. Summary and Future Research Prospective

Based on numerical sensitivity studies using land models with different complexity, conducted from the late 1970s to the 1990s, it has been recognized that, in addition to SST, land surface processes are an important component affecting the West African regional climate, thereby stimulating more extensive research in land/atmosphere interactions for the region. During the past decades, AMMA has conducted extensive large-scale field measurements over the West Africa. Taking the advantage of this unprecedented effort, the ALMIP has produced a valuable analysis of the land surface state and fluxes which has been applied in a number of large-scale African regional studies.

Meanwhile, the latest evidence from satellite data analyses and modeling studies confirm that the West African region has a climate which is particularly sensitive to land surface processes and there is a strong coupling between land surface processes and regional climate. It has been found that, in addition to the sea/land thermal contrast, the development of the West African monsoon is also closely associated with the evolution of a thermal gradient between the Sahara and southern Sahel. The variations of the West African climate system are closely associated with the surface energy partitioning. Mesoscale studies have revealed new understanding of how soil moisture heterogeneity influences the development of convective storms over the course of the diurnal cycle. The strong coupling between convective storms and soil moisture on time scales associated with soil drying (a few days) suggest that considerable progress could be made in short-term prediction of rainfall. All these studies indicate proper land surface process representations and land status initialization would substantially improve predictions and enhance the predictability of West African climate.

Despite this progress, some key scientific issues still remain. Recent observational evidence has supported the notion that there are strong decadal climate variabilities in the Sahel and surrounding areas from the 1950s to the 2000s, not only in precipitation, but also in SST, vegetation cover, land use and land cover (LULC) change, and aerosol types and spatial distributions. Understanding of the complex interactions of the WAM/Sahel precipitation variability associated with SST and land surface and aerosol forcings is still lacking. Further multimodel intercomparison experiments have to be carried out to improve understanding of the possible feedback to SST, land use change, vegetation, and aerosol forcings at seasonal to decadal scales. Given the complexity of the WAM/Sahel precipitation variability associated with these forcings and their interactions, it is imperative to gain basic understanding of the contribution/roles of land surface process relative to other forcings on the regional water cycle of the Sahel, to assess the relative contributions of land and its interaction with other forcings in producing/amplifying the Sahelian seasonal and decadal climate variability, and to understand the mechanisms involved. For instance, Koster et al.'s study [53] and Xue et al.'s [62] study show consistence and discrepancies in soil moisture and VBP effects on the West African monsoon. It is important to further explore the mechanisms behind. Furthermore, there are a few issues that have to be carefully and adequately addressed.

The influence of LULC change on West African drought was the focus of early land/atmosphere interaction studies (see references listed in Section 1). However, due to the controversy in terms of the degree and extent of land degradation there, the assessment of its realistic impact on the West African climate variability is still an outstanding issue. An early study [79] defined the desertification in the Sahel as the expansion of the Sahara Desert. Given the lack of evidence from satellite observations for the expansion of the boundary of the Sahara desert [80], the notion of Sahel land degradation has been discredited even in a very recent study [46]. A LULC change map even shows there were no noticeable LULC change in West Africa since 1870, which may lead to underestimation of its effect in the region [81]. In fact, studies have indicated that land degradation in West Africa is like "patches," which tends to be localized around settlements [82–84]. Based on a ground surveys,

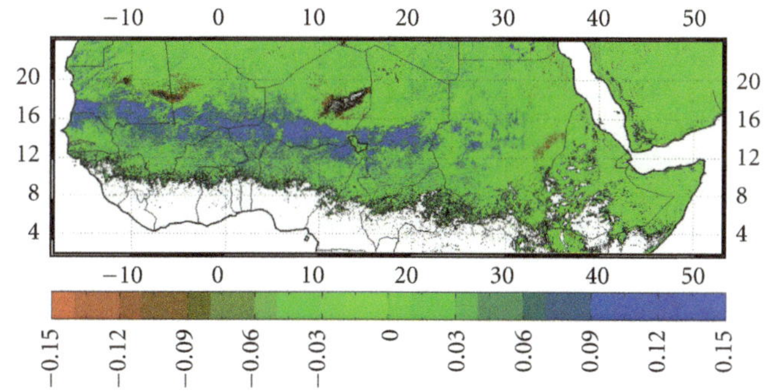

Figure 6: Mean August-September-October broadband surface albedo absolute difference between year 1984 and 2003 over the Sahel region. Missing data are shown in white. Courtesy by Dr. Yves Govaerts of EUMETSAT Based on Govaerts and Lattanzio [51].

Hiernaux et al. [85] found that the overall trend in land use in southwest Niger confirmed the historical increase of the cropped areas since mid-20th century, at an annual rate of 2% from 1994 to 2006. A recently compiled historical land-use data also indicate substantial land-use conversion over West Africa [86], consistent with dramatic population gains across West Africa. Furthermore, more accurate and high-resolution assessments of land condition data start to provide quantitative measurement of land condition change in this region.

For example, Govaerts and lattanzio [51] used the data from the Meteosat spacecraft to compare the derived surface albedo over Sahel from a dry year (1984) and a wet year (2003). They found that, between 12°N and 18°N, the mean August-September-October surface albedo decreased by about 0.06 with a mean zonal difference as large as 0.08 at 16°N between 1984 and 2003, consistent with the partial rainfall recovery in the region. The spatial distribution of surface albedo changes also showed an overall decrease in surface albedo in this period over the Sahel with the largest differences locally exceeding 0.15 (Figure 6). The authors indicate that, although precipitation is identified as a decisive controlling factor in this greening trend [87], human factors, including among other improved land management, might also explain some local trends [88]. According to the United Nation's definition [83], desertification results from various factors, including climate variations and human activities. It could be expected that the great Sahel drought during the 1970s and the1980s should have substantial impact on the land degradation since even a partial rainfall recovery from 1984 to 2003 could cause a change in albedo for about 0.06. The specification of albedo change of 0.1 in a desertification impact study [15] showed a significant impact on the regional climate. However, there is an imperative need to incorporate/identify/distinguish the impact of land degradation caused by anthropogenic land use change and regional climate change. A few studies have tackled this issue. For instance, in an off-line modeling study forced by the NCEP reanalysis, d'Orgeval and Polcher [89] suggested that the impact of land-use changes was much less important than the impact of precipitation changes over the years 1951–2000 in large West African catchments' runoff. More studies with coupled models are necessary to further evaluate this issue, especially since this issue is very relevant to the future climate projection.

A number of studies have explored the climate/vegetation relationship in West Africa based on ground survey/measurements [52, 90]. Rian et al. [90] quantitatively identify strong relationships between vegetation types, LAI, and climate variables (precipitation and surface temperature) and distinguish features in the vegetation/climate transition using the satellite data and field measurements. The study indicates that precipitation differences and the resulting impact on available soil moisture are the main drivers of the land cover transition in this region. The analysis shows the dominant control of summer precipitation on vegetation zonation and structure, with woody savanna being restricted to areas receiving >1000 mm mean annual precipitation, the lightly wooded savanna to areas of >700 mm precipitation, the dense shrub savanna to areas of >500 mm precipitation, and the open and sparse shrub savanna to areas of >200 mm.

The importance of incorporating two-way vegetation/climate interaction has also been demonstrated in modeling studies, some of which focused on the African paleoclimate [91, 92]. Wang et al. [93] investigated the impact of large-scale oceanic forcing and local vegetation feedback on the variability of the Sahel rainfall using a global biosphere-atmosphere model. The study examined the effect of treating vegetation as a dynamic component of the earth climate system rather than a static boundary condition. When vegetation is dynamic, the model realistically reproduces the multidecadal scale fluctuation of rainfall in the Sahel region; keeping vegetation static in the same model results in a rainfall regime characterized by fluctuations at much shorter time scales, indicating that vegetation dynamics acts as a mechanism for persistence of the regional climate. The above-mentioned DVM studies mainly focus on a long-term equilibrium conditions. To study the future climate change and societal planning with enhanced atmospheric CO_2 concentration [94–96], application of DVM for short time scales, for instance interannual to decadal time scale, remains a great challenge. Examining the DVMs with local scale/high resolution satellite data and field data is necessary.

Furthermore, to achieve substantial progress in fully coupled atmospheric modeling at meso to regional scales and to improve seasonal prediction and projection of climate change and its impact, as well as understanding the mechanisms involved, it is evident that effort must be made to improve the representation of land surface and hydrological processes which are specific to this region (such as transpiration from very deep layers in the soil, hydrophobic infiltration effects, significant subgrid lateral water transfer) and to provide numerical weather prediction models with more accurate spatial structure in the initialization of soil moisture. This requires the development of techniques to assimilate satellite data pertaining to the land surface, most notably soil moisture, land surface temperature, and vegetation LAI. ALMIP phase 2 is addressing this issue by evaluating an ensemble of LSM and hydrological models at the local to basin (meso) scale using observational data from the AMMA-CATCH transect [11]. However, the example of the study by Agustí-Panareda et al. [97] provides a cautionary note here. In that study, ALMIP-derived soil moisture was used to initialize a series of forecasts and the output compared with the standard soil moisture initialization based on assimilation of screen-level observations. The authors found that the substantial dry bias in Sahel precipitation in their model was rather insensitive to the soil moisture, suggesting that other processes (notably radiation and convection) were dominating. Only when the description of the basic processes which shape the monsoon system are adequately captured by the models may we expect the full value of additional land surface information to improve forecasts.

Acknowledgments

This research was supported by the US National Science Foundation Grant ATM-0751030. The authors also appreciate Dr. Randy Koster of NASA and Dr. Yves Govaerts of EUMETSAT, Germany, for providing their figures for this paper. The anonymous reviewer provides very constructive and detailed comments/suggestions, which greatly help improving this paper. They really appreciate the reviewer's efforts and contribution.

References

[1] United Nations Office for West Africa, West Africa, http://unowa.unmissions.org/Default.aspx?tabid=793.

[2] M. Hulme, "Validation of large-scale precipitation fields in General Circulation Models," in *Global Precipitations and Climate Change*, M. Desbois and F. Desalmand, Eds., NATO ASI Series, p. 466, Springer, Berlin, Germany, 1994.

[3] S. E. Nicholson, B. Some, and B. Kone, "An analysis of recent rainfall conditions in West Africa, including the rainy seasons of the 1997 El Nino and the 1998 La Nina years," *Journal of Climate*, vol. 13, no. 14, pp. 2628–2640, 2000.

[4] L. Le Barbé, T. Lebel, and D. Tapsoba, "Rainfall variability in West Africa during the years 1950–1990," *Journal of Climate*, vol. 15, no. 2, pp. 187–202, 2002.

[5] A. Dai, P. J. Lamb, K. E. Trenberth, M. Hulme, P. D. Jones, and P. Xie, "The recent Sahel drought is real," *International Journal of Climatology*, vol. 24, no. 11, pp. 1323–1331, 2004.

[6] W. K. M. Lau, K. M. Kim, and M. I. Lee, "Characteristics of diurnal and seasonal cycles in global monsoon systems," *Journal of the Meteorological Society of Japan*, vol. 85, pp. 403–416, 2007.

[7] J. L. Redelsperger, C. D. Thorncroft, A. Diedhiou, T. Lebel, D. J. Parker, and J. Polcher, "African Monsoon Multidisciplinary Analysis: an international research project and field campaign," *Bulletin of the American Meteorological Society*, vol. 88, no. 12, pp. 1739–1746, 2006.

[8] United Nations, "World Population Prospects," United Nations Publication, pp. 801, 2009.

[9] P. J. Lamb, "Large-scale tropical Atlantic surface circulation patterns associated with sub-Saharan weather anomalies," *Tellus*, vol. 30, pp. 240–251, 1978.

[10] C. J. Folland, M. Owen, N. Ward, and A. Colman, "Prediction of seasonal rainfall in the Sahel region using empirical and

dynamical methods," *Journal of Forecasting*, vol. 1, pp. 21–56, 1991.

[11] T. Lebel, B. Cappelaere, S. Galle et al., "AMMA-CATCH studies in the Sahelian region of West-Africa: an overview," *Journal of Hydrology*, vol. 375, no. 1-2, pp. 3–13, 2009.

[12] J. Charney, W. J. Quirk, S. H. Chow, and J. Kornfield, "A comparative study of the effects of albedo change on drought in semi-arid regions," *Journal of the Atmospheric Sciences*, vol. 34, no. 9, pp. 1366–1385, 1977.

[13] K. Laval and L. Picon, "Effect of a change of the surface albedo of the Sahel on climate," *Journal of the Atmospheric Sciences*, vol. 43, no. 21, pp. 2418–2429, 1986.

[14] Y. Xue, K.-N. Liou, and A. Kasahara, "Investigation of the biophysical feedback on the African climate using a two-dimensional model," *Journal of Climate*, vol. 3, pp. 337–352, 1990.

[15] Y. Xue and J. Shukla, "The influence of land surface properties on Sahel climate. Part I: desertification," *Journal of Climate*, vol. 6, no. 12, pp. 2232–2245, 1993.

[16] E. A. B. Eltahir and C. Gong, "Dynamics of wet and dry years in West Africa," *Journal of Climate*, vol. 9, no. 5, pp. 1030–1042, 1996.

[17] C. M. Taylor, E. F. Lambin, N. Stephenne, R. J. Harding, and R. L. H. Essery, "The influence of land use change on climate in the Sahel," *Journal of Climate*, vol. 15, no. 24, pp. 3615–3629, 2002.

[18] N. Zeng, J. D. Neelin, K. M. Lau, and C. J. Tucker, "Enhancement of interdecadal climate variability in the Sahel by vegetation interaction," *Science*, vol. 286, no. 5444, pp. 1537–1540, 1999.

[19] G. Wang and E. A. B. Eltahir, "Role of vegetation dynamics in enhancing the low-frequency variability of the Sahel rainfall," *Water Resources Research*, vol. 36, no. 4, pp. 1013–1021, 2000.

[20] S. Nicholson, "Land surface processes and Sahel climate," *Reviews of Geophysics*, vol. 38, no. 1, pp. 117–139, 2000.

[21] Y. Xue, R. W. A. Hutjes, R. J. Harding et al., "The Sahelian Climate (Chapter A5)," in *Vegetation, Water, Humans and the Climate*, P. Kabat, M. Claussen, P. A. Dirmeyer et al., Eds., pp. 59–77, Springer, Berlin, Germany, 2004.

[22] C. M. Taylor, D. J. Parker, N. Kalthoff et al., "New perspectives on land–atmosphere feedbacks from the African Monsoon Multidisciplinary Analysis," *Atmospheric Science Letters*, vol. 12, pp. 38–44, 2011.

[23] P. Xie and P. A. Arkin, "Global Precipitation: a 17-year monthly analysis based on Gauge observations, satellite estimates, and numerical model outputs," *Bulletin of the American Meteorological Society*, vol. 78, no. 11, pp. 2539–2558, 1997.

[24] J. P. Goutorbe, A. J. Dolman, J. H. C. Gash et al., *HAPEX—Sahel*, Elsevier, Amsterdam, The Netherlands, 1997.

[25] T. Lebel, D. J. Parker, C. Flamant et al., "The AMMA field campaigns: accomplishments and lessons learned," *Atmospheric Science Letters*, vol. 12, no. 1, pp. 123–128, 2011.

[26] A. Boone, P. de Rosnay, G. Balsamo et al., "The AMMA land surface model intercomparison project (ALMIP)," *Bulletin of the American Meteorological Society*, vol. 90, no. 12, pp. 1865–1880, 2009.

[27] G. J. Huffman, R. F. Adler, D. T. Bolvin et al., "The TRMM Multisatellite Precipitation Analysis (TMPA): quasi-global, multiyear, combined-sensor precipitation estimates at fine scales," *Journal of Hydrometeorology*, vol. 8, no. 1, pp. 38–55, 2007.

[28] B. Geiger, C. Meurey, D. Lajas, L. Franchistéguy, D. Carrer, and J. L. Roujean, "Near real-time provision of downwelling shortwave radiation estimates derived from satellite observations," *Meteorological Applications*, vol. 15, no. 3, pp. 411–420, 2008.

[29] P. de Rosnay, M. Drusch, A. Boone et al., "AMMA land surface model intercomparison experiment coupled to the community microwave emission model: ALMIP-MEM," *Journal of Geophysical Research D*, vol. 114, no. 5, Article ID D05108, 2009.

[30] M. Grippa, L. Kergoat, F. Frappart et al., "Land water storage variability over West Africa estimated by Gravity Recovery and Climate Experiment (GRACE) and land surface models," *Water Resources Research*, vol. 47, no. 5, article W05549, 2011.

[31] F. Timouk, L. Kergoat, E. Mougin et al., "Response of surface energy balance to water regime and vegetation development in a Sahelian landscape," *Journal of Hydrology*, vol. 375, no. 1-2, pp. 178–189, 2009.

[32] X. Gao and P. A. Dirmeyer, "A multimodel analysis, validation, and transferability study of global soil wetness products," *Journal of Hydrometeorology*, vol. 7, no. 6, pp. 1218–1236, 2006.

[33] A. A. Boone, I. Poccard-Leclercq, Y. Xue, J. Feng, and P. de Rosnay, "Evaluation of the WAMME model surface fluxes using results from the AMMA land-surface model intercomparison project," *Climate Dynamics*, vol. 35, no. 1, pp. 127–142, 2010.

[34] P. Tulet, M. Mallet, V. Pont et al., "The 7–12 March dust storm over West Africa: mineral dust generation and vertical layering in the atmosphere," *Journal of Geophysical Research*, vol. 113, article D00C08, 13 pages, 2008.

[35] C. Kocha, J. P. Lafore, P. Tulet, and Y. Seity, "High-resolution simulation of a major West African dust-storm: comparison with observations and investigation of dust impact," *Quarterly Journal of the Royal Meteorological Society*, vol. 138, no. 663, pp. 455–470, 2012.

[36] C. Delon, C. Galy-Lacaux, A. Boone et al., "Atmospheric nitrogen budget in Sahelian dry savannas," *Atmospheric Chemistry and Physics*, vol. 10, no. 6, pp. 2691–2708, 2010.

[37] A. Agustí-Panareda, G. Balsamo, and A. Beljaars, "Impact of improved soil moisture on the ECMWF precipitation forecast in West Africa," *Geophysical Research Letters*, vol. 37, no. 20, Article ID L20808, 2010.

[38] F. Guichard, N. Asencio, C. Peugeot et al., "An intercomparison of simulated rainfall and evapotranspiration associated with a mesoscale convective system over West Africa," *Weather and Forecasting*, vol. 25, no. 1, pp. 37–60, 2010.

[39] A. L. Steiner, J. S. Pal, S. A. Rauscher et al., "Land surface coupling in regional climate simulations of the West African monsoon," *Climate Dynamics*, vol. 33, no. 6, pp. 869–892, 2009.

[40] F. Hourdin, F. Guichard, I. Musat et al., "AMMA-Model intercomparison project," *Bulletin of the American Meteorological Society*, vol. 91, no. 1, pp. 95–104, 2010.

[41] R. Meynadier, O. Bock, F. Guichard, A. Boone, P. Roucou, and J. L. Redelsperger, "West African Monsoon water cycle: 1. A hybrid water budget data set," *Journal of Geophysical Research D*, vol. 115, no. 19, Article ID D19106, 2010.

[42] Y. Xue, F. de Sales, W. K. M. Lau et al., "Intercomparison and analyses of the climatology of the West African Monsoon in the West African Monsoon Modeling and Evaluation project (WAMME) first model intercomparison experiment," *Climate Dynamics*, vol. 35, no. 1, pp. 3–27, 2010.

[43] N. Martiny, Y. Richard, and P. Camberlin, "Interannual persistence effects in vegetation dynamics of semi-arid Africa,"

Geophysical Research Letters, vol. 32, no. 24, Article ID L24403, 4 pages, 2005.

[44] X. Zhang, M. A. Friedl, C. B. Schaaf, A. H. Strahler, and Z. Liu, "Monitoring the response of vegetation phenology to precipitation in Africa by coupling MODIS and TRMM instruments," *Journal of Geophysical Research D*, vol. 110, no. 12, Article ID D12103, 14 pages, 2005.

[45] Z. Liu, M. Notaro, J. Kutzbach, and N. Liu, "Assessing global vegetation-climate feedbacks from observations," *Journal of Climate*, vol. 19, no. 5, pp. 787–814, 2006.

[46] S. O. Los, G. P. Weedon, P. R. J. North, J. D. Kaduk, C. M. Taylor, and P. M. Cox, "An observation-based estimate of the strength of rainfall-vegetation interactions in the Sahel," *Geophysical Research Letters*, vol. 33, no. 16, Article ID L16402, 2006.

[47] X. Zeng, P. Rao, R. S. DeFries, and M. C. Hansen, "Interannual variability and decadal trend of global fractional vegetation cover from 1982 to 2000," *Journal of Applied Meteorology*, vol. 42, no. 10, pp. 1525–1530, 2003.

[48] L. Jarlan, Y. M. Tourre, E. Mougin, N. Philippon, and P. Mazzega, "Dominant patterns of AVHRR NDVI interannual variability over the Sahel and linkages with key climate signals (1982–2003)," *Geophysical Research Letters*, vol. 32, no. 4, Article ID L04701, 4 pages, 2005.

[49] N. Philippon, L. Jarlan, N. Martiny, P. Camberlin, and E. Mougin, "Characterization of the interannual and intraseasonal variability of West African vegetation between 1982 and 2002 by means of NOAA AVHRR NDVI data," *Journal of Climate*, vol. 20, no. 7, pp. 1202–1218, 2007.

[50] L. Eklundh and L. Olsson, "Vegetation index trends for the African Sahel 1982–1999," *Geophysical Research Letters*, vol. 30, no. 8, Article ID 1430, 4 pages, 2003.

[51] Y. Govaerts and A. Lattanzio, "Estimation of surface albedo increase during the eighties Sahel drought from Meteosat observations," *Global and Planetary Change*, vol. 64, no. 3-4, pp. 139–145, 2008.

[52] P. Hiernaux, L. Diarra, V. Trichon, E. Mougin, N. Soumaguel, and F. Baup, "Woody plant population dynamics in response to climate changes from 1984 to 2006 in Sahel (Gourma, Mali)," *Journal of Hydrology*, vol. 375, no. 1-2, pp. 103–113, 2009.

[53] R. D. Koster, P. A. Dirmeyer, Z. Guo et al., "Regions of strong coupling between soil moisture and precipitation," *Science*, vol. 305, no. 5687, pp. 1138–1140, 2004.

[54] H. Douville, F. Chauvin, and H. Broqua, "Influence of soil moisture on the Asian and African monsoons. Part I: mean monsoon and daily precipitation," *Journal of Climate*, vol. 14, no. 11, pp. 2381–2403, 2001.

[55] H. Douville, "Influence of soil moisture on the Asian and African monsoons. Part II: interannual variability," *Journal of Climate*, vol. 15, no. 7, pp. 701–720, 2002.

[56] P. A. Dirmeyer, A. J. Dolman, and N. Sato, "The pilot phase of the global soil wetness project," *Bulletin of the American Meteorological Society*, vol. 80, no. 5, pp. 851–878, 1999.

[57] M. Notaro, "Statistical identification of global hot spots in soil moisture feedbacks among IPCC AR4 models," *Journal of Geophysical Research D*, vol. 113, no. 9, Article ID D09101, 2008.

[58] N. Philippon and B. Fontaine, "The relationship between the Sahelian and previous 2nd Guinean rainy seasons: a monsoon regulation by soil wetness?" *Annales Geophysicae*, vol. 20, no. 4, pp. 575–582, 2002.

[59] M. Shinoda and Y. Yamaguchi, "Influence of soil moisture anomaly on temperature in the Sahel: a comparison between wet and dry decades," *Journal of Hydrometeorology*, vol. 4, no. 2, pp. 437–447, 2003.

[60] H. Douville, S. Conil, S. Tyteca, and A. Voldoire, "Soil moisture memory and West African monsoon predictability: artefact or reality?" *Climate Dynamics*, vol. 28, no. 7-8, pp. 723–742, 2007.

[61] B. J. J. M. van den Hurk and E. van Meijgaard, "Diagnosing land-atmosphere interaction from a regional climate model simulation over West Africa," *Journal of Hydrometeorology*, vol. 11, no. 2, pp. 467–481, 2010.

[62] Y. Xue, F. De Sales, R. Vasic et al., "Global and temporal characteristics of seasonal climate/vegetation biophysical process (VBP) interactions," *Journal of Climate*, vol. 23, pp. 1411–1433, 2010.

[63] H. S. Kang, Y. Xue, and G. J. Collatz, "Impact assessment of satellite-derived leaf area index datasets using a general circulation model," *Journal of Climate*, vol. 20, no. 6, pp. 993–1015, 2007.

[64] Y. Xue, H. M. H. Juang, W. P. Li et al., "Role of land surface processes in monsoon development: East Asia and West Africa," *Journal of Geophysical Research D*, vol. 109, no. 3, Article ID D03105, 24 pages, 2004.

[65] C. M. Taylor, "Intraseasonal land-atmospheric coupling in the West African monsoon," *Journal of Climate*, vol. 21, no. 24, pp. 6636–6648, 2008.

[66] S. L. Lavender, C. M. Taylor, and A. J. Matthews, "Coupled land-atmosphere intraseasonal variability of the West African monsoon in a GCM," *Journal of Climate*, vol. 23, no. 21, pp. 5557–5571, 2010.

[67] K. H. Cook, "Generation of the African easterly jet and its role in determining West African precipitation," *Journal of Climate*, vol. 12, no. 5, pp. 1165–1184, 1999.

[68] C. D. Thorncroft and M. Blackburn, "Maintenance of the African easterly jet," *Quarterly Journal of the Royal Meteorological Society*, vol. 125, no. 555, pp. 763–786, 1999.

[69] W. Li, Y. Xue, and I. Poccard, "Numerical investigation of the impact of vegetation indices on the variability of West African summer monsoon," *Journal of the Meteorological Society of Japan*, vol. 85, pp. 363–383, 2007.

[70] M. L. C. Wu, O. Reale, S. D. Schubert, M. J. Suarez, R. D. Koster, and P. J. Pegion, "African easterly jet: structure and maintenance," *Journal of Climate*, vol. 22, no. 17, pp. 4459–4480, 2009.

[71] S. D. Prince, Y. H. Kerr, J. P. Goutorbe et al., "Geographical, biological and remote sensing aspects of the hydrologic atmospheric pilot experiment in the Sahel (HAPEX-Sahel)," *Remote Sensing of Environment*, vol. 51, no. 1, pp. 215–234, 1995.

[72] M. Kohler, N. Kalthoff, and C. Kottmeier, "The impact of soil moisture modifications on CBL characteristics in West Africa: a case-study from the AMMA campaign," *Quarterly Journal of the Royal Meteorological Society*, vol. 136, no. 1, pp. 442–455, 2010.

[73] C. M. Taylor, D. J. Parker, and P. P. Harris, "An observational case study of mesoscale atmospheric circulations induced by soil moisture," *Geophysical Research Letters*, vol. 34, no. 15, Article ID L15801, 2007.

[74] L. Garcia-Carreras, D. J. Parker, C. M. Taylor, C. E. Reeves, and J. G. Murphy, "Impact of mesoscale vegetation heterogeneities on the dynamical and thermodynamic properties of the planetary boundary layer," *Journal of Geophysical Research D*, vol. 115, no. 3, Article ID D03102, 2010.

[75] M. A. Gaertner, M. Domínguez, and M. Garvert, "A modelling case-study of soil moisture-atmosphere coupling," *Quarterly*

Journal of the Royal Meteorological Society, vol. 136, no. 1, pp. 483–495, 2010.

[76] L. Gantner and N. Kalthoff, "Sensitivity of a modelled life cycle of a mesoscale convective system to soil conditions over West Africa," *Quarterly Journal of the Royal Meteorological Society*, vol. 136, no. 1, pp. 471–482, 2010.

[77] C. M. Taylor, A. Gounou, F. Guichard et al., "Frequency of sahelian storm initiation enhanced over mesoscale soil-moisture patterns," *Nature Geoscience*, vol. 4, no. 7, pp. 430–433, 2011.

[78] C. M. Taylor, "Feedbacks on convection from an African wetland," *Geophysical Research Letters*, vol. 37, article L05406, 2010.

[79] H. F. Lamprey, "Report on the desert encroachment reconnaissance in northern Sudan," *Desertification Control Bulletin*, vol. 17, pp. 1–7, 1988.

[80] C. J. Tucker and S. E. Nicholson, "Variations in the size of the Sahara Desert from 1980 to 1997," *Ambio*, vol. 28, no. 7, pp. 587–591, 1999.

[81] A. J. Pitman, N. De Noblet-Ducoudré, F. T. Cruz et al., "Uncertainties in climate responses to past land cover change: first results from the LUCID intercomparison study," *Geophysical Research Letters*, vol. 36, no. 14, Article ID L14814, 2009.

[82] H. E. Dregne and N. T. Chou, "Global desertification dimensions and costs," in *Degradation and Restoration of Arid Lands*, pp. 249–282, Texas Tech University, 1992.

[83] N. Middleton, *The Global Casino. An Introduction to Environmental Issues*, Arnold, 4th edition, 2008.

[84] A. S. Goudie, *The Human Impact on the Natural Environment*, The MIT Press, 6th edition, 2006.

[85] P. Hiernaux, A. Ayantunde, A. Kalilou et al., "Trends in productivity of crops, fallow and rangelands in Southwest Niger: impact of land use, management and variable rainfall," *Journal of Hydrology*, vol. 375, no. 1-2, pp. 65–77, 2009.

[86] G. C. Hurtt, S. Frolking, M. G. Fearon et al., "The underpinnings of land-use history: three centuries of global gridded land-use transitions, wood-harvest activity, and resulting secondary lands," *Global Change Biology*, vol. 12, no. 7, pp. 1208–1229, 2006.

[87] T. Hickler, L. Eklundh, J. W. Seaquist et al., "Precipitation controls Sahel greening trend," *Geophysical Research Letters*, vol. 32, no. 21, Article ID L21415, 4 pages, 2005.

[88] C. Reij, G. Tappan, and A. Belemvire, "Changing land management practices and vegetation on the Central Plateau of Burkina Faso (1968–2002)," *Journal of Arid Environments*, vol. 63, no. 3, pp. 642–659, 2005.

[89] T. d'Orgeval and J. Polcher, "Impacts of precipitation events and land-use changes on West African river discharges during the years 1951–2000," *Climate Dynamics*, vol. 31, no. 2-3, pp. 249–262, 2008.

[90] S. Rian, Y. Xue, G. M. MacDonald et al., "Analysis of climate and vegetation characteristics along the savanna-desert ecotone in mali using MODIS data," *GIScience and Remote Sensing*, vol. 46, no. 4, pp. 424–450, 2009.

[91] S. Levis, G. B. Bonan, and C. Bonfils, "Soil feedback drives the mid-Holocene North African monsoon northward in fully coupled CCSM2 simulations with a dynamic vegetation model," *Climate Dynamics*, vol. 23, no. 7-8, pp. 791–802, 2004.

[92] M. Notaro, Y. Wang, Z. Liu, R. Gallimore, and S. Levis, "Combined statistical and dynamical assessment of simulated vegetation-rainfall interactions in North Africa during the mid-Holocene," *Global Change Biology*, vol. 14, no. 2, pp. 347–368, 2008.

[93] G. Wang, E. A. B. Eltahir, J. A. Foley, D. Pollard, and S. Levis, "Decadal variability of rainfall in the Sahel: results from the coupled GENESIS-IBIS atmosphere-biosphere model," *Climate Dynamics*, vol. 22, no. 6-7, pp. 625–637, 2004.

[94] G. Wang and E. A. B. Eltahir, "Impact of CO_2 concentration changes on the biosphere-atmosphere system of West Africa," *Global Change Biology*, vol. 8, no. 12, pp. 1169–1182, 2002.

[95] C. A. Alo and G. Wang, "Role of dynamic vegetation in regional climate predictions over Western Africa," *Climate Dynamics*, vol. 35, no. 5, pp. 907–922, 2010.

[96] H. Paeth, K. Born, R. Girmes, R. Podzun, and D. Jacob, "Regional climate change in tropical and Northern Africa due to greenhouse forcing and land use changes," *Journal of Climate*, vol. 22, no. 1, pp. 114–132, 2009.

[97] A. Agustí-Panareda, G. Balsamo, and A. Beljaars, "Impact of improved soil moisture on the ECMWF precipitation forecast in West Africa," *Geophysical Research Letters*, vol. 37, no. 20, Article ID L20808, 2010.

Identifying Buried Segments of Active Faults in the Northern Rio Grande Rift Using Aeromagnetic, LiDAR, and Gravity Data, South-Central Colorado, USA

V. J. S. Grauch[1] and C. A. Ruleman[2]

[1] *U. S. Geological Survey, MS 964, Federal Center, Denver, CO 80225, USA*
[2] *U. S. Geological Survey, MS 980, Federal Center, Denver, CO 80225, USA*

Correspondence should be addressed to V. J. S. Grauch; tien@usgs.gov

Academic Editor: Luciano Scarfi

Combined interpretation of aeromagnetic and LiDAR data builds on the strength of the aeromagnetic method to locate normal faults with significant offset under cover and the strength of LiDAR interpretation to identify the age and sense of motion of faults. Each data set helps resolve ambiguities in interpreting the other. In addition, gravity data can be used to infer the sense of motion for totally buried faults inferred solely from aeromagnetic data. Combined interpretation to identify active faults at the northern end of the San Luis Basin of the northern Rio Grande rift has confirmed general aspects of previous geologic mapping but has also provided significant improvements. The interpretation revises and extends mapped fault traces, confirms tectonic versus fluvial origins of steep stream banks, and gains additional information on the nature of active and potentially active partially and totally buried faults. Detailed morphology of surfaces mapped from the LiDAR data helps constrain ages of the faults that displace the deposits. The aeromagnetic data provide additional information about their extents in between discontinuous scarps and suggest that several totally buried, potentially active faults are present on both sides of the valley.

1. Introduction

The combination of aeromagnetic methods and LiDAR (Light Detection And Ranging) mapping has proved useful for identifying active faults in dynamically deforming, compressional terranes in the Pacific Northwest region of USA [1, 2]. Linear aeromagnetic anomalies and gradients corresponding to thrust and tear faults are associated with fault scarps on the ground, located by airborne LiDAR surveys. The scarps are otherwise difficult to identify in the highly populated areas and under widespread, dense vegetative cover.

Recent acquisition of aeromagnetic and airborne LiDAR data over the northern Rio Grande rift in the southwestern USA has afforded the opportunity to test these methods in a less active, extensional basin in a semiarid environment. Farther south, in the central Rio Grande rift, aeromagnetic methods have proved useful for mapping intrasedimentary faults under cover [3], but age of activity cannot be determined from aeromagnetic data alone. Although dense vegetation and population are not as widespread as in the Pacific Northwest, active fault scarps can still be difficult to identify because of widespread eolian cover and dynamic weathering. Thus, combined analysis of high-resolution aeromagnetic data and LiDAR images builds on the strength of the aeromagnetic method to locate normal faults with significant offset under cover and the strength of LiDAR interpretation to identify subtle topographic scarps. Moreover, combining these methods can resolve ambiguities in locating the surface projections of buried faults from aeromagnetic data and in distinguishing tectonic versus fluvial origins of topographic scarps recognized from LiDAR data. Once active faults are identified, information from geologic mapping, trenching, and gravity and seismic-reflection data can provide additional constraints on age, recurrence history, amount of offset, and sense of displacement.

Herein we present a case history of combined interpretation of the recently acquired high-resolution aeromagnetic and LiDAR data from the very northern part of the San Luis Basin, one of the largest of the series of alluvial basins that comprise the northern part of the Rio Grande rift (Figure 1 inset). Although only a few historic earthquakes have been recorded in the northern Rio Grande rift, the rift is still considered active. The potential for seismic hazard is based on evidence of large-magnitude earthquakes (M 6.8–7.1) with recurrence intervals of 10–140 ka [4–10]. Thus, active fault scarps may not be easily recognized because they can be quiescent for long periods of time, subject to erosion. We present several examples that elucidate the strengths and limitations of combining the LiDAR and aeromagnetic data to identify and characterize active faults at basin margins and within the interior of the basin.

2. Geologic and Tectonic Setting

Our study area is located in the northern part of the San Luis Valley, south-central Colorado, within the San Luis Basin of the northern Rio Grande rift (Figure 1). The northwest-elongated valley is bounded on the east by the steep slopes of the Sangre de Cristo Mountains and on the west by low hills at the northeastern edge of the San Juan volcanic field (Figure 1 inset).

The northern San Luis Basin, like other parts of the Rio Grande rift, has been the site of multiple tectonic events that each left an imprint on the geology and underlying structure [11]. Geophysical interpretation of rift structure commonly requires unraveling the similar-looking expressions of older tectonic features from those of the rift. The first of three major prerift events of importance to the region is the Mississippian-Pennsylvanian (Carboniferous) Ancestral Rocky Mountain orogeny. This tectonic event left a regional, north-south uplift that now underlies the central portion of the San Luis Basin. The uplift is evidenced from gravity data, seismic-reflection sections, and deep drilling to the south of the study area [12]. Eventual denudation of the uplift shed detritus into an adjacent basin to the east, accumulating as much as 4 km of clastic sedimentary rocks [13]. These rocks, along with underlying Early Paleozoic marine carbonates and sandstones, unconformably overlie Proterozoic crystalline basement and are exposed at the margins of the valley (Figure 1).

During Late Cretaceous to Eocene, thrust faulting and folding during the Laramide orogeny developed large reverse and low-angle faults, accompanied by strike-slip faulting. The thrusting transported Late Paleozoic rocks kilometers to the east and northeast into deformed and stacked thrust packages that lie mostly within the western front of the Sangre de Cristo Mountains today [14]. Toward the end of the Laramide orogeny, uplifts were again eroded, shedding detritus to the west into a depression now located in the subsurface on the western side of the San Luis Basin [15].

From Late Eocene to Late Oligocene, extensive lava flows and ash-flow eruptions originated from the San Juan volcanic field to the south and west of the study area (inset on Figure 1). The 33.2-Ma Bonanza caldera lies at the western side of the study area [16, 17]. Tertiary volcanic rocks exposed in the northern part of the study area mostly represent Bonanza tuff and precaldera andesite.

Basin subsidence related to Rio Grande rifting overlapped the waning of primarily intermediate to silicic volcanic activity of the San Juan and related volcanic fields in Late Oligocene and the onset of mafic volcanism in the southern San Luis Basin [18]. Subsidence was most active during Miocene time and has diminished since the beginning of the Pliocene [19]. From recent GPS studies, deformation appears to be evenly distributed across the rift instead of being focused at the rift margins [20].

The northern half of the San Luis Basin, in the northern Rio Grande rift, forms an asymmetric, east-tilted half graben, with 6-7 km of total displacement [12, 15]. The Sangre de Cristo fault zone (Figure 1) is commonly cited as the eastern margin of this half graben, extending in length from 79 to 104 km, depending on whether it is defined by orientations of the range front, slip rates, or geomorphic expression [5, 7, 10, 21]. Sediments that filled the Miocene-Pliocene basins consist of poorly consolidated sands, silts, and gravels of the Santa Fe Group. Deposits of this age are found at high elevations to the east and north of the study area, suggesting that early rifting had a different orientation and locus of deposition in this region before Pliocene time ([22] and references therein). Our study area is located where the basin narrows dramatically (Figure 1 inset) and marks the southern end of a poorly understood zone of transition from the strongly east-tilted half-graben of the central San Luis Basin on the south, to a strongly west-tilted half-graben in the next basin to the north [23].

Several periods of Pleistocene glaciation in the region are recorded by deposits of stream and fan alluvium, till, and outwash, which are mainly preserved near the mountain fronts. The relative ages of these deposits are determined from geomorphology and degree of weathering and soil development. From the youngest to the oldest, these deposits are Qfy, Qfi, and Qfo (Figure 1) and are associated with Pinedale, Bull Lake, and pre-Bull Lake glaciations, respectively. Based on regional correlations from previous and recent work, the glaciations in this area correspond to the following age ranges: Pinedale (12–30 ka), Bull Lake (about 120–170 ka), and pre-Bull Lake (170–640 ka) [21, 24–31]. The most recent (<10 ka) deposits in the area are Holocene stream and fan alluvium deposits, unit Qh (Figure 1).

Precipitous, faceted spurs aligned along the range front mark the steeply west-dipping northern Sangre de Cristo fault zone on the east side of the valley (Figure 1). The northwest-trending Villa Grove fault zone, composed of multiple, mainly southwest-facing scarps developed in fan deposits, diverges from the northern Sangre de Cristo fault zone near the southeastern corner of the study area. The zone extends northwestward across the valley for about 10 km, in apparent connection with easterly facing scarps on the west side of the valley near Villa Grove (Figure 1; [10]). Kellogg [32] suggests that much of Rio Grande rift structure is inherited from preexisting Laramide thrust fault geometry, but with reversed sense of motion.

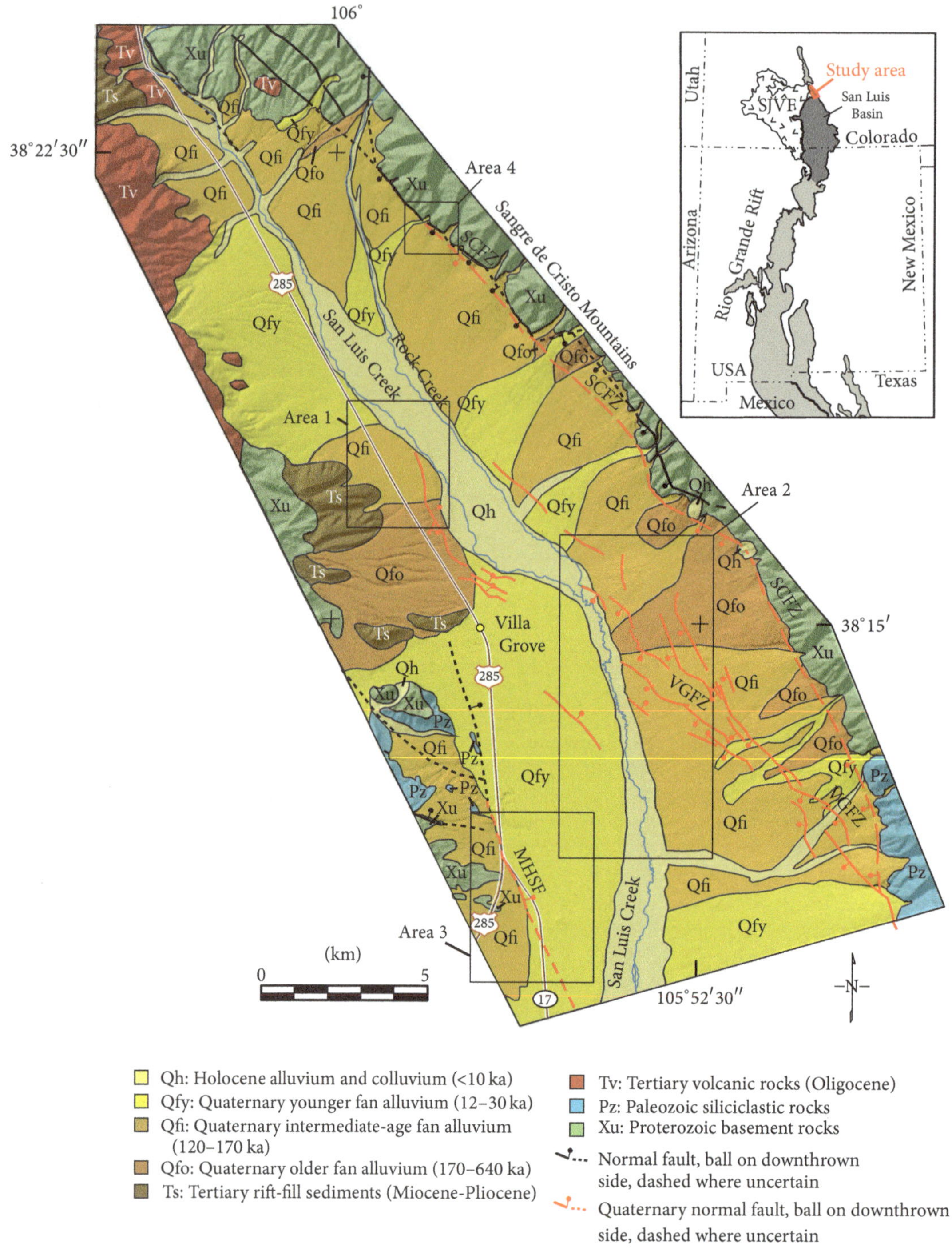

FIGURE 1: Geology of the northern San Luis Valley area from published sources (see text). Modifications and additions to these mapped Quaternary (active) faults and general surficial geology are presented in this study (Figure 15). Underlay of shaded-relief terrain is a graphically smoothed, NW-illuminated hillshade derived from the LiDAR data (Figure 3). MHSF: Mineral Hot Springs fault, SCFZ: Sangre de Cristo fault zone, and VGFZ: Villa Grove fault zone. Boxes locate example Areas 1–4. Inset shows the location of the study area within the San Luis Basin and Rio Grande rift (basins are shaded). SJVF: San Juan volcanic field.

All faults in the study area that displace Quaternary deposits (red lines on Figure 1) are considered active. The potential for seismic hazard is based on evidence of large-magnitude earthquakes (M 6.8–7.1) with recurrence intervals of 10–140 ka [4–10].

3. Data Sources

3.1. Previous Geologic Mapping. The geologic map shown in Figure 1 is generalized from several compilations by others. Geology shown east of 106° longitude and north of 38°22′30″

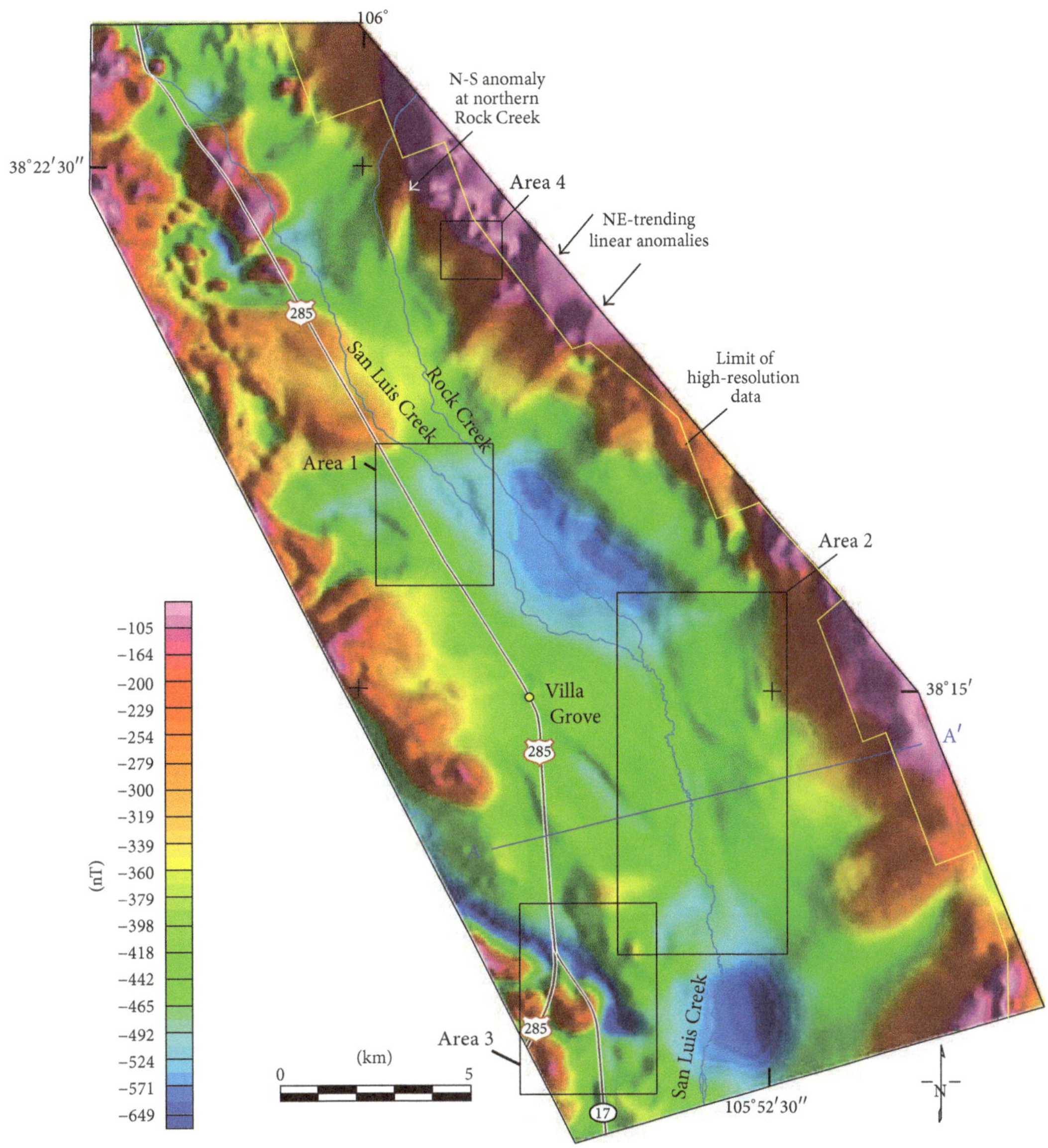

FIGURE 2: Reduced-to-pole (RTP) aeromagnetic data for the study area. Illuminated from the east. Data from the modern high-resolution aeromagnetic survey was merged with older, analog data west of the survey boundary (yellow line), described in text. Boxes locate example Areas 1–4. Model for A-A′ is shown in Figure 16.

latitude was generalized from a geologic compilation for central Colorado undergoing review within the U.S. Geological Survey (USGS) as of 2012. Terry Klein (USGS, retired) led the compilation for this area, which he assembled and edited from previous publications [24, 33–36]. The geology for the remaining part of our study area was generalized from compilations by Colman et al. [24] and Kappa and Wallace [37]. Documentation of ages listed for the surficial units is described in Section 2.

3.2. Aeromagnetic Data. High-resolution aeromagnetic data covering most of the study area (Figure 2) are from a survey flown on contract to the USGS by EON Geosciences, Inc., in October, 2011, over a much larger area of the northern San Luis Basin and regions to the north. The survey was flown by helicopter along lines oriented east northeast (70° east of north) and 150 m apart. Over most of the survey area, the pilot followed a preplanned, smooth flight surface that averaged 100 m above ground over the whole study area but reached as much as 300 m over the mountain front. Survey coverage in the mountainous areas was limited due to regulations restricting low flying over the Sangre de Cristo Wilderness Area. After standard data processing and removal of the 2010 reference field for the Earth for the date of the survey [38], flight-line data were gridded onto a 50 m grid and analytically continued to a more consistent surface 100 m above the ground.

Data for the missing areas over the mountains were obtained from project 4068 from a database developed for North America [39]. The survey for this project was originally flown in 1981 along northeast-southwest lines spaced about 400 m apart at 300 m above ground. Due to numerous problems discovered with the original location data for this survey, the line data were extensively decimated before gridding and no continuation was attempted. Data problems that remain commonly produce linear artifacts oriented along the direction of the flight lines. Nevertheless, the data are useful for providing a more comprehensive picture of the aeromagnetic data for our study area. The data were digitally merged with the high-resolution grid by smoothly blending the overlapping areas.

A standard reduction-to-pole operator was applied to the continued, merged magnetic grid, using a declination and inclination of the presentday Earth's field of 9.2° and 65°, respectively (Figure 2). The underlying assumption in using reduced-to-pole data for interpretation is that magnetizations in the study area are generally collinear with the Earth's field. This assumption is valid for the rift sediments and Paleozoic sedimentary units, which do not carry much remanence. It is likely valid for the Tertiary volcanic rocks and Proterozoic basement rocks that produce large-amplitude anomalies. The Tertiary volcanic rocks formed when Earth field directions were approximately parallel or opposite to those of today. The magnetizations of Proterozoic rocks are likely dominated by the induced component, based on the results of a study just north of our area [40].

3.3. LiDAR Data. LiDAR (Light Detection and Ranging) data were acquired in 2010, using an airborne-laser nominal pulse spacing of 1 meter. Supplemented with ground control positioning, the final dataset has a vertical and horizontal RMSE (root mean square error) of <12.5 cm and <10 cm, respectively. The final full-point cloud was provided as bare-earth coded points in order to generate bare-earth surface raster DEM (Digital Elevation Model) data with a cell size of 1 meter. The required vertical accuracy for the bare-earth DEM was 24.5 cm at a 95% confidence level, but the tested accuracy was 10 cm. The "bare-earth" terrain model was converted to shaded-relief images illuminated from NE, SE, NW, and SW for interpretations used herein (Ted Brandt, USGS, written communication, 2012). The hillshade with NW illumination, decimated for viewing at page size is shown in Figure 3.

3.4. Gravity Data. Gravity data were prepared by Benjamin Drenth (USGS) from principal facts extracted from a gravity-data web portal for the United States [41], supplemented by new data acquired during 2012. Figure 4 shows the irregular data coverage for the study area. The data were checked for consistency, then processed following standard techniques, using a reduction density of 2,670 kg/m^3 for Bouguer and terrain corrections. After gridding at a 1 km interval, a regional field was removed to focus on anomalies of the upper crust. The regional field was computed from an isostatic model using parameters developed by Heywood [42]. The resulting isostatic residual gravity grid is shown in Figure 4.

Measurements of bulk density in nearby areas of the Rio Grande rift indicate that contrasts on the order of 400 kg/m^3 arise from the juxtaposition of bedrock against poorly consolidated Tertiary and Quaternary deposits [40, 43]. Lesser density contrasts of 100–210 kg/m^3 arise from the juxtaposition of Tertiary volcanic rocks or Paleozoic sedimentary rocks against Proterozoic rocks. Significant density contrasts are not typically observed between different facies of the sediments themselves. Thus, gravity lows in the study area are expected to be primarily a function of variations in the overall thickness of rift sediments, implying that the gravity lows generally reflect the shapes of the rift basin.

3.5. Seismic-Reflection Data. Seismic-reflection data were collected along several lines across the southern part of the study area in the 1970s [44, 45] but have limited use for our study. The original data are no longer available; only poor-quality figures of the sections remain. Reflections within the first 0.2 seconds of two-way travel time are insufficiently displayed in the figures to examine shallow rift faults. Davis and Stoughton [45] suggest that ~1.3 km thick sections of complexly folded and faulted Paleozoic sedimentary rocks underlie a fairly shallow rift basin (<500 m thick) underneath the valley. However, geologic interpretations of the sections are ambiguous because no well data are available for comparison.

4. Methods

4.1. Identifying Faults from Aeromagnetic Data. Grauch and Hudson [3] have demonstrated the utility and challenges of using high-resolution aeromagnetic data to interpret shallow normal faults for the central Rio Grande rift. Widespread, partially concealed and totally buried faults that offset Tertiary and younger rift sediments within basins of the central Rio Grande rift are expressed as numerous subtle, northerly trending, linear anomalies. Using field observations, physical property measurements, and magnetic modeling, they found that the aeromagnetic anomalies are caused by the tectonic juxtaposition of magnetically differing strata rather than by chemical processes acting at the fault zone, as widely speculated in earlier work from other areas. Field evidence for chemical processes that produce or destroy magnetic minerals in sufficient volume to create aeromagnetic anomalies from rift sediments is lacking in our study area as well.

Using commonly observed magnetic properties for sediments in the central Rio Grande rift [46], Grauch and Hudson [3] estimate that faults near the surface require at least 30 m of throw to produce the observable aeromagnetic anomalies, indicating that faults with aeromagnetic expression likely have significant vertical displacement. However, not all faults with significant vertical displacement juxtapose layers of differing magnetic properties, even along strike, so that aeromagnetic data may not reveal all faults or the total lengths of individual faults. Moreover, age and sense of motion of aeromagnetically expressed faults cannot be determined without independent information.

A useful approach to mapping the locations of faults involves the computation of the horizontal gradient magnitude (HGM) of the reduced-to-pole aeromagnetic data [3]. The utility and difficulties with the approach are illustrated in four end-member models that represent where magnetic contrasts are developed at simple offsets of layered strata at normal faults (Figure 5). Although developed for faults that offset

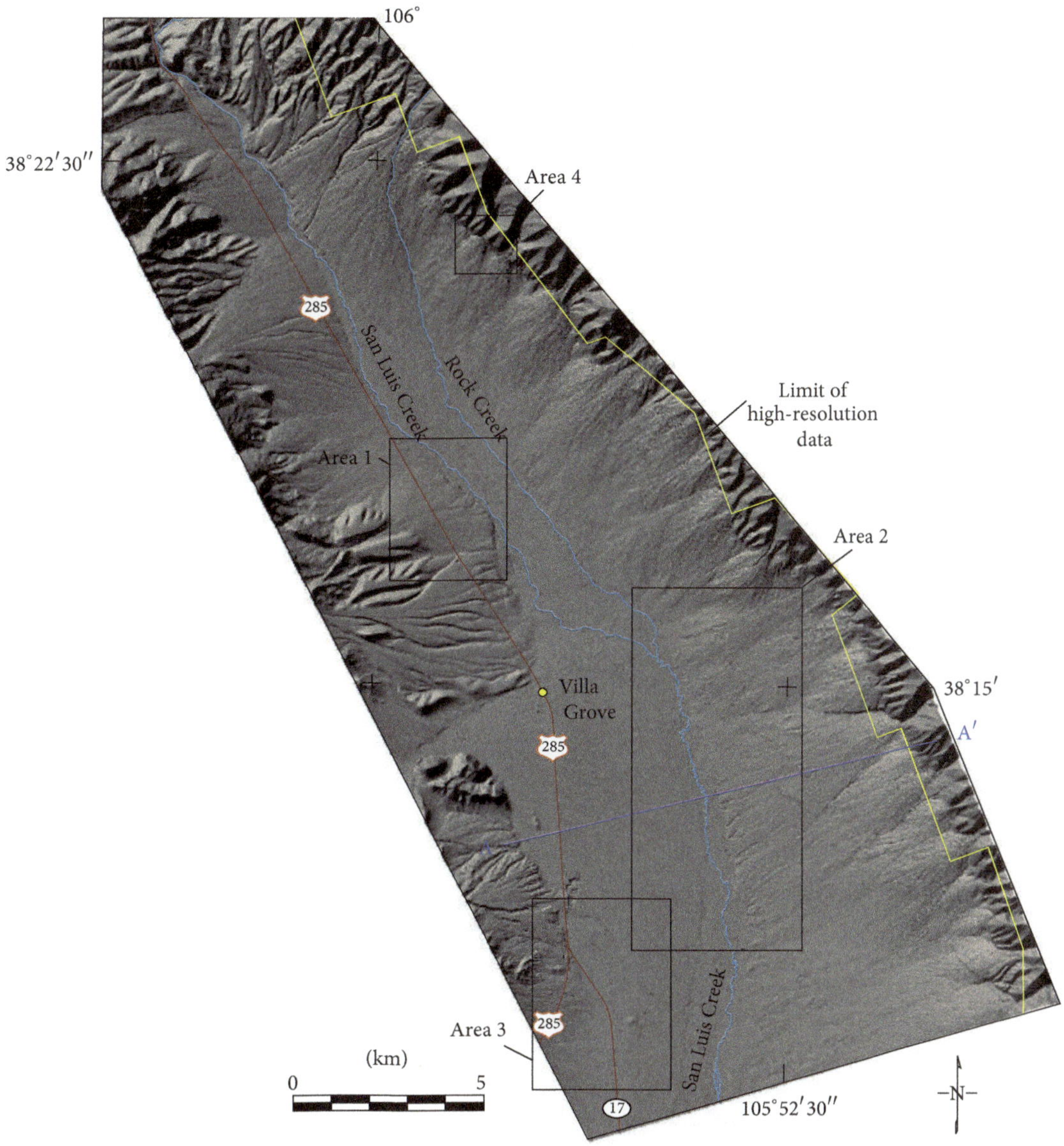

FIGURE 3: Northwest-illuminated hillshade of LiDAR data for the study area, graphically smoothed for viewing at page size. Yellow line indicates the limit of the high-resolution aeromagnetic data. Boxes locate example Areas 1–4. Model for A-A′ is shown in Figure 16.

sediments, the models can be applied to other layered geology as well, such as volcanic layers [47]. Moreover, as discussed by Grauch and Hudson [3, 47], the model layers represent magnetic contrast across the fault, rather than the geologic strata at the fault. The equivalent reduced-to-pole magnetic field and the magnitude of the horizontal gradient (the absolute value of the first derivative in the horizontal direction in the profile case) are computed for each model. The anomalies over two of the models (Figures 5(a) and 5(b)) are conventional fault-type anomalies and produce single HGM peaks for both the reduced-to-pole and magnetic potential computations. The anomaly shapes and multiple HGM peaks for the two other models (Figures 5(c) and 5(d)) are unconventional, but commonly observed. The models demonstrate that the narrowest HGM peak of the reduced-to-pole magnetic data most reliably locates the shallowest projection of the fault. Computing the HGM in two directions produces images where narrow ridges culminate over faults in map view. Although the horizontal gradient is commonly computed from the magnetic potential instead of reduced-to-pole data to avoid multiple HGM peaks over thin bodies [48, 49], this approach does not correctly locate the top projection of the fault for the two unconventional model cases (Figures 5(c) and 5(d)) [3].

Multiple parallel HGM ridges computed from reduced-to-pole aeromagnetic data are commonly observed on maps in the Rio Grande rift south of the study area [3]. Because parallel ridges can indicate multiple parallel normal faults

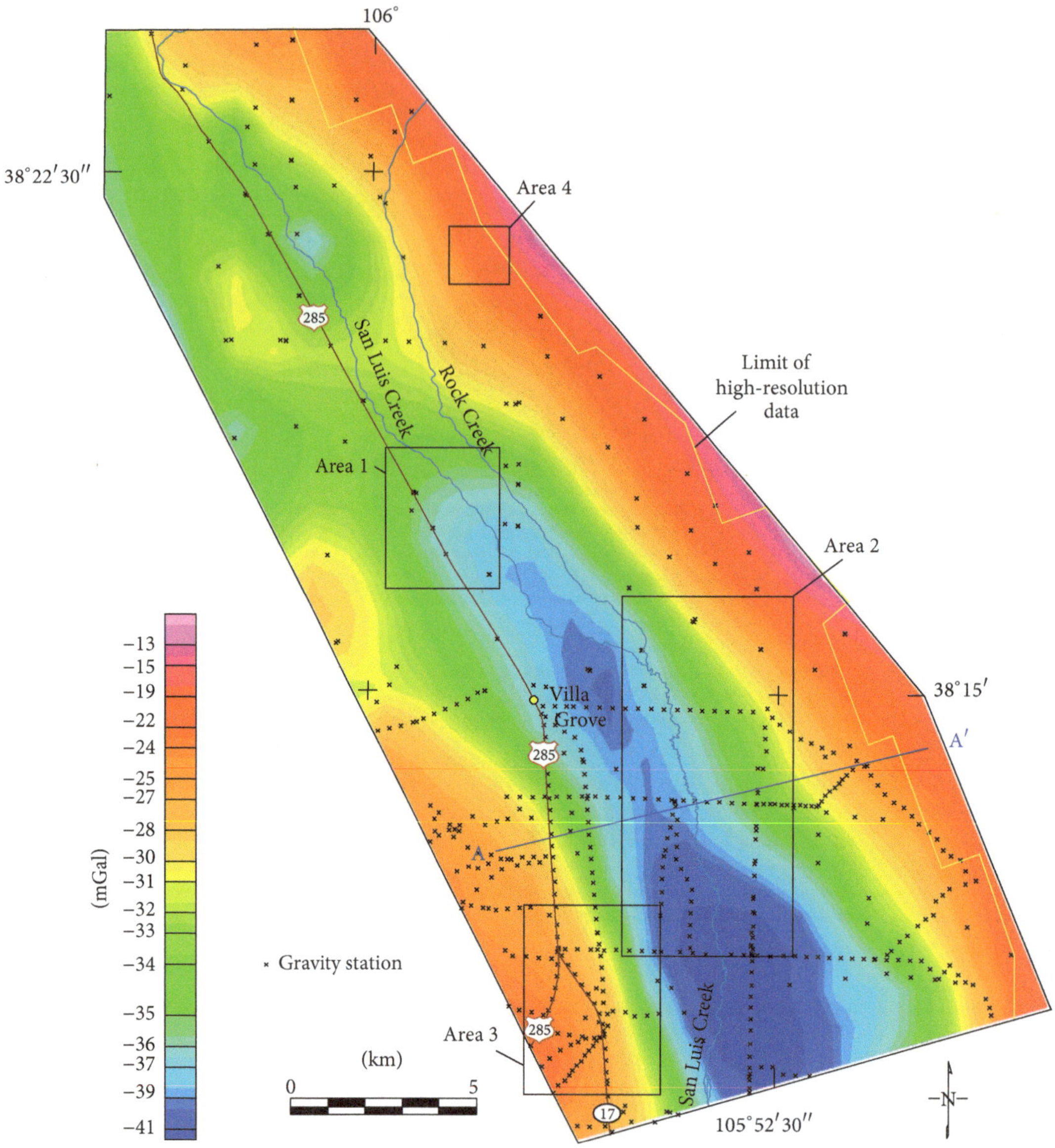

FIGURE 4: Isostatic residual gravity data showing station coverage. Boxes locate example Areas 1–4. Model for A-A′ is shown in Figure 16.

(e.g., Figures 5(a) and 5(b)) or a complicated geometry of juxtaposed layers that produce multiple ridges of variable amplitude (e.g., Figures 5(c) and 5(d)), Grauch and Hudson [3] developed some criteria to distinguish the cause. A complicated geometry is indicated by a pair of HGM ridges, where one ridge is narrower than the other. The narrower ridge (indicating a shallower source) best indicates the location of the fault where it reaches or nears the surface. Similar multiple parallel HGM ridges are observed in a computation of the HGM of the reduced-to-pole aeromagnetic data for our study area (Figure 6). Because of the presence of strong regional gradients that can swamp high-frequency signal from shallow faults, the HGM was computed using the gradient window method [50]. This method removes a best-fit regional trend within a window before computing the HGM across the grid. Afterwards, peaks in the HGM are located in the same positions as those that result from filtering the grid to focus on shallow sources before computing derivatives. However, it is much easier to test different window sizes to focus on removal of different regional gradients. We found a window size of 25×25 grid cells ($1,225\,\text{m} \times 1,225\,\text{m}$) generally shows the best results for the study area, although other window sizes were also inspected (not shown).

Although HGM ridges were important criteria for identifying faults in the aeromagnetic data, other criteria were also used. Abrupt terminations or changes in aeromagnetic anomaly character or in the termination of HGM ridges along a linear edge also indicated faults. Conversely, not all linear HGM ridges represent faults. They may correspond to topographic scarps or other linear geologic features, such as contacts at the eroded edges of tilted blocks, dikes, or paleochannels.

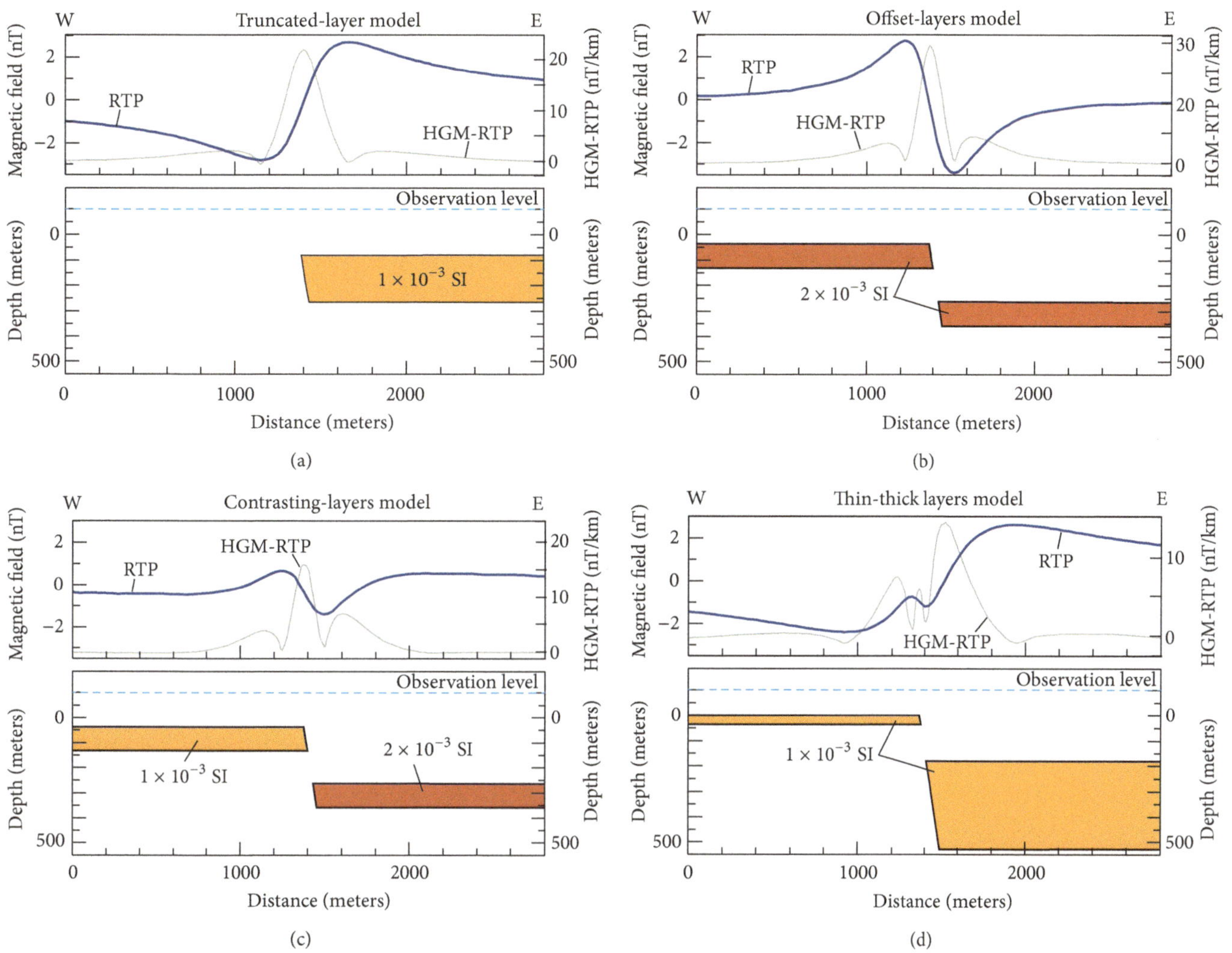

FIGURE 5: Simple 2D geophysical models illustrating the four main types of aeromagnetic signatures associated with intrasedimentary normal faults in the central Rio Grande rift. Model concepts can also be applied to other layered geology, such as volcanic flows. Profiles are computed for the reduced-to-pole magnetic anomaly (bold blue lines) and the horizontal gradient magnitude of the reduced-to-pole anomaly (HGM-RTP—solid gray lines). Magnetic susceptibilities are color coded and labeled.

4.2. Identifying Faults from LIDAR Data. Airborne LiDAR surveys provide extremely detailed digital maps of the elevation of the ground surface over large areas. The detail allows unprecedented views of the geomorphology, leading to the recognition of subtle, gently sloping fault scarps or other topographic expressions of faults that are time consuming to map using conventional methods. As demonstrated in the northern Rio Grande rift, mapping the geomorphology of surficial deposits also leads to identification of their relative ages, with younger deposits showing greater roughness than older ones [21, 31]. Thus, the general age of fault activity and sense of fault motion can be determined from the deposits that are displaced. In addition, digital slope profiles across fault scarps extracted from LiDAR data can be analyzed directly to estimate age of scarp formation [51]. On the other hand, mapping linear topographic scarps from surface evidence alone has limitations. First, fault scarps can be degraded or buried by surficial processes, hindering the mapping of their continuity along strike or their connection to other faults. Such discontinuity of fault scarps is common in the northern Rio Grande rift [24] because of the long intervals between earthquakes (10–140 ka) since early Pleistocene [4, 5, 52]. Second, some fault scarps may represent single-event splays of larger, more active faults that pose greater seismic hazards. Finally, fluvially trimmed stream-terrace deposits are sometimes difficult to distinguish from scarps of tectonic origin. Both produce steep topographic scarps and are easily degraded so that the origin is not evident. The ambiguity has caused debate over several linear topographic scarps within the northern Rio Grande rift [6, 53]. We examine this question in more detail for our study area in following sections.

Initial geologic mapping using LiDAR imagery identified the morphostratigraphic relationships between deposits of differing relative ages. The high-resolution imagery (<1 meter/pixel) permitted greater precision in differentiating surfaces with varying surficial morphologies. Older alluvial fan surfaces have less distinct microtopography (i.e., bar and

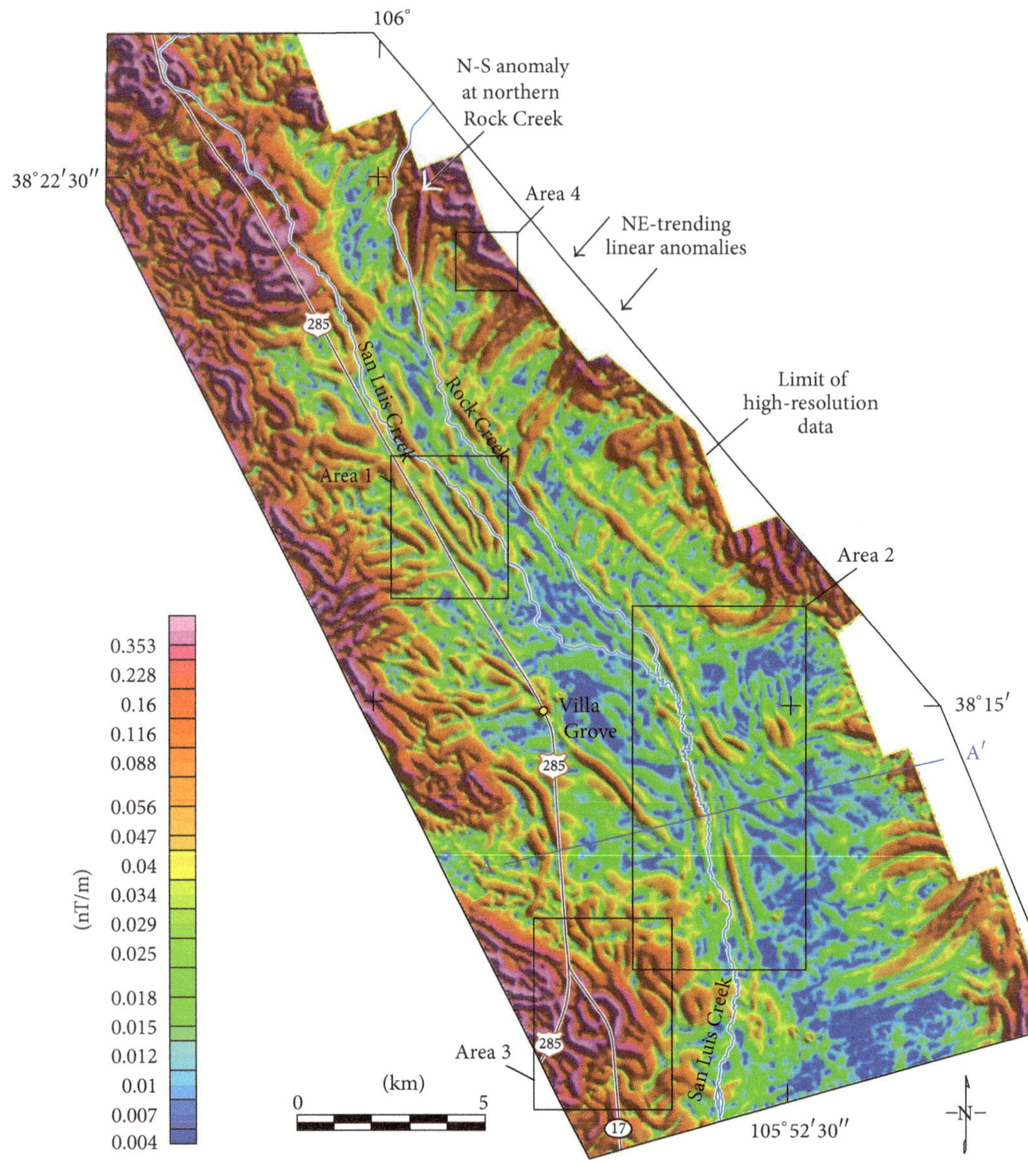

FIGURE 6: Horizontal gradient magnitude (HGM) of reduced-to-pole, high-resolution aeromagnetic data, using the gradient window method (window size = 1,225 × 1,225 m) to focus on gradients caused by shallow sources. HGM image is illuminated from the northeast. Boxes locate example Areas 1–4. Model for A-A′ is shown in Figure 16.

swale topography) and are generally more deeply dissected, indicating abandonment of the original deposition surface. Younger fan surfaces have very distinct bar and swale topography and generally grade to the active axial valley floodplain.

The relative position of these deposits and surfaces of differing ages is directly related to the geomorphic evolution of the region. Bull [54] demonstrated that, along tectonically active range fronts, younger alluvial fans are sequentially stacked at the mountain-piedmont junction. Along inactive range fronts, younger alluvial fans instead incise the fanhead and prograde farther out into the basin.

4.3. Combined Interpretation. Combined interpretation of aeromagnetic and LiDAR data builds on the strength of the aeromagnetic method to locate normal faults with significant offset under cover and the strength of LiDAR interpretation to identify the age and sense of motion of faults. Moreover, ambiguities in interpreting multiple HGM ridges from the aeromagnetic data can be resolved by examining which are associated with scarps in the LiDAR data. Uncertainty in resolving the tectonic versus fluvial origin of scarps next to streams identified by LiDAR data can be resolved by comparing the locations of HGM ridges in the aeromagnetic data.

Combined interpretation was done iteratively, by first identifying or interpreting faults in each dataset separately, then determining similarities and differences and resolving ambiguities by comparing results. The HGM ridges were commonly used to define the buried segments of faults in between scarps identified in the LiDAR data. In some areas, faults inferred from the aeromagnetic data without associated scarps were considered to be potentially active if they were aligned with other active faults and/or followed gravity gradients defining gravity lows. The sense of motion for these inferred faults were assumed to be down into the basin as defined by the gravity low, although antithetic faults are possible.

5. Examples of Combined Interpretation

Examples of combined interpretation for four localities from our study area demonstrate different aspects of how aeromagnetic and LiDAR data were compared to resolve issues specific to the study area. Areas 1–4 are located on Figures 1–4 and 6.

5.1. Area 1 Example. Area 1 is located on the central west side of the valley in a low-relief area where basin-fill and alluvial deposits of several different ages are mapped (Figure 1). Reduced-to-pole aeromagnetic data for Area 1 indicate several narrow, northwest-trending linear anomalies superposed on a broader pattern of highs and lows (Figure 7(a)). The locations of the edges of these anomalies are more apparent as linear ridges in the HGM map (Figure 7(b)). These kinds of patterns are characteristic of shallow faults that displace sediments or rocks of differing magnetic properties, allowing faults to be interpreted along the peaks of the ridges, as described in Section 4.1. However, interpreting the multiple parallel HGM ridges in the central part of the area is uncertain. Recognition of pairs of ridges associated with one fault, where one ridge is narrower than the other (e.g., Figures 5(c) and 5(d)), is ambiguous because there are an odd number of ridges.

Deposits of five different ages for Area 1 are distinguished by variations in morphology of the land surface apparent in the LiDAR hillshade image (Figure 7(c)). They are mapped in Figure 7(d) as Holocene stream alluvium (Qh), Pleistocene glacial fan deposits (from the youngest to the oldest Qfy, Qfi, and Qfo), and Miocene-Pliocene rift fill (Ts). Several subdued, northerly linear breaks in slope cross the deposits (black lines across Qfo, Qfi, and Qh on Figure 7(d)). Scarps are most pronounced within the Holocene alluvium Qh; scarps across the older fans are more difficult to follow. The prominent topographic scarp adjacent to San Luis Creek is ambiguous from LiDAR data alone; it may be a fault scarp, the result of fluvial trimming, or both.

Comparing the HGM map to the LiDAR image provides greater confidence in locating faults (Figure 8). Scarps evident in the LiDAR image partially correspond to four of the multiple HGM ridges (faults A, B, C, and D on Figure 8), indicating these are the HGM ridges that best locate the traces of active faults. Moreover, the east-facing scarps in the LiDAR image suggest the faults are normal and down to east. The eastward offset of the aeromagnetic representation (white line) from the LiDAR scarp (black line) of fault A in the southeastern part of the area probably indicates variability in the source and depth of the magnetically contrasting layers along strike of fault A. In the very southeastern corner of the Area 1, the absence of aeromagnetic expression of fault A, but not its associated scarp, suggests that the fault does not terminate. Instead, the lack of aeromagnetic anomaly likely results from the juxtaposition of strata having similar magnetic properties, which produces no magnetic contrast.

The HGM ridges also help guide connections between scarps on the LiDAR image and suggest that these faults have significant displacement at depth. For example, the HGM ridge for fault A suggests that, from south to north, the fault extends toward a subtle scarp within the stream valley at the northern end of fault A as opposed to following the stream bank to the northwest along the contact between Qfi and Qh. The HGM ridge associated with fault C implies the fault underlying the slope expressed in the LiDAR south of the contact of Qfi and Qfo extends towards the northwest, rather than to the north, where it is tempting to make a connection to the slope associated with fault B (compare Figure 7(d)). The strong northwest-trending HGM ridge near D appears related to the northwest-striking portion of the scarp just south of D and suggests that this fault continues to the northwest underneath Qfi. Fault E is identified aeromagnetically by a break in pattern of HGM ridges that indicate a sharp difference in lithology along a linear contact. Its fault origin is supported by the two nearby scarps determined from the LiDAR data (more evident in hillshade images of different illuminations than shown in Figure 7(c)).

The LiDAR data are important for determining relative age of faults interpreted for Area 1. Fault A has been active most recently, because a scarp is developed in Holocene deposits (Qh). The most recent event for fault B occurred <170 ka but >10 ka (post-Qfi and pre-Qh), because a scarp is apparent in Qfi but not in Qh. The most recent event for fault C occurred <640 ka and >120 ka (post-Qfo and pre-Qfi) and for fault E occurred since 120 ka (post-Qfi). The most recent activity on Fault D, which connects two scarps that displace Qfi, likely occurred <170 ka.

A magnetic model was constructed along a southwest to northeast profile across Area 1 (Figure 9). The model is highly uncertain because of poor constraints on subsurface geology, variable thickness and/or magnetic properties of geologic units in the area, and paucity of gravity data for Area 1 (Figure 4). Moreover, this type of geologic variability tends to produce multiple magnetic sources that are inconsistently related to geologic contacts at depth, making it difficult to interpret the results of standard quantitative techniques, such as depth estimation [3]. Nevertheless, the model helps visualize the multiple sources that are potentially contributing to the observed aeromagnetic anomalies at faults.

Model units were assembled on the basis of geologic concepts, inspection of aeromagnetic maps, and reconnaissance magnetic-susceptibility measurements of outcrops within and near the study area. Magnetic susceptibilities of Holocene stream deposits (Qh) are generally 0.0005 SI or lower, whereas those measured for Quaternary fan material (Qfy, Qfi, and Qfo) range from 0.0010 to 0.0100 SI, with a mode of ~0.0060. The latter values are fairly magnetic for sediments but are within the ranges observed for sediments in the northern Rio Grande rift [46, 47]. We estimate that the collective thickness of the Quaternary deposits generally

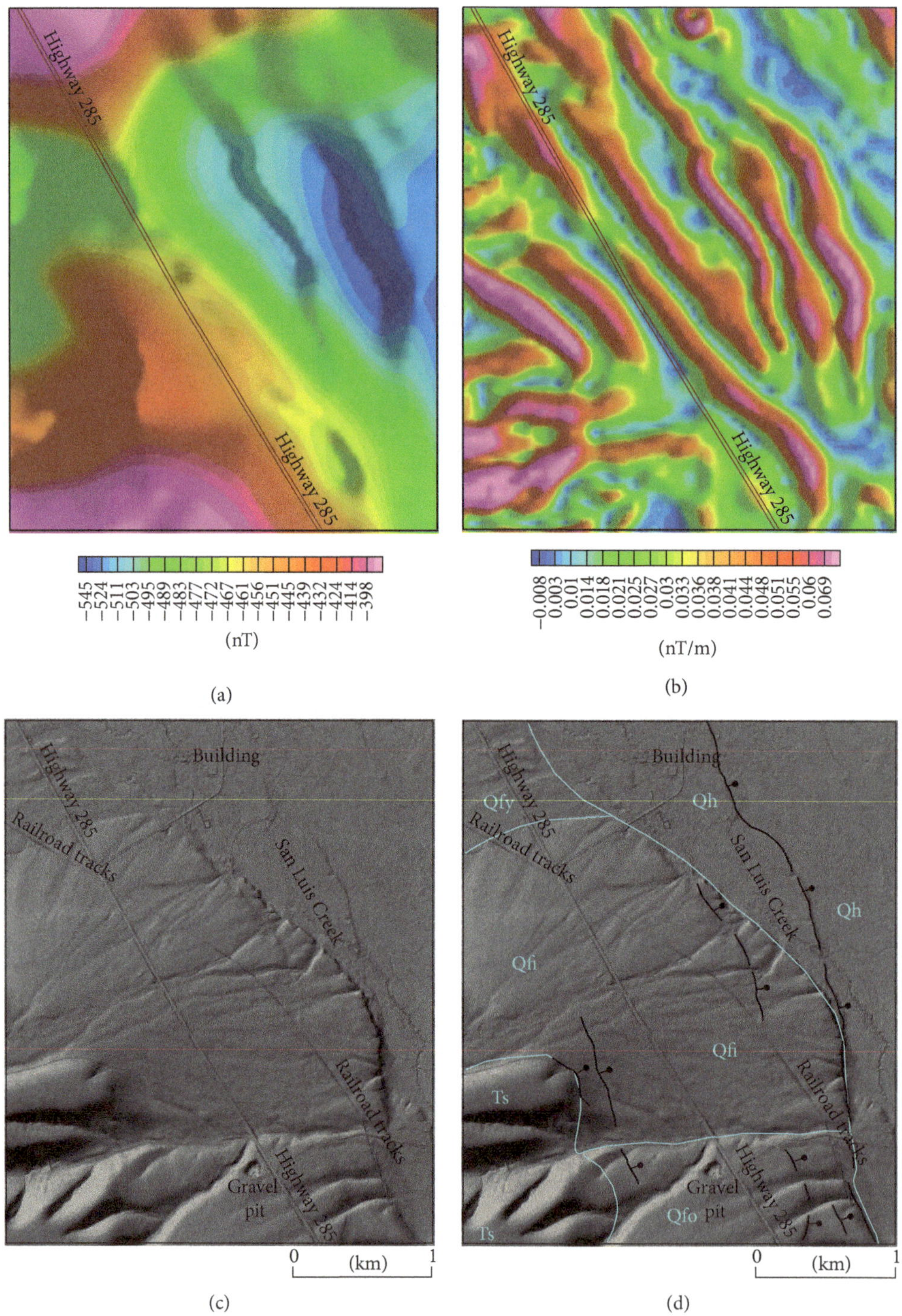

FIGURE 7: Aeromagnetic and LiDAR data for Area 1 (located on Figures 1–4 and 6). (a) Reduced-to-pole (RTP) aeromagnetic data and (b) horizontal-gradient magnitude (HGM) of RTP. NW-illuminated hillshade of LiDAR data, without (c) and with (d) inferred faults and contacts (cyan lines) overlain. Geologic units are explained in Figure 1.

ranges from 10 to ~100 m. Thicknesses, lithologies, and magnetic properties of the underlying Tertiary rift sediments (Santa Fe Group) are highly uncertain. However, similar to observations in the central Rio Grande rift [46], we expect that the sediments are stratified in layers of differing magnetic susceptibilities, with average values ranging from ~0.0005 to 0.0030 SI. How such layers are arranged in the subsurface of our study area is unknown, but the history of tectonic activity [19, 31] suggests that coarse-grained sediments, representing the more magnetic of the sediments, likely lie toward the base of the section and thicken into the deepest parts of the basin. Reconnaissance magnetic-sus-

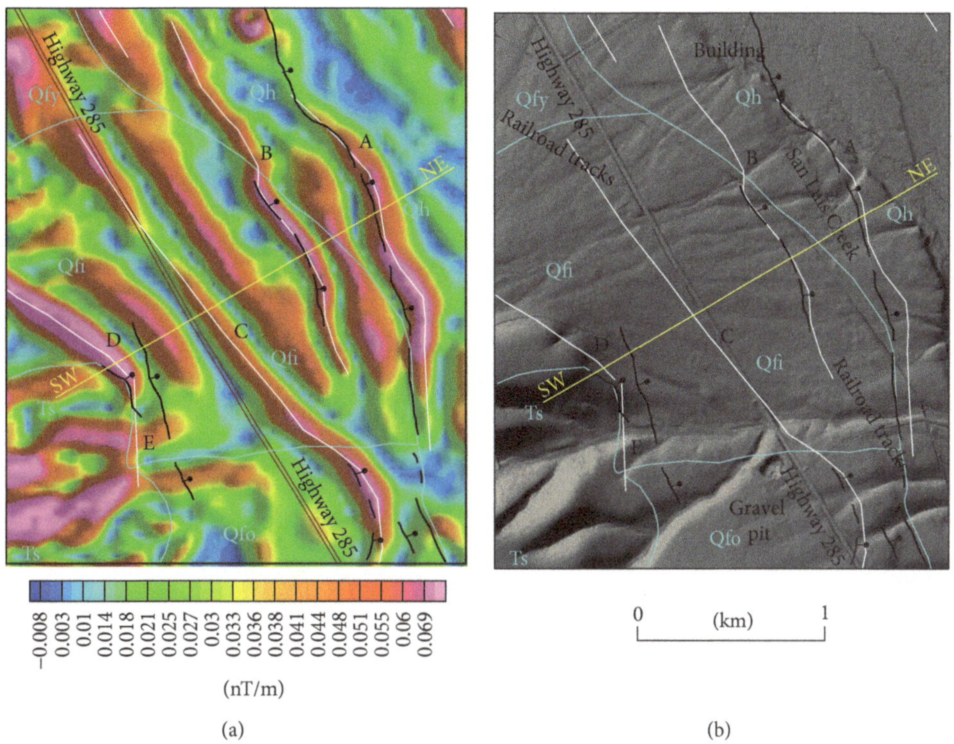

FIGURE 8: Combined interpretation for Area 1 (located on Figures 1–4 and 6). Faults mapped from LiDAR are in black; those interpreted from ridges in HGM image (Figure 6) are in white. Geologic contacts mapped from the LiDAR data are in cyan. Geologic units are explained in Figure 1. Letter labels refer to faults discussed in text. (a) HGM map of the reduced-to-pole aeromagnetic data; (b) NW-illuminated hillshade of LiDAR data. Model for southwest to northeast profile (SW, NE) is shown in Figure 9.

ceptibility measurements of Precambrian rocks in our study area range from <0.0010 to ~0.0250 SI, similar to the range of values found immediately to the north of our area [40, 55]. High-amplitude aeromagnetic anomalies over exposed Proterozoic rocks to the west of the study area suggest that these rocks are fairly magnetic underneath Area 1, so we assign a value of 0.0100 SI.

The model of Figure 9 demonstrates that the juxtapositions of lithologies of differing magnetic properties are the likely sources causing the observed (reduced-to-pole) signal at faults A, B, C, and D for Area 1 (Figure 8). Moreover, multiple HGM peaks in between these inferred faults likely arise from magnetic contrasts at different levels of the faults rather than separate faults. An exception is the HGM peak just to the west of C that does not offset Qfi deposits (Figure 8) and occurs at about 1.5 km distance on the profile (Figure 9). A concealed, major rift fault is modeled at this location, which is required if prerift rocks have the same magnetic properties throughout Area 1. If this interpretation is correct, the concealed fault accommodates much of the vertical displacement at the western basin margin. An alternative model would involve a large magnetic discontinuity within the prerift rocks instead of a rift fault.

To illustrate the relative contributions of different magnetic sources to the overall computed field of the model, the computed field for each of three different groupings of model bodies is also shown on Figure 9. These three curves add together to form the computed field for the whole model. As expected, the computed field of the prerift model units accounts for the broad character of the observed field, with significant contribution to anomalies over faults where the prerift rocks are shallower. The modeled Quaternary units (Qh and Qf) account for most of the character of the short-wavelength anomalies over the faults. The computed field from the Tertiary rift-fill sediments (Ts1 and Ts2) has the least contribution to the character of the whole computed field, so the choice of layers within this model unit does not have significant impact on the model results.

5.2. Area 2 Example. Area 2 is located in the valley and on fan material on the southeastern part of the study area where the swarm of northwesterly normal faults known as the Villa Grove fault zone has been mapped (Figure 1). These faults are quite evident as curvilinear scarps in an enlarged area of the LiDAR image (Figures 10(a) and 10(b)), located on the inset of Figure 10. In contrast to Area 1, there are many more scarps than identifiable HGM ridges in this area (Figure 10(c)). The HGM ridges, except in the northern and western parts of the enlarged area, are also more subdued. Similar to examples shown for Area 1, the differences in HGM amplitudes may be a function of variations in the deposits that are juxtaposed at depth along strike of the faults.

Many of the scarps evident in the LiDAR image are likely secondary features of mass wasting related to more major faults and located in between them [8, 10, 21]. The major faults can be identified by the HGM ridges because the ridges

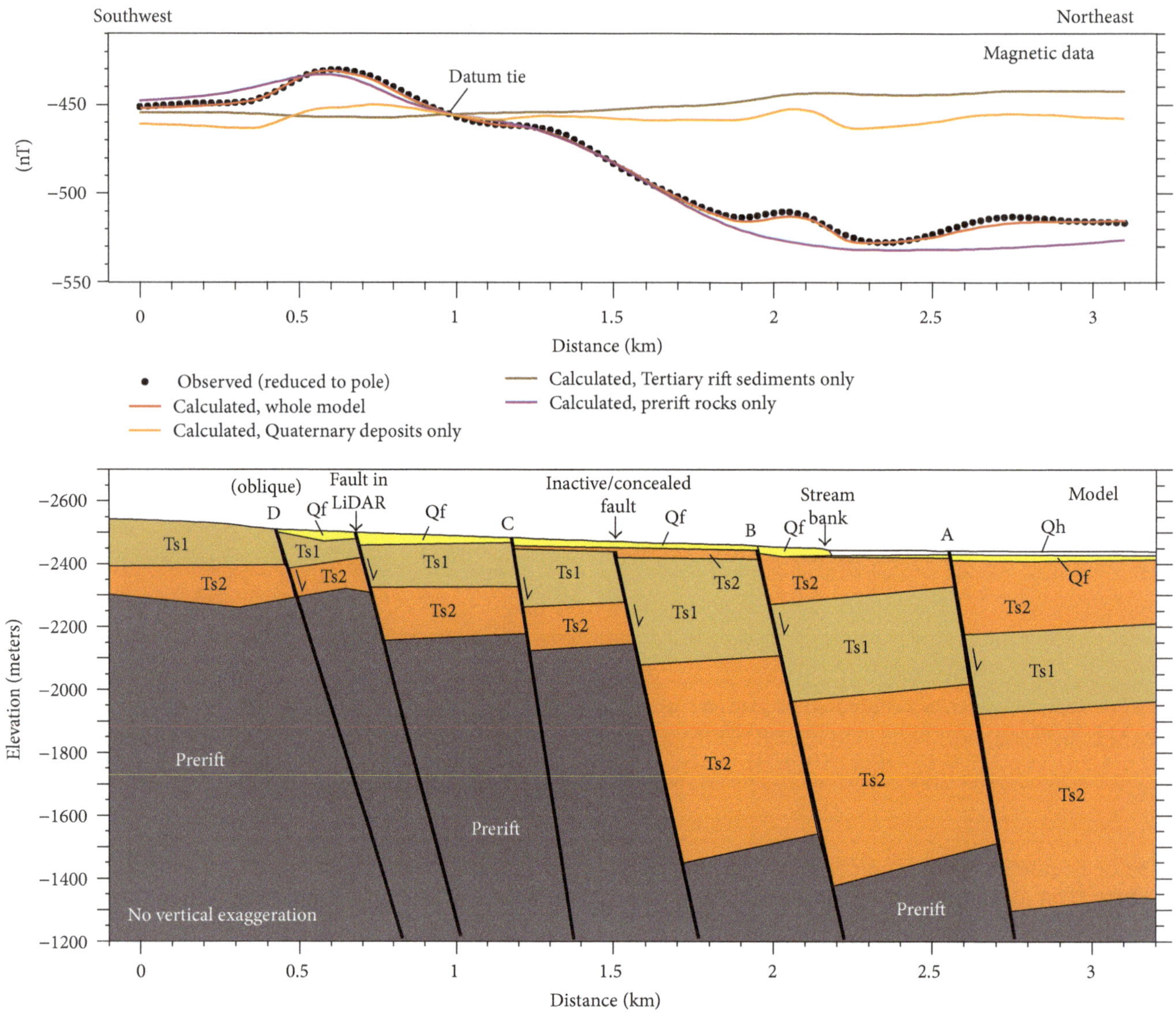

		Label	Magnetic suscept. (SI)	Likely geologic unit(s) from Figure 1
Quaternary deposits	□	Qh	0.0005	Qh
	■ (yellow)	Qf	0.0060	Qfy, Qfi, Qfo undivided
Tertiary rift sediments	■	Ts1	0.0005	Ts (less magnetic)
	■	Ts2	0.0020	Ts (more magnetic)
Prerift rocks	■	Prerift	0.0100	Xu, possibly some Tv

FIGURE 9: 2D magnetic model for the southwest to northeast profile across Area 1 (located on Figure 8). Observed curve was extracted from the reduced-to-pole aeromagnetic grid (Figure 2). Reasoning behind the magnetic model parameters and uncertainties in the model results are discussed in the text. Letter labels along the ground surface refer to the letter labels in Figure 8.

indicate significant offset between materials of differing magnetic properties. For example, a major fault is represented by the prominent scarps in the LiDAR image that follow HGM ridges from A to B to C (Figure 10). Both the LiDAR and HGM images suggest the fault anastomoses at B. Other major faults can be traced by HGM ridges that coincide with scarps that pass northeast of A and southwest of D. All these scarps are southwest facing in the NE-illuminated LiDAR image (scarps are in shadow) and, thus, down to the southwest. The offset to the southwest of the HGM ridge related to the scarp at E is likely the result of magnetic contrast that is downdip from the surface trace. The scarps southwest of the two locations of F are northeast facing (scarps are highlighted) and probably represent antithetic faults down to the

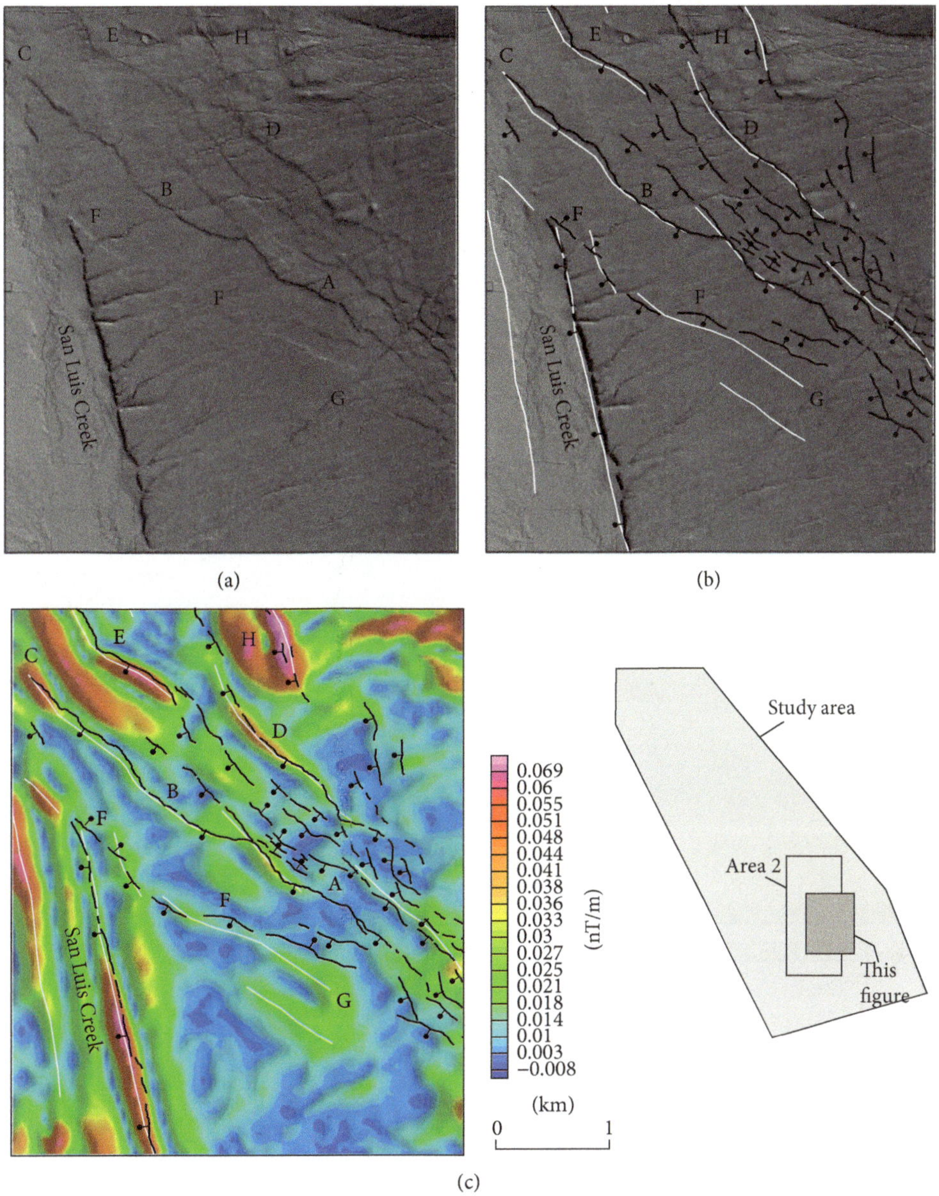

FIGURE 10: Example of combined interpretation for enlarged area related to Area 2 (inset). Faults from the LiDAR interpretation are black and those mapped from HGM image (Figure 6) are white. Letter labels are for discussion purposes. NE-illuminated hillshade of the LiDAR data without (a) and with (b) faults overlain. (c) HGM of reduced-to-pole aeromagnetic data with faults overlain.

northeast. The HGM ridge at G that parallels the northeast-facing scarp to the southwest is not likely associated with the deeper parts of fault F. If it were, it would be offset to the northeast instead. Several northeast-facing scarps northwest of A likely represent secondary antithetic faults as well.

Most of the scarps of the Villa Grove fault zone are well defined as they cross both younger (Qfy) and intermediate-age fan material (Qfi) (Figure 1), indicating the most recent fault activity was fairly young (<15 ka). By contrast, scarps east and west of H, which cross Qfo, are more degraded and discontinuous (Figure 10(a)), implying that these faults may not have ruptured during the most recent event.

The prominent stream bank on the eastern side of San Luis Creek (Figures 10(a) and 10(b)) could be interpreted geomorphologically as a fault scarp or a fluvially trimmed terrace deposit. Ruleman and Machette [10] suggest a tectonic rather than fluvial origin because of the linearity of the topographic scarp and the absence of a paired terrace deposit on the west side of the creek. However, similar geomorphic features south of our study area are still under debate by geologists [53]. The presence of linear HGM ridges coincident with the linear topographic scarp provides support for a tectonic origin. The narrowness of the HGM ridges advocates near-surface, near-vertical boundaries for the magnetic sources [56]; their linearity supports a fault contact rather than a facies change. Near-vertical, distinct magnetic contrasts, required to produce an HGM ridge, are difficult to explain by facies changes caused by depositional

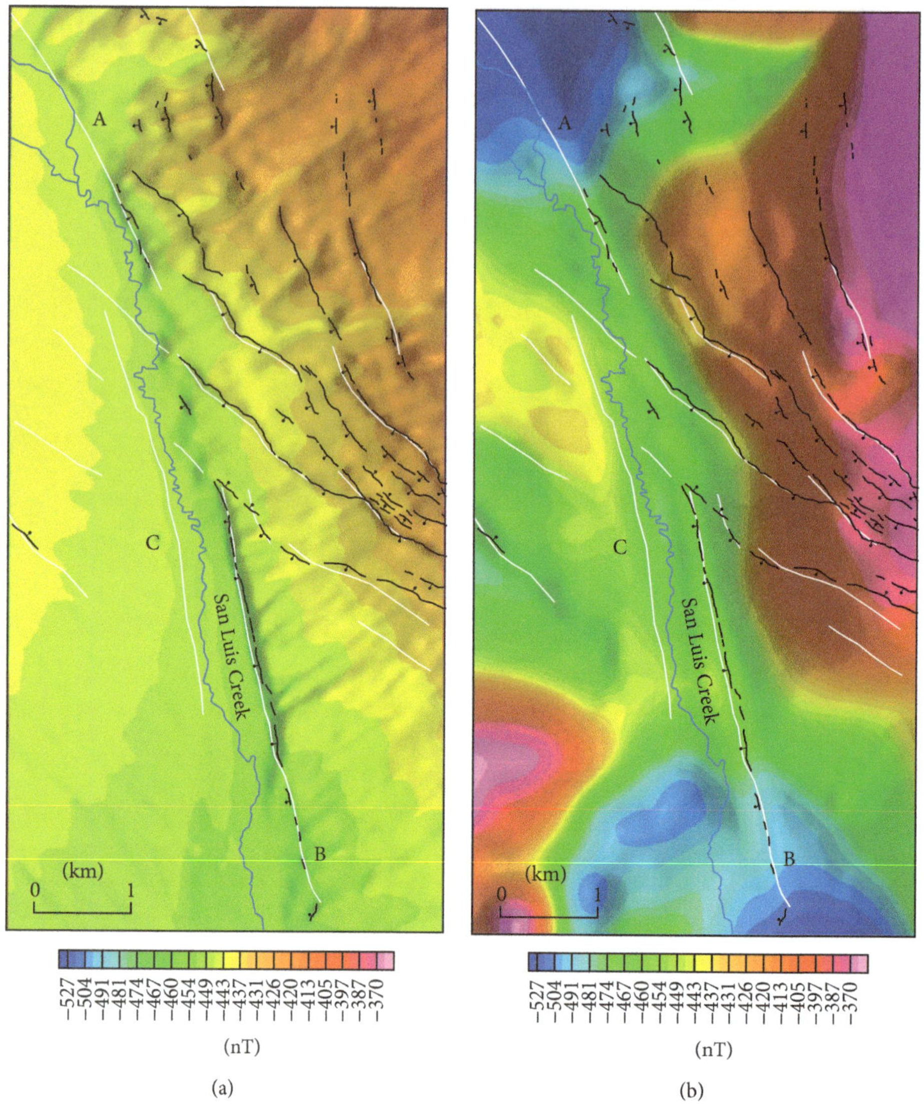

FIGURE 11: Assessment of terrain effects for Area 2. Hypothetical terrain effects in (a) are compared to patterns in the observed RTP aeromagnetic data in (b) to identify anomalies caused solely by topographic relief formed on magnetic Quaternary deposits. Note the lack of correlation along the stream bank on the east side of San Luis Creek. Faults from LiDAR (black) and those from HGM of RTP data (white) are overlain. Letter labels are for discussion purposes. (a) Hypothetical magnetic field computed from digital elevations using uniform magnetic susceptibility of 0.010 SI and height and Earth's field parameters to match the RTP observed data. Digital elevations for the computation were derived from the LiDAR data, gridded at 25-m interval. (b) Observed RTP aeromagnetic data. Northeast illumination and color scales are identical for both images.

processes. Because streams migrate laterally through time as the basin subsides, facies changes in the sedimentary section are likely to develop interfingered, irregular contacts when viewed in cross-section rather than sharp boundaries that extend to a significant depth.

However, linear topographic scarps can produce aeromagnetic anomalies even if they are not fault scarps [3]. To test whether the HGM ridges at the eastern bank of the San Luis Creek are caused by magnetic terrain effects or by underlying sources, we compute a model of hypothetical terrain effects for Area 2 (Figure 11(a)). The hypothetical model uses a digital elevation model and a uniform magnetic susceptibility of 0.010 SI to simulate anomalies due solely to the irregular topographic forms of surficial Quaternary deposits. The computation is accomplished by adding the effects of vertical prisms at each grid interval of the digital elevation model whose top is the average elevation of the surrounding grid points (J. Phillips, unpublished computer program MBOXES).

If all the underlying material in Area 2 were uniformly magnetized with magnetic susceptibility of 0.010 SI, the morphology of the top surface of the surficial deposits alone should produce aeromagnetic anomalies with amplitudes detectable by the aeromagnetic survey, as demonstrated by Figure 11(a). However, comparison to the observed reduced-to-pole aeromagnetic map (Figure 11(b)) shows that (1) an aeromagnetic low is observed over the crest of the stream bank instead (Figure 11(b)), so the stream bank is not a significant source of the observed anomalies, and (2) observed aeromagnetic anomalies generally have greater amplitude than reasonably produced by relief on the terrain alone. Furthermore, HGM ridges follow straighter lines than the stream bank scarps and are continuous across places where the bank has been eroded, such as at A and B on Figure 11. A very linear HGM ridge follows the western stream bank beside C. The magnetic field produced by this stream bank is too subdued to produce the observed aeromagnetic gradient; an additional magnetic source is required at depth. Its similarity in linearity and parallel orientation to the HGM ridge over the eastern stream bank suggests that both HGM ridges are produced by faults in the subsurface.

Thus, we conclude that the HGM ridges along the stream bank originate from magnetic sources that underlie the terrain and the linearity of the ridges suggests a tectonic rather than fluvial origin. The HGM image for the whole study area (Figure 6) suggests that similar faults closely follow San Luis Creek and Rock Creek almost the entire axis of the valley within the study area.

The HGM image for Area 2 can be used to better understand the intersection of the Villa Grove fault zone and the San Luis Creek and Rock Creek fault zones (Figure 12). HGM ridges of the northwest-striking faults of the Villa Grove fault zone clearly cross and interrupt the more northerly HGM ridges related to the San Luis Creek fault (B, C in Figure 12) and the Rock Creek fault zone (A). The Villa Grove fault zone extends west of the San Luis Creek fault zone to D, where it is concealed beneath Holocene material. If the HGM ridge at C is a deeper manifestation of the magnetic contrast expressed at the HGM ridge at A, one could argue that the San Luis Creek fault is displaced down to the southwest relative to the Rock Creek fault at A. The Rock Creek fault zone gradually bends more westerly north of A (Figures 6 and 12), which may also be influenced by the younger Villa Grove fault zone.

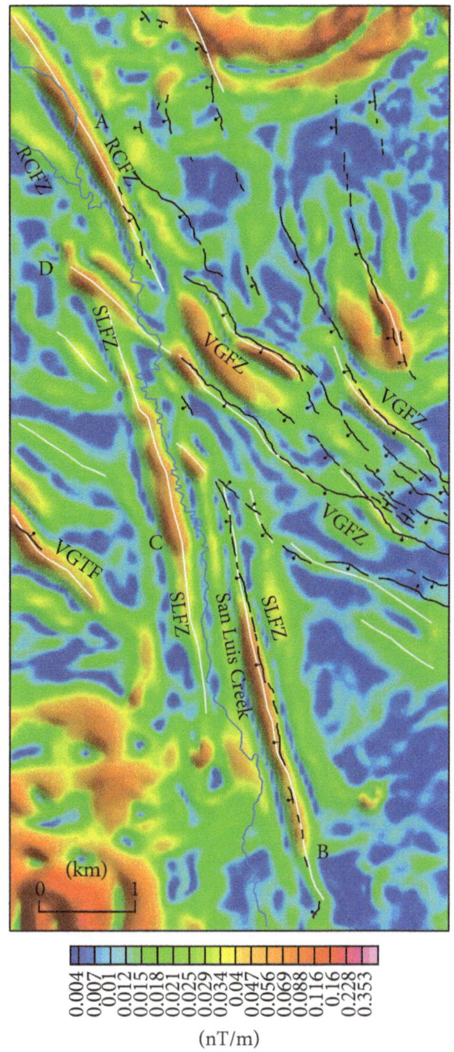

FIGURE 12: Ridges in the HGM of RTP image depicting the crossing of the Villa Grove and San Luis Creek fault zones in Area 2 (located in Figures 1–4 and 6). Letter labels are for discussion purposes. RCFZ: Rock Creek fault zone, SLFZ: San Luis Creek fault zone, VGFZ: Villa Grove fault zone, and VGTF: Villa Grove town fault.

5.3. Area 3 Example. Area 3 is located in the southwestern part of the study area near the western margin of the rift valley (Figure 1). Although the area has low relief, exposures of Proterozoic rocks (Xu) and high-amplitude aeromagnetic anomalies indicate that bedrock is close to the surface over most of the western part of the area (Figure 13). A north-northwest trending normal fault has been previously mapped at the basin margin (red on Figure 13). It passes near the Joyful Journey Hot Springs (previously known as Mineral Hot Springs) and related travertine mounds and is presumed to have an influence on the location of the upwelling thermal waters. The mapped Mineral Hot Springs fault is commonly extended farther south (out of the study area) to connect to the eastern bounding fault of a large intrarift horst concealed within the San Luis basin (e.g., [23]). Several workers argue that a scarp developed in Quaternary deposits (units Qfi and Qfy) along the Mineral Hot Springs fault indicates it is active; others contend that the scarp is an erosional feature (described in Kirkham [6]). Thus, combined interpretation of aeromagnetic and LiDAR data in this area can address this debate and serves as an example of interpreting basin-bounding as well as intrabasin faults from the aeromagnetic data.

Scarps mapped for the Mineral Hot Springs fault are indicated at locations marked A on the LiDAR image (Figure 13(a)), where a linear gradient is superposed on a broader aeromagnetic high (Figure 13(b)). On the north, the aeromagnetic low between locations marked as C (Figure 13(b)) generally corresponds to the westward extension of a normal fault that juxtaposes Proterozoic rocks (Xu) against effectively nonmagnetic Paleozoic sedimentary rocks (Pz). The fault is exposed just west of the northwest corner of Area 3 (Figure 1).

The extension of the mapped Mineral Hot Springs fault inferred southeast of B (dashed red line) is not supported by the aeromagnetic data. Instead, dominant features of the aeromagnetic data are (1) a north-northeast-trending, steep

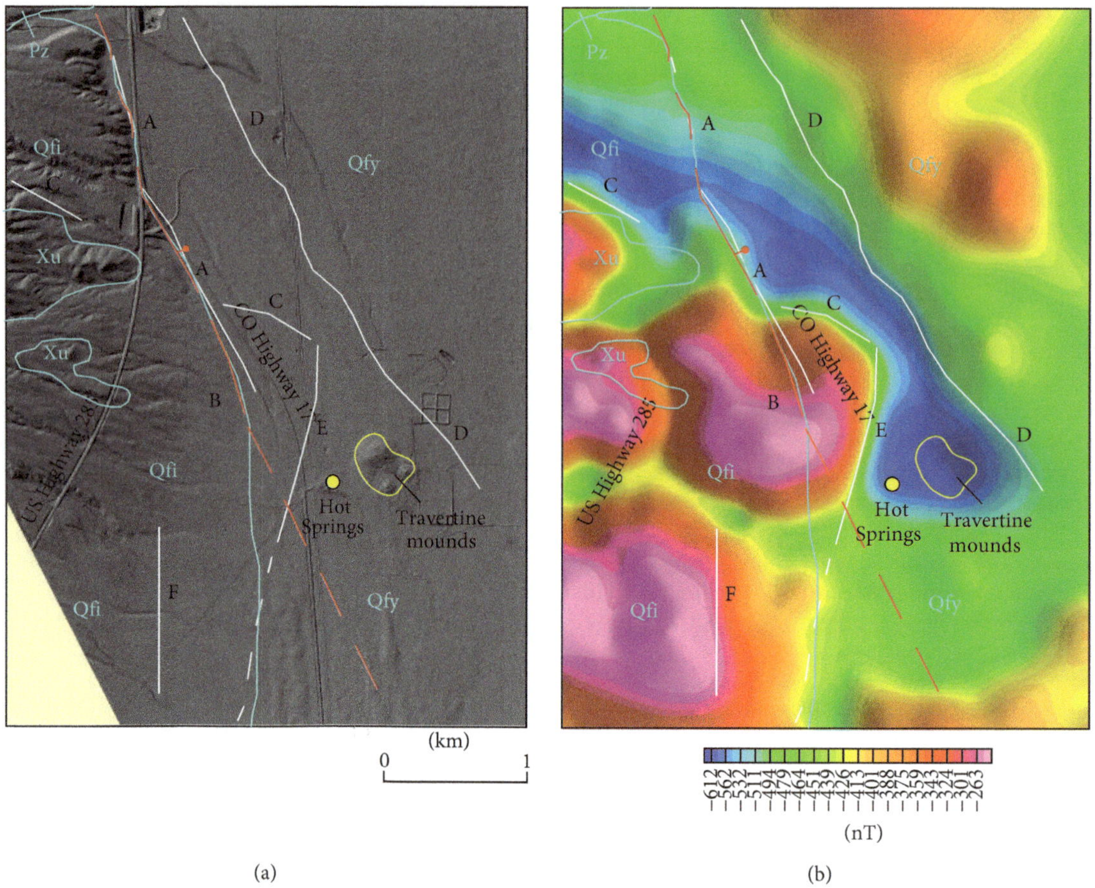

FIGURE 13: Example of combined interpretation from Area 3 (located in Figures 1–4 and 6) suggesting a revision to the location of faults surrounding the active hot springs. Previously mapped Mineral Hot Springs fault [6] is shown as solid and dashed red line. Lines interpreted from the HGM image (Figure 6) are in white. Geologic contacts (cyan lines) and units are from Figure 1. Letter labels are for discussion purposes. (a) NW-illuminated LiDAR image. (b) Reduced-to-pole aeromagnetic data, illuminated from the east.

gradient at E between strong highs on the west and subdued low values on the east that passes west of the hot springs; (2) a northwest-trending linear gradient between low values on the west and moderately high values on the east that passes east of the hot springs (locations D on Figure 13(b)); and (3) a north-south gradient at F. Both features D and E are interpreted as faults, although neither is associated with scarps in the LiDAR data. They surround a triangular shaped aeromagnetic low centered over the hot springs and related travertine mounds, suggesting a geologic association with the active geothermal system. The eastern one (D) is likely rift related, because of its orientation and location at the rift margin that follows trends in the gravity data (Figure 4). If so, fault D has not been active since >12 ka, when unit Qfy was deposited. We interpret the western faults at E and F as inactive, because their strikes do not align with the rift. The aeromagnetic low just east of E may indicate that Proterozoic rocks are juxtaposed against Paleozoic rocks at the inactive fault. The low also could be associated with secondary alteration of crystalline rocks that has destroyed the magnetic minerals. Perhaps thermal waters are currently rising within a fault-bounded hydrogeologic window in the area of the hot springs.

5.4. Area 4 Example. Area 4 is located at the high mountain range front, where the northwest-striking Sangre de Cristo fault follows the base of precipitous facets in the northeastern part of the study area (Figure 1). The LiDAR image clearly shows the fault as thin bands of shadow and highlight near the interface between the steep bedrock slopes and dissected alluvial fans (Figure 14(a)). The high resolution in the LiDAR display of the morphology of the fan surfaces allows detailed mapping of the fault and Quaternary deposits of three different ages (Figure 14(b)). Recurrent displacement along the fault zone is documented east and southeast of A by a sliver of intermediate-age fan material (Qfi) that is stranded and perched on the footwall. This fan was faulted after deposition, younger fan material (Qfy) progressively accumulated across the fault and covered the older fan on the downthrown side, and then both were ruptured during a later event. In the same vicinity (southwest of A), the Sangre de Cristo fault is evident crossing an area mapped as Proterozoic unit Xu. Faulting here has displaced the bedrock and/or colluvium within the footwall, implying that the fault has encroached into the footwall over time rather than stepped farther basinward, as is commonly expected in extensional basins.

In contrast to the detail of the LiDAR image, the aeromagnetic data for Area 4 are dominated by high values on the northeast, which increases almost 300 nanoTeslas over a distance of only about 1.5 km along a northwest-trending gradient (Figure 14(c)). A large component of the high is likely caused by magnetic Proterozoic rocks within the steep terrain, as evidenced by a general correspondence of the bedrock terrain to anomaly shape at locations marked B. However, the northwest-trending gradient does not exactly follow the base of the mountain front (compare the locations of the scarps to the HGM ridges on Figure 14(d)), which would be expected if the gradient were reflecting solely terrain shapes. Differences in Proterozoic rock type likely also contribute to the aeromagnetic expression, as evidenced by a prominent north-northeast trending ridge in the HGM image (C on Figure 14(d)) that does not correspond to slopes in the terrain. Thus, we suspect that the Sangre de Cristo fault zone is partially following older structures, probably related to Laramide thrust faulting or early rift faulting in the Miocene. If so, the structure is somewhat oblique to the extension direction. A relation between Laramide thrust faults and the Sangre de Cristo fault has been noted for this area previously [57]. Linear breaks in slope superposed on the larger regional gradient in the aeromagnetic data are mostly evident in the HGM image (D on Figure 14(d)). They are interpreted as potentially active rift-related faults because they parallel the range-front fault.

Identifying and characterizing active faults are straightforward from the LiDAR data at the range front, where active fan deposition and faulting are clearly expressed in the geomorphology. Isolating the expression of Quaternary faults in the aeromagnetic data is more difficult, because highly magnetic sources in the Proterozoic bedrock dominate the aeromagnetic signal and are the product of multiple events over geologic history. However, using both LiDAR and aeromagnetic data together could provide insights into older, underlying structures and their possible influence on the propagation of seismic energy during an earthquake.

6. Results

Generalized faults identified by combined interpretation of the aeromagnetic and LiDAR data are presented in Figure 15. Faults were classified by the age of their most recent rupture and relations to other active faults. Younger active faults show LiDAR evidence of displacement of the younger Quaternary units, so they ruptured <10 ka (post-Qh) or <30 ka (post-Qfy). Older active faults displace the older units, so they ruptured between 120 ka and 170 ka (post-Qfi) or between 170 ka and 640 ka (post-Qfo) or are cross-cut by younger active faults. Faults identified by the aeromagnetic data only are considered potentially active if (1) they are buried by any of the Quaternary deposits; (2) they align with trends of active faults nearby, or (3) a combination of these. If faults identified solely by aeromagnetic data do not meet the criteria of a potentially active fault, they are considered inactive and concealed (Figure 1). The sense of displacement of normal faults was determined primarily from the orientation of scarp faces apparent from the LiDAR data. They were inferred from gravity data where scarps were not present and the inferred faults followed gravity gradients, although local structures violating this assumption are possible. Modifications to the mapping of Quaternary deposits were recognized while determining the age of faulting. These modifications are depicted generally on Figure 15 as well.

Both the LiDAR and aeromagnetic data (where they cover the range front) suggest the Sangre de Cristo fault is segmented along the northwest strike of the range front and can have multiple discontinuous strands parallel to each other, some of which may be represented by potentially active faults. The Villa Grove fault zone is similar in character as mapped previously, but the aeromagnetic data indicate which of the multiple faults may have experienced the most displacement. The Villa Grove town fault, named herein, suggests that poorly defined faults mapped previously extend for 5–7 km and pass through the town of Villa Grove. The Mineral Hot Springs fault is revised to apply to an inferred fault in the aeromagnetic data underneath Holocene deposits that parallels the previously inferred fault at the stream bank but passes on the east rather than west side of the hot springs. Faults are evident along both San Luis Creek and Rock Creek, supporting a tectonic or fluvially modified tectonic origin for steep scarps located at stream banks. The aeromagnetic data provide additional information about their extents in between discontinuous scarps and suggest several totally buried, potentially active faults are present on both sides of the valley.

7. Discussion

7.1. Two-Dimensional Model. The nature of the rift in relation to the active faults in the valley is further elucidated by a two-dimensional (2D) geophysical model along a roughly east-west profile A-A′ (Figure 16), located on Figures 2–4, 6, and 15. Modeled geologic units were assigned magnetic susceptibilities as described in Section 5.1 regarding the 2D model for Area 1 (Figure 9), except that Holocene unit Qh was grouped with the older Quaternary units because of the more regional scale of this model. Densities were assigned within the ranges of values we measured from similar geologic units in the region and from published sources [40, 43, 55]. As discussed in Section 5.1, thicknesses and/or magnetic susceptibilities of the Tertiary and Precambrian basement rocks are poorly constrained. Thus, we have kept the model as simple as possible. A more complex model could fit the data better but adds greater uncertainty.

The 2D model for A-A′ supports evidence from the gravity map (Figure 4) that the axis of the rift lies between San Luis Creek and the Villa Grove Town fault, with the Villa Grove fault zone and the Mineral Hot Springs fault as the eastern and western margins, respectively. Unlike the east-dipping half graben with large-displacement master fault at the range front that is commonly described for the main San Luis Basin to the south, this basin is located well to the west of the range front. Instead, vertical displacement at the range-front fault is on the order of 100 m. The bedrock surface dips at fairly low angle westward from the range front to the Villa Grove fault zone, which forms a broad zone along the east-

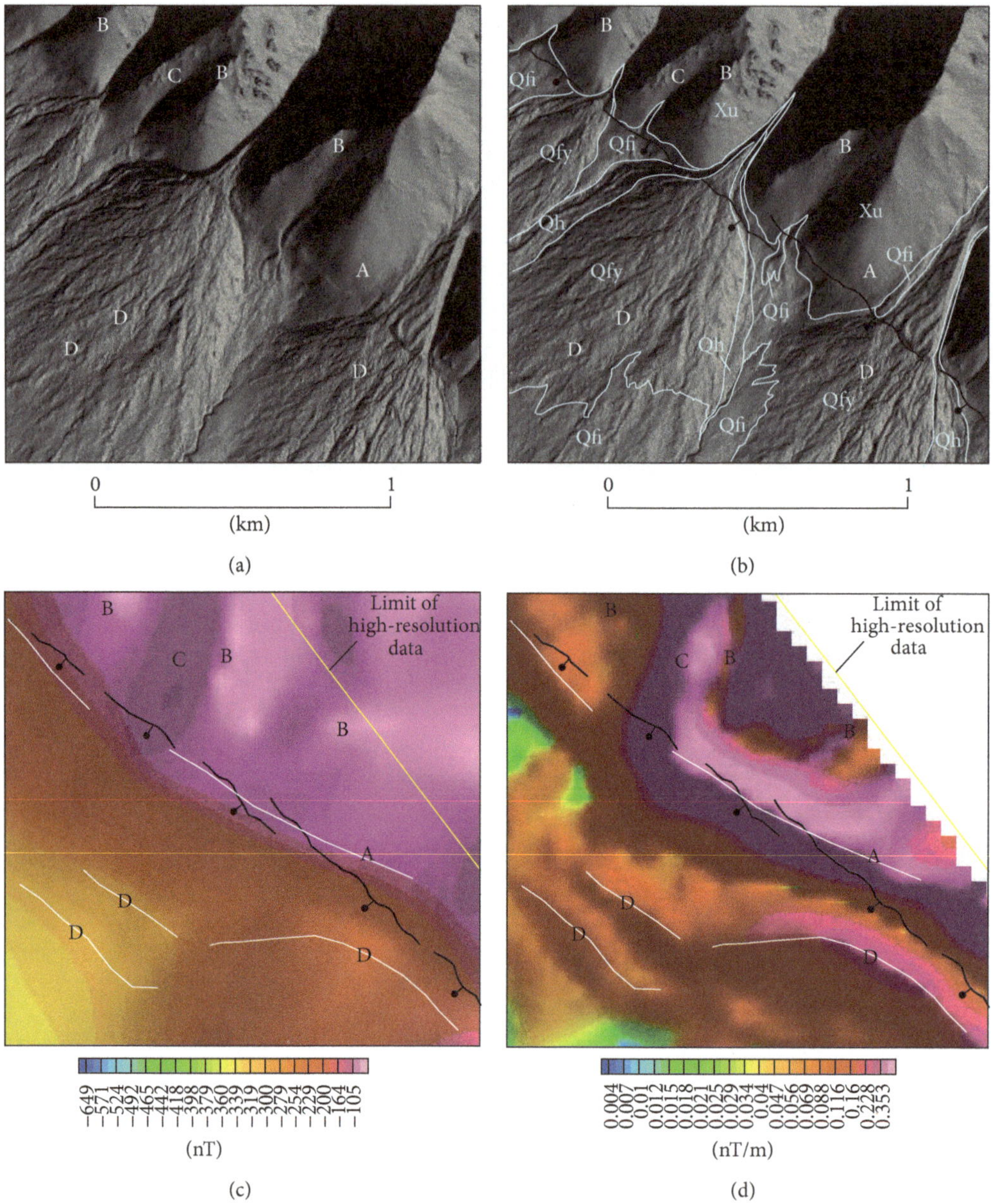

FIGURE 14: Example of combined interpretation from Area 4 (located in Figures 1–4 and 6). Black lines are scarps mapped from the LiDAR image; white lines are faults inferred from the HGM image. Letter labels are for discussion purposes. (a) LiDAR hillshade image, illuminated from the southeast. (b) The same LiDAR image with detailed mapping of Sangre de Cristo fault (black line) and Quaternary deposits (cyan lines). Geologic codes are as in Figure 1. (c) Reduced-to-pole aeromagnetic image illuminated from the east. (d) HGM of reduced-to-pole aeromagnetic data, illuminated from the northeast.

ern margin of the rift basin. Hypotheses to explain the low-angle slope are (1) a relic of previous tectonic activity, such as Laramide thrust faulting; (2) the result of early-rift detachment faulting; or (3) the product of regional tectonic tilting to the west related to early rifting. In any case, the focus of tectonic activity apparently has been concentrated in the vicinity of San Luis Creek, where most of the displacement and basin subsidence has occurred. Tectonic activity at the range front must have started late in rift history, as evidenced by the relatively small displacement across the fault there.

The curve extracted from the HGM grid in the top panel of Figure 16 emphasizes the utility of computing the HGM using the gradient window method to detect the subtle signature of faults in the aeromagnetic data. The HGM curve peaks over subtle gradients caused by shallow sources within the surficial deposits where the smaller gradients can be masked by broad regional magnetic gradients. Individual faults (except the Mineral Hot Springs fault) have even less expression in the gravity profile. The lack of expression is especially noticeable along the broad, almost linear, decrease

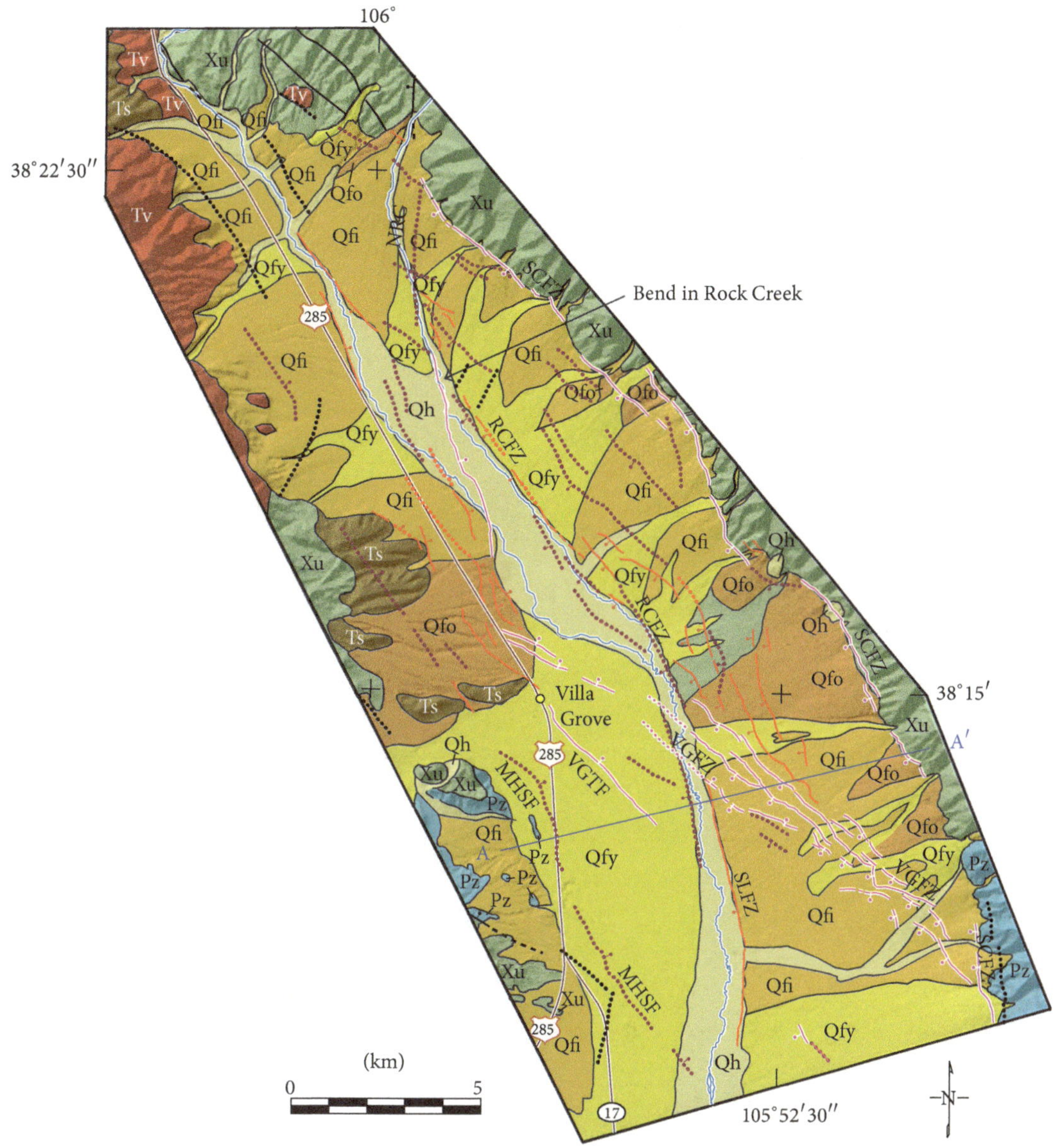

Age of most recent activity

Active (younger): scarps well developed and displace units Qfy (<30 ka) or Qh (<10 ka). Concealed segments inferred from aeromagnetic data

Active (older): scarps degraded and/or displace only units older than Qfy (>12 ka) or Qfi (>120 ka). Concealed segments inferred from aeromagnetic data

Potentially active: no scarps evident, but fault and relation to active faults are inferred from aeromagnetic data. Sense of throw inferred from gravity data.

Inactive: fault geologically mapped or inferred from aeromagnetic data in prerift rocks

Normal fault: ball on downthrown side, dashed where uncertain, dotted where concealed

FIGURE 15: Revised surficial geology and inferred active faults and buried fault segments resulting from combined interpretation of the aeromagnetic and LiDAR data for the study area. MHSF: Mineral Hot Springs fault (revised), NRC: Northern Rock Creek; RCFZ: Rock Creek fault zone, SCFZ: Sangre de Cristo fault zone, SLFZ: San Luis Creek fault zone, VGFZ: Villa Grove fault zone, and VGTF: Villa Grove town fault. Model for A-A′ is shown in Figure 16.

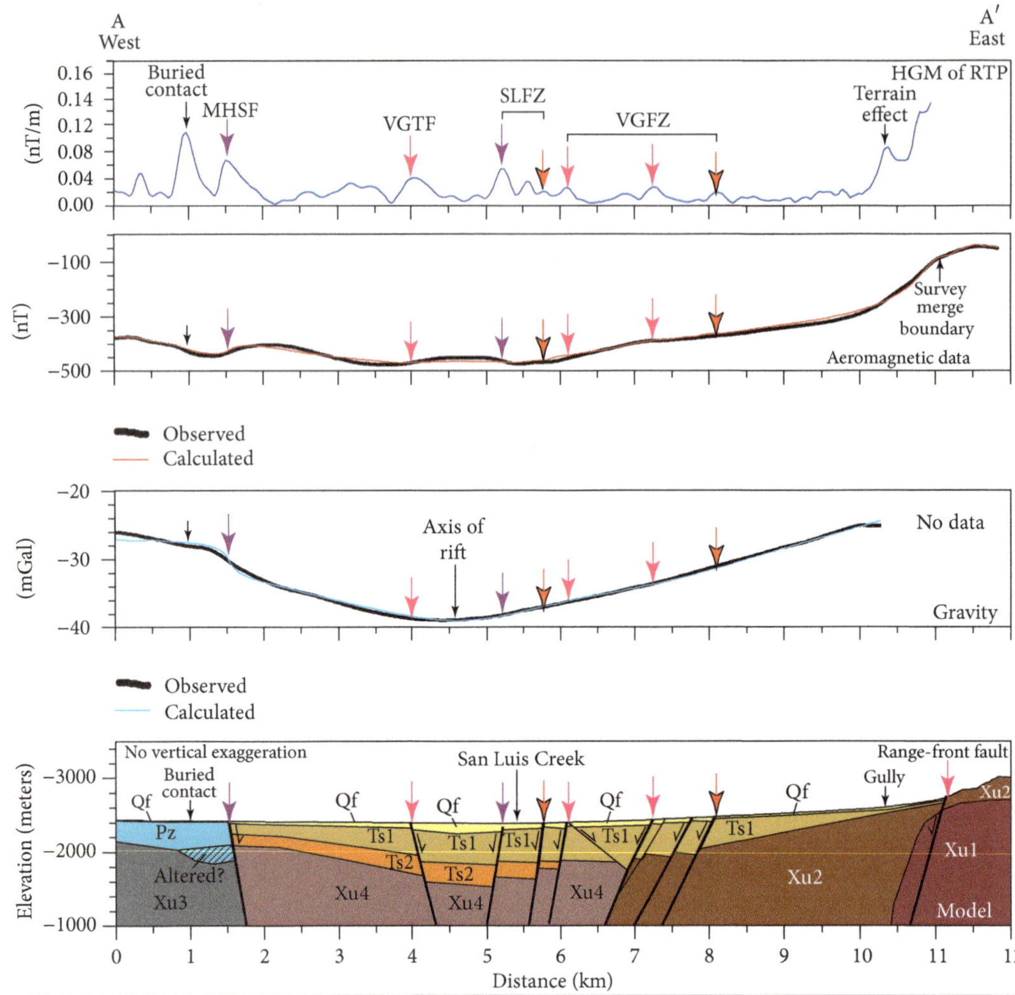

	Label	Magnetic suscept. (SI)	Density (kg/m³)	Likely geologic unit(s) from Figure 1
Rift and younger sediments	Qf	0.0060	2170	Qh, Qfy, Qfi, Qfo undivided
	Ts1	0.0005	2170	Ts (less magnetic)
	Ts2	0.0020	2170	Ts (more magnetic)
Prerift rocks	Pz	0.0000	2670	Pz
	Altered?	0.0000	2670	Pz, or altered Pz or Xu
Precambrian basement rocks	Xu1	0.025	2700	Xu (more magnetic, dense)
	Xu2	0.015	2700	Xu (magnetic, dense)
	Xu3	0.015	2670	Xu (magnetic, less dense)
	Xu4	0.010	2670	Xu (less magnetic, less dense), possibly Tv

FIGURE 16: 2D geophysical model for profile A-A′ (located in Figures 2–4, 6, and 15). Observed curves were extracted from the grids for reduced-to-pole aeromagnetic data (Figure 2), isostatic residual gravity (Figure 4), and HGM of the reduced-to-pole aeromagnetic data (HGM of RTP) (Figure 6). MHSF: Mineral Hot Springs fault (revised), SLFZ: San Luis Creek fault zone, VGFZ: Villa Grove fault zone, and VGTF: Villa Grove town fault. Colored arrows point to the locations of inferred faults from Figure 15, coded in a similar fashion: magenta indicates active (younger), red (with black outline) active (older), and purple potentially active.

of gravity values across the Villa Grove fault zone. These relations support geologic evidence that the Villa Grove fault zone represents a wide zone of small displacement faults, primarily down to the southwest along the eastern margin of the rift [8, 10, 21].

7.2. Fault Patterns in the Valley. The HGM ridges from the aeromagnetic data prominently define faults that closely follow creeks in the valley (Figure 6), confirming previous speculation that steep banks along the eastern sides of Rock Creek and San Luis Creek are tectonic rather than fluvial in origin [10]. In addition to the arguments presented for a tectonic origin in regards to example Area 2, further evidence is provided by the linearity of HGM ridges in other parts of the study area that traverse subdued sections of the stream bank, or entirely underlie the creek itself, as demonstrated north of the confluence of San Luis and Rock Creeks (Figure 6). Thus, we use both the LiDAR and aeromagnetic data together to define the generally north-south San Luis Creek fault zone south of the Villa Grove fault zone, and the northwest-striking Rock Creek fault zone to the north, following Rock Creek upstream to where it bends as it flows from the mountains into the main stream valley (Figure 15).

The Villa Grove town fault should not be considered part of the Villa Grove fault zone, despite previous compilations that list them together (e.g., [8]). A strong signal in the aeromagnetic data (including a strong HGM peak) suggests it has larger displacement than do faults of the Villa Grove fault zone (Figures 2, 6, 12, and 16). Its northeast dip and location on the west side of the gravity low suggest it is part of the western margin of the rift basin (Figure 16). The San Luis Creek fault zone displays multiple splays where it is intersected obliquely by the Villa Grove fault zone and is offset by it (Figure 12), making the San Luis Creek fault an older active system at the eastern margin of the rift basin. It has larger displacement of sediments than the Villa Grove fault zone (as evidenced by stronger aeromagnetic signal), but has little effect on the gravity profile (Figure 16), suggesting that displacement on prerift bedrock associated with these faults is accommodated across multiple faults rather than on one master fault. On the other hand, the western rift margin appears to be focused at the revised Mineral Hot Springs fault (MHSF) (Figure 16). It is associated with a significant drop in gravity values, which signifies a normal offset of Paleozoic and Proterozoic rocks exposed on the west into the rift basin on the east. Decreases in both magnetic and gravity values just west of the Mineral Hot Springs fault were difficult to model geologically. Altered rock that has lost its magnetization is one possible explanation and is compatible with possible explanations for the aeromagnetic low over the hot springs west of the fault to the south (Figure 13).

Fault patterns that generally outline and follow gradients that define gravity lows in the study area support the use of the lows as proxies for defining the shapes of the rift basin (Figure 17). For example, the Villa Grove town fault (VGTF) and the Rock Creek fault zone (RCFZ) nicely outline the eastern and western sides of the gravity low in the center of the study area. The narrowing of this gravity low where it is crossed by profile A-A$'$ before it widens and deepens to the south appears to be controlled by the Villa Grove fault zone (VGFZ). The Villa Grove fault zone also follows the northwest-southeast trend of the broader flanks of the southern gravity low to the intersection with the Sangre de Cristo fault at the southeastern corner of the study area. The widely distributed, small displacement of the faults of the Villa Grove fault zone may explain why its northwestern extension does not appear to affect the shape of the gravity low as it traverses obliquely to meet the northern part of the Villa Grove town fault. However, this northwestward extension parallels the broad bend in San Luis Creek north of the town of Villa Grove that in turn follows the curve of a broad gradient in the aeromagnetic data (Figure 2). The broad aeromagnetic gradient is indicative of large-volume magnetic sources, such as the juxtaposition of underlying Proterozoic and Paleozoic rocks. The similarities between the stream, Quaternary faults, and broad patterns in the aeromagnetic data suggest that the configuration of prerift rocks may have had some influence on the structural patterns.

Whereas the Rock Creek fault zone (RCFZ) follows the eastern margin of the gravity low, the San Luis Creek fault zone (SLFZ) enigmatically traverses across the deepest part of the southern gravity low, apparently crossing the middle of the rift basin rather than following its border. This relation suggests several possibilities: (1) the rift basin was primarily shaped by early rift activity, (2) the gravity low is partially reflecting underlying thickening of Paleozoic rocks, or (3) prerift structure influenced the subsequent location of the fault and the creek. For the first possibility, perhaps the San Luis Creek fault zone originally connected to the revised Mineral Hot Springs fault on the south, forming the tips of an older half graben that has since been widened by the activity on the Villa Grove fault zone. However, contribution to the gravity low from a greater thickness of Paleozoic rocks in this area is also a distinct possibility, as suggested by seismic-reflection sections across the central part of the gravity low [45]. Evaluation of these possibilities is part of future modeling efforts.

7.3. Prerift Structure or Active Fault? The Rock Creek fault zone is intersected at its northern end, at the bend in the creek, by a younger active fault that traverses obliquely across the valley from the west side of San Luis Creek (Figure 15). The younger active fault (A on Figure 8) aligns with the north-northeast orientation of northern Rock Creek (NRC) in the northeast part of the study area (Figure 15). A prominent north-south aeromagnetic anomaly emanating from the range front follows the northern part of Rock Creek (Figure 2). The magnetic source of the anomaly appears to be fault bounded on its eastern side along a north-south linear gradient and faulted along its southern tip, because the anomaly appears terminated. Although the source of this anomaly likely arises from the Proterozoic basement and some of the faulting may be related to the prerift Laramide orogeny, its alignment with an active younger fault suggests it may have a role in active rifting. Several other enigmatic, northeast-trending linear anomalies that abut the range front obliquely on Figure 2 are also likely related to prerift, probably Laramide, structures. However, they appear to be inactive. Alternatively, these anomalies may arise from dikes present within the bedrock. The presence of dikes intruding rift sediments is unlikely because no magmatism subsequent to the prerift San Juan volcanic field has been recognized in the study area (Figure 1).

7.4. Rift Evolution. Implications for the evolution of the Rio Grande rift in the study area are newly apparent from the faults inferred from the combined interpretation (Figure 15). Potentially active faults are considered the oldest and follow northwest trends on either side of the valley. Faults that have apparently not been active since 120 ka (older active faults) are concentrated in the central part of the valley, and they are crossed by faults with more recent activity (younger active faults). An east-migrating sequence of ages is represented in the example for Area 1 (Figure 8), which is not depicted in the generalized age classification of Figure 15: age constraints

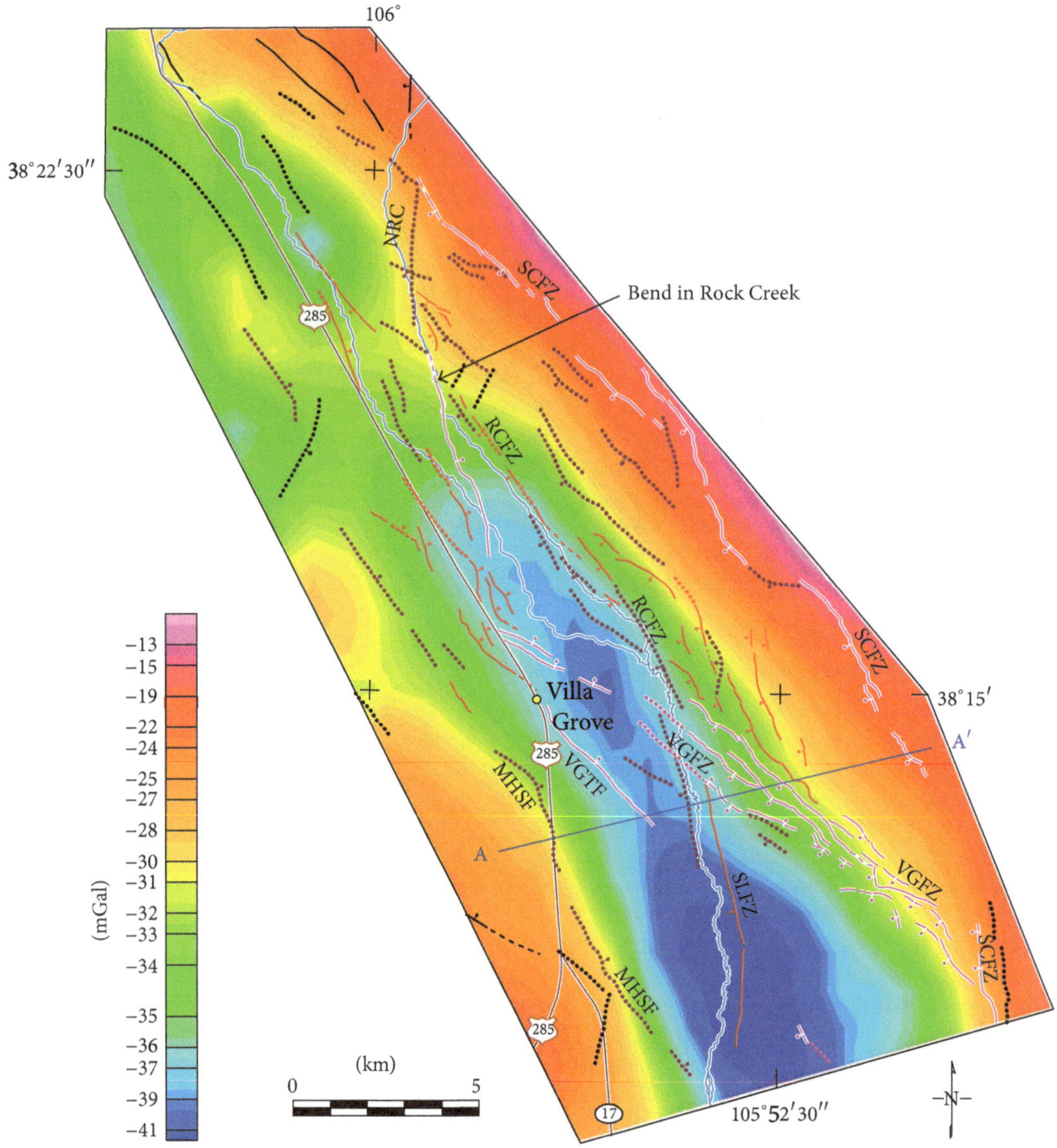

FIGURE 17: Active faults of Figure 15 overlain on the gravity data of Figure 4, a proxy for viewing rift basin shape. MHSF: Mineral Hot Springs fault (revised), NRC: Northern Rock Creek, RCFZ: Rock Creek fault zone, SCFZ: Sangre de Cristo fault zone, SLFZ: San Luis Creek fault zone, VGFZ: Villa Grove fault zone, and VGTF: Villa Grove town fault. Model for A-A′ is shown in Figure 16.

suggest the most recently active fault was on the east side of Area 1 (fault A), preceded by fault B, then C and E (unknown order). Fault D is even older and is classified as potentially active. The oldest interpreted rift fault is concealed, evident from the 2D geophysical model for Area 1 between B and C (Figure 9).

Most recent ruptures on the younger active faults were concentrated on the Sangre de Cristo and Villa Grove fault zones. Although these faults record the most recent ruptures (<30 ka), multiple events recorded on both fault zones suggest they have been active since ~400 ka [5]. Overall, the fault patterns, even those with most recent activity, show an influence from older structure, yet to be fully understood.

8. Conclusions

Combined interpretation of aeromagnetic and LiDAR data builds on the strength of the aeromagnetic method to locate normal faults with significant offset under cover and the strength of using LiDAR images to identify the age and sense of motion of faults. Each dataset helps resolve ambiguities in interpreting the other. In addition, gravity data can be used to infer the sense of motion for totally buried faults inferred solely from aeromagnetic data.

The most common benefits from combining the methods are as follows:

(i) results from aeromagnetic interpretation help guide the proper connection of fault traces between discontinuous scarps apparent in the LiDAR data and help resolve tectonic versus fluvial origin for topographic scarps next to streams;

(ii) results from LiDAR interpretation help locate the surface fault trace from multiple or ambiguous peaks in aeromagnetic gradient data;

(iii) identification of fault scarps from the LiDAR data provide evidence that an inferred fault is active;

(iv) orientations of fault scarp faces from the LiDAR data suggest the sense of displacement for inferred faults;

(v) patterns of active rift faults or locations of basin margins determined from gravity data provide guidance for determining the relative age of completely buried, aeromagnetically inferred faults from their locations and strike directions.

Combined interpretation to identify active faults in the northern San Luis Basin of the northern Rio Grande rift has confirmed general aspects of previous geologic mapping but has also provided significant improvements. The interpretation revises and extends mapped fault traces, confirms tectonic versus fluvial origins of steep stream banks, and gains additional information on the nature of active and potentially active partially and totally buried faults. In general, tectonics appears to have a large control on the geomorphology. Tectonic activity was concentrated in the central part of the valley throughout most of rift history. Activity at the range front began fairly late and has produced little displacement so far (~100 m).

Detailed morphology of surfaces mapped from the LiDAR data helps constrain ages of the faults that displace the deposits. The aeromagnetic data provide additional information about their extents in between discontinuous fault scarps and suggest several totally buried, potentially active faults are present on both sides of the valley, surrounding gravity lows that likely represent the shape of the rift basin. An enigmatic linear aeromagnetic anomaly in line with a younger active fault suggests that prerift structure may have a role in rifting.

Acknowledgments

The authors are grateful to Ted Brandt (USGS) who developed the LiDAR images and Ben Drenth (USGS) who prepared the gravity data and shared his preliminary modeling with them. They appreciate reviews by Jonathan Glen (USGS) and two journal reviewers, which greatly improved the paper. This project was supported by the National Cooperative Geologic Mapping Program. Aeromagnetic data were partially supported by the Colorado Geological Survey, USGS Geothermal Program, and the Task Force for Stability and Business Operations. LiDAR acquisition was part of a coalition of Federal and State partners, coordinated by Carol Giffin (USGS). Any use of trade, product, or firm names in this publication is for descriptive purposes only and does not imply endorsement by the U.S. Government. Authors have no affiliation with nor financial interest in EONGeosciences company.

References

[1] R. J. Blakely, B. L. Sherrod, J. F. Hughes, M. L. Anderson, R. E. Wells, and C. S. Weaver, "Saddle Mountain fault deformation zone, Olympic Peninsula, Washington: Western boundary of the Seattle uplift," *Geosphere*, vol. 5, no. 2, pp. 105–125, 2009.

[2] R. J. Blakely, B. L. Sherrod, C. S. Weaver et al., "Connecting the Yakima fold and thrust belt to active faults in the Puget Lowland, Washington," *Journal of Geophysical Research B*, vol. 116, no. 7, Article ID B07105, 2011.

[3] V. J. S. Grauch and M. R. Hudson, "Guides to understanding the aeromagnetic expression of faults in sedimentary basins: lessons learned from the central Rio Grande Rift, New Mexico," *Geosphere*, vol. 3, no. 6, pp. 596–623, 2007.

[4] R. M. Kirkham and W. P. Rogers, "Earthquake potential in Colorado," Colorado Geological Survey Bulletin 43, 1981, http://geosurvey.state.co.us/pubs/online/Documents/1981%20Bulletin%2043.pdf.

[5] J. P. McCalpin, "Quaternary tectonics of the Sangre de Cristo and Villa Grove fault zones," in *Contributions to Colorado Seismicity and Tectonics, a 1986 Update*, W. P. Rogers and R. M. Kirkham, Eds., Colorado Geological Survey Special Publication 28, pp. 59–64, 1986.

[6] R. M. Kirkham, Compiler, "Fault number 2320, Mineral Hot Springs fault," in *Quaternary fault and fold database of the United States*, U.S. Geological Survey, 1998, http://earthquakes.usgs.gov/regional/qfaults.

[7] R. M. Kirkham, Compiler, "Fault number 2321b, Northern Sangre de Cristo fault, Crestone section," in *Quaternary fault and fold database of the United States*, U.S. Geological Survey, 1998, http://earthquakes.usgs.gov/regional/qfaults.

[8] R. M. Kirkham, Compiler, "Fault number 2319, Villa Grove fault zone," in *Quaternary fault and fold database of the United States*, U.S. Geological Survey, 1998, http://earthquakes.usgs.gov/regional/qfaults.

[9] Colorado Earthquake Hazards Mitigation Council, *Colorado Earthquake Hazards*, Colorado Geological Survey, 2008, http://geosurvey.state.co.us/hazards/Earthquakes/Documents/Earthquake_Map_2008.pdf.

[10] C. Ruleman and M. N. Machette, "An overview of the Sangre de Cristo fault system and new insights to interactions between Quaternary faults in the northern Rio Grande rift," in *2007 Rocky Mountain Section Friends of the Pleistocene Field Trip—Quaternary Geology of the San Luis Basin of Colorado and New Mexico, September 7–9, 2007*, M. N. Machette, M. M. Coates, and M. L. Johnson, Eds., U.S. Geological Survey Open-File Report 2007-1193, pp. 187–197, 2007, http://pubs.usgs.gov/of/2007/1193/pdf/OF07-1193_ChJ.pdf.

[11] G. R. Keller and W. S. Baldridge, "The Rio Grande rift: a geological and geophysical overview," *Rocky Mountain Geology*, vol. 34, no. 1, pp. 121–130, 1999.

[12] C. F. Kluth and C. H. Schaftenaar, "Depth and geometry of the northern Rio Grande rift in the San Luis Basin, south-central Colorado," in *Basins of the Rio Grande Rift: Structure, Stratigraphy, and Tectonic Setting*, G. R. Keller and S. M. Cather, Eds., Geological Society of America Special Paper 291, pp. 27–37, 1994.

[13] D. A. Lindsey, R. F. Clark, and S. J. Soulliere, "Minturn and Sangre de Cristo Formations of southern Colorado: a prograding fan delta and alluvial fan sequence shed from the Ancestral Rocky Mountains," in *Paleotectonics and Sedimentation*, J. A. Peterson, Ed., American Association of Petroleum Geologists Memoir 41, pp. 541–561, 1986.

[14] D. A. Lindsey, B. R. Johnson, and P. A. M. Andriessen, "Laramide and Neogene structure of the northern Sangre de Cristo Range, South-Central Colorado," in *Rocky Mountain Foreland Basins and Uplifts*, J. D. Lowell, Ed., pp. 219–228, Rocky Mountain Association of Geologists, Denver, Colo, USA, 1983.

[15] B. S. Brister and R. R. Gries, "Tertiary stratigraphy and tectonic development of the Alamosa basin (northern San Luis Basin), Rio Grande rift, south-central Colorado," in *Basins of the Rio Grande Rift: Structure, Stratigraphy, and Tectonic Setting*, G. R. Keller and S. M. Cather, Eds., Geological Society of America Special Paper 291, pp. 39–58, 1994.

[16] R. J. Varga and B. M. Smith, "Evolution of the early Oligocene Bonanza caldera, northeast San Juan volcanic field, Colorado," *Journal of Geophysical Research*, vol. 89, no. 10, pp. 8679–8694, 1984.

[17] P. W. Lipman and W. C. McIntosh, "Erosionally exhumed fill and floor of a ring-fault-bounded plate-subsidence caldera: 33. 2-Ma Bonanza, Southwest Colorado," *Eos Transactions of the American Geophysical Union*, vol. 90, no. 52, 2009, Fall Meeting Supplement, Abstract V23C-2080.

[18] R. A. Thompson, C. M. Johnson, and H. H. Mehnert, "Oligocene basaltic volcanism of the northern Rio Grande Rift: San Luis Hills, Colorado," *Journal of Geophysical Research*, vol. 96, no. 8, pp. 13577–13592, 1991.

[19] C. E. Chapin and S. Cather, "Tectonic setting of the axial basins of the northern and central Rio Grande rift," in *Basins of the Rio Grande Rift: Structure, Stratigraphy, and Tectonic Setting*, G. R. Keller and S. M. Cather, Eds., Geological Society of America Special Paper 291, pp. 5–23, 1994.

[20] H. T. Berglund, A. F. Sheehan, M. H. Murray et al., "Distributed deformation across the Rio Grande Rift, Great Plains, and Colorado Plateau," *Geology*, vol. 40, no. 1, pp. 23–26, 2012.

[21] J. McCalpin, "Quaternary geology and neotectonics of the west flank of the Northern Sangre de Cristo Mountains, South-Central Colorado," *Colorado School of Mines Quarterly*, vol. 77, no. 3, p. 97, 1982.

[22] J. R. Shannon and J. P. McCalpin, "Geologic map of the Maysville quadrangle, Chaffee County, Colorado," Colorado Geological Survey Open-File Report 06-10, 2009, http://geosurveystore.state.co.us/p-935-maysville-quadrangle-chaffee-county-colorado.aspx.

[23] O. Tweto, "The Rio Grande rift system in Colorado," in *Rio Grande Rift: Tectonics and Magmatism*, R. E. Riecker, Ed., pp. 33–56, American Geophysical Union, Washington, DC, USA, 1979.

[24] S. M. Colman, J. P. McCalpin, D. A. Ostenaa, and R. M. Kirkham, "Map showing upper Cenozoic rocks and deposits and Quaternary faults, Rio Grande rift, South-Central Colorado," U.S. Geological Survey Miscellaneous Investigations Map I-1594, 2 sheets, scale 1:125, 000, 1985, http://pubs.er.usgs.gov/publication/i1594.

[25] M. N. Machette, R. A. Thompson, and B. J. Drenth, "Geologic map of the San Luis quadrangle, Costilla County, Colorado," U.S. Geological Survey Scientific Investigations Map 2963, 1 sheet, scale 1:24, 000, 2008, http://pubs.usgs.gov/sim/2963/.

[26] K. A. Brugger, "Cosmogenic ^{10}Be and ^{36}Cl ages from Late Pleistocene terminal moraine complexes in the Taylor River drainage basin, central Colorado, USA," *Quaternary Science Reviews*, vol. 26, no. 3-4, pp. 494–499, 2007.

[27] K. S. Kellogg, R. R. Shroba, B. Bryant, and W. R. Premo, "Geologic map of the Denver West 30′ x 60′ quadrangle, north-central Colorado," U.S. Geological Survey Scientific Investigations Map 3000, 2008, http://pubs.usgs.gov/sim/3000/.

[28] J. P. Briner, "Moraine pebbles and boulders yield indistinguishable ^{10}Be ages: a case study from Colorado, USA," *Quaternary Geochronology*, vol. 4, no. 4, pp. 299–305, 2009.

[29] C. A. Ruleman, R. G. Bohannon, B. Bryant, R. R. Shroba, and W. R. Premo, "Geologic map of the Bailey 30′ x 60′ Quadrangle, Colorado," U.S. Geological Survey Scientific Investigations Map 3156, scale 1:100, 000, 2011, http://pubs.usgs.gov/sim/3156/.

[30] N. E. Young, J. P. Briner, E. M. Leonard, J. M. Licciardi, and K. Lee, "Assessing climatic and nonclimatic forcing of Pinedale glaciation and deglaciation in the western United States," *Geology*, vol. 39, no. 2, pp. 171–174, 2011.

[31] C. A. Ruleman, R. A. Thompson, R. R. Shroba et al., "Late Miocene-Pleistocene evolution of a Rio Grande rift subbasin, Sunshine Valley-Costilla Plain, San Luis Basin, New Mexico and Colorado," in *New Perspectives on Basins of the Rio Grande Rift: From Tectonics to Groundwater*, M. R. Hudson and V. J. S. Grauch, Eds., Geological Society of America Special Paper 494, pp. 47–73, 2013.

[32] K. S. Kellogg, "Neogene basins of the northern Rio Grande rift: partitioning and asymmetry inherited from Laramide and older uplifts," *Tectonophysics*, vol. 305, no. 1–3, pp. 141–152, 1999.

[33] D. H. Knepper Jr., *Tectonic analysis of the Rio Grande rift zone, central Colorado [Ph.D. thesis]*, Colorado School of Mines, Golden, Colo, USA, 1974.

[34] D. A. Lindsey and S. J. Soulliere, "Geologic map and sections of the Valley View Hot Springs Quadrangle, Custer and Saguache counties, Colorado," U.S. Geological Survey Miscellaneous Field Studies Map, MF-1942, scale 1:24, 000, 1987.

[35] G. R. Scott, R. B. Taylor, R. C. Epis, and R. A. Wobus, "Geologic map of the Pueblo 1 x 2 quadrangle, South-Central Colorado," U.S. Geological Survey Miscellaneous Investigations Map I-1022, scale 1:250, 000, 1978, http://pubs.er.usgs.gov/publication/i1022.

[36] R. E. Van Alstine, "Geologic map of the Bonanza NE quadrangle, Chaffee and Saguache Counties, Colorado," U.S. Geological Survey Open-File Report 75-53, scale 1:62, 500, 1975, http://ngmdb.usgs.gov/ngm-bin/ILView.pl?sid=15201_1.sid&vtype=b&sfact=1.5.

[37] J. A. Cappa and C. A. Wallace, "Saguache county, geology and mineral resources," Colorado Geological Survey Resource Series 44, CD-ROM, 2007, http://geosurveystore.state.co.us/p-1026-saguache-county-geology-and-mineral-resources.aspx.

[38] C. C. Finlay, S. Maus, C. D. Beggan et al., "International geomagnetic reference field: the eleventh generation," *Geophysical Journal International*, vol. 183, no. 3, pp. 1216–1230, 2010.

[39] U. S. Geological Survey and National Geophysical Data Center, "Digital aeromagnetic datasets for the conterminous United States and Hawaii—a companion to the North American magnetic anomaly map," U.S. Geological Survey Open-File Report 02-361, version 1. 0, project 4068, 2002, http://pubs.usgs.gov/of/2002/ofr-02-361/.

[40] J. E. Case and R. F. Sikora, "Geologic interpretation of gravity and magnetic data in the Salida region, Colorado," U.S. Geological Survey Open-File Report 84-372, 1984, http://pubs.usgs.gov/of/1984/0372/report.pdf.

[41] Pan-American Center for Earth and Environmental Studies (PACES), "Gravity data portal for the United States," University

of Texas, El Paso, Tex, USA, 2009, http://irpsrvgis00.utep.edu/repositorywebsite/.

[42] C. E. Heywood, "Isostatic residual gravity anomalies of New Mexico," U. S. Geological Survey, Water-Resources Investigations Report 91-4065, 1992, http://pubs.er.usgs.gov/publication/wri914065.

[43] V. J. S. Grauch and S. D. Connell, "New perspectives on the geometry of the Albuquerque Basin, Rio Grande Rift, New Mexico—insights from geophysical models of rift-fill thickness," in *New Perspectives on Rio Grande Rift Basins: From Tectonics to Groundwater*, M. R. Hudson and V. J. S. Grauch, Eds., Geological Society of America Special Paper 494, pp. 427–462, 2013.

[44] D. Stoughton, *Interpretation of seismic reflection data from the San Luis Valley, south-central Colorado [Unpublished PhD thesis T-1960]*, Colorado School of Mines, Golden, Colo, USA, 1977.

[45] T. L. Davis and D. Stoughton, "Interpretation of seismic reflection data from the northern San Luis Valley, South-Central Colorado," in *Rio Grande Rift: Tectonics and Magmatism*, R. E. Riecker, Ed., pp. 185–194, American Geophysical Union, Washington, DC, USA, 1979.

[46] M. R. Hudson, V. J. S. Grauch, and S. A. Minor, "Rock magnetic characterization of faulted sediments with associated magnetic anomalies in the Albuquerque Basin, Rio Grande rift, New Mexico," *Geological Society of America Bulletin*, vol. 120, no. 5-6, pp. 641–658, 2008.

[47] V. J. S. Grauch and M. R. Hudson, "Aeromagnetic anomalies over faulted strata," *The Leading Edge*, vol. 30, no. 11, pp. 1242–1252, 2011.

[48] L. Cordell and V. J. S. Grauch, "Mapping basement magnetization zones from aeromagnetic data in the San Juan Basin, New Mexico," in *The Utility of Regional Gravity and Magnetic Anomaly Maps*, W. J. Hinze, Ed., pp. 181–197, Society of Exploration Geophysicists, Tulsa, Okla, USA, 1985.

[49] R. J. Blakely and R. W. Simpson, "Approximating edges of source bodies from magnetic or gravity anomalies," *Geophysics*, vol. 51, no. 7, pp. 1494–1498, 1986.

[50] V. J. S. Grauch and C. S. Johnston, "Gradient window method: a simple way to separate regional from local horizontal gradients in gridded potential-field data," *Society of Exploration Geophysicists Technical Program Expanded Abstracts*, pp. 762–765, 2002.

[51] R. C. Bucknam and R. E. Anderson, "Estimation of fault-scarp ages from scarp height-slope-angle relationships," *Geology*, vol. 7, pp. 11–14, 1979.

[52] M. N. Machette, "Contrasts between short- and long-term records of seismicity in the Rio Grande rift—important implications for seismic hazard assessments in areas of slow extension," in *Western States Seismic Policy Council Proceedings Volume, Basin and Range Province Seismic Hazards Summit*, W. R. Lund, Ed., Utah Geological Survey Miscellaneous Publication 98–2, pp. 84–95, Utah Geological Survey, 1998.

[53] A. Valdez, "Stop A8—closed basin overflow and origin of Hansen Bluff," in *2007 Rocky Mountain Section Friends of the Pleistocene Field Trip—Quaternary Geology of the San Luis Basin of Colorado and New Mexico, September 7–9, 2007*, M. N. Machette, M. M. Coates, and M. L. Johnson, Eds., U.S. Geological Survey Open-File Report 2007-1193, pp. 38–39, 2007, http://pubs.usgs.gov/of/2007/1193/pdf/OF07-1193_ChA.pdf.

[54] W. B. Bull, "The alluvial fan environment," *Progress in Physical Geography*, vol. 1, no. 2, pp. 222–270, 1977.

[55] O. Tweto and J. E. Case, *Gravity and Magnetic Features As Related to Geology in the Leadville 30-Minute Quadrangle, Colorado*, U.S. Geological Survey Professional Paper 726-C, 1972.

[56] J. D. Phillips, R. O. Hansen, and R. J. Blakely, "The use of curvature in potential-field interpretation," *Exploration Geophysics*, vol. 38, no. 2, pp. 111–119, 2007.

[57] B. R. Johnson, D. A. Lindsey, R. M. Bruce, and S. J. Soulliere, "Reconnaissance geologic map of the Sangre de Cristo Wilderness Study Area, South-Central Colorado," U.S. Geological Survey Miscellaneous Field Studies Map MF-1635-B, scale 1:62, 500, 1987.

Permissions

The contributors of this book come from diverse backgrounds, making this book a truly international effort. This book will bring forth new frontiers with its revolutionizing research information and detailed analysis of the nascent developments around the world.

We would like to thank all the contributing authors for lending their expertise to make the book truly unique. They have played a crucial role in the development of this book. Without their invaluable contributions this book wouldn't have been possible. They have made vital efforts to compile up to date information on the varied aspects of this subject to make this book a valuable addition to the collection of many professionals and students.

This book was conceptualized with the vision of imparting up-to-date information and advanced data in this field. To ensure the same, a matchless editorial board was set up. Every individual on the board went through rigorous rounds of assessment to prove their worth. After which they invested a large part of their time researching and compiling the most relevant data for our readers. Conferences and sessions were held from time to time between the editorial board and the contributing authors to present the data in the most comprehensible form. The editorial team has worked tirelessly to provide valuable and valid information to help people across the globe.

Every chapter published in this book has been scrutinized by our experts. Their significance has been extensively debated. The topics covered herein carry significant findings which will fuel the growth of the discipline. They may even be implemented as practical applications or may be referred to as a beginning point for another development. Chapters in this book were first published by Hindawi Publishing Corporation; hereby published with permission under the Creative Commons Attribution License or equivalent.

The editorial board has been involved in producing this book since its inception. They have spent rigorous hours researching and exploring the diverse topics which have resulted in the successful publishing of this book. They have passed on their knowledge of decades through this book. To expedite this challenging task, the publisher supported the team at every step. A small team of assistant editors was also appointed to further simplify the editing procedure and attain best results for the readers.

Our editorial team has been hand-picked from every corner of the world. Their multi-ethnicity adds dynamic inputs to the discussions which result in innovative outcomes. These outcomes are then further discussed with the researchers and contributors who give their valuable feedback and opinion regarding the same. The feedback is then collaborated with the researches and they are edited in a comprehensive manner to aid the understanding of the subject.

Apart from the editorial board, the designing team has also invested a significant amount of their time in understanding the subject and creating the most relevant covers. They scrutinized every image to scout for the most suitable representation of the subject and create an appropriate cover for the book.

The publishing team has been involved in this book since its early stages. They were actively engaged in every process, be it collecting the data, connecting with the contributors or procuring relevant information. The team has been an ardent support to the editorial, designing and production team. Their endless efforts to recruit the best for this project, has resulted in the accomplishment of this book. They are a veteran in the field of academics and their pool of knowledge is as vast as their experience in printing. Their expertise and guidance has proved useful at every step. Their uncompromising quality standards have made this book an exceptional effort. Their encouragement from time to time has been an inspiration for everyone.

The publisher and the editorial board hope that this book will prove to be a valuable piece of knowledge for researchers, students, practitioners and scholars across the globe.

List of Contributors

Cesar E. Valladares
Institute for Scientific Research, Boston College, Chestnut Hill, MA 02467, USA

Matthew A. Hei
Plasma Physics Division, Naval Research Laboratory, Washington, DC 20375, USA

Tengfei Ma and Zhongliang Wu
Institute of Geophysics, China Earthquake Administration, Beijing 100081, China

G. Colangelo
Department of Infrastructure and Civil Protection, Basilicata Region, C.so Garibaldi 139, 85100 Potenza, Italy

A. Perrone
Institute of Methodologies for Environmental Analysis, CNR, C.da S. Loja, 85050 Tito Scalo, Italy

Guiling Wang and Clement A. Alo
Department of Civil and Environmental Engineering, University of Connecticut, 261 Glenbrook Road, Storrs, CT 06269, USA

Ibrahima Kalil Kante, Daouda Badiane, Saïdou Moustapha Sall, Abdoulaye Deme and Arona Diedhiou
Laboratoire de Physique de l'Atmosph`ere et de l'Oc´ean Sim´eon Fongang, Ecole Sup´erieure Polytechnique, Universit´e Cheikh, Anta Diop Dakar-S´en´egal, BP: 5085, Dakar, Senegal

Ibrahima Kalil Kante
Direction Nationale de la M´et´eorologie de Guin´ee-Conakry, BP: 566 Rep., Guinea

Arona Diedhiou
IRD/LTHE, BP 53, 38041 Grenoble Cedex 9, France

Javed N. Malik, Ashutosh Kumar, Sravanthi Satuluri, Bishuddhakshya Puhan, and Asmita Mohanty
Department of Civil Engineering, Indian Institute of Technology Kanpur, Kanpur 208016, India

Masashi Kawamura
Department of Earth Sciences and Graduate Institute of Geophysics, National Central University, Jhongda Road no. 300, Taoyuan, Jhongli 32001, Taiwan

Takeshi Kudo
General Education Division, College of Engineering, Chubu University, Matsumoto-cho 1200, Aichi, Kasugai 487-8501, Japan

Koshun Yamaoka
Research Center for Seismology, Volcanology and Disaster Mitigation, Graduate School of Environmental Studies, Nagoya University, Chikusa-ku, Nagoya 464-8601, Japan

A. J. Komkoua Mbienda, D. A. Vondou and F. Mkankam Kamga
Laboratory of EnvironmentalModeling and Atmospheric Physics, Department of Physics, Faculty of Science, University of Yaounde 1, P.O. Box 812, Yaounde, Cameroon

C. Tchawoua
Laboratory of Mechanics, Department of Physics, Faculty of Science, University of Yaounde 1, P.O. Box 812, Yaounde, Cameroon

D. Tapete
Department of Earth Sciences, University of Florence, Via G. La Pira 4, 50121 Florence, Italy

F. Cigna
British Geological Survey, Natural Environment Research Council, Nicker Hill, Keyworth, Nottingham NG12 5GG, UK

Nana A. K. Browne
Dep ylla2artment of Physics, University of Cape Coast, Cape Coast, Ghana

Mouhamadou B. Sylla
Earth System Physics (ESP) Section, International Centre for Theoretical Physics (ICTP), 34151 Trieste, Italy

Tada-nori Goto
Graduate School of Engineering, Kyoto University, C1-1-113, Kyodai-Katsura, Nishikyo, Kyoto 615-8540, Japan

Kazuya Kondo, Rina Ito, Keisuke Esaki and Yasuo Oouchi
Department of Environmental Earth Sciences, Aichi University of Education, 1 Hirosawa, Igaya-cho, Kariya, Aichi 448-8542, Japan

Yutaka Abe and Maki Tsujimura
Graduate School of Life and Environmental Sciences, University of Tsukuba, 1-1-1 Tennodai,

Tsukuba 305-8572, Japan

Frédéric Ouattara
Ecole Normale Sup´erieure de l'Universit´e de Koudougou, P.O. Box 376, Koudougou, Burkina Faso

Doua Allain Gnabahou
Lyc´ee Provincial de Koudougou, Direction R´egional du Centre Ouest, P.O. Box 39, Koudougou, Burkina Faso

Christine Amory Mazaudier
LPP-UPMC, CNRS, 4 Avenue de Neptune, 94107 Saint Maur des Foss´es, Paris, France

R. R. Ilma and M. C. Kelley
School of Electrical and Computer Engineering, Cornell University, Ithaca, NY 14853, USA

C. A. Gonzales
IBM Thomas J. Watson Research Center, Yorktown Heights, NY 10598, USA

L. Jouniaux
Institut de Physique du Globe de Strasbourg, UdS-CNRS UMR 7516, Universit´e de Strasbourg, 5 rue Ren´e Descartes, 67084 Strasbourg, France

C. Bordes
Laboratoire des Fuides Complexes et leurs R´eservoirs, UPPA-CNRS UMR5150, Universit´e de Pau et des Pays de l'Adour,64013 Pau, France

Antonella Guida, Antonello Pagliuca, and Alessandro Tranquillino Minerva
Department of European and Mediterranean Cultures, School of Architecture, University of Basilicata,
Via Lazazzera, 75100 Matera, Italy

Mojtaba Naghavi and Zaher Hossein Shomali
Institute of Geophysics, University of Tehran, Tehran, 14155-6466, Iran

Mehdi Zare
International Institute of Earthquake Engineering and Seismology (IIEES), Tehran, 19395-3913, Iran

Iginio Marson
Department of Civil Engineering and Architecture, University of Trieste, Via Valerio 6, 34127 Trieste, Italy

Baljeet Singh
Department of Mathematics, Post Graduate Government College, Sector 11, Chandigarh 160 011, India

Ritu Sindhu and Jagdish Singh
Department of Mathematics, Maharshi Dayanand University, Rohtak, 124 001 Haryana, India

Shemer Keydar, Benjamin Medvedev and Michael Ezersky
Project Department, The Geophysical Institute of Israel, Lod 71100, Israel

Benjamin Medvedev
Schlumberger, Via dell'Unione Europea, 4 SanDonatoMilanese, Milan, Italy

Abdallah Al-Zoubi and Emad Akkawi
Engineering Faculty, Al-Balqa' Applied University, Salt 19117, Jordan

Y. Otsuka
Solar-Terrestrial Environment Laboratory, Nagoya University, Nagoya, Japan

Giacomo Di Giacomo and Giuseppe Scardozzi
Istituto per i Beni Archeologici e Monumentali (IBAM), Consiglio Nazionale delle Ricerche (CNR), Strada per Monteroni c/o Campus Universitario, 73100 Lecce, Italy

Aaron V. Wandler, Thomas L. Davis and Paritosh K. Singh
Department of Geophysics, Colorado School of Mines, 1500 Illinois Street, Golden, CO 80401, USA

P. W. J. Glover and E. Walker
Department of Geology and Engineering Geology, Laval University, Qu´ebec City, QC, Canada G1K 7P4

J. Ruel and E. Tardif
Department of Mechanical Engineering, Laval University, Qu´ebec City, QC, Canada G1K 7P4

Javier Sanchez-Rojas
Institut f¨ur Geowissenschaften Abteilung Geophysik, Christian-Albrechts-Universit¨at zu Kiel, Otto-Hahn-Platz 1,24118 Kiel, Germany
Departamento de Geof´ısica, Fundaci´on Venezolana de Investigaciones Sismol´ogicas FUNVISIS, 1070 Caracas, Venezuela

Mahmoud Abd El-Gelil
Geomatics Division, Faculty of Engineering & the Built Environment, University of Cape Town, Cape Town 7700, South Africa

Spiros Pagiatakis
Department of Earth and Space Science and Engineering, York University, Toronto, ON, Canada M3J 1P3

Ahmed El-Rabbany
Department of Civil Engineering, Ryerson University, Toronto, ON, Canada M5B 2K3

Yongkang Xue
Department of Geography, University of California, Los Angeles, CA 90095, USA

Yongkang Xue
Department of Atmospheric and Oceanic Sciences, University of California, Los Angeles, CA 90095, USA

Aaron Boone
Meteo-France, CNRM, Toulouse, France

Christopher M. Taylor
Centre for Ecology & Hydrology, Crowmarsh Gifford, Wallingford, Oxon OX10 8BB, UK

V. J. S. Grauch
U. S. Geological Survey, MS 964, Federal Center, Denver, CO 80225, USA

C. A. Ruleman
U. S. Geological Survey, MS 980, Federal Center, Denver, CO 80225, USA